ALLE ZEIT WACH
SJ
1842

Springer Series in Wood Science

Editor: T.E. Timell

M.H. Zimmermann
Xylem Structure and the Ascent of Sap (1983)

J.F. Siau
Transport Processes in Wood (1984)

R.R. Archer
Growth Stresses and Strains in Trees (1986)

W.E. Hillis
Heartwood and Tree Exudates (1987)

S. Carlquist
Comparative Wood Anatomy (1988)

L.W. Roberts / P.B. Gahan / R. Aloni
Vascular Differentiation and Plant Growth Regulators (1988)

C. Skaar
Wood-Water Relations (1988)

J.M. Harris
Spiral Grain and Wave Phenomena in Wood Formation (1989)

B.J. Zobel / J.P. van Buijtenen
Wood Variation (1989)

P. Hakkila
Utilization of Residual Forest Biomass (1989)

J.W. Rowe
Natural Products of Woody Plants (1989)

K.-E.L. Eriksson / R.A. Blanchette / P. Ander
Microbial and Enzymatic Degradation of Wood and Wood Components (1990)

R.A. Blanchette / A.R. Biggs
Defense Mechanisms of Woody Plants Against Fungi (1992)

S.Y. Lin / C.W. Dence
Methods in Lignin Chemistry (1992)

G. Torgovnikov
Dielectric Properties of Wood and Wood-Based Materials (1993)

Stephen Y. Lin Carlton W. Dence (Eds.)

Methods in Lignin Chemistry

With 190 Figures

Springer-Verlag
Berlin Heidelberg New York London Paris
Tokyo Hong Kong Barcelona Budapest

Dr. STEPHEN Y. LIN
LignoTech USA Inc.
100 Highway 51 South
Rothschild, WI 54474, USA

Professor Emeritus Dr. CARLTON W. DENCE
600 Black Walnut Court
Hendersonville, NC 28792, USA

Series Editor:

T.E. TIMELL
State University of New York
College of Environmental Science and Forestry
Syracuse, NY 13210, USA

Cover: Transverse section of *Pinus lambertiana* wood. Courtesy of Dr. Carl de Zeeuw, SUNY College of Environmental Science and Forestry, Syracuse, New York

ISBN-13:978-3-642-74067-1 e-ISBN-13:978-3-642-74065-7
DOI: 10.1007/978-3-642-74065-7

Library of Congress Cataloging-in-Publication Data. Methods in lignin chemistry / Stephen Y. Lin, Carlton W. Dence (eds.). p. cm. – (Springer series in wood science) Includes bibliographical references and index. ISBN-13:978-3-642-74067-1 1. Lignin.
I. Lin, Stephen Y. II. Dence, Carlton W. III. Series. TS933.L5M47 1992 661'.802–dc20 92-6677

Softcover reprint of the hardcover 1st edition 1992

Typesetting: Best-set Typesetter Ltd., Hong Kong, and K+V Fotosatz GmbH, Beerfelden
31/3145-5 4 3 2 1 0 – Printed on acid-free paper

Dedication

As this book neared completion, we learned of the sudden and unexpected death of Dr. Kyosti V. Sarkanen, Professor of Wood Chemistry at the University of Washington. Just prior to his passing, Dr. Sarkanen had reviewed several portions of this book and was in the process of co-authoring the section on ozonolysis. Ironically, these activities coincided with the culmination of a professional career that spanned nearly five decades during which he established himself as a dominant figure in the field of wood and, in particular, lignin chemistry.

It is not our intention to echo here the numerous eulogies marking his passing or even to attempt to highlight his scientific accomplishments. It suffices to say that those of us who were fortunate to know him personally as a teacher, researcher, adviser, colleague, or co-worker are fully aware of the enormous impact he made on the development of wood science; others who come to know him indirectly through his published work have the opportunity to gauge his uniqueness for themselves. As co-editors, we deem it a privilege to dedicate this book to the memory of Dr. Sarkanen as a fitting tribute to a colleague whose research contributions and scientific insight provided the basis for the shaping and refining of many of the concepts expressed within its covers.

CARLTON W. DENCE and STEPHEN Y. LIN

Preface

The structural complexity of lignin has continually challenged the ingenuity of researchers to develop suitable methods for its characterization prior to and following a wide variety of chemical, biological, and physical treatments. Initially, activity along these lines was fueled by a desire to interpret technical delignification (i.e., pulping) processes in terms of accompanying structural changes in the lignin. Subsequently, increasingly wide ranging, in-depth investigations on the structure and reactivity of lignin exposed the inadequacy of many of the methods currently in use and underscored the ever-continuing need to develop new methods capable of solving the unique analytical problems associated with lignin. Characteristically, such methods should be selective, sensitive, suitable for quantitative measurements, and capable of being applied directly to, and without destruction of, the lignin or lignocellulose sample. One notable example of the headway being made in reaching this objective is the relatively recent development and refinement of methods based on the use of sophisticated instrumentation, e.g., ^{1}H- and ^{13}C-NMR spectroscopy. Although the utility of many of these and other recently developed methods described in this book has yet to be fully and satisfactorily exploited, we believe that progress already made in this direction will continue and most likely accelerate.

The decision to produce this book was prompted mainly by the acknowledged need for an up-to-date, single source compilation of lignin methodology. Hitherto, this need was, in part, satisfied by B. L. Browning's two-volume set of books, *Methods of Wood Chemistry*, which was published nearly 30 years ago. The recognition that these books had, in part, become outdated was a major consideration in arriving at our decision to undertake to assemble a book reflecting the current status of lignin methodology. While we envision *Methods in Lignin Chemistry* as broadly serving the needs of professionals in academia, the pulp and paper industry, and allied industries, we also anticipate that the format adopted here for presenting the information will be particularly useful to those beginning a career in wood and/or lignin chemistry and to those with no previous background in or experience with lignin or lignocellulosics and whose interest is projected to be on a one-time-only, short-term basis.

The core element common to each chapter in this book is a detailed, step-by-step description of what, in the judgment of the author(s), is the preferred method for conducting a specific analysis or operation. This particular format was selected to enable the reader to use the book essentially as a laboratory manual thereby obviating the need to search for and evaluate methods scattered throughout the literature. The methods included are a blend of the traditional and the contemporary. In some instances, what appeared to be promising analytical methods were bypassed in favor of more widely accepted methods on the grounds that they had not been field-tested sufficiently to authenticate their reliability. The description of the preferred method is augmented by supporting information including, for example, a survey of alternative methods, a discussion of the principle underlying the preferred method, the advantages and disadvantages of the method, the type and significance of information provided by the analysis, and the presentation of typical analytical data.

From the onset of this undertaking it was obvious to us that the range of topics we wished to include virtually necessitated our drawing on the expertise of a considerable number of individuals well versed in the principles and applications of the selected methodology. We were also aware that the adoption of a multi-author format entailed risks that might impact negatively on the uniformity, general quality, and timetable for completion of the book. Without question, some of these concerns did prove to be justified. On the other hand, when this book is viewed in its totality, we believe that the overall quality and usefulness of its contents vindicates our decision to proceed as we did. With this in mind, we commend the international team of authors for service well performed in helping to make *Methods in Lignin Chemistry* a reality. Others who played important roles in its creation are acknowledged below.

STEPHEN Y. LIN
CARLTON W. DENCE

Acknowledgments

We are indebted to the following persons who generously contributed their time to review and recommend revision for several chapters in this book: F. Agblevor, L. Amos, S. Black, J. Boxell, H.L. Chum, R. Evans, D.A.I. Goring, M. Himmel, D.K. Johnson, L. Landucci, J. Obst, S. Omori, M.G. Paice, K.V. Sarkanen, and T.P. Schultz. We also acknowledge with gratitude the special contributions of T.E. Timell, Series Editor, who, in addition to reviewing and editing each manuscript, advised us on matters related specifically to organization, language, and style conventions during the entire period in which material for the book was being assembled.

Our sincere appreciation is also expressed to Sarah Remon for her valued service in retyping promptly and accurately many of the edited and revised manuscripts.

Finally, we individually wish to thank G.A. Rasmussen of LignoTech USA, Inc. (SYL) for his support and encouragement in this undertaking and the SUNY College of Environmental Science and Forestry (CWD) for permission to discharge attendant editorial responsibilities concurrently with customary academic duties.

Rib Mountain, WI
Hendersonville, NC
August 1992

STEPHEN Y. LIN
CARLTON W. DENCE

Contents

Chapter 8 Determination of Molecular Weight, Size, and Distribution ... 485

List of Contributors

ABE I., Faculty of Agriculture, Department of Forestry, Mie University, Mie, 514, Japan

AGARWAL U. P., Forest Products Laboratory, USDA, Forest Service, One Gifford Pinchot Drive, Madison, WI 53705-2398, USA

ANDERSON C. D., Department of Chemistry, Pacific Lutheran University, Tacoma, WA 98447, USA

ATALLA R. H., Forest Products Laboratory, USDA, Forest Service, One Gifford Pinchot Drive, Madison, WI 53705-2398, USA

BEATSON R. P., Canfor Research and Development Centre, Canadian Forest Products Ltd., 101, 1750 West 75th Avenue, Vancouver, BC, Canada V6P 6G2

BOND J. S., The Institute of Paper Science and Technology, Inc., 575 14th Street NW, Atlanta, GA 30318, USA

CHANG H.-m., Department of Wood and Paper Science, North Carolina State University, Box 8005, Raleigh, NC 27695-8005, USA

CHEN C.-L., Department of Wood and Paper Science, North Carolina State University, Box 8005, Raleigh, NC 27695-8005, USA

CHIANG V. L., Institute of Wood Research, Michigan Technological University, Houghton, MI 49931-1295, USA

DENCE C. W., 600 Black Walnut Court, Hendersonville, NC 28792, USA

DONALDSON L. A., Forest Research Institute, Ministry of Forestry, Private Bag, Rotorua, New Zealand

FAIX O., Federal Research Center for Forestry, Leuschnerstr. 91, W-2050 Hamburg 80, FRG

FUKAZAWA K., Department of Forest Products, Faculty of Agriculture, Hokkaido University, Kita 9, Nishi 9, KitaKu, Sapporo 060, Japan

FUNAOKA M., Faculty of Agriculture, Department of Forestry, Mie University, Mie, 514, Japan

GELLERSTEDT G., Royal Institute of Technology, Drottning Kristinas Vag 53, S-10044 Stockholm, Sweden

HATAKEYAMA H., Industrial Products Research Institute, Agency of Ind. Science and Technology, 1-1-4, Higashi, Tsukuba, Ibaraki, Japan

HON D. N.-S., Clemson University, 125 Lehotsky Hall, Clemson, SC 29631-1003, USA

ISLAM A., College of Forest Resources, University of Washington, Seattle, WA 98195, USA

LAI Y.-Z., SUNY College of Environmental Science and Forestry, 1 Forestry Drive, Syracuse, NY 13210, USA

LAPIERRE C., Laboratoire de Biochimie, Institut National Agronomique, F-78850 Thiverval-Grignon, France

LEARY G. J., Chemistry Division, Department Scientific/Ind. Res., Private Bag, Petone, New Zealand

LEWIS N. G., Institute of Biological Chemistry, 467 Clark Hall, Washington State University, Pullman, WA 99164-6340, USA

LIN S. Y., LignoTech USA Inc., 100 Highway 51 South, Rothschild, WI 54474, USA

LUNDQUIST K., Department of Organic Chemistry, Chalmers University of Tech., S-41296 Göteborg, Sweden

MEIER D., Federal Research Center for Forestry, Leuschnerstr. 91, W-2000 Hamburg 80, FRG

MESHITSUKA G., Faculty of Agriculture, Department of Forest Products, University of Tokyo, Tokyo, 113, Japan

MONTIES B., Laboratoire de Biochimie, INRA, Institut National Agronomique, Paris, Grignon, F-78850 Thiverval-Grignon, France

NAKANO J., Fukui Institute of Technology, 3-6-1, Gakuen, Fukui City 910, Japan

NEWMAN R. H., Chemistry Division, D.S.I.R., Private Bag, Petone, New Zealand

PLA F., Laboratoire Des Sciences Du, Genie Chimique, University de Nancy, E.N.S.I.C., 1, rue Grandville BP 451, 54001, Nancy Cedex, France

ROBERT D., Centre d'Etudes Nucléaires de Grenoble, DRF-Laboratoires de Chimie, 85X, F-38041 Grenoble, France

ROLANDO C., Department de Chimie, Ecole Normale Superieure, 24 rue Lhamond, F-75231 Paris, France

SAKA S., Department of Wood Science and Technology, Faculty of Agriculture, Kyoto University, Kitashirawaka Oiwake-Cho, Sakyo-ku, Kyoto 606, Japan

SAKAKIBARA A., Department of Forest Products, Faculty of Agriculture, Hokkaido University, Kita 9, Nishi 9, KitaKu, Sapporo 060, Japan

SARKANEN K. V., College of Forest Resources, University of Washington, Seattle, WA 98195, USA

Chapter 1
Introduction

1 Introduction

C.W. DENCE and S.Y. LIN

1.1 General Structural Features of Lignin

The development of lignin methodology can be credited in large measure to the recognized need to define more accurately and precisely the structure of lignin as present in plant tissue and in isolated form. Thus, as a result of the sustained application of standard and uniquely adapted analytical procedures and of increasingly sophisticated instrumentation, the concept of lignin has been progressively refined to a degree where, at present, lignin structure and morphology are considered to be adequately, if not perfectly, defined.

According to the widely accepted concept, lignin may be defined as an amorphous, polyphenolic material arising from an enzyme-mediated dehydrogenative polymerization of three phenylpropanoid monomers, coniferyl (1), sinapyl (2) and p-coumaryl (3) alcohols. The conventions followed throughout this book for designating the carbon atoms in phenylpropanoid units are illustrated by the accompanying figure.

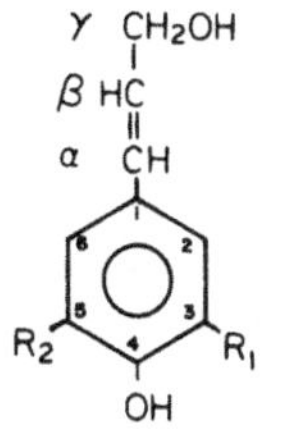

(1) $R_1 = OCH_3$, $R_2 = H$
(2) $R_1 = R_2 = OCH_3$
(3) $R_1 = R_2 = H$

The biosynthesis process, which consists essentially of radical coupling reactions, sometimes followed by the addition of water or of primary, secondary, and phenolic hydroxyl groups to quinonemethide intermediates, leads to the formation of a three-dimensional polymer which lacks the regular and ordered repeating units found in other natural polymers such as cellulose and proteins. For this reason, lignin is viewed not as a constitutionally defined compound, but as a composite of physically and chemically heterogeneous materials whose structure may be represented by models such as those proposed for spruce and beech lignins (Figs. 1.1 and 1.2, respectively). These models should not be regarded as depicting the structural formulas for lignin in the usual sense, but as vehicles for illustrating the types and linkage modes of the constituent structural elements and the proportions in which they are believed to occur in lignin.

Springer Series in Wood Science
Methods in Lignin Chemistry
(Edited by S.Y. Lin and C.W. Dence)

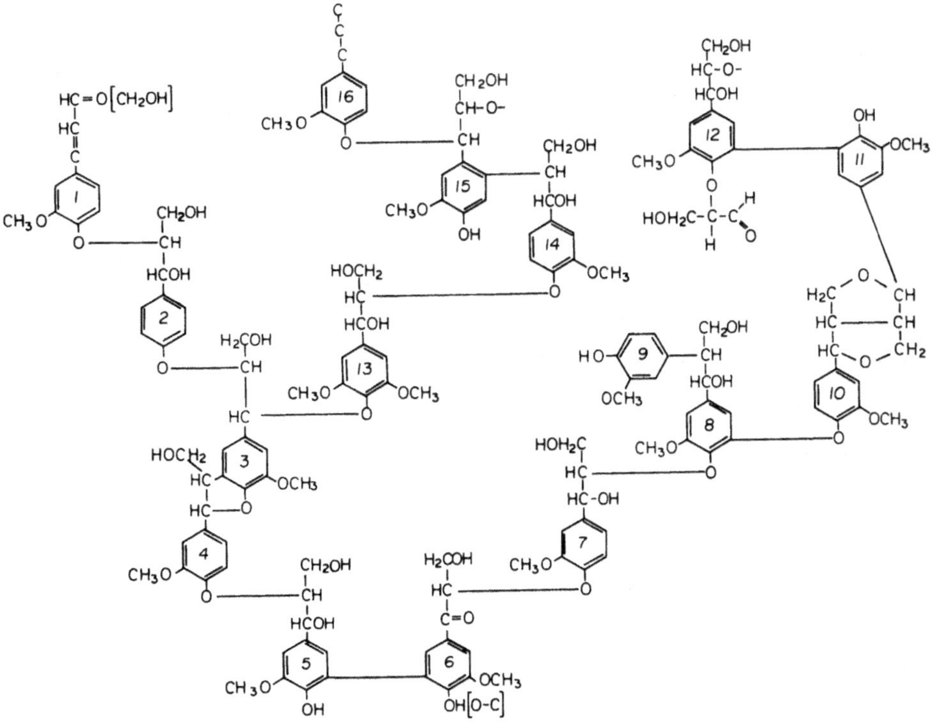

Fig. 1.1. Structural model of spruce lignin. (Adler 1977)

In normal softwood lignin, usually referred to as guaiacyl lignin, the structural elements are derived principally from coniferyl alcohol, (1) (more than 95% of the total number of structural units), with the remainder consisting mainly of p-coumaryl alcohol-type units (Fig. 1.1, unit 2) and trace amounts of sinapyl alcohol-derived units (Fig. 1.1, unit 13).

Normal hardwood lignins, termed guaiacyl-syringyl lignins, are comprised of coniferyl alcohol- and sinapyl alcohol-derived units in varying ratios (Fig. 1.2). In lignins of this type, the methoxyl content per phenylpropanoid unit typically is in the range 1.2 to 1.5 (Sarkanen and Hergert 1971).

Grass lignins are also classified as guaiacyl-syringyl lignins. However, unlike hardwood (angiosperm) lignins, grass lignins additionally contain small but significant amounts of structural elements derived from p-coumaryl alcohol, (3). Grass lignins also contain p-coumaric acid, (4), and ferulic acid (5) residues attached to the core lignin through ester linkages. Compression wood, the reaction wood of conifers, contains a p-hydroxylphenyl-guaiacyl lignin (Erickson et al. 1973b, Yasuda and Sakakibara 1975).

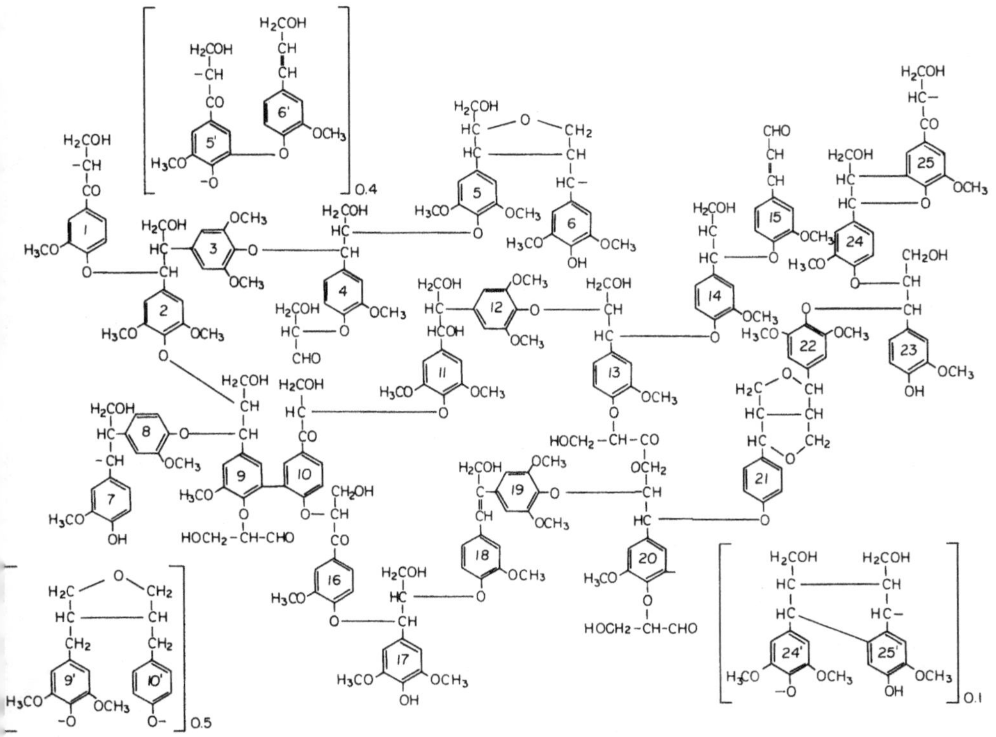

Fig. 1.2. Structural model of beech lignin. (Nimz 1974)

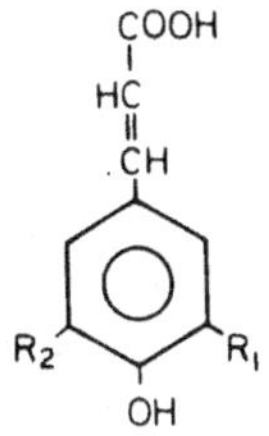

(4) $R_1 = R_2 = H$
(5) $R_1 = OCH_3$, $R_2 = H$

Although not indicated in the model structures shown in Figs. 1.1 and 1.2, there is ample evidence supporting the concept that lignin is occasionally linked with polysaccharides, primarily through arabinose, xylose, and galactose units in the hemicellulose components (Eriksson et al. 1980, Koshijima et al. 1989). The stability of such linkages has been cited as a reason for the difficulty encountered in removing the residual lignin in the final stages of alkaline pulping (Minor 1986).

The structural elements comprising lignin are linked by carbon-carbon and ether bonds. Units that are trifunctionally linked to adjacent units represent branching sites which give rise to the network structure characteristic of lignin (see Fig. 1.1, units 3, 8, and 15). As the interunit carbon-carbon linkages are difficult to rupture without excessively fragmenting the carbon skeleton of the

Table 1.1. Types and frequencies of interunitary linkages in softwood and hardwood lignins (number of linkages per 100 C_9-units)

Linkages	Erickson et al. (1973a)[a]	Nimz (1974)[b]	Unit no. in Fig. 1.1
β-0-4	49-51	65	1/2, 4/5, 6/7, 7/8, 13/1
α-0-4	6-8		3/13, 15/16, 3/4
β-5	9-15	6	3/4
β-1	2	15	8/9
5-5	9.5	2.3	5/6, 11/12
4-0-5	3.5	1.5	8/10
β-β	2	5.5	10/11

[a] Softwood (spruce) lignin.
[b] Hardwood (beech) lignin.

lignin, solvolysis of the ether linkages is often utilized as the best approach for degrading lignin to structures which retain some characteristic features of the original lignin. The types and frequencies of several of the more prominent interunit lignin linkages are summarized in Table 1.1.

Of the functional groups attached to the basic phenylpropanoid skeleton, those having the greatest impact on the reactivity of lignin include phenolic hydroxyl, benzylic hydroxyl, and carbonyl groups. Although relatively unreactive, methoxyl groups serve the useful function of providing an approximate measure of the phenylpropanoid content of a particular lignin. The frequency of these groups, reported throughout Chapter 7.6, may vary according to the morphological location of the lignin, wood species, and method of isolation.

1.2 Variations in Concentration and Composition of Lignin in Xylem

Lignin contents generally vary from 24 to 33% in normal softwoods, from 19 to 28% in normal, temperate-zone hardwoods (Sarkanen and Hergert 1971), and from 26 to 35% in tropical hardwoods (Pettersen 1984); in nonwoody fiber sources, the lignin contents are typically in the 11 to 27% range (Bagby et al. 1971) (see Chap. 2.2). In compression wood, the average lignin content is as high as 35–40%, but tension wood, the reaction wood of the angiosperms, contains only 15–20% lignin.

Both the amount and the nature of the lignin in wood can vary within wide limits. Juvenile wood at the center of a tree stem has a higher lignin content than mature wood, and earlywood has a higher proportion of lignin than latewood. The composition of lignin is also influenced by its age such that, for example, in mature poplar wood the syringyl-guaiacyl ratio is high but decreases in the younger xylem and in the phloem (Sarkanen and Hergert 1971). The distribution of lignin within individual cells has been determined principally by UV microscopy (Scott et al. 1969) and by electron microscopy combined with energy-dispersive X-ray analysis, entailing either bromination (Saka et al. 1978) or mercurization (Westermark et al. 1988) of the lignin. As illustrated in Fig. 1.3 for spruce tracheids, lignin is distributed relatively uniformly across the secondary wall, but its concentration increases abruptly at the boundary

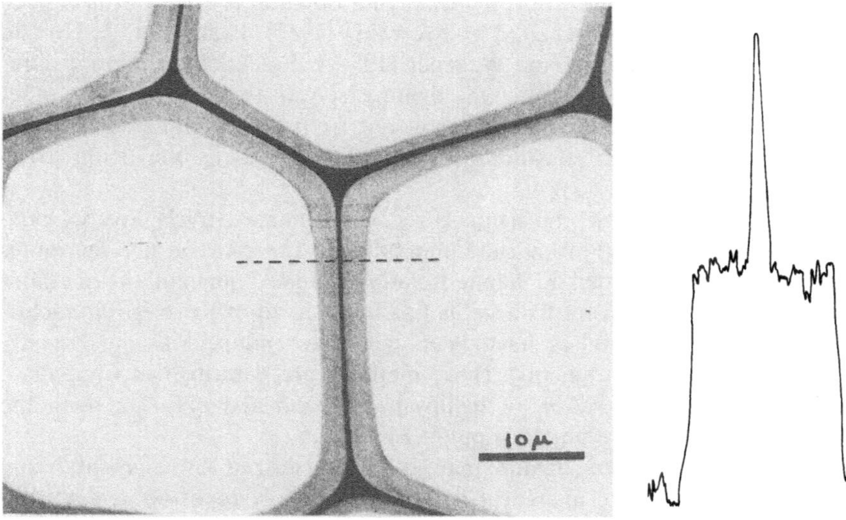

Fig. 1.3. Ultraviolet micrograph of transverse section of black spruce tracheid with the microdensitometer trace taken across *dotted line*

between the secondary wall and the compound middle lamella (M + P). The average lignin concentration in the secondary wall is 2–25% in both softwood and hardwood and varies between 50 and 100% in M + P. Because the middle lamella accounts for only 5–15% of the wood volume, 60–80% of the total lignin is located within the secondary wall (Fergus et al. 1969, Fergus and Goring 1970b). In compression wood tracheids, the outer portion of the S_2 layer has a lignin concentration of 60%, and only 10% of the total lignin is located in M + P (Wood and Goring 1971). By contrast, the innermost layer of tension wood fibers, the so-called G-layer, is unlignified and consists largely of cellulose (Norberg and Meier 1966).

The nature of the lignin also varies on the microscopic level. In softwood, the lignin in the compound middle lamella has a higher content of p-hydroxyphenyl units and contains fewer free phenolic hydroxyl groups than that in the secondary wall (Whiting and Goring 1982). Hardwood lignins are even more variable. The lignin in the secondary walls of fibers and ray cells has a high content of syringyl units, while the compound middle lamella contains a guaiacyl-syringyl lignin. The vessels, in contrast, contain a guaiacyl lignin in both the secondary wall and the M + P (Fergus and Goring 1970a, Musha and Goring 1975). It should be added, however, that while these results have been confirmed more recently by some investigators, they have been challenged by others (Obst 1982, Westermark 1985).

1.3 Isolation and Recovery of Lignin

Several of the analytical procedures described in this book stipulate that the lignin first be isolated from a plant substance to help reduce or eliminate interference from nonlignin constituents. Lignin isolation procedures have been classified, described, and evaluated by Browning (1967), Pearl (1967a), Lai and Sarkanen (1971), and Fengel and Wegener (1984a). Essentially, the procedures are classified according to whether the lignin is selectively removed from other plant constituents and subsequently recovered from the resulting solution or, conversely, the nonlignin constituents are removed, leaving the lignin as an insoluble residue (Chap. 2.2).

In an ideal procedure, the lignin is recovered quantitatively, free of nonlignin contaminants, and chemically unmodified. In spite of the enormous amount of effort expended in lignin isolation studies, none of the methods currently available is universally accepted as fulfilling all of these requirements. However, several procedures have been generally endorsed as most nearly meeting the above requirements. These methods are described in Chapter 3, which deals with the isolation of lignins from wood and pulp and with the recovery of lignin from commercial pulping liquors.

Isolated and purified lignins may be characterized by elemental and methoxyl analyses. These analytical data are commonly reported as empirical formulae calculated on a C_9-basis. Thus, the formula of a typical softwood milled wood lignin (MWL) is reported as $C_9H_{8.3}O_{2.7}(OCH_3)_{0.97}$ and that of a typical hardwood MWL as $C_9H_{8.7}O_{2.9}(OCH_3)_{1.58}$. These formulas serve to lend emphasis to the fact that the monomeric units comprising lignins have C_9 carbon skeletons and to underscore the significance of the methoxyl group for both identifying the lignin type and for indicating the contributions of component monomer units. Empirical formulas also find use in calculating the unit weight of an isolated lignin. However, in situations where the lignin side chain undergoes extensive chemical degradation as, for example, in kraft pulping, the practice of reporting the empirical formula on a C_9-basis may not be justified.

1.4 Physical Properties of Lignin

As an amorphous polymer, lignin behaves as a thermoplastic material, undergoing a glass transition at temperatures which vary widely depending on the method of isolation, sorbed water, and heat treatment (Goring 1971, Hatekeyama et al. 1972). Thermal softening is an important property since it affects lignin dissolution in chemical pulping and fiber separation in mechanical pulping.

Because it is not possible to isolate lignin quantitatively from plant materials without chemical or mechanical degradation, its true molecular weight is not known. In the isolation process, lignin is more or less randomly degraded to fragments of varying size. This property of isolated lignins, termed polydispersity, is commonly measured as the ratio of weight and number average molecular weights, $\bar{M}w/\bar{M}n$. Polydispersity ranging from about 2.3 to 3.5 appears to be typical of milled wood lignins (Wayman and Obiaga 1974, Chang et al. 1975, Faix et al. 1981) and kraft lignins (Marton and Marton 1964). By contrast, the polydispersity of lignosulfonates is much greater, with the $\bar{M}w/\bar{M}n$ ratio in the range of 6 to 8 (Lin 1990).

The various methods used to determine number- and weight-average molecular weights have been outlined and critically evaluated by Goring (1962, 1971) and Pearl (1967b); procedures for several methods currently in use are described in detail in Chapter 8. Reported molecular weight values cover the range 10^2 to 10^6, reflecting the lignin source and isolation process, degree of purification, fractionation history, and method of measurement. Weight- and number-average molecular weights of approximately 20000 and 8000, respectively, have been reported for a spruce milled wood lignin (Wayman and Obiaga 1974). For technical lignins, molecular weights ranging approximately from 10^2 to 10^4 have been obtained for lignosulfonates (Goring 1971, Lin and Detroit 1981), and generally lower average values for kraft lignin (Marton and Marton 1964). Based on a very limited number of results, hardwood lignins appear to have a slightly lower molecular weight than do softwood lignins (Goring 1971).

The ability of a solvent to dissolve lignin is a function of the cohesive energy density and hydrogen-bonding capacity of the solvent (Schuerch 1952). In a single solvent or mixture of solvents, these properties are expressed as $\Delta\mu$, the infrared shift in the O-D bond when the solvent is mixed with CH_3OD, and the Hildebrand solubility parameter; δ. Solvents having a $\Delta\mu$ in excess of 0.14 micron and δ values of around 10–11 $(cal/cc)^{1/2}$ are considered good lignin solvents. Solvents meeting these requirements include dioxane, acetone, methyl cellosolve, pyridine, and dimethylsulfoxide (DMSO) (Fengel and Wegener 1984b). Other physical properties of lignin have been comprehensively reviewed by Goring (1971).

1.5 Structure-Modifying Reactions

Lignin has been subjected to a vast array of reactions in fundamental and applied studies aimed, for example, at defining its structure, separating it from associated polysaccharides, and evaluating its potential as a source of chemicals. To reduce the total number of reported lignin-modifying reactions to a manageable level, the following description has been restricted to a survey in which only the main types of structural units and functional groups formed in the more typical and universally applied lignin-modifying reactions are identified and discussed. The reactions singled out for discussion have been arbitrarily classified according to whether the propanoid side chain or the aromatic ring is totally or principally involved in a particular transformation. Reactions forming the basis for analytical procedures have been excluded from consideration.

1.5.1 Modification of the Aromatic Ring

1.5.1.1 Electrophilic Substitution

Of the processes encompassing this general reaction type, chlorination has received by far the most attention because of its widespread, albeit currently diminishing, use in delignifying chemical pulps in multistage bleaching

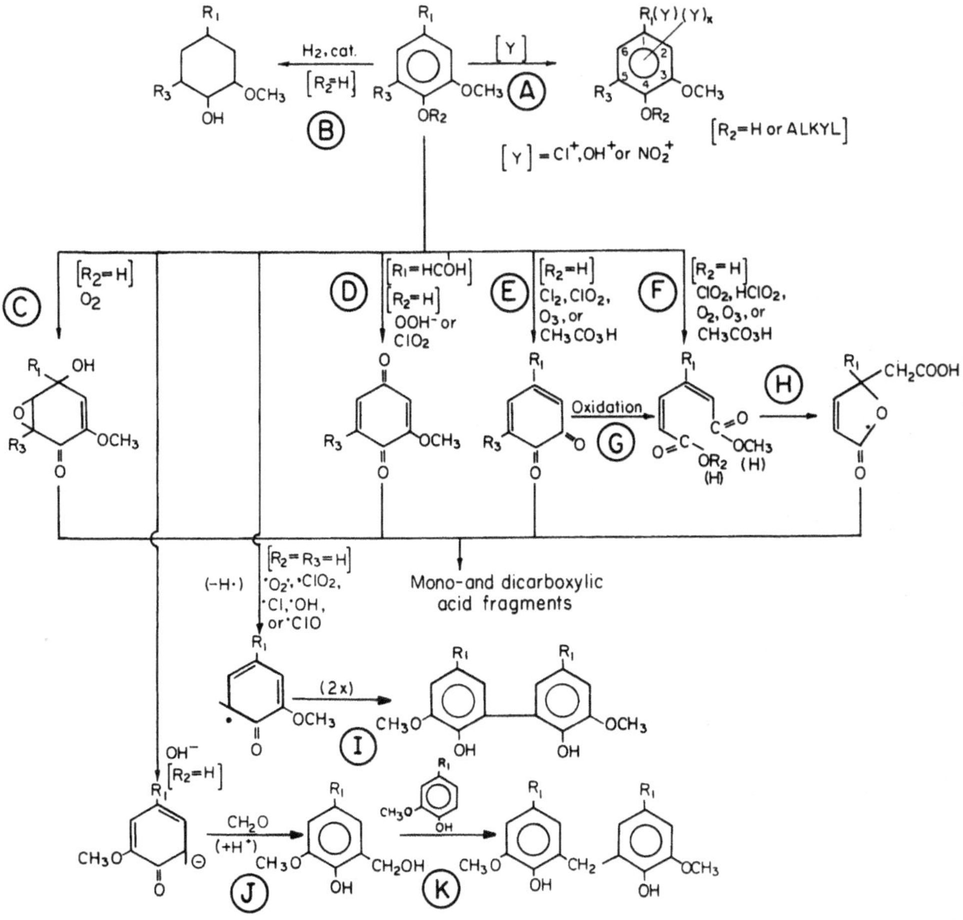

Fig. 1.4. Aromatic ring modification reactions

sequences. In such processes, elemental chlorine, applied to the pulp in the form of a dilute aqueous solution, rapidly reacts with lignin in consequence of which the aromatic ring is nonuniformly substituted with chlorine mainly at the 5- and 6-positions and at the 1-position through side chain displacement (A, Fig. 1.4) (Dence 1971). Although functioning primarily as oxidants, hypochlorous acid, hypochlorite anion, chlorous acid, and chlorine dioxide may also chlorinate the aromatic units in lignin although to a much lower extent than does elemental chlorine.

Other processes in which the aromatic ring is modified via electrophilic substitution reactions are nitration and ozonation. The electrophilic substitution of nitro groups on the aromatic moiety of lignin and accompanying reactions have been reviewed by Dence (1971). A nitration process consisting of the pretreatment of chemical pulp with nitrogen dioxide followed by treatment with oxygen, the so-called Prenox Process, recently has been proposed as an alternative for delignification with chlorine-containing chemicals (Andersson and Samuelson 1985). As one of several competing processes, electrophilic attack of ozone on lignin ultimately leads to ring hydroxylation (Gierer 1982a,b).

1.5.1.2 Conversion of Aromatic Rings to Nonaromatic Cyclic Structures

On treatment with oxidants such as chlorine, hypochlorite anion, chlorine dioxide, oxygen, hydrogen peroxide, and peroxy acids, the aromatic nuclei in lignin typically are converted to o- and p-quinonoid structures and oxirane derivatives of quinols (E, D, C, resp., Fig. 1.4). It should be noted that the conversion of aromatic nuclei to o-quinonoid rings is accompanied by loss of the methoxyl group as methanol and that conversion to p-quinonoid groups in many cases leads to displacement of the side chain.

Because of their relatively high reactivity, the foregoing structures often appear as transient intermediates in a series of reactions rather than as end products. Several of the documented reactions of these intermediates include rearomatization (e.g., reduction to catechol and hydroquinone derivatives), further oxidation to mono- and dicarboxylic acids (see below), benzylic acid rearrangements (Corbett 1966, Corbett and Fooks 1967), cycloaddition reactions (Teuber et al. 1966), and various condensation reactions (Erdtman and Granath 1954). The last-named processes, which are accelerated in acidic and basic media, often give rise to structurally complex and poorly defined materials.

Aromatic rings in lignin may be converted to cyclohexanol derivatives by catalytic hydrogenation (B, Fig. 1.4). Conversions of this type require highly active catalysts such as copper-chromium oxide and high temperatures (250 °C) and pressures (200–350° atm.) (Harris and Adkins 1938). Similar reduction of aromatic to saturated rings has been achieved using as reductants sodium in liquid ammonia (Shorygina et al. 1949) and hydrogen in the presence of a soluble catalyst (Eachus and Dence 1975).

1.5.1.3 Conversion of Cyclic to Acyclic Structures

On oxidative treatment, the aromatic rings may be converted directly to acyclic structures, e.g., muconic acid derivatives or indirectly through secondary reactions by oxidative splitting of o-quinonoid rings (F, Fig. 1.4). Under alkaline conditions, the muconic acid monomethyl ester derivatives are hydrolyzed to the corresponding muconic acid dianions with release of methanol; under acidic conditions, muconic acids may undergo ring closure via lactonization (H, Fig. 1.4) (Gierer and Imsgard 1977). With further oxidation, cyclic and acyclic structures alike are transformed to mixtures of mono- and dicarboxylic acid fragments. As a consequence of such reactions, oxidized lignocellulosics generally contain carboxylic acid fragments attached to the lignin network.

1.5.1.4 Ring Coupling and Condensation Reactions

Oxidants, which are themselves free radicals (e.g., ClO_2 and O_2), or which are convertible to such species (e.g., conversion of peroxides to OH·), abstract hydrogen atoms from phenolic hydroxyl groups thereby generating phenoxy radicals. Coupling of the proper mesomeric forms of such radicals ultimately leads to diphenyl structures, as is illustrated by I in Fig. 1.4. Either or both of the aromatic rings in such structures may undergo further oxidation, yielding quinonoid and carboxylic acid derivatives analogous in structure to those produced from monomeric units.

In alkaline media, phenolic units may react with formaldehyde, forming methylol derivatives that condense with themselves or with another phenol (J, Fig. 1.4). This formaldehyde condensation reaction forms the basis for using technical lignins in the production of adhesives.

1.5.2 Modification of the Propanoid Side Chain

1.5.2.1 Cleavage of Ether Bonds

Ether linkages at the α- and β-positions are the most abundant functional groups on the propanoid side chain of lignin. Under acidic conditions, these linkages undergo solvolytic cleavage forming initially the corresponding secondary alcohols (L_1, Fig. 1.5) (Wallis 1971). α-Ethers, i.e., benzyl ethers, are especially labile in these reactions. When the solvolysis is performed under strongly acidic conditions such as exist during acidolysis (Chap. 6.1), the hydroxyl-substituted side chain undergoes a combination of dehydration reactions and allylic rearrangements which ultimately leads to the introduction of carbonyl, ethylene, and carboxyl groups, and fragmentation of the side chain (Wallis 1971).

α-Ether linkages are also cleaved in alkali-promoted reactions when the side chain is attached to a phenolic ring. Nonphenolic α-ethers are cleaved provided a hydroxyl group is present at the adjacent β-position. Cleavage of β-ether linkages in phenolic and nonphenolic units similarly occurs in strongly alkaline media at elevated temperature (e.g., under soda pulping conditions) when a hydroxyl or carbonyl group is present in the adjacent α- or γ-position. As observed for acid-promoted cleavage of ether bonds, arylglycerol derivatives are formed (L_2, Fig. 1.5) which, unlike their behavior in acidic media, are stable in alkali (Wallis 1971, Gierer 1982a).

The alkali-promoted cleavage of α- and β-ether linkages, a pivotal step in alkaline pulping processes, is mainly responsible for the fragmentation and dissolution of lignin. Addition of bisulfide ion to the aqueous alkaline (sodium hydroxide) medium, as is the case in kraft pulping, enhances the rate and extent of β-aryl ether cleavage in phenolic units (Gierer 1980, 1982a) and hence the rate and extent of delignification (Rydholm 1965). Under pulping conditions, the primary intermediate is a p-hydroxycinnamyl alcohol derivative (reaction M, Fig. 1.5) (Gierer and Lindeberg 1980). This intermediate undergoes further reactions under pulping conditions ultimately yielding fragments with side chains

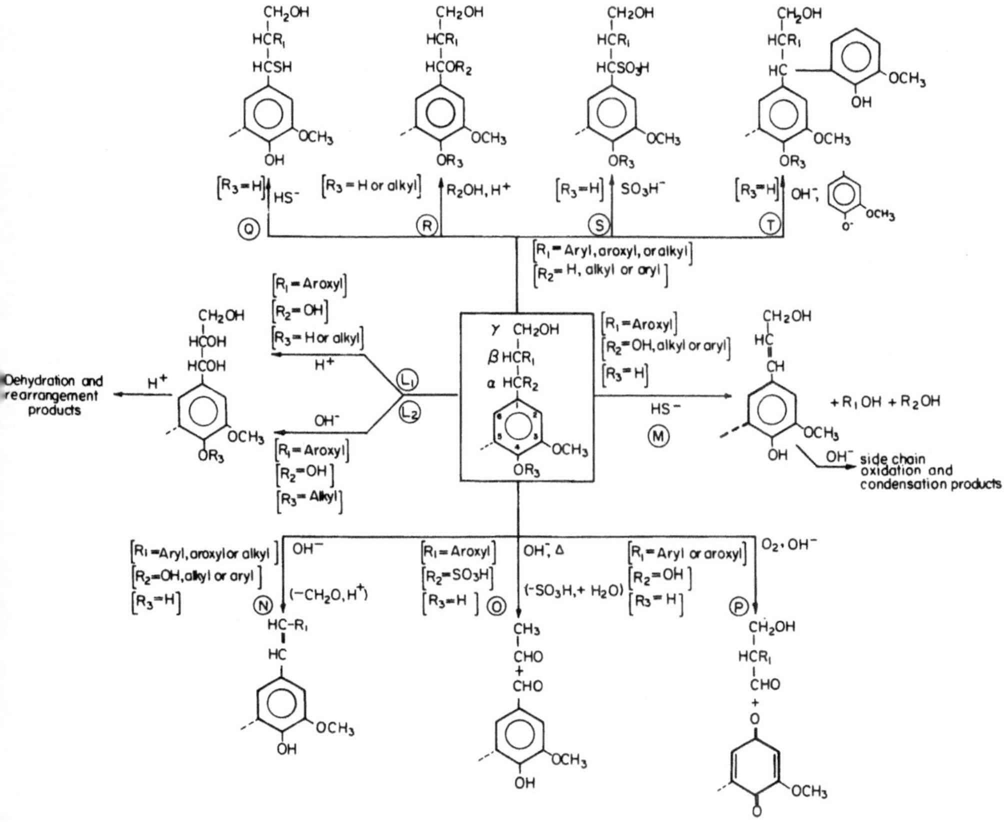

Fig. 1.5. Representative side chain ether cleavage, carbon-carbon fragmentations, and substitution reactions

comprised of 1 to 3 carbon atoms bearing carbonyl and ethylenic substituents (Marton 1971, Yamasaki et al. 1981).

1.5.2.2 Cleavage of Carbon-Carbon Bonds

Under appropriate conditions, the propanoid side chain in lignin may be ruptured at each of the three possible sites (1-α, α-β, or β-γ, Fig. 1.5), giving rise in the process to three-, two-, and one-carbon fragments, respectively. In contrast to the extensive lignin fragmentation resulting from the rupture of interunitary ether linkages, the effect of carbon-carbon bond cleavage is relatively minor in this respect (Gierer 1980).

Depending on the type of substituents on the propanoid side chain, the reagent(s) involved in the treatment, and the reaction conditions (e.g., acidity or alkalinity, temperature), carbon-carbon fragmentation of the propanoid side chain may occur in a variety of laboratory treatments and technical processes, for example, in the brightening of high-yield pulps with H_2O_2 (Gierer 1982b), in neutral sulfite, soda and kraft pulping (Gierer 1982a), in the bleaching of chemical pulps with Cl_2, ClO_2, OCl^-, and O_2, in chemical degradation methods, including many of those described in Chapter 6, in microbial degradation (Kirk and Farrell 1987), and in photooxidation (Gellerstedt and Pettersson 1975, 1977). The cleavage of carbon-carbon bonds in these cases is frequently the result of retrograde aldol condensation reactions or of an attack by an appropriate electrophilic (oxidizing) species. The fragments generated in such reactions are mainly aldehydes, ketones, and carboxylic acids.

Representative side chain carbon-carbon fragmentation reactions are illustrated in Fig. 1.5. Cleavage of the β-γ bond, typified by reaction N in the figure, results in the loss of the terminal methylol group as formaldehyde. This reaction is especially significant in soda pulping as the resulting vinyl aryl ether bond is relatively stable in alkali. When the β-carbon is attached to an aryl group via a carbon-carbon linkage, the elimination reaction leads to the formation of stilbenoid (i.e., 1,2-diphenylethene) derivatives.

As an example of α-β carbon bond cleavage may be cited the fragmentation arising from the alkaline hydrolysis of guaiacylglycerol-α-guaiacoxy-α-sulfonic acid-type compounds (reaction O, Fig. 1.5) (Kratzl et al. 1974). This reaction is the basis for the commercial manufacture of vanillin from spent sulfite liquor.

Finally, 1-α-bond cleavage is illustrated by the so-called Dakin-like reaction (P, Fig. 1.5) (Kratzl et al. 1974). In this reaction the side chain is displaced from a phenolic ring through attack by molecular oxygen under alkaline conditions.

1.5.2.3 Substitution Reactions

As the benzyl (α) carbon is the most reactive site on the propanoid side chain (Allan 1971), it is not surprising that many substitution reactions are documented as occurring at this position and at the vinylogous γ-carbon atoms of p-hydroxycinnamyl alcohol units. The variety of substitution reactions occurring at the benzyl carbon atom has been thoroughly reviewed by Allan (1971).

Typically, substitution reactions occur by attack of a nucleophilic reagent on a benzyl carbon present in the form of a carbonium ion or a methine group in a quinonemethide structure. Several representative substitution reactions are illustrated in Fig. 1.5. At moderate temperatures (~100 °C) and under mildly alkaline conditions, benzylic hydroxyl groups in phenolic units are converted to thiols by reaction with bisulfide (Q, Fig. 1.5). At higher temperatures and alkalinities, e.g., under kraft pulping conditions, the mercaptide group undergoes a series of transformations in which the sulfur is ultimately eliminated.

In a reversal of the ether cleavage reactions described above, benzyl alcohols and ethers may be transformed to alkyl or aryl ethers via acid-catalyzed etherification or transetherification, respectively, by reaction with the appropriate alcohol or phenol (reaction R, Fig. 1.5).

The conversion of a benzyl alcohol or ether to a sulfonic acid group is illustrated by reaction S in Fig. 1.5. This transformation is among the most important side chain modification reactions as it is central to the solubilization of lignin in the sulfite pulping process (Gellerstedt 1976). The γ-position is also sulfonated in phenylpropane-α-carbonyl-β-aryl ether structures (Gellerstedt 1976) as is the γ-carbon atom in cinnamyl alcohol-type structures (Glennie 1971). Sulfonation at the benzyl carbon atom is pH-dependent and as the pH is increased, the substitution becomes increasingly restricted to phenolic lignin units. Under neutral sulfite pulping conditions, β-aroxyl groups in phenolic lignin units undergo nucleophilic displacement by sulfite ions with concomitant introduction of a sulfonic acid group in the β-position (Gierer 1982a).

Benzyl and γ-carbon atoms frequently are involved in acid- and base-catalyzed condensation reactions (Gierer et al. 1976, Gierer 1980, Gierer and Lindeberg 1978, Glennie 1961). As illustrated by reaction T (Fig. 1.5), the base-catalyzed condensation reaction of a cyclohexadienone anion with a benzyl alcohol or benzyl ether group in a phenolic lignin unit gives rise to a diphenylmethane structure. Condensation reactions of the type illustrated in the figure are important in chemical pulping processes, where they may have an adverse effect on lignin solubilization.

1.5.2.4 Formation and Elimination of Multiple Bond Functionalities

The multiple bond structures of importance in side chain modification reactions are carbonyl and ethylenic groups and the processes frequently involved in the formation and elimination of such groups are reduction, hydrogenation, oxidation, and dehydration, here arbitrarily restricted to reactions in which no accompanying carbon-carbon fragmentation occurs.

Reduction of ring-conjugated carbonyl groups to the corresponding primary and secondary alcohols is generally achieved by reaction with sodium borohydride (reactions U_1 and V_1, Fig. 1.6) (Adler and Marton 1959). The determination of carbonyl groups in lignin is described in detail in Chapter 7.4. Ring-conjugated ethylenic groups may be converted to their saturated counterparts by hydrogenation over Pd-$BaSO_4$ (reaction W_1, Fig. 1.6) (Adler and Marton 1961). When the ethylenic group is further conjugated with a formyl group, as in cinnamaldehyde structures, hydrogenation of the ethylene group and reduction of the aldehyde group both occur (reaction W_2, Fig. 1.6). The determination of ethylenic groups is described in Chapter 7.3.

Reagents generating multiple bond structures on lignin side chain sites frequently tend to be nonspecific in their action with the result that more than one structural modification may occur in any single reaction. In selecting reactions to illustrate multiple-bond formation in Fig. 1.6, an attempt was made to include examples in which competing and serial reactions are minimal. Thus, for example, conversion of benzylic hydroxyl to α-keto groups has been achieved by chemical oxidation with 2,3-dichloro-5,6-dicyano-1,4-benzoquinone (DDQ) (Becker and Adler 1961, Gierer and Norén 1982), permanganate and pyridinium dichromate (Ljunggren and Olsson 1984), and through bio-oxidation with white-rot fungi (Chua et al. 1982) (reaction U_2, Fig. 1.6).

In an alkali-promoted dehydration reaction, β-1 phenolic lignin units may give rise to ring-conjugated ethylenic groups, as indicated by the results of a

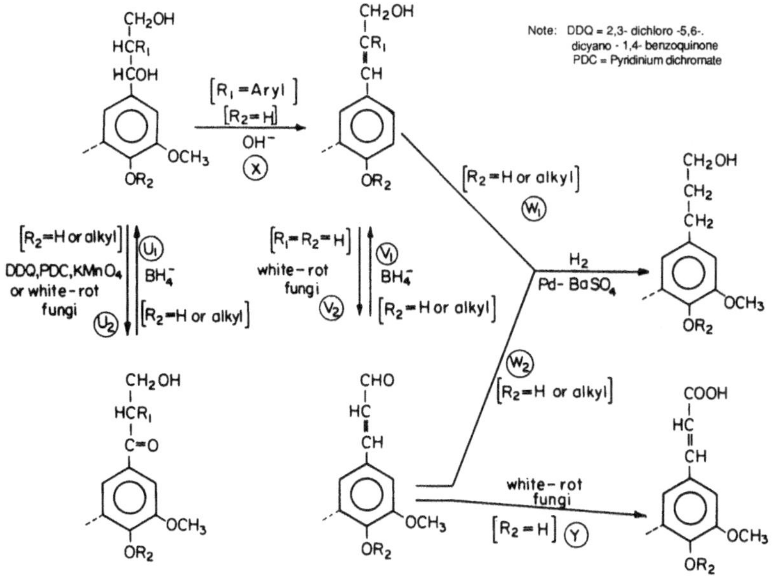

Fig. 1.6. Reactions illustrating formation and elimination of multiple bond side chain functional groups

lignin model compound study (reaction X, Fig. 1.6) (Nonni and Dence 1988). Sequential bio-oxidation mediated by certain white-rot fungi convert cinnamyl alcohol to cinnamic acid derivatives through cinnamaldehyde intermediates (reactions V_2, Y, Fig. 1.6) (Christman and Oglesby 1971).

Reactions that involve the formation and elimination of multiple-bond functional groups may significantly influence the color of residual lignin in bleached and unbleached pulps. Thus, ethylenic and carbonyl groups conjugated with phenolic or quinonoid structures may contribute to color as components of chromophore or leucochromophore systems.

1.5.3 Future Analytical Needs

The above brief survey of the main lignin-modifying reactions is based mostly on our knowledge of the structural changes occurring in treatments of lignin model compounds with various chemicals. These changes may or may not represent transformations occurring in polymeric lignins. Our lack of vital information about lignins themselves, which represents a significant hindrance to the advancement of many industrial technologies requiring a precise understanding of lignin chemistry, may be due to the complexity of the lignin polymers and the inadequacy of analytical techniques available for characterizing solid-state composite materials. Thus, it is reasonable to conclude that there is still an imperative need for analytical methods suitable for the characterization of solid, polymeric lignins when present in wood or in an isolated form. It will continue to be a major challenge for lignin chemists to devise and apply improved methodology to lignin chemistry in the search for solutions to both fundamental and applied lignin-related problems.

References

Adler E, Marton J (1959) Zur Kenntnis der Carbonylgruppen im Lignin I. Acta Chem Scand 13:75–96

Adler E, Marton J (1961) Carbonyl groups in lignin II. Catalytic hydrogenation of model compounds containing aryl carbinol, aryl carbinol ether, ethylene and carbonyl groups. Acta Chem Scand 15:357–369

Adler E (1977) Lignin chemistry – past, present and future. Wood Sci Technol 11:169–218

Allan GG (1971) Modification reactions. In: Sarkanen KV, Ludwig CH (eds) Lignins. Occurrence, formation, structure and reactions. Wiley-Interscience, New York, 511–573

Andersson S-I, Samuelson O (1985) Formation of soluble compounds during pretreatment of kraft pulp with nitrogen dioxide. Sven Papperstidn 88:R102–R104

Bagby MO, Nelson GH, Helman EG, Clark EF (1971) Determination of lignin in non-wood plant fiber sources. Tappi 54:1876–1878

Becker H-D, Adler E (1961) Zur Oxydation mit Chinonen. Acta Chem Scand 15:218–219

Browning BL (1967) Methods of wood chemistry, Vol. II. Interscience, New York, 717–746

Chang H-m, Allan GG (1971) Oxidation. In: Sarkanen KV, Ludwig CH (eds) Lignins. Occurrence, formation, structure and reactions. Wiley-Interscience, New York, 433–485

Chang H-m, Cowling EB, Brown W, Adler E, Miksche G (1975) Comparative studies on cellulolytic enzyme lignin and milled wood lignin of sweetgum and spruce. Holzforschung 29:153–159

Christman RF, Oglesby RT (1971) Microbiological degradation and the formation of humus. In: Sarkanen KV, Ludwig CH (eds) Lignins. Occurrence, formation, structure and reactions. Wiley-Interscience, New York, 769–795

Chua MSG, Chen C-L, Chang H-m, Kirk TK (1982) ^{13}C NMR spectroscopic study of spruce lignin degraded by *Phanerochaete crysosporium* I. New structures. Holzforschung 36: 165–172

Corbett JF (1966) The chemistry of hydroxyquinones. Part I. The reaction of 2-hydroxybenzoquinones with alkaline hydrogen peroxide. J Chem Soc (C):2308–2311

Corbett JF, Fooks AG (1967) The chemistry of hydroxy-quinones. Part III. The reactions of 2,5-dihydroxy-benzoquinones with alkali. J Chem Soc (C):1909–1913

Dence CW (1971) Halogenation and nitration. In: Sarkanen KV, Ludwig CH (eds) Lignins. Occurrence, formation, structure and reactions. Wiley-Interscience, New York, 373–432

Eachus SW, Dence CW (1975) Hydrogenation of lignin model compounds in the presence of a homogeneous catalyst. Holzforschung 29:41–48

Erdtman H, Granath M (1954) Studies on humic acids. V. The reaction of p-benzoquinone with alkali. Acta Chem Scand 8:811–816

Erickson M, Larsson S, Miksche GE (1973a) Gas-chromatographische Analyse von Ligninoxydationsprodukten. VII. Zur Struktur des Lignins der Fichte. Acta Chem Scand 27:903–904

Erickson M, Larsson S, Miksche GE (1973b) Zur Struktur des Lignins des Druckholzes von Pinus mugo. Acta Chem Scand 27:1673–1678

Erickson Ö, Goring DAI, Lindgren BO (1980) Structural studies on the chemical bonds between lignins and carbohydrates in spruce wood. Wood Sci Technol 14:267–279

Fengel D, Wegener G (1984a) Wood: chemistry, ultrastructure, reactions. Walter de Gruyter, New York, 49–53

Fengel D, Wegener G (1984b) Wood: chemistry, ultrastructure reactions. Walter de Gruyter, New York, 157 pp

Fergus BJ, Goring DAI (1970a) The location of guaiacyl and syringyl lignins in birch xylem tissue. Holzforschung 24:113–116

Fergus BJ, Goring DAI (1970b) The distribution of lignin in birch wood as determined by ultraviolet microscopy. Holzforschung 24:118–124

Fergus BJ, Procter AR, Scott JAN, Goring DAI (1969) The distribution of lignin in sprucewood as determined by ultraviolet microscopy. Wood Sci Technol 3:117–138

Gellerstedt G (1976) The reactions of lignin during sulfite pulping. Sven Papperstidn 79:537–543

Gellerstedt G, Pettersson E-L (1975) Light-induced oxidation of lignin. The behavior of structural units containing a ring-conjugated double bond. Acta Chem Scand B29: 1005–1010

Gellerstedt G Pettersson E-L (1977) Light-induced oxidation of lignin. Part 2. The oxidative degradation of aromatic rings. Sven Papperstidn 80:15–21

Gierer J (1970) The reactions of lignin during pulping. A description and comparison of conventional pulping processes. Sven Papperstidn 73:571–596

Gierer J (1980) Chemical aspects of kraft pulping. Wood Sci Technol 14:241–266

Gierer J (1982a) The chemistry of delignification – a general concept. Part 1. Holzforschung 36:43–51

Gierer J (1982b) The chemistry of delignification – a general concept. Part 2. Holzforschung 36:55–64

Gierer J, Imsgard F (1977) The reactions of lignins with oxygen and hydrogen peroxide in alkaline media. Sven Papperstidn 80:510–518

Gierer J, Imsgard F, Pettersson I (1976) Possible condensation and polymerization reactions of lignin fragments during alkaline pulping processes. Appl Polym Symp 28:1195–1211

Gierer J, Lindeberg O (1978) Reactions of lignin during sulfate pulping. Part XV. The behavior of intermediary coniferyl alcohol structures. Acta Chem Scand B32:577–587

Gierer J, Lindeberg O (1980) Reactions of lignin during sulfate pulping. Part XIX. Isolation and identification of new dimers from a spent sulfate liquor. Acta Chem Scand B34:161–170

Gierer J, Norén I (1982) Oxidative pretreatment of pine wood to facilitate delignification during kraft pulping. Holzforschung 36:123–130

Glennie DW (1966) Chemical structure of lignin sulfonates III. Some reactions of monomeric lignin sulfonates. Tappi 49:237–243

Glennie DW (1971) Reactions in sulfite pulping. In: Sarkanen KV, Ludwig CH (eds) Lignins. Occurrence, formation, structure and reactions. Wiley-Interscience, New York, 597–637

Goring DAI (1962) The physical chemistry of lignin. Pure Appl Chem 5:233–254

Goring DAI (1971) Polymer properties of lignin and lignin derivatives. In: Sarkanen KV, Ludwig CH (eds) Lignins. Occurrence, formation,, structure and reactions. Wiley-Interscience, New York, 695–768

Harris EE, Adkins H (1938) Reactions of hydrogen with lignin. Pap Trade J 107(20):38–40

Hatakeyama H, Kubota K, Nakano J (1972) Thermal analysis of lignin by differential scanning calorimetry. Cellul Chem Technol 6:521–529

Kirk TK, Farrell RL (1987) Enzymatic "combustion": the microbial degradation of lignin. Annu Rev Microbiol 41:465–505

Koshijima T, Watanabe T, Yaku F (1989) Structure and properties of the lignin-carbohydrate complex polymer as an amphipathic substance. In: Glasser WG, Sarkanen S (eds) Lignin: properties and material. ACS Symp Ser 397:11–28

Kratzl K, Claus P, Lonsky W, Gratzl JS (1974) Model studies on reactions occurring in oxidations of lignin with molecular oxygen in alkaline media. Wood Sci Technol 8:35–49

Lai Y-Z, Sarkanen KV (1971) Isolation and structural studies. In: Sarkanen KV, Ludwig CH (eds) Lignins. Occurrence, formation, structure and reactions. Wiley-Interscience, New York, 165–189

Lin SY (1990) Lignin. In: Ullmann's Encyclopedia Ind Chem, 5th edn, Vol. 15. VCH, Weinheim, FRG, 305–315

Lin SY, Detroit WJ (1981) Chemical heterogeneity of technical lignins – its significance in lignin utilization. Proc Int Symp Wood Pulp Chem (Stockholm) SPCI Rep 38, 4:44–52

Ljunggren S, Olsson A (1984) The specificity in oxidation of some lignin and carbohydrate models and pine wood shavings with permanganate and pyridinium dichromate before the kraft pulping process. Holzforschung 38:91–99

Lundquist K (1976) Low molecular weight lignin hydrolysis products. Appl Polym Symp 28:1393–1407
Marton J (1971) Reactions in alkaline pulping. In: Sarkanen KV, Ludwig CH (eds) Lignins. Occurrence, formation, structure and reactions. Wiley-Interscience, New York, 639–654
Marton J, Marton T (1964) Molecular weight of kraft lignins. Tappi 47:471–476
Minor JL (1986) Chemical linkage of polysaccharides to residual lignin in loblolly pine kraft pulp. J Wood Chem Technol 6:185–201
Musha Y, Goring DAI (1975) Distribution of syringyl and guaiacyl moieties in hardwoods as indicated by ultraviolet microscopy. Wood Sci Technol 9:45–58
Nimz HH (1974) Beech lignin – proposal of a constitutional scheme. Angew Chem Int Ed 13:313–321
Nonni AJ, Dence CW (1988) The reactions of alkaline hydrogen peroxide with lignin model dimers. Part 3: 1,2-Diaryl-1,3-propanediols. Holzforschung 42:37–46
Norberg PH, Meier H (1966) Physical and chemical properties of the gelatinous layer in tension wood fibres of aspen (Populus tremula L). Holzforschung 20:174–178
Obst JR (1982) Guaiacyl and syringyl lignin composition in hardwood cell components. Holzforschung 36:143–152
Pearl IA (1967a) The chemistry of lignins. Marcel Dekker, New York, 7–37
Pearl IA (1967b) The chemistry of lignins. Marcel Dekker, New York, 250–258
Pettersen RC (1984) The chemical composition of wood. In: Rowell RM (ed) The chemistry of solid wood. Adv Chem Ser 207:57–126
Rydholm SA (1965) Pulping processes. Interscience, New York, 576–578, 590
Saka S, Thomas RJ, Gratzl JS (1978) Lignin distribution determined by energy-dispersive analysis of X rays. Tappi 61(1):73–76
Sarkanen KV, Hergert HL (1971) Classification and distribution. In: Sarkanen KV, Ludwig CH (eds) Lignins. Occurrence, formation, structure and reactions. Wiley-Interscience, New York, 43–94
Schuerch C (1952) The solvent properties of liquids and their relation to the solubility, swelling, isolation and fractionation of lignin. J Am Chem Soc 74:5061–5067
Scott JAN, Procter AR, Fergus BJ, Goring DAI (1969) The application of ultraviolet microscopy to the distribution of lignin in wood. Description and validity of the technique. Wood Sci Technol 3:73–92
Shorygina NN, Kafeli TYa, Samechkina AF (1949) Cleavage of lignin by sodium in liquid ammonia. J Gen Chem USSR 19:1558–1566
Teuber HJ, Heinrich P, Dietrich M (1966) Structure of 3-hydroxy-1,2-benzoquinone dimers. Ann Chem 696:64–71
Wallis AFA (1971) Solvolysis by acids and bases. In: Sarkanen KV, Ludwig CH (eds) Lignins. Occurrence, formation, structure and reactions. Wiley-Interscience, New York, 345–392
Wayman M, Obiaga TI (1974) Molecular weights of milled-wood lignin. Tappi 54(4):123–126
Westermark U (1985) The occurrence of p-hydroxyphenylpropane units in the middle-lamella lignin of spruce (Picea abies). Wood Sci Technol 19:223–232
Westermark U, Lidbrandt O, Eriksson I (1988) Lignin distribution in spruce (Picea abies) determined by mercurization with SEM-EDXA technique. Wood Sci Technol 22:243–250
Whiting P, Goring DAI (1982) Chemical characterization of tissue fractions from the middle lamella and secondary wall of black spruce tracheids. Wood Sci Technol 16:261–267
Wood JR, Goring DAI (1971) The distribution of lignin in stem wood and branch wood of Douglas fir. Pulp Pap Mag Can 72:T95–T102
Yamasaki T, Hosoya S, Chen C-L, Gratzl JS, Chang H-m (1981) Characterization of residual lignin in kraft pulp. Proc Int Symp Wood Pulp Chem (Stockholm) SPCI Rep 38, 2:34–42
Yasuda S, Sakakibara A (1975) Hydrogenolysis of protolignin in compression wood. IV. Isolation of a diphenyl ether and three dimeric compounds with carbon to carbon linkage. Mokuzai Gakkaishi 23:383–387

Chapter 2
Detection and Determination

2.1 The Detection of Lignin

J. NAKANO and G. MESHITSUKA

2.1.1 Introduction

The characteristic color-forming responses of lignified tissue and some lignin preparations on treatment with certain organic and inorganic reagents were recognized as early as the 19th century when Runge (1834) noted that spruce wood turned greenish blue after treatment with phenol and hydrochloric acid. Color reactions have since found widespread use in the detection and identification of lignin and several have also been adapted for use in the quantitative determination of lignin. Recently, color reactions have been proposed for the detection of specific groups within the lignin molecule.

2.1.2 Reagents for the Detection of Lignin

More than 150 color reactions have been proposed for the detection of lignin, most of them by botanists and plant physiologists. Reagents used in these reactions may be classified under the following headings: aliphatic compounds, phenolic compounds, aromatic amines, heterocyclic compounds, and inorganic chemicals (Brauns 1952). Detection of lignin based on coloration produced by dyestuffs is excluded from consideration in covering the subject matter in this section.

2.1.2.1 Aliphatic Compounds

Lignified tissues give color reactions with several alcohols and ketones in the presence of small amounts of mineral acids (Brauns 1952). For example, methanol- or acetone-hydrochloric acid forms a red color on contact with spruce wood; amyl alcohol-sulfuric acid gives a blue color and methylheptenone, $(CH_3)_2C{=}CHCH_2CH_2COCH_3$, a purple-red color when applied to spruce wood.

2.1.2.2 Phenols and Aromatic Amines

A large number of phenols and aromatic amines exhibit color reactions with lignified tissue and certain lignin preparations (Brauns 1952, Brauns and Brauns

Springer Series in Wood Science
Methods in Lignin Chemistry
(Edited by S.Y. Lin and C.W. Dence)

Table 2.1.1. Color reactions of lignin with phenols and amines

Phenol	Color	Amine	Color
Phenol	Bluish green	Aniline	Yellow
o-, m-Cresol	Bright blue	o-Nitroaniline	Yellow
p-Cresol	Olive-green	m, p-Nitroaniline	Orange
o-, m-Nitrophenol	Yellow	o-, m-, p-Chloroaniline	Orange-yellow
p-Nitrophenol	Olive-yellow	Sulfanilic acid	Yellow-orange
Hydroquinone	Olive	o-Phenylenediamine	Orange-brown
Resorcinol	Green-violet	m-Phenylenediamine	Yellow
Phloroglucinol	Violet-red	p-Phenylenediamine	Orange-red
1-Naphthol	Greenish blue	Benzidine	Orange
2-Naphthol	Rose	Quinoline	Yellow

1960). A sampling of the colors produced in the reaction of various phenols and amines with lignin is presented in Table 2.1.1. Phenolic compounds are applied as acidic aqueous solutions and aromatic amines as aqueous solutions of the corresponding hydrochloride or sulfate in the presence of acids. Of the phenols, phloroglucinol is the most important, and its solution in hydrochloric acid is known as the Wiesner reagent.

2.1.2.3 Heterocyclic Compounds

The heterocyclic compounds commonly used in the detection of lignin include furane, pyrrole, and indole derivatives. Furane-type reagents give a green color, pyrroles a red color, and indoles a red shade on reaction with lignified tissue. On the other hand, no color is produced when the oxygen atom in the furane ring is replaced by sulfur as, for example, in thiophene. A comprehensive list of heterocyclic compounds giving rise to colored products on reaction with lignin has been compiled by Brauns (1952).

2.1.2.4 Inorganic Reagents

Of the inorganic reagents used to detect lignin, those constituting the Cross and Bevan and the Mäule detection systems are the most noteworthy because of their value in differentiating woods from gymnosperms and angiosperms. The Mäule test is based on successive treatments of the wood with aqueous potassium permanganate, hydrochloric acid, and ammonia. In hardwoods, an intense purple-red color is generated whereas for softwoods, the color is a brownish shade. The Cross and Bevan test involves successive treatments of the wood with chlorine and aqueous sodium sulfite. The color changes for hardwoods and softwoods are similar to those obtained in the Mäule reaction. The colors formed in the reaction of lignin with several of the most commonly used inorganic reagents are summarized in Table 2.1.2.

Table 2.1.2. Color reactions with inorganic reagents

Reagent	Color
Conc. hydrochloric acid, conc. sulfuric acid	Green
Ferric sulfate-potassium ferrocyanide	Deep blue
Chlorine-sodium sulfite	Yellow-brown, purple-red
Potassium permanganate-hydrochloric acid-ammonia	Yellow-brown, purple-red
Vanadium pentoxide/phosphoric acid	Yellow-brown
Cobalt thiocyanate	Deep blue
Bisulfite/conc. sulfuric acid	Red

2.1.2.5 Unclassified Reagents

In addition to the reactions referred to above, other color reactions, which cannot be classified under the headings previously listed, have found use in the detection of lignin. These include principally the formation of o-quinones (reddish brown) on oxidation of lignin with potassium nitrosodisulfonate (Fremy's salt) (Adler and Lundquist 1961), quinonemethides (yellow) by successive treatments of the lignified tissue with hydrogen bromide in chloroform and sodium bicarbonate, and indophenols (blue) by displacement of the lignin side chain on reaction with quinonemonochlorimide (Gierer 1954, 1956). It is noteworthy that these reactions were investigated in connection with studies aimed at the detection and determination of specific structural units in lignin.

2.1.3 Color-Forming Reaction Sequences

The sequence leading to the formation of color in the reactions of phloroglucinol-hydrochloric acid with lignified tissue (Wiesner reaction) has been clarified by Adler et al. (1948). These investigators implicated coniferaldehyde (1) units in lignin as the groups responsible for color formation based on the similarity of visible absorption spectra of the phloroglucinol/HCl-treated lignin and that of the coniferaldehyde condensation product, (2).

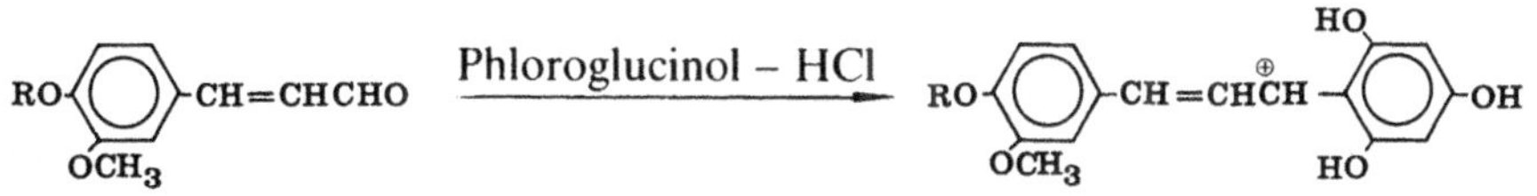

(1) R = H or alkyl (2) $\lambda_{max} \sim 550$ nm

This reaction has been used to determine the amount of cinnamaldehyde units in native lignin (see Table 2.1.1).

Following Adler's identification of coniferaldehyde units as the structural type giving rise to color in the Wiesner reaction, many other color reactions were re-examined and found attributable to the same structural unit. These include the reaction of lignified tissue with methanol-hydrochloric acid (Brauns and Brauns 1960), with concentrated mineral acids and hydrogen sulfide-sulfuric acids (Brauns and Brauns 1960), with indoxyl (Harada and Nikuni 1950a,b), and with barbituric and thiobarbituric acids (Hachihama and Jyodai 1946). The orange color produced by the reaction of lignin with Schiff's reagent (Nakamura and Kitaura 1957) and the reddening of lignin with human urine (Harada and Nikuni 1950b) are also ascribed to coniferaldehyde structures (1). In the latter case, enzymatically liberated indoxyl (3) is believed to react with coniferaldehye-type structures as follows (Brauns and Brauns 1960):

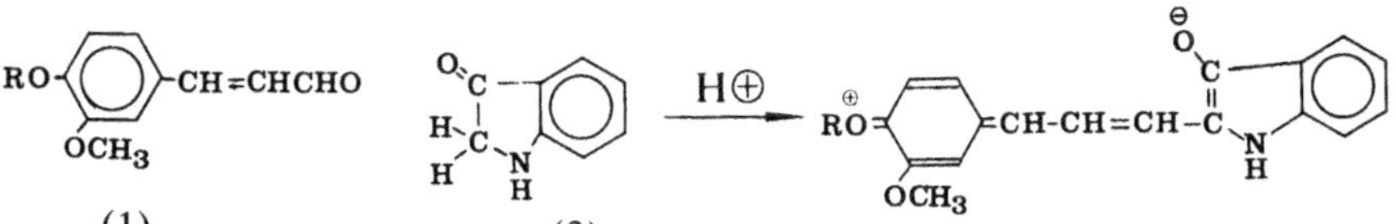

(1) (3)

The color-forming reactions in the Mäule and the Cross and Bevan tests have been assumed to be identical in that the chlorination of lignin was considered an essential step in each case. In the Mäule reaction, potassium permanganate is reduced by the wood to manganese dioxide, which is deposited on the fiber; the latter reacts with hydrochloric acid, forming chlorine which reacts with the lignin. However, a positive color reaction was obtained when a hardwood was treated with sulfuric acid instead of hydrochloric acid following treatment with permanganate. Moreover, the color formed in the Mäule reaction is more stable than that formed in the Cross and Bevan reaction. Collectively, these observations suggest that the essential color-forming step may be somewhat different in each case.

The steps comprising the Mäule reaction may be portrayed as follows (Meshitsuka and Nakano 1978, Iiyama and Pant 1988):

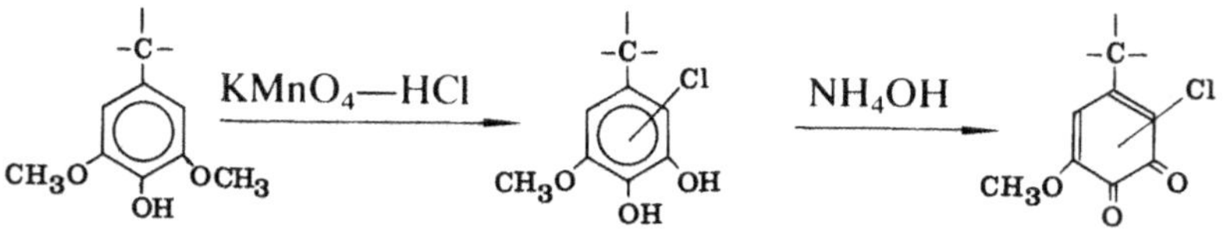

(4) (5) (6)

λ_{max} 515–517 nm

The syringyl nucleus (4) in hardwood lignin is converted to the methoxycatechol structure (5) by treatment with potassium permanganate and hydrochloric acid and thence to a methoxy-o-quinone (6) following reaction with ammonium hydroxide.

The steps in the Cross and Bevan reaction may be depicted as follows (Migita and Nakano 1955):

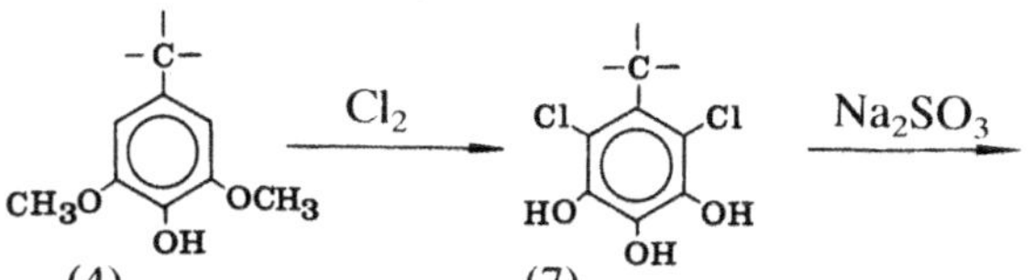

Reddish purple color
$\lambda_{max} \sim 540$ nm

(4) (7)

The 2,6-dichloropyrogallol structure (7), which is produced by chlorination of the syringyl nucleus (3), is thereafter transformed to a colored material of unknown structure on reaction with sodium sulfite.

Lignin also gives color reactions with other inorganic reagents, but most of these are not specific to lignins. For instance, the color formed in the reaction of lignin with vanadium pentoxide in phosphoric acid and with cobalt thiocyanate is due, respectively, to the presence of a phenolic hydroxyl group and the formation of an aromatic ring – Co^{2+} π complex (Nakano et al. 1960).

Several color reactions form the basis for both the detection and estimation of specific structural elements in lignin. For example, coniferyl alcohol-type units (8) ultimately may be transformed to the reddish purple chromophore (10) through the series of reactions outlined below (Lindgren and Mikawa 1957).

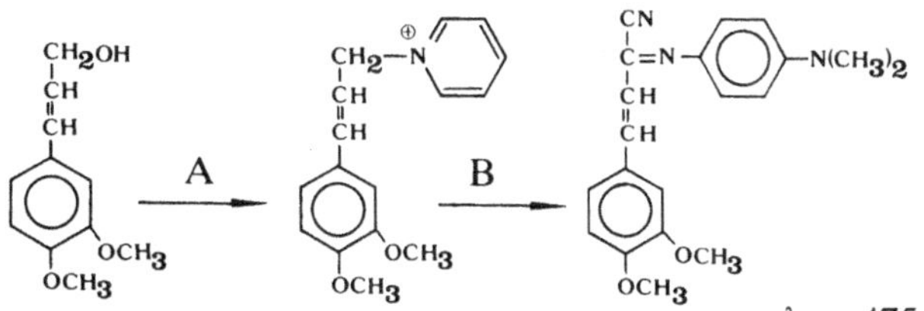

λ_{max} 475 nm

(8) (9) (10)

A: Tosyl chloride in pyridine
B: Nitrosodimethylaniline-potassium cyanide

By means of the procedure developed by Lindgren and Mikawa (1957), the coniferyl alcohol content of lignin in softwood slices was estimated to be 2/100 phenylpropane units.

p-Hydroxybenzyl alcohols (11) and p-hydroxybenzyl alkyl and aryl ethers (12) may be converted to quinonemethides (yellow) through successive treatments with hydrogen bromide in chloroform and pyridine or sodium bicarbonate (Adler and Stenemur 1956):

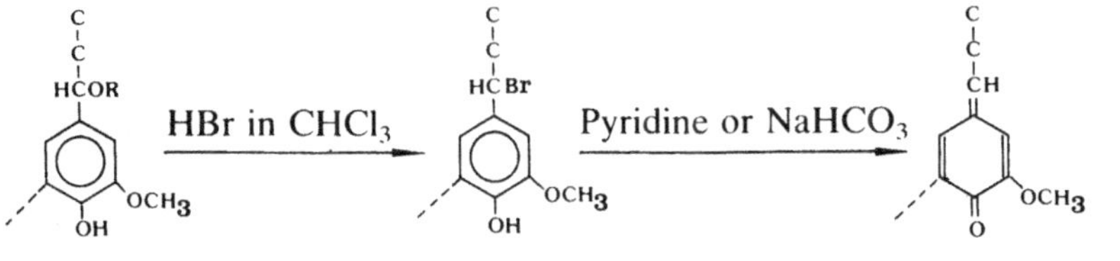

(11) R = H (13) (14) λ_{max} 320–340 nm

(12) R = Alkyl or aryl

The absorption spectra of (14) and wood slices treated according to the above sequence are similar, thereby providing evidence that structures (11) and (12) are present in native lignin. Quinone monochlorimide (15) reacts only with p-hydroxybenzyl alcohols (11), forming indophenol (blue) as shown below (Gierer 1954, 1956).

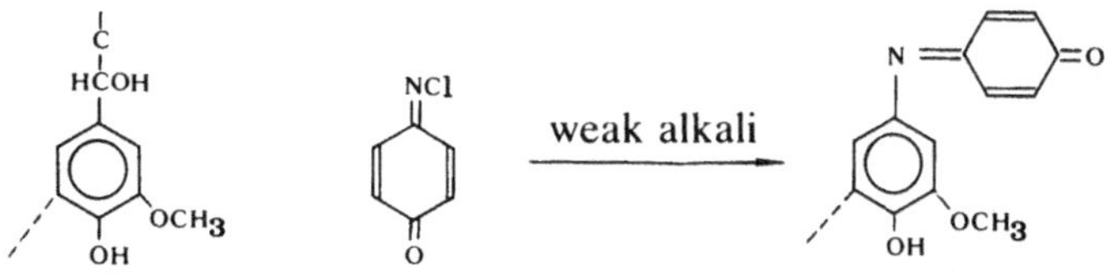

(11) (15) (16) $\lambda_{max} \sim 640\,nm$

Colorimetric analysis based on the above indophenol sequence indicates the presence of 0.12–0.14 and 0.07–0.10 p-hydroxybenzyl alcohol units/C_6–C_3 in spruce Brauns native lignin and spruce milled wood lignin, respectively (Gierer 1954, 1956).

Uncondensed phenolic units in lignin have been estimated by a color reaction based on oxidation with potassium nitrosodisulfonate (Fremy's salt) (Adler and Lundquist 1961) as illustrated below:

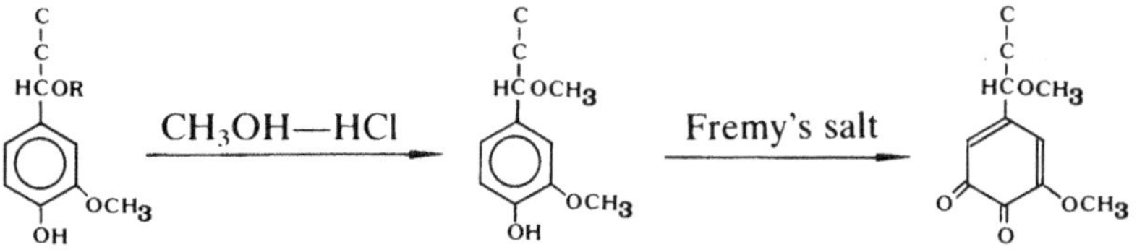

(17) R = H (19) (20)

(18) R = alkyl or aryl

To eliminate interfering side reactions caused by units having carbonyl and hydroxyl groups at the α-position, the lignin is reduced with borohydride and alkylated with methanolic hydrochloric acid prior to oxidation to convert these groups to o-quinones. Following a procedure based on the above reaction, spruce milled-wood lignin was found to contain 15–18% of uncondensed phenolic units (Adler and Lundquist 1961).

2.1.4 Procedures for the Detection of Lignin

The procedures described below comprise a representative sampling of test methods in which use of the principal types of color-forming reagents is illustrated. Prior to performing the color reaction, the lignified material should be extracted in a Soxhlet apparatus with ethanol-benzene (1:2 v/v) and then with water to remove extractives that otherwise might interfere with the test.

The reactions are usually performed by treating the extracted sample in a test tube or on a porcelain spot test plate. When an organic reagent is part of the test, the reaction is conducted with a 1% solution of the reagent in 3 M hydrochloric acid or sulfuric acid at room temperature. The colors formed generally are stable for several hours.

2.1.4.1 Color Reaction with Phloroglucinol-Hydrochloric Acid

The reagent is prepared by mixing 50 ml of a 2% solution of phloroglucinol in 95% ethanol and 25 ml of concentrated hydrochoric acid. Since the reagent is not very stable, the phloroglucinol solution should be stored in the dark in a tightly stoppered brown bottle and mixed with the hydrochloric acid immediately before use. When the reagent is applied dropwise to a wood slice or to wood meal, a reddish violet color appears instantaneously. The test is quite sensitive to native lignin or, in other words, to coniferaldehyde structures in lignin.

When this color reaction is applied to isolated lignin, such as milled wood and dioxane lignins, 4 ml of the phloroglucinol-hydrochloric acid reagent is added to a solution of 0.1 mg of lignin in 1 ml of dioxane. After a short period, the color intensity is measured spectrophotometrically at 550 nm. Using this procedure, the relative amount of chromogenic structures in different lignin samples may be evaluated by comparing color intensities at exactly the same reaction time.

2.1.4.2 The Mäule Color Reaction

The procedure for this test consists basically of three sequential treatments of the lignified tissue with 1% potassium permanganate, 3% hydrochloric acid (Note 1), and concentrated ammonium hydroxide. A small amount of wood meal or wood slices is placed on a watchglass or microscope slide and immersed in a 1% potassium permanganate solution for 5 min at room temperature. The solution is discarded and the sample is washed twice with water and then treated with a 3% hydrochloric acid solution until the color turns from black or dark brown to light brown. If necessary, additional hydrochloric acid may be added to facilitate discharge of the dark color. The hydrochloric acid solution is discarded, and the sample is washed twice with water. On subsequent treatment with concentrated ammonium hydroxide, a red-purple color develops for hardwoods and some tannins, and a brown color for softwoods.

When isolated lignins, such as milled wood lignin and hydrochloric acid lignin, are subjected to the Mäule reaction, the colored solutions may be collected by filtration immediately after the reaction and their absorption spectra measured. Typical visible spectra for milled wood lignin often after undergoing the Mäule reaction are shown in Fig. 2.1.1.

2.1.4.3 The Cross and Bevan Color Reaction

A chlorine solution is prepared by bubbling chlorine gas into ice-cold water. This solution is added to wood meal or wood slices in a small glass-stoppered Erlenmeyer flask. After 10 min, the solution is discarded and the sample is rinsed twice with water. On subsequent treatment with 4% sodium sulfite solution, color appears which is similar to that observed in the Mäule reaction,

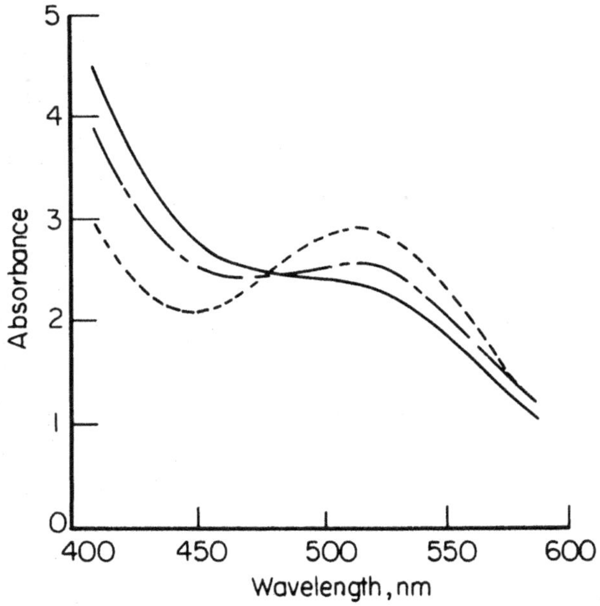

Fig. 2.1.1. Absorption spectra of beech milled wood lignin after Mäule reaction (1% $KMnO_4$ followed by 3% HCl for 1 min (——), 5 min (—·—·-) and 20 min (......)

that is, a purple red color (λ_{max} ~ 510 nm) for hardwoods and a brown or brownish yellow color for softwoods.

2.1.4.4 Color Reaction with Tosyl Chloride/Pyridine, p-Nitrosodimethylaniline/Potassium Cyanide

Freshly prepared wood slices are treated with a solution of tosyl chloride (2 g) in pyridine (20 ml) for 24 h at room temperature. The wood slices are washed successively with benzene, ethanol, dilute hydrochloric acid, and water, and then treated with a solution of p-nitrosodimethylaniline (1.1 g) and potassium cyanide (0.9 g) in 85% aqueous ethanol (20 ml) for 1.5 h. The wood slices turn reddish purple (λ_{max} ~ 475 nm) and, after they are washed successively with ethanol and benzene, the color is recorded while the material is still wet with benzene.

2.1.4.5 Color Reaction with Quinone Monochlorimide

Wood meal (3 mg) is suspended in 2 ml of ethanol, and 0.2 ml of an ethanolic solution of quinone monochlorimide (1 mg) and 1 ml of an 0.04 M sodium hydroxide solution are added. The suspension is shaken for 1 h at room temperature. After removing the solution by filtration, the residual wood is washed with ethanol, and the blue color which forms is recorded while the material is still wet with ethanol.

For an isolated lignin, a 20-mg sample is dissolved in 50 ml of ethanol or methyl cellosolve. To 2 ml of this solution are added the same amounts of quinone monochlorimide and sodium hydroxide as were used in the above procedure. The solution is allowed to stand for 1 h at room temperature. Two ml of the solution is diluted to 25 ml with methyl cellosolve or ethanol and the absorption spectrum (λ_{max} 640 nm) is recorded spectrophotometrically.

2.1.4.6 Color Reaction with Potassium Nitrosodisulfonate

Modification of the lignin: 5 g of the lignin sample (Note 2) is dissolved in 50 ml of 0.1 M sodium hydroxide solution and 3 g sodium borohydride is added. The reaction is continued for several days or preferably for at least 1 week to insure complete reduction of the carbonyl groups. The reaction mixture is carefully acidified with acetic acid and the lignin is recovered by centrifugation. Two g of the reduced lignin is reacted with 50 ml of a methanol-dioxane solution (1 : 1 v/v) of 0.15 M hydorchloric acid for 48 h at 24 °C. The methylated lignin is recovered after evaporation of the solvent.

Preparation of Fremy's salt: a freshly prepared sodium bisulfite solution (100 ml) is added to a mixture of sodium nitrite (345 g) and ice (200 g) in a 1-l beaker. On addition of 20 ml of glacial acetic acid, the reaction mixture turns black. Concentrated ammonium hydroxide (25 ml) is added and the mixture is cooled in an ice bath. A 0.2 M solution (400 ml) of potassium permanganate is added dropwise over 1 h and the black precipitate which forms is immediately removed by filtration. A saturated solution of potassium chloride (250 ml) is added slowly over a 45-min period while the filtrate is gently stirred in an ice water bath. The precipitate that forms is collected by filtration and washed successively with a saturated solution of potassium chloride and methanol, each containing 5% of concentrated ammonium hydroxide and, finally, with acetone. Fremy's salt should be stored in a desiccator over calcium oxide. A small amount of ammonium carbonate in a desiccator is also very effective in protecting Fremy's salt against decomposition.

Treatment of lignin with Fremy's salt: one g of the reduced and methylated lignin is dissolved in 50 ml of a dioxane-water (1:1, v/v) mixture. Fremy's salt (1 g) is dissolved in 100 ml of an ice-cold dioxane-water mixture (1:1, v/v) containing 0.3 g of sodium acetate. This solution (violet color) is added to the lignin solution and the mixture is maintained at 0–5 °C for 0.5 h while being stirred gently. The oxidized lignin is purified by dialysis against water or by gel permeation. The lignin is reddish brown; its visible difference spectrum (measured against original lignin as reference material) shows a broad maximum at approximately 475 nm.

2.1.4.7 Color Reaction Based on the Formation of Nitrosophenols (Pearl-Benson Method)

The procedure for the Pearl-Benson method is described in detail in Chapter 2.2.4. This method has been used primarily to detect and estimate quantitatively lignosulfonates present in low (ppm) concentrations in receiving waters. The principal drawback of the method is its lack of specificity for phenols derived solely from lignin.

Note 1. A dilute sulfuric acid solution (~4%) may be used in place of a hydrochloric acid solution although the color formed for a hardwood is less distinct.

Note 2. The test may also be performed on lignocellulosic materials, for example, groundwood pulp (Singh 1966).

References

Adler E, Björkquist J, Häggroth S (1948) Über die Ursache der Farbreaktionen des Holzes. Acta Chem Scand 2:93–94

Adler E, Lundquist K (1961) Estimation of "uncondensed" phenolic units in spruce lignin. Acta Chem Scand 15:223–224

Adler E, Stenemur B (1956) Lignin model studies – quinonemethides. Chem Ber 89:291–303

Brauns FE (1952) The chemistry of lignin. Academic Press, New York, 24–45

Brauns FE, Brauns DE (1960) The chemistry of lignin: Supplement volume. Academic Press, New York, 33–59

Gierer J (1954) Die Reaktion von Chinomonochlorimid mit Lignin. Acta Chem Scand 8: 1319–1331

Gierer J (1956) The reaction of quinone monochlorimide with lignin. II. Isolation and identification of dyes formed. Chem Ber 80:257–262

Hachihama Y, Jyodai S (1946) The chemistry of lignin. Nippon Hyorinsha, Tokyo, 42 pp

Harada T, Nikuni Z (1950a) Mechanism of the color reaction between lignin and indoxyl. Mem Inst Sci Ind Res Osaka Univ 7:163–164

Harada T, Nikuni Z (1950b) A new bacterium in urine which makes lignin red. VI. Mechanism of the color reaction. Structure of lignin. J Agric Chem Soc Jpn 23:415–421; Chem Abstr 45:10578

Iiyama K, Pant R (1988) The mechanism of the Mäule colour reaction. Introduction of methylated syringyl nuclei into softwood lignin. Wood Sci Technol 22:167–175

Lindgren BO, Mikawa H (1957) The presence of cinnamyl alcohol groups in lignin. Acta Chem Scand 11:826–835

Meshitsuka G, Nakano J (1978) Studies on the mechanism of lignin color reaction. XII. Mäule color reaction. Mokuzai Gakkaishi 24:563–568

Migita N, Nakano J (1955) On the mechanism of color reaction of lignin. IV. Cross-Bevan color reaction of hardwood lignin. J Jpn For Soc 37:26–34

Nakamura T, Kitaura S (1957) Lignin color reactions. Ind Eng Chem 49:1388

Nakano J, Ishizu A, Sasaki A, Migita N (1960) On the mechanism of color reaction in lignin. IX. Color reactions of lignin with inorganic reagents. Mokuzai Gakkaishi 8:144–147

Runge FE (1834) Über einige Produkte der Steinkohlendestillation. J Prakt Chem 1:24–32

Singh RP (1966) Bleaching of groundwood pulp. Investigation of methods based on chemical modification of woodpulp. TAPPI 49:281–286

2.2 The Determination of Lignin

C.W. Dence

2.2.1 Introduction

The determination of lignin is an analysis that is performed routinely for characterizing lignocellulosic material, for assessing the effects of chemical, physical, and biological treatments of wood and pulp, for monitoring effluents in wood processing industries and, in the case of chemical pulps, for estimating bleach chemical requirements. The general approaches followed in this determination, together with their associated procedural variations, have been outlined and thoroughly detailed by several authors (Brauns 1952, Brauns and Brauns 1960, Browning 1967a, Pearl 1967). Therefore, the content of this section is confined mainly to a description and discussion of the preferred and/or widely accepted methods for determining lignin when present in a variety of representative lignified materials.

In considering the various methods proposed for the determination of lignin, it is important to bear in mind that none can be regarded as totally satisfactory. This situation arises from the fact that lignin has yet to be isolated in a pure or unmodified condition. As a result there is no definitive structural formula of lignin currently available upon which to base the development of a rational and quantitative measurement or to assess accurately the validity of the methods currently available. In addition, many of the methods which are commonly applied to the determination of lignin have fundamental defects which seriously undermine their accuracy. For these reasons, values reported in the literature may not always provide a reliable indication of the "true" lignin content of a material.

In organizing the subject matter for this chapter, it was found expedient to divide the presentation into two main subsections, one dealing with the analytical methods applicable to lignin when initially present as a component of a solid mixture and the other focusing on the determination of lignin in solution.

2.2.2 Direct Methods

2.2.2.1 Acid Hydrolysis of Wood and Pulp

A substantial proportion of the methods falling into this category consist of the hydrolysis and solubilization of the carbohydrate component of the lignified material, leaving the lignin as a residue which is determined gravimetrically.

Springer Series in Wood Science
Methods in Lignin Chemistry
(Edited by S.Y. Lin and C.W. Dence)

Interfering extraneous materials are removed prior to the hydrolysis step by extraction with an appropriate solvent or combination of solvents. Hydrolysis of the polysaccharides is catalyzed by a variety of strong mineral acids, applied either individually or in combinations. The range of possibilities has been surveyed by Browning (1967a) and Pearl (1967), who also included detailed descriptions of many of the proposed methods.

The most widely applied variation of the acid hydrolysis approach employs sulfuric acid to promote carbohydrate hydrolysis. The lignin isolated in this treatment is referred to as acid-insoluble lignin, or more commonly as Klason lignin, in recognition of the wood chemist who pioneered the development of the isolation method. In its present modified form the method consists basically of two steps: treatment of the lignified material with cold (10–15 °C) 72% sulfuric acid for a specified time at 20 °C followed by dilution of the acid to 3.0% and boiling to complete the hydrolysis. The procedure described below is essentially the Official Test Method T222 om-83 of TAPPI (Technical Association of the Pulp and Paper Industry 1983) for the determination of lignin in wood and pulp.

2.2.2.2 Determination of Acid-Insoluble (Klason) Lignin in Wood and Pulp

Reagents

Sulfuric acid, 72% H_2SO_4 solution, 24 ± 0.1 N, sp, gr. 1.6338 at 20°/4° is prepared by carefully pouring 665 ml of concentrated H_2SO_4 (95.5 to 96.5%, sp. gr. 1.84) into 300 ml of water and, after cooling, diluting to 1000 ml. The strength is adjusted to 24 ± 0.1 N after titration with standard alkali, or measurement of specific gravity. A variation of 0.1% in the strength of acid at this concentration causes a change of 0.0012 in specific gravity. The acid is cooled to 10–15 °C before use.

Ethanol-benzene mixture is prepared by mixing one volume of approximately 95% ethanol and two volumes of C.P. benzene.

Sample Preparation

The wood specimen is ground to pass a 0.4 mm (40-mesh) screen. A sample of the wood meal equivalent to about 5 g on an oven-dry basis is placed in a coarse porosity alundum or fritted glass extraction thimble and extracted successively in a Soxhlet apparatus with ethanol-benzene for 6–8 h and 95% ethanol for 4 h. Between extractions the wood meal is transferred to a Büchner funnel, where the excess solvent is removed by suction and washing with ethanol.

At the conclusion of the extraction with 95% ethanol, the wood meal is once again transferred to the Büchner funnel and the excess solvent removed by suction and washing with distilled water. The wood is then transferred to a 1000-ml Erlenmeyer flask, 500 ml of boiling distilled water is added and the flask is heated in a boiling water bath for 1 h. After the hot water extraction, the wood meal is filtered on a Büchner funnel, washed with 500 ml of boiling distilled water and allowed to air dry.

The moisture content of the extracted wood is determined by drying a 2-g sample to constant weight in an oven at 105 ± 3 °C. The moisture content of the wood is calculated to the nearest 0.1%.

To prepare a pulp for Klason lignin analysis, a sample equivalent to approximately 10 g oven-dry is disintegrated in a blender or ground to pass a 10- to 20-mesh screen. Pulps which do not contain coarse fibers or shives and which can be dispersed in sulfuric acid readily may be used without prior disintegration.

Groundwood and high-yield pulps containing a significant amount of resins are extracted with ethanol-benzene: a sample equivalent to 10 g of oven-dry pulp is cut into pieces approximately 10 mm square and placed in a coarse porosity alundum or fritted glass extraction thimble and extracted with ethanol-benzene in a Soxhlet apparatus for 4–5 h (minimum of 24 siphoning cycles). The pulp is filtered on a Büchner funnel and washed with ethanol and hot water. The moisture content of an air-dried sample is determined as described above.

Procedure

Duplicate samples of extracted wood meal (1.0 ± 0.1 g, oven-dry basis) (Note 1) or pulp (2.0 ± 0.1 g, oven-dry basis) (Note 2), weighed to the nearest 0.1 mg, are transferred to 100-ml beakers; cold (10–15 °C) 72% sulfuric acid (15 ml for wood and 40 ml for pulp) is added gradually in small increments while stirring and macerating the material with a glass rod. Following dispersion of the wood or pulp in the acid, the mixture is allowed to stand for 2 h in a water bath maintained at 20 ± 1 °C while being stirred frequently. The contents of the beaker are transferred to an Erlenmeyer flask containing 300 to 400 ml of water. Additional water is used to rinse the beaker and to adjust the total volume of the solution to 575 ml for the wood sample and 1540 ml for the pulp sample (H_2SO_4 concentration of 3%). The solution is boiled for 4 h (Note 3) maintaining constant volume by use of a reflux condenser (Note 4) or by frequent addition of hot water. The insoluble residue (lignin) is allowed to settle until the supernatant solution is clear (Note 5). Without disturbing the precipitate, the supernatant solution is decanted or siphoned off through a tared filtering crucible (Note 6). The lignin is then transferred quantitatively to the filter using hot water and a rubber policeman. The lignin is washed free of acid with hot water and dried to constant weight in an oven at 105 ± 3 °C. The lignin content is calculated from the oven-dry weight of the residue and reported as the average of two determinations to the nearest 0.1%.

Note 1. Whiting et al. (1981) and Effland (1977) have modified the standard procedure so as to accommodate situations where only small amounts (1–10 mg and 200–300 mg, resp.) are available for analysis.

Note 2. Groundwood and other high-yield pulps having high lignin contents may be regarded as being the same as wood and the same procedure used for wood samples may be followed.

Note 3. In a modification developed by Yoshihara et al. (1984), this step is replaced by an autoclave treatment at 121 °C for 0.5 h as a means of shortening the analysis, thereby making it more suitable for routine analysis.

Note 4. When the concentration of acid-soluble lignin in the filtrate is to be determined, a reflux condenser should not be used to maintain constant volume during boiling.

Note 5. If acid-soluble lignin is to be determined, a sample of the undiluted supernatant solution or filtrate is removed.

Note 6. Various types of filtering crucibles may be used provided that the filtration is reasonably fast and all of the lignin is retained on the filter resulting in a clear filtrate. Glass filtering crucibles (30-ml capacity) with a sintered glass disc of a fine (F) or medium (M) porosity can be used to filter lignin from wood and most pulp samples. Lignin in low-yield sulfite pulp forms a fine dispersion which often clogs the pores of the sintered glass discs and slows the filtration. A disc of a glass fiber paper or long-fiber asbestos, fitted in the crucible, facilitates the filtration. Alundum or porous porcelain crucibles, with a mat of glass fibers, may also be used. Crucibles should be ignited before weighing if the lignin residue is to be ashed. Glass crucibles cannot be used in such instances.

The precision of the method as applied to wood and pulp is shown by the repeatability and reproducibility values listed in Table 2.2.1. In the present context, repeatability is defined as a limit within which agreement may be expected 95% of the time between two test results obtained under essentially the same conditions and from the same homogeneous sample of material; reproducibility is defined as a limit within which agreement may be expected 95% of the time between two test results obtained in different, qualified laboratories for the same homogeneous sample of material.

Discussion of Method

The acid-insoluble lignin contents of a variety of representative lignified materials are recorded in Table 2.2.2. These values reflect generally recognized trends

Table 2.2.1. Precision of data

Material	*Lignin content* (%)	*Repeatability*	*Reproducibility*
Wood, normal	19–30	0.34	0.79
Sulfate (kraft) pulps	2.6–19.1	0.17	0.45
Sulfite pulps	6.5–28.0	0.48	1.04

Table 2.2.2. Acid-insoluble (Klason) lignin contents of representative lignified materials

Material	Lignin (%)	Reference
Wood		
Softwoods		
Norway spruce normal wood	28.6	Timell 1986
Norway spruce compression wood	38.8	"
Larch sapwood	26.0	"
Larch heartwood	28.6	"
Pine earlywood	28.8	"
Pine latewood	27.4	"
Hardwoods		
White birch normal wood	22.0	Timell 1969
White birch tension wood	16.1	"
Birch fibers	12.6	Obst 1982
Birch ray cells	26.7	"
Nonwood Fiber		
Bagasse	19.6	Bagby et al. 1971
Bamboo	22.2	"
Wheat straw	17.0	"
Kenaf	10.9	"
Sorghum	7.9	"
Pulp		
Pine kraft	4.8 (32.0)[a]	Kyrklund and Strandell 1969
Birch kraft	5.0 (31.4)	"
Spruce kraft	2.8 (13.0)	"
Birch acid sulfite	3.2 (16.0)	"
Spruce bisulfite (Mg base)	9.9 (51.1)	"
Birch bisulfite (Mg base)	4.0 (20.4)	"
Scots pine kraft (>50 mesh)	5.2	Lindström and Glad-
" (<300 mesh)	23.4	Nordmark 1978
Norway spruce TMP (>150 mesh)	23.5	Robinson 1977
Norway spruce TMP (<150 mesh)	35.8	"

[a] Value in parenthesis is kappa number corresponding to lignin content.

based on source (woods compared to annual plants), botanical classification (gymnosperms or softwoods compared to angiosperms or hardwoods), anatomical differences (normal compared to reaction wood, earlywood compared to latewood, sapwood compared to heartwood), and in the case of pulps, the process used to delignify the wood (acid sulfite, bisulfite, kraft), and on fiber classification (fines compared to fiber fractions).

The procedure as described above is applicable to the determination of acid-insoluble lignin in wood and in all grades of unbleached pulps. In semi-bleached pulp the lignin content should not be less than approximately 1% to provide enough lignin (approximately 20 mg) for an accurate weighing. The method is not applicable to bleached chemical pulps where the amounts of lignin are too small to be weighed accurately.

In addition to the foregoing analytical constraints, there is a fundamental consideration (often overlooked) which further restricts the range of applicability of the method. The designation of the residue isolated by the Klason procedure as "lignin" should not negate the need for this material to have the structural features and reactivity generally recognized as being unique to lignin (Kratzl and Billek 1957). Thus, a situation may be envisioned where the lignin has been modified to such a degree, for example in the bleaching of chemical pulp, that its characteristic properties totally or partially disappear and yet is still capable of yielding a residue after being treated with 72% H_2SO_4 according to the Klason procedure.

The factors affecting the determination of acid-insoluble lignin have been discussed in detail by Brauns (1952) and hence will be touched on only briefly in this section. The yield and composition of the lignin isolated in the Klason procedure are affected by the concentration of sulfuric used to promote carbohydrate hydrolysis and by the time and temperature of treatment. At sulfuric acid concentrations below 65%, the carbohydrates are not completely hydrolyzed in a reasonable length of time; at sulfuric acid concentrations above 80% the risk is incurred of the carbohydrates being converted to partially insoluble degradation products through humification reactions (Lai and Sarkanen 1971). Incomplete carbohydrate hydrolysis may also result if the reaction time is too short or the temperature is too low. On the other hand, prolongation of the reaction or too high a temperature may also lead to falsely high lignin contents as a consequence of carbohydrate products condensing with the lignin.

The pre-extraction of wood or pulp is routinely performed to remove extraneous substances which would otherwise interfere with the determination. However, as emphasized by Lai and Sarkanen (1971), any advantage gained through inclusion of a pre-extraction step may be offset as a result of extractants such as hot water and ethanol removing a portion of the lignin. According to Brauns (1952), the hot water extraction may be omitted in the case of chemical pulps and the ethanol-benzene extraction may be disregarded for pulps produced under alkaline conditions.

The filtrate from the second hydrolysis stage in the determination of acid-insoluble lignin may contain material whose properties and behavior closely approximate those of lignin. Such material is referred to as acid-soluble lignin. In some instances the amount of acid-soluble lignin is sufficiently high that it must be independently determined and added to the acid-insoluble lignin content as a correction factor to provide a more accurate value for the total lignin content of the wood or pulp. As stated in TAPPI Official Test Method T222 om-83 (1983), the amount of soluble lignin in softwoods and in sulfate (kraft) pulps is small (0.2 to 0.5%). In hardwoods, nonwood fibers, and in sulfite pulps the acid-soluble lignin content is approximately 3 to 5% and in semi-bleached pulps acid-soluble lignin may amount to as much as one-half or more of the total lignin content. Several authors (Stewart et al. 1951, Swan 1965, Lai and Sarkanen 1971) have reported acid-soluble lignin contents for hardwood species considerably higher than those stated above (3–5%). In one type of approach followed for the determination of acid-soluble lignin, the material is ultimately recovered from the acid hydrolyzate of the acid-insoluble lignin determination and weighed. More commonly, however, ultraviolet spectroscopy has been employed for the determination of acid-soluble lignin and a method of this type is described below. The method consists essentially of measuring the absorbance of the decantate or filtrate from the second stage of the acid-insoluble lignin determination at a wavelength (205 nm) characteristic of lignin and relating this value to lignin concentration through application of Beer's Law.

2.2.2.3 Determination of Acid-Soluble Lignin in Wood and Pulp

Procedure

The following procedure is essentially that described as part of TAPPI Useful Method UM-250 (TAPPI 1985b). A sample of the clear filtrate (or decantate) from the second stage acid hydrolyzate in the acid-insoluble lignin determination is placed in a fused silica absorption cell having a 10-mm light path. The absorbance at 205 nm is measured using 3% H_2SO_4 as a blank or reference solution. If the absorbance is higher than 0.7, the filtrate is diluted in a volumetric flask with 3% H_2SO_4 to obtain an absorbance reading in the range 0.2 to 0.7 and the diluted filtrate is used as a test specimen for absorbance measurement.

The lignin content in the filtrate is calculated from the following expression of Beer's Law:

$$\text{Lignin (g/l)} = \frac{\text{Absorbance (A)}}{\text{b (light path in cm)} \times \text{a (absorptivity in l g}^{-1}\,\text{cm}^{-1})}$$

$$b = 1;\ a = 110\ \text{l g}^{-1}\,\text{cm}^{-1}$$

Hence

$$\text{Lignin (g/l), B} = \frac{A}{110}, \text{ in undiluted filtrate}$$

$$\text{Lignin (g/l), B} = \frac{A}{110}D, \text{ in diluted filtrate}$$

$$\text{where } D = \frac{V_D}{V_O};\ V_D = \text{volume of diluted filtrate and}$$

$$V_O = \text{volume of original filtrate taken}$$

The acid-soluble lignin content in the wood or pulp sample is calculated from the expression

$$\text{Lignin, \%} = \frac{B \times V \times 100}{1000 \times W},$$

where V = total volume of the filtrate, that is, 575 ml for wood and 1540 ml for pulps, W = oven-dry weight of wood or pulp sample, g.

Discussion of Method

The acid hydrolysis of wood and pulp gives rise to carbohydrate degradation products such as furfural and hydroxymethylfurfural which absorb strongly at 280 nm, the wavelength commonly used to monitor lignin in solution. Although interference from these carbohydrate degradation products may be reduced by boiling rather than refluxing the second stage acid hydrolysis mixture, the absorbance of acid-soluble lignin is instead conventionally measured at 205 nm where the risk of interference from the furan aldehydes is considerably less. However, after re-evaluating the acid-soluble lignin determination, Maekawa et al. (1989) concluded that while 205 nm is the preferred wavelength for measuring the absorbance of the Klason lignin filtrates from hardwoods, the most suitable wavelength for softwoods and bamboo is the actual absorption wavelength maximum in the 200 nm region.

As is the case with all determinations of lignin based as UV spectroscopy, the reliability of the absorptivity value, a, used to calculate the lignin content is a critical factor to consider when assessing the reliability of such data. Since the absorptivity value of lignin may vary to a greater or lesser degree depending on the source and previous history of the wood or pulp sample, ideally it would be desirable to determine the absorptivity for each sample being analyzed. As such an undertaking is not practicable, a value reported in the literature for a standard lignin preparation, such as milled wood lignin, is commonly used in calculating lignin content. In the calculation of acid-soluble lignin described above, the absorptivity value 110 $l\,g^{-1}\,cm^{-1}$ is an average of several reported values.

Another factor affecting the reliability of the acid-soluble lignin determination is the partial hydrolysis of ester groups in the lignin leading, in the case of aspen wood, to vanillic, syringic and, especially, p-hydroxybenzoic acids (Pearl and Busche 1960). Although small, some loss of methoxyl as methanol from lignin has been reported (Bland 1956) to result from hydrolysis with sulfuric acid.

2.2.2.4 Acid Hydrolysis of Annual Plants, Developing Wood, and Forage

Methods proposed for the determination of lignin in annual plants, developing wood, and forage are essentially adaptations of the sulfuric acid hydrolysis procedure described above for the determination of acid-insoluble lignin in wood and pulp. These methods, referenced by Bagby et al. (1971), differ mainly in the concentration of the sulfuric acid applied to the material and in the time and temperature of the treatment. The method described below is that developed by Ellis (1949) and adapted as an official method by the Association of Official Agricultural Chemists. It consists of a pre-extraction of the sample with ethanol-benzene followed by successive treatments with a proteinolytic enzyme, dilute sulfuric acid, and finally 72% sulfuric acid. This is the preferred method in cases where the protein content of the material is high (for example in young and developing tissue); for mature plants having a protein content in the order of 2.5% of less, the TAPPI standard method described above is routinely used.

2.2.2.5 Determination of Acid-Insoluble Lignin in Annual Plants, Developing Wood, and Forage

Procedure

The sample (1 g), previously ground to pass a 40-mesh sieve, is extracted with a 95% ethanol-benzene mixture (1:2, v/v) for 4 h in a Soxhlet extractor or comparable apparatus, washed with ethanol and ether, and dried. The sample is transferred to a 250-ml wide-mouth Erlenmeyer flask, pepsin solution (40 ml, 1% solution in 0.1 N hydrochloric acid) is added, taking precautions to insure the residue being uniformly wetted, and the mixture is incubated at 40 °C overnight. Hot water (20–30 ml) is added and the mixture is filtered with the aid of a filter stick (Note 1). The washing is repeated twice; the residue is washed into the flask by forcing 5% sulfuric acid (7–8 ml) downward through the filter stick with air pressure. The stick is washed further with 5% sulfuric acid and finally the volume is brought to 150 ml with the acid. The mixture is refluxed for 1 h, the acid is removed by filtration, and the residue is washed with three portions (30 ml) of hot water, then with two portions (20 ml) of ethanol and finally with two portions (15 ml) of ether. The residue is dried by pulling air through the filter stick and then transferred to the flask by tapping and brushing. Sulfuric acid (20 ml, 72%) at 20 °C is added to the residue and kept at 20 °C for 2 h with occasional stirring. Water (125 ml) is added, the residue is filtered, washed once with hot water (20 ml) and filtered again. The residue is washed from the filter stick and refluxed with dilute sulfuric acid (150 ml, 3%) for 2 h. The residue is allowed to settle and is filtered through a tared Gooch crucible with an asbestos pad and washed with hot water until acid-free. The crucible is dried at 105–110 °C and the lignin is determined by the loss in weight at 600 °C (Note 2).

Note 1. A glass filtering crucible containing a 30-mm diameter medium porosity fritted glass disc may be used in place of the filter stick. In this situation the filtered residue is removed through the careful use of a spatula and subsequent brushing. Any remaining traces of the residue are removed by rinsing the crucible with 5% sulfuric acid. Diatomaceous earth or any other filter aid is pre-ashed before being added to the crucible.

Note 2. When the silica content of the test specimen is sufficiently low to justify omission of the ashing stage, the fritted glass filtering crucible used in the pretreatment may be used to filter the acid-insoluble lignin. In such instances, the weight of any filter aid used must be deducted from the weighed residue to obtain the actual weight of the lignin.

Discussion of Method

Annual plants, developing wood, and forage contain proteinaceous material in amounts varying with their age. Although soluble in sulfuric acid under the conditions of the acid-insoluble lignin determination, the proteins condense with lignin, thereby adding to its weight. To reduce the amount of this interference, the protein is first removed by pretreatment with proteolytic enzymes such as pepsin. However, as pointed out by Lai and Sarkanen (1971), there is no assurance that such a pretreatment effects the complete elimination of protein and, as a result, acid-insoluble lignin contents may require correction based on nitrogen analysis.

Of similar concern in the analysis is the possible error originating in the formation of carbohydrate degradation products, such as furfural and hydroxymethylfurfural, which may also condense with the lignin under acidic conditions and thus yield falsely high results. To reduce this possibility, the sample is pretreated with 5% sulfuric acid as described above to partially remove the hemicelluloses which give rise to the interfering compounds.

The ash content of most woods is low enough (approximately 0.5% or less) to be neglected in the determination of acid-insoluble lignin. On the other hand, the silica content of nonwoody materials is often sufficiently high to require a correction for ash when calculating acid-insoluble lignin contents. In grasses, for example, the siliceous ash content may be as high as 5–10% (Browning 1967b) and in recognition of this fact, the procedure described above specifies that an ash correction be made.

The acid-soluble lignin contents of wheat straw and bamboo have been reported as 2 and 1.5%, respectively (Schöning and Johansson 1965). In such cases, where an accurate value for total lignin content is required, acid-soluble lignin must be determined and the value applied as a correction to the acid-insoluble lignin content.

2.2.3 Indirect Methods

In contrast to the approach utilized in the direct determination of lignin, indirect methods do not involve the isolation of a lignin residue. Instead, the lignin content may be calculated as the difference between 100% and the polysaccharide content of an extractive-free sample or, more commonly, by measuring some characteristic structural functionality, property, or chemical response and relating the result to concentration.

2.2.3.1 Spectrophotometric Methods

Methods falling into this category commonly depend on the capacity of lignin to absorb radiation corresponding to various regions in the electromagnetic spectrum, thereby producing a characteristic response pattern in the form of a spectrum. Under proper conditions the magnitude of the response at a selected frequency is proportional to the concentration of the lignin, thus making the procedure adaptable to quantitative measurement.

The lignin content of wood or pulp has been measured by incorporating a finely ground sample (0.1 to 2.0 mg) of the material in a potassium chloride pellet and measuring the absorbance at 210 or 280 nm (Bolker and Somerville 1962). In a related approach, a UV-microspectrophotometric method, developed by Goring and co-workers, has been applied mainly to determine the distribution of lignin in the various cell wall regions of softwoods (Fergus et al. 1969) and hardwoods (Fergus and Goring 1970). In this procedure the absorbance is measured either by recording the microscope image photographically and subjecting the negative to densitometric analysis directly or by recording the light intensity microphotometrically.

The UV-microspectrophotometric method has been critically reviewed by Boutelje and Jonsson (1980). The analytical uncertainties associated with sample preparation have been addressed by Boutelje and Eriksson (1982), who were able to simplify this phase of the procedure by combining UV-microscopy with interference microscopy. Although UV-microscopy generally has been found suitable for determining the distribution of lignin in the morphological regions of softwood, difficulty has been encountered in attempting to extend the method to hardwoods because of the differences in the individual absorptivities of the guaiacyl and syringyl units comprising hardwood lignin.

Kolboe and Ellefsen (1962) and Michell et al. (1965) provided preliminary results indicating the feasibility of employing infrared spectroscopy to determine the lignin content of finely ground wood and pulp samples embedded in potassium chloride. Further development and refinement of this technique have led to methods for determination of lignin based on multiple internal reflectance infrared spectrometry (Marton and Sparks 1967) and diffuse reflectance Fourier transform spectrometry (Schultz et al. 1985). Lignin contents have also estimated by ^{13}C CP/MAS/NMR spectrometry (Haw et al. 1984, Hemmingson and Newman 1985) (see Chap. 4.5).

The methods listed above have the advantage of being applicable to the determination of lignin in the solid state and, being nondestructive, they obviate the risk of concurrent chemical modification. In addition, these analyses required very small amounts of sample.

A procedure supplementing the UV-microscopic technique outlined above has been developed by Saka et al. (1978). In this procedure the lignin is brominated and the bromine uptake, which is proportional to the lignin content, is determined by a combination of scanning or transmission electron microscopy (SEM or TEM) and energy-dispersive X-ray analysis (EDXA). As in the case of the UV-microscopic technique, the SEM-EDXA method has been employed mainly to determine the distribution of lignin in the various morphological regions of wood and pulp. The procedure and its applications are more fully discussed in Chap. 4.4.

A number of spectral methods for the determination of lignin consist essentially of totally dissolving the sample in a suitable solvent and measuring the UV-absorbance of the solution at a wavelength characteristic for lignin, usually 280 nm. Among the solvents used to dissolve the lignocellulosic material are sulfuric acid (Giertz 1945), phosphoric acid (Bethge et al. 1952), nitric acid (Bartunek 1959, Henriksen and Kesler 1970), sodium chlorite solution (Schadenböck and Prey 1972), cadoxene (cadmium oxide in ethylenediamine) (Sjöström and Enström 1966), and acetyl bromide in acetic acid (Johnson et al. 1961, Iiyama and Wallis 1988a).

Of the above methods, the acetyl bromide method appears to have earned the most widespread acceptance. This recognition can be credited to the following advantages of the procedure: it is rapid and simple, adaptable to small (milligram size) samples, does not require a correction for acid-soluble lignin, provides precise absorbance values from which the lignin content can be calculated, involves minimal chemical modification of the sample, and is accompanied by diminished interference from nonlignin products.

The acetyl bromide procedure has been modified by Marton (1967) to make it suitable for the analysis of unbleached softwood pulp and uncoated paper samples. Morrison (1972a,b) has adapted the procedure for the determination of lignin in grasses and legumes and Bagby et al. (1973) have modified the basic procedure for use in the determination of lignin in fibrous materials derived from kenaf. The method described below is based on that developed originally by Johnson et al. (1961) and modified by Iiyama and Wallis (1988a).

2.2.3.2 Determination of Lignin in Wood and Pulp by the Acetyl Bromide Method

Reagents

The reagents used for the analysis are ultra pure (EP) grade acetyl bromide (Tokyo Chemical Industries), 2 M sodium hydroxide, acetic acid (BDH Analar grade or equivalent) and analytical grade 70% perchloric acid. The acetyl bromide should be stored under nitrogen when not being used (Wu and Wilson 1967).

Procedure

Using a Wiley mill, wood samples are ground and the 40–60 mesh fraction is Soxhlet-extracted with ethanol:benzene (1:2, v/v). Pulp samples are used without size reduction and are solvent exchanged with acetone before drying at 60 °C at 20 mm (Note 1).

The samples (5 mg for woods and 10–15 mg for pulps) (Note 2) are placed in glass reaction bottles (15 ml) with solutions of 25% (w/w) acetyl bromide in acetic acid (5 ml) containing perchloric acid (70%, 0.2 ml) (Note 3). The bottles are sealed with PTPE-coated silicon caps (Note 4) and placed in an oven at 70 °C for 30 min. During the digestion, the bottle is swirled gently at 10-min intervals. After digestion, the solution is transferred to a 100-ml volumetric flask containing 2 M sodium hydroxide (10 ml) (Note 5) and acetic acid (25 ml). The bottle is rinsed, and the solution is made up to 100 ml with acetic acid (Note 6). The UV absorption spectrum is measured against a blank solution which is run in conjunction with the sample (Note 7). The lignin content is measured from the absorbance at 280 mm using the following expression:

$$\text{Lignin \%} = \frac{100(A_s - A_b)V}{aW} - B,$$

where A_s and A_b = absorbance of sample and blank, resp., V = volume of solution, l, a = absorptivity of lignin standard, $l\,g^{-1}\,cm^{-1}$ (Note 8), W = weight of sample, g, B = correction factor for pulp samples only (Marton 1967); B = 1.70 for kraft pulps and 1.38 for sulfite pulps

Note 1. Pulp samples dried directly from the water-wet state require Wiley-milling to insure their dissolution in the acetyl bromide reagent.

Note 2. Samples ranging in size from 2 to 105 mg and containing from 0.6 to 21.3 mg of lignin have been successfully examined (Johnson et al. 1961). According to these authors, sample size is not necessarily restricted to these limits and hence, in theory, the method should be applicable to materials containing even smaller amounts of lignin. In practice, however, sample size may affect the analytical result (see Discussion section below).

Note 3. The addition of perchloric acid accelerates dissolution of the material and permits coarser samples to be analyzed.

Note 4. In this step care should be taken to avoid moisture contamination of the sample (van Zyl 1978).

Note 5. The amount of 2 M sodium hydroxide added is double that specified in the original method (Johnson et al. 1961). This procedural modification effectively removes the interfering 267 nm peak present in the spectrum of the digested sample and eliminates the need to add hydroxylamine for this purpose as was the case in the original procedure.

Note 6. According to Johnson et al. (1961), the time lapse between the transfer step and final dilution with acetic acid must be standardized and kept short (5 min or less), since loss of absorptivity results from extended standing in the water-rich mixture.

Note 7. Using freshly prepared acetyl bromide solution made up in analytical grade acetic acid, the absorbances at 280 nm of the digested samples were constant after standing at 20 °C for periods up to 30 min. This is contrary to the findings of van Zyl (1978) and Cho (1984), who observed that when freshly prepared reagents were used, the absorbances were not stable over this period.

Note 8. An absorptivity value of $20.09\,l\,g^{-1}\,cm^{-1}$ for both softwood and hardwood samples was used by Iiyama and Wallis (1988a) in their modification of the Johnson et al. (1961) procedure.

Discussion of Method

The sample size specified in the various modifications of the acetyl bromide procedure is not uniform. Ideally, the sample size is dictated by the need to obtain a solution which, after final dilution, has an absorbance reading between 0.3 and 0.7. Alternatively, the final volume may be adjusted so as to provide an absorbance reading falling within this range. However, as reported by Wu and Wilson (1967), sample size, in combination with dilution and absorbance, has an important bearing on the analytical result. For example, in the determination of lignin in Douglas-fir wood, statistical treatment of the data revealed that more than half of the 0.5% variation in these replicate determinations was attributable to the sample weight alone or in combination with the other two variables.

The use of a proper absorptivity value is of paramount importance in arriving at an accurate figure for the lignin content. Although a number of such values have been reported for various reference materials (Table 2.2.3), the preferred approach is to determine the actual absorptivity at 280 nm of the lignin in the material under investigation by measuring the absorbance of the lignin isolated from the source material, for example, milled wood lignin or enzyme

Table 2.2.3. Selected absorptivity values for representative reference materials

Material	Absorptivity at 280 nm, $l\,g^{-1}\,cm^{-1}$	Reference
Spruce milled wood lignin (MWL)	21.9	Hardell et al. 1980
Pine MWL	21.65[a]	Marton 1967
Spruce MWL	19.8[b]	Sjöström and Enström 1966
Spruce "native lignin"	20.9[a]	Johnson et al. 1961
Spruce "native lignin"	18.8	Hess 1952
Spruce dioxane lignin	19.8[b]	Sjöström and Enström 1966
Pine dioxane lignin	17.8[b]	"
Birch dioxane lignin	13.1[b]	"
Aspen dioxane lignin	12.2[c]	Bolker and Somerville 1962
Spruee enzyme lignin	19.1[d]	Johnson et al. 1961
Douglas-fir Klason lignin	25.0[e]	Bolker and Somerville 1962
Kenaf lignin	34.9[a,f]	Bagby et al. 1973
Nine softwoods	23.3[a,f,g] (av.)	Johnson et al. 1961
Nine hardwoods	23.6[a,f,g] (av.)	" "
Softwood and hardwoods	20.0	Iiyama and Wallis 1988a
Pine kraft lignin	25.0[a]	Marton 1967
Spruce kraft lignin	18.7	McNaughton et al. 1967
Hardwood kraft lignin	26.1[a]	Marton 1967
Purified spruce sodium lignosulfonate	14.7	Yean and Goring 1965
Birch sodium lignosulfonate	11.0	Sjöström and Enström 1966
Spruce sodium lignosulfonate	12.7	Lorås and Loschbrandt 1956
Spruce sodium lignosulfonate	120[h]	" "
Birch acid-soluble lignin	14.5	Swan 1965

[a] Measured in acetyl bromide.
[b] Measured in cadoxene.
[c] Measured in dioxane.
[d] Corrected for carbohydrates.
[e] Measured as suspension in methyl cellosolve.
[f] Calculation based on Klason lignin content of material.
[g] Variation from average, ± 4.5%.
[h] Absorptivity at 205 nm.

lignin or of the lignin in the whole (lignocellulose) material after its lignin content has been determined by some standard procedure, such as, Klason lignin analysis. The absorptivity is then calculated from the absorbance and the lignin concentration by application of the Beer-Lambert relationship.

The approach followed by Iiyama and Wallis (1988a,b) was to measure the absorbance of several isolated softwood and hardwood lignins after reaction with acetyl bromide according to their modification of the original Johnson et al. (1961) procedure and to calculate the corresponding absorptivity values. The values obtained for P. *radiata* and E. *regnans* milled wood lignins, steam-exploded hardwood lignin, and Indulin-AT ranged from 18.0 to 20.9 $l\,g^{-1}\,cm^{-1}$. On the other hand, absorptivity values for eucalypt and mixed hardwood kraft lignins varied widely.

The lignin contents of P. *radiata* and E. *regnans* woods and pulps determined by the standard Klason method and by the conventional and modified acetyl bromide methods are listed in Table 2.2.4. The lignin contents determined using the modified procedure generally agree better with those obtained by the Klason method than do results obtained using the conventional acetyl bromide method.

Table 2.2.4. Lignin contents of P. *radiata* and *Eucalyptus* woods and pulps obtained by various methods

Sample	Standard method			Acetyl bromide method	
	Klason lignin %	Acid-soluble lignin %	Total lignin %	Conventional[a] (Johnson et al. 1961) %	Modified[b] (Iiyama and Wallis 1988b) %
P. *radiata* WM[c]	28.1	0.5	28.6	26.3	29.0
E. *regnans* WM[c]	23.0	4.8	27.8	24.1	27.5
P. *radiata* RMP[d]	28.1	0.4	28.5	25.6	28.3
Mixed eucalypt CSP[e]	22.4	5.3	27.7	24.1	27.6
E. *regnans* EP[f]	33.4	2.4	35.8	–[j]	37.6
E. *regnans* AEEP[g]	15.9	1.4	17.3	–[j]	17.9
P. *radiata* BP[h]	1.7	3.5	5.2	4.6	5.4
P. *radiata* KP1[i]	9.1	0.4	9.5	10.0[k]	8.9
P. *radiata* KP2[i]	5.0	0.4	5.4	6.4[k]	5.4
E. *regnans* KP1[i]	8.0	1.8	9.8	6.1[k]	10.7
E. *regnans* KP2[i]	3.3	1.1	4.4	4.1[k]	5.6

[a] Absorptivity 23.3 $l\,g^{-1}\,cm^{-1}$ (softwood), 23.6 $l\,g^{-1}\,cm^{-1}$ (hardwood).
[b] Absorptivity 20.0 $l\,g^{-1}\,cm^{-1}$ for both softwood and hardwood.
[c] Wood meal.
[d] Refiner mechanical pulp.
[e] Cold soda pulp.
[f] Exploded pulp.
[g] Alkali-extracted exploded pulp.
[h] Bisulfite pulp.
[i] Kraft pulp.
[j] Sample not dissolved after 50 min.
[k] Digestion time 50–55 min.

The acetyl bromide method has been applied less frequently to pulp than to wood. It has, however, been successfully applied by Marton (1967) to the determination of lignin in softwood kraft and bisulfite pulps with kappa numbers in the 25–120 range but to hardwood pulps with "inconclusive" results. Using their modified acetyl bromide procedures, Iiyama and Wallis (1988a,b) determined the lignin contents of several pine and eucalypt bisulfite and kraft pulps and compared the results with those obtained using the Klason method. The results for the pine kraft and bisulfite pulps agreed reasonably well while the lignin contents of eucalypt kraft pulps were 10–20% higher than those obtained by the Klason method (Table 2.2.4). Using a modification of the original Johnson et al. (1961) procedure, Bagby et al. (1973) determined the lignin content of kraft pulp made from kenaf fiber. Both Marton and Bagby et al. report relative standard deviations for replicate determinations of approximately ±2%.

Marton (1967) found that, with the exception of bleached pine kraft pulp, bleached pulps dissolved incompletely in the acetyl bromide/acetic acid solution. Iiyama and Wallis (1988a) noted (Table 2.2.4) that several unbleached eucalypt pulps dissolved incompletely or only after prolonging the digestion time beyond the standard 30 min.

In those instances where the sample has undergone extensive chemical reaction such as, for example, oxidation during bleaching, the spectral absorption characteristics of the lignin may have been altered to such a degree as to make measurement of absorbance at 280 nm meaningless. Under these circumstances neither the acetyl bromide method nor any other spectrophotometric method represents a viable analytical approach.

2.2.3.3 Methods Based on Oxidant Consumption

These methods, which are used exclusively in the analysis of unbleached pulps, were developed to provide a simple and rapid means of estimating residual lignin content for quality control purposes in commercial pulping processes. The procedures are all based on the common principle that the lignin will consume the applied oxidant at a much higher rate than the carbohydrate component of the pulp; hence oxidant consumption under carefully specified conditions can be regarded as a measure of lignin concentration in the pulp. Lignin concentration determined by such procedures is usually expressed as the amount of oxidant consumed per unit weight of pulp (e.g., Roe chlorine number or kappa number). These numbers may be converted to acid-insoluble (Klason) lignin or other lignin values by application of appropriate, empirically determined conversion factors (Table 2.2.5).

The two oxidants most commonly used in these procedures are chlorine and potassium permanganate. Chlorine reacts rapidly and extensively with lignin in unbleached pulp at room temperature. In such situations chlorine consumption not only provides a method for estimating lignin content but is also an indicator of the amount of delignification in the pulping process, of the total bleach requirement, and especially of the bleach requirement in the chlorination stage of a conventional multi-stage bleach sequence.

Table 2.2.5. Relationships for the interconversion of lignin content and kappa, permanganate, and chlorine (hypo) numbers

Relationship	Reference
Kraft pulps	
%Klason lignin = kappa number × 0.15	Kyrklund and Strandell 1969, Tasman 1959
Sulfite pulps	
%Klason lignin = kappa number × 0.165	Lorås and Loschbrandt 1961
%Total lignin = kappa number × 0.252	Kyrklund and Palenius 1964
%Klason lignin = kappa number × 0.187	Kyrklund and Strandell 1969
%Total lignin = chlorine (hypo) number × 0.90	Kyrklund and Strandell 1969
Kraft pulps	
%Klason lignin = Roe chlorine number × 0.84	Brauns 1952
Sulfite and bisulfite pulps	
Kappa number = chlorine number × 4.82	Kyrklund and Strandell 1969
Kraft and polysulfide pulps	
Kappa number = chlorine number × 5.91	Kyrklund and Strandell 1969
Kappa number = chlorine number × 5.0	Tasman 1959
Kraft pulps	
Permanganate number = Roe chlorine number × 2.87	Brauns 1952
Chlorine (hypo) number = Roe chlorine number × 1.052	Kyrklund and Strandell 1967
Softwood pulps	
Log kappa number = 0.958 + 0.0253 permanganate number	Hatton 1975
Hardwood pulps	
Log kappa number = 0.727 + 0.0421 permanganate number	Hatton 1975

The Roe chlorine number determination, which consists essentially of measuring the uptake of gaseous chlorine by a known weight of unbleached pulp in a specially constructed apparatus, was modified by Johansson (1935) and later adopted by the Technical Association of the Pulp and Paper Industry (TAPPI) as Standard Method T202 ts-66. This method has in turn been superseded by the simpler hypo number determination which was developed by Colombo et al. (1962) and McLean (1965) and eventually adopted by TAPPI as revised Official Test Method T253 om-86. In this procedure the pulp is reacted with acidified sodium or calcium hypochlorite under rigidly controlled conditions and the chlorine consumption is measured titrimetrically. In essence, the hypo number is the modified Tingle number described by Trout et al. (1953) and the modified chlorine number detailed by Kyrklund and Strandell (1967, 1969). In contrast to the standard permanganate number test described below, the hypo number determination is applicable to all types of pulp irrespective of species, pulping process, yield, or degree of delignification. The chlorine number, i.e., hypo number, is a linear function of the Roe number, the ratio of chlorine number to Roe number being 1.052.

In a related approach, Ni et al. (1990) have developed a procedure for determining lignin in softwood kraft pulp based on the conversion of methoxyl groups in the lignin to methanol when the pulp is chlorinated under a specified set of conditions. Pending further development, this method, termed "Methanol number", is viewed by the authors as a possible alternative to the Kappa number test.

As is the case for both the chlorine and hypo number determinations, the permanganate number test is based on the principle that the oxidant, in this case potassium permanganate, readily oxidizes lignin, whereas the carbohydrate constituents are relatively unreactive. Several adaptations of the basic procedure have been made, but all have the common feature of being based on the addition of an excess of 0.1 N $KMnO_4$ to an aqueous suspension of the pulp sample and measurement of the amount consumed by titration of the residual permanganate at the end of a prescribed reaction time. The permanganate number is conventionally reported as the number of milliliters of 0.1 N $KMnO_4$ consumed by 1 g of oven-dry pulp under the test conditions. As do the chlorine and hypo numbers, the permanganate number affords a measure of the lignin remaining in unbleached and semi-bleached pulps, thus providing an estimate of the bleach demand of the pulp in subsequent bleaching stages.

The permanganate number is affected by the size of the pulp sample and the amount of permanganate applied. Thus, for pulps having both low and high lignin contents, discontinuities in the permanganate-lignin relationship are introduced at each point where either sample weight or permanganate volume is changed. This problem has been resolved by modifying the procedure according to the suggestions made by Tasman and Berzins (1957a,b) which include adjustment of the sample size to ensure that approximately half of the applied permanganate is consumed and correcting the volume of permanganate actually consumed by the sample so that it corresponds to exactly 50% consumption of the applied amount. This corrected procedure has been termed "kappa number" to distinguish it from other permanganate determinations. The kappa number determination has been adopted as a standard procedure by the technical pulp and paper organizations of several countries and specifically by TAPPI as Historical Method T236 hm-85 in the form detailed below.

2.2.3.4 Determination of the Kappa Number of Pulp

Reagents

Potassium permanganate solution, standardized 0.1000 ± 0.0005 N $KMnO_4$; sodium thiosulfate solution, approximately 0.2 N $Na_2S_2O_3$, normality known with an accuracy of ±0.0005 N; 1.0 N potassium iodide solution; 4 N sulfuric acid; 0.2% starch indicator solution.

Sample Preparation

Air-dried sheets are torn into small pieces to provide a total weight of 3–4 g. Screened slush sheets are mixed and 3–4 g (dry weight) are made into a pad on a Büchner funnel, taking care to avoid any loss of fibers. The pad is air-dried and torn into small pieces. An unscreened pulp sample should be screened prior to the test to remove shives and knots if the pulp is normally screened before bleaching and other processing.

Procedure

Prior to weighing, the test samples are conditioned for at least 20 min in the atmosphere near the balance. A pulp sample which will consume approximately 50% of the potassium permanganate solution is weighed out to the nearest 0.001 g (Note 1). The consumption must be between 30 and 70%. At the same time a second sample is weighed and its moisture content determined.

The test sample is disintegrated in 500 ml or less of distilled water until free of fiber clots and undispersed fiber bundles. Methods of disintegration which involve extensive cutting of the fibers should be avoided. The disintegrated test specimen is transferred to a 2000-ml beaker using sufficient distilled water (25.0 ± 0.2 °C) for rinsing to bring the total volume to 795 ml. The beaker is placed in a constant temperature bath to maintain the temperature at 25.0 ± 0.2 °C (Note 2) during the entire reaction. The suspension is stirred continuously (Note 3) so as to produce a vortex about 25-mm deep but not fast enough to introduce air into the mixture.

Potassium permanganate (100.0 ± 0.1 ml) and 100 ml of the sulfuric acid solution are pipeted into a 250-ml beaker. The mixture is brought to 25 °C quickly and added immediately to the pulp suspension while simultaneously starting a stopwatch. The beaker is rinsed out using not more than 5 ml of distilled water and the washings are added to the reaction mixture. The total volume should be 1000 ± 5 ml. At the end of exactly 10 min, the reaction is terminated by adding 20 ml of the potassium iodide solution. Immediately after mixing, but without filtering out the fibers, the free iodine is titrated (Note 4) with the sodium thiosulfate solution, adding a few drops of the starch indicator toward the end of the reaction. A blank is carried out using the same method as above but omitting the pulp.

The kappa number is calculated as follows:

$$K = \frac{p \times f}{W} \quad \text{and} \quad p = \frac{(b - a)N}{0.1},$$

where K = kappa number, f = factor for correction to a 50% permanganate consumption, depending on the volume of p (Table 2.2.6), W = weight of

Table 2.2.6. Factors (f) to correct for different percentages of permanganate used

P	+ 0	1	2	3	4	5	6	7	8	9
30	0.958	0.960	0.962	0.964	0.966	0.968	0.970	0.973	0.975	0.977
40	0.979	0.981	0.983	0.985	0.987	0.989	0.991	0.994	0.996	0.998
50	1.000	1.002	1.004	1.006	1.009	1.011	1.013	1.015	1.017	1.019
60	1.022	1.024	1.026	1.028	1.030	1.033	1.035	1.037	1.039	1.042
70	1.044									

moisture-free pulp in the specimen, g, p = amount of 0.1 N permanganate actually consumed by the sample, ml, b = amount of thiosulfate consumed in the blank determination, ml, a = amount of thiosulfate consumed by the sample, ml, N = normality of the thiosulfate.

Kappa numbers under 100 are reported to the nearest 0.1 and those over 100 are reported to the nearest whole number.

Note 1. For pulps having ash contents higher than 2–3% (as is, for example, the case with rice straw pulps), the weight of the pulp sample should be increased accordingly (Srinivasa Rao et al. 1983).

Note 2. When the reaction is performed at a temperature of between 20 and 30 °C other than 25 °C, the kappa number is corrected using the following relationship:

$$K = pf[1 + 0.0013(25 - t)]$$

where t = actual reaction temperature in °C and K, p, and f are as identified above

Note 3. A propeller-type stirrer made of glass or other noncorrosive material may be used; a plastic- or glass-covered stirring bar used in conjunction with a magnetic stirrer is also satisfactory.

Note 4. The titration should be performed as rapidly as possible to avoid volatilization of the liberated iodine. In this connection, Germer and Maiorova (1972) have recommended that only half of the prescribed amount of permanganate be added to the blank, thereby reducing the amount of liberated iodine and the risk of its volatilizing.

The repeatability of the measurement is high for semi-bleached pulps but decreases with increasing kappa numbers and becomes fairly constant for kappa numbers above 20. With semi-bleached pulps with kappa numbers below 20, the method will have a minimum repeatability of 0.2 kappa number. Repeatability decreases systematically from 13% for kappa numbers below 2 to 2.3% for kappa numbers between 10 and 20. The average repeatability is 3.8% for kappa numbers between 2.5 and 20. With unbleached pulps having kappa numbers above 20, the repeatability varies unsystematically between 0.9 and 1.5%. The average repeatability is 1.2% for kappa numbers between 20 and 190.

Discussion of Method

The kappa number determination is applicable to all types and grades of semichemical, unbleached, and semi-bleached pulps obtained in yields under 60% and for pulps in yields of up to 70% if the pulp has been well screened prior to the test. Typical values for representative pulps are listed in Table 2.2.7. In

Table 2.2.7. Typical kappa numbers for representative pulps

Pulp	Kappa no. range
Kraft (bleachable grade)	
Softwood	25–35
Hardwood	14–18
Kraft softwood liner	
Primary	80–95
Secondary	40–50
Neutral sulfite semi-chemical (softwood)	80–100
Bisulfite (softwood)	30–50
Acid sulfite	
Softwood	16–22
Hardwood	14–20
Kraft (chlorinated and alkali-extracted[a])	
Softwood	5–8
Hardwood	3–6
Acid sulfite (chlorinated and alkali-extracted[a])	
Softwood	3–5
Hardwood	2–4

[a] When oxygen is added the kappa number values are reduced by an additional 1 to 3 units.

cases where the history of the pulp includes an oxidative treatment, the kappa number determination may not reflect the "true" lignin content. For example, when a pulp consumes oxidant without accompanying loss of lignin, then consumption of permanganate in a kappa number determination will perforce be less because the "oxidant demand" of the lignin has been partially satisfied in the prior oxidative treatment.

Jayme and Jerratsch (1960) have modified the basic kappa number determination mainly through the addition of two permanganate preoxidation steps to make the method suitable for semi-chemical pulping having yields in excess of 70%. Another variation of the standard kappa number determination is the so-called micro-kappa number developed by Berzins (1966). This procedure is applicable to very small samples and to semi-bleached pulps with low permanganate consumption (lignin contents approximately 0.15%). This method has been adopted by TAPPI and designated as Useful Method UM-246 (1985a).

Berzins (1965) has further modified the standard kappa number determination by reducing the reaction time from 10 to 5 min, thus providing a procedure that is useful for routine quality control where rapid test results are required. The values obtained using this method may be converted to standard kappa numbers through application of correction factors provided by the author. The rapid kappa number test has been adopted by TAPPI as a useful method, UM 245 (1985b). Other variations of the standard kappa number procedure have been outlined by Browning (1967a). More recently, a near-infrared spectroscopic method has been developed for estimating kappa number in unbleached pulps (Birkett and Gambino 1989).

2.2.4 Determination of Lignin in Solution

The dissolved lignin sources whose lignin contents are most frequently sought include pulping and bleaching liquors and ground, surface, and sea waters containing discharges from pulp and paper mills and other wood-processing industries.

One analytical strategy that has been employed in situations such as these is simply to evaporate the sample to dryness and determine the lignin content of the residue by one of the several methods available for the determination of lignin in solid materials. This approach obviously is not acceptable for the on-line monitoring of lignin concentrations during pulping and bleaching and, for this and other reasons, dissolved lignin is more commonly determined by direct analysis of the solution. The methods used to determine lignin in solution have been reviewed by Pearl (1967), Browning (1967a), and Goldschmid (1971) and may be classified according to whether the procedure is based on derivative formation, light absorbance, or modification of the lignin by a specific chemical reaction.

Derivative formation is based on the reaction of lignosulfonates with organic bases to produce a precipitate which is determined gravimetrically or turbidimetrically. A number of organic bases used to precipitate lignosulfonates have been listed by Browning (1967a) together with a detailed procedure for performing the analysis. Since the lignosulfonate is usually not precipitated quantitatively, a correction factor is applied in calculating the weight of lignosulfonic acid. The method has been applied to the determination of lignin when present in concentrations of upwards of 50 ppm or <1 ppm using trypaflavine or "surfen", [(1,3-bis(4-amino-2-methyl-6-quinolyl) urea], respectively, as precipitant.

UV absorption spectrophotometry has been employed to determine the concentration of lignin in sulfite (Patterson et al. 1951, Joyce and Kleinert 1957) and kraft (Kleinert and Joyce 1957a, Alén and Hartus 1988) pulping liquors and after discharge into receiving waters. The methodology for such measurements has been described previously in connection with the determination of acid-soluble lignin. As was stressed on that occasion, the selection of an appropriate wavelength at which to measure absorbance and an absorptivity value for calculating lignin content are important considerations in such an analysis. The effects of interfering substances (SO_2 and carbohydrate degradation products) on the absorbance measurements at 205 and 280 nm have been discussed by Browning (1967a), Pearl (1967), and Goldschmid (1971). Representative absorptivity values are listed in Table 2.2.3. The lower concentration limits for the measurement of lignin in sulfite and kraft pulping liquors at 205 nm are in the 5–10 ppm range (Kleinert and Joyce 1957a,b). In situations where lignin concentration is very low, for example in natural waters, impurities interfering with absorbance measurements at 280 nm may require removal by solvent extraction (Bilikova 1973, Jurkiewicz 1977).

Another spectroscopic method for the determination of lignin in solution is that utilizing fluorescence spectrophotometry. This technique is reportedly capable of detecting lignin in concentrations two to three orders of magnitude less than are required for absorption spectroscopy (Bublitz and Meng 1978). This greater sensitivity requires that pulping effluents be diluted 10^3 to 10^4 times in order to reduce lignin concentration to a level where it is linearly related to fluorescence intensity. Fluorescence spectroscopy has been examined specifically an a means of monitoring delignification in sulfite (Bublitz 1981) and kraft (Bublitz 1981, Paul et al. 1985) pulping effluents, and presumably is applicable to the determination of lignin present as a contaminant of natural waters provided the background fluorescence of nonlignin materials is sufficiently low. Where it is not, the interfering materials may possibly be removed by ether extraction prior to performing the fluorimetric analysis (Bogomolov et al. 1986).

Measurement of total organic carbon (TOC) in pulping liquors has been related to lignin concentration and hence provides a means of monitoring delignification (Genco et al. 1984). In this approach a liquor sample is combusted to CO_2, which is measured gravimetrically after adsorption onto Dryerite and Ascarite/magnesium perchlorate beds or by infrared spectrometry.

Of the methods for determining lignin in solution based on a specific chemical reaction, that involving nitrosation, the so-called Pearl-Benson method, has found the widest application. In this procedure, reaction of the phenolic units in lignin with acidified sodium nitrite leads to the formation of a nitrosophenol which, upon addition of alkali, is tautomerized to an intensely colored quinone mono-oxime. The absorbance of the latter structure is measured at 430 nm and related to lignin concentration by calibration with a standard lignin. The procedure described below is essentially that developed by Barnes et al. (1963), who modified the original Pearl-Benson method (Pearl and Benson 1940) to improve its sensitivity.

2.2.4.1 Determination of Dissolved Lignin by the Modified Pearl-Benson (Nitrosation) Method

Reagents

10% Acetic Acid. Pipet 100 ml of reagent grade acetic acid into a 1000-ml volumetric flask and dilute to the mark with distilled water.

10% Sodium Nitrite. Weigh 10 g of reagent grade $NaNO_2$ to the nearest 0.02 g, transfer to a 100-ml volumetric flask, and add distilled water to bring final solution to volume. If the solution contains any suspended particles or sediment, filter through a coarse sintered glass filter. The clear solution is stored in a glass-stoppered bottle.

2N Ammonium Hydroxide. Dilute 135 ml of concentrated ammonium hydroxide (28% as NH_3) to 1000 ml in a volumetric flask. This solution is stored in a Pyrex or borosilicate glass bottle and should be used only as long as it remains free from sediment when mixed with the sample of dissolved lignin in the procedure used to test reagent purity (Note 1).

Sample Preparation

The water sample is allowed to stand for 12 to 24 h at room temperature (20–25 °C) after which the clear supernate is pipeted or decanted off. Samples which remain turbid after standing (i.e., those samples which show absorbance greater than 0.10 for a 10-cm path at 600 nm) should be filtered through a type AA Millipore filter (Millipore Filter Co., Bedford, MA).

Procedure

A 50-ml (Note 2) sample of water (settled or filtered through type AA Millipore filter) is placed in a 125-ml glass-stoppered Erlenmeyer flask and 1 ml each of the acetic acid and sodium nitrite solutions are added from a buret or pipet. The solution is mixed by swirling and allowed to stand for 15 min at room temperature. Two ml of the ammonium hydroxide solution is added from a buret or pipet and mixed, and the solution is allowed to stand for 10 min at room temperature. The absorbance (A_T) is read at 430 nm in a cell with a 10-cm light path against a distilled water reference.

The blank absorbance (A_B) in determined in the same way except that the reagents are added to the water sample in the order: acetic acid, ammonium hydroxide, sodium nitrite. Colored nitro compounds are not formed in the basic solution and thus the blank absorbance is due to the reagents and the substances originally present in the water. The net absorbance (A_S) is calculated as the difference between A_T and A_B.

Lignin concentration is calculated from the Beer-Lambert equation:

$$C = \frac{A_S}{ab},$$

Where a = absorptivity, $cm^{-1} ppm^{-1}$; A_S = net absorbance, C = concentration of lignin (10% solids basis), ppm, and b = length of light path, cm. The absorptivity is determined either on the same liquor as that present in the water sample or preferably from "standard" solids recovered from the liquor known or suspected of being the source of the lignin in the water sample.

In the recommended procedure for determining absorptivity, the "standard" solids are weighed out and then dissolved in fresh or sea water (Note 3) to make 1000 ml of a stock standard solution containing 1000 ppm of the liquor (10% solids basis). Ten ml of the stock solution is then pipeted into a 1000-ml volumetric flask, diluted to the mark with fresh or salt water to provide a secondary solution containing 100 ppm of the liquor (10% solids basis). Suitably sized aliquots of this secondary stock solution are further diluted to provide solutions containing liquor concentrations in the 5–50 ppm range (10% solids basis). These solutions are nitrosated according to the standard procedure and the absorbance at 430 nm measured. From the slope of the line (m) obtained by plotting absorbance against the concentration of the liquor in ppm, the average absorptivity is calculated from the equation:

$a = m/b$, $cm^{-1} ppm^{-1}$,
where b = length of light path in cm.

Absorptivities determined in this manner for calcium lignosulfonates are $a = 0.000247\,cm^{-1}\,ppm^{-1}$ (fresh water) and $a = 0.000242\,cm^{-1}\,ppm^{-1}$ (sea water) (Barnes et al. 1963).

Note 1. The reagents are tested for purity by using distilled water instead of the water sample and determining A_T, A_B and A_S by the above procedure with both normal volume of reagents and twice normal volume. If A_S (twice normal volumes) − A_S (normal volumes) is greater than 0.004 (10-cm light path), new reagent solutions should be prepared.

Note 2. The amounts of water sample and reagents may be increased proportionately if required.

Note 3. When using "standard" lignins such as black liquor solids which are insoluble in water, the material may be dissolved in the smallest amount of glacial acetic acid necessary to maintain the solid in solution after dilution with fresh or salt water.

The results of the analysis of several water samples indicate a standard deviation in the absorbance (A_S) corresponding to ±3–6 ppm spent sulfite liquor at 10% total solids content (Felicetta and McCarthy 1963).

Discussion of Method

Although developed originally and used principally for the determination of lignosulfonates in sea water, the applicability of the Pearl-Benson method has been widened to include the determination of not only lignosulfonates but of other lignins as well in fresh water. In its present form the sensitivity of the method is such that, in the absence of any extraneous absorbance, lignosulfonate concentrations as low as 0.2 to 0.5 ppm can be determined (Goldschmid 1971).

The principal drawback of the method is its lack of specificity for phenols having the characteristic lignin structure. Thus, other phenolic impurities in water (e.g., tannins) would most likely be nitrosated under the conditions specified in the procedure and thereby contribute the absorbance reading. In addition, certain nitrogen-containing and inorganic substances commonly found in fresh- and sea water are also known to react with nitrous acid (Felicetta and McCarthy (1963). However, in comparison to other colorimetric and to UV spectrophotometric procedures, the Pearl-Benson method has been found to be less affected by interfering impurities (Jurkiewicz 1977).

As previously emphasized in the discussion of quantitative methods employing UV spectroscopy, the accuracy of the results is strongly dependent on the selection or determination of an absorptivity value that reflects the spectral characteristics of the lignin actually being measured.

References

Alén R, Hartus T (1988) UV spectrophotometric determination of lignin from alkaline pulping liquors. Cellul Chem Technol 22:613–618

Bagby MO, Cunningham RL, Maloney RL (1973) Ultraviolet spectral determination of lignin. Tappi 56(4):162–163

Bagby MO, Nelson GH, Helman EG, Clark TF (1971) Determination of lignin in non-wood plant fiber sources. Tappi 54:1876–1878

Barnes CA, Collias EE, Felicetta VF, Goldschmid O, Hrutfiord BF, Livingston A, McCarthy JL, Toombs GL, Waldichuk M, Westley R (1963) A standardized Pearl-Benson, or nitroso, method recommended for the examination of spent sulfite liquor or sulfite waste liquor concentration in waters. Tappi 46:347–351

Bartunek R (1959) A nonhydrolytic spectrophotometric method for determining lignin in high-purity pulps. Tappi 42:553–556

Berzins V (1965) A rapid procedure for the determination of kappa number. Tappi 48:15–20

Berzins V (1966) Micro-kappa numbers. Pulp Pap Mag Can 67:T206–T208

Bethge PO, Gran G, Ohlsson K-E (1952) Determination of lignin in chemical wood pulp. I. Principles and methods. Sven Papperstidn 55:44–48

Bilikova A (1973) Verification of analytical methods for the determination of lignin in water. Vodni Hospod 23(2):44; Abstr Bull Inst Pap Chem 48:1245

Birkett MD, Gambino MJT (1989) Estimation of pulp kappa number with near-infrared spectroscopy. Tappi J 72(9):193–197

Bland DE (1956) Spectrophotometric determination of lignin in eucalypt soda pulp. Appita 10:287–295

Bogomolov BD, Pivovarova VA, Gorbunova OF (1986) Determination of lignin and ether-soluble substances in aqueous solutions by fluorimetry. Khim Drev (Riga) 2:54–57

Bolker HI, Somerville NG (1962) Ultraviolet spectroscopic studies of lignins in the solid state. I. Isolated lignin preparations. Tappi 45:826–829

Boutelje JB, Eriksson I (1982) A UV-microscopy study of lignin in middle lamella fragments from fibers of mechanical pulp of spruce. Sven Papperstidn 85:R39–R42

Boutelje JB, Jonsson U (1980) Ultraviolet microscope photometry of pulp fibers. UV-absorbance and its relationship to chlorine number, kapp number and lignin content. Cellul Chem Technol 14:53–67

Brauns FE (1952) The chemistry of lignin. Academic Press, New York, 151–177

Brauns FE, Brauns DA (1960) The chemistry of lignin: supplement volume. Academic Press, New York, 127–167

Browning BL (1967a) Methods of wood chemistry Vol. II. Interscience, New York, 785–823

Browning BL (1967b) Methods of wood chemistry Vol. I. Wiley-Interscience, New York, 14

Bublitz WJ (1981) Fluorescence of pulping liquors: a tool for digester control? Tappi 64(6):73–76

Bublitz WJ, Meng TY (1978) The fluorometric behavior of pulping waste liquors: a valuable tool for lignin and pulping research. Tappi 61(2):27–30

Cho NS (1984) Spectrophotometric determination of lignin in woody samples. J Tappik 16(1): 3–9

Colombo P, Corbetta D, Pirotta A, Ruffini G (1962) Chlorine number as a method for evaluation of lignin content of a pulp. Pulp Pap Mag Can 63:T126–T140

Effland MJ (1977) Modified procedure to determine acid-insoluble lignin in wood and pulp. Tappi 60(10):143–144

Ellis GH (1949) Report on lignin and cellulose in plants. J Assoc Off Agric Chem 32:287–291

Felicetta VF, McCarthy JL (1963) Spent sulfite liquor. X. The Pearl-Benson, or nitroso, method for the estimation of spent sulfite liquor concentration in waters. Tappi 46:337–347

Fergus BJ, Goring DAI (1970) The distribution of lignin in birchwood as determined by ultraviolet microscopy. Holzforschung 24:118–124

Fergus BJ, Procter AR, Scott JAN, Goring DAI (1969) The distribution of lignin in sprucewood as determined by ultraviolet microscopy. Wood Sci Technol 3:117–138

Genco JM, Hassler JC, Busayasakul N (1984) Total organic carbon as an indicator of wood delignification. Tappi 67(7):92–95
Germer ZI, Maiorova ID (1972) [Proposed] changes in the method of determining the degree of delignification of pulp. Bumazh Prom 8:23–24; Abstr Bull Inst Pap Chem 43: 8400
Giertz HW (1945) Bestamning av restlignin. En kolorimetrisk method for knotroll av blekningsforloppet (Determination of residual lignin. A colorimetric method for control of bleaching). Sven Papperstidn 48:485–489
Goldschmid O (1971) Ultraviolet spectra. In: Sarkanen KV, Ludwig CH (eds) Lignins. Occurrence, formation, structure and reactions. Wiley-Interscience, New York, 160–263
Hardell H-L, Leary GJ, Stoll M, Westermark U (1980) Variation in lignin structure in defined morphological parts of spruce. Sven Papperstidn 83:44–49
Hatton JV (1975) Kappa number-permanganate number relationships for softwoods and hardwoods. Tappi 58(10):150–151
Haw JF, Maciel GE, Schroeder HA (1984) Carbon-13 nuclear magnetic resonance spectrometric study of wood and wood pulping with cross polarization and magic-angle spinning. Anal Chem 56:1323–1329
Hemmingson JA, Newman RH (1985) A CP/MAS ^{13}C NMR study of the effect of steam explosion processes on wood composition and structure. J Wood Chem Technol 5: 159–188
Henriksen A, Kesler RB (1970) The Nu-number, a measure of lignin in pulp. Tappi 53: 1131–1140
Hess CL (1952) An investigation of the homogeneity of isolated native black spruce lignin. Tappi 35:312–320
Iiyama K, Wallis AFA (1988a) An improved acetyl bromide procedure for determining lignin in woods and wood pulps. Wood Sci Technol 22:271–280
Iiyama K, Wallis AFB (1988b) Spectrophotometric determination of lignin in polyphenol-containing eucalyptus woods by the acetyl bromide method. Appita 41:442–446
Jayme G, Jerratsch H-J (1960) Die K_J-Methode, eine modifizierte Kappa-Zahl-Bestimmung für Halbzellstoffe in über 70% Ausbeute (The K_J-method, a modified kappa-number determination for semichemical pulps of more than 70% yield). Papier 14:718–723
Johansson D (1935) Testing methods for evaluation of unbleached chemical pulp. Pap Trade J 101(13):101–104
Johnson DB, Moore WE, Zank LC (1961) The spectrophotometric determination of lignin in small wood samples. Tappi 44:793–798
Joyce CS, Kleinert TN (1957) Short wavelength ultraviolet absorption of various lignin and related substances. II. Lignin determination in sulfite pulping liquors. Pulp Pap Mag Can 58(6):131–134
Jurkiewicz S (1977) Problem of the effect of interfering factors in the quantitative determination of lignin compounds in aqueous solutions. Przegl Papier 33(1):30–35; Abstr Bull Inst Pap Chem 48:1245
Kleinert TN, Joyce CS (1957a) Short wavelength absorption of various lignins and related substances. IV. Lignin determination in sulfate pulping liquors. Pulp Pap Mag Can 58(11):147–152
Kleinert TN, Joyce CS (1957b) Short wavelength absorption of various lignins and related substances. III. The flow-micro cuvette for continuous measurement of lignin concentration in sulfite cooking liquors. Pulp Pap Mag Can 58(7):215–219
Kolboe S, Ellefsen O (1962) Infrared investigations of lignin. A discussion of some recent results. Tappi 45:163–166
Kratzl K, Billek G (1957) Synthesis and testing of lignin precursors. Tappi 40:269–285
Kyrklund B, Palenius I (1964) The total lignin content of sulphite pulps and its relation to the kappa number. Pap Puu 46:513–520
Kyrklund B, Strandell G (1967) A modified chlorine number for evaluation of cooking degree of high-yield pulps. Pap Puu 49:99–106
Kyrklund B, Strandell G (1969) Applicability of the chlorine number for evaluation of the lignin content of pulp. Pap Puu 51:299–305

Lai Y-Z, Sarkanen KV (1971) Isolation and structural studies. In: Sarkanen KV, Ludwig CH (eds) Lignins. Occurrence, formation, structure and reactions. Wiley-Interscience, New York, 190–195

Leithe W (1973) The analysis of organic pollutants in water and waste water. Ann Arbor Science, Ann Arbor Michigan, 191–193

Lindström T, Glad-Nordmark G (1978) Chemical characterization of the fines from unbleached kraft pulp. Sven Papperstidn 81:489–492

Lorås V, Loschbrandt F (1956) Gravimetric and absorptiometric determination of lignin in pulp and wood. Nor Skogind 10:1–12

Lorås V, Loschbrandt F (1961) Determination of lignin in sulfite pulps. Nor Skogind 15:302–309

Maekawa F, Ichizawa T, Koshijima T (1989) An evaluation of the acid-soluble lignin determination in analysis of lignin by the sulfuric acid method. J Wood Chem Technol 9:549–567

Marton J (1967) Determination of lignin in small pulp and paper samples using the acetyl bromide method. Tappi 50:335–337

Marton J, Sparks HE (1967) Determination of lignin in pulp and paper by infrared multiple internal reflectance. Tappi 50:363–368

McLean JD (1965) Hypo number (chlorine number using sodium hypochlorite solution). Pulp Pap Mag Can 66 T103–T106

McNaughton JG, Yean WQ, Goring DAI (1967) Macromolecular properties of kraft lignins from spruce made soluble by a continuous flow process. Tappi 50:548–553

Michell AJ, Watson AJ, Higgins HG (1965) An infrared spectroscopic study of delignification of Eucalyptus regnans. Tappi 48:520–532

Morrison IM (1972a) A semimicro method for the determination of lignin and its use in predicting the digestibility of forage crops. J Sci Food Agric 23:455–463

Morrison IM (1972b) Improvements in the acetyl bromide technique to determine lignin and digestibility and its application to legumes. J Sci Food Agric 23:1463–1469

Ni Y, Kubes GJ, Van Heiningen ARP (1990) Methanol number: a fast method to determine the lignin content of pulp. J Pulp Pap Sci 16:J83–J86

Obst JR (1982) Guaiacyl and syringyl lignin composition in hardwood cell components. Holzforschung 36:143–152

Patterson RF, Keays JL, Hart JS, Strapp RK, Luner P (1951) The spectrophotometric determination of lignin in sulfite cooking liquor. Pulp Pap Mag Can 52(12):105–111

Paul SK, Oswal N, Maheswari S (1985) Monitoring the progress of pulping by fluorometric analysis of spent sulfate liquor. Ind Pulp Pap 40:11–16

Pearl IA (1967) The chemistry of lignin. Marcel Dekker, New York, 32–58

Pearl IA, Benson HK (1940) The determination of lignin in sulfite pulping liquors. Pap Trade J 111(18):35–36

Pearl IA, Busche LR (1960) Studies on the chemistry of aspenwood. XI. The Klason lignin determination as applied to aspenwood with special reference to acid-soluble lignin. Tappi 43:961–970

Robinson DH (1977) A comprehensive study of the bleaching of mechanical pulps with hydrogen peroxide and sodium dithionite. MS thesis SUNY College of Environmental Science and Forestry, Syracuse, NY, 98 pp

Saka S, Thomas RJ, Gratzl JS (1978) Lignin distribution. Determination by energy-dispersive analysis of X-rays. Tappi 61(1):73–76

Schadenböck W, Prey V (1972) Eine neue quantitative Lignin-Bestimmung mit Hilfe der UV-Spektrophotometrie (A new quantitative lignin determination by ultraviolet spectrophotometry). Papier 26:117–118

Schöning AG, Johansson G (1965) Absorptiometric determination of acid-soluble lignin in semichemical bisulfite pulps and in some woods and plants. Sven Papperstidn 68:607–613

Schultz TP, Templeteon MC, McGinnis GD (1985) Rapid determination of lignocellulose by diffuse reflectance Fourier transform infrared spectrometry. Anal Chem 57:2867–2869

Sjöström E, Enström B (1966) Spectrophotometric determination of the residual lignin in pulp after dissolution in cadoxene. Sven Papperstidn 69:469–476

Srinivasa Rao Y, Shukla RN, Maheshwari S, Kulkarni AY, Jivendra Y (1983) Determination of permanganate and kappa number of pulp with high ash content. IPPTA 20(3):55–56
Stewart CM, Foster OH, Cohen WE, Leslie RT, Watson AJ (1951) The selection and preparation of a reporducible wood sample for chemical studies. Appita 5:267–311
Swan B (1965) Isolation of acid-soluble lignin from the Klason lignin determination. Sven Papperstidn 68:791–795
Tasman JE, Berzins V (1957a) The permanganate consumption of pulp materials. I. Development of a basic procedure. Tappi 40:691–694
Tasman JE, Berzins V (1957b) The permanganate consumption of pulp materials. II. The KAPPA number. Tappi 40:695–699
Tasman JE (1959) The permanganate consumption of pulp materials. IV. The kappa correction coefficient. Pulp Pap Mog Can 60(No. C):231–235
Technical Association of the Pulp and Paper Industry, Atlanta, (1983) Acid-insoluble lignin in wood and pulp. Official Test Method T-222 om-8
Technical Association of the Pulp and Paper Industry, Atlanta, (1985a) Micro kappa number. Useful Method UM 246
Technical Association of the Pulp and Paper Industry, Atlanta, (1985b) Acid-soluble lignin in wood and pulp. Useful Method UM 250
Timell TE (1969) The chemical composition of tension wood. Sven Papperstidn 72:173–181
Timell TE (1986) Compression wood in gymosperms, Vol. 1. Springer, Berlin Heidelberg New York Tokyo, 294, 295, 299, 311
Trout PE, Franckowiak PA, Ham JL, Browning BL, Bublitz LO (1953) Routine determination of lignin in semichemical pulps by a modified Tingle procedure. Tappi 36:123–126
van Zyl JD (1978) Notes on the spectrophotometric determination of lignin in wood samples. Wood Sci Technol 12:251–259
Whiting P, Favis BD, St. Germain FGT, Goring DAI (1981) Fractional separation of middle lamella and secondary wall tissue from spruce wood. J Wood Chem Technol 1:29–42
Wu Y-T, Wilson JW (1967) Lignification within coniferous growth zones. Pulp Pap Mag Can 68:T159–T164, T-171
Yean WQ, Goring DAI (1965) The molecular weight of lignosulfonates from morphologically different subdivisions of wood structure. Sven Papperstidn 68:787–790
Yoshihara K, Kobayashi T, Fujii T, Akamatsu I (1984) A novel modification of Klason lignin quantitative method. J Jpn Tappi 38(4):86–95

Chapter 3
Isolation and Purification

3.1 Wood

K. Lundquist

3.1.1 Introduction

Several different methods have been used for isolation of lignin from wood. Only three of them, however, have been found to give a chemically essentially unchanged lignin. Extraction with organic solvents was suggested by Brauns (1939, 1952) and Brauns and Brauns (1960), and lignin samples obtained in this way are here referred to as Brauns lignins. Björkman introduced the use of a preliminary, extensive grinding, followed by solvent extraction, to give what is here referred to as milled wood lignin (MWL) (Björkman 1956, 1957, Björkman and Person 1957). Later, it was found that the amount of solubilized lignin could be increased if the finely ground wood meal was treated with hydrolytic enzymes for removal of associated polysaccharides prior to solvent extraction (Pew 1957, Pew and Weyna 1962, Chang et al. 1975). Such preparations are here referred to as cellulolytic enzyme lignins (CEL). A critical review of methods for the isolation of lignins has been published by Lai and Sarkanen (1971). The subject is also discussed in the monographs by Brauns (1952), Brauns and Brauns (1960), and Fengel and Wegener (1984).

A procedure for the isolation and purification of MWL is given below; it follows largely the standard procedure given by Björkman (1956), but the purification steps are modified according to Lundquist et al. (1977).

3.1.2 Preferred Method for Isolation and Purification of Lignin

In the selection of starting material, one should be aware that there may be a variation in lignin composition depending on the part of a tree from which the wood is taken. Therefore, sapwood free of reaction wood should be used for the preparation of a reference lignin from a particular species. Size reduction and removal of extractives from the wood can be accomplished according to the directions given by Björkman (1956).

3.1.2.1 Isolation

Grinding

The wood sample (dried in vacuo over P_2O_5) is suspended in toluene (purified by distilling over sodium) and milled in a modified National Bureau of Standards

Springer Series in Wood Science
Methods in Lignin Chemistry
(Edited by S.Y. Lin and C.W. Dence)

mill for 48 h according to Björkman (1956). Each milling jar is charged with 5–6 g of wood. The toluene should cover the milling balls (to about 80% of the volume of the vessel). The temperature in the milling jars should be below 35 °C. This can be accomplished by regulating the air temperature in combination with the use of fans. The milling in a Lampén mill (Björkman 1956) is omitted (Björkman 1957, Björkman and Person 1957).

Extraction
The toluene is carefully removed by centrifugation and the residue is extracted (occasional stirring) with dioxane-water (96:4, v/v) (5–10 ml/g of milled wood) for 24 h (reagent grade dioxane purified by distilling over sodium should be used). The extract is centrifuged off and replaced with a fresh dioxane-water mixture. Removal of extract and addition of fresh dioxane-water mixture is repeated the following day. After 2 days the extract is centrifuged off and combined with the two other extracts. Solvents are removed by film evaporation.

3.1.2.2 Purification

For simplicity, the purification of lignin from 1 g of the product obtained in the preceding step is described. The sample is dissolved in 28 ml of pyridine-acetic acid-water (9:1:4, v/v/v) and the solution is extracted with 36 ml of chloroform. The mixture is centrifuged (an efficient centrifuge is required, particularly when hardwood lignins are purified) until the organic layer is completely clear. The organic layer is removed by a syringe to avoid contamination with the aqueous layer. The aqueous layer from 64 ml of a mixture of pyridine-acetic acid-water-chloroform (9:1:4:18, v/v/v/v) is added to the organic extract. Chloroform (36 ml) is added and the mixture is shaken. The organic layer is separated using the technique described above, and solvents are removed by film evaporation (temperature <35 °C). If complete removal of pyridine is desired, this can be achieved by repeated addition and removal (film evaporation) of ethanol (Adler et al. 1987). The residue is dissolved in 10–20 ml of 1,2-dichloroethane-ethanol (2:1, v/v) and the solution is added dropwise to magnetically stirred ether (250 ml). The precipitated lignin is centrifuged off, washed twice with ether, and dried in vacuo over P_2O_5 and KOH.

3.1.3 Discussion

3.1.3.1 Modifications of and Comments on the Preferred Method

There is a risk for solvolytic reactions during the removal of the extractives. Comparatively mild treatments such as Soxhlet extraction with ether (or dichloromethane) and acetone, complemented with leaching at room temperature with water, should be considered when it is particularly important to avoid such reactions. An alternative mode of milling has been described by Brownell (1965). This entails the use of a porcelain ball mill. While this procedure requires a longer grinding time, the fact that much larger amounts of wood can be used is a distinct advantage. This type of ball mill is also less expensive and more readily available than the mill originally designed by Björkman (1956).

The purification procedure can be modified by omitting the last addition of chloroform (Lundquist and Kirk 1980). This results in a somewhat less efficient removal of carbohydrates but separation of the organic layer is easier. The precipitation in ether included in the preferred method removes residual extractives and other contaminants but also causes losses of ether-soluble lignin constituents. Purification procedures involving gel permeation chromatography (Kirk and Lundquist 1970, Bardet et al. 1985) should be considered when there is an interest in retaining the latter type of materials. Ether solubles from a preparation of spruce MWL have been examined (Lundquist et al. 1977). Hydrophilic constituents of dioxane-water extracts from milled spruce wood have recently been investigated (Lundquist et al. 1990).

The extraction steps in the preferred method replace a precipitation in water from an acetic acid solution included in the original procedure (Björkman 1956). It is noteworthy that precipitation in water may result in the formation of a colloid if no acid or electrolyte is present. The extraction steps in the preferred method sometimes require that an efficient centrifuge be available. Purification according to the preferred method provides products with a low carbohydrate content, namely 3.7% in birch lignin (Lundquist et al. 1979) and 0.2–0.3% in spruce lignin (Lundquist et al. 1980), and it is possible to perform the purification within 1 day. The methoxyl contents of spruce lignin (representative of softwood lignins) are of the order of 16%. In birch lignin (representative of hardwood lignins), the methoxyl contents are in the range 21–22%.

3.1.3.2 Milled Wood Lignin as a Representative of the Lignin in Wood

Milling is expected to cause chemical changes in the lignin but to what extent this occurs is largely unknown (Lai and Sarkanen 1971). A slight increase in carbonyl content during milling in the presence of oxygen has been reported (Björkman and Person 1957, Chang et al. 1975). The effect of milling on lignocellulosic materials has been discussed in general terms by Hon (1983).

The yield of the MWL obtained according to the preferred method usually constitutes about 25% of the Klason lignin. It has been suggested that MWL originates from the middle lamella or at least that lignin from this morphological region is over-represented in the samples (Lai and Sarkanen 1971). Whiting and Goring (1981), in contrast, have more recently found that most of the MWL has its origin in the secondary wall of spruce tracheids. Lee et al. (1981) found that milled wood lignins successively isolated from a birch wood sample have decreasing amounts of guaiacyl units. This is in accordance with current concepts regarding the distribution of guaiacyl and syringyl units in lignins located in different morphological regions of hardwoods since middle lamella lignin has been shown to be comparatively rich in guaiacyl units (Eom et al. 1987). It appears from what has been said above that the problems regarding the morphological origin of MWL are as yet unresolved. However, it seems that MWL obtained in yields as high as 25% of the Klason lignin are fairly representative of the lignin in the wood examined. To obtain milled wood lignins in high yield (~50% of the Klason lignin), very extensive milling is required. Such a lignin would be more representative of the total lignin in wood with respect to morphological origin but chemical modification of the lignin during milling might be fairly severe.

3.1.3.3 Milled Wood Lignin Compared with Brauns Lignin

Lai and Sarkanen (1971) have discussed the properties of Brauns lignin as well as MWL in some detail. Brauns lignin is obtained in very low yield; it has a low molecular weight and a high phenol content, and it is sometimes contaminated with extractives. The latter drawback may be overcome by the use of improved purification procedures involving gel permeation chromatography (Kirk and Lundquist 1970, Kirk and Chang 1974). Parenthetically, it may be mentioned that the syringyl content of a Brauns lignin obtained from aspen was found to be higher than that of a MWL obtained from the same species (Bardet et al. 1985 and unpublished data; the similarities between Brauns lignin and lignin in lignin carbohydrate complexes noted by Košiková et al. (1977) are also of interest in this context).

3.1.3.4 Milled Wood Lignin Compared with Cellulolytic Enzyme Lignin

The step involving enzymic hydrolysis of carbohydrates makes it possible to obtain CEL in amounts which correspond to a large percentage of the Klason lignin. It should be noted that the preparation of CEL as well as MWL requires an initial grinding step. Comparative studies of milled wood lignins and cellulolytic enzyme lignins from softwoods as well as from hardwoods indicate that the number of conjugated carbonyl groups is somewhat larger in MWL (Chang et al. 1975, Bezúch and Polčin 1978, Fišerová and Šutý 1980, Lapierre et al. 1982, 1983). For hardwood lignins, it has been shown that the syringyl content is significantly higher in CEL than in MWL (Lapierre et al. 1982, 1983, 1985). The isolation of CEL has also been studied by Wayman and Obiaga (1974), Polčin and Bezúch (1978), and Glasser and Barnett (1979).

It is evident that CEL, in certain respects, is more representative of the total lignin in the wood than is MWL. Part of the CEL could, however, be described as lignin-carbohydrate complexes. A drawback is that the procedure required for the preparation of CEL is rather tedious. On the basis of an evaluation of the drawbacks and advantages of various methods for the isolation of lignins, the preparation of MWL lignin under standardized conditions has been selected here as the preferred method because of its simplicity and reproducibility. It should, however, be emphasized that cellulolytic enzyme lignins are of great value in lignin research. The enzymic treatment involved in the preparation of CEL efficiently solubilizes the major portion of the wood lignin. Specific procedures for the preparation of cellulolytic enzyme lignins are given in papers referred to in this section (e.g. in the paper by Chang et al. 1975). A procedure for the isolation of lignin from pulp using cellulolytic enzymes is described in Chapter 3.2.2.

References

Adler E, Brunow G, Lundquist K (1987) Investigation of the acid-catalyzed alkylation of lignins by means of NMR spectroscopic methods. Holzforschung 41:199–207

Bardet M, Lundquist K, Robert D (1985) On the reactions and degradation of the lignin during steam hydrolysis of aspen wood. Sven Papperstidn 88:R61–67

Bezúch B, Polčin J (1978) The structural study of milled wood and enzymatically isolated lignin fractions from spruce wood. Cellul Chem Technol 12:473–482

Björkman A (1956) Studies on finely divided wood Part 1. Extraction of lignin with neutral solvents. Sven Papperstidn 59:477–485

Björkman A (1957) Studies on finely divided wood Part 5. The effect of milling. Sven Papperstidn 60:329–335

Björkman A, Person B (1957) Studies on finely divided wood Part 2. The properties of lignins extracted with neutral solvents from softwoods and hardwoods. Sven Papperstidn 60: 158–169

Brauns FE (1939) Native lignin I. Its isolation and methylation. J Am Chem Soc 61:2120–2127

Brauns FE (1952) The chemistry of lignin. Academic Press, New York, 49–132, 742–744

Brauns FE, Brauns DA (1960) The chemistry of lignin, supplement volume. Academic Press, New York, 62–126

Brownell HH (1965) Isolation of milled wood lignin and lignin-carbohydrate complex. Tappi 48:513–518

Chang H-m, Cowling EB, Brown W, Adler E, Miksche G (1975) Comparative studies on cellulolytic enzyme lignin and milled wood lignin of sweetgum and spruce. Holzforschung 29:153–159

Eom TJ, Meshitsuka G, Ishizu A, Nakano J (1987) Chemical characteristics of lignin in the differentiating xylem of a hardwood III. Mokuzai Gakkaishi 33:716–723

Fengel D, Wegener G (1984) Wood chemistry, ultrastructure and reactions. de Gruyter, Berlin, 49–55

Fišerová M, Šutý L (1980) Comparative studies on milled wood lignins and cellulolytic enzyme lignin of beech wood (Fagus silvatica L.). Cellul Chem Technol 14:243–252

Glasser WG, Barnett CA (1979) The structure of lignins in pulps II. A comparative evaluation of isolation methods. Holzforschung 33:78–86

Hon D N-S (1983) Mechanochemical reactions of lignocellulosic materials. Appl Polym Symp 37:461–481

Kirk TK, Chang H-m (1974) Decomposition of lignin by white-rot fungi I. Isolation of heavily degraded lignins from decayed spruce. Holzforschung 28:217–222

Kirk TK, Lundquist K (1970) Comparison of sound and white-rotted sapwood of sweetgum with respect to properties of the lignin and composition of extractives. Sven Papperstidn 73:294–306

Košiková B, Polčin J, Joniak D (1977) NMR Studies on the lignin-saccharidic complexes II. Hydrolysis of the lignin-saccharidic complex isolated from acetylated beechwood. Holzforschung 31:191–193

Lai YZ, Sarkanen KV (1971) Isolation and structural studies. In: Sarkanen KV, Ludwig CH (eds) Lignins – occurrence, formation, structure and reactions. Wiley, New York, 165–240

Lapierre C, Lallemand JY, Monties B (1982) Evidence of poplar lignin heterogeneity by combination of ^{13}C NMR and ^{1}H NMR spectroscopy. Holzforschung 36:275–282

Lapierre C, Rolando C, Monties B (1983) Characterization of poplar lignins acidolysis products: Capillary gas-liquid and liquid-liquid chromatography of monomeric compounds. Holzforschung 37:189–198

Lapierre C, Monties B, Rolando C (1985) Thioacidolysis of lignin: comparison with acidolysis. J Wood Chem Technol 5:277–292
Lee ZZ, Meshitsuka G, Cho NS, Nakano J (1981) Characteristics of milled wood lignins isolated with different milling times. Mokuzai Gakkaishi 27:671–677
Lundquist K, Kirk TK (1980) Fractionation-purification of an industrial kraft lignin. Tappi 63(1):80–82
Lundquist K, Ohlsson B, Simonson R (1977) Isolation of lignin by means of liquid-liquid extraction. Sven Papperstidn 80:143–144
Lundquist K, Simonson R, Tingsvik K (1979) On the occurrence of carbohydrates in milled wood lignin preparations. Sven Papperstidn 82:272–275
Lundquist K, Simonson R, Tingsvik K (1980) Studies on lignin carbohydrate linkages in milled wood lignin preparations. Sven Papperstidn 83:452–454
Lundquist K, Simonson R, Tingsvik K (1990) On the composition of dioxane-water extracts of milled spruce wood: characterization of hydrophilic constituents. Nord Pulp Pap Res J 5:107–113 and 199
Pew JC (1957) Properties of powdered wood and isolation of lignin by cellulytic enzymes. Tappi 40:553–558
Pew JC, Weyna P (1962) Fine grinding, enzyme digestion, and the lignin-cellulose bond in wood. Tappi 45:247–256
Polčin J, Bezúch B (1978) Enzymic isolation of lignin from wood and pulps. Wood Sci Technol 12:149–158
Wayman M, Obiaga TI (1974) Molecular weights of milled-wood lignins. Tappi 57(4):123–126
Whiting P, Goring DAI (1981) The morphological origin of milled wood lignin. Sven Papperstidn 84:R120–R122

3.2 Isolation of Lignin from Pulp

H.-m. Chang

3.2.1 Introduction

Isolation of residual lignin from a chemical pulp, especially from a bleachable grade, has been a difficult task since the residual lignin content in pulp is relatively low and is probably linked chemically to the carbohydrates. A method for the quantitative isolation of residual lignin in chemical pulp for subsequent structural studies was developed in the early 1980s (Yamasaki et al. 1981a,b). The procedure is based on selective hydrolysis and dissolution of the carbohydrates in pulp by commercially available cellulolytic enzymes leaving lignin as an insoluble residue. The method was adapted from the enzymatic method first reported by Pew 1957, Pew and Weyna (1962) and later adapted by Chang et al. (1975) for the isolation of lignin from wood. The main difference between the two procedures is that chemical pulp, unlike wood, is digestible by cellulolytic enzymes without ball-milling. Consequently, this method provides an ideal method for the isolation of the residual lignin in chemical pulps without structural modification.

This enzymatic procedure may be used to isolate residual lignin from chemical pulps obtained in various yields for structural characterization. However, for higher yield chemical pulps such as linerboard grade, the residual lignin recovered after cellulolytic enzyme treatment is only partially soluble in common lignin solvents such as aqueous dioxane, dimethylformamide, dimethylsulfoxide, and 1 M NaOH (Chang 1972, unpubl. results). This limitation makes it difficult to fully characterize the residual lignin in high yield pulps. On the other hand, the residual lignin in bleachable grade pulps is almost completely soluble in the above solvents. The enzymatic procedure may also be used to isolate residual lignin from semi-bleached pulps such as oxygen-bleached pulp (Yamasaki et al. 1981b) and chlorinated and alkali-extracted pulp (Jiang et al. 1987). In these cases, a portion of the residual lignin becomes solubilized during the enzyme treatment and must be precipitated from the filtrate by acidification. Furthermore, cellulolytic enzymes tend to be strongly adsorbed on the residual lignin in semi-bleached pulp and hence must be removed from it.

Springer Series in Wood Science
Methods in Lignin Chemistry
(Edited by S.Y. Lin and C.W. Dence)

3.2.2 Procedures

3.2.2.1 Preparation of Pulp Samples and Enzyme Solution

The pulp to be treated with cellulolytic enzymes should first be screened in a laboratory flat screen to remove any shives. The screened pulp is beaten either in a laboratory Hollander beater (TAPPI Standard Method T 200 os-70) or in a PFI mill (TAPPI Standard Method T 248 cm-85), depending on the amount of pulp to be treated, to a final Canadian Standard Freeness of 200–300 ml. The beating operation helps convert shives into fibers and improves the digestibility of the pulp towards cellulolytic enzymes. The pulp is dewatered to a consistency of approximately 25% and stored in a polyethylene bag after fluffing.

A pH 4.5 acetate buffer solution is prepared by dissolving 6.56 g of sodium acetate and 4.64 ml of acetic acid in 4 l of distilled water. An enzyme stock solution is prepared by dissolving 10 g of commercially available crude cellulase enzymes (such as Cellulases TV Concentrate from Miles Labs., Inc. or Onozuka SS 1500 Cellulase from Japan Biochemicals Corp.) in 100 ml of the acetate buffer solution. The enzyme solution is then filtered through a fine porous glass filter to remove any insoluble material.

3.2.2.2 Isolation of Residual Lignins in Unbleached Pulps

In each of four 500-ml Erlenmeyer flasks with screw caps, a 20-g (OD) sample of pulp at a consistency of approximately 25% is treated with 12 ml of the enzyme solution at 5% consistency using the acetate buffer solution to dilute the pulp slurry to a final volume of 400 ml. After incubation in a shaking water bath under agitation at 45 °C for 48 h, the hydrolyzates are removed by centrifugation. The residues from two flasks are combined and placed in one flask and subjected to a second enzyme treatment by adding 12 ml of the enzyme solution and using the acetate buffer solution to bring the final volume to 400 ml. After incubation for 48 h under the same conditions as before, the residues from the two flasks are collected by centrifugation and combined into one flask for a third enzyme treatment under the same conditions. The fourth enzyme treatment is carried out using the same procedure as the third, except that only 6 ml instead of 12 ml of the enzyme solution is used. The residue from the fourth treatment is washed twice with dilute hydrochloric acid solution (pH 2.5) and then freeze-dried from a water suspension to give a water-insoluble residual lignin.

3.2.2.3 Isolation of Residual Lignins from Semi-Bleached Pulps

The residual lignin in a semi-bleached pulp such as oxygen-bleached pulps or chlorinated and alkali extracted pulps can be isolated using the same procedure as described previously. However, during the successive enzymatic hydrolyses, some residual lignin in the semi-bleached pulp becomes soluble in the acetate buffer solution. The water-soluble residual lignin can be isolated by combining the hydrolyzates from the four enzyme treatments and acidifying to pH 2.5 with dilute hydrochloric acid. The precipitated lignin is recovered by centrifugation, washed twice with dilute hydrochloric acid (pH 2.5), and finally freeze-dried. Thus, two residual lignin fractions are isolated from the semi-bleached pulps: one as an insoluble residue after enzymatic treatment and the other precipitated from the combined hydrolyzates by acidification.

3.2.2.4 Purification of the Residual Lignins

All residual lignins isolated by the enzymatic method are contaminated with enzymes to varying extents. This is especially true for the residual lignins isolated from semi-bleached pulps (Jiang et al. 1987). The contamination can be largely removed by the following procedure:

The residual lignin is dissolved in 1 M NaOH at a lignin concentration of 1%, and kept at room temperature under a nitrogen atmosphere and with magnetic stirring for 2 h. The solution is then centrifuged and the precipitate is washed with 1 M NaOH and discarded. The washings and supernatant are combined and acifified to pH 2.5 with HCl. The precipitate is collected by centrifugation and redissolved in 2 M NaOH to provide a lignin concentration of 1%. The solution is stirred at 70 °C under a nitrogen atmosphere for 48 h in a shaking water bath. The solution is then neutralized with acetic acid and transferred to a dialysis tubing having a molecular weight cutoff of 1000. The solution is dialyzed against distilled water for 48 h, dilute acetic acid (pH 3.0) for 24 h, and distilled water again for another 24 h. During the dialysis, the distilled water or the dilute acetic acid is changed regularly. The purified residual lignin is recovered by freeze-drying the dialyzed solution.

3.2.3 Concluding Remarks

All residual lignins isolated by the enzymatic procedure contain some carbohydrates of both cellulose and hemicellulose origin. These carbohydrates cannot be removed by either prolonged and repetitive enzymatic treatments or purification procedures commonly used for milled wood lignins (Yamasaki et al. 1981a) and are believed to be linked covalently to lignin. Typical carbohydrate contents of residual lignins isolated from unbleached and semi-bleached pulps are in the range of 3–7% (Yamasaki et al. 1981a, Jiang et al. 1987).

Contamination by enzymes is a main drawback of the procedure, especially in the isolation of residual lignin from semi-bleached pulps. The contaminants cannot be completely removed by the purification procedure described above. Consequently, it is important to determine the nitrogen content of the residual lignin. Typical values of nitrogen content are 2.5% and 7.3% before purification and 0.6% and 2.3% after purification for residual lignins isolated from an unbleached and a semi-bleached southern pine kraft pulp, respectively (Jiang et al. 1987).

References

Chang H-m, Cowling EB, Brown W, Adler E, Miksche G (1975) Comparative studies on cellulolytic enzyme lignin and milled wood lignin of sweetgum and spruce. Holzforschung 29:153–159

Jiang J-E, Chang H-m, Bhattacharjee SS, Kwoh DLW (1987) Characterization of residual lignins isolated from unbleached and semi-bleached softwood kraft pulps. J Wood Chem Technol 7:81–96

Pew JC (1957) Properties of powdered wood and isolation of lignin by cellulytic enzymes. TAPPI 40:553–558

Pew JC, Weyna P (1962) Fine grinding, enzyme digestion, and the lignin-cellulose bond in wood. Tappi 45:247–256

Yamasaki T, Hosoya S, Chen C-L, Gratzl JS, Chang, H-m (1981a) Characterization of residual lignin in kraft pulp. Proc Int Symp Wood Pulp Chem. Stockholm, Sweden, Vol. II:34–42

Yamasaki T, Hosoya S, Chen C-L, Gratzl JS, Chang H-m (1981b) Properties of residual lignin in kraft pulp. Proc 26th Lignin Symp, Oct. 29–30, Tsukuba, Japan, 45–48

3.3 Commercial Spent Pulping Liquors

S.Y. LIN

3.3.1 Introduction

Several techniques are available for the isolation and purification of lignin from commercial spent pulping liquors generated in the kraft and sulfite pulping processes. The techniques may be based on the differences between lignin and contaminants in solubility and molecular weight, or a combination of both. Extraction with various solvents and precipitation of lignin by lowering the pH of spent pulping liquors are examples of fractionation based on solubility. Ultrafiltration and gel permeation separate lignin and contaminants primarily according to their molecular weight/size.

The criteria used for selecting methods for the isolation of lignin from the spent pulping liquors are: (1) the lignin should be isolated in a quantitative yield, (2) the isolated lignin should be free from contaminants, and (3) the procedure should be simple and easy to perform. These criteria are satisfied when kraft lignin is isolated by acidification of concentrated black liquor with a mineral acid, and filtration and washing of the precipitated lignin, or when sulfite lignin is recovered from a spent sulfite liquor by treatment with a long-chain alkyl amine and extraction of the lignosulfonate-amine complex with a suitable organic solvent.

3.3.2 Principle

3.3.2.1 Kraft (Sulfate) Lignin

Kraft lignin has been conventionally recovered from black liquor in the laboratory and on an industrial scale by acid precipitation and heat coagulation. Indeed, the principle was patented by W.J. Hough as long ago as 1910. In his patent, Hough stated that, "My invention consists in neutralizing the spent liquors with an acid, whereupon the resin and lignin are precipitated. . . . It is preferable that the solution be quite hot before filtering, as the precipitated resin and lignin become more solid when the solution is in this condition, the filtration being thus greatly facilitated." Neutralization of the phenolic hydroxyl groups in kraft lignin occurs in the pH range of about 10.5–11, resulting in precipitation of the lignin in a concentrated liquor. The word "concentrated" is significant since only a small quantity of kraft lignin can be precipitated from black liquor when the concentration is low. For example, kraft lignin remains soluble when a

Springer Series in Wood Science
Methods in Lignin Chemistry
(Edited by S.Y. Lin and C.W. Dence)

typical black liquor is neutralized. The amount of lignin precipitated increases at higher concentrations and lower pH of the liquor. In essence, the isolation procedure consists of three steps, namely, acidification, filtration, and washing.

Acidification of black liquor is normally done with a mineral acid, such as sulfuric or hydrochloric acid, under good agitation. To prevent localized acidification and ensure uniform precipitation, too high an acid concentration should be avoided. A concentration of between 5 and 20% is preferred. The black liquor may be acidified in one step to pH 2 to precipitate practically all lignin, or it may be acidified stepwise to various pH values to achieve fractional precipitation. In either case, the resultant slurry is heated to coagulate the lignin particles for ease of filtration. The filtration step is critical to obtaining a pure lignin. If "channeling" occurs on filtration (which is always performed under suction), the filter cake may contain an undesirably large amount of occluded liquor and, as a result, retain inorganic salts and organic contaminants. A remedy is to reslurry the lignin in acidic water and repeat the filtration step. Alternatively, the filter cake may be washed several times with water at pH 2 until the ash content of the isolated kraft lignin is reduced to less than 2%.

3.3.2.2 Lignosulfonate (Sulfite Lignin)

As a first step, spent sulfite liquor is treated with a long-chain alkyl amine to form a water-insoluble lignosulfonic acid-amine complex which is then extracted with solvents such as are shown in Table 3.3.3 to separate it from sugars, oligosaccharides, inorganic salts, and other nonlignin contaminants. The alcoholic solution of the complex can be extracted with alkaline water (e.g., NaOH, KOH or NH_4OH solution) to recover the sodium, potassium, or ammonium salt of the lignosulfonate in the aqueous phase and the alkylamine in the alcohol phase. Essentially, the procedure consists of three steps:

Step 1 Complex formation

$$\text{Lignin-SO}_3^-\text{H} + \text{R}_1\text{R}_2\text{R}_3\text{NH Cl} \rightarrow \underset{\text{(Complex)}}{\text{Lignin-SO}_3^-\overset{+}{\text{N}}\text{H R}_1\text{R}_2\text{R}_3} + \text{HCl}$$

Step 2 Extraction of complex with organic solvent
Step 3 Regeneration of lignosulfonate

$$\begin{aligned}&\text{Lignin-SO}_3^-\ \overset{+}{\text{N}}\text{HR}_1\text{R}_2\text{R}_3 + \text{NaOH (or other alkali)} \rightarrow\\&\qquad \text{Lignin-SO}_3\text{Na} + \text{NR}_1\text{R}_2\text{R}_3 + \text{H}_2\text{O},\end{aligned}$$

where R_1, R_2, R_3 are hydrogen or alkyl groups.

3.3.3 Methods

3.3.3.1 Isolation Procedure for Kraft Lignin

The following steps can be used to isolate fractions of kraft lignin from black liquor.

1. Heat 40% black liquor to 90–95 °C in a hot water bath under rapid stirring.
2. Acidify the hot liquor carefully with 20% (by weight) sulfuric acid to pH 11 and cool to 40 °C in an ice bath.
3. Filter the acidified liquor through a Whatman No. 4 filter paper. Recover the filter cake.
4. Reslurry the filter cake in distilled water and under good stirring acidify the slurry to pH 2 with 20% sulfuric acid. Centrifuge off the precipitated lignin. After several washings with water at pH 2, air dry the recovered lignin.
5. Acidify the filtrate in Step 3 to pH 10.5 and repeat Steps 3 and 4 to recover a second fraction of lignin.
6. Repeat Steps 2 to 4 with appropriate H_2SO_4 addition to recover lignin fractions at pH 9.5, 8.5, 8, and 2.

The kraft lignin fractions thus obtained are heterogeneous with regard to their chemical properties (see Sect. 3.3.4 for analytical data). When the black liquor contains fibrous materials, it is filtered through a Whatman No. 2 paper prior to the acidification step.

3.3.3.2 Isolation Procedure for Lignosulfonates

1. Preparation of Liquid Ion Exchanger, Amine Hydrochloride

Dissolve dodecylamine in octyl alcohol to make a 20% (by weight) solution, which is then shaken with 1 M HCl (using an amount slightly in excess of an equivalent amount of dodecylamine) for 10 min. The organic phase is washed twice with distilled water to remove excess hydrochloric acid.

2. Formation and Extraction of the Lignosulfonate-Amine Complex

Mix equal volumes of the liquid ion exchanger and spent sulfite liquor and stir at 50 °C for 30 min. The amount of spent sulfite liquor is calculated to give the same equivalent amount of sulfonic acid groups as the dodecylamine in the ion exchanger. Alternatively, the equivalent amount of the ion exchanger can be added in portions to effect fractionation of the lignosulfonate. After mixing, carefully pour the mixture into a separatory funnel and allow 2 h for the aqueous and organic phases to separate. Keep the mixture at 50–60 °C to facilitate phase separation. Remove the organic phase for back extraction.

3. Back Extraction

To the organic phase from Step 2 add an equal volume of distilled water. Stir the mixture and adjust the pH to 9 with sodium hydroxide (or other alkali hydroxides if necessary). Pour the mixture into a separatory funnel and allow 24 h for the phases to separate while keeping the temperature at 50–60 °C. Recover sodium lignosulfonate (or other metal salts) from the aqueous phase.

3.3.4 Composition of Isolated Lignin

3.3.4.1 Kraft Lignin Fractions

Following the steps outlined in Section 3.3.3.1, six kraft lignin fractions, having the yields and composition shown in Table 3.3.1 (Lin 1978), are isolated from a typical Kamyr black liquor of Southern pine. The bulk of the kraft lignin (about 80%) is recovered at pH 10.5 and has a uniform chemical composition. The rest of the lignin, which is isolated at pH 9.5 or below, accounts for 20% of the total kraft lignin in the liquor. The low pH kraft lignin is heterogeneous with respect to chemical functionalities such as methoxyl and carboxyl groups – the lower the pH of precipitation, the higher is the carboxyl content and the lower is the methoxyl content. In a fractionation study of spruce wood black liquor, Wada et al. (1962) observed a similar heterogeneity of kraft lignin fractions with respect to methoxyl, phenolic, and carboxyl contents.

3.3.4.2 Lignosulfonate Fractions

Fractionation of a spent sulfite liquor from pulping of spruce wood according to Section 3.3.3.2 by stepwise addition of dodecylamine yields five fractions of lignosulfonate. The yields and composition of the fractions are shown in Table 3.3.2. As is the total lignin in black liquor; lignosulfonate in the spent sulfite liquor is heterogeneous with respect to chemical composition. The fraction that is extracted first with the amine has lower sulfonic acid and sodium contents and a higher methoxyl content than those that are extracted later. Generally, lignosulfonates of high molecular weight and low sulfonic acid content are insolubilized first when less than an equivalent amount of alkylamine is used.

Table 3.3.1. Analytical data for kraft lignin fractions isolated from pine Kamyr black liquor. (S.Y. Lin 1978, unpubl. data)

Fraction	pH of precip.	Yield (% of total)	Ash (%)	Acetone solubility (%)	Methoxyl (%)	Carboxyl (%)
1	11	55	0.9	45	13.1	1.8
2	10.5	26	0.5	74	13.1	2.3
3	9.5	5	0.6	84	12.0	9.0
4	8.5	3	0.3	86	11.7	9.1
5	8	3	0.9	86	9.5	9.4
6	2	5	1.5	88	9.8	13.9

Table 3.3.2. Analytical data for lignosulfonate fractions isolated from spruce spent sulfite liquor. (S.Y. Lin 1978, unpubl. data)

Fraction	Quantity of amine[a] (% Eq.)	Yield (% of total)	Sulfonic sulfur (%)	Methoxyl (%)	Sodium (%)
1	5	6	7.2	10.9	6.5
2	15	20	6.9	11.4	6.7
3	25	30	7.2	10.6	7.5
4	25	23	9.1	9.8	8.9
5	30	18	9.3	5.2	14.8

[a] % Equivalents of octylamine on the basis of total sulfonic sulfur content of the spent sulfite liquor.

3.3.5 Discussion

In the above subsection, the heterogeneity of kraft lignin and lignosulfonate in spent pulping liquors is demonstrated by the analytical data of lignin fractions obtained by fractional precipitation. The nonuniformity of industrial lignins should be kept in mind when isolating lignins for research uses. In kraft lignin, the chemical composition and molecular weight of the isolated material depend, to a great extent, on the pH at which the sample is obtained. About 20% of low molecular weight kraft lignin in the black liquor has high phenolic hydroxyl and carboxyl contents and a low methoxyl content. Contamination with such lignin significantly alters the analytical data for the bulk of the kraft lignin in the pulping liquor. However, to understand how the kraft pulping process changes lignin structure, it is necessary to isolate quantitatively all lignin in the liquor. Failure to include the 20% of the lignin having a lower molecular weight, as is commonly done in the laboratory when lignin is recovered by acidification of black liquor to pH 10.5, results in inaccurate information on the true chemical composition of the whole kraft lignin in a black liquor sample.

The purity of kraft lignin is measured by its ash content. For example, an ash content of less than 1% is indicative of "high purity". When the ash content exceeds 2%, the isolated kraft lignin should be washed repeatedly with water at pH 2 until a desirable ash level is obtained.

Two problems are usually encountered in isolating lignosulfonates by the alkylamine method, namely, formation of an emulsion and a foam. These problems occur in the extraction step of the lignosulfonic acid-amine complex. Lignosulfonic acid (particularly in softwood spent sulfite liquors) is a good emulsion stabilizer for an alkyl alcohol-water mixture and thus tends to keep the two phases (aqueous and organic) from separating completely in a reasonably short time. An incomplete phase separation results in a less than quantitative recovery of lignosulfonate in the pulping liquor. These problems can be solved by proper selection of the alkyl alcohol and amine used in the isolation procedure. In the past, numerous studies have been undertaken to examine the effects of solvent and amine on the degree of extraction and fractionation (Felicetta and McCarthy 1957, Harris and Hogan 1957, Eisenbraun 1963, Forss

Table 3.3.3. Amines and solvents used in the extraction of lignosulfonates

Amine	Solvent	Reference
Dodecylamine	Amyl alcohol	Harris and Hogan (1957)
Dicyclohexylamine	Butanol	Felicetta and McCarthy (1957)
Tri-n-hexylamine	Butanol	Eisenbraun (1963)
Dodecylamine	Butanol	Forss and Pirhonen (1974)
Dodecylamine, dioctylamine, trioctylamine	Butanol, pentanol, methyl isobutyl ketone, 1,2-dichloroethane, cyclohexane	Kontturi and Sundholm (1986)

and Pirhonen 1974). Recently, this issue has been discussed in detail by Kontturi and Sundholm (1986). In Table 3.3.3 are listed the amines and solvents used in the previous extraction studies.

The purity of isolated lignosulfonate is measured by the amounts of reducing sugars and inorganic salts (e.g., sodium sulfate and sulfite) it contains. Although ultraviolet light absorptivity has been often used to measure the purity of lignosulfonate, its accuracy and reliability is open to question. The absorptivity changes with fractionation, chemical modification of the lignosulfonate, and pulping conditions.

In addition to the procedures described in this Chapter, there are many other isolation and purification methods available both for laboratory use and industrial processing. However, few of these methods meet the criteria set forth in the beginning of this section. For example, ultrafiltration is an economic process for production of high molecular weight lignosulfonate, yet the process is neither selective enough to remove completely sugars and other impurities from the lignosulfonate, nor quantitative enough to allow recovery of all (or practically all) lignosulfonate from a spent sulfite liquor. Gel permeation chromatography and ion exclusion techniques are useful for isolating kraft lignin and lignosulfonate, but they also suffer from the same shortcomings.

References

Ball FJ (1967) U.S. Patent No. 3312643

Eisenbraun EW (1963) The separation and fractionation of lignosulfonic acids from spent sulfite liquor with tri-n-hexylamine in organic solvents. Tappi 46:104–107

Felicetta VF, McCarthy JL (1957) Tappi 40:851–866

Forss KG, Pirhonen IMJ (1974) U.S. Patent No. 3825526

Harris EE, Hogan D (1957) Isolation and fractionation of lignosulfonates in spent sulfite liquors. Ind Eng Chem 9:1393

Hough WJ (1910) U.S. Patent No. 949324

Kontturi AK, Sundholm G (1986) The extraction and fractionation of lignosulfonates with long chain aliphatic amines. Acta Chem Scand 40:121–125

Wada S, Iwamida T, Iijima R, Yabe K (1962) Fractional precipitation of kraft black liquor from spruce wood. Chem High Polymers (Kabunshi Kagaku) 19:699–705

Chapter 4

Characterization in Solid State

4.1 Fourier Transform Infrared Spectroscopy

O. FAIX

4.1.1 Principles and Instrumental Techniques

4.1.1.1 Introduction

Since the early 1950s, IR spectroscopy has been a routine analytical tool for lignin chemists. In the past, spectra were recorded using the so-called dispersive technique, i.e., with grating-type or prism instruments. In the last decade, Fourier transform infrared (FTIR) spectrometers have become increasingly available for routine laboratory work.

4.1.1.2 Scope of FTIR Spectroscopic Applications

FTIR spectroscopy may be applied to good advantage in such specialized areas as micro analysis where high sensitivity is required, in the analysis of aqueous solutions or dark, solid state samples that require the use of special reflectance techniques, in investigations placing emphasis on quantitative evaluation, and in experiments where analysis time is a limiting factor, e.g., in process or quality control measurements.

4.1.1.3 Description of a Standard FTIR Instrument

In FTIR spectroscopy, an interference wave interacts with the sample in contrast to a dispersive instrument where the interacting energy assumes a well-defined wavelength range. The interference wave is produced in an interferometer (Fig. 4.1.1), the most common of which is the Michelson interferometer. A computer is used to control the interferometer, to collect and store data, and to perform the Fourier transformation. In addition, the computer performs post-spectroscopic operations such as spectral presentation, resolution enhancement, calibration, and calculation of correlation equations.

A collimated light beam from the IR source is directed to the Michelson interferometer where it is divided by the beam splitter. One half of the beam is reflected from a fixed mirror and the other half from a moving mirror. The two light beams recombine after returning from the mirrors and give rise to a reconstructed beam which is optically an interference wave. The interference light beam passes through the sample and is modified by its interaction with the

Springer Series in Wood Science
Methods in Lignin Chemistry
(Edited by S.Y. Lin and C.W. Dence)

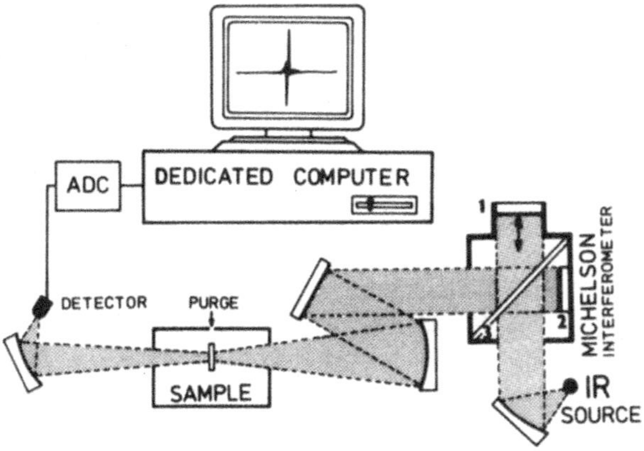

Fig. 4.1.1. Schematic diagram of a FTIR instrument: *1* moving mirror; *2* fixed mirror; *3* beam splitter made of KBr or CsI as supporter coated with a thin germanium layer. *ADC* analog-digital-converter

sample. Usually, a deuterated triglycine sulfate (DTGS) pyroelectric detector receives the modified light. The analog signals arriving at the detector are digitized by an analog-digital-converter (ADC) and stored in the computer. Laser technology is indispensable for good performance in this procedure. A laser beam undergoes the same change of optical path in the interferometer as the IR beam. Hence, it can serve as a reference for the position of the mirror during the scan. The laser beam also governs the collection of data points as a function of the movement of the mirror.

A complete IR spectrum can be reconstructed after only one cycle of mirror movement by use of the fast Fourier transform algorithm (Cooley and Tukey 1965). If time is not a limiting factor, additional scans are usually made. After each addition, an average interferogram is calculated. An interferogram is an electrical signal depicted as a function of the time delay (retardation) of the two beams in the interferometer. The addition procedure improves the signal-to-noise ratio, since the noise level is inversely related to the square root of the number of scans. The average interferogram is then prepared through further processing, such as zero filling, phase correction, and apodization. The application of these mathematical techniques is necessary for the compensation of instrumental artifacts.

Fourier transformation is a decoding procedure by which the signal is converted from the so-called time domain, where the signal is a function of retardation, to the frequence domain, where it is a function of frequency, according to the formula:

$$S(\nu) = 2 \int_0^\infty I(x) \cos(2\pi\bar{\nu}x)\, dx,$$

where x is the retardation; I(x), the interferogram; $\bar{\nu}$ is any wavenumber, and S(ν), the spectrum.

4.1.1.4 Advantages of FTIR Spectroscopy

Signal-to-Noise Ratio and Linearity

FTIR spectroscopy has a high signal-to-noise ratio; this is the most important advantage of this type of spectroscopy, caused partly because all spectral elements (wavelengths) are observed simultaneously during the whole recording time. By contrast, in grating spectrometers only one resolution element can be examined at a time. The signal-to-noise ratio improves with the square root of the observation time of each resolution element, $\bar{\nu}$. The signal-to-noise advantage that the interferometer offers for a spectral region S between $\bar{\nu}_1$ and $\bar{\nu}_2$ is equal to $(S/\Delta\nu)^{1/2}$. This is called multiplex advantage. Another reason for the better signal-to-noise ratio is the high optical throughput in FTIR spectrometers, which is about 200 times higher than for dispersive IR spectroscopy. This is referred to as the throughput advantage. The combined multiplex and throughput advantages results in a high dynamic range of linearity, making FTIR spectroscopy most suitable for quantitative work.

Accuracy

The high accuracy in frequency, often referred to as laser reference advantage, is a result of using laser technology for coordinating the mirror movement and data recording. This is, however, more important for gas phase investigations than for studies of lignins and lignocellulosics. Nevertheless, high accuracy is also indispensable for data handling, such as in spectral subtraction.

Data Handling Facility

The advent of computer-controlled FTIR spectrometers has accelerated the use of software for spectral data manipulation, which is particularly beneficial in the study of lignin and lignocellulosics. Well-designed routine programs permit the following operations: (1) base line correction and normalization of spectra, a prerequisite for intra- and interlaboratory comparisons of spectra and for the use of automatic library search programs; (2) subtraction of interfering absorbances, which is important for analyzing the composite spectra of lignocellulosics; (3) expansion of co-ordinates, which is useful for magnifying spectral details of interest; (4) application of mathematical resolution techniques such as derivative spectroscopy (first and second derivatives), deconvolution, and band-fitting (band shape analysis); and (5) statistical analysis such as partial least squares (PLS), principal components analysis (PCA), and other methods often included with the manufacturer's software, that are very useful quantitative or semi-quantitative tools for analysis of multiple components.

Mechanical Simplicity

The mechanical simplicity of an FTIR spectrometer contrasts sharply with its theoretical complexity. The only moving part is the mirror, which rides on air or mechanical bearings to facilitate maintenance and improve reliability. The round cross-section of the FTIR beam is especially advantageous in situations where a circular attenuated total reflectance (CATR) accessory for liquids is used. For heat-sensitive compounds it is important to note that the sample is not heated significantly during the FTIR measurement.

There is a large body of literature on FTIR spectroscopy including, for example, Bracewell (1965), Horlick (1968), Bell (1972), Griffiths (1975, 1983), Ferraro and Basile (1978), Koenig (1981), Griffiths and de Haseth (1986), Perkins (1986, 1987), Mackenzie (1988). Cameron and Moffatt (1984), and Gillette et al. (1985) have explained the basic mathematical concepts of deconvolution, derivation and smoothing in FTIR spectroscopy.

4.1.2 Method

4.1.2.1 FTIR Spectroscopy in Practice

At present, the manufacturers listed below deliver reliable instruments with satisfactory working software. The type of instruments and models are not listed because the models change frequently and the software is updated almost annually. Users may select from a variety of beam splitters, detectors, computer memory, and software. The entire spectral range, from the far infrared through the near infrared, is covered by various FTIR instruments. Instrument suppliers include Analect (USA), Bomem (Canada), Bruker (FRG), Bio-Rad, Digilab Division (USA), Mattson (USA), Jasco (Japan), Nicolet (USA), and Perkin-Elmer (UK). Accessory manufacturers include Spectra-Tech Inc. (USA), Specac (UK, USA), Harrick Scientific Corp. (USA), and AAB SPEC (USA).

Figure 4.1.2 shows several elements of FTIR spectroscopic measurements with which one needs to become familiarized before proceeding, namely the interferogram, the appearance of the single beam spectrum composed of instrumental artifacts and the spectrum of air in the mid-IR range using a DTGS detector, and the so-called "100% line" between 2200 and 2000 cm^{-1}. The size and shape of interferogram is optimized by adjusting the instrument. The maximum intensity of the interferogram is called the centerburst. Ideally, it should be matched to the analog-digital-computer (ADC) and should have a peak form. The centerburst optimization should be repeated if accessories with high energy consumption (e.g., reflectance techniques, micro sampling) are used. The single-beam spectrum provides, besides background interference data specific to the instrument, constant information on the humidity and CO_2 content of the air. Using a standard DTGS detector, the response at 4000 cm^{-1} should be 20% of that at 1800 cm^{-1}. The magnified section of a transmission spectrum of an empty beam (100% line) recorded with 8 cm^{-1} resolution between 2200 and 2000 cm^{-1} provides information on the noise contributed by

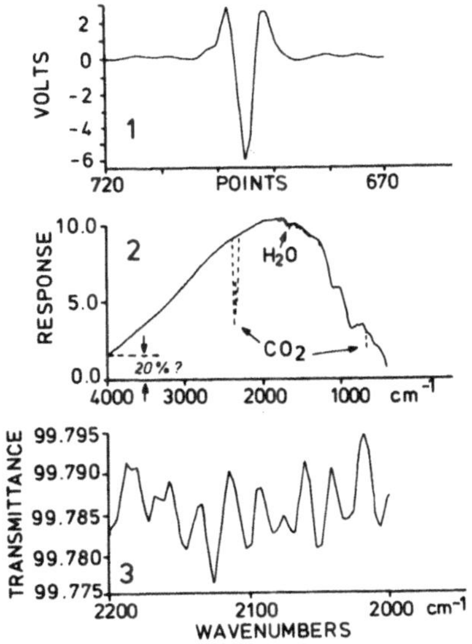

Fig. 4.1.2. Important elements of a FTIR spectrospcopy. *1* Interferogram. *2* Single-beam spectrum. *3* 100% line

the instrument. Because of the complexity of the adjustment procedure, which varies from instrument to instrument, consultation of the operating manual is always necessary.

4.1.2.2 Procedures

Lignin IR spectra are easy to obtain routinely. As a rule of thumb, one should use the KBr transmission technique (TR) for routine work wherever possible. Reflection techniques, e.g., attenuated total reflectance (ATR) or diffuse reflectance (DR), are methods for dealing with special problems, such as liquids or high consistency pastes, and surface investigations in the solid state. The same is true for photoacoustic (PA) methods. TR, DR, and microsampling techniques are addressed below because of their importance in lignin and wood chemistry. ATR and PA methods are less used for the characterization of pure lignins. Details of the PA technique have been discussed by Rosencwaig (1981), Graham et al. (1985), Harbour et al. (1985), Kuo et al. (1988) and Oelichmann and Hauser (1988).

Sample Preparation in the Transmission Mode

The most common way to obtain IR spectra for quantiative evaluation is to embed the sample in an alkali halide pellet. Dry, spectroscopy-grade potassium bromide or potassium chloride are suitable embedding media. (Drying procedure: 120 °C in oven for at least 1 h, followed by cooling in a desiccator over

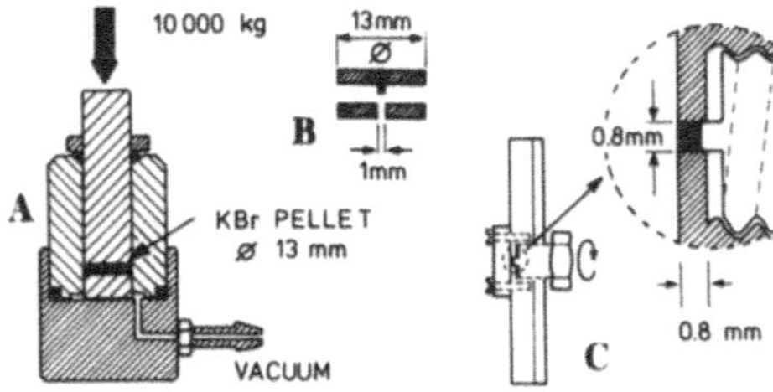

Fig. 4.1.3A–C. Preparation of alkali halide pellets. **A** Press for preparation of standard (macro) pellets. **B** Adapter for preparation of micro pellets. **C** Adapter and sample holder for micro pellet

P_2O_5). A standard 13-mm diameter pellet is prepared by pressing a 1-2 mg sample in 350 mg of KBr in an evacuated die (Fig. 4.1.3A).

The IR spectrum of a blank pellet is recorded and stored in the computer memory as background. The spectrum of the blank reveals whether the effort to remove water has been successful. Water has signals at about 1625 and 3700 cm^{-1}. The finer the powder, the greater the amount of water that can be adsorbed in a given time. A detailed water/KBr interaction has been provided by Malhotra et al. (1989). The KBr should not be ground in a mortar or milled for an unnecessarily long time to avoid water uptake. The use of a glove box with dry air purge is highly recommended in special cases. However, for routine runs, a small amount of water in the pellets can be tolerated if both sample and blank pellets are prepared in exactly the same way. In such cases, the water peak is cancelled out by subtracting the spectrum of the blank from that of the sample. A heated die also helps to avoid water in the pellet.

The band resolution in the carbonyl stretching region can be improved relative to that obtained in the KBr technique through use of the paraffin or fluorocarbon oil-mull method, also known as the Nujol technique (Hergert 1971). In this procedure, the sample is milled uniformly in an agate mortar together with a few drops of oil. The spectrum is then run as an oil suspension sandwiched between salt plates especially constructed for this purpose. This technique is recommended only for situations in which the carbonyl regions are of interest since the C—H bonds in the hydrocarbons of the mulling medium cause spectral interference.

Diffuse Reflectance (DR)

Infrared light is partly reflected from the surface and partly transmitted to a depth of a few micrometers in powdered material. The light that enters the material is absorbed or scattered. The re-emitted light has components from the surface and layers close to the surface. This phenomenon, in which the light is reflected in many different directions, is termed diffuse reflectance (DR). In an ideal case, the energy distribution of the diffusely reflected light is homogeneous irrespective of the reflection angle. Another mode of reflection is the regular, mirror-like reflection normally observed with smooth surfaces and described as specular reflection. When measuring diffuse reflectance, the specular component of reflection should be reduced as much as possible since these two reflection modes obey different laws.

Low levels of reflected energy, even from micro samples, can be measured by FTIR instruments. This method is termed diffuse reflectance infrared Fourier transform (DRIFT) spectroscopy. A reflectometer design with hemiellipsoidal mirrors sliding back for sample positioning is very convenient (Fig. 4.1.4).

For DRIFT studies, a wood wafer, paper sheet, or milled wood sample dispersed in KBr (or KCl) is placed in a cup at the focal point of the concave, ellipsoidal mirror so that the incident light is focused on the sample. The scattered light coming from the sample is collected from the concave mirror and directed by a suitable mirror system to the detector of the FTIR instrument. The pressure used for smoothing the sample has to be adjusted so that reproducible results can be obtained (Yeboah et al. 1984). The contribution of specular reflectance can be diminished by reducing the particle size and by increasing the sample dilution. For powder samples, as indicated above, the diluent is KBr or KCl. Good results are normally obtained with alkali halide powders that contain 1–2% of sample. In certain cases, the sample concentration may be increased up to 10%.

When measuring paper sheets or wood surfaces, the so-called Blocker device (Spectra-Tech Inc.) must be used to eliminate the specular light. Surface quality often influences the results. For example, DRIFT spectra are slightly different when taken on the top side or the wire side of a paper sheet.

DRIFT spectroscopy is of special interest in studying wood surface phenomena, for example, the effects of oxidation, photooxidation, and weathering. Moreover, it is advantageous for the analysis of highly absorbing materials

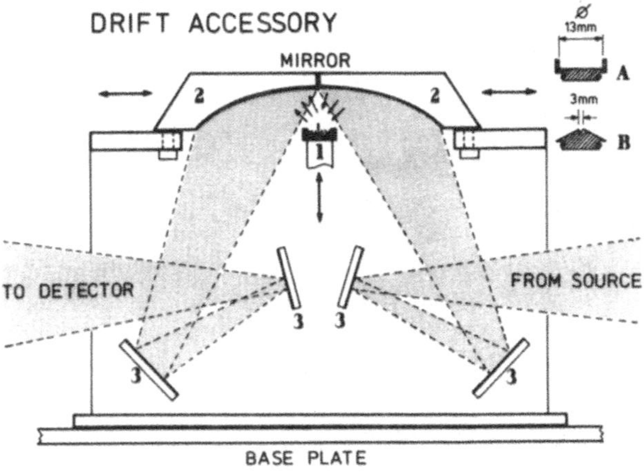

Fig. 4.1.4. Diffuse reflectance accessory (Spectra-Tech Inc.). *1* Movable stage with Blocker device to diminish the specular component of reflectance; *2* two movable parts of the hemiellipsoidal mirror. The mirror opens for sample positioning; *3* mirror system to reflect the IR beam from the source to the detector; *A* macro sample holder; *B* micro sample holder

such as charcoal or technical lignins. DRIFT is also well suited for infrared microsampling in GPC and HPLC analyses.

A general description of DRIFT has been provided by Ferraro and Rein (1985). The application of DRIFT to wood, pulp, and lignin chemistry has been discussed by Schultz et al. (1985a,b), Schultz and Glasser (1986), Yang et al. (1986), Berben et al. (1987), Grandmaison et al. (1987), Faix and Németh (1988), Hauser and Oelichmann (1988), Michell (1988a,b), Michell et al. (1989) and Ostmeyer et al. (1989).

Micro Techniques

The need for micro sampling often arises from the morphological and chemical inhomogeneity of wood. The sample size for DRIFT can be reduced to nanograms by using a micro cup (see Fig. 4.1.4). Manufacturers of accessories also offer devices for the preparation of KBr discs ranging down to 1 mm diameter (see Fig. 4.1.3 B and C). Micro pellets are usually prepared in a small agate mortar as described above for the preparation of macro pellets. Diamond anvil cells are useful for micro analysis of wood fibers. In these cells, sample thickness may be reduced by the application of pressure. Optimal adjustment of the instrument is especially important in micro analysis. Sample positioning has to be optimized in the "set up" mode of the instrument. This is done by observing the interferogram while adjusting the sample holder. As is true for all accessories, the highest interferogram response indicates optimal positioning. Incorrectly oriented samples may result in irregularities such as vignetting and interference fringes.

For very small samples (in the picogram range, or with disc diameters below 0.5 mm), the use of a FTIR microscope is required. A FTIR microscope is schematically shown in Fig. 4.1.5 for Bio-Rad Digilab Model UMA 300 A. The microscope permits visual inspection of the sample for selecting a spot of

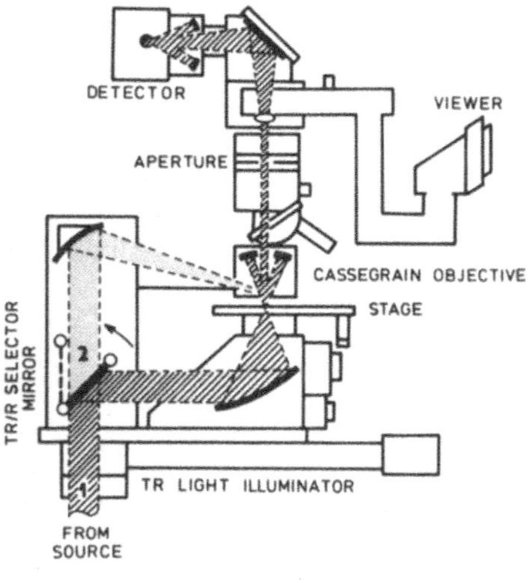

Fig. 4.1.5. FTIR microscope UMA 300 A (Bio-Rad, Digilab) operating in the transmission *1* or in the reflectance *2* mode

interest. The spectrum is recorded only from this part of the object by focusing the infrared beam on the sample with the switchable objective. After being transmitted through the sample, the IR beam is collected by a Cassegrain mirror objective. The Cassegrain objective works on the principle of reflection since the standard microscopic transmission objective of glass or quartz, used for the visible-range measurements, cannot be used in IR spectroscopy. A photoconductivity detector made of mercury cadmium-telluride (MCT) with a small collecting area collects the transmitted IR beam.

The resolution, R, depends on the numerical aperture, NA, and the wavelength, λ, according to the formula

$$R = \frac{\lambda(\text{in } \mu m)}{NA}.$$

R is approximately between 4 and 30 μm in the mid-IR range (2.5–20 μm) when NA = 0.70. Sometimes it is possible to obtain IR spectra of good quality for samples with dimensions below the theoretical limits. The high diffraction effects of Cassegrain objectives may require the use of two apertures, one below and the other above the sample, to minimize the diffraction in the transmission mode.

Spectra of good quality normally can be obtained in a reasonably short measuring time (less than 5 min) with disc diameters ranging from 20–30 μm. FTIR microspectroscopy is described by Fuller and Griffiths (1980), Bartick (1985), Krishnan (1988), Schiering et al. (1988), Sommer et al. (1988) and Sweeney (1989).

4.1.3 Lignin Characterization in the Mid-Infrared Region (4000–500 cm^{-1})

4.1.3.1 Band Assignment and Classification of Lignin IR Spectra

Table 4.1.1 lists band assignments for the FTIR spectra of three different milled wood lignins, namely guaiacyl-type (G), guaiacyl/syringyl-type (GS), and hydroxyphenyl/guaiacyl/syringyl-type (HGS). Only a few group frequencies and bands are uneqivocally assigned, and many bands can be interpreted in various ways. The O—H, C—H, and C=O stretching modes above 1600 cm^{-1} and the aromatic skeletal vibration around 1510 cm^{-1} are "pure" bands, whereas the 1600 cm^{-1} aromatic skeletal vibration band is a superimposed band that has been broadened by the C=O stretching mode (Hergert 1971). Interpretation of the bands below 1430 cm^{-1} is more difficult. Here, nearly all bands are complex, with contributions from various vibration modes.

The band wavenumbers in Table 4.1.1 were obtained at 4 cm^{-1} resolution. The intensity values shown in the parentheses were obtained from baseline-corrected and standardized spectra (Figs. 4.1.6 and 4.1.7). As shown in Fig. 4.1.7, most IR bands are common to the three main lignin types although their intensities may differ. In the wavenumber range 2700–3700 cm^{-1}, the band intensities of all lignin types are similar (Table 4.1.1, Fig. 4.1.6).

A semi-quantitative comparison of band intensities can be used for lignin classification as first described by Kawamura and Higuchi (1964a,b).

Table 4.1.1. Band assignments in the mid-infrared region of nonderivatized G, GS, and HGS-type milled wood lignins. Band intensities (*in parenthesis*) were measured from the zero line of baseline-corrected and normalized spectra. Intensity of the highest band equals 100

Wavenumbers (cm^{-1}): Baseline (Abs. = 0)	Wavenumbers (cm^{-1}): Range of maxima	No. in Fig. 4.1.7	Band origin, short comments	Spruce MWL cm^{-1}	Spruce MWL % A	Beech MWL cm^{-1}	Beech MWL % A	Bamboo MWL cm^{-1}	Bamboo MWL % A
from 3695 to 3042	3412–3460	1	O—H stretch	3412	(58)	3460	(49)	3428	(45)
from 3042 to 2782	3000–2842	2	C—H stretch in methyl and methylene groups	3000 2937 2879 2840	(5) (24) (15) (12)	3000 2940 2880 2840	(7) (22) (12) (12)	3002 2942 2879 2840	(6) (22) (11) (11)
from 1800	1738–1709	3	C=O stretch in unconjugated ketones, carbonyls and in ester groups (frequently of carbohydrate origin); conjugated aldehydes and carboxylic acids absorb around and below 1700 cm^{-1}	1722	(11)	1735	(18)	1709	(45)
	1655–1675	4	C=O stretch; in conjugated p-subst. aryl ketones; strong electronegative substituents lower the wavenumber	1663	(29)	1658	(23)		
	1593–1605	5	aromatic skeletal vibrations plus C=O stretch; S > G; G condensed > G etherified	1596	(46)	1593	(54)	1601	(75)
	1505–1515	6	aromatic skeletal vibrations; G > S	1510	(95)	1505	(60)	1511	(77)
	1460–1470	7	C—H deformations; asym. in $—CH_3$ and $—CH_2—$	1464	(60)	1462	(63)	1462	(68)
	1422–1430	8	aromatic skeletal vibrations combined with C—H in-plane deform.	1423	(53)	1422	(53)	1423	(56)

1365–1370	9	aliphatic C—H stretch in CH_3, not in OMe; phen. OH	1367	(33)	1367	(27)		
1325–1330	10	S ring plus G ring condensed; (i.e., G ring substituted in pos. 5)	1326	(38)	1329	(48)	1329	(57)
1266–1270	11	G ring plus C=O stretch	1269	(100)	1266	(48)	1267	(80)
1221–1230	12	C—C plus C—O plus C=O stretch; G condensed > G etherified	1221	(70)	1227	(67)	1229	(81)
1166	13	typical for HGS lignins; C=O in ester groups (conj.)					1166	(71)
1140	14	aromatic C—H in-plane deformation; typical for G units; whereby G condensed > G etherified	1140	(78)				
1128–1125	15	aromatic C—H in-plane deformation (typical for S units); plus secondary alcohols plus C=O stretch			1126	(100)	1127	(100)
1086	16	C—O deformation in *secondary alcohols* and aliphatic ethers	1086	(45)				
1030–1035	17	aromatic C—H in-plane deformation, G > S; plus C—O deform. in primary alcohols; plus C=O stretch (unconj.)	1032	(76)	1033	(54)	1032	(58)
966–990	18	—HC=CH— out-of-plane deform., (trans)						
915–925	19	C—H out-of-plane; aromatic	919	(5)	925	(20)		
853–858	20	C—H out-of-plane in position 2, 5, and 6 of G units	858	(11)				
834–835	21	C—H out-of-plane in position 2 and 6 of S, and in all positions of H units			835	(10)	834	(26)
817–832	22	C—H out-of-plane in positions 2, 5, and 6 of G units			817	(8)		
to 780								

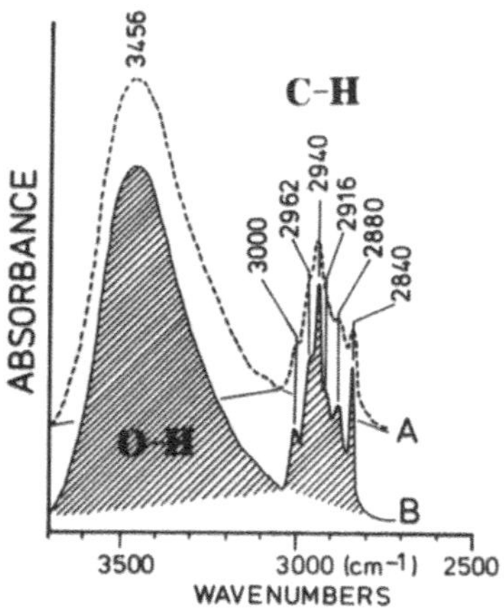

Fig. 4.1.6. O—H and C—H stretching bands of European beech (*Fagus sylvatica*) MWL. *A* 4 cm^{-1} resolution; *B* spectrum after mathematical resolution enhancement using deconvolution technique. (Experimental conditions described in legend to Fig. 4.1.11)

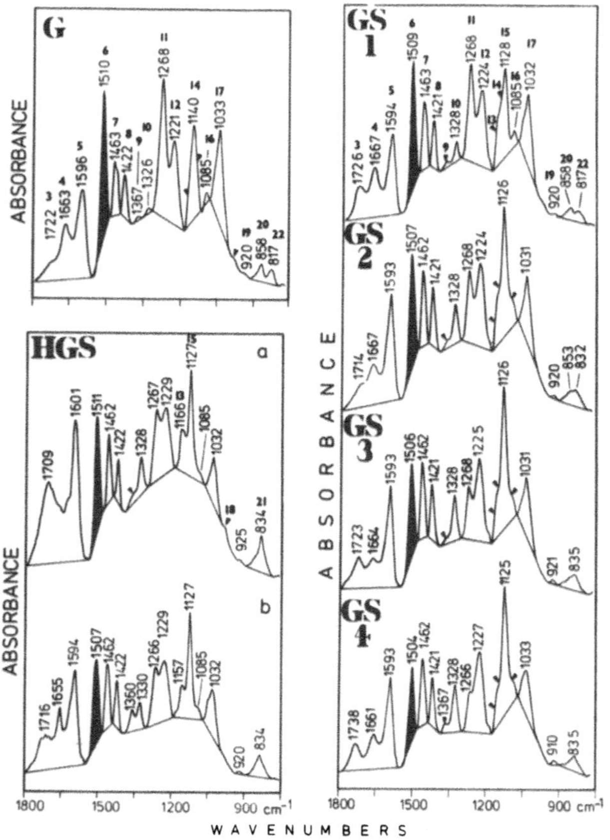

Fig. 4.1.7. Baseline-corrected and normalized FTIR spectra illustrating spectral types G, GS 1-GS 4, and HGS. *G* Norway spruce (*Picea abies*); *GS 1* (*Gnetum venosum*); *GS 2* lauan (*Shorea polysperma*); *GS 3* dabema (*Piptadeniastrum africanum*); *GS 4* birch (*Betula* sp.); HGS *a* bamboo (*Bambusa* sp.); *b* barley (*Hordeum vulgare*). (Instrument: FTS 40 Bio-Rad, Digilab, 4 cm^{-1} resolution, 32 scans, KBr pellet technique). According to Faix 1991

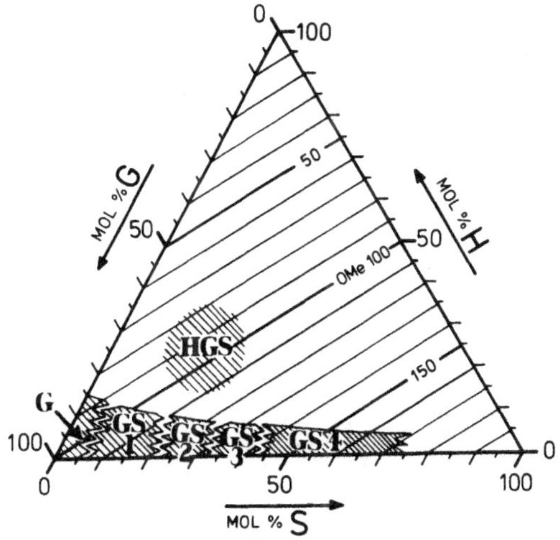

Fig. 4.1.8. Triangular coordinate system displaying the three types of lignin (*G*, *GS*, and HGS) in realtion to the composition of the lignins (*H*, *G*, and *S* units expressed as mol % and the methoxyl contents as the number/100 phenylpropane units). According to Faix 1991

In Fig. 4.1.7, the GS-type is subdivided into four subgroups based on the ratio of the 1509/1462 and 1268/1224 bands. The spectral differences are mostly the result of variation in the H, G, and S ratios in lignin. For simplicity, the three lignin types and the four subgroups, GS1–GS4, are described by a triangular coordinate system (Fig. 4.1.8) based on the contents of the basic lignin building units (i.e., G, S, and H units). In agreement with earlier studies (Sarkanen and Ludwig 1971), most wood lignins belong to the GS type containing trace amounts of H units. The S content of GS-type lignins may vary from a few percent to 70%. The HGS type is isolated from the other two types and has an H content of 15 to 20% and an S content of about 20%. This subject has been discussed in greater detail by Faix (1991).

4.1.3.2 Influence of Carbonyl Groups

As shown in the above section, the ratio of H, G, and S units exerts a major influence on the characteristics of lignin IR spectra. Carbonyl groups (bands 3 and 4) also show dominant effects, particularly in bands 5, 11, 12, 13, and 17. The effects usually are eliminated by reduction, as indicated by the so-called $NaBH_4$-reduction difference spectrum shown in Fig. 4.1.9. Sodium borohydride reduction decreases the band intensity at 1724–1741 cm^{-1} (all samples), 1668 cm^{-1} (*Picea abies*), 1602 cm^{-1} (bamboo), 1593 cm^{-1} (*Picea abies*), 1270–1230 cm^{-1} (all samples), and 1170–1127 cm^{-1} (all samples), resulting in positive peaks in the difference spectra. The bands at 3500, 3200, 2930, 1610–1620, 1516, 1202, 1100–1110, and 1014 cm^{-1} increase in intensity, resulting in the negative peaks. Figure 4.1.10 is a compilation of C=O group assignments for lignin model compounds. Unsaturated aliphatic esters are not reduced by

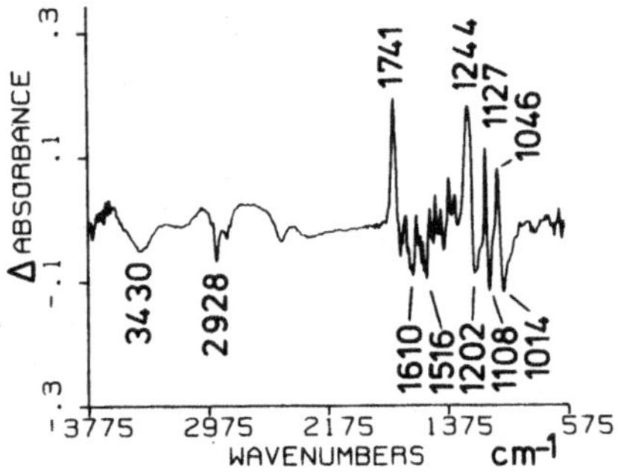

Fig. 4.1.9. Difference spectrum of oak (*Quercus obtusata*) obtained by subtracting the spectrum of MWL from the spectrum of the same MWL after $NaBH_4$ reduction. (Instrument: Nicolet 20SX D, $4\,cm^{-1}$ resolution, 32 scans, KBr pellet technique)

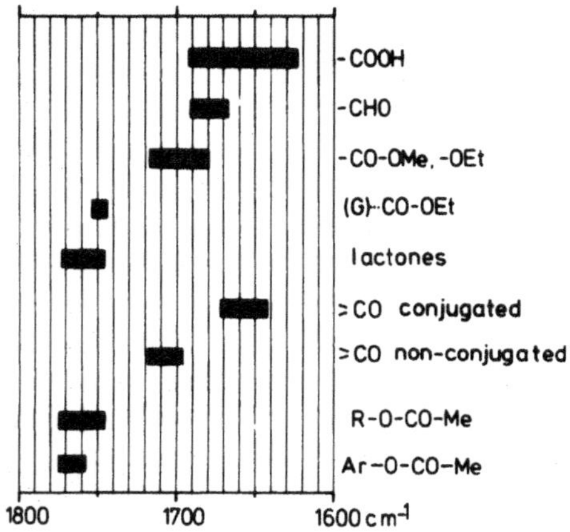

Fig. 4.1.10. Range of carbonyl group vibrations observed in lignin model compounds

$NaBH_4$ and the carboxyl bands are shifted to $1590-1560\,cm^{-1}$ upon conversion to their sodium salts. Water present in KBr pellets gives rise to a signal at $1625\,cm^{-1}$, which may interfere with the conjugated C=O band. A high signal intensity in the $1700-1760\,cm^{-1}$ range often indicates the presence of hemicellulose in milled wood lignin. Phenolic acid esters, especially in grass lignins, may also contribute to the signal in this range.

4.1.3.3 Mathematical Resolution of Spectra

In the spectra of macromolecules, the overlapping bands cannot be resolved by instrumental manipulation alone. Instead, mathematical techniques such as deconvolution, derivative spectroscopy, and band fitting are used for improving the spectral quality.

Deconvolution

Deconvolution is manipulation of the interferogram with an exponential weighting function. It may be viewed as the inverse of smoothing. In deconvoluted spectra, shoulders and points of inflection become peaks, and narrow bands are sharpened. For theoretical reasons, few advantages can be expected from the very broad composite carbonyl bands of lignin at 1630–1750 cm^{-1}, and no effect can be seen on the OH band (Fig. 4.1.6). Moreover, the spectra frequently show background noise in the carbonyl region due to traces of vapor in the instrument and adsorbed water in the pellet. This is a serious limitation because a high signal-to-noise ratio is a prerequisite for successful deconvolution.

Deconvolution provides a better indication of the approximate number of individual components contributing to a superimposed band. Figure 4.1.11 shows the spectra of three milled wood lignins before and after deconvolution (Faix and Beinfhoff 1988). Table 4.1.2 lists the wavenumbers and intensities of bands after deconvolution. The band intensities change when a different band width is used for the deconvolution, but the band positions (wavenumbers) remain constant.

Derivative Spectroscopy

First derivative IR spectra of a softwood and hardwood lignin are shown in Fig. 4.1.12. They display 20–22 positive and negative peaks instead of the 12 to 14 maxima appearing in the original spectra. Broad bands with low intensity are

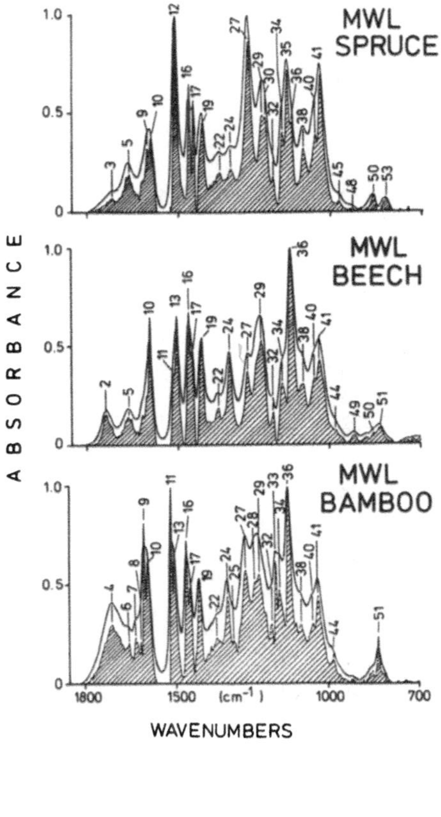

Fig. 4.1.11. FTIR spectra of spruce, beech, and bamboo MWL before and after deconvolution (Faix and Beinhoff 1988). (Instrument): Bio-Rad, Digilab FTS 40, resolution 4 cm^{-1}, 64 scans, bandwith for deconvolution 15 cm^{-1}, KBr pellet technique). By courtesy of Marcel Dekker Inc.

Table 4.1.2. Wavenumbers and intensities of IR bands after enhanced resolution obtained by deconvolution; bandwith for deconvolution, 15 cm^{-1}. According to Faix and Beinhoff 1988, by courtesy of Marcel Dekker Inc.

Band #			MWLs						Remarks
4.1.7	4.1.11	x cm^{-1}	Spruce cm^{-1}	% A	Beech cm^{-1}	% A	Bamboo cm^{-1}	% A	
	1	1771							in GS DHPs
3	2	1737	1737	(4)	1737	(17*)	1737	(18)	
	3	1720	1718	(7*)	1718	(10)	1718	(25)	
	4	1710	1710	(5)	1710	(5)	1710	(30*)	
4	5	1664	1664	(19*)	1664	(16*)	1664	(19)	
	6	1654					1655	(21*)	
	7	1634					1634	(24*)	
5	8	1614	1616	(16)	1617	(16)	1616	(21*)	
	9	1604	1604	(33*)	1604	(26)	1605	(82*)	
	10	1593	1593	(37*)	1592	(67*)	1593	(62*)	
6	11	1516			1515	(38*)	1515	(100*)	
	12	1511	1511	(100*)					
	13	1505			1504	(67*)	1504	(65*)	
	14	1498							in G DHPs
	15	1493	1493	(12)					
7	16	1465	1465	(66*)	1465	(69*)	1465	(73*)	
	17	1453	1451	(58*)	1455	(48*)	1453	(49*)	
8	18	1428	1427	(28)			1428	(15)	
	19	1422	1421	(46*)	1421	(56*)	1422	(52*)	
9	20	1384	1384	(16)			1384	(19*)	
	21	1373							in DHPs
	22	1367	1366	(20*)	1368	(18*)	1365	(23*)	
	23	1344							in DHPs
10	24	1329	1328	(22*)	1328	(46*)	1329	(44*)	
	25	1311					1311	(21*)	
	26	1290					1290	(29)	
11	27	1270	1269	(87*)	1270	(38*)	1269	(58*)	
12	28	1239			1240	(47)	1237	(52*)	
	29	1225	1224	(49*)	1224	(54*)	1225	(56*)	
	30	1214	1215	(48*)					
	31	1204					1204	(37*)	
	32	1186	1189	(18*)	1185	(17*)	1184	(31*)	
13	33	1170					1170	(63*)	
	34	1157	1158	(41*)	1154	(32*)	1158	(48*)	
14	35	1143	1141	(72*)					
15	36	1128	1128	(46*)	1128	(100*)	1127	(100*)	
	37	1104					1104	(31*)	
16	38	1086	1089	(33*)	1086	(32*)	1088	(30*)	
	39	1057							in DHPs
17	40	1048	1048	(39*)	1048	(34*)	1048	(31*)	
	41	1032	1031	(70*)	1032	(44*)	1031	(45*)	
	42	1020	1020	(24)	1020	(34)	1020	(24)	

Table 4.1.2. (continued)

Band #			MWLs						Remarks
4.1.7	4.1.11	x cm^{-1}	Spruce cm^{-1}	% A	Beech cm^{-1}	% A	Bamboo cm^{-1}	% A	
18	43	991							in GS DHPs
	44	982			981	(8*)	984	(12*)	
	45	972	972	(6*)					
	46	966							in G DHPs
	47	945							in G DHPs
19	48	926	927	(2*)					
	49	920			918	(6*)	922	(2*)	
20	50	854	856	(10*)	854	(7*)	854	(7*)	
21	51	831			834	(9*)	834	(24*)	
22	52	822							in G DHPs
	53	816	816	(7*)			815	(5*)	
	54	704							in GS DHPs

Bands labeled with * show a maximum. A = Absorbance of normalized spectra. "A" for the highest band equals 100. (Instrument: Bio-Rad, Digilab FTS 40)

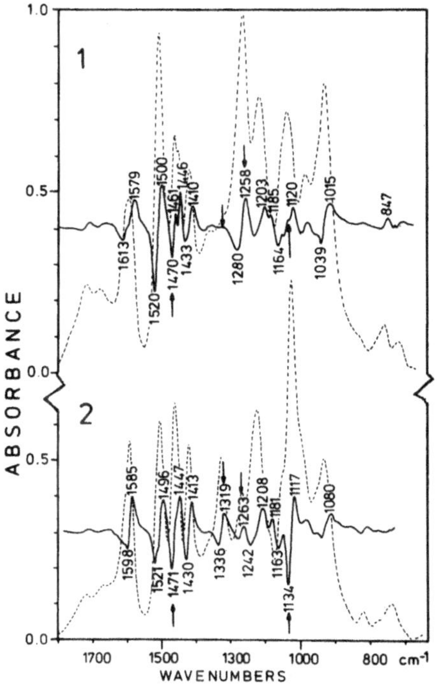

Fig. 4.1.12. IR spectra (*dotted line*) and first derivative IR spectra of two Willstätter lignins. *1* Bald cypress (*Taxodium distichum*), example of a G lignin; *2* tulip tree (*Liriodendron tulipifera*), example of a pronounced hardwood (GS 4) lignin

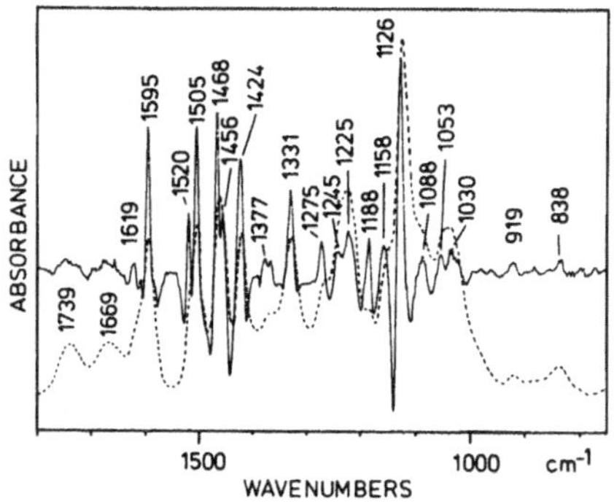

Fig. 4.1.13. IR spectrum (*dotted line*) and second derivative IR spectrum of tulip tree (*Liriodendron tulipifera* sapwood MWL. (Resolution 4 cm^{-1}, 64 scans, KBr pellet technique)

suppressed and the disturbing effect of a tilted baseline disappears. A disadvantage is that the resulting spectra display no similarity with the original. Despite this, first derivative spectra can be evaluated quantitatively. The peaks labeled with an arrow, for example, clearly highlight the differences between G and GS lignins. However, information on the carbonyl groups is lost in the derivatization. Figure 4.1.13 displays the original and the second derivative spectra of a hardwood MWL. At least 26 positive major peaks are visible in full conformity with the results of the deconvolution data presented in Table 4.1.2. Deconvolution and second derivative spectra give exactly the same wave numbers for the bands arising from the increase resolution. The use of second derivative IR spectroscopy for characterization of celluloses and technical lignins has been reported by Michell (1988d).

Band Fitting

Band fitting (curve fitting, or band shape analysis) is conveniently performed with commercially available software packages which fit, interactively or automatically, Gaussian or Lorentzian line shapes (or their combinations) to an unknown band profile. The applicability of this technique to lignin chemistry is not known at present.

4.1.3.4 Effect of Acetylation

Derivatization of lignin often discloses hidden features of IR spectra. A difference spectrum of acetylated lauan (*Shorea polysperma*) MWL, shown in Fig. 4.1.14, demonstrates the effects of acetylation. The OH band at 3410–3460 cm^{-1} disappears upon acetylation; acetoxy bands appear at 1765–1772 cm^{-1} (shoulder, aromatic), 1743 cm^{-1} (aliphatic), 1371 cm^{-1}, as do peaks in a cluster with maxima at 1217, 1225–1227, and 1198 cm^{-1}. It is noteworthy

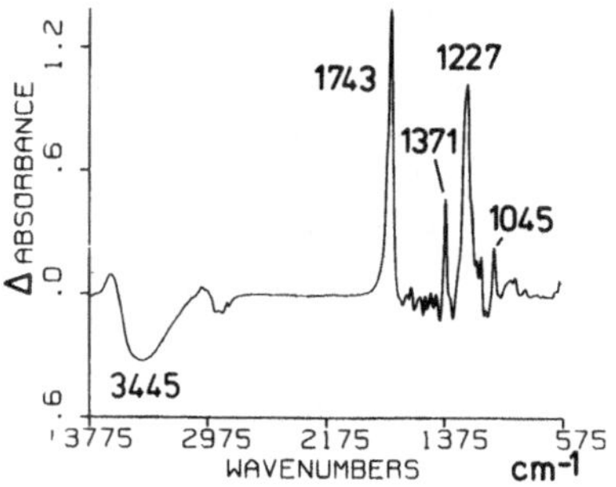

Fig. 4.1.14. Difference spectrum obtained by subtracting the spectrum of acetylated red lauan (*Shorea polysperma*) MWL from the spectrum of the same MWL before acetylation. (Resolution 4 cm^{-1}, 64 scans, KBr pellet technique)

that a sub-band of the aromatic band at 1518 cm^{-1} also disappears upon acetylation. (This effect can be seen only in the spectrum of the acetylated lignin).

In the spectra of acetylated organosolv lignins (i.e., lignin solubilized through partial splitting of its interunit ether linkages by alcohols), additional maxima appear at 1600, 1131, and 1044 cm^{-1}. The ratio of the acetoxy bands 1765/1743 cm^{-1} can be used for estimating the ratio of phenolic/aliphatic OH groups, as illustrated in Fig. 4.1.15. In this case, the pure acetoxy groups from acetylated lignins are shown before and after deconvolution. The increase in the phenolic hydroxyl group content of the organosolv lignin resulting from β-0-4 splitting reactions is clearly shown. The usefulness of FTIR spectroscopy for the structural analysis of acetylated lignins has been documented by Schultz and Glasser (1986).

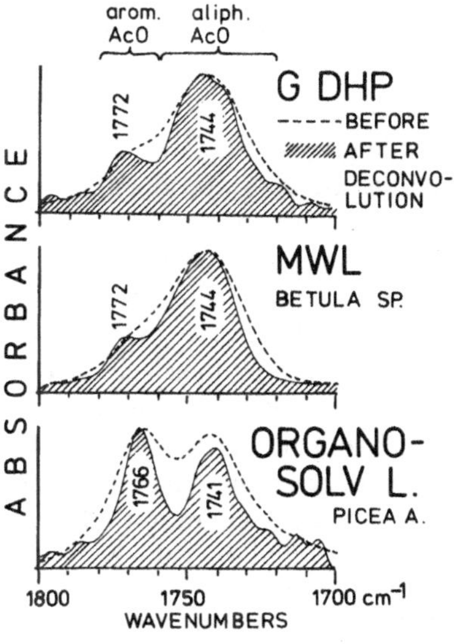

Fig. 4.1.15. Acetyl bands of acetylated lignins before and after deconvolution (Faix 1987). Acetylation in Ac_2O/pyridine. (Conditions described in legend to Fig. 4.1.11 caption)

4.1.3.5 Quantitative Evaluation

FTIR spectroscopy furnishes rather comprehensive data for quantitative work, even though the overlapping of bands occasionally is a limiting factor. The procedure for quantitative evaluation consists of several phases as described below.

Generation of IR Data Sets

IR spectra of lignin can be presented in three forms, namely as raw spectra, as baseline-corrected raw spectra, and as baseline-corrected normalized spectra. The band heights are determined from either the absolute zero line or from a baseline. In the latter case, there are again two possibilities: measurement of the heights of two or more bands from one common baseline or from the construction of an individual baseline for each band constructed by connecting the left side minimum with the right side minimum of the particular band. If an "individual" baseline is constructed, the integrated area below this band can be used for quantification. Better interlaboratory comparability is achieved with data taken from baseline-corrected and normalized spectra, provided that the baseline correction is always made in the same way. Even with the normalized spectra, it is advantageous to select an internal standard band for calculating the relative band intensities of all absorption bands. The aromatic skeletal vibration band at 1505–1510 cm^{-1} is conventionally used as a reference band, but the bands at 2900, 1463, or 1420 cm^{-1} also may be used. Schultz et al. (1985a, and Schultz and Glasser 1986), for example, selected band numbers 5, 6 and 8 in Table 4.1.1 as the internal standards for generating IR data sets whereas Zavarin et al. (1982) used the 1420 cm^{-1} band for internal standardization. Nevertheless, the 1505–1510 cm^{-1} band should be used, if possible, to increase comparability.

The choice of the reference band (or bands), however, influences less the results than the reproducibility of the baseline construction and the correct reading of the maxima. Modern software in FTIR spectroscopy are useful in automatic and reproducible location of baselines, maxima, integral values, and baseline-corrected absorbances.

Reproducibility problems may arise when tangential baselines (for example on the right side of the 1033 cm^{-1} band) have to be constructed. As a rule, data sets generated on the basis of height measurements are as reliable and reproducible as those based on integral values. The reproducibility of relative band intensities in the transmission mode is, for most of the bands, about ±0.5 to 1%. An exception is the 1600 cm^{-1} band which, because it is strongly influenced by humidity, exhibits poor reproducibility (±10%).

Sample preparation influences DRIFT spectra, as already pointed out. For example, dilution with KBr affects mainly the carbonyl region and the wave-number range between 1157 and 1046 cm^{-1}. Prolonged milling of the wood in KBr causes changes in the 1450–1046 cm^{-1} region (Fig. 4.1.16) as a result of decreased particle size. The use of KCl and NaCl as diluting media helps avoid noise in the carbonyl region resulting from moisture.

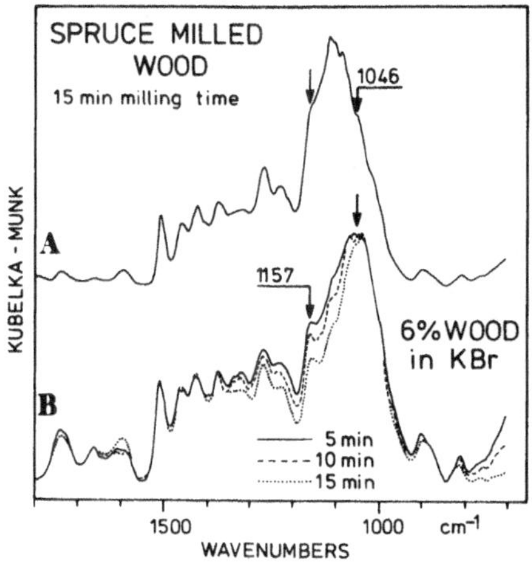

Fig. 4.1.16. DRIFT spectra of spruce milled wood. *A* Without dilution with KBr; *B* diluted with KBr (6% conc.) after 5, 10, and 15 min milling time in a vibratory ball mill

Calibration

The most important multivariate calibration techniques include multiple linear regression (MLR), principal component regression (PCR), and the partial least squares (PLS) method. All of these statistical techniques are used to arrive at a model that relates the IR data set (R) to the analytical data set (C). They are, in the given order, increasingly reliable with regard to mathematical stability and less sensitive to statistically irrelevant information. Under ideal conditions (i.e., when all bands show linear responses and there are no interfering signals and co-linearities between the data), the above techniques give similar results. In practice, however, the PLS and PCR techniques have essential advantages (Beebe and Kowalski 1987).

The curve-fitting calibration technique is a process whereby sample spectra are modeled as linear combinations of constituent spectra. This procedure is better suited for the analysis of a mixture of known components than for complex biopolymers such as lignin. Nevertheless, ongoing successful work throughout the world on the calibration of FTIR data with those of wet chemistry demonstrates that FTIR spectroscopy can be a powerful tool for quantitative (or at least semi-quantitative) lignin analysis.

For additional information on multivariate calibration techniques, the reader should consult Antoon et al. (1979), Gillette (1983), Gillette et al. (1985), Sjöström et al. (1983), Frank et al. (1984), Naes and Martens (1984), Beebe and Kowalski (1987), Hirschfeld (1987), Compton et al. (1987), and Schultz et al. (1989).

For lack of space, the quantitative IR spectral evaluation of lignin cannct be dealt with in detail here. However, the literature surveyed below may help the reader to access this exciting new area of lignin chemistry. Sarkanen et al. (1967a,b) were the first to report on the quantification of IR spectroscopic data.

They observed a linear relationship when the relative IR band intensities of several $NaBH_4$-reduced hardwood and softwood MWLs were plotted against their OMe contents. Later, several authors such as Faix and Schweers (1974), Kawamura et al. (1974, 1977), Lai and Sarkanen (1975), Salud and Faix (1980), Yasuda and Sakakibara (1981), Obst (1982), and Chum et al. (1984) reported on quantitatively evaluated IR spectra. Recently, the advent of FTIR spectroscopy has led to an increasing number of publications dealing with quantitatively evaluated IR spectra and their correlation with wet chemistry data, for example Schultz et al. (1985a,b), Faix (1986, 1987), Schultz and Glasser (1986), Schultz and Nicholas (1987), Abott et al. (1988), Obst et al. (1989), and Owen and Thomas (1989). Schultz and Glasser (1986) and Grandmaison et al. (1987) have provided examples of quantitative data evaluation in DRIFT spectroscopy.

4.1.4 Lignin Spectra in the Near-Infrared Region (4000–10 000 cm^{-1})

FTIR spectra in the near-IR region consist entirely of overtones and combinations of primary bands within the mid-IR region. For macromolecules or complex mixtures, the excessive overlapping of bands produces a diffuse absorption continuum with few characteristic features, making unequivocal band assignment practically impossible. Thus, this spectral range has a limited use in qualitative analysis. Even so, a major asset of near-IR analysis is the ease with which reproducible spectra can be obtained by reflectance and transreflectance (a combination of transmission and reflectance) techniques in every state of aggregation without complicated sample preparation.

Figure 4.1.17a and b show near-IR spectra of milled softwoods and hardwoods and milled wood lignins isolated from the same wood. The envelopes of nonresolved overlapping bands originate mainly from the 1st, 2nd, and 3rd overtones of various types of C—H and O—H vibrations. There are only minor, but significant, differences between the spectra that are useful for quantitative analysis. Here, derivative spectroscopy is particularly beneficial for eliminating the tilted baselines and accentuating these small differences. Acetylation of milled wood lignin causes significant changes in the spectra (Fig. 4.1.17c).

Osborne and Fern (1986) have presented detailed band assignments in the near-IR range, and Wetzel (1983) and William and Norris (1987) have summarized the salient points of near-IR spectroscopy. The latter authors emphasize applications for the agricultural and food industries. Birkett et al. (1989) have reported the successful kappa number determination of pulps using spectra in the near-IR range.

4.1.5 Lignin Spectra in the Far-Infrared Region (50–500 cm^{-1})

Very little information is available on lignin spectroscopy in this range. The spectra of beech and bamboo MWLs (Fig. 4.1.18) again show only diffuse bands originating from aromatic ring deformation vibrations. Sulfur-containing technical lignins display more pronounced features because of C—S vibrations.

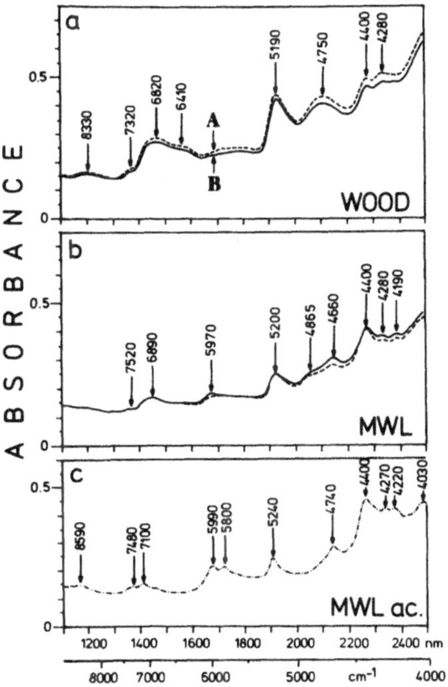

Fig. 4.1.17a–c. Near-infrared spectra. (**a**) Milled wood and MWL from common wingnut (*Pterocarya fraxinifolia*), hardwood lignin. (**b**) Milled wood and MWL from dawn redwood, (*Metasequoia glyptostroboides*). (**c**) Acetylated Norway spruce (*Picea abies*) MWL. (Instrument: Technikon Infralyzer 500; courtesy of E. Ignatzek, Bran & Luebbe, Hamburg)

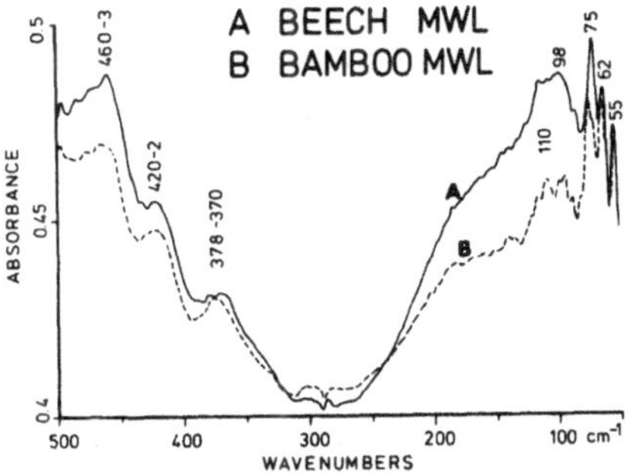

Fig. 4.1.18. Infrared spectrum of European beech (*Fagus sylvatica* L.) and bamboo (*Bambusa* sp.) MWLs in the near infrared range. (Instrument: Nicolet 7000; Resolution: 8 cm^{-1}, 128 scans; courtesy of J. Mink, Budapest)

4.1.6 Concluding Remarks

FTIR spectroscopy is a versatile, rapid, and reliable technique for lignin characterization. Baseline-corrected and normalized spectra facilitate interlaboratory comparisons. Using this analytical technique, the p-hydroxyphenyl, guaiacyl and syringyl units, methoxyl groups, carbonyl groups, and the ratio of phenolic OH to aliphatic OH groups can be determined. The DRIFT technique is well suited for wood surface characterization and lignin determination in wood and lignocellulosics in general, although its reproducibility is poor. Mathematical resolution techniques are useful for revealing hidden spectral infromation. However, FTIR spectroscopy in the near- and far-IR ranges needs further development.

References

Abott TP, Palmer DM, Gordon SH, Bagby MO (1988) Solid state analysis of plant polymers by FTIR. J Wood Chem Technol 8:351–357

Antoon MK, D'Esposito L, Koenig JL (1979) Factor analysis applied to Fourier transform infrared spectra. Appl Spectrosc 33:351–357

Bartick EG (1985) Microscopy/infrared spectroscopy for routine sample size. Appl Spectrosc 39:885–890

Beebe KR, Kowalski BR (1987) An introduction to multivariate calibration and analysis. Anal Chem 59:1007A–1017A

Bell RJ (1972) Introductory Fourier transform spectroscopy. Academic Press, New York, 382 pp

Berben SA, Rademacher JP, Sell LO, Easty DB (1987) Estimation of lignin in wood pulp by diffuse reflectance Fourier-transform infrared spectrometry. Tappi J 70(11):129–133

Birkett M, Gambino M, Meyer JH, Egers D (1989) Estimation of kappa number of pulps by near-infrared spectroscopy. Tappi J 72(9):193–197

Bracewell R (1965) The Fourier transformation and its applications. McGraw-Hill, New York, 381 pp

Cameron DG, Moffatt DJ (1984) Deconvolution, derivation, and smoothing of spectra using Fourier transforms. Test Eval 76:83

Chang H-m, Sarkanen KV (1973) Species variation in lignins. Effect of species on the rate of kraft delignification. Tappi 56:132–134

Chum HL, Ratcliff, Schroeder HA, Sopher DW (1984) Electrochemistry of biomass-derived materials. Characterization, fractionation, and reductive electrolysis of ethanol-extracted explosively depressurized aspen lignin. J Wood Chem Technol 4:505–532

Compton DAC, Young JR, Kollar RG, Mooney JR, Grasselli JG (1987) In: McClure GL (ed) Computerized quantitative infrared analysis. ASTM, Philadelphia, 36–57

Cooley JW, Tukey JW (1965) An algorithm for the machine calculation of complex Fourier series. Math Comput 19:297–301

Faix O (1986). Investigations on lignin polymer models (DHP's) by FTIR spectroscopy. Holzforschung 40:273–280

Faix O (1987) Quantitative FTIR-spektroskopische Untersuchungen an Ligninen und Ligninmodellsubstanzen. Habilitation Thesis, University of Hamburg

Faix O (1991) Classification of lignins from different botanical origins by FTIR spectroscopy. Holzforschung 45 (Suppl, Sept):21–27

Faix O, Beinhoff O (1988) FTIR spectra of milled wood lignins and lignin polymer models (DHPs) with enhanced resolution obtained by deconvolution. J Wood Chem Technol 8:505–522

Faix O, Németh K (1988) Monitoring of wood photodegradation by DRIFT-spectroscopy. Holz Roh- Werkst 46:112

Faix O, Patt R, Beinhoff O (1987) Grundlagen und Anwendung von FTIR-Spektroskopie bei der Herstellung und Analyse von Zellstoffen. Papier 41:657–663

Faix O, Schweers W (1974) Vergleichende Untersuchungen an Polymermodellen des Lignins (DHPs) verschiedener Zusammensetzungen. 3 Mitt. IR-spektroskopische Untersuchungen. Holzforschung 28:50–54

Ferraro JR, Basile LJ (1978) Fourier transform infrared: application to national problems. In: Ferraro JR, Basile LJ (eds) Fourier transform infrared spectroscopy – applications to chemical systems, Vol. 4. Academic Press, New York, 275–302

Ferraro JR, Rein AJ (1985) Application of diffuse reflectance spectroscopy in the far-infrared region. In: Ferraro JR, Basile LJ (eds) Fourier transform infrared spectroscopy – applications to chemical systems, Vol. 4. Academic Press, New York, 244–282

Frank IE, Feikema J, Constantine N, Kowalski BR (1984) Prediction of product quality from spectral data using the partial least squares method. J Chem Inf Comput Sci 24:20–24

Fuller MP, Griffiths PR (1980) Infrared microsampling by diffuse reflectance Fourier transform spectrometry. Appl Spectrosc 34:533–539

Gillette PC (1983) Factor analysis for separation of pure component spectra from mixture spectra. Anal Chem 55:630–633

Gillette PC, Lando JB, Koening JL (1985) A survey of infrared spectral data processing techniques. In: Ferraro JR, Basile LJ (eds) Fourier transform infrared spectroscopy – applications to chemical systems, Vol. 4. Academic Press, New York, 1–47

Graham JA, Grim WM III, Fateley WG (1985) Fourier transform infrared photoacoustic spectroscopy of condensed-phase samples, In: Ferraro JR, Basile LJ (eds) Fourier transform infrared) spectroscopy – applications to chemical systems, Vol. 4. Academic Press, New York, 345–392

Grandmaison JL, Thibault J, Kaliaguine S, Chantal PD (1987) Fourier transform infrared spectrometry and thermogravimetry of partially converted lignocellulosic materials. Anal Chem 59:2153–2157

Green DW, Reedy GT (1978) Matrix-isolation studies with Fourier transform infrared. In: Ferraro JR, Basile LJ (eds) Fourier transform infrared spectroscopy – applications to chemical systems, Vol. 1. Academic Press, New York, 1–59

Griffiths PR (1975) Chemical infrared Fourier transform spectroscopy. Wiley, New York, 340 pp

Griffiths PR (1983) Fourier transform infrared spectrometry. Science 222:297–302

Griffiths PR, de Haseth JA (1986) Fourier transform infrared spectrometry. Wiley, New York, 672 pp

Harbour JR, Hopper MA, Marchessault RH, Dobbin CJ, Anczurowski E (1985) Photoacoustic spectroscopy of cellulose, paper and wood. J Pulp Pap Sci 11:J42–J47

Hauser M, Oelichmann J (1988) A critical comparison of solid sample preparation techniques in infrared spectroscopy. Microchim Acta (Wien), Spec. Issue, I:39–43

Hergert HL (1971) Infared spectra. In: Sarkanen KV, Ludwig CH (eds) Lignins. Occurrence, formation, structure and reactions. Wiley-Interscience, New York, 267–293

Hirschfeld T (1987) In: McClure GL (ed) Computerized quantitative infrared analysis. ASTM, Philadelphia, 169–179

Horlick G (1968) Introduction to Fourier transform spectroscopy. Appl Spectrosc 22:617–626

Kawamura L, Higuchi T (1964a) Comparative studies of milled wood lignins from different taxonomical origins by infrared spectroscopy. In: Grenoble Symposium 1964, Chimie et biochimie de la lignine, de la cellulose et des hemicelluloses. Les Imprimeries Reunies de Chambery, Chambéry, 439–456

Kawamura I, Higuchi T (1964b) Studies on the properties of lignins of plants in various taxonomical positions II. On the I.R. absorption spectra of lignins. Mokuzai Gakkaishi 10:200–206

Kawamura I, Shinoda Y, Nonomura S (1974) The comparison of relative intensities of IR absorption bands of MWL of various woods from tropical and temperate zones. Mokuzai Gakkaishi 20:15–20

Kawamura I, Shinoda Y, Ai TV, Tanada T (1977) Chemical properties of lignin of Eyrthrina wood. Mokuzai Gakkaishi 23:400–404

Koenig JL (1981) Fourier transform infrared spectroscopy of chemical systems. Acc Chem Res 15:171–178

Krishnan K (1988) Characterization of semiconductor silicon using the FT-TR microsampling techniques. In: Messerschmidt RG, Harthcock MA (eds) Infrared microspectroscopy: theory and applications. Marcel Dekker, New York, 139–151

Kuo MI, McClelland JF, Luo S, Chien PL, Walker RD, Hse CY (1988) Applications of infrared photoacoustic spectroscopy for wood samples. Wood Fiber Sci 20:132–145

Lai Y-Z, Sarkanen KV (1975) Structrual variation in dehydrogenation polymers of coniferyl alcohol. Cellul Chem Technol 9:239–245

Mackenzie MW (1988) Advances in applied Fourier transform infrared spectroscopy. Wiley, New York, 350 pp

Malhotra VM, Jasty S, Mu R (1989) FT-IR spectra of water in microporous KBr pellets and water's desorption kinetics. Appl Spectrosc 43:638–645

Michell AJ (1988a) Note on a technique for obtaining infrared spectra of treated wood surfaces. Wood Fiber Sci 20:272–276

Michell AJ (1988b) Infrared spectroscopy transformed – new applications in wood and pulping chemistry. Appita 41:375–380

Michell AJ (1988c) Usefulness of Fourier-transform infrared difference spectroscopy for studying the reactions of wood during pulping. Cellul Chem Technol 22:105–113

Michell AJ (1988d) Second derivative F.T.-I.R. spectra of celluloses I and II and related mono- and oligosaccharides. Carbohydr Res 173:185–195

Michell AJ, Garland CP, Nelson PJ (1989) Diffuse-reflectance infrared Fourier transform (DRIFT) spectroscopic study of bleaching and yellowing of eucalypt cold soda pulp. J Wood Chem Technol 9:85–103

Naes T, Martens H (1984) Multivariate calibration II. Chemometric methods. Trends Anal Chem 3:266–271

Obst JR (1982) Guaiacyl and syringyl lignin composition in hardwood cell components. Holzforschung 36:143–152

Obst JR, McMillan NJ, Blanchette RA, Christensen DJ, Crawford DM, Kuster TA, Landucci LL, Faix O, Newman RH, Pettersen RC, Schwandt VH, Weselowsky MF (1989) Proc 5th Int Symp Wood Pulp Chem, Raleigh, NC, Poster Sessions, 289–308, Tappi press, Athanta, GA and Geological Survey of Canada, Bulletin 403 (1991):123–146

Oelichmann J, Hauser M (1988) Feststoffuntersuchungen in Infrarot-Spektroskopie. Perkin Elmer's Applied Infrared Spectroscopy, 24, 7700 Überlingen, FRG

Osborne BG, Fearn T (1986) Near infrared spectroscopy in food analysis. Langma Scientific and Wiley, New York, 200 pp

Ostmeyer JG, Elder TJ, Winandy JE (1989) Spectroscopic analysis of southern pine treated with chromated copper arsenate II. Diffuse-reflectance Fourier-transform infrared spectroscopy (DRIFT). J Wood Chem Technol 9:105–122

Owen NL, Thomas DW (1989) Infrared studies of "hard" and "soft" woods. Appl Spectrosc 43:451–455

Painter P, Starsinic M, Coleman M (1985) Determination of functional groups in coal by Fourier transform interferometry. In: Ferraro JR, Basile LJ (eds) Fourier transform infrared spectroscopy – applications to chemical systems, Vol. 4. Academic Press, New York, 169–241

Pakdel H, Grandmaison JL, Roy C (1989) Analysis of wood vacuum pyrolysis solid residues by diffuse reflectance infrared Fourier transform spectrometry. Can J Chem 67:310–314

Perkins WD (1986) Fourier transform-infrared spectroscopy. Part I. Instrumentation. J Chem Educ 63:A5–A10

Perkins WD (1987) Fourier transform-infrared spectroscopy. Part II. Advantages of FT-IR. J Chem Educ 64:A269–A271

Rosencwaig A (1981) Photoacoustics and photoacoustic spectroscopy. Wiley, New York, 324 pp

Salud EC, Faix O (1980) The isolation and characterization of lignins of Shorea species. Holzforschung 34:113–121

Sarkanen KV, Chang H-m, Ericsson B (1967a) Species variation in lignins. Conifer lignins. Tappi 50:538–587

Sarkanen KV, Chang H-m, Allan GG (1967b) Species variation in lignins. III. Hardwood lignins. Tappi 50:587–590

Sarkanen KV, Hergert HL (1971) Classification and distribution. In: Sarkanen KV, Ludwig CH (eds) Lignins. Occurrence, formation, structure and reactions. Wiley-Interscience, New York, 43–94

Sarkanen KV, Ludwig CH (1971) Definitions and nomenclature. In: Sarakanen KV, Ludwig CH (eds) Lignins. Occurrence, formation, structure and reactions. Wiley-Interscience, New York, 1–18

Schiering DW, Oelichmann J, Rau A (1988) Prinzipien und Anwendungen der Infrarot-Mikroskopie. Perkin Elmer's Applied Infrared Spectroscopy, 7700 Überlingen, FRG

Schultz TP, Templeton MC, McGinnis GD (1985a) Rapid determination of lignocellulose by diffuse reflectance Fourier transform infrared spectrometry. Anal Chem 57:2867–2869

Schultz TP, McGinnis GD, Bertran MS (1985b) Estimation of cellulose crystallinity using Fourier transform-infrared spectroscopy dynamic thermogravimetry. J Wood Chem Technol 5:543–557

Schultz TP, Glasser WG (1986) Quantitative structural analysis of lignin by diffuse reflectance Fourier transform infrared spectromety. Holzforschung 40 (Suppl):37–44

Schultz TP, Nicholas DD (1987) Fourier transform infrared spectrometry. Detection of incipient brown rot decay in wood. Int Analyst 41:35–39

Sjöström M, Wold, S, Lindberg W, Persson J-A, Martens H (1983) A multivariate calibration problem in analytical chemistry solved by partial least-squares models in latent variables. Anal Chim Acta 150:61–70

Sommer AJ, Lang PL, Miller BS, Katon JE (1988) Application of molecular microspectroscopy to paper chemistry. Prac Spectrosc 6 (Infrared Microspectrosc): 245–258

Sweeney KM (1989) FTIR microscopy of pulp and paper samples. Tappi J 72(2):171–174

Wetzel DL (1983) Near-infrared reflectance analysis. Anal Chem 55:1165A–1176A

Williams P, Norris K (1987) Near-infrared technology in the agricultural and food industries. Am Assoc of Cereal Chemists, St. Paul, MN, 330 pp

Yang PW, Mantsch HH, Baudais F (1986) A critical evaluation of three types of diffuse reflectance infrared accessories. Appl Spectrosc 40:974–978

Yasuda T, Sakakibara A (1981) Hydrogenolysis of protolignin in compression wood. V. Isolation of two trimeric compounds with lactone ring. Holzforschung 35:183–187

Yeboah SA, Wang S-H, Griffiths PR (1984) Effect of pressure on diffuse reflectance infrared spectra of compressed powders. Appl Spectrosc 38:259–264

Zavarin E, Nguyen C, Romero E (1982) Preparation of enzymatically liberated lignin from naturally brown-rotted wood. J Wood Chem Technol 2:343–37

4.2 Ultraviolet Microscopy

K. Fukazawa

4.2.1 Introduction

Lignin in wood absorbs ultraviolet light much more strongly than the polysaccharide components. It is not surprising, therefore, that many authors have used UV microscopy to study lignin in wood. Lange (1954) determined the lignin content in the middle lamella and the secondary wall of spruce tracheids from UV photomicrographs. Wardrop (1957) utilized UV photomicrographs for detecting the beginning of lignification and its subsequent progress in differentiating xylem. Both Lange (1954) and Wergin (1965) elucidated the unusual lignin distribution in compression wood by means of UV photomicrographs. Jayme and Torgersen (1967) reported on the topochemistry of delignification in sulfite and sulfate pulping of spruce wood using a Leitz UV microscope.

More recently, Goring and his co-workers (Fergus et al. 1969, Fergus and Goring 1969, 1970a,b, Procter et al. 1967, Scott et al. 1969, Scott and Goring 1970a,b, Wood and Goring 1971, 1973, Wood et al. 1972, Musha and Goring 1975a,b, Yang and Goring 1978a,b, 1980) established quantitative methods for determination of the lignin distribution in xylem cells by UV microscopy. A variety of methods and equipment were used in these studies. They analyzed lignin quantitatively using mostly densitometric analysis of UV photomicrograph negatives. Tissue volumes were also calculated from the densitometric line profile. Subsequently, an image-analyzing system was introduced by Fukazawa (1974), Fukazawa and Imagawa (1981, 1983), and Takano et al. (1983) to reduce the error derived from densitometric analyses.

Boutelje and Jonsson (1980) used UV microscope photometry for the study of lignin in pulp fibers. They discussed the errors associated with illumination, UV absorption by substances other than lignin, and the varying specific volume of the fiber wall, as well as limitations of the technique when the Zeiss UV microscope photometer was used. They also developed a technique for lignin analysis using fragments from mechanical pulp (Boutelje and Eriksson 1982, 1984). UV spectra, taken with a computerized Zeiss UV microspectrophotometer, have been reported for cell walls degraded by enzymatic hydrolysis (Fujii et al. 1987), milled wood lignin embedded with epoxy resin (Fujii 1987), and the morphological elements of hardwoods (Takabe et al. 1987).

UV microscopic techniques have become increasingly precise over the years. In the present section an outline will be presented of the type of information which can be obtained by UV microscopy, together with a discussion of experimental techniques and the limitations of the methods.

Springer Series in Wood Science
Methods in Lignin Chemistry
(Edited by S.Y. Lin and C.W. Dence)

4.2.2 Information Obtained by UV Microscopy

4.2.2.1 Photomicrographs

The UV microscope provides an image of the lignin in a particular area of a woody tissue. Such an image is useful in determining the microscopic distribution of the lignin in the cell wall (Fig. 4.2.1). For example, unusual patterns of lignin distribution are seen clearly in the tissue of compression wood (Wood and Goring 1971, Fukazawa 1974), and in the delignification of CTMP fibers with ozone (Fig. 4.2.2) (Kojima et al. 1988a,b). The progress of lignification in tracheids (Wardrop 1957, Imagawa et al. 1976) and in the phloem cells (Fig. 4.2.3) (Imagawa 1981) have been followed by UV microscopy.

The resolving power of a UV microscope is better than that of an optical microscope, but much lower than that of an electron microscope and is given by the expression

$$d\ (\mu m) = (\text{wavelength} / \text{NA}) \times 0.61, \qquad (1)$$

where d = the minimum resolvable distance in μm, NA = the numerical aperture.

UV microscopes are supplied mainly by two optical companies, Zeiss and Leitz, in the Federal Republic of Germany. The major difference between the Zeiss and the Leitz instruments lies in their optical systems. The Leitz microscope uses quartz reflecting objectives, while Zeiss uses quartz refractive objectives. Both objectives are achromatic for a wide portion of the UV and visible light range, with no shift in focus accompanying the change of wave-

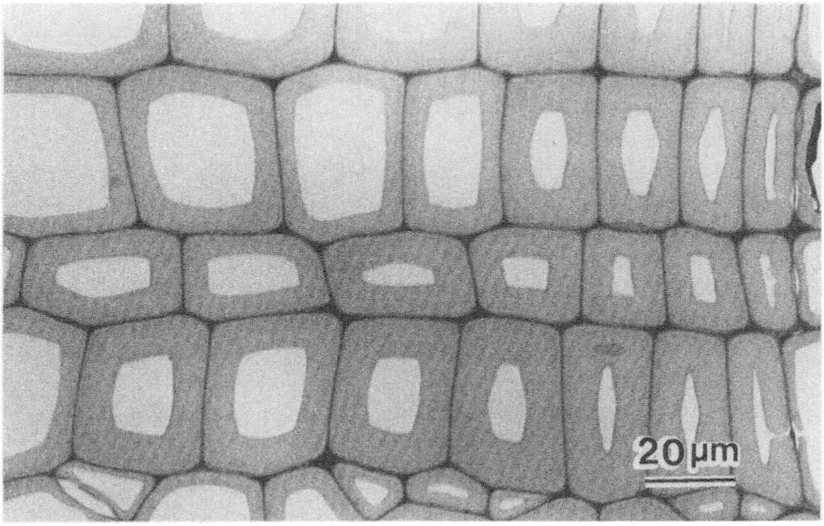

Fig. 4.2.1. A UV photomicrograph of *Picea glehnii* latewood taken at 280 nm. (Takano et al. 1983)

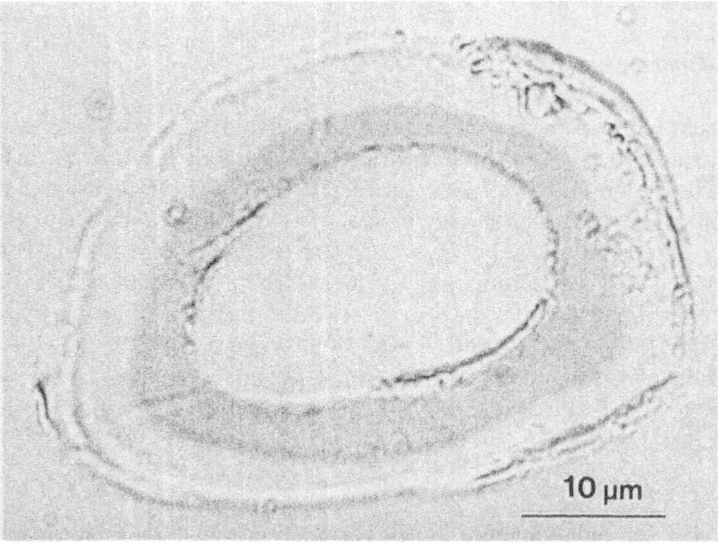

Fig. 4.2.2. A UV photomicrograph of a transverse section of CTMP fiber treated with ozone for 30 min showing the spread of delignification area to the middle part of secondary wall. (Kojima et al. 1988a)

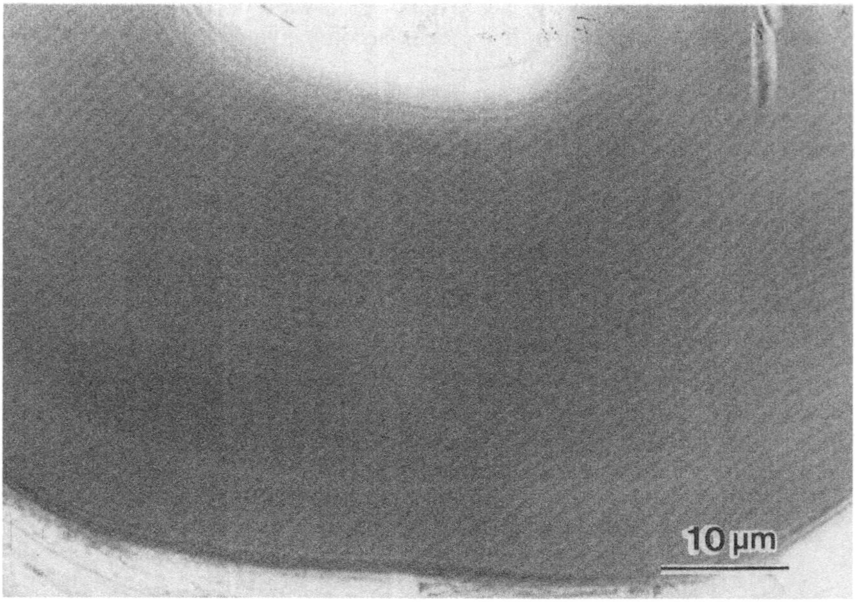

Fig. 4.2.3. A UV photomicrogrpah of thickened secondary wall of *Larix leptolepis* sclerenchyma cell. (Imagawa 1981)

length. However, the resolving power of refractive objectives is higher than that of reflecting types because the former have higher numerical apertures. For example, ULTRAFLUARS 100x glycerine immersion refractive lens (Zeiss) has an NA of 1.25, while the Leitz reflecting objectives Q Glyc. 300x has an NA of 0.85.

4.2.2.2 Lignin Concentration

When the proper conditions are met, UV microscopy can provide a quantitative measure of the lignin concentration in a microscopic area of wood tissue according to the Beer-Lambert law:

$$A = D \cdot c \cdot d, \qquad (2)$$

where A = UV absorbance, D = extinction (absorption) coefficient ($l\,g^{-1}$ cm^{-1}), c = concentration ($g\,l^{-1}$), d = section thickness (cm).

If the extinction coefficient and the section thickness remain constant, the lignin concentration is proportional to UV absorbance. Normal softwood lignin is composed mostly of guaiacyl units, and a constant value of D across the various morphological regions in the cell wall can be reasonably assumed (Fergus et al. 1969). With hardwoods, however, the method has a serious limitation in that hardwood lignins often contain more than one type of monomer unit, namely, guaiacyl, syringyl, and p-hydroxyphenyl. These monomer units have different absorptivities. Therefore, unless the composition of the lignin in the area under study is known, the absorptivity of the lignin is indeterminate and its concentration cannot be calculated. For example, the ratio of the guaiacyl to syringyl lignins in birch is approximately 3:1 (Fergus and Goring 1970a,b), and thus the calculation of lignin concentration becomes more complicated. This point is discussed in more detail in a later section.

4.2.2.3 UV Absorption Spectra

By measuring light absorption over a range of wavelengths, the UV spectrum of a wood section can be obtained. UV spectra differ depending on the type of lignin and are sometimes used to identify the monomer composition in a particular area of wood tissue. Guaiacyl, syringyl, and p-hydroxyphenyl units have different absorption maxima in their UV spectra. According to Fergus and Goring (1970a,b), UV absorbance maxima for guaiacyl and syringyl model compounds occur at 280 and 270 nm, respectively.

Takabe et al. (1987) recorded the UV absorption spectra within the various cell elements and at various differentiating stages for several hardwood species. Figure 4.2.4a to e shows UV absorption spectra of various cell wall layers of hybrid poplar during differentiation. The secondary wall of vessel elements (V-SW) exhibits an absorption maximum in the 274–276 nm range at the initial stage of lignification. The absorption maximum is shifted to 280 nm at a later stage of differentiation (Fig. 4.2.4a). This indicates that a p-hydroxyphenyl lignin is formed during the initial lignification, while guaiacyl lignin is deposited

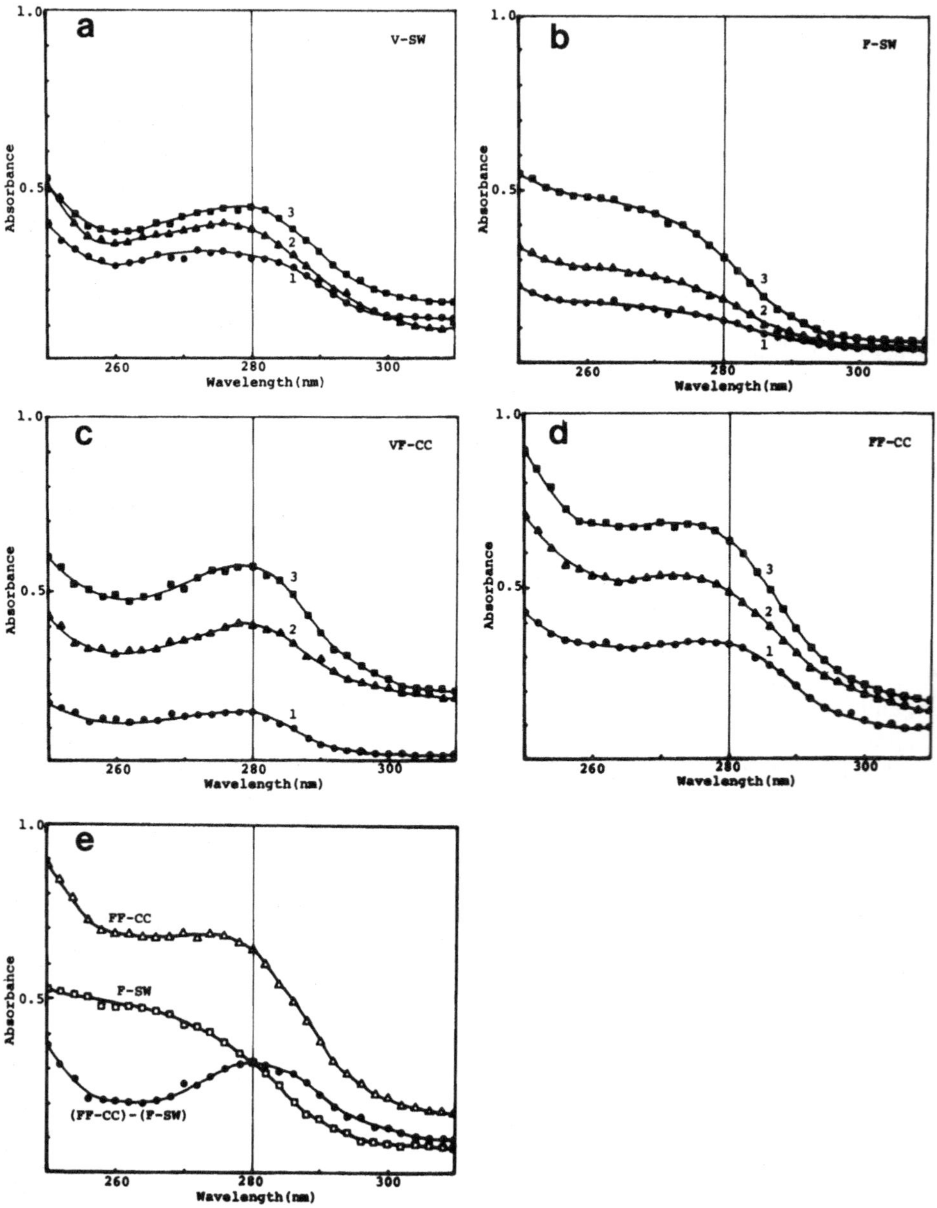

Fig. 4.2.4. a–d Changes in the UV absorption spectra of various cell wall layers of hybrid poplar during differentiation, showing the initial (*1*), middle (*2*), and later (*3*) stages of lignification. *V-SW* Secondary wall of vessel; *F-SW* secondary wall of fiber; *FF-CC* cell corner of middle lamella between fibers; *VF-CC* cell corner of middle lamella between vessel and fiber. **e** UV absorption spectra of mature fiber secondary wall (*F-SW*) and mature cell corner of middle lamella between fibers (*FF-CC*) obtained from the same poplar fiber. The UV spectrum shown by (*FF-CC*)-(*F-SW*) is the difference spectrum between the cell corner middle lamella and fiber secondary wall. (Takabe et al. 1987)

later. The secondary wall of the fibers (F-SW) shows no peak in the UV absorption spectra at any stage of the lignification (Fig. 4.2.4b), indicating the availability of a continuous supply of p-coumaryl alcohol during the lignification of F-SW. The cell corner middle lamella between the fibers (FF-CC) has a spectrum that is different from that of the fiber wall (Fig. 4.2.4d). The difference spectrum, obtained by subtracting the F-SW spectrum from the FF-CC spectrum, shows a distinctive maximum at 280 nm (Fig. 4.2.4e), suggesting that, at the cell corner, a considerable amount of guaiacyl lignin is deposited in addition to the p-hydroxyphenyl lignin. The lignins in the cell corner region between wood fibers and vessel elements (VF-CC) are mainly composed of guaiacyl units at all stages of lignification (Fig. 4.2.4c).

UV absorption spectra may also be used to estimate the phenolic hydroxyl content in lignin by application of the so-called $\Delta\varepsilon$ method (Aulin-Erdtman 1954). The increase in UV absorption at 300 nm ($\Delta\varepsilon$), as the pH of wood tissue is changed from 6 to 12, measures the concentration of ionized phenolic groups in the lignin. Yang and Goring (1978a,b, 1980) found that there is a greater concentration of phenolic hydroxyl groups in the secondary wall lignin than in the middle lamella lignin. Similarly, Boutelje and Eriksson (1982, 1984), using small, flake-like fragments of middle lamella and secondary wall from mechanical pulp, found that the phenolic hydroxyl content of secondary wall lignin is 1.5 times that of middle lamella lignin.

Limitations in the use of spectral data arise because of the complex nature of lignin and the difficulty in relating UV absorption data to lignin structure. Also, the precision of the results obtained with wood sections is somewhat lower than that obtained when measuring the UV spectra of model compounds in solution.

4.2.3 Experimental Methods

4.2.3.1 Specimen Preparation

Small wood chips are employed in UV microscopic measurements. If partially delignified wood is to be studied, care must be taken to conduct the delignification in such a manner that the fibers are not dispersed before the specimen is embedded. In the quantitative analysis of lignin, the wood chips should be solvent-extracted to remove extraneous materials. For *Larix leptolepis*, Imagawa and Fukazawa (1978) recommend sequential extraction with n-hexane at room temperature for 5 days, alcohol-benzene (1:2, v/v) in a Soxhlet extractor for 48 h, and acetone-water (1:1, v/v) at room temperature for 4 days.

After the extraction, the chips are embedded in a suitable resin using a capsule or a rubber bed. Embedding media commonly used for electron microscopy are also suitable for UV microscopy. However, the well-known Epon 812 formulation of Luft (1961) shows increasing absorbance at lower wavelengths (Wood and Goring 1973, Boutelje and Jonsson 1980). Instead, the following formulation is recommended for low absorbance:

	Component	*Weight %*
Epon 812	Mixture of di- and triglycidyl ethers of glycerol	30.9
ERL 4206	Vinyl cylohexene diepoxide	14.1
NSA	Nonenyl succinic anhydride	55.0
S-1	Dimethylaminoethanol	1.0

Removal of the epoxy resin from wood sections is sometimes necessary, especially for the determination of low concentrations of lignin (Imagawa et al. 1976). The preparation of sections for UV microscopy requires much practice and skill. Sectioning is usually carried out with a conventional ultramicrotome using a diamond knife. Defects such as thickness variation in successive sections, unequal thickness within a section, wrinkle or chatter deformity, are usually caused by a dull knife or by mismatching knife angle, cutting speed, block hardness, and cutting surface width.

Section thickness is an important parameter. The absorbance is higher in thick sections than in thin ones and thus the contrast between light and shade in the UV microscopic image is greater in thick sections. However, the increase of section thickness introduces errors caused by light scattering, nonparallel illumination, and diffraction. After a detailed study of the various factors involved, Scott et al. (1969) recommended a section thickness of $0.5\,\mu$m. The upper limit for section thickness for UV microscopy is about $1-2\,\mu$m (Boutelje and Jonsson 1980, Fujii 1987). For quantitative work it is necessary to know the section thickness accurately. This value may be estimated from the interference colors produced by the sections or from measurements with an interference microscope (Boutelje and Eriksson 1982, 1984).

4.2.3.2 Photography

UV photomicrography is an important part of UV microscopy. The Leitz UV microscope shown in Fig. 4.2.5 provides excellent photomicrographs. The instrument consists of a linear mirror monochromator, quartz reflection optics, an image converter, and a camera. For focusing at a particular UV wavelength, the image converter is used. The image from UV light is deflected to a photocathode, and a visible image is produced by electrons from the photocathode striking a fluorescent screen of the image converter. With the ULTRAFLUARS objectives (Zeiss), which has a range of UV to visible light, focusing is achieved with visible light from a halogen lamp or blue light from the monochromator.

Kodak Spectrum Analysis film 5367, No. 1, which has an optimum range of spectral sensitivity from 250 to 440 nm was used by Scott et al. (1969). This film provides strong contrast, high resolving power (200 lines/mm), and fine granularity (Eastman Kodak 1987). However, it is difficult to handle because of its delicate surface. In addition, it must be bought on special order and is expensive. Conventional films, such as Ilford EP4 (ASA 125) or Fuji Neopan SS (ASA 100), may also be used for image recording. Exposure time for ordinary film is about 30 s at 280 nm, when using a 5-nm band width and a 100x ULTRAFLUAR objective, and about 2–3 s when using a 32x ULTRAFLUAR

Fig. 4.2.5. A Leitz UV microscope used by the Goring and associates (PAPRICAN)

objective. A linear relationship between the silver density of the negatives and the logarithm of the exposure time should be confirmed for each wavelength in the UV region. Long exposures should be avoided to prevent degradation of the lignin. For example, exposure of a 0.5-μm transverse wood section to UV light for 2 h causes the absorbance at 280 nm to decrease to less than 25% of its original value (Scott and Goring 1970b).

To obtain sufficient sensitivity for measuring low lignin concentrations in pulp fibers, Wood and Goring (1973) used wavelengths below 240 nm, a high-intensity light source (1000 W xenon-mercury lamp), and a special film with a low gelatin content (Kodak Special Film 101-01). With these adaptations, the investigators were able to study the later stages of delignification.

4.2.3.3 Quantitative Analysis

The density of UV photomicrograph negatives is analyzed with a micro-densitometer such as the Joyce-Loebl. This instrument measures successively the density of a small area with a photomultiplier as the negative moves on the scanning stage. For converting silver density to lignin concentration, the methods developed by Scott et al. (1969) and Fergus et al. (1969) may be used.

The UV microscope may also be used as a microphotometer in which the transmittance from selected areas of the image is measured directly. The light intensity through a particular part of the specimen is measured by a photomultiplier. A diaphragm aperture is used to restrict the light from the selected area of the magnified image of the specimen. Flare light errors, known as the S-V effect (Schwarzschild-Villiger effect), are eliminated by inserting a pin hole diaphragm in the monochromatic beam and forming a reduced image of the pin hole on the specimen surface. For a Zeiss UMSP 80, the minimum measuring diameter in the object plane is $0.5\,\mu m$, which is suitable for measurement in the S2 layer and cell corner region of the middle lamella, but is not too large to fit the tangential or radial middle lamella without overlap.

The image analysis of UV photomicrographs is carried out by means of a microdensitometer using an X-Y scanning stage or a TV camera. However, the absorbance value obtained may be influenced by the methods of film development or by unevenness of illumination. It is difficult to eliminate the latter completely from the optical system of a UV microscope.

Rather than attempting an image analysis on photomicrographs, it is much better to use a modern system with a scanning stage and image analyzer built into the microscope. The Zeiss UMSP 80 microscope photometer, fitted with an X-Y scanning stage and with a minimum increment of $0.25\,\mu m$ (Fig. 4.2.6), has the advantage of permitting a random selection of measurement values and automatic recording of UV absorbance in an isodensity area. The computer program supplied by Zeiss, termed APAMOS (Automatic Photometric Analysis of Microscopic Objects by Scanning), can supply the following information:

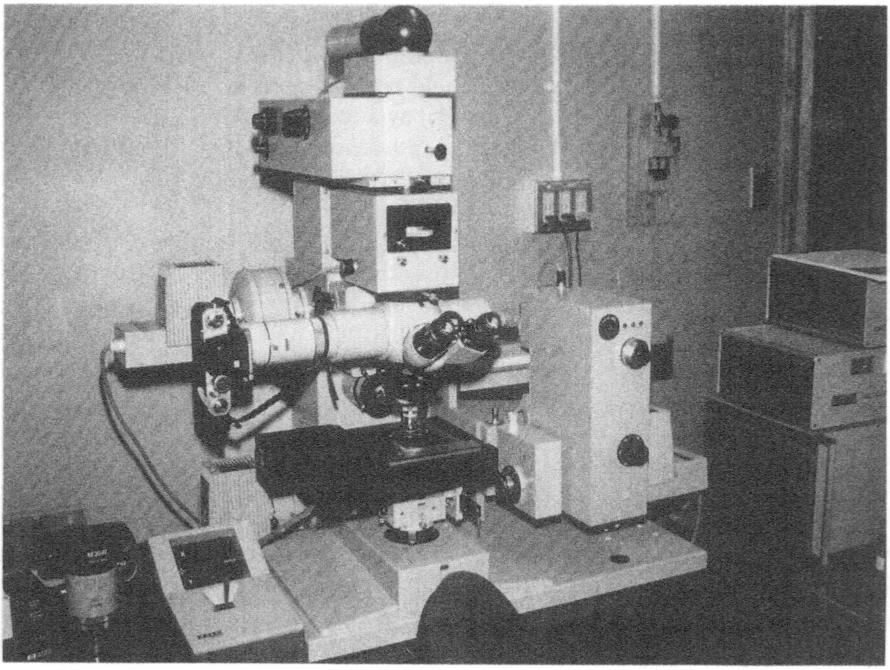

Fig. 4.2.6. A Zeiss universal microspectrophotometer system, Model UMSP 80

1. Printouts of data in their geometric arrangement.
2. Horizontal or vertical line profiles.
3. Plots of isodensity areas or intensity histograms.
4. Computed values of mean absorbance and accumulated area of measurement.

Another basic program for the Zeiss UMSP 80 is LAMBDA-SCAN, which is useful for spectral analysis. The light intensities at fixed wavelength intervals are measured automatically and printed out as a spectrum. The bandwidth can be precisely set to within 1 nm and the spectra corrected for the influence of ambient light.

The Leitz UVM microscope photometer system comes also with the PLUG computer program for single value measurements, and the SPECTRA program for spectral measurements. However, the latest Leitz MPV 3 system is not designed for UV light and its use as a UV microscope photometer requires special accessories.

It is also possible to obtain UV absorption spectra from a series of UV photomicrographs taken at intervals of 2.5 nm between 265 and 290 nm and developed under the same conditions (Fergus and Goring 1970a). However, such methods are tedious and less precise.

In the past, researchers have used different UV microscopic methods for the determination of lignin concentration. Two of the procedures are briefly described for softwoods.

Steps in the calculation of softwood lignin content based on the UV absorbance of isodensity areas (Fukazawa and Imagawa 1981, Takano et al. 1983) are as follows:

1. Determine the Klason lignin content of wood.
2. Obtain isodensity histogram showing cell wall areas of identical UV absorbance or transmittance.
3. Calculate the lignin content of each isodensity area using the established relationship between the mean absorbance and Klason lignin content of the wood.
4. Plot a cumulative curve of area versus lignin concentration.
5. Divide the cumulative curve into regions of secondary wall and middle lamella on the basis of lignin concentration.
6. Determine the average lignin concentrations in the two regions from the cumulative curve.

Steps in the calculation of lignin content based on the assumed absorptivity of softwood of lignin (Fergus et al. 1969) are as follows:

1. Assume for softwood lignin that

   $$D_{280} = 15.61\,g^{-1}\,cm^{-1}.$$

2. Determine the the UV absorbance of the secondary wall.
3. Calculate the lignin content in the secondary wall by use of Eq. (2) (in g/cm^3).
4. Convert the lignin content in g/cm^3 to g/g by multiplying by the specific volume of the water-swollen cell wall (e.g., 1.07 cm^3/g for spruce wood).

4.2.4 Concluding Remarks

In addition to UV microscopy, there are other microscopic techniques for investigating the topochemistry of lignin in wood. These methods include fluorescence and electron microscopy. The combination of these techniques with UV microscopy should give the best results. However, when a composite technique is employed, precautions must be taken. For example, in measurements of lignin concentration by bromination/UV microscopy, a correction factor is needed to account for the difference in the reactivities of middle lamella and secondary wall lignins toward bromine (Saka et al. 1982) (see Chap. 4.4).

UV microscopy is an established technique for determining the concentration and chemical structure of lignin in different layers of the cell wall. The chemistry of morphogenesis during the progress of cell differentiation has been elucidated by the use of UV microspectroscopy. However, in qualitative and quantitative analyses involving the use of UV microscopy, many less-than-precise assumptions are made regarding the structure and light absorptivity of lignin. Until these features of the cell wall or portion of the cell wall under investigation are better known, the UV microscopic technique will always be imperfect.

References

Aulin-Erdtman G (1954) Spectrographic contributions to lignin chemistry V. Sven Papperstidn 57:745–760

Boutelje JB, Eriksson I (1982) An UV-microscopy study of lignin in middle lamella fragments from fibers of mechanical pulp of spruce. Sven Papperstidn 85:R39–42

Boutelje JB, Eriksson I (1984) Analysis of lignin in fragments from thermomechanical spruce pulp by ultraviolet microscopy. Holzforschung 38:249–252

Boutelje JB, Jonsson U (1980) Ultraviolet microscope photometry of pulp fibers. Cellul Chem Technol 14:53–67

Eastman Kodak Co. (1987) Scientific imaging with Kodak films and plates. Publ No. P-315, Rochester, New York

Fergus BJ, Goring DAI (1969) The topochemistry of delignification in kraft and neutral sulphite pulping of birch wood. Pulp Pap Mag Can 70:T-314

Fergus BJ, Goring DAI (1970a) The location of guaiacyl and syringyl lignins in birch xylem tissue. Holzforschung 24:113–117

Fergus BJ, Goring DAI (1970b) The distribution of lignin in birch wood as determined by ultraviolet microscopy. Holzforschung 24:118–124

Fergus BJ, Procter AR, Scott JAN, Goring DAI (1969) The distribution of lignin in sprucewood as determined by ultraviolet microscopy. Wood Sci Technol 3:117–138

Fujii T (1987) Ultraviolet microspectophotometry of MWL embedded with epxy-resin. Proc 32nd Jpn Lignin Symp, Fukuoka, Japan, 21–24

Fujii T, Shimizu K, Yamaguchi A (1987) Enzymatic saccharification on ultrathin sections and ultraviolet spectra of Japanese hardwood and softwoods. Mokuzai Gakkaishi 33:400–407

Fukazawa K (1974) The distribution of lignin in compression- and lateral-wood of Abies sachalinensis using ultraviolet microscopy. Res Bull Coll Exp For Hokkaido Univ 31:87–114

Fukazawa K, Imagawa H (1981) Quantitative analysis of lignin using an UV microscopic image analyzer. Variation within one growth increment. Wood Sci Technol 15:45–55

Fukazawa K, Imagawa H (1983) Ultraviolet and fluorescence microscopic studies of lignin. 1983 Proc Int Symp Wood Pulp Chem May 23–27, 1983, Tsukuba, Japan, TAPPI I:20–23

Imagawa H (1981) Study on the seasonal development of the secondary phloem in Larix leptolepis. Res Bull Coll Exp For Hokkaido Univ 38:31–44

Imagawa H, Fukazawa K (1978) Influence of wood extractives on UV microscopy for lignin determination. Mokuzai Gakkaishi 24:583–586

Imagawa H, Fukazawa K, Ishida S (1976) Study on the lignification in tracheids of Japanese larch, Larix leptolepis Gord. Res Bull Coll Exp For Hokkaido Univ 33:127–138

Jayme G, Torgersen HF (1967) Topochemie der Delignifizierung beim Aufschluss von Fichtenholz nach dem Sulfit- und Sulfatverfahren Teil 1 Ultraviolett-mikroskopische Untersuchungen an teilweise delignifiziertem Fichtenholz. Holzforschung 21:110–116

Kojima Y, Yoon SY, Kayama T (1988a) Delignification from the cell wall of wood fibers with ozone. Mokuzai Gakkaishi 34:697–701

Kojima Y, Yoon SY, Kayama T (1988b) A study of production of CTMP from hardwood. Part 3. Characterization of pulps produced by CTMP-O_3 process. J Jpn Tappi 42:953–962

Lange PW (1954) The distribution of lignin in the cell wall of normal and reaction wood from spruce and a few hardwoods. Sven Papperstidn 57:525–532

Luft JH (1961) Improvements in epoxy resin embedding method. J Biophys Biochem Cytol 9:409–414

Musha Y, Goring DAI (1975a) Cell dimensions and their relationship to the chemical nature of the lignin from the wood of broad-leaved trees. Can J For Res 5:259–268

Musha Y, Goring DAI (1975b) Distribution of syringyl and guaiacyl moieties in hardwoods as indicated by ultraviolet microscopy. Wood Sci Technol 9:45–58

Procter AR, Yean WQ, Goring DAI (1967) The topochemistry of delignification in kraft and sulphite pulping of spruce wood. Pulp Pap Mag Can 68:T45–T53

Saka S, Whiting P, Fukazawa K, Goring DAI (1982) Comparative studies on lignin distribution by UV microscopy and bromination combined with EDXA. Wood Sci Technol 16:269–277

Scott JAN, Goring DAI (1970a) Lignin concentration in the S3 layer of softwoods. Cellul Chem Technol 4:83–93

Scott JAN, Goring DAI (1970b) Photolysis of wood microsections in the ultraviolet microscopy. Wood Sci Technol 4:237–239

Scott JAN, Procter AR, Fergus BJ, Goring DAI (1969) The application of ultraviolet microscopy to the distribution of lignin in wood. Description and validity of the technique. Wood Sci Technol 3:73–92

Takabe K, Fukazawa K, Terashima N, Fukushima K (1987) Lignification process of hardwoods xylem elements by UV microspectrophotometry. Proc 32nd Jpn Lignin Symp, Fukuoka, Japan, 1–4

Takano T, Fukazawa K, Ishida S (1983) Within-a-ring variation of lignin in Picea glehnii by UV microscopic image analysis. Res Bull Coll Exp For Hokkaido Univ 40:709–722

Wardrop AB (1957) The phase of lignification in the differentiation of wood fibers. Tappi 40:225

Wergin W (1965) Über Entstehung und Aufbau von Reaktionsholzzellen. 4. Mitt.: Nachweise der Ligninverteilung in den Zellwänden des Druckholzes durch Untersuchungen im UV-Licht. Flora 156:322–331

Wood JR, Ahlgren PA, Goring DAI (1972) Topochemistry in the chlorite delignification of spruce wood. Sven Papperstidn 15:15–19

Wood JR, Goring DAI (1971) The distribution of lignin in stem wood and branch wood of Douglas fir. Pulp Pap Mag Can 72:T95–T102

Wood JR, Goring DAI (1973) Ultraviolet microscopy at wavelengths below 240 nm. J Microscopy 100:105–111

Yang JM, Goring DAI (1978a) A comparison of the concentration of free phenolic hydroxyl groups in the secondary wall and middle lamella regions of softwoods. Trans Tech Sect CPPA 4:2–5

Yang JM, Goring DAI (1978b) The topochemistry of the reaction of chlorine dioxide with lignin in black spruce wood. Holzforschung 32:185–188

Yang JM, Goring DAI (1980) The phenolic hydroxyl content of lignin in spruce wood. Can J Chem 58:2411–2414

4.3 Interference Microscopy

L.A. Donaldson

4.3.1 Introduction

Quantitative Jamin-Lebedeff interference microscopy can be used to measure the distribution of lignin in sections of wood or pulp fibers. The technique is fully quantitative, accurate, and can be calibrated with any desired solid state lignin standard. Resolution is adequate to measure lignin concentration in the cell corner middle lamella (ccml) and S2 regions of the specimen (Boutelje 1972, Donaldson 1985a), while a modification of the basic technique allows measurements to be made in other cell wall regions (Donaldson 1987). Interference microscopy is ideally suited to the study of lignification during cell development, lignin removal during pulping, and any other application where information on lignin concentration at the cellular level is required.

4.3.2 Principle

Interference microscopy according to Jamin (1868) and Lebedeff (1930) produces interference by splitting a single beam of polarized light into two beams which are in phase (Fig. 4.3.1). One beam then passes through the specimen while the other passes through the reference medium in which the specimen is immersed. The two beams travel over different optical path lengths because of the difference in optical density between the specimen and the reference medium. The two beams, now out of phase, are recombined and interfere with each other to produce an image with color or intensity contrast between the specimen and background. By manipulating the image contrast, it is possible to measure the refractive index of the specimen and hence obtain information on the chemical composition of the sample.

Although several techniques are available for the determination of refractive index, only the double embedding procedure has been used with wood specimens (Boutelje 1972, Donaldson 1985a). This procedure involves measuring optical path difference (o.p.d.) successively in two different reference media. Selection of reference media depends on a number of criteria:

1. The medium should not react chemically or physically with the specimen, i.e., it should not extract or swell the specimen.
2. The two media should have different refractive indices and should have refractive indices different from that of the specimen. If the medium and the

Springer Series in Wood Science
Methods in Lignin Chemistry
(Edited by S.Y. Lin and C.W. Dence)

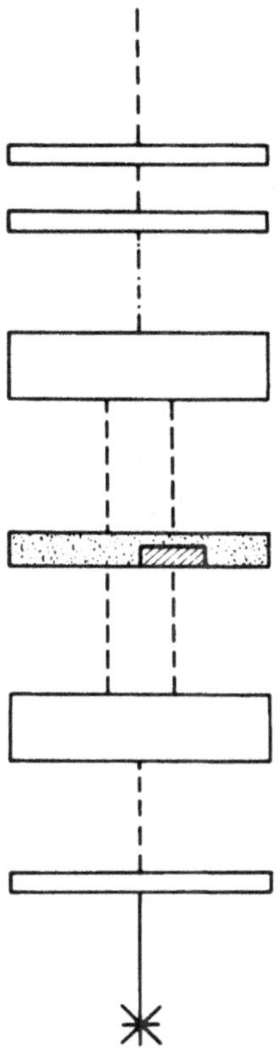

Fig. 4.3.1. A diagramatic representation of the optical path for the Zeiss Photomicroscope II (Carl Zeiss), equipped for quantitative interference microscopy

specimen have the same refractive index, the specimen will be invisible, i.e., o.p.d. = 0.

3. The first medium should preferably be volatile to facilitate its removal prior to remounting in the second medium.

The basic measurement made with the microscope is analyzer rotation in degrees. Angular readings are converted to o.p.d. values using the following equation:

$$\Delta\delta = (180 - a) \times 3.034, \tag{1}$$

where $\Delta\delta$ = o.p.d. (nm), a = analyzer rotation (°), 3.034 = a constant for de Sénarmont compensation (546.1/180).

(Equations given here apply to a Zeiss Photomicroscope II equipped for interference microscopy. For other optical systems different equations may apply.)

For the double embedding procedure, refractive index can be calculated from o.p.d. values using the following equation:

$$N_o = \frac{(\Delta\delta_1 N_2) - (\Delta\delta_2 N_1)}{\Delta\delta_1 - \Delta\delta_2}, \quad (2)$$

where N_o = refractive index of specimen, N_1 = refractive index of medium 1, N_2 = refractive index of medium 2, $\Delta\delta_1$ = o.p.d. in medium 1 (nm), $\Delta\delta_2$ = o.p.d. in medium 2 (nm).

More detailed information on the theory involved in these calculations has been provided by Ross (1967) and Tolansky (1968).

To calculate lignin concentration from the refractive index of the specimen, it is necessary to use calibration values. Refractive index varies in a linear fashion with change in concentration (Barer 1953). By determining refractive indices for specimens with 0% lignin and 100% lignin it is possible to calculate the lignin concentration of an intermediate specimen using the following equation:

$$L = \frac{N_o - N_a}{N_b - N_a} \times 100, \quad (3)$$

where L = lignin concentration in specimen (%), N_o = refractive index of specimen N_a = refractive index at 0% lignin, N_b = refractive index at 100% lignin.

4.3.3 Method

4.3.3.1 Specimen Preparation

Each wood specimen to be examined is divided into two matched samples. The first sample is dissected into small blocks, usually by cutting a thick (100 µm) radial section with a sledge microtome followed by further dissection with a razor blade. Because orientation of the sample is important at latter stages, care must be taken to prepare blocks in such a way that the transverse plane can readily be determined. The simplest way of doing this is to cut blocks 1 mm wide by 2 mm in length taking care to make the edges perpendicular and parallel to the fiber axis, respectively. These blocks are dehydrated in an ethanol series and then extracted in a 2:1 (v/v) mixture of benzene and ethanol for 3 days to remove extractives. After washing in ethanol and acetone, the blocks are embedded in Spurr resin (Spurr 1969). These blocks are designated as whole wood.

The second sample is dissected into matchstick-sized pieces and delignified in a 1:1 (v/v) mixture of glacial acetic acid and hydrogen peroxide (30% w/v) for 6–8 h at 80–90 °C until fully delignified. As a check that delignification is complete, the following criteria may be used: complete dissolution of the middle lamella, absence of $KMnO_4$ staining, and absence of phloroglucinol staining. (see Chap. 2.1). Chlorite delignification (Uprichard 1965) may be used as an alternative approach and gives comparable results (Donaldson 1985a).

After exhaustive washing in water, the delignified material is dehydrated in acetone, dissected into shives (groups of adhering fibers) and embedded in Spurr resin. This material is designated as holocellulose.

Embedded material is sectioned in the transverse plane at a thickness of 2 μm using glass knives. Section thickness is an important consideration in determining the o.p.d value which must lie within the range of the interference system being used. When the de Sénarmont compensator is used (Fig. 4.3.1), this range is 546.1 nm. The maximum section thickness, for example, of lignin with refractive index of 1.604 in a medium with a refractive index of 1.474 (glycerol) would be 4.2 μm [see Eq. (4)]. Care must be taken to ensure that sections of both whole wood and holocellulose are cut at the same orientation, preferably within 4° of each other (Donaldson 1985a). This precaution is necessary to avoid error resulting from measurement of refractive index at different orientations in the birefringent S2 layer of the cell wall. Orientation is not an important consideration for measurements in the nonbirefringent middle lamella (Hermans 1946, Donaldson 1985a).

Slides containing sections are washed in a saturated solution of sodium hydroxide in ethanol for 5 min to remove the embedding plastic from the sections. After careful washing in ethanol and drying, the sections are ready for measurement.

4.3.3.2 Measurement of Refractive Index

The section is first mounted in ethanol. The instrument is adjusted to give the maximum extinction position (m.e.p.) of the background (Fig. 4.3.2) by tilting the crystal plate of the beam splitter in the condenser (Fig. 4.3.1). Measurements of analyzer rotation are made by rotating the analyzer filter until the area of the specimen to be measured reaches its m.e.p. (Fig. 4.3.2).

The angle is read off a vernier scale on the control knob. This procedure is repeated several times and an average value is calculated. This value is converted to an o.p.d. value using Eq. (1). The coverslip is then removed and the ethanol allowed to evaporate. The section is remounted in glycerol and the same point on the same section is remeasured. Refractive index can be calculated using Eq. (2) where N_1 and N_2 are 1.362 and 1.474 for ethanol and glycerol, respectively. All measurements are made at 20 °C unless otherwise stated.

4.3.3.3 Calculation of Lignin Concentration from Refractive Index

To calculate lignin concentration from refractive index of the specimen, calibration values are required. For measurements in the secondary wall, value N_a in Eq. (3) is the refractive index of the S2 region in a matched specimen of holocellulose. This value varies from specimen to specimen and must be measured each time a new specimen is examined (Donaldson 1985a). The only exception is for earlywood and latewood from the same growth ring. For measurements in the middle lamella, N_a has a constant value of 1.516 which

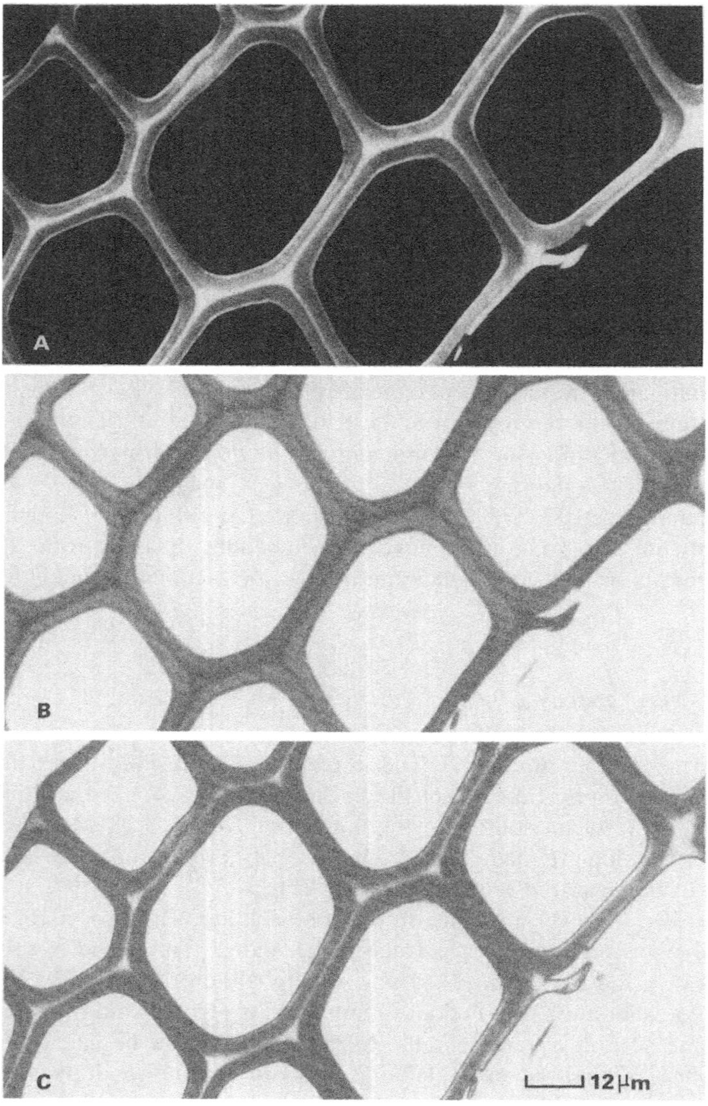

Fig. 4.3.2A–C. A series of interference micrographs showing the maximum extinction positions of (**A**) the background, (**B**) the cell corner middle lamella of *Pinus radiata* earlywood, and (**C**) the S2 region of the secondary wall (interference photomicrograph)

is the refractive index of the unlignified middle lamella measured either in differentiating secondary xylem or in primary xylem where the middle lamella does not lignify (Donaldson 1985a). The value for N_b is 1.604 based on the refractive index of the middle lamella in wood which has been hydrolyzed with sulfuric acid (comparable to Klason lignin) (Donaldson 1985a).

4.3.3.4 Measurement of Porosity

An alternative approach to measurement of lignin distribution is to assess variation in the porosity of wood which has been hydrolyzed to form a lignin skeleton (Donaldson 1987). Hydrolysis with sulfuric acid is particularly useful because of the swelling that occurs. This magnifies areas of the cell wall such as the S3 layer or the compound middle lamella which would otherwise be too small to measure.

Wood which has been hydrolyzed in 72% sulfuric acid for 3 h at 70°C is washed exhaustively in water, dehydrated in acetone, and embedded in Spurr resin. Measurements of refractive index are made in the same way as described above. In addition, an assessment of porosity is carried out as follows: For sections mounted in glycerol, o.p.d. is measured in the regions to be compared, e.g., S3 and S2. These measurements are then converted to effective thickness values using the following equation:

$$t = \frac{\Delta\delta}{N_o - N_m}, \tag{4}$$

where t = effective thickness (nm), N_o = refractive index of specimen, N_m = refractive index of medium (glycerol), $\Delta\delta$ = o.p.d. in glycerol (nm).

Relative porosity can then be determined by comparing the effective thickness of two regions, e.g., S3 and S2, using the following equation:

$$P = (1 - t_1/t_2) \times 100, \tag{5}$$

where P = relative porosity (%). t_1 = effective thickness of region 1, e.g., S3 (nm). t_2 = effective thickness of region 2, e.g., S2 (nm).

4.3.3.5 Calculation of Lignin Concentration from Porosity

The first step is to determine a correction value for swelling which varies from region to region. This is done by measuring the width of each region on matched samples of whole wood and hydrolyzed wood by examination with the scanning electron microscope (Donaldson 1987). A swelling correction based on the linear component of volumetric swelling can be obtained using the following equation:

$$s = \sqrt[3]{\frac{w_1}{w_2}}, \tag{6}$$

where s = swelling correction, w_1 = width of region in hydrolyzed wood (μm), w_2 = width of region in whole wood (μm).

Relative lignin concentration can then be calculated using the following equation:

$$RL = \frac{t_1 s_1}{t_2 s_2}, \tag{7}$$

where RL = relative lignin concentration, t_1 = effective thickness of region 1, e.g., S3 (nm), t_2 = effective thickness of region 2, e.g., S2 (nm), s_1 = swelling correction region 1, e.g., S3, s_2 = swelling correction region 2, e.g., S2.

Relative lignin concentration can be converted to an actual concentration by determining the lignin concentration of one region using the conventional method, e.g., S2, and multiplying by RL to give L for the unknown region, e.g., S3.

4.3.4 Discussion

Three quantitative approaches to measure lignin distribution are interference microscopy, UV microscopy, and bromination/EDXA. The choice of method probably depends more on the availability of equipment than any other factor since each method can be applied to a similar range of specimen types.

4.3.4.1 Comparison with Other Techniques

UV microscopy has been used most extensively due to the pioneering work of Goring and others (Fergus et al. 1969, Scott et al. 1969). No specific comparison has been made between UV microscopy and interference microscopy. Interference microscopy gives consistently lower values than UV microscopy when compared on a w/w basis (Table 4.3.1).

Table 4.3.1. Comparison of lignin distribution (% w/w) in softwoods determined by interference microscopy and UV microscopy. All values for interference microscopy have been converted from v/v to w/w using density values for lignin and carbohydrate of 1.335 and 1.521. (Stamm and Sanders 1966)

Genus	Method	S2	ccml	Reference
Picea	IM	EW 18	82 (79)[a]	Boutelje (1972)
		LW 16	82 (79)	
Pinus		EW-	70 (65)	
		LW 20	85 (82)	
Pinus	IM	EW 19	81 (79)	Donaldson (1985b)
		LW 19	78 (75)	
Picea	UV	EW 23	85	Fergus et al. (1969)
		LW 22	100	
Pseudotsuga (1971)	UV	EW 25	83	Wood and Goring
		LW 23	88	

IM = interference microscopy.
UV = ultraviolet microscopy.
EW = earlywood.
LW = latewood.
ccml = cell corner middle lamella.
[a] Values in parentheses have been recalculated with N_a = 1.516 and N_b = 1.604.

UV microscopy (Chap. 4.2) has a number of advantages over interference microscopy, including automated measurement (Fukazawa and Imagawa 1981), and the ability to measure the guaiacyl/syringyl ratio in hardwood lignin (Fergus and Goring 1970, Musha and Goring 1975). Interference microscopy has not yet been applied to hardwoods and it is not known if there is a change in refractive index with lignin type.

UV microscopy, however, also has a number of disadvantages including the need for 0.5-μm sections of uniform thickness and the need to measure the proportions of different cell wall layers to convert relative lignin concentration to true lignin concentration using estimates of lignin content based on Klason lignin analysis (Fergus et al. 1969, Scott et al. 1969). Interference microscopy is the only technique which directly estimates lignin concentration independent of differences in section thickness between the secondary wall and middle lamella, and without the need to measure lignin content and cell wall morphology.

Bromination/EDXA (Chap. 4.4) has several disadvantages when compared to the other two techniques. Only relative lignin concentration can be determined directly by this method and a correction factor of 1.7 is required to obtain agreement between techniques (Saka et al. 1982, Donaldson and Ryan 1987). When bromination/EDXA and interference microscopy are used on matched samples, agreement is poor (Donaldson and Ryan 1987) as shown in Table 4.3.2. Not only is there poor quantitative agreement between the two methods, but EDXA data are also much more variable within each specimen.

4.3.4.2 Types of Specimen that Can Be Examined

Interference microscopy can be applied to the study of lignin distribution in wood, in differentiating xylem, and in chips during pulping. The only constraints are that the material can be embedded and sectioned, matched samples are available for calibration, and the specimen can be oriented to obtain transverse sections.

The most common difficulties arise from the second constraint. For example, a pulp sample containing wood from different sources cannot be examined by interference microscopy because the value of N_a will vary from fiber to fiber, making accurate calibration impossible.

4.3.4.3 Accuracy and Precision

The accuracy of interference microscopy depends on the calibration standards used. For measurements in the secondary wall, these values are 1.604 for N_b and a range of values for N_a depending on the specimen. The variation in N_a arises presumably from variation in microfibril angle (Hermans 1946). The need to measure N_a for each sample is one disadvantage of this technique that increases the amount of work necessary to perform the measurement. If N_a is not measured for each sample, an error of $\pm 11\%$ is introduced compared to the usual error of $\pm 2\%$ (Donaldson 1985a). N_a does not vary between earlywood

Table 4.3.2. Relative lignin concentration (ccml/S2) for two matched samples determined by interference microscopy and bromination/EDXA. (Donaldson and Ryan 1987)

Tree no.	Site[a] no.	Interference microscopy	EDXA
1	1	3.38	1.98 (3.37)[b]
	2	3.38	1.83 (3.11)
	3	3.43	1.91 (3.25)
	4	3.27	2.02 (3.43)
	5	3.48	2.59 (4.40)
	6	–	1.52 (2.58)
	7	–	1.80 (3.06)
	8	–	1.64 (2.79)
	9	–	2.22 (3.77)
	10	–	1.99 (3.38)
	$\bar{x}$	3.39 ± 0.14[c]	3.31 ± 0.51
2	1	4.38	1.35 (2.30)
	2	4.38	1.72 (2.92)
	3	4.00	1.20 (2.04)
	4	4.18	0.84 (1.43)
	5	4.23	0.93 (1.58)
	6	–	1.06 (1.80)
	7	–	1.23 (2.09)
	8	–	1.65 (2.81)
	9	–	0.98 (1.67)
	10	–	1.06 (1.80)
	$\bar{x}$	4.23 ± 0.28	2.04 ± 0.51

[a] Sites are not matched.
[b] With a 1.7 correction factor (Saka et al. 1982).
[c] Least significant difference at p = 0.05.

and latewood in a single growth ring but does vary from corewood to outerwood in a single tree as well as between trees (Donaldson 1985b).

Although N_b varies slightly between the secondary wall and middle lamella when measured in hydrolyzed wood, a single value of 1.604 is recommended for both regions (Donaldson 1985a). It is assumed that the lower value in the S2 region (1.596) is due to contamination with carbohydrates, although this has not been experimentally verified. Recent work (Donaldson, unpubl.) has shown that, in compression wood, areas of middle lamella adjacent to intercellular spaces can have refractive indices of 1.601–1.604, suggesting that if one assumes this layer is pure lignin, 1.604 may be very close to the refractive index of native lignin, thus confirming its suitability as a calibration value. Frey (1959) measured the refractive indices of brown rot lignin and ultrasonic lignin, obtaining values of 1.601 and 1.605, respectively. It appears that the method of purification does not greatly influence the value of N_b.

Accuracy of measurements in the middle lamella depends not only on N_b, but also on the value of N_a for this region. Donaldson (1985a) measured the refractive index of the middle lamella in differentiating tracheids before lignification and in primary xylem, where the middle lamella is unlignified, and obtained values of approximately 1.516. Further work (Donaldson 1991) has since confirmed that a value of 1.516 is appropriate to all trees (at least for *Pinus radiata*). Boutelje (1972) used the refractive index of pectin (1.504) to calibrate measurements in the middle lamella leading to an error of 2–3% in lignin concentration (Donaldson 1985a).

Interference microscopy typically yields values of 20–22% v/v (standard deviation, sd = 1.378) for the secondary wall, and 68–91%* v/v (sd = 5.932) for the cell corner middle lamella (ccml) region (Donaldson 1985b)* (recalculated using N_a = 1.516). The precision of these data depends primarily on the ability of the operator to accurately locate the m.e.p. of both the background and the specimen. In practice this depends on operator experience and on the size of the area to be measured. For example, it is much easier to determine the m.e.p. of a large area of background (±0.1° of analyzer rotation) than of a small area of specimen (±0.5° of analyzer rotation). This uncertainty has little effect on the result because a mean value is used, thus greatly increasing the precision. The difference in sd between secondary wall and ccml data noted above can be attributed to this source of error. In a later report. Donaldson (1986) indicated sd values of 0.812 and 0.775 for secondary wall and ccml data which demonstrates the importance of operator experience.

It is worth discussing some aspects of precision that relate to the way in which data are processed. Typically, m.e.p. is estimated as the average of ten measurements on a single site. These values often deviate from a normal distribution and should not be used as replicates (sample sizes of 100 or more are normally distributed). To obtain replicates, independent determinations of o.p.d. are made on different sections, using the mean value in each case.

For measurements in the secondary wall, data will consist of replicates of values for N_o and N_a. There are two alternatives for calculating replicates of L: either use the mean of N_a, or rank N_o and N_a according to magnitude and use the resulting pairs to calculate replicate values of L. If N_o and N_a are randomly assigned in pairs, the variation in L is unnecessarily enhanced. Donaldson (1985a) recommended the ranking procedure because it incorporates variation from both sets of measurements and yields the least variation among replicates. This problem does not arise for measurements in the middle lamella, where N_a is a constant.

References

Barer R (1953) Determination of dry mass, thickness, solid and water concentration in living cells. Nature 172:1098

Boutelje JB (1972) Calculation of lignin concentration and porosity of cell wall regions by interference microscopy. Sven Papperstidn 75:683–686

Donaldson LA (1985a) Critical assessment of interference microscopy as a technique for measuring lignin distribution in cell walls. N Z J For Sci 15:349–360

Donaldson LA (1985b) Within- and between-tree variation in lignin concentration in the tracheid cell wall of Pinus radiata. N Z J For Sci 15:361–369

Donaldson LA (1986) Between-tree variation in lignin concentration in Pinus radiata tracheids with growth rate, stem eccentricity, site, and silvicultural treatment. N Z J For Sci 16:118–123

Donaldson LA (1987) S3 lignin concentration in radiata pine tracheids. Wood Sci Technol 21:227–234
Donaldson LA (1991) Seasonal changes in lignin distribution during tracheid development in Pinus radiata D. Don. Wood Sci Technol 25:15–24
Donaldson LA, Ryan KG (1987) A comparison of relative lignin concentration as determined by interference microscopy and bromination/EDXA. Wood Sci Technol 21:303–309
Fergus BJ, Goring DAI (1970) The location of guaiacyl and syringyl lignins in birch xylem tissue. Holzforschung 24:113–117
Fergus BJ, Procter AR, Scott JAN, Goring DAI (1969) The distribution of lignin in sprucewood as determined by UV microscopy. Wood Sci Technol 3:117–138
Frey HP (1959) Über die Einlagerung des Lignins in der Zellwand. Holz Roh-Werkst 17: 313–381
Fuykazawa K, Imagawa H (1981) Quantitative analysis of lignin using an UV microscopic image analyser. Variation within one growth increment. Wood Sci Technol 15:45–55
Hermans PH (1946) Contribution to the physics of cellulose fibres. Elsevier, Amsterdam, 221 pp
Jamin MJ (1868) Sur un réfracteur differential pour la lumiére polarisèe construit par Lutz. C R Acad Sci 67:814–816
Lebedeff AA (1930) Interféromètre à polarisation et ses applications. Rev Opt 9:385–413
Musha Y, Goring DAI (1975) Distribution of syringyl and guaiacyl moieties in hardwoods as indicated by ultraviolet microscopy. Wood Sci Technol 9:45–58
Ross KFA (1967) Phase contrast and interference microscopy for cell biologists. Edward Arnold, London, 238 pp
Saka S, Whiting P, Fukazawa K, Goring DAI (1982) Comparative studies of lignin distribution by UV microscopy and bromination combined with EDAX. Wood Sci Technol 16: 269–277
Scott JAN, Procter AR, Fergus BJ, Goring DAI (1969) The application of UV microscopy to the distribution of lignin in wood. Description and validity of the technique. Wood Sci Technol 3:73–92
Spurr AR (1969) A low viscosity epoxy resin embedding medium for electron microscopy. J Ultrastruct Res 26:31–43
Stamm AJ, Sanders HT (1966) Specific gravity of the wood substance of loblolly pine as affected by chemical composition. Tappi 49:397–400
Tolansky S (1968) Interference microscopy for the biologist. Charles C. Thomas, Springfield, Illinois, USA, 166 pp
Uprichard JM (1965) The alpha-cellulose content of wood by the chlorite procedure. Appita 19:36–39
Wood JR, Goring DAI (1971) The distribution of lignin in stem wood and branch wood of Douglas-fir. Pulp Pap Mag Can 72(3):T95–T102

4.4 Electron Microscopy

S. Saka

4.4.1 Introduction

Many staining techniques have been used in combination with light microscopy to study the distribution of lignin in wood (Brauns 1952). However, since the advent of electron microscopy, considerable efforts have been made to obtain a better understanding of lignin morphology in wood cell walls. One of the techniques in electron microscopy for detection of lignin involves the use of potassium permanganate ($KMnO_4$) stains (Crocker 1921, Hepler et al. 1970, Bland et al. 1971, Parham 1974, Kutscha and Schwarzmann 1975, Saka et al. 1979). Although some doubts have been expressed as to its specificity for lignin (Hoffmann and Parameswaran 1976, Kishi et al. 1982), the $KMnO_4$ staining technique provides many excellent details of the ultrastructural features of lignin in wood.

Another method in electron microscopy involves preparation of lignin skeletons by removal of carbohydrates with brown rot fungi (Meier 1955) or concentrated hydrofluoric acid (Sachs et al. 1963, Côté et al. 1968, Bentum et al. 1969, Parham and Côté 1971, Timell 1973, Fujii et al. 1981). Condensation reactions may alter the lignin itself, but the overall results using this method are in good agreement with those obtained by $KMnO_4$ staining (Parham 1974).

Although useful for indicating the location of lignin in wood cell walls, these methods provide only qualitative estimation of lignin distribution. For quantitative assay, ultraviolet (UV) microscopy (Fergus et al. 1969, Scott et al. 1969, Fergus and Goring 1970a,b, Wood and Goring 1971) (see Chap. 4.2) or interference microscopy (Lange and Kjaer 1957, Boutelje 1972) (see Chap. 4.3) has usually been the methods of choice in the past. Lately, a quantitative electron microscopy method has been developed that involves bromination of lignin in a nonaqueous system, followed by measurements of bromine concentration by transmission or scanning electron microscopy coupled with energy dispersive X-ray analysis (TEM-EDXA or SEM-EDXA). By knowing the relative reactivity of bromine with different types of lignin, the distribution of lignin in the various morphological regions of wood can be determined quantitatively.

Springer Series in Wood Science
Methods in Lignin Chemistry
(Edited by S.Y. Lin and C.W. Dence)

4.4.2 Principles

4.4.2.1 Glossary

A glossary of symbols and descriptions used in this section is found below.

I_i	Bromine concentration in morphological region i
V_i	Tissue volume percentage of morphological region i
W	Lignin content of the whole wood
S	Secondary wall
CML	Compound middle lamella
ML	Middle lamella
CC	Middle lamella at the cell corner
C_i	Lignin concentration at morphological region i
C_{sy}	Concentration of the syringyl lignin
C_{gu}	Concentration of the guaiacyl lignin
Ki	Correction factor for bromine reactivity in morphological region i
UV microscopy	Ultraviolet microscopy
A_i	UV absorbance of a 0.5 μm section in morphological region i
TEM	Transmission electron microscopy
SEM	Scanning electron microscopy
EDXA	Energy dispersive X-ray analysis

4.4.2.2 Principles of SEM-EDXA and TEM-EDXA

One of the limitations of electron microscopy is that it allows only a visual characterization of the cell wall structure. Once electron microscopy is combined with energy dispersive X-ray analysis (EDXA), the system becomes a powerful analytical tool for elucidating the fine details of lignin morphology in wood.

In an electron microscope, the incident electron beam impinges upon the specimen and excites the emission of electrons (Fig. 4.4.1). The emitted electrons are utilized in scanning electron microscopy (SEM) to create a surface image of the specimen on a cathode-ray tube, in contrast to transmission electron microscopy (TEM) which uses transmitted electrons in visualizing the image. Also emitted from the specimen under electron bombardment are X-rays whose energies are characteristic of each element (Fig. 4.4.1). In the EDXA system, a silicon or other type of detector is utilized to capture and measure these X-ray photons. An attached computer stores and sorts the measured energies and displays the various energy peaks on a cathode ray tube.

Generally, detection is limited to elements with an atomic number higher than neon. This limitation is advantageous in wood research since it eliminates the interference from the elements carbon and oxygen present in wood. Satisfactory application of this technique to lignin detection requires the incorporation of an element such as bromine reacting specifically with lignin and having an appropriate atomic number.

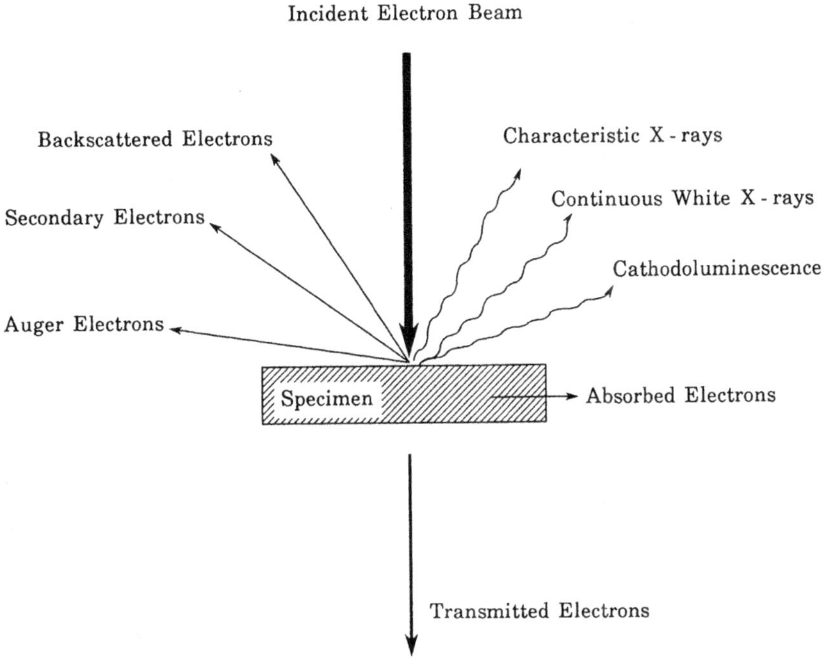

Fig. 4.4.1. Types of signals occurring when incident electrons impinge upon the specimen

As applied to bromine, the X-rays generated when this element is bombarded with electrons are manifested in emission spectra as lines designated Br-K_α, Br-K_β, and Br-L (more precisely, L_α) which correspond to energy levels of 11.9, 13.3, and 1.48 keV, respectively (Russ 1974). These characteristic lines for bromine can be used in EDXA analysis.

4.4.3 Methods

4.4.3.1 Specimen Thickness

Thick or bulky specimens are commonly used in scanning electron microscopy. However, as shown in Fig. 4.4.2, there are two disadvantages in using such thick specimens: (1) a low spatial resolution, and (2) interferences from the effects of backscattering, absorption and fluorescence (Russ 1974). The poor spatial resolution can be remedied by using thin specimens (Fig. 4.4.2). The use of cross-sections of 0.5 μm thickness, for example, would permit analysis of the compound middle lamella in cell walls as a separate entity (Saka and Thomas 1982). The second problem can also be minimized or eliminated by using thin specimens (Russ 1974, Saka and Thomas 1982).

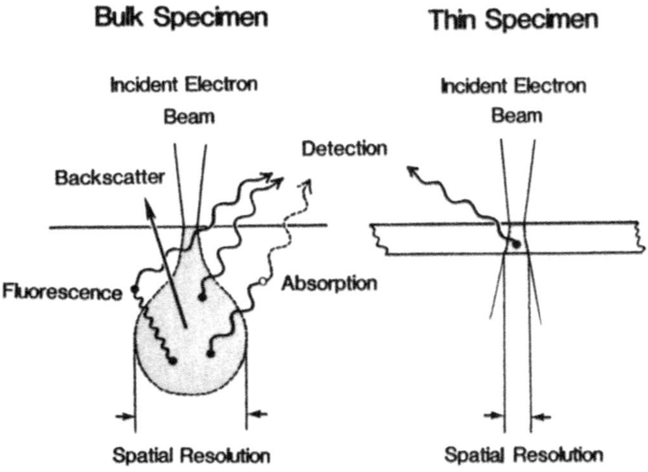

Fig. 4.4.2. Comparison of bulk and thin specimens in spatial resolution of X-rays generated and induced effects of incident electron beam. (Saka and Thomas 1982)

4.4.3.2 Procedure for Bromination-EDXA

Prior to bromination, small sticks of wood (1.5 mm radial; 1.5 mm tangential; 5 mm longitudinal) are extracted overnight with benzene-ethanol (2:1, v/v) in a Soxhlet extractor. The extracted sample is dehydrated sequentially with ethanol and water mixtures with increasing ethanol concentration up to 100%, and finally with chloroform. The sample is brominated at room temperature in a nonaqueous system with a solution of bromine in chloroform (0.3 ml in 20 ml). The bromine solution is slowly dropped into a 125-ml Erlenmeyer flask that contains 70 ml of chloroform and 1 g of wood sticks with stirring. After stirring for 30 min, the brominated wood sticks are washed several times with an excess of chloroform at room temperature to remove unreacted bromine and then with ethanol.

The brominated sticks are embedded in Spurr epoxy resin (Spurr 1969) and sectioned with a diamond knife or glass knife mounted on an ultramicrotome to give cross sections of 0.15 μm thickness for TEM-EDXA or 0.5 μm thickness for SEM-EDXA. The sections are then placed on a carbon-coated collodion film on the specimen support grid. The specimens are carbon-coated again to avoid charging.

The brominated specimens may be studied by a SEM-EDXA or a TEM-EDXA system. Since the intensity of the X-ray emission is directly proportional to the illuminating current, it is essential, for quantitative assay, to ascertain the current during the analysis. A Faraday cup (Howitt et al. 1976) in conjunction with an electrometer (e.g., Keithley 616 Digital Electrometer) may be used. The illuminating current should preferably be in the range of 10^{-9}–10^{-10} amperes.

Other variables, such as the distance between X-ray detector and specimen, tilting angle of specimen holder, specimen height, and take-off angle, should be

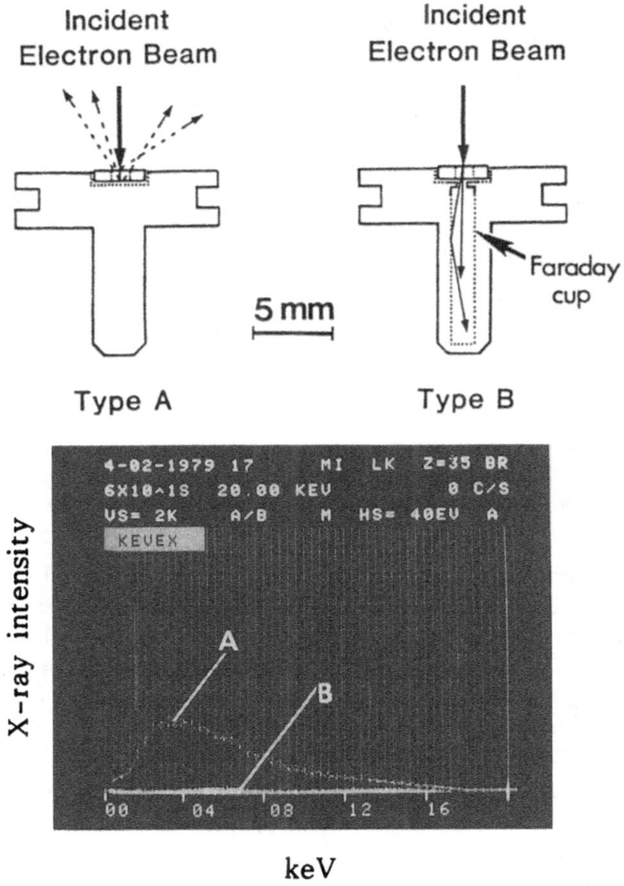

Fig. 4.4.3. The effect of two types of carbon specimen holders for carbon grids on continuous white X-ray emission. (Saka and Thomas 1982).

held constant throughout the entire study. The Br-L (1.42–1.56 keV) or the Br-K_{α} (11.80–12.02 keV) line of X-rays can be used for analysis.

Thin sections may readily be examined with TEM-EDXA or SEM-EDXA in the transmission mode (STEM-EDXA). However, to use conventional SEM-EDXA, some modifications of the equipment are required. As shown in Fig. 4.4.3, the electrons transmitted through a thin specimen impinge upon the carbon plate underneath. This causes the emission of a strong continuous spectrum of white X-rays. To minimize this problem, a carbon grid holder with a Faraday cup should be used. If the depth of the cylinder in the Faraday cup is more than 20 times the size of aperture, the electrons can be effectively collected (Grubb 1971, Howitt et al. 1976), thus reducing a large portion of the white X-rays emitted.

There are three analytical techniques available, namely, X-ray distribution mapping, line scan analysis, and point analysis. In most cases, only point analysis is needed for quantitative assay of bromine distribution. In this pro-

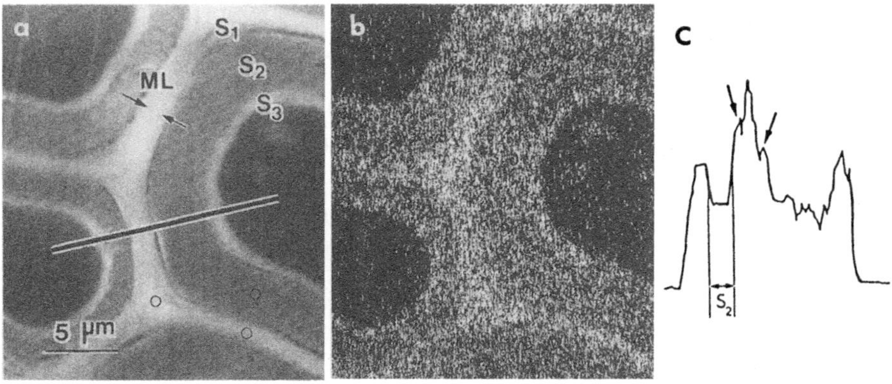

Fig. 4.4.4a–c. **a** Scanning electron micrograph of brominated loblolly pine tracheids (0.5 μm section). *Open cricles* indicate locations of point analysis. **b** Distribution map of Br-L X-rays taken from the same area as in the SEM micrograph. **c** The distribution of bromine taken along the line across the double cell walls. *Arrows* indicate S_1 layer. (Saka et al. 1981)

cedure, using SEM or TEM image as the positional reference (Fig. 4.4.4a), the electron beam is kept stationary for a suitable period of time in a designated morphological region to obtain a spectrum from which the bromine concentration is calculated. X-ray distribution mapping (Fig. 4.4.4b) and line scan analysis (Fig. 4.4.4c), on the other hand, are convenient methods for visualizing the distribution of bromine across the cell walls.

Figure 4.4.5 shows the spectra obtained by a 60-s point analysis of the various morphological regions shown as circles in Fig. 4.4.4a. The backgrounds shown in the spectra are almost identical, but the Br-L (1.42–1.56 keV) peak from the middle lamella at the cell corner (CC) is higher than that from the compound middle lamella (CML) which, in turn, is higher than that for the secondary wall (S). Since white X-rays are emitted in proportion to the weight of the sample, the peak intensity of bromine relative to background should be a measure of bromine concentration. However, the background includes not only the continuous white X-rays from the specimen but also those from the surroundings because of electron backscattering. The latter cannot be ignored (Fig. 4.4.3). Therefore, a quantitative determination of bromine concentration should be based on the total counts of Br-L or Br-K_α X-rays after background subtraction.

4.4.4 Lignin Distribution in Cell Walls

4.4.4.1 Softwoods

To determine the distribution of lignin from bromine analysis, it is essential to know the reactivity of bromine toward lignin in different morphological regions. For black spruce (*Picea mariana*), Douglas-fir (*Pseudotsuga menziesii*) and loblolly pine (*Pinus taeda*), the lignin in the secondary wall of tracheids is 1.7

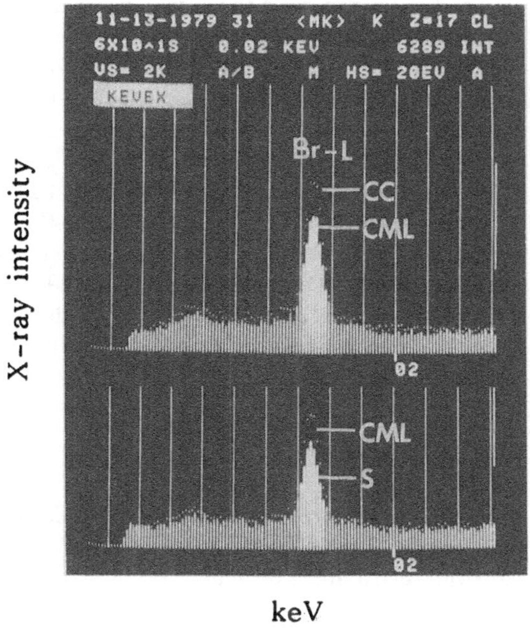

Fig. 4.4.5. Comparison of the spectra from three different morphological regions. *Upper* CC and CML; *lower* CML and S

times more reactive toward bromination than the middle lamella lignin (Saka et al. 1982). By applying this value as a correction to the bromine concentration (I_i) measured by EDXA, relative lignin concentration can be estimated. In addition, a quantitative estimation of lignin distribution requires determination of the tissue volume percentage of different morphological regions (V_i) and the lignin content of the whole wood (W). The volume fraction is obtained from electron micrographs while the lignin content of the whole wood is determined from the sum of Klason and acid-soluble lignins.

The lignin content of ray cells is generally assumed to be 50% higher than that of whole wood (W) (Fergus et al. 1969) while its volume percentage is about 4.5% of the total wood volume (Back 1960). The lignin content of the tracheid tissue is then $W(1 - 1.5 \times 0.045) = 0.933W$. Assuming uniform density and swelling throughout the tracheid cell wall, then

$$C_s\left(V_s + k_{ML}\frac{I_{ML}}{I_s}V_{ML} + k_{ML}\frac{I_{cc}}{I_s}V_{cc}\right) = 0.933W, \qquad (1)$$

where C_s = the weight concentration of lignin in g/g of dry tracheid secondary wall, k_{ML} = a correction factor for bromine reactivity in middle lamella lignin.

The substitution in Eq. (1) of the data for V_i, k_{ML} I_{ML}/I_s, and I_{cc}/I_s gives the weight concentration of lignin in the secondary wall (C_s) from which the lignin concentration of the tracheids in different morphological regions can be calculated.

Table 4.4.1 gives the distribution of lignin in Douglas-fir (*Pseudotsuga menziesii*) earlywood tracheids (Saka and Thomas 1982). The overall lignin content (W) of the earlywood portion is 0.285 (g/g). The Klason and acid-soluble lignin determinations should be made on wood taken near the location from which the sections are collected for EDXA measurements.

The value of k_{ML} is likely to be in the vicinity of 1.7 for mature normal softwoods (Saka et al. 1982). However, compression (Westermark 1985) and juvenile woods (Donaldson and Ryan 1987) are reported to have a different correction factor. This may be due to the fact that such abnormal softwoods contain different types of lignins in both the secondary wall and middle lamella regions.

4.4.4.2 Hardwoods

By combining the bromination-EDXA technique with ultraviolet (UV) microscopy, it is possible to determine not only the distribution of lignin but also the distribution of guaiacyl and syringyl residues in various morphological regions (Saka et al. 1988, Saka and Goring 1988).

4.4.4.2.1 Distribution of Lignin

The lignin concentration of the S_2 layer of the tracheid secondary wall (C_{S_2}) obtained by the method described in the previous section is used as a calibration standard for the Br-K_α X-ray counts (I_{S_2}) for a wood section of 0.15 μm thickness. The lignin in the softwood tracheid secondary wall is assumed to be entirely of the guaiacyl type (Whiting and Goring 1982). Thus, the counts of Br-K_α X-rays for a guaiacyl lignin in the section is I_{S_2}/C_{S_2}. The bromine uptake by syringyl lignin is approximately 1.2 times higher than that of guaiacyl lignin (Saka et al. 1988). Thus, the counts of Br-K_α X-rays for pure syringyl lignin in the section is $1.2 I_{S_2}/C_{S_2}$.

Table 4.4.1. The distribution of lignin in Douglas-fir earlywood tracheids as determined by bromination with SEM-EDXA. (Saka and Thomas 1982)

Morphological region	Tissue volume percentage (V_i)	Relative lignin conc. $\left(= k_i \frac{I_i}{I_s}\right)^a$	Lignin conc. (g/g) $\left(= k_i \frac{I_i}{I_s} C_s\right)^b$
S	84	1.00	0.21
ML	12	1.7 × 1.40	0.50
CC	4	1.7 × 2.10	0.75

[a] k_i = 1 for secondary walls and 1.7 for middle lamella regions.
[b] C_s = 0.21 g/g using an earlywood lignin content of 0.285 g/g.

Assuming the total X-ray counts of Br-K_{α} is the sum of the X-ray intensities from the brominated syringyl and guaiacyl lignins, then

$$I = \frac{I_{S_2}}{C_{S_2}}(1.2\, C_{sy} + C_{gu}), \tag{2}$$

where I = intensity of Br-K_{α} X-rays for a 0.15-μm-thick wood section, C_{sy} = concentration of the syringyl lignin (g/g), C_{gu} = concentration of the guaiacyl lignin (g/g).

For the same tracheid secondary wall in softwoods in which its lignin concentration is known to be C_{S_2}, the UV absorbance (A_{S_2}) of a wood section of 0.5 μm thickness is determined by UV microscopy. The UV absorptivity of guaiacyl lignin in the section is then calculated as A_{S_2}/C_{S_2}. The guaiacyl lignin absorptivity is 3.5 times that of the syringyl lignin absorptivity at 280 nm (Fergus and Goring 1970a). Thus, the UV absorptivity of syringyl lignin in the section is $A_{S_2}/(3.5C_{S_2})$. Assuming the total UV absorbance of the wood section is the sum of the UV absorbances of the syringyl and guaiacyl lignins, then

$$A = \frac{A_{S_2}}{C_{S_2}}\left(\frac{C_{sy}}{3.5} + C_{gu}\right), \tag{3}$$

where A = UV absorbance of a 0.5-μm section.

From Eqs. (2) and (3), the concentration of the syringyl and guaiacyl lignins in the wood cells are calculated from the expressions

$$C_{sy} = (1.09C_{S_2}/I_{S_2})I - (1.09C_{S_2}/A_{S_2})A \tag{4}$$

$$C_{gu} = (1.31C_{S_2}/A_{S_2})A - (0.313C_{S_2}/I_{S_2})I, \tag{5}$$

and the total concentration of lignin (C_{total}) is given by

$$C_{total} = C_{sy} + C_{gu}. \tag{6}$$

To avoid the effect of the sample variation, care must be taken to prepare sections for EDXA and UV microscopy as nearly contiguous as possible. The same area of a particular cell wall may then be examined by both techniques to collect the data for I and A in each morphological region.

Prior to the EDXA measurements on a hardwood specimen, the values of C_{S_2}/I_{S_2} and C_{S_2}/A_{S_2} in Eqs. (4) and (5) must be determined. The following is an example of their determination for a given set of conditions. For a 0.15-μm section of the black spruce (*Picea mariana*) tracheid secondary wall, the Br-K_{α} X-ray counts are determined to be 464. Since the lignin concentration of its secondary wall is 0.225 g/g, the value of C_{S_2}/I_{S_2} is 4.85×10^{-4} (= 0.225/464). The same tracheid secondary wall has a UV absorbance of 0.125 in a 0.5-μm section (Fergus et al. 1969). Therefore, the value of C_{S_2}/A_{S_2} is 1.8 (= 0.225/0.125). Substitution in Eqs. (4), (5) and (6) of these values and measurement of (I) and (A) give the concentration of lignin for each morphological region of the hardwood.

Table 4.4.2. The distribution of lignin in white birch as determined by TEM-EDXA coupled with UV microscopy. (Saka and Goring 1988)

Element	Morphological region	Tissue volume (%)	Br-K_{α} X-ray counts[a]	UV absorbance[b]	Lignin conc. (g/g)	Guaiacyl:syringyl
Fiber	S_1	11.4	347	–	0.14	–
	S_2	58.5	350	0.030	0.14	12:88
	S_3	3.5	298	–	0.12	–
	ML	5.2	756	–	0.36	–
	CC (F/F)[c]	2.4	949	0.236	0.45	91:9
Vessel	S_1	1.6	546	–	0.26	–
	S_2	4.3	543	0.131	0.26	88:12
	S_3	2.3	564	–	0.27	–
	ML	0.8	861	–	0.40	–
	CC (F/V)[d]	≃0	1250	0.277	0.58	80:20
Ray Parenchyma	S	8.0	277	0.043	0.12	49:51
	ML	2.0	767	–	0.38	–
	CC (F/R)[e]	≃0	959	0.262	0.47	100:0
	CC (R/R)[f]	≃0	868	0.209	0.41	88:12

[a] Section thickness = 0.15 μm.
[b] Section thickness = 0.5 μm, λ = 280 nm.
[c] Fiber/Fiber.
[d] Fiber/Vessel.
[e] Fiber/Ray.
[f] Ray/Ray.

Table 4.4.2 shows the experimental data for Br-K_{α} X-ray counts (I), UV absorbance (A) and the distribution of lignin for different morphological regions of white birch (*Betula papyrifera*) wood. Because of the limited resolution of UV microscopy, the UV absorbance of the S_1, S_3 layers and the middle lamella between cell corners (ML) cannot be measured. Instead, the guaiacyl/syringyl ratios are assumed to be the same as in the S_2 layer and the cell corner middle lamella (CC), respectively.

4.4.4.2.2 Distribution of Guaiacyl and Syringyl Lignins

The substitution in Eqs. (4) and (5) of the data presented in Table 4.4.2 gives the concentrations of both syringyl and guaiacyl units in a particular morphological region. Table 4.4.2 also shows the ratio of these units in different morphological regions. The calculations do not take into account the possible presence of *p*-hydroxybenzoic acid residues in the hardwood lignin. The poplar species are known to contain this moiety (Musha and Goring 1975). Therefore, further studies will be necessary before the EDXA combined with UV technique can be applied to such woods.

4.4.5 Comparisons of Bromination-EDXA with Other Methods

For softwoods, the distribution of lignin in black spruce (*Picea mariana*) obtained by the bromination-EDXA technique and UV microscopy has been compared. By applying the correction factor of 1.7 for bromination-EDXA, a good agreement was achieved between the results obtained by the two methods (Saka et al. 1982).

For white birch (*Betula papyrifera*), the results obtained by the bromination-EDXA combined with UV technique (Saka et al. 1988, Saka and Goring 1988) confirm, in general, the previous data obtained by UV microscopy alone (Fergus and Goring 1970a,b). This agreement indicates that any errors caused by specimen degradation in the electron beam were small (Mary et al. 1986). However, some discrepancies are found for the middle lamella and the ray parenchyma.

Recently, a new method of EDXA involving lignin mercurization has been reported (Eriksson et al. 1988, Westermark et al. 1988). Some of the results obtained using this method differ from those obtained by the bromination-EDXA with UV technique. The mercurization of lignin is claimed to be independent of the lignin type, in contrast to bromination. However, quantification of the lignin distribution by mercurization still requires a correction of the data because of differences in the molecular weight of the syringyl and guaiacyl moieties in hardwood lignins.

References

Back E (1960) Tracheidal and parenchymatous cells in Picea abies (Karst.) pulpwood and their behavior in sulphite pulping. Sven Papperstidn 63:695–698

Bentum ALK, Côté WA Jr, Day AC, Timell TE (1969) Distribution of lignin in normal and tension wood. Wood Sci Technol 3:213–218

Bland DE, Foster RC, Logan AF (1971) The mechanism of permanganate and osmium tetroxide fixation and distribution of lignin in the cell wall of Pinus radiata. Holzforschung 25:137–143

Boutelje JB (1972) Calculation of lignin concentration and porosity of cell wall regions by interference microscopy. Sven Papperstidn 75:683–686

Brauns FE (1952) The chemistry of lignin. Academic Press, New York, 24–48

Côté WA Jr, Day AC, Timell TE (1968) Studies on compression wood VII. Distribution of lignin in normal and compression wood of tamarack Larix laricina (Du Roi, K Koch). Wood Sci Technol 2:13–37

Crocker EC (1921) An experimental study of the significance of "lignin" color reactions. Ind Eng Chem 13:625–627

Donaldson LA, Ryan KG (1987) A comparison of relative lignin concentration as determined by interference microscopy and bromination/EDXA. Wood Sci Technol 21:303–309

Erikkson I, Lidbrandt O, Westermark U (1988) Lignin distribution in birch (Betula verrucosa) as determined by mercurization with SEM- and TEM-EDXA. Wood Sci Technol 22: 251–257

Fergus BJ, Goring DAI (1970a) The location of guaiacyl and syringyl lignins in birch xylem tissue. Holzforschung 24:113–117

Fergus BJ, Goring DAI (1970b) The distribution of lignin in sprucewood as determined by ultraviolet microscopy. Holzforschung 24:118–124

Fergus BJ, Procter AR, Scott JAN, Goring DAI (1969) The distribution of lignin in sprucewood as determined by ultraviolet microscopy. Wood Sci Technol 3:117–138
Fujii T, Harada H, Saiki H (1981) Ultrastructure of "amorphous layer" in xylem parenchyma cell wall of angiosperm species. Mokuzai Gakkaishi 27:149–156
Grubb DT (1971) The calibration of beam measurement devices in various electron microscopes using an efficient Faraday cup. J Phys E Sci Instrum 4:222–224
Hepler PK, Fosket DE, Newcomb EH (1970) Lignification during secondary wall formation in Coleus: an electron microscope study. Am J Bot 57:85–96
Hoffmann P, Parameswaran N (1976) On the ultrastructural localization of hemicelluloses within delignified tracheids of spruce. Holzforschung 30:62–70
Howitt DG, Thomas G, Toutolmin W (1976) Comparisons of the measurement of beam-current densities in an electron microscope using a Faraday cup and solid-state detector. J Appl Phys 47:1694–1696
Kishi K, Harada H, Saiki H (1982) The distribution of lignin in vessel walls after treatments on ultrathin sections. Bull Kyoto Univ For 54:209–216
Kutscha NP, Schwarzmann JM (1975) The lignification sequence in normal wood of balsam fir. Holzforschung 29:79–84
Lange PW, Kjaer A (1957) Kvantitativ kemisk analys av cellväggens delar i ved-och cellulosafibrer med användning av interferensmikroskopi. Nor Skogind 11:425–432
Mary M, Revol JF, Goring DAI (1986) Mass loss of wood and its components during transmission electron microscopy. J Appl Polym Sci 31:957–963
Meier H (1955) Über den Zellwandabbau durch Holzvermorschungpilze und die submikroskopische Struktur von Fichtentracheiden und Birkenholzfasern. Holz Roh-Werkst 13:323–338
Musha Y, Goring DAI (1975) Distribution of syringyl and guaiacyl moieties in hardwoods as indicated by ultraviolet microscopy. Wood Sci Technol 9:45–58
Parham RA (1974) Distribution of lignin in kraft pulp as determined by electron microscopy. Wood Sci Technol 6:305–315
Parham RA, Côté WA Jr (1971) Distribution of lignin in normal and compression wood of Pinus taeda L. Wood Sci Technol 5:49–62
Russ JC (1974) X-ray microanalysis in the biological sciences. J Submicrosc Cytol 6:55–79
Sachs IB, Clark IT, Pew JC (1963) Investigation of lignin distribution in the cell wall of certain woods. J Polym Sci Part C 2:203–212
Saka S, Goring DAI (1988) The distribution of lignin in white birch wood as determined by bromination with TEM-EDXA. Holzforschung 42:149–153
Saka S, Hosoya S, St-Germain FGT, Goring DAI (1988) A comparison of the bromination of syringyl- and guaiacyl-type lignins. Holzforschung 42:79–83
Saka S, Thomas RJ (1982) Evaluation of the quantitative assay of lignin distribution by SEM-EDXA technique. Wood Sci Technol 16:1–18
Saka S, Thomas RJ, Gratzl JS (1979) Lignin distribution in soda-oxygen and kraft fibers as determined by conventional electron microscopy. Wood Fiber 11:99–108
Saka S, Thomas RJ, Gratzl JS (1981) Lignin distribution in Douglas-fir and loblolly pine as determined by energy dispersive X-ray analysis. Proc ISWPC, Stockholm, Vol. 1 SPCI Report No 38 35–42
Saka S, Whiting P, Fukazawa K, Goring DAI (1982) Comparative studies on lignin distribution by UV microscopy and bromination combined with EDXA. Wood Sci Technol 16: 269–277
Scott JAN, Procter AR, Fergus BJ, Goring DAI (1969) The application of ultraviolet microscopy to the distribution of lignin in wood. Description and validity of the technique. Wood Sci Technol 3:73–92
Spurr AR (1969) A low-viscosity epoxy embedding medium for electron microscopy. J Ultrastruct Res 26:31–43
Timell TE (1973) Studies on opposite wood in conifers III. Distribution of lignin. Wood Sci Technol 7:163–172
Westermark U (1985) Bromination of different morphological parts of spruce wood (Picea abies Karst.). Wood Sci Technol 19:323–328

Westermark U, Lidbrandt O, Eriksson I (1988) Lignin distribution in spruce (Picea abies) determined by mercurization with SEM-EDXA technique. Wood Sci Technol 22:243–250

Whiting P, Goring DAI (1982) Chemical characterization of tissue fractions from the middle lamella and secondary wall of black spruce tracheids. Wood Sci Technol 16:261–267

Wood JR, Goring DAI (1971) The distribution of lignin in stem wood and branch wood of Douglas-fir. Pulp Pap Mag Can 72:T95–T102

4.5 Cross Polarization/Magic Angle Spinning Nuclear Magnetic Resonance (CP/MAS NMR) Spectroscopy

G.J. LEARY and R.H. NEWMAN

4.5.1 Introduction

Standard techniques for solution-state ^{13}C NMR seldom yield useful spectra when applied to solid samples as the signals are usually too weak and broad. However, the problems with weak signals can be largely overcome by using cross polarization pulse sequences (Pines et al. 1973), and the broad linewidths can be sharpened by spinning samples at the "magic angle" of 54.7° relative to the applied magnetic field (Schaefer and Stejskal 1976). The first CP/MAS spectrum of solid wood showed a resonance band assigned to lignin (Schaefer and Stejskal 1976), but there was a gap of several years before resolution improved sufficiently to reveal detailed structural information about solid lignin (Bartuska et al. 1980). The number of research groups using the technique then increased rapidly as commercial CP/MAS NMR equipment became available and wood scientists began to recognize the advantages of being able to characterize lignin in situ.

Figure 4.5.1a shows a typical CP/MAS NMR spectrum of solid lignin. This sample was extracted from steam-exploded *Eucalyptus regnans* wood. The linewidths are broader than in the solution spectrum (Fig. 4.51b) but it is still possible to resolve signals separated by only a few parts per million. Signal areas are roughly proportional to the amount of carbon contained in each type of functional group, giving NMR an advantage over IR and other spectroscopic techniques which require independent calibration of oscillator strengths for each signal. Publications describing CP/MAS NMR studies of lignin have been reviewed (Newman 1989).

4.5.2 Principle of CP/MAS NMR

4.5.2.1 Cross Polarization

Cross polarization (CP) pulse sequences are designed to transfer magnetic polarization from abundant nuclei, e.g., protons, to rare nuclei, e.g., ^{13}C nuclei at 1.1% natural isotopic abundance. The result is an enhancement of the NMR signal from the rare nucleus. The principle is sketched here; more detailed explanations may be found in books by Mehring (1983) and by Gerstein and Dybowski (1985).

Springer Series in Wood Science
Methods in Lignin Chemistry
(Edited by S.Y. Lin and C.W. Dence)

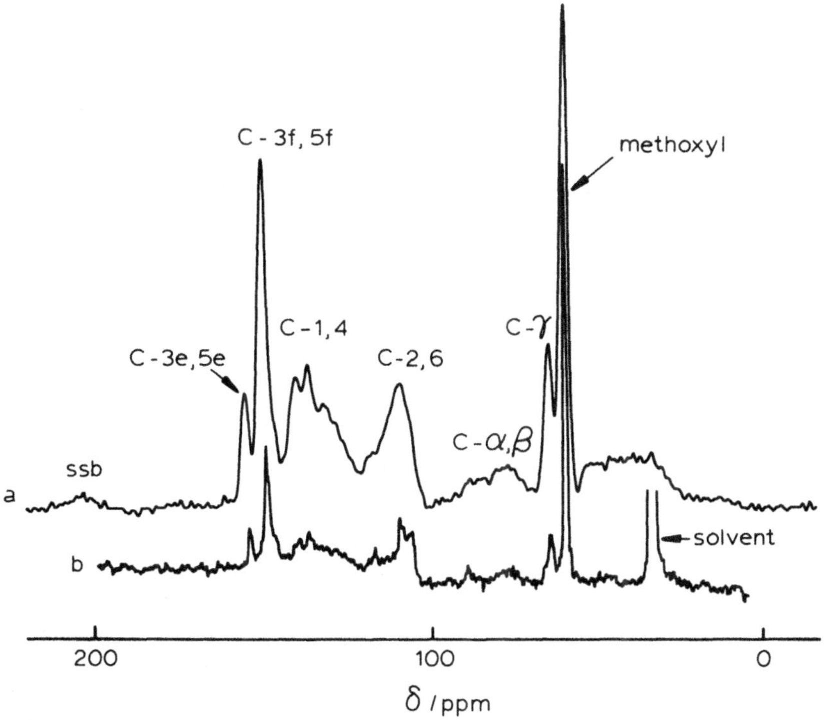

Fig. 4.5.1. ^{13}C 50 MHz NMR spectra of lignin extracted from steam-exploded *E. regnans* wood: *a* CP/MAS spectrum run with MAS at 5.0 kHz, 15 min of data averaging and digital resolution enhancement; *b* solution NMR spectrum after 2 h of data averaging. Both spectra run on a Varian XL-200 spectrometer. Assignments show carbons labeled *1* to *6* around the aromatic ring and from *α* to *γ* along the sidechain of a C_9 lignin unit. Abbreviations *e* and *f* refer to units etherified or free (not etherified) through the oxygen on C-4. A spinning sideband is labeled *ssb*. The sharpest signals in the CP/MAS NMR spectrum are at 153 ppm (*C-3e, 5e*), 148 ppm (*C-3f, 5f*), 62 ppm (*C-γ*), and 56 ppm (methoxyl)

A generalized CP pulse sequence is shown in Fig. 4.5.2, with vertical displacements indicating transmitter power and the horizontal axis indicating elapsed time. A CP pulse sequence begins with a pulse delay, t_d, to allow recovery of proton polarization along the static field of the NMR magnet. This can be achieved if t_d greatly exceeds $T_1(H)$, the time constant for recovery of equilibrium polarization.

The first radiofrequency (RF) pulse is at the proton NMR frequency and is of a duration t_p chosen for a 90° displacement of the magnetization vector away from the static field. At the end of t_p, the proton transmitter is left on but the oscillation is phase-shifted through 90°. The RF field now oscillates in synchronization with precession of the proton magnetization vector; instead of displacing the vector further from the static field, it counteracts any tendency for the vector to drift away from synchronized precession. The magnetization is then said to be "spin locked".

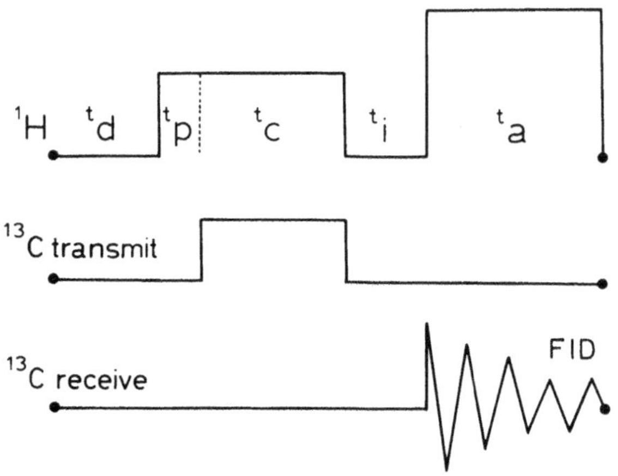

Fig. 4.5.2. Representation of a generalized pulse sequence for CP/MAS NMR

While the proton magnetization remains spin locked, a second RF transmitter is switched on at the ^{13}C NMR frequency for a contact time t_c. Any ^{13}C magnetization that builds up during the contact time is conserved through spin locking to the second RF field. Once this is switched off, the receiver is switched on to acquire data over a period of t_a.

The processes involved in CP NMR can be explained by analogy with a hydraulic model (Fig. 4.5.3) in which the thermal equilibrium population of spins aligned along the static magnetic field is represented by the main reservoir labeled "lattice". The applied RF pulse is represented by a "pump" and the resultant component of magnetization moved into the plane perpendicular to the

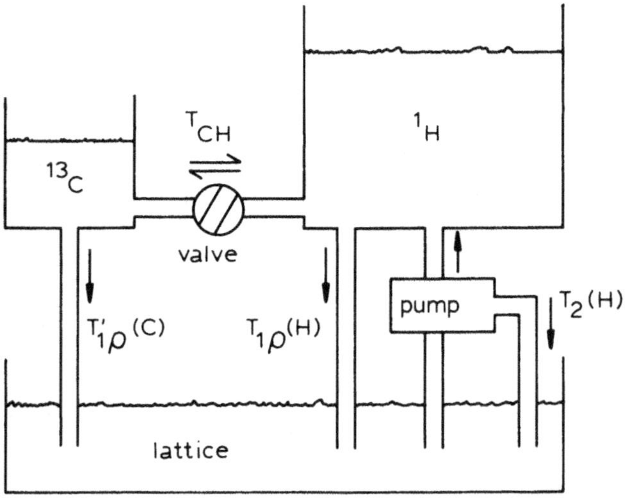

Fig. 4.5.3. Hydraulic model for transfer of spin polarization during the CP/MAS NMR process

static field is represented by the reservoir "H". In practice the "pump" is not totally efficient because spin-spin interactions gradually destroy the component of perpendicular magnetization. The time constant, $T_2(H)$, for spin-spin relaxation in lignin is typically of similar magnitude to t_p (Newman 1987, Tekely and Vignon 1987).

Once the magnetization has become "spin locked", the effects of $T_2(H)$ relaxation are mostly countered by the RF field, but there is still some loss of magnetization represented by a leak from the ^1H reservoir. The time constant for this leak is $T_{1\rho}(H)$.

Switching on the ^{13}C RF transmitter is represented by opening the valve between the reservoirs ^1H and ^{13}C. The relative powers of the proton and ^{13}C RF transmitters are adjusted to maximize interactions between the two types of precessing nuclei. Polarization can then be transferred between neighboring nuclei through spin "flip-flop" processes. Optimization is achieved when the Hartmann-Hahn condition is met, i.e., the ^1H and ^{13}C RF field strengths are in a ratio set close to 1:4 (Pines et al. 1973). Magnetization is then transferred with a time constant T_{CH}.

Spin locking of the ^{13}C nuclei is not totally effective and magnetization is gradually lost with time constant $T'_{1\rho}(C)$. This loss is represented by a leak in the ^{13}C reservoir in Fig. 4.5.3. It is commonly assumed that the loss is too slow to have a significant effect on signal strengths.

At the end of t_c the ^{13}C receiver is switched on to monitor the precessing magnetization through the oscillating voltage induced in the RF coil. This corresponds to measuring the depth in the reservoir ^{13}C. The oscillating voltage is digitized and processed by Fourier transformation to provide a spectrum.

The hydraulic model of Fig. 4.5.3 can be described by a pair of simultaneous differential equations with a solution in the form of a double exponential expression (Schaefer et al. 1981):

$$S = \left(\frac{KNT_{1\rho}(H)}{T_{1\rho}(H) - T_{CH}}\right)\{\exp[-t_c/T_{1\rho}(H)]-\exp[-t_c/T_{CH}]\}, \qquad (1)$$

where S is the strength of the signal from N ^{13}C nuclei, and K is a constant based on the sensitivity of a particular spectrometer. Despite the simplifying assumptions, Eq. (1) provides a useful basis for exploring the consequences of adjusting the contact time t_c. Fig. 4.5.4 shows experimental data for a signal at a chemical shift of 149 parts per million (ppm) assigned to nonprotonated carbon in lignin, along with a best-fit curve based on Eq. (1) with $T_{CH} = 0.38$ ms and $T_{1\rho}(H) = 4.4$ ms. The broken line shows a second curve predicted for a CH group with the same value of $T_{1\rho}(H)$ but with T_{CH} reduced to 0.05 ms, i.e., to a typical value for protonated carbon as estimated in the same set of experiments. Contact times shorter than 1 ms would result in underestimation of nonprotonated carbon because of slower transfer from the ^1H to ^{13}C reservoirs of Fig. 4.5.3. Contact times longer than 1 ms would result in overestimation of nonprotonated carbon because of slower transfer from the ^{13}C reservoir back to the ^1H reservoir as the latter becomes depleted by $T_{1\rho}(H)$. The consequences of variation of t_c have been discussed in more detail by Hatfield et al. (1987).

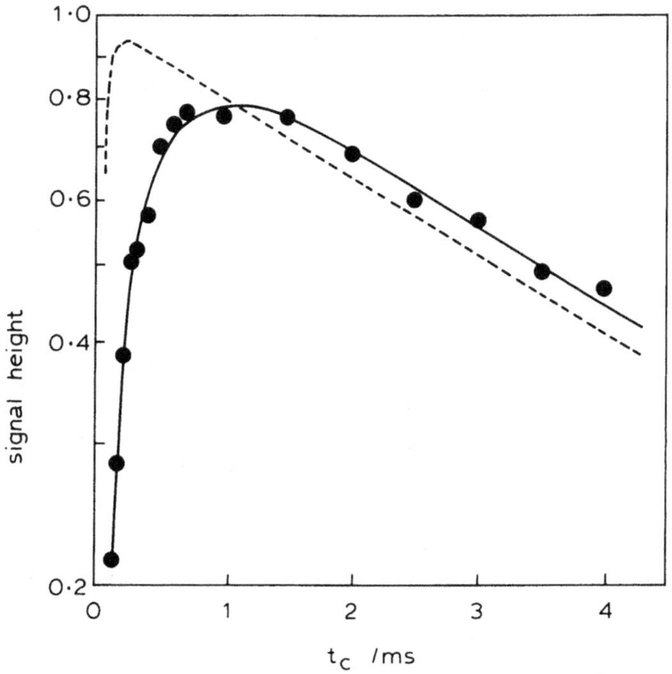

Fig. 4.5.4. Signal heights as a function of the contact time t_c for the lignin signal at 149 ppm in *Pinus radiata* wood (*circles*) along with a best-fit curve (*solid line*) based on Eq. (1) and a simulated curve (*broken line*) for a CH group

4.5.2.2 Magic Angle Spinning

Solution NMR spectra are routinely sharpened by spinning the NMR tube. The spectra of solids can likewise be sharpened, but the orientation of the spinning axis is as important as fast spinning.

The precise ^{13}C NMR frequency depends on the orientation of each molecule relative to the static magnetic field. Spinning the sample therefore modulates the resonance condition, resulting in a centerband signal at the averaged frequency and spinning sideband (SSB) signals at harmonics of the spinning frequency. It can be shown that the residual width of the centerband signal is proportional to a factor $(1-3\cos^2\theta)$, where θ is the angle between the spinning axis and the static magnetic field. The centerband can therefore be collapsed by selecting the magic angle of $\theta = \cos^{-1}(1/3)^{1/2}$.

The strength of the SSB signals can be reduced if the spinner frequency is sufficient to place the first-order SSB signals beyond the range of chemical shifts associated with different orientations of the sample. This range is about ±100 ppm for CH groups in aromatic rings (Veeman 1984). The target spinning frequency would therefore be 100 Hz for every MHz of the ^{13}C NMR frequency. This would not eliminate the SSB signals entirely; in fact, the graphs published by Herzfeld and Berger (1980) show that the first-order SSB signals would still be about one-tenth the height of the centerband signal and that 22% of the total signal would be spread out in the SSB pattern.

Such performance was beyond the capabilities of the NMR spectrometers used in the first few years of CP/MAS NMR studies of lignin. The problem of SSB strength was at first avoided by using low-field spectrometers, but this was unsatisfactory because of the poor sensitivity associated with low fields. The problem was later partly overcome by SSB-suppression pulse sequences (Dixon et al. 1981, Barron et al. 1985), or by applying correction factors to the centerband strength (Hemmingson and Newman 1985).

4.5.3 Description of Methods for CP/MAS NMR Spectrocopy

4.5.3.1 Choice of Spectrometer

Line broadening effects can be divided into two categories: those that are independent of the magnetic field and those that broaden the line in proportion to the strength of the field applied. There is an advantage in the use of a high-field spectrometer if the former effects are significant by comparison with the latter, as in the case of lignin (Fig. 4.5.5). An NMR spectrum run at 75 MHz (Fig. 4.5.5b) showed better separation of the C-γ and methoxyl signals compared with a spectrum of the same sample run at 25 MHz (Fig. 4.5.5a).

A high-field NMR spectrometer also provides better signal-to-noise ratios because the proton spins are more strongly polarized. The two spectra in Fig. 4.5.5 involved similar numbers of averaged transients, yet the high-field NMR spectrum (Fig. 4.5.5b) shows far less noise. Improved signal-to-noise ratios can be traded against even better resolution through the use of digital resolution enhancement as in Fig. 4.5.1.

The disadvantage of a high-field spectrometer is the requirement for faster magic-angle spinning as discussed in Section 4.5.2.2. The SSB signals from aromatic carbon are prominent in Fig. 4.5.5b, across the ranges 40 to 90 ppm and 170 to 220 ppm. The authors use a spectrometer operating at 50 MHz as a compromise between good signal-to-noise ratios and freedom from spinning sidebands.

4.5.3.2 Sample Preparation

Most MAS rotors require between 0.1 and 0.5 g of sample. Samples should be ground to sufficiently small size for uniform packing in the MAS rotor. Wiley milling of wood to 40 or 60 mesh has been commonly used, but ball-milled wood is undesirable because the severe physical damage to cellulose fibers broadens NMR signals (Kolodziejski et al. 1982).

Moistening the sample assists proton spin-lattice relaxation. Experiments in the authors' laboratory gave a value of $T_1(H) = 0.05$ s for a sample of *Pinus radiata* wood containing 45% moisture; this rose to 0.75 s when the sample was

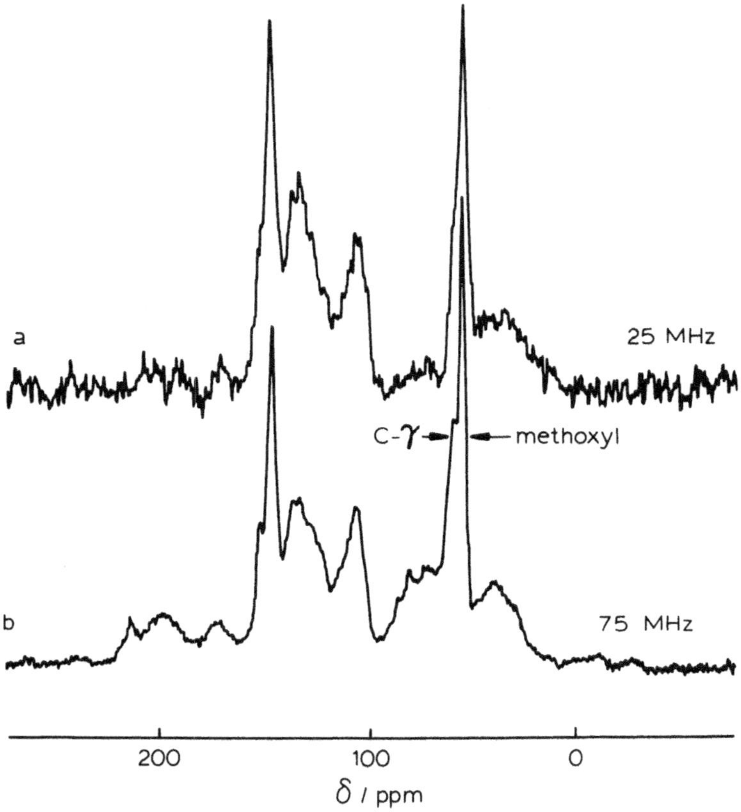

Fig. 4.5.5. Improvement in CP/MAS NMR results through the use of a stronger magnetic field. Spectra of *E. regnans* lignin (as in Fig. 1) run at two ^{13}C NMR frequencies and plotted without resolution enhancement. *a* t_d = 1 s, t_p = 3.5 μs, t_c = 1 ms, t_a = 50 ms, MAS at 4.1 kHz and 996 transients averaged on a Bruker MSL-100 spectrometer; *b* t_d = 0.5 s, t_p = 4.8 μs, t_c = 1 ms, t_a = 12 ms, MAS at 5.0 kHz, and 1148 transients averaged on a Bruker AM-300 spectrometer

oven-dried, and returned to 0.05 s when the sample was remoistened. This meant that the pulse delay had to be 25 times longer for the oven-dried sample, wasting considerable spectrometer time. Moisture contents as high as 45% are difficult to retain during MAS operation, so a compromise is recommended.

Moistening the sample has the additional advantage of sharpening line shapes for carbohydrate signals in the CP/MAS NMR spectra of wood, reducing interference with lignin signals (Willis and Herring 1986, Sterk et al. 1987). It can also improve the efficiency of cross-polarization dynamics (Willis and Herring 1986).

4.5.3.3 Determination of Lignin in Wood and Pulp

Radiofrequency (RF) power output levels should be adjusted for a Hartmann-Hahn match. The pulse sequence illustrated in Fig. 4.5.2 should be used with the following parameters:

1. t_d = 2 s (seconds) if $T_1(H)$ is unknown, or shorter provided $t_d > 3T_1(H)$. The value of $T_1(H)$ can be estimated in a preliminary experiment by starting with t_d = 2 s and halving t_d in successive short runs until signal suppression becomes obvious.
2. t_p adjusted for a 90° displacement of magnetization. The actual value will depend on RF power and can be estimated by varying t_p until null signals are obtained for multiples of 180° displacements. Typical values of t_p are a few μs (microseconds).
3. t_c = 1 ms (millisecond) for relatively weak proton RF field strengths (5 to 10 Gauss) or preferably t_c = 2 ms and a stronger field (>10 Gauss) if the RF power output is available.
4. t_a = 50 ms if this is within the permissible duty cycle of the spectrometer, or t_a > 15 ms otherwise. Shorter values can result in noticeable broadening of signals.
5. $t_i = 0$.

Between 10^3 and 10^4 transients should be averaged to provide an adequate signal-to-noise ratio.

The lignin content can be estimated from the spectrum by measuring the fractional area, A, of a band of signals between chemical shifts of 141 and 159 parts per million (ppm) relative to the total area of the spectrum. This band is assigned to C-3 and C-4 of guaiacyl units or C-3 and C-5 of syringyl units; the fact that it is associated with just two carbon atoms of each structural unit means that the method can be used for both softwoods and hardwoods.

This procedure is not valid if the wood contains tannins since these polyphenolic substances contribute to area A. The presence of tannins can be tested for examining the spectrum for a signal at 145 ppm. A signal will appear at this chemical shift regardless of whether the tannin is a "condensed tannin" (procyanidin polymer) or "hydrolyzable tannin" (based on gallic or ellagic acids) (Morgan and Newman 1987). An example of this test is provided in Fig. 4.5.6; the spectrum of oak (*Quercus robur*) wood was obtained with t_d = 0.6 s, t_p = 5.5 μs, t_c = 2 ms, t_a = 15 ms, 5 kHz MAS, and 10^4 transients averaged. Tannins are clearly present.

The area A can be determined, e.g., by drawing a baseline and vertical lines at 141 and 159 ppm and then cutting out and weighing the areas enclosed. Area A must then be corrected for SSB strengths and converted from mole fraction to weight fraction. The method summarized here follows procedures described in more detail by Hemmingson and Newman (1985) and Leary et al. (1986).

First, we need to assume an average number (m) of methoxyl groups per lignin unit. The result will not be particularly sensitive to errors in the choice of m, but it is advisable to be consistent in the use of a single value of m for a series of samples. Reasonable choices are m = 1 for the predominantly guaiacyl lignins of softwoods and m = 1.5 for the mixed guaiacyl/syringyl lignins of hardwoods. We then calculate the fraction, F, of carbon contained in lignin:

$$F = Ak(9 + m)/2, \tag{2}$$

where k is a correction factor for signal strength lost in SSB signals (see Table 4.5.1). The weight percentage, C, of carbon in the lignin is also dependent on a chosen value of m:

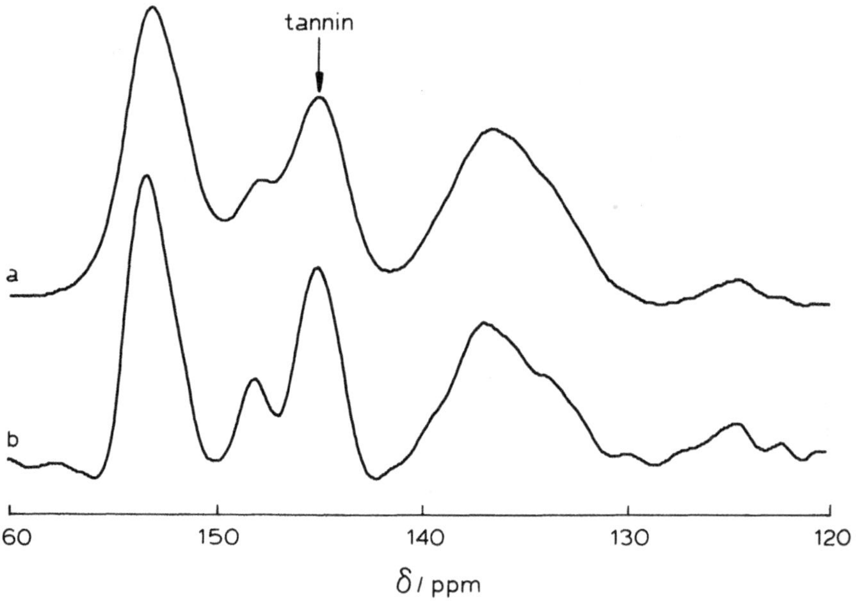

Fig. 4.5.6. Portion of a CP/MAS NMR spectrum of oak wood, showing a signal assigned to tannin; *a* without and *b* with digital resolution enhancement

Table 4.5.1. Correction factors (k) for spinning sidebands, to be used in lignin determination by CP/MAS NMR. The sideband separation (in ppm) is obtained by dividing the MAS frequency by the ^{13}C NMR frequency

Sideband separation	k
50	1.57
60	1.33
80	1.19
100	1.12
120	1.08
140	1.06

$$C = 100(108 + 12\,m)/(157.3 + 30\,m). \quad (3)$$

The results of evaluating Eqs. (2) and (3) are now combined to provide the lignin content, L, as a percentage (by weight) of dry wood:

$$L = 4440F/\{C - (C - 44.4)F\}. \quad (4)$$

The method can be further refined by allowing for weak SSB signals cast into area A from cellulose and C-2,6 of syringyl groups (Leary et al. 1986). Naturally, it would be more desirable to choose a spinning speed sufficient to cast these SSB signals well clear of the region 141 to 159 ppm.

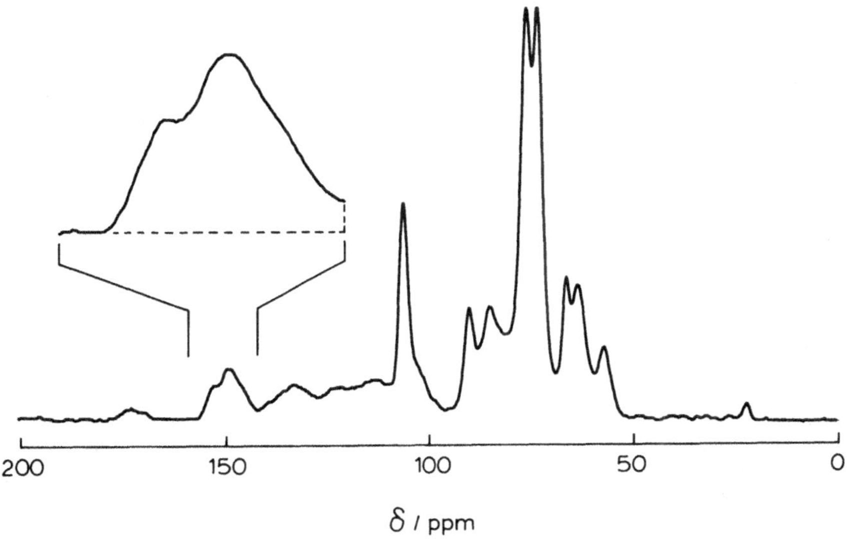

Fig. 4.5.7. CP/MAS NMR spectrum of *Pinus radiata* wood. The area of the enlarged region is used for determination of the lignin content

An example of a lignin-determination experiment is provided in Fig. 4.5.7. The sample was *Pinus radiata* wood, run with t_d = 0.2 s, t_p = 5.5 μs, t_c = 2 ms and t_a = 20 ms. The sample was spun at 5 kHz, and transients from 2000 contacts were averaged. The area between 141 and 159 ppm accounted for a fraction 0.0571 of the total spectrum. Since the sample was a softwood, a value m = 1 was chosen, yielding [from Eq. (2)] F = 0.320 and [from Eq. (3)] C = 64.1. The lignin content [from Eq. (4)] was therefore estimated at 25% of the dry weight of wood. A Klason lignin determination gave the same result for this sample.

4.5.3.4 Interrupted Decoupling

The CP/MAS NMR spectra of lignin and wood can be simplified by suppressing signals from CH and CH_2 groups in the lignin or associated carbohydrates. The signals that remain include a band assigned to nonprotonated carbon atoms in lignin aromatic rings. This band stretches across chemical shifts from about 130 to 165 ppm and features within the band can provide useful information about substitution patterns around the aromatic rings.

The pulse sequences used for this purpose are known as "interrupted decoupling" (Gerasimowicz et al. 1984, Haw et al. 1984, Barron et al. 1985, Maciel et al. 1985), "delayed decoupling" (Manders 1987) or "dipolar dephasing" (Hatcher 1987). All involve an interval, t_i, inserted in the pulse sequence between the end of the contact time and the beginning of data acquisition as shown in Fig. 4.5.2. Both ^{13}C and proton RF transmitters are switched off for the duration of t_i, and strong static dipolar interactions between ^{13}C

nuclei and directly bonded protons cause rapid destruction of the precessing magnetization for CH and CH_2 groups.

Some of the pulse sequences involve additional refocusing pulses in the ^{13}C channel, the proton channel, or both channels. This discussion and examples presented here are confined to the simplest of interrupted-decoupling sequences, as illustrated in Fig. 4.5.2, since the reader may not have access to the computer code for the more sophisticated versions.

The operating parameters listed in Section 4.5.3.3 are appropriate for interrupted decoupling also, except that:

1. The requirement $t_d > 3T_1(H)$ can be relaxed to $t_d > T_1(H)$ since it is unlikely that $T_1(H)$ will vary significantly between different units of the lignin macromolecules.
2. The delay t_i should be adjusted for a null signal in the chemical shift region around 75 ppm, i.e., for the strongest of the signals assigned to cellulose. The optimal value of t_i can be sensitive to variations in MAS frequency, but is typically between 40 and 80 μs.
3. The number of averaged transients may need to be increased to 10^4 or 10^5 for an adequate signal-to-noise ratio.

Interrupted-decoupling experiments provide the spectra required for the analytical procedures described in the following two sections.

4.5.3.5 Estimation of Syringyl/Guaiacyl Ratios

A typical CP/MAS NMR spectrum of a hardwood shows a sharp signal at 153 ppm and a broader band of signals centered on a chemical shift of about 149 ppm. These signals are assigned primarily to syringyl units (C-3 and C-5) and guaiacyl units (C-3 and C-4), respectively. The signals are not clearly resolved, but it is possible to estimate a ratio of the frequency of syringyl and guaiacyl units by CP/MAS NMR. The method described here is derived from that used by Manders (1987) and is based on two assumptions: that the spectrum of a typical softwood resembles the guaiacyl component in the spectrum of a hardwood, and that only guaiacyl groups contribute signals in the region 146 to 148 ppm. Two spectra are required, a spectrum, X, from which a syringyl/guaiacyl ratio is to be determined, and a spectrum, Y, of a typical softwood. The symbols X and Y represent arrays of numbers, i.e., signal strengths as a function of chemical shift.

First, the vertical scales of the spectra are normalized to the same area across the chemical-shift range 141 to 159 ppm, after which linear combinations of spectra are generated. The coefficient z in the combination $(1 - z)X - zY$ is adjusted until the signal strength in the region of 146 to 148 ppm is brought down to the baseline. The syringyl/guaiacyl ratio becomes simply $1/z$.

This procedure is illustrated in Fig. 4.5.8. All three spectra were run with $t_d = 0.3$ s, $t_p = 5.5\,\mu s$, $t_c = 2$ ms, $t_a = 50$ ms, 5 kHz MAS, and between 21×10^3 and 36×10^3 transients averaged. The spectrum of wood from spruce (*Picea abies*) was chosen as typical of a softwood. The results were syringyl/guaiacyl

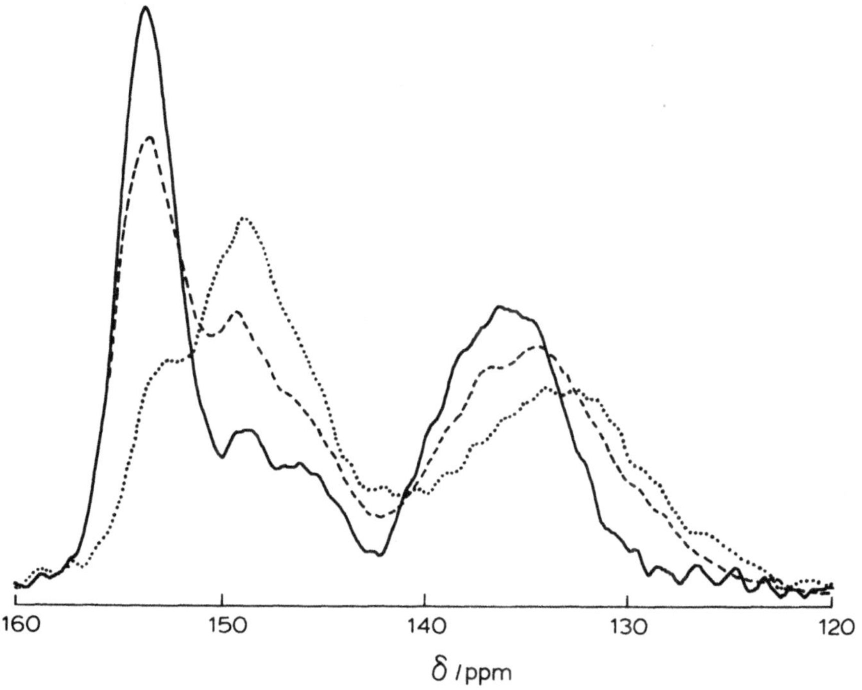

Fig. 4.5.8. Interrupted-decoupling CP/MAS NMR spectra of wood from *Eucalyptus obligua* (*solid line*), *Pseudowintera axillaris* (*broken line*) and *Picea abies* (*dotted line*)

ratios of 1.3 for the hardwood *Eucalyptus obliqua* and 0.7 for the endemic New Zealand hardwood *Pseudowintera axillaris*.

4.5.3.6 Estimation of the Degree of Etherification

Interrupted-decoupling CP/MAS NMR spectra of softwoods show a band of signals from about 140 to 156 ppm, with some evidence of partly resolved structure. Variations in the shape of this band are attributed to variations in the abundance of β-O-4 ether linkages. The degree of etherification can be estimated by curve fitting. The procedure described here follows Leary et al. (1987).

Simulated NMR spectra are generated by summing four Gaussian lineshapes. Two peaks of equal intensity at 152.6 and 148.4 ppm represent contributions from etherified units and two peaks of equal intensity at 148.4 and 145.0 ppm represent contributions from nonetherified units. The width of each component should be about 210 Hz at half height, although some variation may be observed between spectrometers. Relative contributions are adjusted until the result fits the experimental spectrum across the range 143 to 156 ppm. The procedure is illustrated in Fig. 4.5.9 for spruce wood (spectrum as dotted in Fig. 4.5.8). The best fit was obtained for 60% etherification.

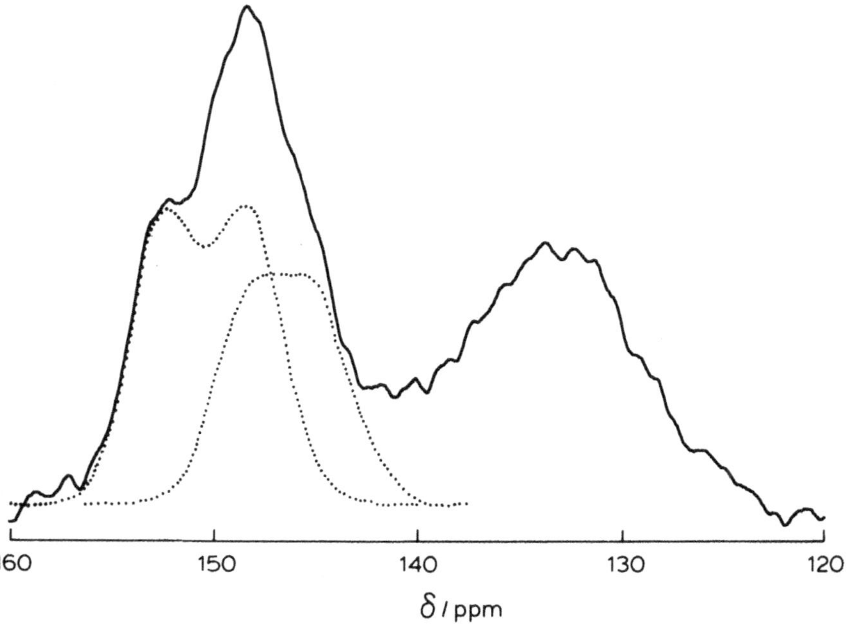

Fig. 4.5.9. Interrupted-decoupling CP/MAS NMR spectrum of wood from *Picea abies*. The *dotted lines* show components of a simulated spectrum based on contributions from etherified (*left*) and nonetherified (*right*) lignin units

The degree of etherification of a hardwood lignin cannot be estimated as easily. The presence of both syringyl and guaiacyl units adds to the number of peaks required for curve fitting and it is particularly difficult to distinguish a signal at 148 ppm (assigned to C-3 and C-5 in nonetherified syringyl units) from a broader band at the same chemical shift (assigned to guaiacyl units). Relative changes in the degree of etherification can, however, be estimated from a shift in signal intensity across a vertical line drawn at about 151 ppm (Leary et al. 1986).

4.5.3.7 Resolution Enhancement

Digital resolution enhancement is sometimes used to improve the appearance of CP/MAS NMR spectra. The method described here follows that used by Leary et al. (1986). The number of averaged transients must be increased beyond normal requirements because signal-to-noise ratio is traded against resolution. The averaged data table is multiplied by an exponentially increasing function (time constant 3 ms) and a Gaussian decay curve (time constant 6 ms) and then Fourier-transformed. Examples are shown in Fig. 4.5.1a and Fig. 4.5.6b. Comparison of the two curves in Fig. 4.5.6 reveals an improvement in the separation of a tannin signal at 145 ppm from a lignin signal at 148 ppm.

4.5.4 Discussion

The reliability of any attempt at determination or characterization of lignin by NMR is dependent on fortuitous similarities in the various spin relaxation parameters. For example, values of $T_2(H)$ for lignins tend to be longer than those for cellulose or hemicellulose (Newman 1987, Tekely and Vignon 1987), but the differences are too small to be important for typical values of t_p. Values of $T_{1\rho}(H)$ are sometimes indistinguishable for various components of a sample, and even when differences have been observed, the values are so long that the differences would have little effect on the results (Gerasimowicz et al. 1984, Haw et al. 1984, Maciel et al. 1985).

These fortuitous similarities should not be taken for granted, and it is important that the NMR methods should be tested against other techniques. The method for determination of lignin contents described in Section 4.5.3.3 has provided results in good agreement with Klason lignin contents for softwoods. The NMR results for hardwoods were genrally 4% higher, but this could be explained by inclusion of acid-soluble lignins in the NMR determinations (Leary et al. 1986).

Himmelsbach et al. (1983) have used a method similar to that described in Section 4.5.3.3, but based their interpretation on the NMR spectra of known standards. This approach is, of course, just as dependent on fortuitous circumstances, i.e., similarities in spin relaxation parameters between the standards and the samples. Other methods have been proposed for determination of lignin by NMR but are designed for less general use. Haw et al. (1984) integrated over all signals assigned to aromatic carbon in softwoods; but this method cannot be used for hardwoods because of the coincidence of signals for C-1 of cellulose and C-2,6 of syringyl lignin units. Willis et al. (1986) used signal heights rather than areas for the band at 148–150 ppm and used samples of known lignin content for calibration. This method would not be reliable if the structure of the lignin differed between samples.

4.5.5 Future Developments

Recent literature contains no reference to NMR of lignin in solid samples spun faster than 4.0 kHz (Newman 1989), yet commercial MAS probes are now available for routine spinning at 5 kHz or more. Most of the illustrations used in this chapter were generated by MAS at 5 kHz. The faster MAS probes should allow spectroscopists to use stronger static magnetic fields, improving the dispersion of NMR signals without spreading too much signal strength into spinning sidebands. The use of stronger fields will in turn provide better signal-to-noise ratios after shorter data-averaging times.

The introduction of faster MAS probes will require modifications of procedures for lignin analysis. In particular, the "interrupted decoupling" pulse sequence depends on dipolar interactions which become averaged out by high-speed spinning. Accordingly, modified sequences are being developed to ensure that spectroscopists can take advantage of the new MAS technology.

Acknowledgment. The authors thank Charles Rodger and Sandra Mooibroek of Bruker Spectrospin (Canada) Ltd for running the spectra shown in Fig. 4.5.5.

References

Barron PF, Frost PL, Doimo L, Kennedy MJ (1985) ^{13}C-CP/MAS NMR examination of some Australian hardwoods and their chemical and biochemical residues. J Macromol Sci Chem A22:303–322

Bartuska VJ, Maciel GE, Bolkers HI, Fleming BI (1980) Structural studies of lignin isolation procedures by ^{13}C NMR. Holzforschung 34:214–217

Dixon WT, Schaefer J, Sefcik MD, Stejskal EO, McKay RA (1981) Quantitative chemical composition of materials such as humic soils, lignins, and coals by high-resolution carbon-13 NMR. J Magn Reson 45:173–176

Gerasimowicz WV, Hicks KB, Pfeffer PE (1984) Evidence for the existence of associated lignin-carbohydrate polymers as revealed by carbon-13 CPMAS solid-state NMR spectroscopy. Macromolecules 17:2597–2603

Gerstein BC, Dybowski CR (1985) Transient techniques in NMR of solids: an introduction to theory and practice. Academic Press, Orlando, 295 pp

Hatcher PG (1987) Chemical structural studies of natural lignin by dipolar dephased solid-state ^{13}C nuclear magnetic resonance. Org Geochem 11:31–39

Hatfield GR, Maciel GE, Erbatur O, Erbatur G (1987) Qualitative and quantitative analysis of solid lignin samples by carbon-13 nuclear magnetic resonance spectrometry. Anal Chem 59:172–179

Haw JF, Maciel GE, Schroeder HA (1984) Carbon-13 nuclear magnetic resonance spectrometric study of of wood pulping with cross polarization and magic-angle spinning. Anal Chem 56:1323–1329

Hemmingson JA, Newman RH (1985) A CP/MAS ^{13}C NMR study of the effect of steam explosion processes on wood composition and structure. J Wood Chem Technol 5: 159–188

Herzfeld J, Berger AE (1980) Sideband intensities in NMR spectra of samples spinning at the magic angle. J Chem Phys 73:6021–6030

Himmelsbach DS, Barton FE, Windham WR (1983) Comparison of carbohydrate, lignin, and protein ratios between grass species by cross polarization-magic angle spinning carbon-13 nuclear magnetic resonance. J Agric Food Chem 31:401–404

Kolodziejski W, Frye JS, Maciel GE (1982) Carbon-13 nuclear magnetic resonance spectrometry with cross polarization and magic-angle spinning for analysis of lodgepole pine wood. Anal Chem 54:1419–1424

Leary GJ, Morgan KR, Newman RH (1987) Solid state carbon-13 nuclear magnetic resonance study of Pinus radiata wood. Appita 40:181–184

Leary GJ, Newman RH, Morgan KR (1986) A carbon-13 nuclear magnetic resonance study of chemical processes involved in the isolation of Klason lignin. Holzforschung 40:267–272

Maciel GE, Haw JF, Smith DH, Gabrielson BC, Hatfield GR (1985) Carbon-13 nuclear magnetic resonance of herbaceous plants and their components, using cross polarization and magic-angle spinning. J Agric Food Chem 33:185–191

Manders WF (1987) Solid-state ^{13}C NMR determination of the syringyl/guaiacyl ratio in hardwoods. Holzforschung 41:13–18

Mehring M (1983) Principles of high resolution NMR in solids. 2nd edn. Springer, Berlin Heidelberg New York, 342 pp

Morgan KR, Newman RH (1987) Estimation of the tannin content of eucalypts and other hardwoods by carbon-13 nuclear magnetic resonance. Appita 40:450–454

Newman RH (1987) Effects of finite preparation pulse power on carbon-13 cross-polarization NMR spectra of heterogeneous samples. J Magn Reson 72:337–340

Newman RH (1989) Carbon-13 NMR studies of lignin in solid samples – a review. Chemistry Division Report, DSIR, New Zealand.

Pines A, Gibby MG, Waugh JS (1973) Proton-enhanced NMR of dilute spins in solids. J Chem Phys 59:569–589

Schaefer J, Stejskal EO (1976) Carbon-13 NMR of polymers spinning at the magic angle. J Am Chem Soc 98:1031–1032

Schaefer J, Stejskal EO, Sefcik MD, McKay RA (1981) Applications of high-resolution ^{13}C and ^{15}N NMR of solids. Philos Trans R Soc Lond A299:593–608

Sterk H, Sattler W, Esterbauer H (1987) Investigations of lignocellulosic materials by the carbon-13 NMR CP-MAS method. Carbohydr Res 164:85–95

Tekely P, Vignon MR (1987) Proton T_1 and T_2 relaxation times of wood components using ^{13}C CP/MAS NMR. J Polym Sci Part C: Polym Lett 25:257–261

Veeman WS (1984) Carbon-13 chemical shift anisotropy. Prog NMR Spectrosc 16:193–235

Willis JM, Herring FG (1986) Effect of water on the ^{13}C CP/MAS NMR spectrum of white spruce wood. Macromolecules 20:1554–1556

Willis JM, Jagannathan NR, Herring FG (1986) The potential of ^{13}C CP/MAS NMR in the study of kraft pulping kinetics. J Wood Chem Technol 6:249–267

4.6 Raman Spectroscopy

R.H. ATALLA, U.P. AGARWAL, and J.S. BOND

4.6.1 Introduction

Raman scattering was first observed in 1928 and was used to investigate the vibrational states of many molecules in the 1930s. Initially, spectroscopic methods based on the phenomenon were used in research on the structure of relatively simple molecules. Over the past 20 years, however, the development of laser sources and new generations of monochromators and detectors has made possible the application of Raman spectroscopy to the solution of many problems of technological interest.

In many industrial laboratories, Raman spectrscopy is routinely used, together with infrared spectroscopy, for acquisition of vibrational spectra. Raman spectrometer systems for routine analytical applications are commercially available. An important expansion of the potential of the technique has arisen from the use of the microprobe, which permits acquisition of spectra from domains as small as one micron.

Application of Raman spectroscopy to lignin analytical chemistry is relatively new, and only limited information has been obtained. However, the technique offers several potential advantages and, though it is complementary to infrared spectroscopy, it gives information that is not accessible with the latter alone.

4.6.2 Principle

The phenomena underlying Raman spectroscopy can be described by comparison with infrared spectroscopy as shown schematically in Fig. 4.6.1. The primary event in infrared absorption is the transition of a molecule from a ground state (M) to a vibrationally excited state (M*) by absorption of an infrared photon with energy equal to the difference between the energies of the ground and the excited states. The reverse process, infrared emission, occurs when a molecule in the excited state (M*) emits a photon during the transition to a ground state (M). In infrared spectroscopy, one derives information by measuring the frequencies of infrared photons that a molecule absorbs and interpreting these frequencies in terms of the characteristic vibrational motions of the molecule. In complex molecules, some of the frequencies are associated with functional groups that have characteristic localized modes of vibration.

Springer Series in Wood Science
Methods in Lignin Chemistry
(Edited by S.Y. Lin and C.W. Dence)

As is also shown in Fig. 4.6.1, the same transitions between molecular vibrational states (M) and (M*) can result in Raman scattering. A key difference between the Raman and infrared processes is that, in the former process, the photons involved are not absorbed or emitted but rather shifted in frequency by an amount corresponding to the energy of the particular vibrational transition. In the Stokes process, which is the parallel of absorption, the scattered photons are shifted to lower frequencies as the molecules abstract energy from the exciting photons; in the anti-Stokes process, which is parallel to emission, the scattered photons are shifted to higher frequencies as they pick up the energy released by the molecules in the course of transitions to the ground state. In addition, a substantial number of the scattered photons are not shifted in frequency. The process which gives rise to these photons is known as Rayleigh scattering. This scattering arises from density variations and optical heterogenieties and is many orders of magnitude more intense than Raman scattering.

Figure 4.6.1 depicts another major difference between Raman scattering and infrared processes. To be active in the infrared spectra, transitions must have a change in the molecular dipole associated with them. For Raman activity, in contrast, the change has to be in the polarizability of the molecule. These two molecular characteristics are qualitatively inversely related.

A Raman spectrum is obtained by exposure of a sample to a monochromatic source of exciting photons and measurement of the frequencies of the scattered light. Because the intensity of the Raman scattered component is much lower than the Rayleigh scattered component, a highly selective monochromator and a very sensitive detector are required.

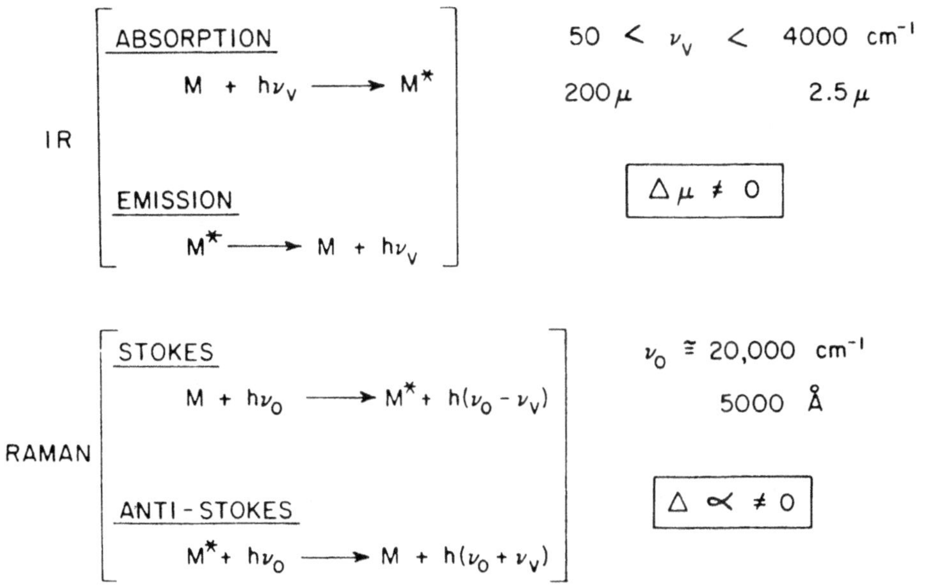

Fig. 4.6.1. A comparison of infrared and Raman phenomena; μ = dipole moment, α = polarizability, ν_v = vibrational frequency, ν_0 = exciting frequency

The exciting photons are typically of much higher energies than those of the fundamental vibrations of most chemical bonds or systems of bonds, usually by a factor ranging from about 6 for O—H and C—H bonds to about 200 for bonds between very heavy atoms, as for example in I_2. The 514.5 and 488 nm lines from an argon ion laser are often used as exciting frequencies.

4.6.3 Method

4.6.3.1 Raman System

A typical Raman system consists of the following basic components: (1) an excitation source, usually a laser; (2) optics for sample illumination; (3) a double or triple monochromator; and (4) a signal processing system consisting of a detector, an amplifier, and an output device. A diagram showing various components of the Raman spectrometer is shown in Fig. 4.6.2.

A number of stages are involved in the acquisition of Raman spectrum. A sample is mounted in the sample chamber and laser light is focused on it with the help of a lens. Generally, liquids and solids are sampled in a Pyrex capillary tube. The scattered light is collected using another lens and is focused at the entrance slit of the monochromator. Monochromator slit widths are set for desired spectral resolution. The monochromator effectively rejects stray light and serves as a dispersing element for incoming radiation. The light leaving the exit slit of the monochromator is collected and focused on the surface of a detector. This optical signal is converted to an electrical signal within the detector and further manipulated using detector electronics. Such a signal is stored in computer memory for each predetermined frequency interval. In a

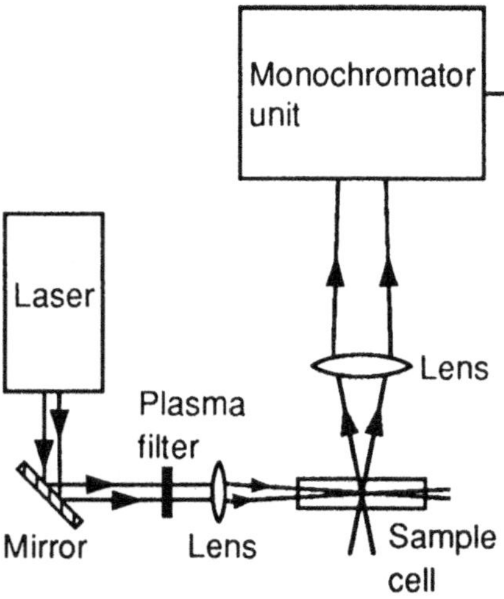

Fig. 4.6.2. Schematic of a laser Raman system

conventional Raman system using a photomultiplier tube (PMT) detector, light intensity at various frequencies is measured by scanning the monochromator. A plot of signal intensity against wavenumber constitutes a Raman spectrum.

There are various ways in which Raman systems can be designed depending upon the nature of the information desired. They can be classified as macro- and micro-systems according to whether the information is to represent the sample at the macro level or at a particular micro-domain of the sample. For homogeneous samples, information from the two systems is identical. In contrast, this information is different for heterogeneous samples. For example, a spectrum acquired over a large number of woody tracheids indicates an average over-all morphological feature and is typical of spectra acquired in the macromode. On the other hand, spectra characteristic of the middle lamella, the secondary wall, or other morphologically distinct features of woody tissue require the microsampling mode. Raman spectra obtained by using both types of systems will be considered. It is also possible to carry out time-resolved measurements of Raman spectra. This has been greatly facilitated by the availability of pulsed lasers and gatable multichannel detectors.

The macromode spectra described here are acquired with an Instruments SA Jobin Yvon Ramanor HG.2S system. Sample excitation is done with either argon or krypton ion lasers. This scanning spectrometer has a thermoelectrically cooled PMT detector and is fitted with a modified Nachet 400 microscope accessory for Raman microprobe work. The microprobe is capable of providing information from domains as small as $1\,\mu$ in diameter.

Recent advances in solid state detectors have led to more efficient laser Raman spectrometers. These spectrometers are based on multichannel detectors (MCD) and are at least an order of magnitude faster than the systems based on PMT detector. However, at present these systems do not match the resolution capabilities of the scanning systems. The Spex Triplemate 1877 A is a new generation instrument which can be fitted with an intensified diode array consisting of 1024 elements. Figure 4.6.3 shows spectra obtained with these two types of detectors.

4.6.3.2 Scattering Geometries

In a Raman experiment, light scattered from a laser excited sample is collected and analyzed. Focusing lenses are used both for sample excitation and light collection purposes. Generally, there are two geometries in which a sample can be studied (Fig. 4.6.4). In the 90° geometry, the laser beam direction and the axis of the collection lens are at 90° to each other, but in the backscattering or the 180° scattering geometry they are coincident.

The 90-degree scattering geometry is frequently used in the macromode work, whereas back scattering geometry is usually the choice for microprobe experiments. Specimen sampling needs are somewhat different for these two types of geometries.

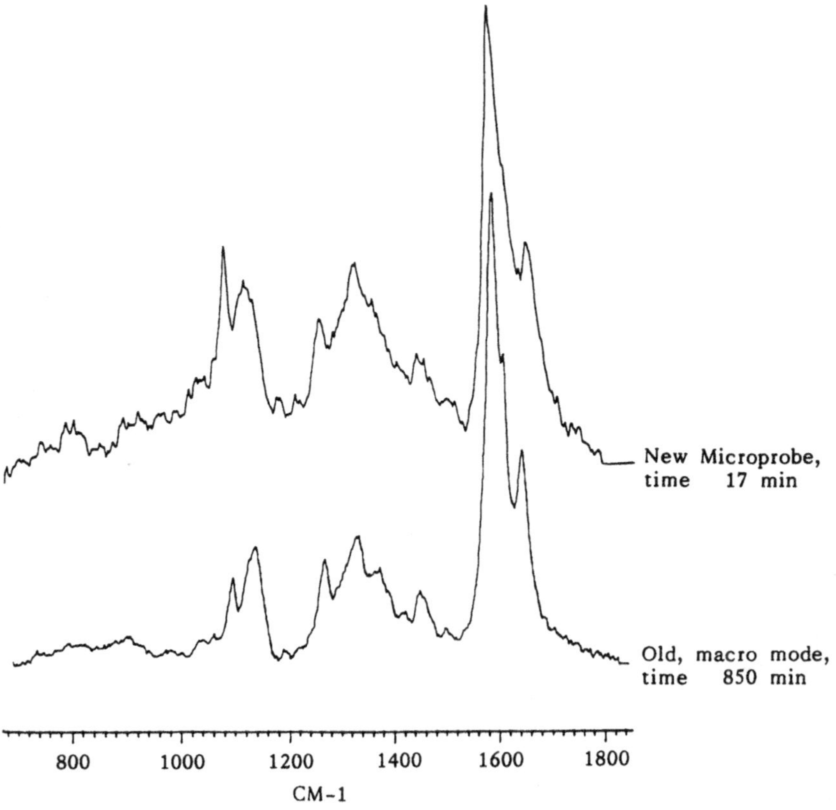

Fig. 4.6.3. A comparison of spectra recorded on old and new Raman microprobes

4.6.3.3 Specimens and Sampling

Laser-induced fluorescence from lignin-containing samples is a major hindrance to obtaining reasonably good spectra. Certain lignin structural units and/or lignin radicals are evidently the major contributors to the problem. Two sampling procedures are effective in significantly reducing the fluorescence of lignocellulosics.

Water Immersion Technique

The technique of water immersion involves sampling lignocellulosics in an aqueous environment. Although originally developed for Raman microprobe investigations of woody tissue, this technique has given much better quality spectra in the conventional macromode studies of lignin-containing samples. The sampling procedure for microprobe studies of woody tissue is briefly described below.

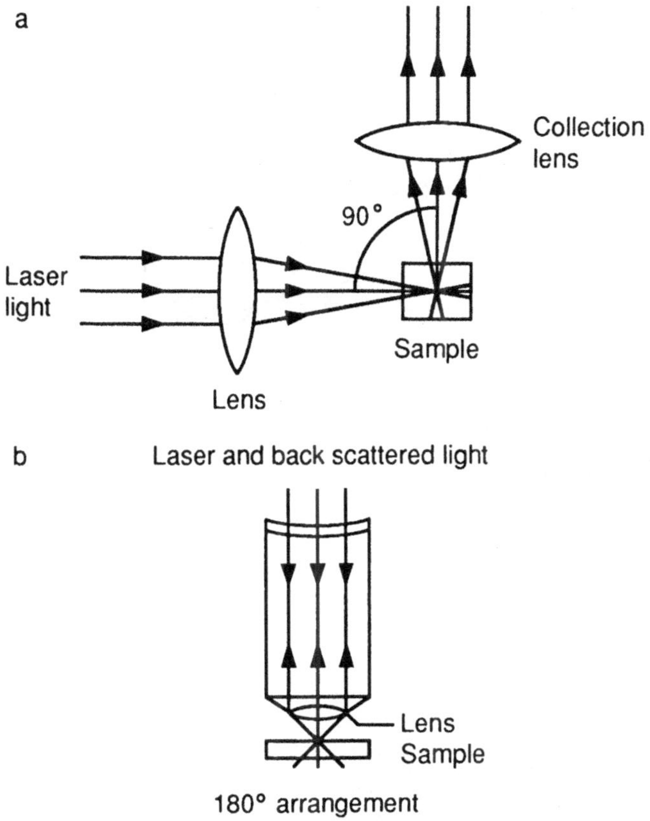

Fig. 4.6.4. **a** 90° and **b** 180° (backscattering) geometries in Raman spectroscopy

Sections of woody tissue, 30μ in thickness, are extracted with a toluene/ethanol mixture (2:1, v/v) and repeatedly washed with distilled water. A flat-base Pyrex glass beaker is attached to a slide in a manner that allows transmitted light to pass through. On the inner side of the bottom, another portion of a glass slide is attached. A wet cell wall section is sandwiched between the slide portion and a cover glass through the center of which a small hole has been drilled; the cover glass holds the section immobile while D_2O is slowly added to the beaker until the sample is completely immersed. The beaker is then mounted on the stage of the microprobe and the water immersion objective is lowered into the D_2O. To avoid evaporation or exchange with atmospheric moisture, a thin layer of mineral oil is placed on the surface of the D_2O after the microprobe objective is submerged in the medium. This sampling device is shown in Fig. 4.6.5 (Atalla and Agarwal 1986).

Figure 4.6.6 shows improvement in the Raman signal-to-background (RS/B) ratio and elimination of Raman features of the cover glass (cf. Figs. 6.6.6b and 6.6.6c). One shortcoming associated with the immersion technique is the intensity reduction of certain lignin features with time.

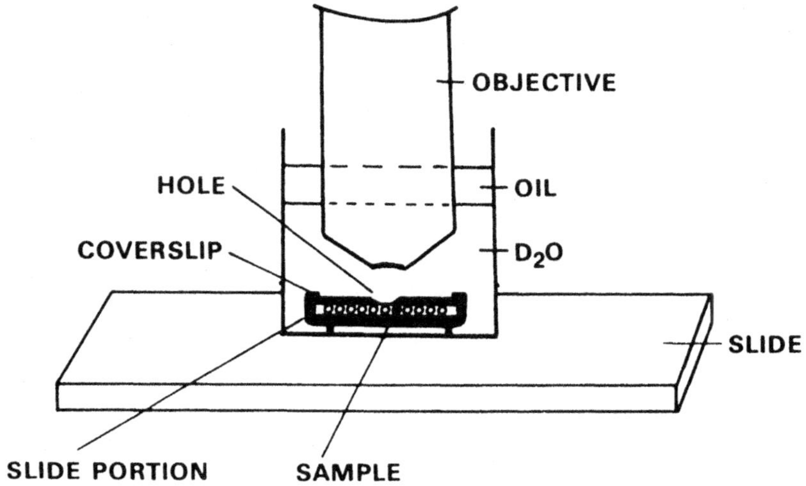

Fig. 4.6.5. Water immersion technique for sampling woody tissue sections in a microprobe experiment

Oxygen Flushing Technique

Molecular oxygen is an effective quencher for lignocellulose fluorescence (Agarwal and Atalla 1986a). This effect is now routinely used to study lignin samples by conventional Raman spectroscopy, i.e., in the macromode with 90-degree scattering geometry. Lignin-containing solid samples are sandwiched between two cover slips and inserted in a specially constructed cell designed to allow gaseous flushing of the sample. Using a lens, the laser beam is focused on the sample and oxygen is flushed through the cell. The fluorescence part of the total signal declines rapidly, thereby resulting in improved RS/B ratios. A variation of the method is to modify the sample cell so that it can withstand a higher oxygen pressure. A Raman spectrum of woody tissue obtained by using the technique is shown in Fig. 4.6.7a. In this particular instance, 50 psi oxygen pressure has been used. (Care must be exercised when exposing organic compounds to focused laser radiation under high pressure oxygen. In one instance, an explosion occurred during the study of a photoyellowed handsheet specimen.) The residual background in Fig. 4.6.7a, underlying the Raman features in the spectrum, can be removed by subtracting a mathematically constructed curve that approximates the background shape (Fig. 4.6.7b). The result of a computer subtraction is shown in Fig. 4.6.7c and d, which differ only with respect to full scale intensity values.

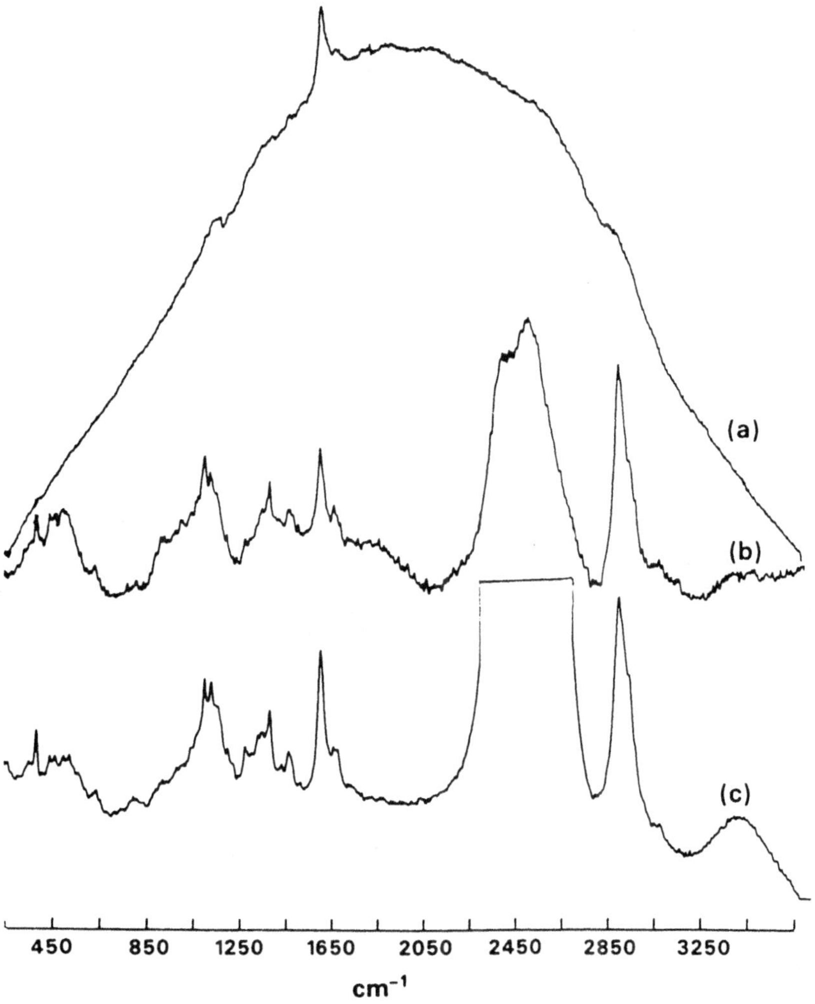

Fig. 4.6.6. Raman microprobe spectra of wood cell wall in a transverse section: *a* air-dried sample; *b* wet sample sandwiched between glass slide and a cover glass; *c* deuterated water immersed sample (a cover glass with a hole was used)

4.6.4 Spectral Information

4.6.4.1 Identification of Native Lignin Features

Wood and pulp samples contain, in addition to lignin, cellulose, and hemicelluloses. As a result, lignin spectral features need to be distinguished from those of the other main components. There are a number of complementary ways to derive this information. First, there are regions of the spectrum of

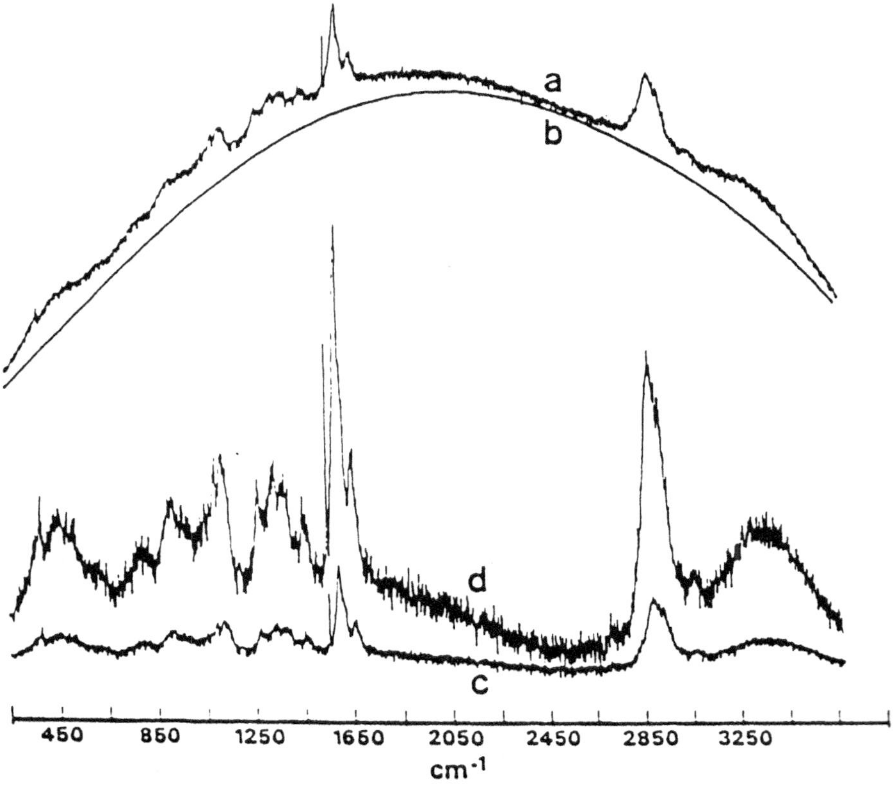

Fig. 4.6.7. *a* Raman spectrum of a wood section obtained under 50 psi oxygen; *b* a mathematically constructed background; *c* difference of *a* and *b* at the same intensity scale; *d* spectrum *c* at lower intensity scale

wood where cellulose and hemicelluloses do not contribute and only features attributable to lignin are seen. This is the case, for example, in the 1600 cm^{-1} region where only aromatic ring stretching vibrations are observed. However, there are other regions of the Raman spectrum where all components have bands, and the interpretation is therefore somewhat complex. To identify lignin bands in these regions, the Raman spectrum of a completely delignified wood sample is subtracted from the native wood spectrum. The difference spectrum is considered to represent the native lignin. This information, along with that obtained directly from native wood spectra, is then interpreted using Raman studies of lignin models (Ehrhardt 1984). In addition, infrared assignments for various lignins and their models reported in literature (Hergert 1971) have been used in arriving at the most reasonable interpretation of the lignin spectra. Raman band positions and assignments are shown in Table 4.6.1 (Ehrhardt 1984, Atalla, unpubl. 1989).

Table 4.6.1. Prominent softwood lignin Raman bands and assignments (see Fig. 4.6.7 and text)

Wavenumber (cm^{-1})[a]	Assignment
3070 w	Aromatic C—H stretching
2900	Aliphatic C—H stretching
1654 ms	C=C stretching and (C=O stretching?)
1620 sh	C=C stretching
1595 vs	Aromatic ring stretching
1455 m	Aliphatic C—H deformation
1396 sh	H—C=O bending
1375 sh	C—H deformation or C—C—H bending
1334 w	?
1271 m	C—O stretching of aromatic C—OCH_3
1139 m	Aromatic C—H in plane deformation (guaiacyl type)

[a] m = medium, ms = medium to strong, sh = shoulder, vs = very strong, w = weak.

4.6.4.2 Quantitative Analysis of Lignin in Wood

In a Raman spectrum, band heights are proportional to sample concentration (Schmid and Brosa 1971). In the Raman spectroscopic analysis of lignin (Fig. 4.6.8), band height at 1595 cm^{-1} is used as the indicator of lignin content. The cellulose peak at 1098 cm^{-1} is the reference and serves as an internal standard. In Fig. 4.6.9, relative lignin peak intensity of acid chlorite-delignified southern pine wood meal is plotted against its Klason lignin content. The good correlation indicates that Raman spectroscopy is suitable for quantifying lignin in wood pulps after a correlation factor has been predetermined for such materials.

In Fig. 4.6.9, the enhanced peak height of the lignin band for the untreated wood (data points in top right-hand corner) arises in part from aromatic ring-conjugated structures in lignin. Chemical compounds with such structures are known to have enhanced Raman scattering (Schmid and Brosa 1971). Experiments carried out with certain lignin model compounds confirm the presence of such effects (Atalla, unpubl. 1989).

The intense lignin peak in Fig. 4.6.8 is also due to a pre-resonance Raman effect. The pre-resonance Raman effect observed with some chemical compounds (Long 1977) has also been seen when Raman studies of wood samples are carried out using a number of different excitation frequencies (Atalla, unpubl. 1989). The 1595 cm^{-1} band is reduced to less than half its intensity when excitation is changed from 514.4 to 647.1 nm. The pre-resonance Raman effect in wood, when it is excited at 514.5 nm, may be attributed to phenoxy radicals and other lignin structures capable of light absorption at that wavelength (Tripathi and Schuler 1984, Schmid and Brosa 1971).

4.6.4.3 Orientation and Composition Studies of Lignin in Woody Tissue

Developments in Raman spectroscopy and wood sampling techniques have made it possible to carry out in situ studies on the organization of lignin and

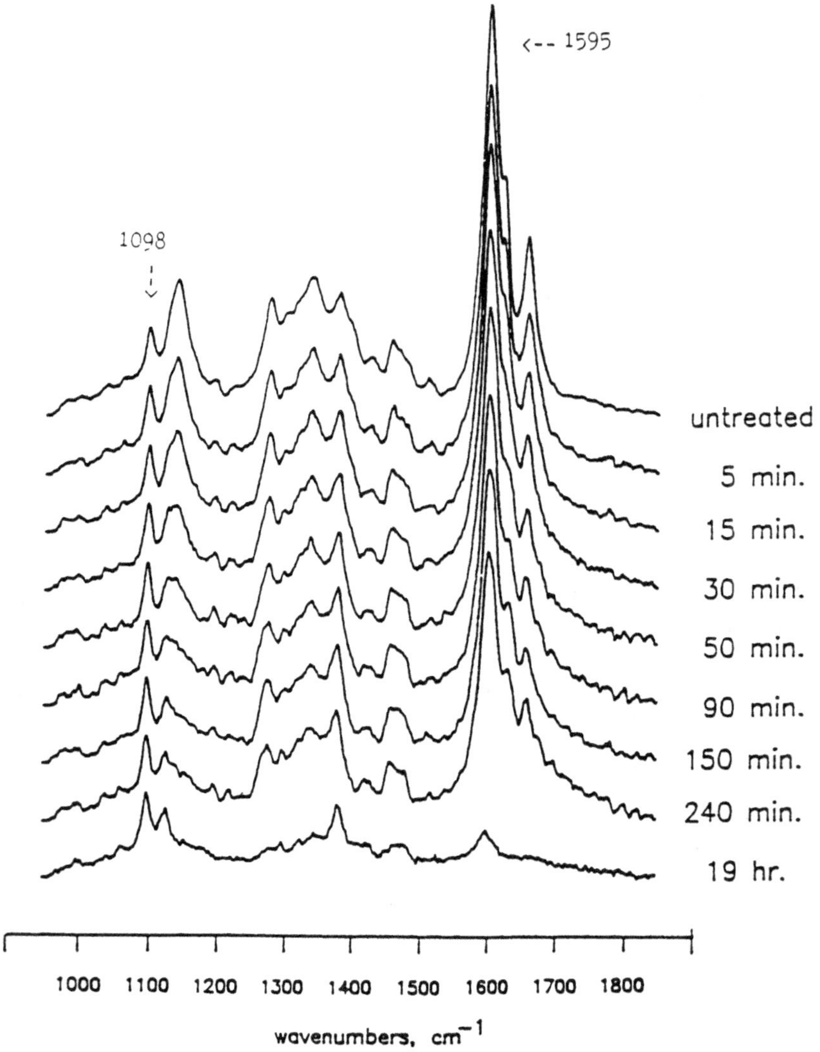

Fig. 4.6.8. Raman spectra of southern pine wood meal before and after various durations of acid chlorite treatment

other constituents in wood. The Raman microprobe can examine regions of woody tissue as small as 1 μ in diameter. Analyses of Raman spectra, obtained in various scattering geometries, suggest that lignin is oriented at the molecular level (Atalla and Agarwal 1985, Agarwal and Atalla 1986b). One set of such spectra, obtained on a longitudinal section, is shown in Fig. 4.6.10. The 1595 cm^{-1} lignin band intensity changes when the electric vector (ev) of the exciting laser radiation is rotated with respect to the cell wall orientation. This change indicates lignin organization within the cell wall.

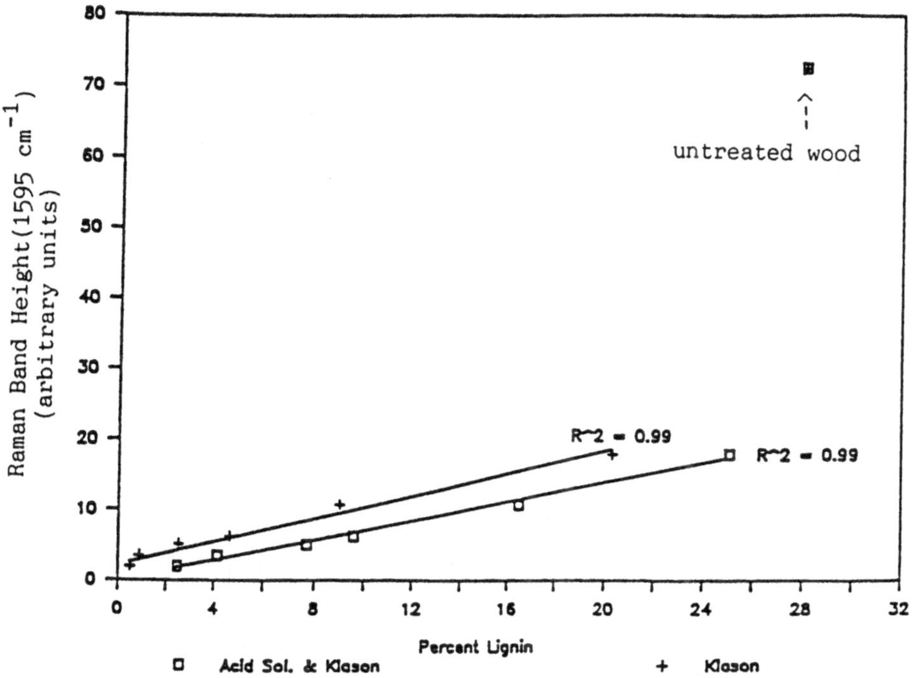

Fig. 4.6.9. Correlation of the 1595 cm^{-1} band height with lignin content of Southern pine wood meal after acid chlorite treatments. *Data points in top right-hand corner* correspond to lignin in untreated wood

4.6.5 Discussion

At this time, Raman spectroscopic studies of lignin have just begun and the capability of the technique has not been fully realized. Work done so far indicates a great potential for the technique. Moreover, Raman spectroscopy is by no means a fully matured branch of spectroscopy and further development in this field is expected.

4.6.5.1 Comparison With Other Techniques

The nature of the information that can be obtained from Raman experiments, namely, vibrational frequencies and band intensities, is similar but not identical to that of infrared spectroscopy. Therefore, it is appropriate to compare these two techniques. As mentioned previously, for a molecular vibration to be IR active, a change in dipole moment accompanying the vibrational transition is

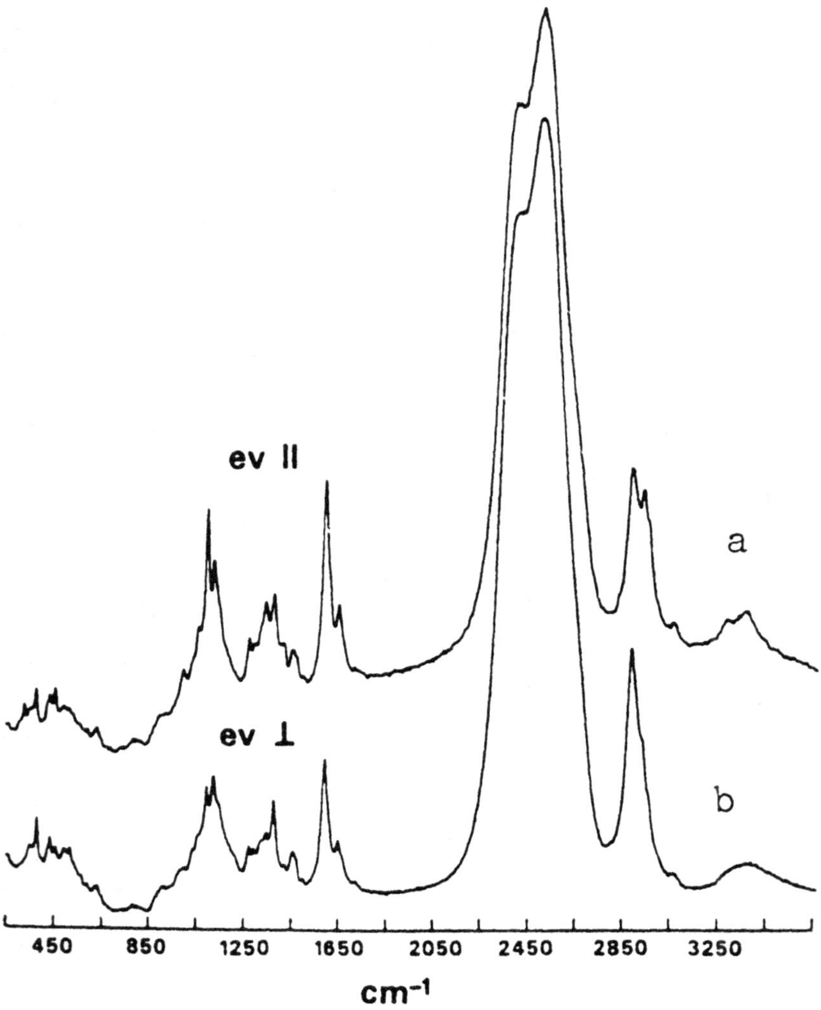

Fig. 4.6.10. Polarized Raman spectra from approximately 1 μ domain of the secondary wall in a longitudinal section of woody tissue: *a* electric vector parallel to the cell wall; *b* electric vector perpendicular to the wall

needed. On the other hand, a vibration is Raman active whenever a change in molecular polarizability occurs during the transition. In light of this, polar bond systems with a high dipole show up strongly in IR, whereas bond systems with highly covalent character are quite easily seen in Raman. Water, as a consequence, is a weak scatterer in Raman spectroscopy but strongly absorbs in IR. Indeed, water is used as a part of the sampling procedure for some of the experiments. Because of the difference in selection rules between these two types of spectroscopic transitions, the information obtained from the two types of spectra is complementary.

Ordinary Raman scattering is an inefficient process, and in general it is less sensitive than IR absorption. However, in certain lignin samples, conjugation, resonance, or pre-resonance Raman effects can arise, and for particular vibrational modes a higher level of sensitivity can be achieved (Long 1977, Schmid and Brosa 1971). In lignins, the latter effect can be induced by proper laser frequency selection.

Another advantage of Raman spectroscopy is ease of working with heterogeneous samples. In IR spectroscopy this is difficult because of Rayleigh scatter of infrared photons. The degree of this scatter depends upon differences in the refractive indices at optical heterogeneities. Because the refractive index varies with wavelength in regions of strong IR absorption, it is difficult to separate the extinction due to molecular absorption from the extinction caused by Rayleigh scattering. For Raman spectral measurements, in contrast, discrimination against Rayleigh scattering is relatively simple.

There are additional advantages in using Raman spectroscopy to study lignin. Since it is possible to use a number of excitation wavelengths, choice of a proper frequency allows selective lignin excitation. Time-resolved Raman studies of excited lignin and its photomodifications can be carried out using pulsed laser based Raman spectrometers. This capability opens up a new field for lignin research. IR spectroscopy, in contrast, does not have this possibility, as the frequencies involved are far removed from those which cause electronic excitation.

4.6.5.2 Future Development

Since Raman studies of lignin in heterogeneous lignocellulosic samples can now be carried out, various topics in lignin research will greatly benefit from Raman spectroscopic measurements. Using conventional instrumentation, samples under the category "difficult" are those that are highly colored. With these, the problem is of high levels of fluorescence and/or thermal degradation caused by excessive absorption of laser radiation. In the future, however, this may be remedied as 1064 nm excitation-based laser FT-Raman instruments become increasingly available.

References

Agarwal UP, Atalla RH (1986a) Oxygen sensitive background in the Raman spectra of woody tissue. In: Peticolas WL, Hudson B (eds) Xth Int Conf Raman Spectrosc, Eugene, Oregon, Paper 14–16. Univ Oregon, Eugene

Agarwal UP, Atalla RH (1986b) In situ Raman microprobe studies of plant cell walls: macromolecular organization and compositional variability in the secondary wall of Picea mariana (Mill.) B.S.P. Planta 169:325–332

Atalla RH, Agarwal UP (1985) Raman microprobe evidence for lignin orientation in cell walls of native woody tissue. Science 227:636–638

Atalla RH, Agarwal UP (1986) Recording Raman spectra from plant cell walls. J Raman Spectrosc 17:229–231

Ehrhardt SM (1984) An investigation of vibrational spectra of lignin model compounds, Ph.D. thesis dissertation. Institute of Paper Chemistry, Appleton, Wisconsin
Hergert HL (1971) Infrared spectra. In: Sarkanen KV, Ludwig CH (eds) Lignins. Occurrence, formation, structure and reactions. Wiley-Interscience, New York, 267–297
Long DA (1977) Raman spectroscopy. McGraw-Hill, New York, 253 pp
Schmid ED, Brosa B (1971) Raman-Intensität und Konjugation I. Substituentenabhängigkeit der Raman-Intensitaten der 1600 cm^{-1} Ringschwingungen monosubstituierter Benzolderivate. Ber Bunseng Phys Chem 75:1334–1343
Tripathi GNR, Schuler RH (1984) The resonance Raman spectrum of phenoxyl radical. J Chem Phys 81:113–121

4.7 Pyrolysis-Gas Chromatography-Mass Spectrometry

D. Meier and O. Faix

4.7.1 Introduction

Pyrolysis is the transformation of a nonvolatile compound into a volatile degradation mixture by heat in the absence of oxygen. A rate of heating to the final temperature in the millisecond range is typical for analytical pyrolysis, in contrast to the slow heating rates employed for other thermal characterization techniques, e.g., thermogravimetry (TG) and differential scanning calorimetry (DSC), that are in the range of minutes or hours. Simple sample preparation (drying and milling), rapid analysis times (from minutes up to 1.5 h) and small sample size (1 to 100 μg) are the key features of analytical pyrolysis. Comprehensive books on special applications of pyrolysis are available (Irwin 1982, Meuzelaar et al. 1982, Voorhees 1984, Liebman and Levy 1985, Linskens and Jackson 1986).

The analysis of lignin normally includes tedious and time-consuming chemical degradation and product isolation procedures. However, analytical pyrolysis (Py) combined with high resolution capillary gas chromatography (GC) and/or mass spectrometry (MS) eliminates the need to isolate lignin. In pyrolysis, molecules break down by heat at specific points having low bonding energies. The stable, volatile degradation fragments thus formed can provide useful information about the whole molecule. More detailed information is obtained when fewer undesirable secondary reactions occur and when fragment yields are higher.

When lignocellulosic material is pyrolyzed, the degradation products from polysaccharides and lignin can be separated by gas chromatography and identified by mass spectrometry. Under appropriate conditions, pyrolysis of lignins, even in the presence of carbohydrates, gives rise to a characteristic product pattern of substituted monomeric phenols in which the lignin propanoid side chain has been split off completely or shortened to one or two carbons. New double bonds are created in the side chains through pyrolytic dehydrogenation. The peaks of lignin origin in the pyrolysis GC chromatogram, or pyrogram, are clearly recognizable because of the abundance of molecular ions. Lignin mass spectra are characteristically different from those of carbohydrate pyrolysis products, thus making it possible to analyze lignin in lignocellulosic materials without first removing the carbohydrates.

Several applications of the Py-GC-MS technique are listed below:

Springer Series in Wood Science
Methods in Lignin Chemistry
(Edited by S.Y. Lin and C.W. Dence)

1. Lignin classification based on the quantities of the three phenylpropanoid units, namely 4-hydroxyphenylpropane (H), guaiacylpropane (G), and syringylpropane (S), formed in the pyrolysis of lignocellulosic materials
2. Microanalysis of lignin in wood cells and cell fragments of different morphological origin
3. Analysis of residual lignin in pulps
4. Fingerprinting and identification of technical lignins from various pulping processes

4.7.2 Principle and Methods of Analytical Pyrolysis

4.7.2.1 Combination of Pyrolysis with Other Instruments

Analytical pyrolysis is always performed in combination with one or more systems for detecting the pyrolysis products. When the pyrolysis system is separated from the analytical instrument, i.e., the pyrolysis and analysis are performed in two steps, the approach is termed "off-line" (Faix et al. 1987). Contrarily, in the "on-line" approach, the pyrolysis unit and the analytical instrument are directly coupled. Pyrolysis may be combined with a monitoring instrument, preferably a mass or FTIR spectrometer (Py-MS, Py-FTIR) (Metzger 1979, Genuit and Boon 1985, Haw and Schultz 1985, Faix et al. 1988b). These relatively rapid techniques are most suitable for fingerprinting and classification studies when used in conjunction with modern statistical pattern recognition methods (Hartley and Haverkamp 1984, Windig et al. 1984). Alternatively, pyrolysis may be combined with gas chromatography using a detection system based on FID, MS, or FTIR. One advantage of this methodology is that the pyrolysis products can be separated for individual identification based on their retention times, and mass or IR spectral patterns. This method is more time-consuming (1–1.5 h) because of the GC separation step. The less expensive Py-GC configuration is often used for routine work after the system has been calibrated with authentic compounds or by GC-MS.

4.7.2.2 Pyrolysis Techniques

Rapid heating of the sample to the maximum pyrolysis temperature is of paramount importance to obtain optimum thermal fragmentation and to avoid secondary reactions. The selected pyrolysis temperature should be high enough to ensure the complete thermal breakdown of the macromolecule. To avoid secondary recombination reactions, the fragmentation products are removed rapidly from the hot zone of the pyrolyzer. This is normally done with the aid of an inert gas (e.g., helium) which is usually used as a carrier in GC and MS analyses. Figure 4.7.1 shows a temperature-time profile together with the terminology of pyrolysis. The influence of pyrolysis parameters on the results have been described by Ericsson (1980, 1985).

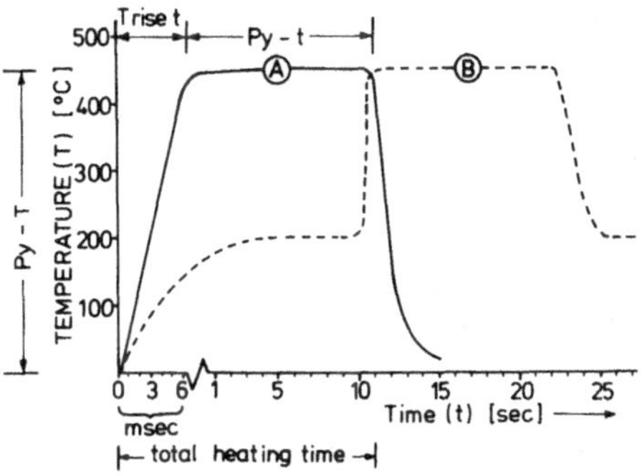

Fig. 4.7.1. Temperature-time profile of a Pyroprobe instrument: *A* ideal pyrolysis with the ribbon probe; *B* coil probe with quartz boat and 200 °C interface temperature. Explanation of terms: *Py-T* pyrolysis temperature; *Py-t* pyrolysis time; *T rise t* temperature-rise time

Pyrolysis may be performed using continuous-mode or pulse-mode instruments. In the first instance, the material must be introduced rapidly into the furnace at a predetermined temperature, which is maintained throughout the pyrolysis. Common modern pulse-mode instruments allow very rapid sample heating within a specified time. There are three different types of pulse-mode pyrolyzers:

1. Curie-point pyrolyzer (high frequency inductive heating of a ferromagnetic carrier).
2. A resistance-heated system (platinum ribbon or coil).
3. Laser pyrolyzer.

Most of the pyrolysis experiments in the field of natural polymer characterization involve the use of Curie-point instruments or resistance-heating apparatus. Both methods are described in the following sections. Laser pyrolyzers are not yet common in biopolymer analysis and will not be discussed here.

4.7.2.2.1 Induction Heating (Curie-Point Pyrolysis)

Curie-point pyrolysis employs high-frequency (HF) inductive heating of a ferromagnetic wire ($d \leqslant 1$ mm). A diagram of a Curie-point pyrolyzer designed for Py-GC is shown in Fig. 4.7.2.

A dry sample or a sample suspension in water or methanol is deposited directly on the surface of the wire which, for suspensions, must then be dried. A glass liner with a ferromagnetic wire is placed in an induction coil which in turn is connected to a high-frequency power supply. The relatively low temperature of the reaction tube helps minimize unwanted secondary reactions during pyrolysis. Additionally, fewer volatile, high boiling pyrolysis products condense on the glass wall, thus preventing contamination of the system. Furthermore, the glass liner reduces the dead volume of the system.

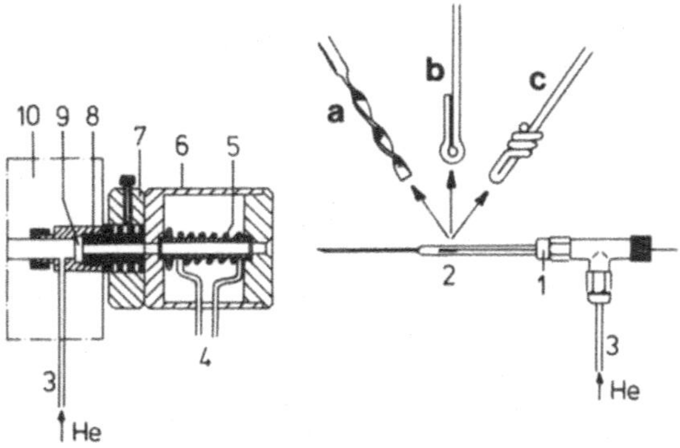

Fig. 4.7.2. Schematic diagram of a Curie-point pyrolyzer (Fischer, Germany). Note the possible modifications of the wire tip (*a*, *b*, and *c*) for solid samples. Pyrolysis glass injector (*1*), ferromagnetic wire (*2*), carrier gas inlet (*3*), impulse cable from power generator (*4*), induction coil (*5*), aluminum box (*6*), adapter for GC injector (*7*), GC inlet (*8*), GC septum (*9*), GC oven (*10*)

After switching on the HF generator, the wire is inductively heated to its Curie-point temperature, where a transition from ferromagnetism to paramagnetism occurs. When the dimensions of the ferromagnetic wire and the field strength are properly matched, the equilibrium temperature of the wire stabilizes close to the Curie-point temperature. Highly reproducible temperatures of up to 1128 °C can be attained in less than 200 milliseconds (ms). The final temperature in a Curie-point system is selected by using an appropriate alloy of ferromagnetic materials. It should be noted, however, that alloys of the same Curie-point temperature may have different temperature-time profiles. The temperature-rise time depends on the diameter of the wire, its composition, the strength of the HF field, and the field frequency and typically is 100 ms. The total heating time is normally in the range of 1 to 10 s. The temperature-time profile, an important parameter for reproducing pyrograms, can be easily standardized by the Curie-point temperature which is the equilibrium temperature of the wire material. Figure 4.7.3 shows the temperature-rise time and the equilibrium temperature of three ferromagnetic materials.

Apparatus Sources

Several Curie-point pyrolyzers are obtainable from Fischer (Germany), Scientific Glass Engineering (Australia), Horizon Instruments (England) and Japan Analytical Industries. In addition, laboratory-constructed models have been described in the literature (Saiz-Jimenez and De Leeuw 1984, 1986, Boon et al. 1987, Genuit et al. 1987, Pouwels and Boon 1987, Saiz-Jimenez et al. 1987).

Experimental Considerations

Clean wires should be used in pyrolysis experiments. A preferred cleaning method is to wash the wires with solvent and heat them overnight at 600 to 700 °C, or preheat the wires with the system immediately prior to performing the

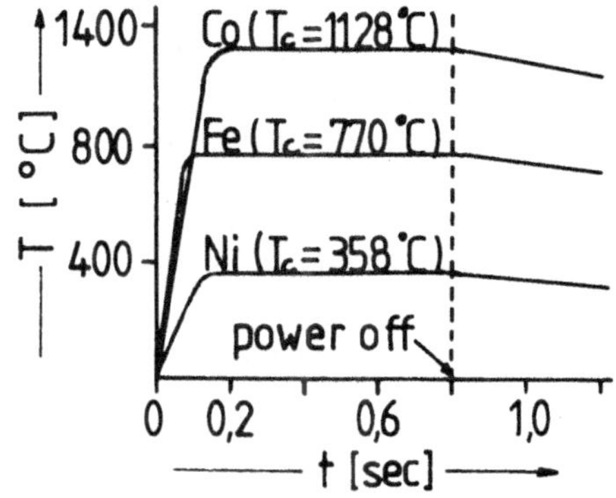

Fig. 4.7.3. Temperature-time profiles and Curie-point temperatures (T_c) for pure Ni, Fe, and Co wires. (Tromp 1987)

experiment. Direct flame cleaning is not recommended because oxidation may change the heat transfer characteristics of the metal. The smaller the sample, the more significant solvent impurities become. Therefore, the solvent used in wire washing should be free of impurities.

Sample loading has a significant effect on data reproducibility. The position of the ferromagnetic wire in the induction coil and sample coating of the wire should be properly controlled to reproduce the sample heating rate. A microliter syringe is positioned over the rotating wire to obtain uniform coating of the wire with the lignin solution. A hairdryer or an infrared lamp can be used to evaporate the solvent. An insoluble lignin sample is placed directly on the surface of the wire. In this case, the geometry of the wire tip is modified according to Fig. 4.7.2a,b, and c.

Generally, samples should be similarly prepared to minimize changes in the pyrogram appearance. After being coated, the wire is carefully inserted into the glass liner which is placed in the holder. Typical Curie-point temperatures used for lignin pyrolysis are 510 °C (Saiz-Jimenez and de Leeuw 1984, 1986) and 650 °C (Pouwels and Boon 1987). A lower temperature, 450 °C, may be used to reduce secondary fragmentation reactions (Faix et al. 1988a).

4.7.2.2.2 Resistance Heating

Heated-filament pyrolyzers are often used to analyze lignins (Kratzl et al. 1965, Lindberg et al. 1982, Obst 1983, Gardner et al. 1985, Faix et al. 1987, 1991, Funazukuri et al. 1987, Salo et al. 1989). In this type of analyzer, electric current is passed through a resistance ribbon or coiled wire, both made of platinum. The dissipation of power increases the temperature of the conductor. Heat-up and pyrolysis times are selected from an instrument control. Characteristic parameters of this type of pyrolyzer have been described by Wells et al. (1980) and Wampler and Levy (1987).

Apparatus Sources

The most widely used system is the Pyroprobe (Chemical Data Systems, USA). The system is designed for varying the heating rate from 0.1 to 20 °C per ms linearly over the initial temperature-rise period. Typical temperature-rise times without ramp control are, for the ribbon probe, 8 ms to 600 °C and 17 ms to 1000 °C. For the coil filament, where the sample is held in a quartz tube or boat positioned within the coil, the heating rate (to 600 °C in 600 ms) is lower owing to the higher thermal mass of the system. Another system, Pyrola, is marketed by Pyrolab in Sweden. It is designed for qualitative, quantitative, and kinetic analyses and is equipped with a process and control unit.

Experimental Considerations in the Use of a Pyroprobe Analyzer

Attachment of the Pyrolysis System. The attachment of a pyrolyzer to a GC system should be made so that minimum dead volume remains in the system. Dead volume can be tested for by injection of methane into the GC column; a tailing methane peak indicates the existence of dead volume. Such voids drastically reduce resolution and may also trap polar or more volatile fragments. The system should also be tested for contamination from previous runs by firing the pyrolyzer without sample. Generally, such a blank run should be made from time to time to ensure the absence of memory effects. A typical configuration of the so-called "on-line" approach is presented in Fig. 4.7.4.

Precleaning. Precleaning of the probe is performed by heating it at 1000 °C in air before pyrolysis using the "clean" mode of the control unit. Quartz tubes or boats for the coil probe may be cleaned in a Bunsen burner or muffle furnace.

Ribbon Probe Compared to Quartz Tubes or Boats. Positioning of a wood or lignin sample is of paramount importance. Soluble lignins may be applied directly to the surface of the ribbon probe with a microliter syringe. A solution (2 μl) of 10 mg/μl concentration is transferred to the ribbon probe. The solvent is evaporated either by air-drying or by heating the ribbon at an appropriate temperature.

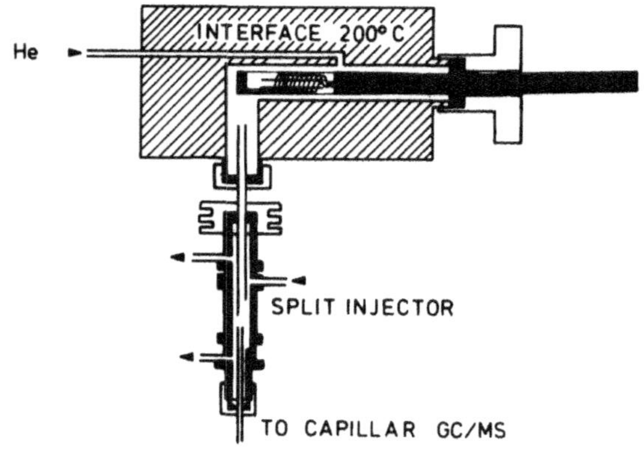

Fig. 4.7.4. Schematic diagram of Pyroprobe interface with coil probe connected to an GC injector ("on-line" approach)

For solid lignin samples, a quartz tube or a quartz boat with coil filaments is used. The boat is preferred for ease in sample weighing and positioning. Dry lignin samples are pressed against the boat walls so that they are not blown away by the carrier gas stream. Quartz wool plugs (see Fig. 4.7.5) that allow a free flow of carrier gas may also be used.

The sample is secured inside the tube or boat which is carefully slid into the heating coil. The probe can be connected either directly to a horizontal GC injection port or first into the heated interface of the Pyroprobe system (see Fig. 4.7.4) that is connected with the GC injection port. The interface temperature is set typically between 200 and 250 °C. After sample insertion but before triggering the pyrolysis, a period of 5 to 10 s is required to eliminate oxygen from the system.

Quantification of Pyrolysis Products. The pyrolysis must be performed separately ("off-line" approach, see Fig. 4.7.5), and the products are trapped completely so that an internal standard can be added. Accurate quantification of the pyrolysis products is achieved using a gas chromatograph equipped with a flame ionization detector (Faix et al. 1987). Fluoranthene is a suitable internal standard as it elutes from the column after the last detectable lignin pyrolysis product (trans-sinapyl alcohol).

4.7.2.3 Sample Preparation, Size, Geometry, and Contamination: General Considerations

If possible, samples of both extractive-free wood and its corresponding milled wood lignin (MWL) should be used for Py-GC-MS characterization of plant lignins. The peaks in pyrograms originating from polysaccharides and lignin can be clearly distinguished by comparing the pyrograms of the two samples. In general, the carbohydrate "impurities" of a lignin sample can be recognized easily with Py-GC-MS or Py-GC-FID. A low thermal gradient throughout the sample is important for obtaining a uniform and reproducible pyrolysis. For this reason, a small sample size (5–50 μg) is recommended.

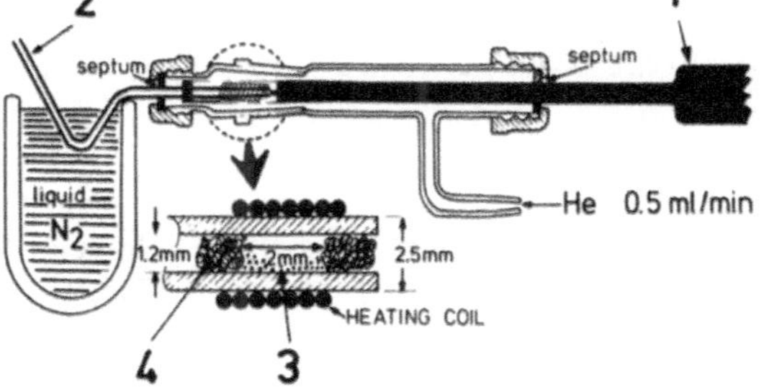

Fig. 4.7.5. Schematic diagram of the "off-line" approach equipment. Coil probe (*1*), quartz tube (*2*), sample (*3*), silanized quartz wool (*4*)

Sufficient heat transfer is achieved through good sample contact with the heating element of the pyrolyzer. Thus, a lignin is preferably dissolved and applied directly to the surface of the heating element. When the sample is insoluble, its shape and particle size become important parameters. To obtain reproducible results, wood samples should be homogenized in a vibratory ball mill and subsequently dried completely.

One of the most common causes of non-reproducible results is contamination of sample or pyrolysis equipment with unidentified foreign material such as plasticizers contained in solvents used for MWL preparation and from plastic bottles used to store samples.

4.7.3 Gas Chromatography

Because of the outstanding performance of modern quartz capillary columns, the gas chromatographic separation of lignin pyrolysis products is easily achieved. Generally, nonpolar or medium polarity columns are used; the selection of the stationary phase is less important than for conventional packed columns. For the separation of pyrolysis products from wood or lignin a modified dimethylpolysiloxane liquid phase (e.g., SE-54 or OV-1701) should be used.

4.7.3.1 Operating Conditions

In the "on-line" approach, helium is recommended as the carrier gas to avoid the risk of an explosion in the pyrolysis system. Before pyrolysis, the GC system is optimized with respect to linear gas flow velocity in the column and methane peak shape (indication of dead volumes). Column connections should be made as specified by the column manufacturer.

Excellent separation of pyrolysis products from lignin and carbohydrates can be obtained using a column packed with chemically bonded OV-1701 (86% dimethyl-14% cyanopropylpolysiloxane). Figure 4.7.6 shows two pyrograms of beech wood that illustrate the influence of film thickness using a DB-1701 column (J & W Scientific, USA, distributed by Carlo Erba Instruments, Italy); pyrogram A was obtained with a 0.25-μm film and pyrogram B with a 1-μm film. Low volatility products are separated better on a thick film although the peaks are broader. Typical operating conditions for obtaining pyrograms are:

Pyrolysis		*Gas Chromatography*	
Sample weight	90–155 μg	Injector temp.	250 °C
Temperature-rise-time	ca. 400 ms	Detector temp.	250 °C
Pyrolysis time	10 s	Column dimensions	30 m × 0.25 mm
Final temp.	450 °C	Carrier gas velocity	38 cm/s
Interface temp.	200 °C	Oven program	45 °C, 4 min 4 °C/min to 240 °C, 39 °C/min to 280 °C

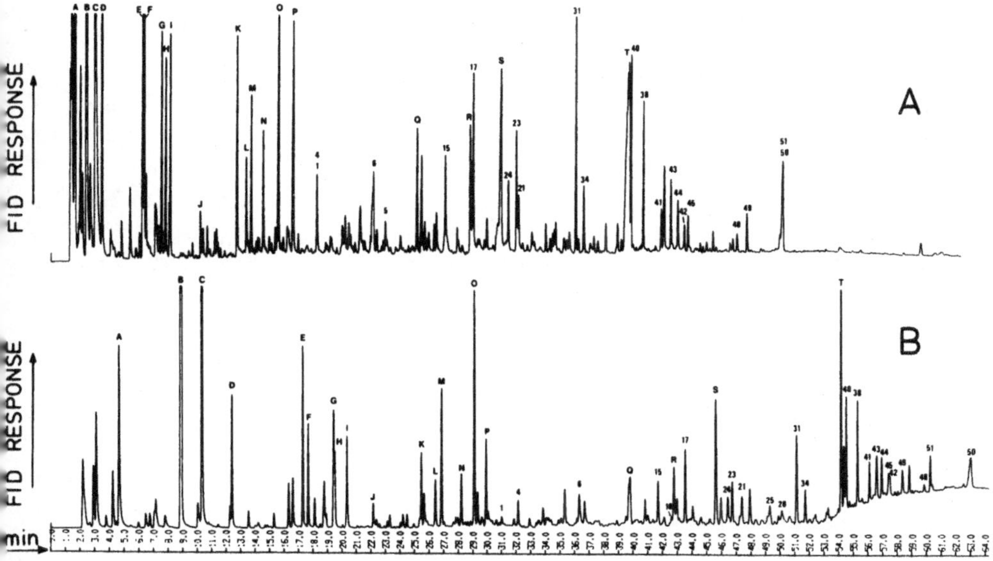

Fig. 4.7.6. Influence of film thickness (A = 0.25 μm, B = 1.0 μm) on the retention of beech wood pyrolysis products. Column medium, chemically bonded OV-1701. Peaks *labeled with letters* are carbohydrate-derived products; *numbered peaks* correspond to lignin degradation products (see Table 4.7.1)

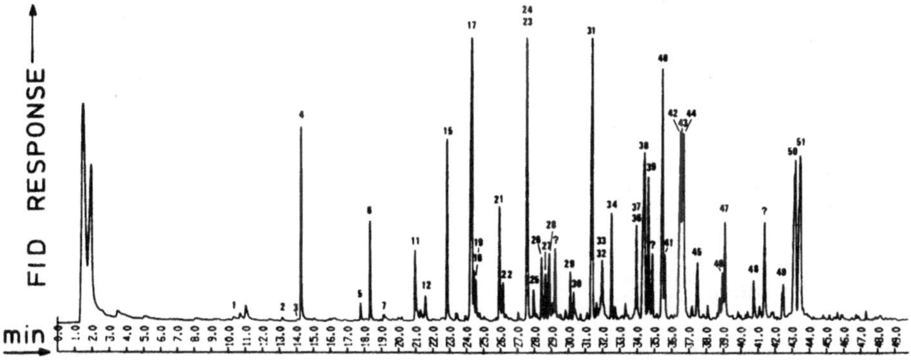

Fig. 4.7.7. Pyrogram of beech MWL on chemically bonded SE 54 (for *peak numbers* see Table 4.7.1)

Good separations can also be obtained with an SE 54 column (95% dimethyl-5% diphenylpolysiloxane). The pyrogram of a beech MWL is shown in Fig. 4.7.7. The separation was obtained with an equivalent DB-5 column under the following conditions:

Pyrolysis		*Gas Chromatography*	
Sample weight	65 μg	Injector temp.	250 °C
Temperature-rise time	ca. 400 ms	Detector temp.	250 °C
Pyrolysis time	10 s	Column dimensions	30 m × 0.25 mm
Final temp.	450 °C	Carrier gas velocity	38 cm/s
Interface temp.	200 °C	Oven program	45 °C, 4 min 4 °C/min to 290 °C

When the pyrolysis temperature is raised to 620 °C, about 20% more products are obtained as the result of the occurrence of undesirable secondary reactions (see Faix et al. 1987).

Table 4.7.1 lists phenolic lignin-derived pyrolysis products according to their relative retention times on DB-1701 (0.25- and 1.0-μm films) and DB-5 (0.25-μm films) quartz capillary columns.

4.7.4 Mass Spectrometry

The combination of a gas chromatograph with a mass spectrometer provides one of the most specific and sensitive means of analyzing and identifying the numerous pyrolysis products from lignin (see Chap. 9.1). The mass spectrometer acts as a detector and records the mass spectrum of each compound eluting from a GC column. This information, together with the retention time, allows unequivocal product identification to be made in most instances.

4.7.4.1 Instrumentation

Basically, mass spectrometers can be classified as magnetic sector and quadrupole instruments. For analytical pyrolysis, all instruments are useful provided the scan rate is fast enough (at least 1 scan/s) to record the narrow peaks eluting from the high-resolution GC column.

4.7.4.2 Ionization

The most common ionization technique is electron impact (EI) ionization at 70 eV. As applied to lignin pyrolysis products, this mode of ionization produces molecular and typical fragment ions in sufficient abundance to permit structure assignments to be made.

4.7.4.3 Data Handling

A complex lignin pyrogram can be analyzed only with a powerful data system that stores thousands of acquired mass spectra from a single analysis. Usually,

Table 4.7.1. Retention times of phenolic lignin pyrolysis products relative to guaiacol (1.0) separated on columns with differing liquid phases

No.	Name	Relative retention time		
		DB-5	DB-1701 0.25 μm	DB-1701 1.0 μm
1	Phenol	0.71	0.98	0.96
2	Phenol, 2-methyl-	0.92	1.06	n.d.
3	Phenol, 3 + 4-methyl-	0.98	1.13	n.d.
4	Guaiacol (G)	1.0	1.0	1.0
5	Phenol, dimethyl-	1.24	1.21	n.d.
6	Guaiacol, 4-methyl-	1.28	1.18	1.1
7	Catechol	1.32	1.49	n.d.
8	Phenol, 4-vinyl-	1.36	1.42	1.28
9	Phenol, 4-allyl-	1.45	1.49	n.d.
10	Catechol, 3-methyl-	1.45	n.d.	n.d.
11	Catechol, 3-methoxy-	1.46	1.41	n.d.
12	Guaiacol, 4-ethyl-	1.5	1.32	1.24
13	Catechol, 4-methyl-	1.56	1.55	1.39
14	Phenol, 4-propenyl-(cis)	1.59	1.54	n.d.
15	Guaiacol, 4-vinyl-	1.59	1.42	1.3
16	Phenol, 4-propenyl-(trans)	1.67	1.6	n.d.
17	Syringol (S)	1.7	1.51	1.35
18	Eugenol	1.7	1.45	1.33
19	Guaiacol, 4-propyl-	1.71	1.51	1.33
20	Benzaldehyde, 4-hydroxy-	1.74	1.82	1.57
21	Vanillin	1.81	1.66	1.47
22	Isoeugenol (cis)	1.82	1.54	n.d.
23	Syringol, 4-methyl-	1.92	1.65	1.46
24	Isoeugenol (trans)	1.92	1.62	1.44
25	Homovanillin	1.93	1.74	1.54
26	G—CH=C=CH2?	1.98	1.67	n.d.
27	G—CH=C=CH2?	1.99	1.68	n.d.
28	Acetoguaiacone	2.01	1.77	1.56
29	Syringol, 4-ethyl-	2.1	1.75	1.53
30	Guaiacyl acetone	2.11	1.85	n.d.
31	Syringol, 4-vinyl-	2.19	1.84	1.59
32	Propioguaiacone	2.22	1.89	1.64
33	G—CO—CH=CH_2	2.23	1.91	1.66
34	Syringol, 4-allyl-	2.26	1.86	1.61
35	Syringol, 4-propyl-	2.28	1.86	1.61
36	Syringol, 4-propenyl-(cis)	2.36	1.93	n.d.
37	Dihydroconiferyl alcohol	2.36	2.02	1.71
38	Syringaldehyde	2.4	2.06	1.72
39	Coniferyl alcohol (cis)	2.41	2.06	n.d.
40	Syringol, 4-propenyl-(trans)	2.47	2.01	1.7
41	Homosyringaldehyde	2.48	2.11	1.75
42	Coniferaldehyde	2.53	2.19	1.79
43	Acetosyringone	2.54	2.14	1.76
44	Coniferyl alcohol (trans)	2.55	2.17	1.77
45	Syringyl acetone	2.61	2.2	1.79
46	Propiosyringone	2.71	2.24	1.82
47	S—CO—CH=CH_2	2.72	2.26	1.83
48	Dihydrosinapyl alcohol	2.84	2.35	1.86
49	Sinapyl alcohol (cis)	2.97	2.38	1.83
50	Sinapaldehyde	3.01	2.51	1.97
51	Sinapyl alcohol (trans)	3.03	2.5	1.88

the total ion current (TIC) is reconstructed to form a chromatogram similar to the response of the flame ionization detector. Occasionally, however, it is helpful to select single, typical masses for the reconstruction of mass chromatograms. Figure 4.7.8 is an example of this kind of evaluation. Here, typical mass fragments from the p-hydroxyphenyl (H)-, guaiacyl (G)- and syringyl (S)-propane units are selected and reintegrated to provide information about the H-G-S ratios. Table 4.7.2 contains the ion masses of typical compounds from the three types of lignin units.

4.7.5 Evaluation of Pyrograms and Spectra

Thc pyrograms of plants or lignins can be evaluated through use of the similarity index (Salo et al. 1989), as defined by the following equation:

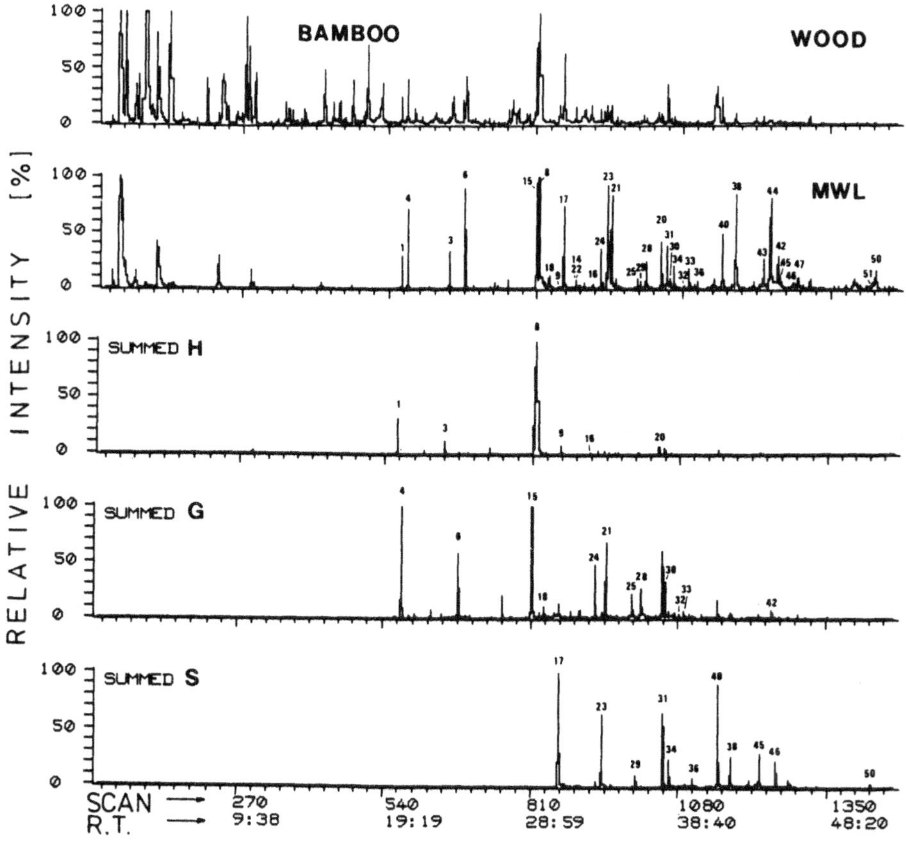

Fig. 4.7.8. Total ion chromatograms (TIC) of bamboo wood, bamboo MWL and mass chromatograms of H, G, and S units. Selected ion masses are listed in Table 4.7.2

Table 4.7.2. Typical parent ion masses for pyrolysis products from H, G, and S lignin units (see Fig. 4.7.8)

H[a]		G[a]		S[a]	
m/z	Substance	m/z	Substance	m/z	Substance
94	Phenol	124	Guaiacol	154	Syringol
107	Alkylphenols	135	Vinylguaiacol	165	Vinylsyringol
108	Methylphenols	137	Ethylguaiacol, homovanillin, coniferyl alcohol	167	Ethylsyringol, syringylacetone, homosyringaldehyde
120	Vinylphenol	138	Methylguaiacol	168	Methylsyringol
121	Alkylphenols	151	Vanillin, acetoguaiacone, propioguaiacone	181	Syringaldehyde, acetosyringone, propiosyringone
134	Allyl- + propenyl-phenols	164	Allyl- + propenyl-guaiacol	194	Allyl- + propenyl-syringol
148	Cumarylaldehyde	178	Coniferylaldehyde	208	Sinapaldehyde

[a] H = p-hydroxyphenylpropane; G = guaiacylpropane; S = syringylpropane.

$$\text{Similarity Index (SI) [\%]} = 100 - \frac{\sum_{n=1}^{k}\left(\mathrm{RRT}_n^a \times \mathrm{Ar}_n^a - \mathrm{RRT}_n^b \times \mathrm{Ar}_n^b\right)}{\sum_{n=1}^{k}\left(\mathrm{RRT}_n^a \times \mathrm{Ar}_n^a\right) + \sum_{n=1}^{k}\left(\mathrm{RRT}_n^b \times \mathrm{Ar}_n^b\right)} \times 100$$

where: Ar_n^a = normalized peak area of the nth peak in pyrogram a, Ar_n^b = normalized peak area of the nth peak in pyrogram b, RRT_n^a = relative retention time of the nth peak in pyrogram a, RRT_n^b = relative retention time of the nth peak in pyrogram b, k = number of peaks in the chromatogram.

The SI values can be used for comparing the pyrogram of an unknown sample and that of a known substance, and for determining the reproducibility of the pyrolysis system. Within the same laboratory, a similarity index of 95% for two consecutive Py-GC runs is possible using the Pyroprobe system. However, comparison of reproducibility as reported by different laboratories may not be justified, since individual groups have their own pyrolysis techniques.

The mass spectra of lignin pyrolysis products show several main characteristic masses; for example, mass 137 is typical of guaiacyl fragments having at least two carbon atoms in the alkyl side chain, and mass 167 is typical of syringyl residues. Mass fragments 151 and 181 are indicative of α-carbonyl-containing guaiacyl and syringyl groups, respectively. Figure 4.7.9 shows possible fragmentation pathways accounting for the formation of mass 137 and 151 ion peaks from guaiacyl units.

Additional fragmentation patterns of lignin model compounds have been reported by Kováčik et al. 1969, 1980, Kováčik and Škamla 1969, and Faix et al.

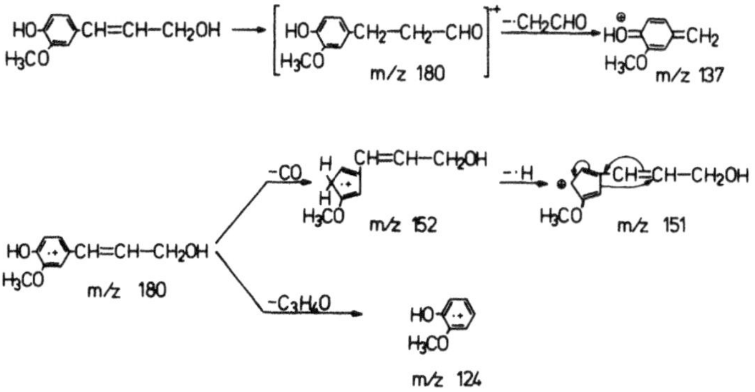

Fig. 4.7.9. Fragmentation patterns of coniferyl alcohol. (Kováčik et al. 1969)

1988a. The mass spectra of 51 important phenolic lignin pyrolysis products are presented in Fig. 4.7.10.

4.7.6 Utility of the Method

Recently, many useful applications of analytical pyrolysis for the analysis of diverse lignin samples have been reported. The method has proven to be extremely useful for a quick comparison of lignins of different origins. Based on the peak areas of H-, G-, and S-derived pyrolysis products in the pyrogram, lignins can be easily classified as G- or G/S-type. The results of such a comparison are shown in Fig. 4.7.11. Pyrograms of MWLs from spruce, teak, beech and whitewood show clearly the increase in S-derived pyrolysis products.

Saiz-Jimenez et al. (1987) applied analytical pyrolysis-gas chromatography to recent and buried alder, oak, and spruce woods. In comparison with the native woods, the yields of pyrolysis products from the guaiacyl and syringyl lignins were significantly higher in the alder and oak woods, while for spruce, this increase was less significant.

Faix et al. (1991) studied the residual lignin in wood after degradation by different white-rot fungi. Py-GC-FID revealed that the lignin moiety of the samples was modified more extensively than the polysaccharide fraction. By normalization of the pyrograms and calculation of the congruence indices, they found that the guaiacyl moiety had changed more than the syringyl moiety. Further, the guaiacyl/syringyl ratio had increased slightly.

Decayed wood was also investigated by Mulder et al. (1990). They used Py-MS and Py-GC-MS for studying the changes of lignin in the samples. Py-MS data revealed that after decay by a white-rot fungus, no mass peaks indicative of lignin were present. According to the relative intensities of mass peaks, the relative abundance of syringyl compounds decreased faster than that of the coniferyl compounds. On the other hand, brown-rot fungi did not degrade the lignin. However, the pyrolysis data showed an increase in oxygenated lignin pyrolysis products.

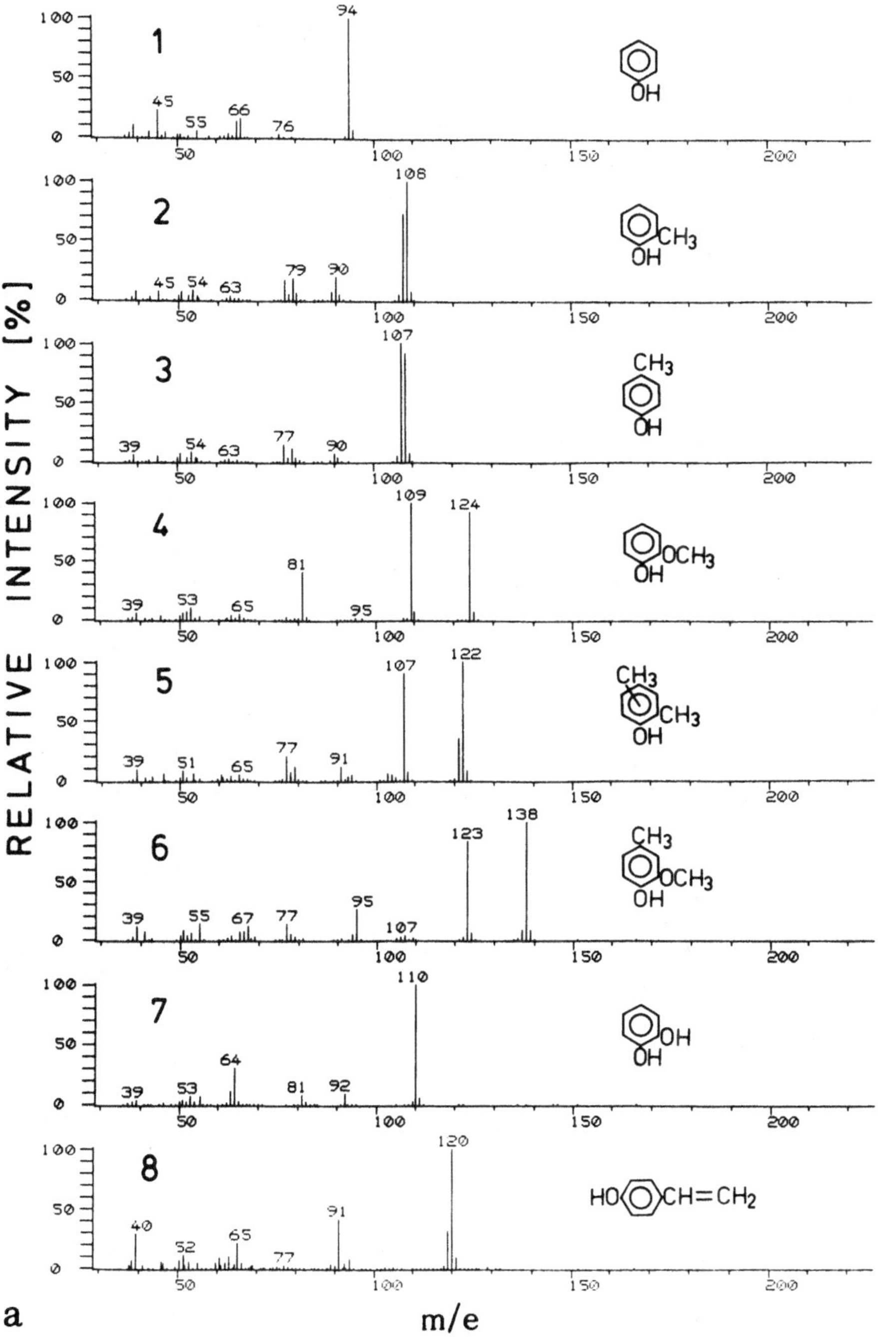

a

Fig. 4.7.10. a–f (pp 191–196). Electron impact (EI) mass spectra of lignin pyrolysis products. *Numbers* refer to Table 4.7.1. Spectra recorded at 70 eV, source temp. 170 °C

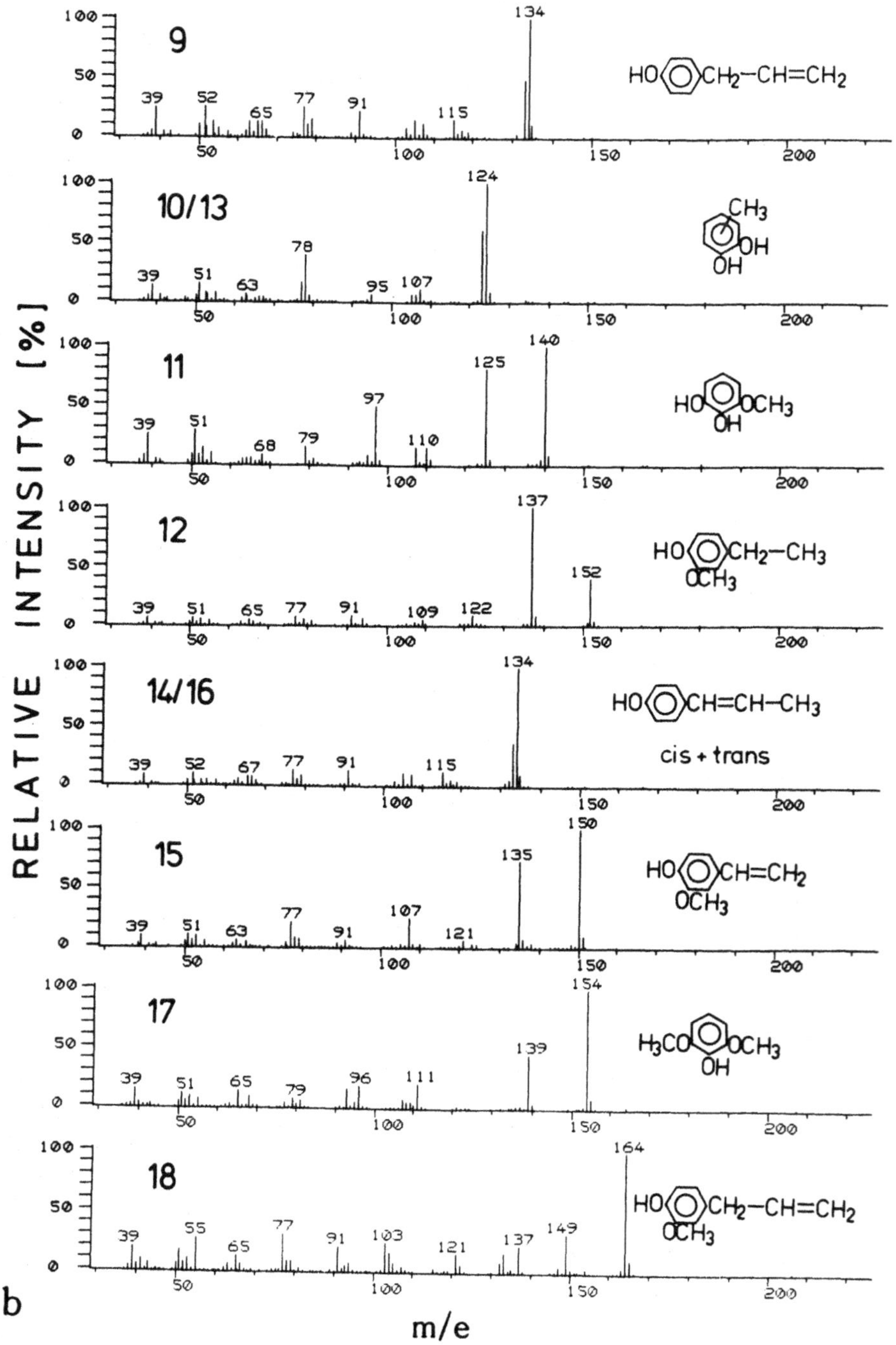

Fig. 4.7.10b

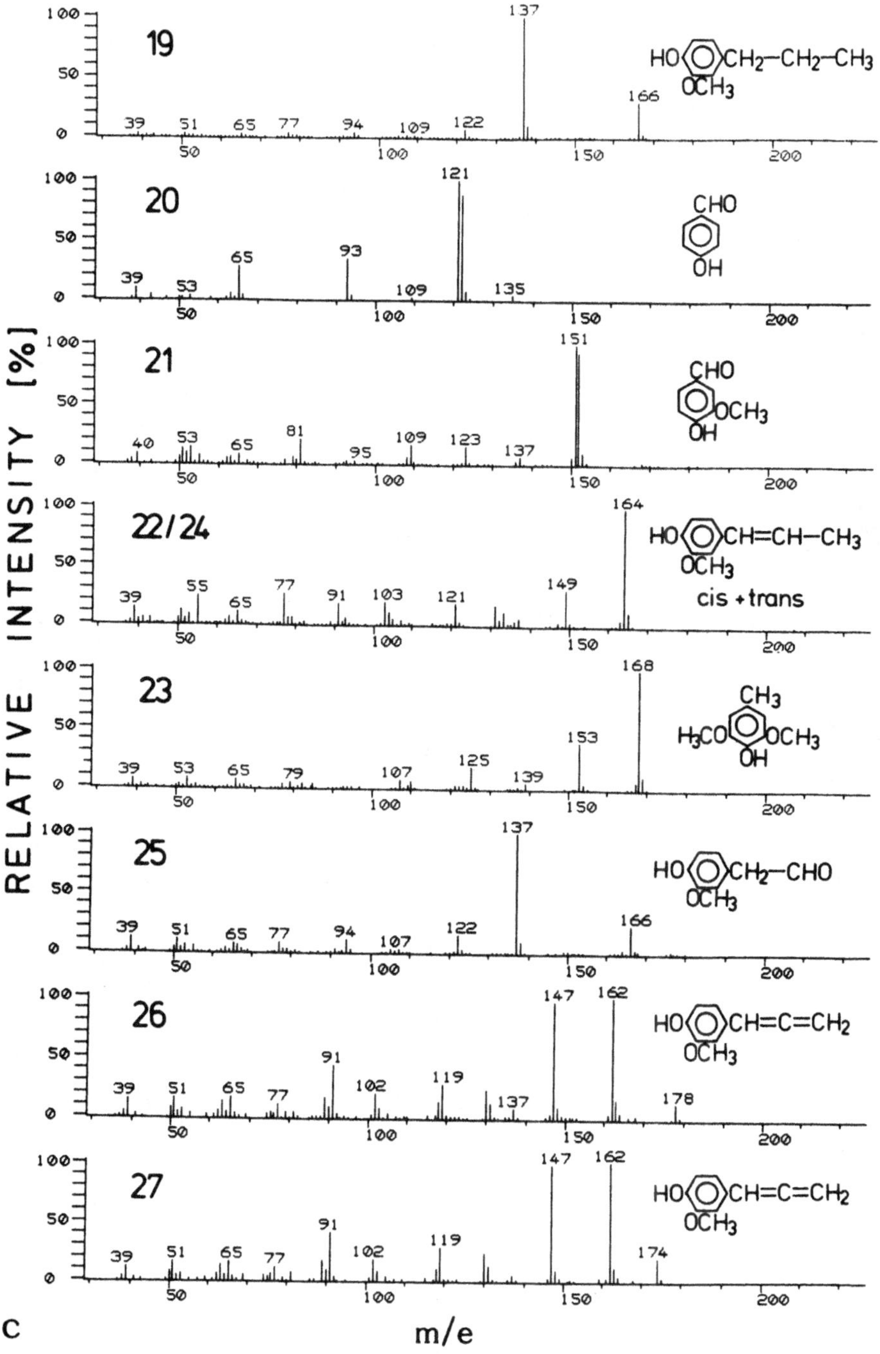
RELATIVE INTENSITY [%]
19
HO CH2–CH2–CH3
OCH3
137
166
20
CHO
OH
121
93
65
21
CHO
OCH3
OH
151
22/24
HO CH=CH–CH3
OCH3
cis + trans
164
149
23
CH3
H3CO OCH3
OH
168
153
25
HO CH2–CHO
OCH3
137
166
26
HO CH=C=CH2
OCH3
147
162
178
27
HO CH=C=CH2
OCH3
147
162
174
C
m/e

Fig. 4.7.10c

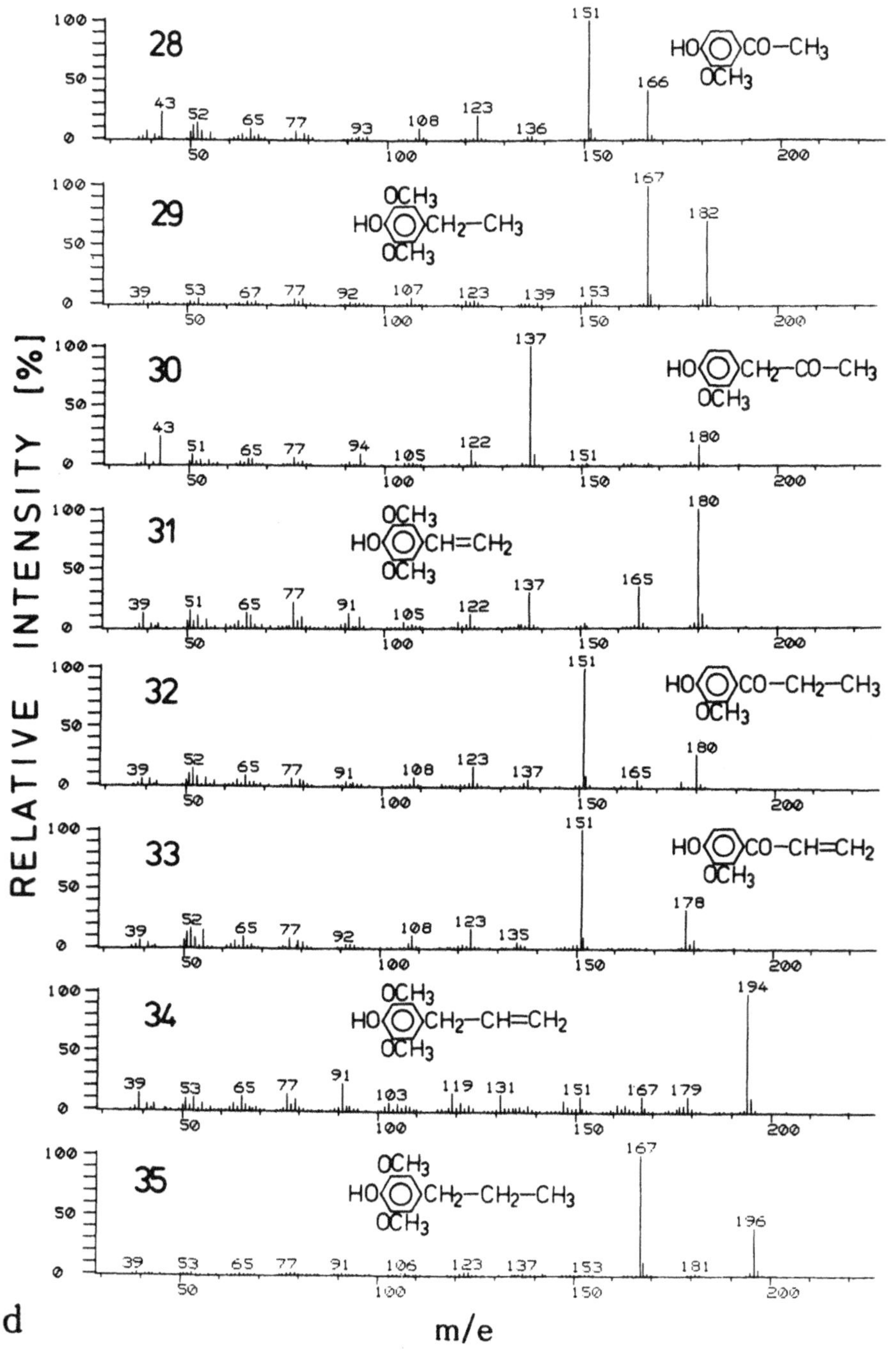
RELATIVE INTENSITY [%]
m/e
d
28
HO CO—CH3 OCH3
151 166 43 52 65 77 93 108 123 136
29
OCH3 HO CH2—CH3 OCH3
167 182 39 53 67 77 92 107 123 139 153
30
HO CH2—CO—CH3 OCH3
137 43 51 65 77 94 105 122 151 180
31
OCH3 HO CH=CH2 OCH3
180 165 137 39 51 65 77 91 105 122
32
HO CO—CH2—CH3 OCH3
151 180 39 52 65 77 91 108 123 137 165
33
HO CO—CH=CH2 OCH3
151 178 39 52 65 77 92 108 123 135
34
OCH3 HO CH2—CH=CH2 OCH3
194 39 53 65 77 91 103 119 131 151 167 179
35
OCH3 HO CH2—CH2—CH3 OCH3
167 196 39 53 65 77 91 106 123 137 153 181
0 50 100
50 100 150 200

Fig. 4.7.10d

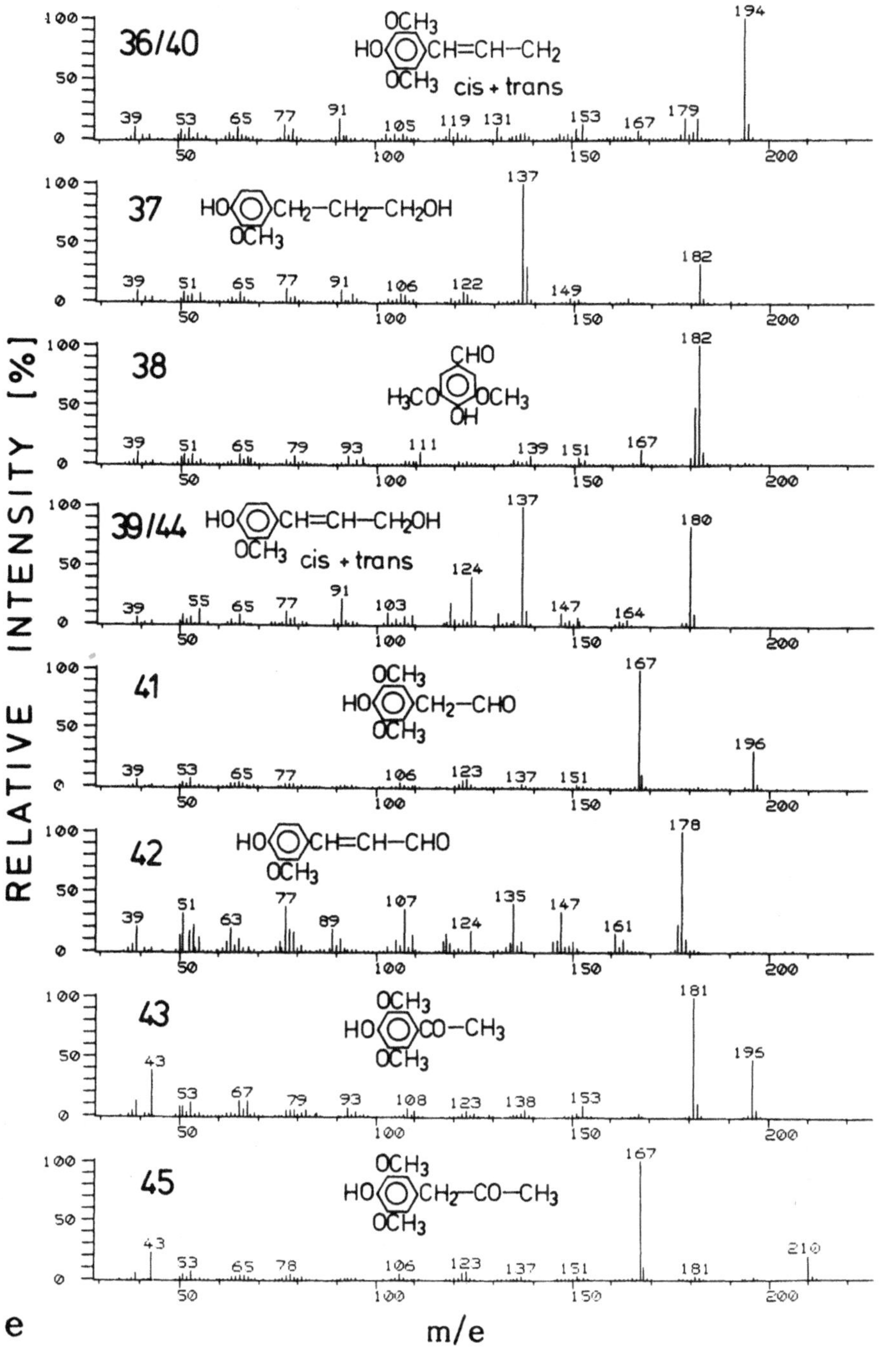
RELATIVE INTENSITY [%]
36/40
$HO-C_6H_2(OCH_3)_2-CH=CH-CH_3$ cis + trans
194
37
$HO-C_6H_3(OCH_3)-CH_2-CH_2-CH_2OH$
137
182
38
CHO
H_3CO OCH_3
OH
182
39/44
$HO-C_6H_3(OCH_3)-CH=CH-CH_2OH$ cis + trans
137
180
41
$HO-C_6H_2(OCH_3)_2-CH_2-CHO$
167
196
42
$HO-C_6H_3(OCH_3)-CH=CH-CHO$
178
43
$HO-C_6H_2(OCH_3)_2-CO-CH_3$
181
196
45
$HO-C_6H_2(OCH_3)_2-CH_2-CO-CH_3$
167
210
e
m/e

Fig. 4.7.10e

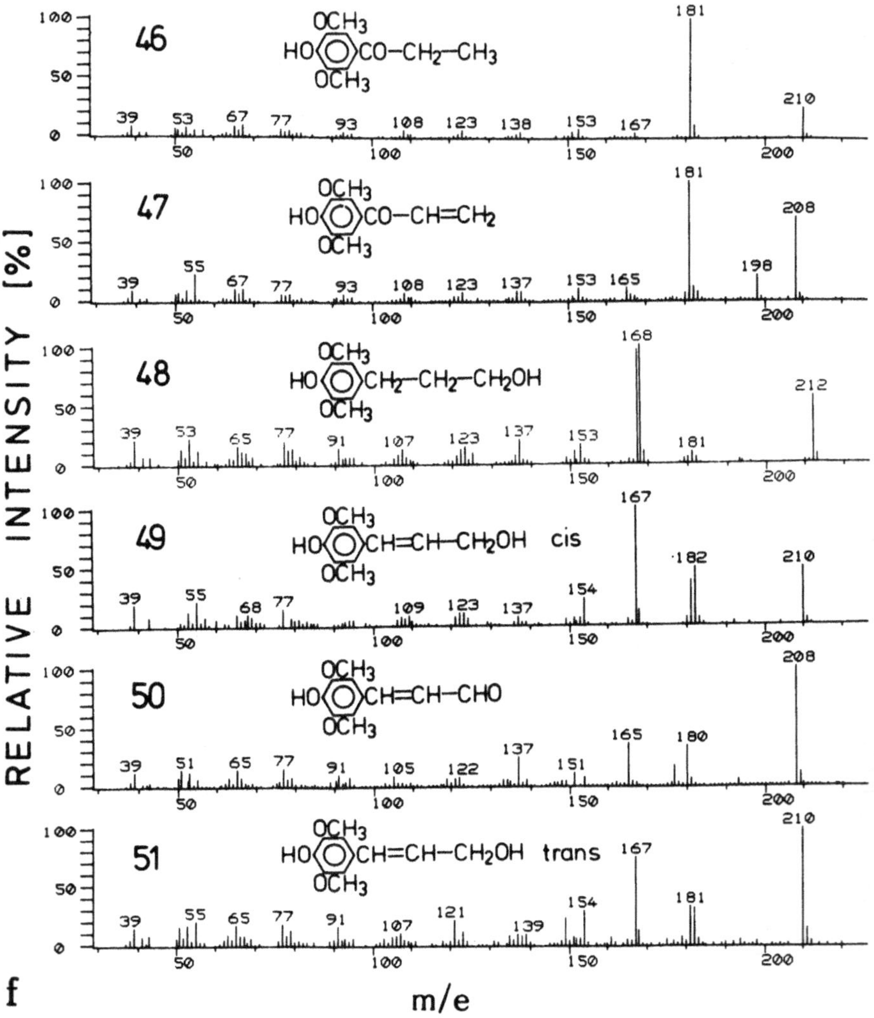

Fig. 4.7.10f

A catalog of degradation products from the analytical pyrolysis of lignin has been published recently (Faix et al. 1990a,b). The compilation comprises retention time data and the complete mass spectra of 82 compounds.

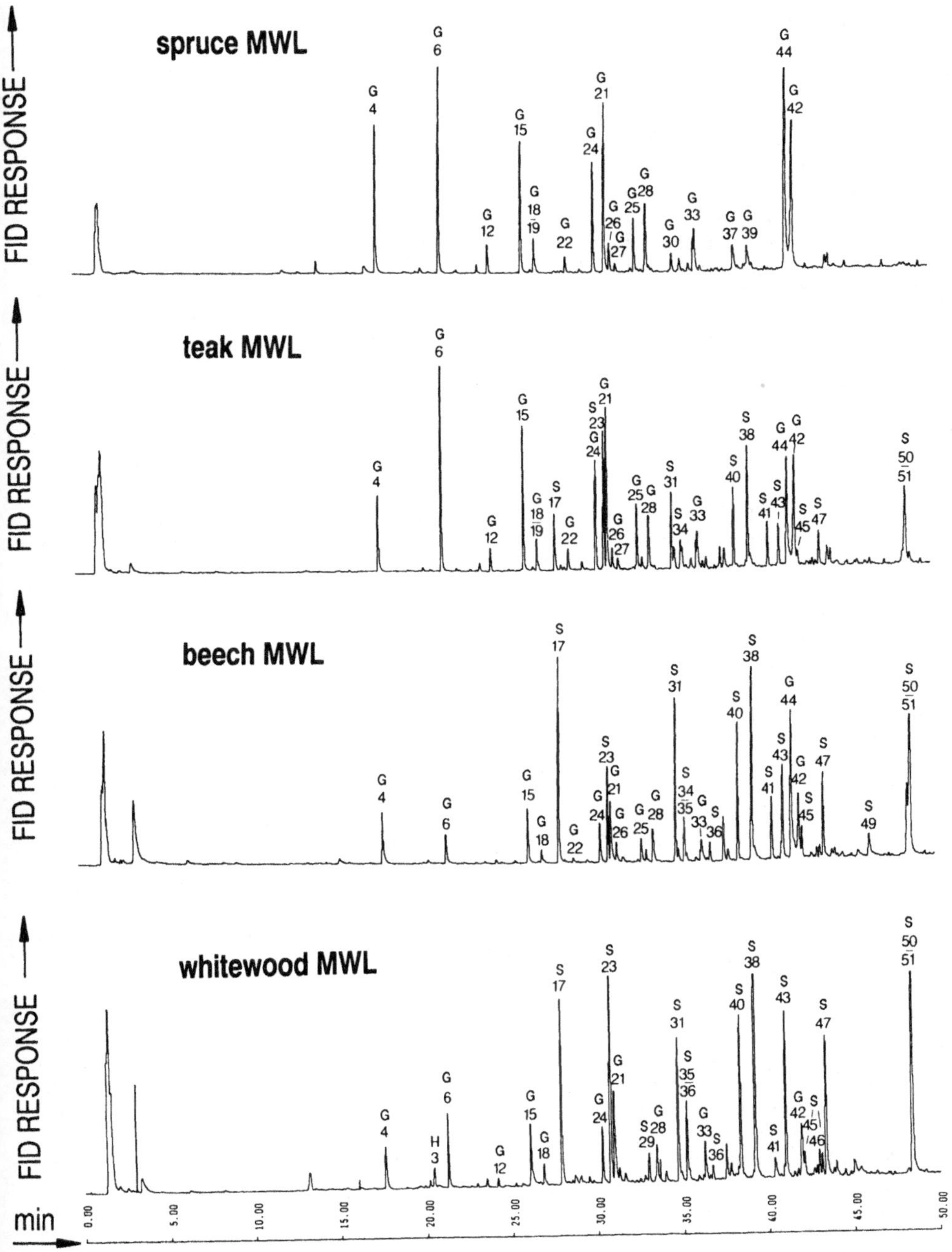

Fig. 4.7.11. Comparison of pyrograms from Py-GCFID experiments. *H*, *G*, and *S* indicate p-hydroxyphenyl, guaiacyl, and syringyl units. Compounds corresponding to *numbered peaks* are listed in Table 4.7.1

References

Boon JJ, Pouwels AD, Eijkel GB (1987) Pyrolysis high-resolution gas chromatography-mass spectrometry studies on beech wood: capillary high-resolution mass spectrometry of a beech lignin fraction. Biochem Soc Trans 15:170–174

Danielson ND, Glajch JL, Rogers LB (1978) Pyrolysis gas chromatography of enzymes. J Chromatogr Sci 16:455–461

Ericsson I (1980) Determination of the temperature-time profile of filament pyrolyzers. J Anal Appl Pyrolysis 2:187–194

Ericsson I (1985) Influence of pyrolysis parameters on results in pyrolysis-gas chromatography. J Anal Appl Pyrolysis 8:73–86

Faix O, Meier D, Grobe I (1987) Studies on isolated lignins and lignins in woody materials by pyrolysis-gas chromatography-mass spectrometry and off-line pyrolysis-gas chromatography with flame ionization detection. J Anal Appl Pyrolysis 11:403–416

Faix O, Meier D, Grobe I (1988a) Pyrolysis-gas chromatography-mass spectrometry of two trimeric lignin model compounds with alkyl-aryl structure. J Anal Appl Pyrolysis 14:135–148

Faix O, Jakab E, Till F, Székely T (1988b) Study on low mass thermal degradation products of milled wood lignins by thermogravimetry-mass-spectrometry. Wood Sci Technol 22:323–334

Faix O, Meier D, Fortmann I (1990a) Thermal degradation products of wood. Gas chromatographic separation and mass spectrometric characterization of monomeric lignin derived products. Holz Roh- Werkst 48:281–285

Faix O, Meier D, Fortmann I (1990b) Thermal degradation of wood. A collection of electron-impact (EI) mass spectra of monomeric lignin derived products. Holz Roh- Werkst 48:351–354

Faix O, Bremer J, Schmidt O, Stevanovic T (1991) Monitoring of chemical changes in white-rot degraded beech wood by pyrolysis-gas chromatography and Fourier-transform infrared spectroscopy. J Anal Appl Pyrolysis 21:147–162

Funazukuri T, Hudgins RR, Silveston PL (1987) Product distribution for flash pyrolysis of cellulose in a coil pyrolyzer. J Anal Appl Pyrolysis 10:225–249

Gardner DJ, Schultz TP, McGinnis GD (1985) The pyrolytic behavior of selected lignin preparations. J Wood Chem Technol 5:85–110

Genuit W, Boon JJ (1985) Pyrolysis-gas chromatography-photoionization-mass spectrometry, a new approach in the analysis of macromolecular materials. J Anal Appl Pyrolysis 8:25–40

Genuit W, Boon JJ, Faix O (1987) Characterization of beech milled wood lignin by pyrolysis-gas chromatography-photoionization mass spectrometry. Anal Chem 59:508–512

Hartley RD, Haverkamp J (1984) Pyrolysis-mass spectrometry of the phenolic constituents of plant cell walls. J Sci Food Agric 35:14–20

Haw JF, Schultz TP (1985) Carbon-13 CP/MAS NMR and FT-IR study of low-temperature lignin pyrolysis. Holzforschung 39:289–296

Irwin WJ (1982) Analytical pyrolysis – a comprehensive guide. Chromatographic Science, Vol. 22. Marcel Dekker, New York, 578 pp

Kováčik V, Mihalov V, Brezny R (1980) Mass spectrometry of lignin model substances III. Structure determination of β-linked dimers by mass spectrometry. Cellul Chem Technol 14:233–241

Kováčik V, Škamla J (1969) Massenspektrometrie einiger Modellsubstanzen des Lignins II. Chem Ber 102:3623–3631

Kováčik V, Škamla J, Joniak D, Kosiková B (1969) Massenspektrometrie einiger Modellsubstanzen des Lignins I. Chem Ber 102:1513–1522

Kratzl K, Czepel H, Gratzl J (1965) Gaschromatographische Studien zur Pyrolyse des Lignins. Holz Roh Werkst 23:237–240

Lattimer RP, Schur KM (1985) Quantitative analysis of rubber triblends by pyrolysis-mass spectrometry. J Anal Appl Pyrolysis 8:95–107

Liebman SA, Levy EJ (1985) Pyrolysis and GC in polymer analysis. Marcel Dekker, New York, 557 pp

Lindberg JJ, Raisanen S, Niemi R (1982) GC-MS studies on the pyrolysis of wood and lignin. Gov Rep Announce Index (U.S.) 1984 84:163–173

Linskens HF, Jackson JF (1986) Gas chromatography/mass spectrometry. Springer, Berlin Heidelberg New York Tokyo, 304 pp

Martin F, Saiz-Jimenez C, Gonzalez-Vila FJ (1979) Pyrolysis-gas chromatography-mass spectrometry of lignins. Holzforschung 33:210–212

Metzger J (1979) Chemische Ionisations-Massenspektrometrie von Ligninen. Fresenius Z Anal Chem 295:45–46

Meuzelaar HLC, Haverkamp J, Hileman FD (1982) Pyrolysis mass spectrometry of recent and fossil biomaterials – compendium and atlas. Elsevier, Amsterdam, 293 pp

Mulder MM, Pureveen JBM, Boon JJ, Martínez AT (1990) An analytical pyrolysis mass spectrometric study of Eucryphia cordifolia wood decayed by white-rot and brown-rot fungi. J Anal Appl Pyrolysis 19:175–191

Müller M (1980) Beitrag zum Einsatz der Curie-Punkt-Pyrolyse-Massenspektrometrie in der Analytik von Naturstoffen. Dissertation ETH 6743, Zürich

Obst JR (1983) Analytical pyrolysis of hardwood and softwood lignins and its use in lignin-type determination of hardwood vessel elements. J Wood Sci Technol 3:377–397

Pouwels AD, Boon JJ (1987) Analysis of lignin and chlorolignin residues in a beech xylan fraction by pyrolysis gas chromatography mass spectrometry. J Wood Sci Technol 7:197–213

Saiz-Jimenez C, DeLeeuw JW (1984) Pyrolysis-gas chromatography-mass spectrometry of isolated, synthetic and degraded lignins. Org Geochem 6:417–422

Saiz-Jimenez C, DeLeeuw JW (1986) Lignin pyrolysis products: their structure and their significance as biomarkers. Adv Org Geochem 1985, Org Geochem 10:869–876

Saiz-Jimenez S, Boon JJ, Hedges JI, Hessels JKS, DeLeeuw JW (1987) Chemical characterization of recent and buried woods by analytical pyrolysis. Comparison of pyrolysis data with ^{13}C NMR and wet chemical data. J Anal Appl Pyrolysis 11:437–450

Salo S, Niemelä S, Elomaa M, Lindberg JJ (1989) Identification of lignins by pyrolytical methods. Holzforschung 43:257–261

Tromp PJJ (1987) Coal pyrolysis – basic phenomena relevant to conversion processes. Dissertation, University Amsterdam, FOM-Institute, Amsterdam

Voorhees KJ (1984) Analytical pyrolysis – techniques and applications. Butterworths, London, 486 pp

Wampler TP, Levy EJ (1987) Reproducibility in pyrolysis: recent developments. J Anal Appl Pyrolysis 12:75–82

Wells G, Voorhees KJ, Futrell JH (1980) Heating profile curves for resistively heated filament pyrolyzers. Anal Chem 52:1782–1784

Wieten G, Meuzelaar HLC, Haverkamp J (1984) Analytical pyrolysis in clinical and pharmaceutical microbiology. In: Odham G, Larsson L, Mardh PA (eds) Gas chromatography/mass spectrometry – applications in microbiology. Plenum, New York, 335–358

Windig W, Meuzelaar HLC, Shafizadeh F, Kelsey RG (1984) Biochemical analysis of wood and wood products by pyrolysis-mass spectrometry and multivariate analysis. J Anal Appl Pyrolysis 6:233–250

4.8 Thermal Analysis

H. Hatakeyama

4.8.1 Introduction

Thermal analysis is an important technique for determining the physical and chemical properties of polymeric materials. It may be defined as a set of methods used to measure the physical or chemical changes of substances as a function of temperature. Some common thermal analysis techniques are given in Table 4.8.1.

Thermogravimetry (TG), differential thermal analysis (DTA), and differential scanning calorimetry (DSC) are the most frequently used techniques in lignin chemistry, although thermomechanical analysis (TMA) has also been used effectively in the analysis of thermal properties of lignin (Goring 1963). In this section, the principles of TG, DTA, and DSC, and their application to lignin are described.

4.8.2 General Principles and Techniques of Thermal Analysis

4.8.2.1 Thermogravimetry

Thermogravimetry is a technique that measures the weight change of a sample as a function of temperature or time (time is suitable only when thermal analysis is performed at specified constant temperature increments). The solid or liquid sample is heated or cooled at a selected rate or isothermally maintained at a fixed temperature. TG is used to measure degradation, oxidation, reduction, evaporation, sublimation, and other heat-related changes occurring in polymers.

4.8.2.1.1 Apparatus

As shown in Fig. 4.8.1, a TG apparatus typically consists of a (1) recording balance, (2) furnace, (3) furnace temperature controller, (4) temperature programmer, (5) temperature detector, and (6) recorder. Most commercial models, e.g., Perkin-Elmer Model TGA-7, TA Instruments 2950 TGA, Mettler TC11/TG50, Seiko Instruments SSC 5200 TG-DTA 220, and Shimadzu TGA-50, consist of these basic components.

A thermobalance can be one of three types according to the position of the furnace in relation to the recording balance: (1) below balance, (2) above

Springer Series in Wood Science
Methods in Lignin Chemistry
(Edited by S.Y. Lin and C.W. Dence)

Table 4.8.1. Frequently used thermal analysis techniques

Techniques	Parameter measured
Thermogravimetry (TG)	Mass
Differential Thermal Analysis (DTA)	Temperature
Differential Scanning Calorimetry (DSC)	Enthalpy
Thermomechanical Analysis (TMA)	Deformation
Thermooptometry	Light
Thermosoniometry	Sound
Thermoelectrometry	Current or resistance

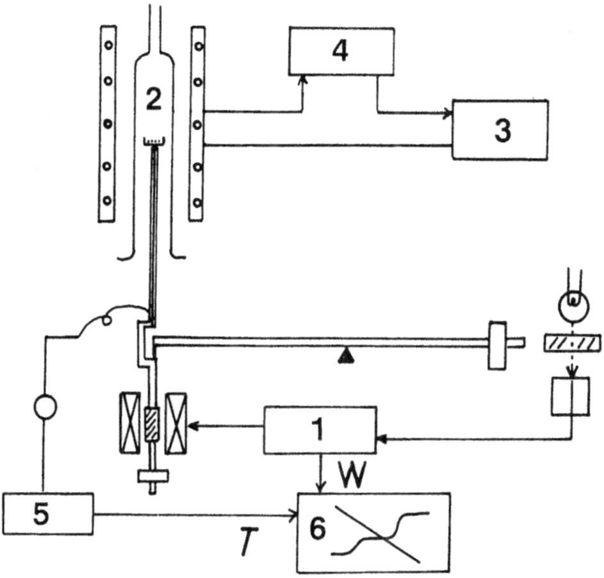

Fig. 4.8.1. Schematic diagram of a TG apparatus: recording balance (*1*), furnace (*2*), furnace temperature (*3*), temperature programmer (*4*), temperature detector (*5*), and recorder (*6*)

balance and (3) parallel to balance (Fig. 4.8.2). In a Type 1 apparatus, a sample is suspended in the furnace from under the beam of the balance. This type is suitable for a comparatively large sample where the buoyancy effect is small. Nevertheless, a measurement error may be introduced in the recording balance because of the influence of heat from the furnace and gas from the decomposed sample. In a Type 2 apparatus, the sample is supported with a rod on the balance, and is more severely influenced by buoyancy than in a Type 1 configuration. However, the influence of heat from the furnace is smaller, and the collection of the gas from the decomposed sample is easier. In a Type 3 apparatus, the buoyancy of the sample and the thermal expansion of the balance beams are large, but the effect of heat from the furnace is small. None of the above three arrangements is ideal for meeting the conditions of low sample

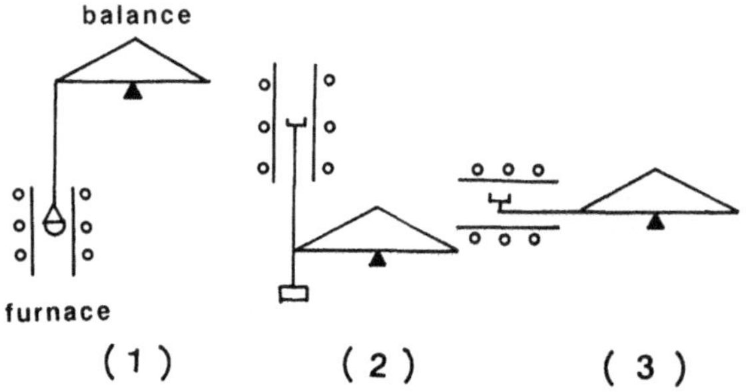

Fig. 4.8.2. Furnace positions in TG apparatus: (*1*) below, (*2*) above, and (*3*) parallel to balance

buoyancy, small thermal expansion of balance beams, and minimal heat effect on the balance.

4.8.2.1.2 TG Curve

As shown in Fig. 4.8.3, a TG curve is recorded by plotting the weight (W) or weight ratio (W_t/W_o), where W_o is the original weight and W_t, the weight after time t, of a sample against temperature (T) or time (if the heating rate is constant). The following information can be obtained from a TG curve: the temperature (or time) at which (1) weight loss begins (T_1), (2) one-half of the weight loss occurs (T_2), and (3) complete weight loss occurs (T_3).

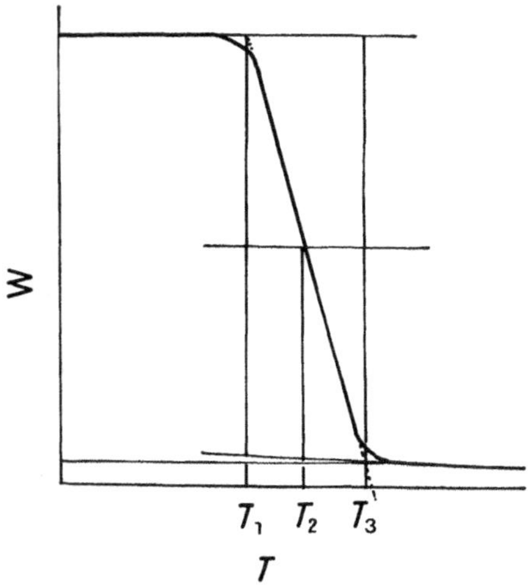

Fig. 4.8.3. Typical TG curve: temperatures corresponding to beginning weight loss (T_1), half weight loss (T_2), and complete weight loss (T_3)

4.8.2.1.3 Factors Affecting TG Measurements

Temperature
The difference between the actual temperature of a sample and the temperature indicated on TG recording paper is highly dependent on the geometrical relationship of the sample, furnace, and temperature sensor. If the sensor is not directly attached to the sample holder, the temperature indicated may be profoundly influenced by ambient conditions and heating rate; in such instances, a temperature calibration is usually required.

Sample Weight
Some potential problems exist in determining accurately the sample weight in a TG apparatus because of the buoyancy effect. Thus, the aerodynamic forces of flow gas, the condensation of gaseous products on the cooler parts of the sample suspension system, and sample loss from the holder caused by an excessive gas flow rate may introduce errors in the determination of sample weight loss.

Sample Parameters
The major factors that affect thermal analytical data are related to the shape, size, and weight of the sample. Samples are preferably in a powder or particle form, smaller than 0.1 mm in diameter. If water or volatile solvents are present, the sample diameter should be smaller than 0.05 mm. However, an exceedingly small sample size may create a buoyancy effect and should be avoided.

4.8.2.2 DTA and DSC

DTA measures, as a function of temperature or time (if the heating rate is constant), the difference between the temperature of a sample and a reference material in the furnace. In contrast to DTA, DSC measures, as a function of temperature, the energy input required to keep a sample and a reference material at the same temperature. Therefore, DSC provides more direct information than DTA on the enthalpy changes associated with the phase transition occurring on heating or cooling.

4.8.2.2.1 Apparatus

As shown in Figs. 4.8.4 and 4.8.5, typical DTA and DSC apparatus consist of the following basic components: (1) furnace or heating device, (2) differential temperature detector, (3) amplifier, and (4) recorder. There are many modifications of this basic design. However, all instruments measure the difference in temperature (DTA) or enthalpy (DSC) between a sample and an inert reference material.

Most commercial DTA and DSC models are patterned after those shown in the figures. Representative commercially available models include Perkin-Elmer DSC 7 and DTA 1700, TA Instruments 2910 DSC and DTA, Mettler TC11/

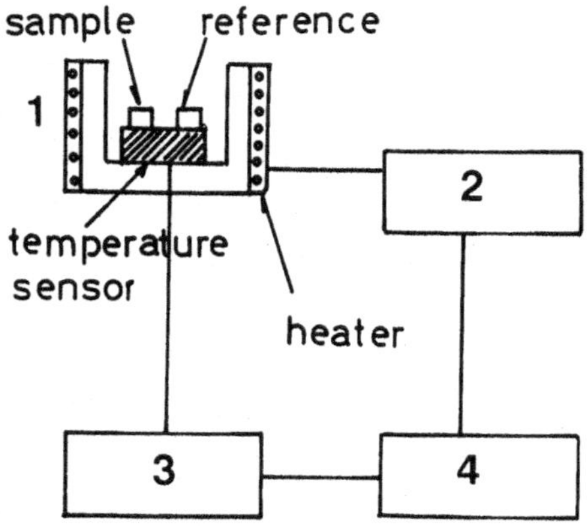

Fig. 4.8.4. Schematic diagram of a DTA apparatus: furnace (*1*), furnace temperature programmer (*2*), amplifier (*3*), and recorder (*4*)

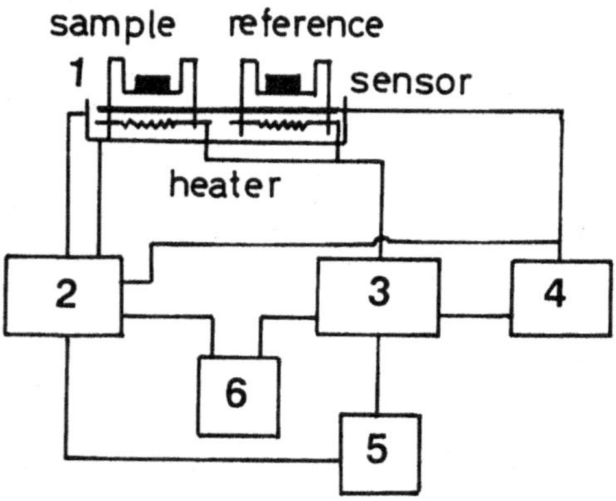

Fig. 4.8.5. Schematic diagram of a DSC apparatus: furnace (*1*), temperature programmer (*2*), temperature controller (*3*), amplifier (*4*), baseline controller (*5*), and recorder (*6*)

DSC30, Seiko Instruments SSC 5200, DSC 220, and TG-DTA 220, and Shimadzu DSC-50 and DTA-50.

Three kinds of sample holders are available for DTA and DSC (Fig. 4.8.6). Type 1 holders are commonly used for a DTA apparatus. In this type, the sample and reference holders are placed on the same metal block and heated by the same heater; the temperature difference between the two holders is indicated by a thermocouple. Type 2 holders are generally used in a quantitative DTA (heat-flux DSC) apparatus. Both sample and reference holders are maintained at the same temperature by two individual internal heaters, which, in turn, are heated by the same main heater. The temperature difference between the sample and the reference material is measured by a thermocouple placed outside of the holders. Type 3 holders are customarily used for a power-compensated DSC apparatus. This apparatus has separate heaters for heating the sample and reference holders thus maintaining the sample and the reference

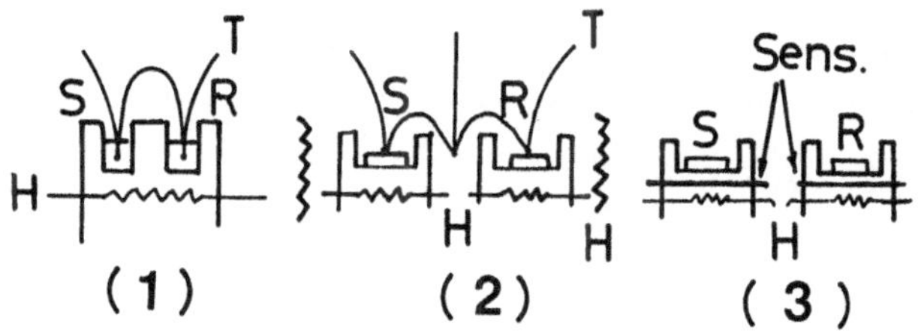

Fig. 4.8.6. Arrangements of sample holder (*S*) and reference holder (*R*) in the heater of a DTA or DSC apparatus: *1* DTA; *2* heat-flux DSC; *3* power-compensated DSC

material at the same temperature. The amount of heat required to maintain isothermal conditions is recorded as a function of time or temperature.

4.8.2.2.2 DTA and DSC Curves

DTA and DSC curves are commonly recorded by plotting the temperature or enthalpy differences, respectively, against temperature (T) or time (t).

Figure 4.8.7 shows typical DSC curves of a crystalline polymer containing amorphous regions. From the DSC curves, the following phase transition temperatures can be obtained as indicated by the arrows: melting peak temperature in the heating run (T_m), crystallization peak temperature in the cooling run (T_c), and, after quenching the polymer from the melting stage, the glass transition temperature (T_g), cold crystallization peak temperature (T_{cc}), and

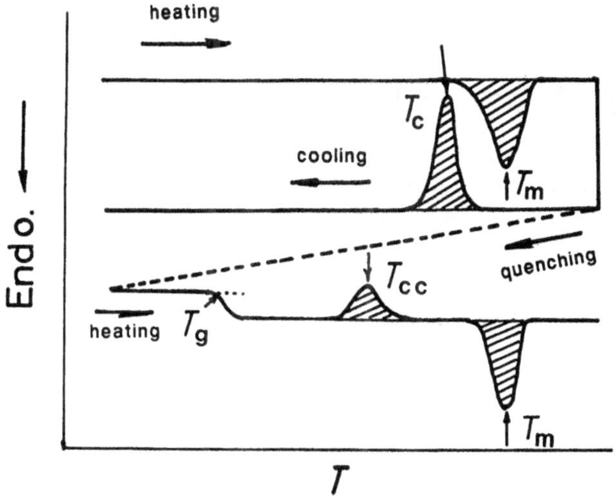

Fig. 4.8.7. Typical DSC curves of a crystalline polymer: melting temperature (T_m), crystallization temperature (T_c), glass transition temperature (T_g), and cold crystallization temperature (T_{cc})

melting peak temperature in the reheating run. The enthalpy of each transition can be calculated from the area under the curve using the area of the melting peak of indium (enthalpy, 28.59 J/g) as the reference.

4.8.2.2.3 Factors Affecting DTA and DSC Measurements

To obtain accurate data from DTA and DSC measurements, attention should be paid to the following parameters:

Sample Pans
Different sample pans are used in DTA and DSC measurements for samples of various shapes and physical states. The sample pans are of two types: open pans for solid samples and sealed pans for liquids or volatile samples. The materials used for making sample pans are mostly aluminum, silver, gold, stainless steel, and carbon. When aluminum and silver sample pans are used for samples containing water, the sample pans should be pretreated with water in a small pressure vessel to avoid reactions between the pans and water during subsequent heating.

Sample Parameters
Sample size and weight, packing density, heat capacity, and heat conductivity are among the sample parameters that may impact the outcome of a thermal analysis. It is important that a small sample be used and that the sample pan fits the sample shape.

Temperature
The temperature can be measured in the furnace, in the reference material holder, or in the sample holder by placing a thermocouple or sensor in the center of the furnace or in the holders. However, experience indicates that the best choice is to place the thermocouple or sensor in the sample holder. Temperature calibration of the apparatus using a standard material is necessary and recommended. A standard material that has a transition in a similar temperature range as the sample should be used for calibration. The scanning rate is also a factor that affects the shape of DTA and DSC curves. In general, a low scanning rate (0.1–2 K/min) improves the resolution, while higher heating rates (>20 K/min) yield larger peaks of poor resolution. Heating rates of 5–10 K/min are commonly used.

4.8.3 Applications of Thermal Analysis to Lignin

Lignin is a complex amorphous polymer composed of a cross-linked and branched network phenylpropane units. Thermal analyses such as TG, DTA, and DSC are useful for elucidating the physicochemical properties of lignin.

4.8.3.1 Kinetic Study of the Thermal Degradation of Lignin

The random degradation and reaction kinetics of high-molecular weight polymers can be determined by an approximate integral method (Ozawa 1965, Hirose and Hatakeyama 1986). Generally, the fractional weight, W, of a reacting material can be expressed as a function of the fraction of a structural quantity which is represented by x, i.e.,

$$W = f(x), \tag{1}$$

where x changes according to the ordinary kinetic equation:

$$-dx/dt = A\,[\exp(-\Delta E/RT)]g(x). \tag{2}$$

By integration,

$$-\int_{x_0}^{x} dx/g(x) = A\int_{t_0}^{t} \exp(-\Delta E/RT)dt, \tag{3}$$

where x_o = the value of x at $t=t_o$, T = absolute temperature, K, R = gas constant, t = time, A = pre-exponential factor, ΔE = activation energy.

Assuming that the temperature of the sample is raised at a constant rate, k, and that the reaction is very slow at low temperature, the change in x is given by the following equation:

$$-\int_{o}^{x} dx/g(x) = A\tau, \tag{4}$$

where

$$\tau = (1/k)\int_{o}^{T} \exp(-\Delta E/RT)dT. \tag{5}$$

τ is expressed by using function, p

$$\tau = (\Delta E/R)p(\Delta E/RT). \tag{6}$$

$p(\Delta E/RT)$ can be approximated:

$$\log p(\Delta E/RT) = -2.315 - 0.4567(\Delta E/RT). \tag{7}$$

For a given value of W, the left side of Eq. (4) is constant, and the following equation is obtained:

$$(A/k_1)(\Delta E/R)p(\Delta E/RT_1) = (A/k_2)(\Delta E/R)p(\Delta E/RT_2) = \ldots \tag{8}$$

Using Eq. (6), the following relationship can be derived:

$$-\log k_1 - 0.4567(\Delta E/RT_1) = -\log k_2 - 0.4567(\Delta E/RT_2) = \ldots \tag{9}$$

Thus, plots of log k against the reciprocal absolute temperature for a given value of W is a straight line, the slope of which is the activation energy.

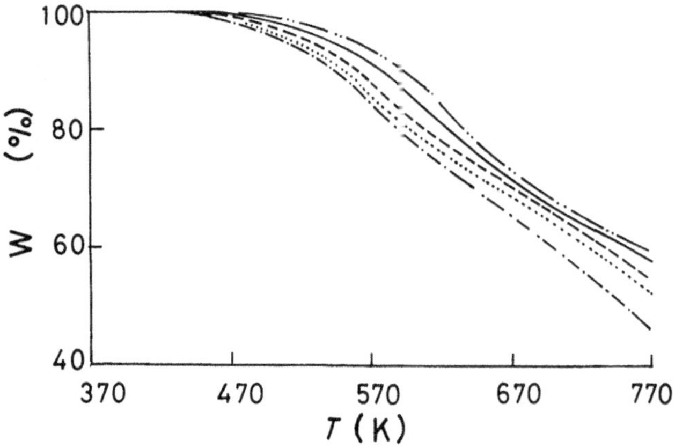

Fig. 4.8.8. TG curves of dioxane lign n heating rate: -.- 2 K/min,...5 K/min, --- 10 K/min, — 20 K/min, -..- 30 K/min. (Hirose and Hatakeyama 1986)

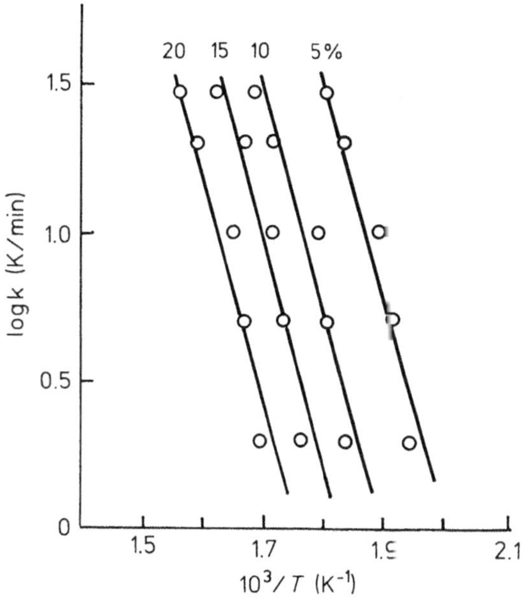

Fig. 4.8.9. Logarithmic plots of heating rate (k) against reciprocal absolute temperature. (Hirose and Hatakeyama 1986)

Figure 4.8.8 shows the TG curves of softwood dioxane lignin at various heating rates, k (Hirose and Hatakeyama 1986). Applying Eq. (7) to the TG curves, the linear relationship shown in Fig. 4.8.9 is obtained. The activation energy calculated from the data is 136 kJ/mol.

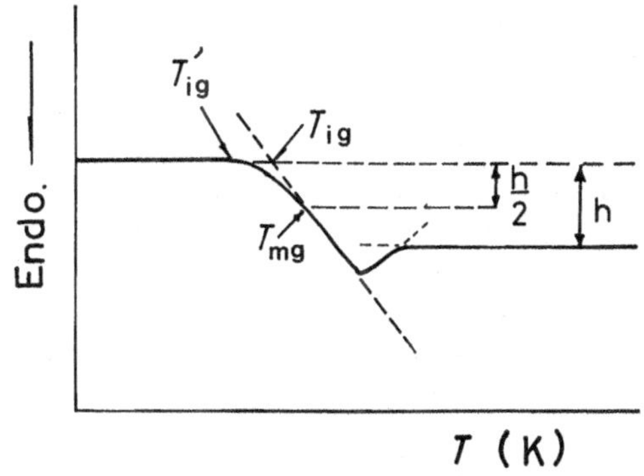

Fig. 4.8.10. Schematic DSC curve illustrating the determination of glass transition temperature (T_g)

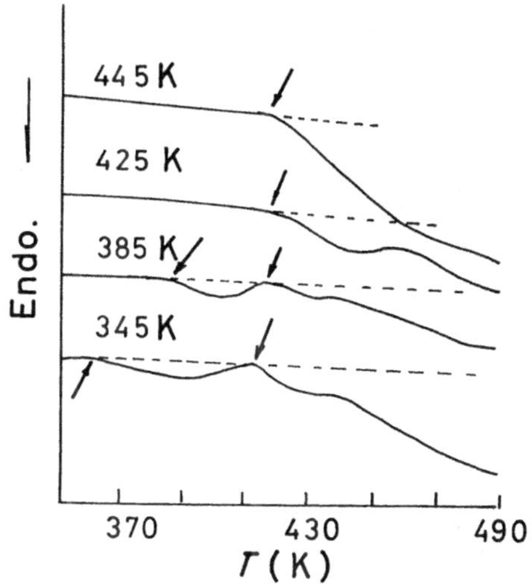

Fig. 4.8.11. DSC heating curves of milled wood lignin preheated for 30 min at varying temperature. (Hatakeyama et al. 1972)

4.8.3.2 Glass Transition of Lignin

The following temperatures, illustrated by the DSC curve in Fig. 4.8.10, are used to determine the glass transition temperature (T_g):

T_{ig}' = onset temperature
T_{ig} = extrapolated onset temperature
T_{mg} = midpoint temperature

T_{ig}', T_{ig}, and T_{mg} are used by the Committee on Standardization of ICTA (Nakamura et al. 1988) to determine the T_g of typical amorphous polymers such as polystyrene and lignin. Generally, T_{ig} is used as the T_g value.

As an amorphous polymer, lignin undergoes chain segment motion upon heating. This motion, a glass transition, is characteristic of all amorphous polymers, and is indicated by an endothermic shift in the DTA or DSC curves. This glass transition is accompanied by abrupt changes in free volume, heat capacity, and thermal expansion coefficient.

The T_g of lignin is affected by such factors as the presence of low molecular weight contaminants (including water and solvents), molecular weight, thermal history, cross-linking, and pressure. Therefore, a typical T_g value for any particular lignin cannot be precisely reported. However, the values of T_g of dioxane lignin and milled wood lignin are generally around 440 and 420 K, respectively.

Figure 4.8.11 shows the DSC curves of MWL pretreated at various temperatures for 30 min under nitrogen before the room temperature DSC run (Hatakeyama et al. 1972). Two T_g's (shown by arrows) were observed when the treatment temperature was lower than 405 K. At the treatment temperatures above 405 K, the two T_g's merge into one. The low-temperature T_g increases with increasing treatment temperature, but the high-temperature T_g increases to a maximum at about 415 K and then decreases slightly with increasing treatment temperature. This observation suggests that the glassy state of lignin is non-equilibrium and that molecular rearrangement takes place depending on the time and temperature of the heat treatment.

The T_g values of lignin also change markedly with the breaking of hydrogen bonds between hydroxyl groups. Hydrogen bonds are easily broken in the presence of a small amount of water and/or by blocking the hydroxyl groups by

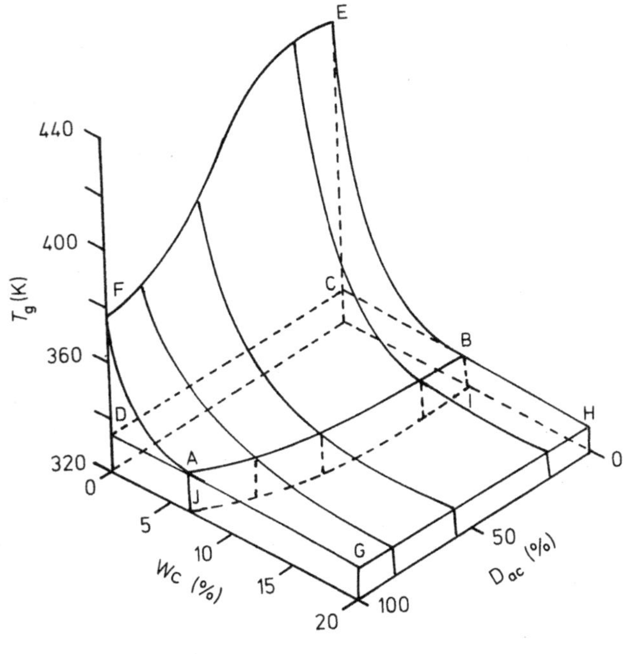

Fig. 4.8.12. Diagram showing the effect of degree of acetylation and water content on the glass transition temperature, T_g, of dioxane lignin. (Hatakeyama and Hatakeyama 1983)

acetylation or methylation. Figure 4.8.12 shows a three-dimensional diagram for acetylated dioxane lignin (ADL) in which T_g is plotted against the water content (W_c) and degree of acetylation (D_{ac}) of dioxane lignin (DL) (Hatakeyama and Hatakeyama 1983). In this figure, point E represents the T_g of DL and point F, the T_g of ADL with 100% relative degree of acetylation. The T_g of DL decreases rapidly with an increase in water content (W_c) as do the values of the acetylated DLs. The slope of the T_g curve varies with the degree of acetylation. Line AB shows the leveling off of the decrease in the T_g values of DL and the various ADLs with increasing water content (W_c). Plane EFGH is intersected by plane ABIJ. Section ABEF of the plane shows the rapid decrease in the T_g values of DL and the ADLs with increasing W_c, reflecting the breaking of hydrogen bonds by water. The ABGH section of the plane is almost flat and represents the region in which there is no further breaking of bonds. The excess water molecules in the DL and ADLs remain as bulk water at W_cs above 8–10%. This excess water has a plasticizing effect on the lignins with the result that T_g values decrease gradually with increasing W_c.

For synthetic polymers composed of 4-hydroxylphenyl, guaiacyl, and syringyl units and for polyurethanes containing lignin, the T_g values are affected by factors such as structure, molecular flexibility, and molecular weight of the other components of the polymers (Hatakeyama and Kanetsuna 1989, Yoshida et al. 1990).

4.8.3.3 Heat Capacity of Lignin

The molar heat capacity (C_p) of a pure compound at constant pressure is defined as

$$C_p(J/g \cdot K) = dq/dT, \tag{10}$$

where q is heat at constant pressure and T the absolute temperature (Moelwyn-Hughes 1961). C_p is measured by the following procedure (Hatakeyama and Kanetsuna 1989):

1. Adjust the sensitivity, heating rate, baseline, and other parameters of the DSC apparatus. Place the empty sample pans in the sample and reference holders. The initial and end temperatures (T_i and T_e) (Fig. 4.8.13) for scanning are set and the sample and reference holders are maintained at T_i for a suitable time to obtain the straight line I. Scanning at a heating rate of 5 or 10 K/min is conducted from T_i to T_e to obtain DSC curve II. The sample and reference holders are maintained at T_e for some time to obtain the straight line, III.
2. The reference material (e.g., sapphire or Al_2O_3) is placed in the reference pan which is placed in the reference holder, and scanning from T_i to T_e is carried out to obtain curve IV.
3. Scanning from T_i to T_e is performed with the sample in the sample pan to obtain curve V. If the three curves, II, IV, and V, do not meet at T_i and T_e, adjust the apparatus to obtain the correct baseline and repeat the procedure.
4. C_p can be calculated using the following equation:

$$C_p = (l_x/l_s) \times (W_r/W_s)C_{ps}, \tag{11}$$

where C_p = molar heat capacity of the sample, C_{ps} = molar heat capacity of

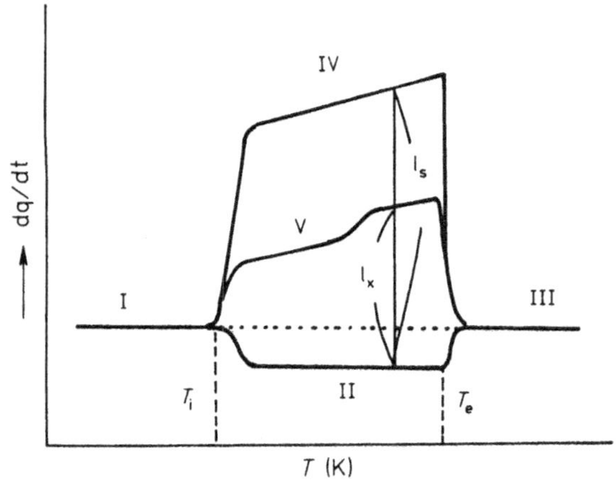

Fig. 4.8.13. Schematic DSC curves illustrating calculation of heat capacity

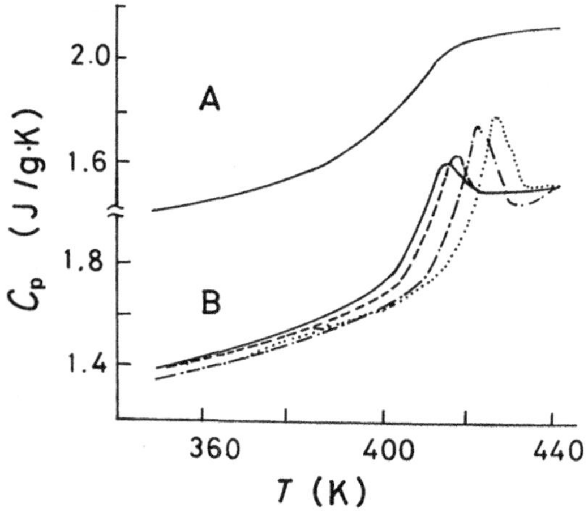

Fig. 4.8.14. Heat capacity (C_p) curves of quenched (cooled rapidly) and annealed (heat-treated) dioxane lignin. *A* quenched from 423 K to 310 K; *B* annealed at 395 K for — 60 min, --- 100 min, -.- 940 min, ... 1300 min. (Hatakeyama et al. 1982)

the reference material, W_s = reference material weight, l_x and l_s = the values shown in Fig. 4.8.13.

Because of the temperature fluctuation of sample holders in the initial period of scanning, the use of data from the initial stage for calculating heat capacity should be avoided.

The C_p values for the glassy and transition regions reflect the structural change in the amorphous state of lignin and are dependent on the temperature of measurement. A typical C_p value for dioxane lignin at 350 K is 1.342 J/(g · K). Figure 4.8.14 shows C_p curves of dioxane lignin as a function of temperature under various treatment conditions (Hatakeyama et al. 1982). The appearance of a C_p peak in the temperature range slightly higher than T_g in the heat-treated samples indicates that relaxation of the molecular chain occurs during annealing at 395 K. The C_p peak resulting from the annealing is related to the enthalpy of relaxation and increases with annealing time, suggesting that a rearrangement of local order takes place in the structure of the amorphous lignin.

4.8.4 Conclusions

Thermal analysis is capable of providing accurate information on the phase transition temperatures, degradation temperatures, heat capacity, and enthalpy of transition of polymers using comparatively simple DTA, DSC, and TG instruments. The measurement time is short compared with other techniques, such as viscoelastic measurement and nuclear magnetic resonance spectroscopy. Moreover, any kind of material, e.g., powders, flakes, films, fibers, and liquids, may be used. The required amount of sample is small, normally in the range of several milligrams.

Thermal analysis is suitable for studying the physical properties of lignin and its derivatives. In such studies, it provides information on molecular arrangements, phase transitions, and the interaction between lignin and low molecular weight substances such as water. In the application area, thermal analysis may be used to measure the softening temperature corresponding to the glass transition, degradation temperature, and the durability of lignin and related materials.

A weakness of thermal analysis is that it provides only general information on the molecular arrangements of polymers although it may contribute very precise data on transition and degradation temperatures. As a result, it is difficult to estimate the higher order structure of polymers on the basis of thermal analysis data alone.

References

Goring DAI (1963) Thermal softening of lignin, hemicellulose and cellulose. Pulp Pap Mag Can 64:T-517–527

Hatakeyama H, Hirose S, Hatakeyama T (1989) High-performance polymers from lignin degradation products. In: Glasser WG, Sarkanen S (eds) Lignin. Properties and materials. ACS Symposium Series 397. American Chemical Society, Washington, DC, 205–218

Hatakeyama H, Kubota K, Nakano J (1972) Thermal analysis of lignin by differential scanning calorimetry. Cellul Chem Technol 6:521–529

Hatakeyama T, Hatakeyama H (1983) Differential scanning calorimetric studies on bound water in 1,4-dioxane acidolysis lignin. Makromol Chem 184:1265–1274

Hatakeyama T, Kanetsuna H (1989) Thermal analysis of polymer samples by the round robin method. Part III. Heat capacity measurement by DSC. Thermochim Acta 146:311–316

Hatakeyama T, Nakamura K, Hatakeyama H (1982) Studies on heat capacity of cellulose and lignin by differential scanning calorimetry. Polymer 23:1801–1804

Hirose S, Hatakeyama H (1986) A kinetic study on lignin pyrolysis using the integral method. Mokuzai Gakkaishi 32:621–625

Moelwyn-Hughes EA (1961) Physical chemistry. Pergamon, Oxford, 258–296

Nakamura S, Todoki M, Nakamura K, Kanetsuna H (1988) Thermal analysis of polymer samples by a round robin method. Part I. Reprodubility of melting, crystallization and glass transition temperatures. Thermochim Acta 136:163–178

Ozawa T (1965) A new method of analyzing thermogravimetric data. Bull Chem Soc Jpn 88:1881–1886

Yoshida H, Mörck R, Kringstad KP, Hatakeyama H (1990) Kraft lignin in polyurethanes. II. Effects of the molecular weight of kraft lignin on the properties of polyurethanes from a kraft lignin-polyether triol-polymeric MDI system. J Appl Sci 40:1819–1932

Chapter 5

Characterization in Solution: Spectroscopic Methods

5.1 Ultraviolet Spectrophotometry

S.Y. Lin

5.1.1 Introduction

Ultraviolet spectrophotometry is among the most convenient and useful methods for the quantitative and qualitative analyses of lignin in solution. Because of its aromatic nature, lignin strongly absorbs ultraviolet light and exhibits characteristic maxima in the ultraviolet light region. The location and intensity of the maxima depend on the type of lignin, its chemical modifications, and on the solvent used in the photometric measurements. References relating to the variety of interpretations of ultraviolet light absorption spectra of lignin and its derivatives are published extensively in reviews (Brauns 1952, Brauns and Brauns 1960, Goldschmid 1971, Fengel and Wegener 1984).

An ultraviolet spectrum is a plot of the wavelength or frequency of absorption against the absorption intensity (transmittance or absorbance). In lignin spectroscopy, the term "difference spectrum" refers to a spectrum which is obtained by subtraction of the absorption curves of two components. A $\Delta\varepsilon$- or a ΔD-curve is commonly used in lignin chemistry. The $\Delta\varepsilon$- and ΔD-curves are obtained by plotting against wavelength the change in absorptivity, ε, in $l\,mol^{-1}cm^{-1}$ or D in $l\,g^{-1}cm^{-1}$ of a compound before and after a chemical treatment. For example, in $NaBH_4$ reduction, the difference spectrum is termed a $NaBH_4$ reduction $\Delta\varepsilon$- or ΔD-curve. Such a spectrum is useful for identification of structures having $NaBH_4$-reducible α-carbonyl groups. An ionization $\Delta\varepsilon$- or ΔD-curve is used for measurement of free phenolic hydroxyl groups in lignin.

Second derivative ultraviolet spectroscopy is a simple yet powerful technique for highlighting the fine structure of spectral curves. It involves calculating the second derivative of a spectrum with respect to wavelength and plotting the derivative rather than the spectrum itself. Second derivative absorbance spectra of lignin model compounds, milled wood lignins and lignosulfonates have been published (Lin 1982).

5.1.2 Principle of Measurement

The ultraviolet portion of the electromagnetic spectrum suitable for lignin spectroscopy extends from 200 to 380 nm. Energy absorbed in this region produces changes in the electronic energy of a lignin molecule resulting from transitions of valence electrons in the molecule. These transitions consist of the

Springer Series in Wood Science
Methods in Lignin Chemistry
(Edited by S.Y. Lin and C.W. Dence)

excitation of an electron from a filled molecular orbital (usually a bonding π-orbital) to the next higher energy orbital (an antibonding π-orbital), designated as π-π^*.

The relationship between the energy absorbed (ΔE) in an electronic transition and the frequency (ν) or wavelength (λ) of radiation producing the transition is

$$\Delta E = h\nu = hc/\lambda,$$

where h is Planck's constant and c is the velocity of light. The principal characteristics of an absorption band are its position and intensity. The position of absorption corresponds to the wavelength of radiation whose energy is equal to that required for an electronic transition. The intensity of absorption is related to two factors: the probability of interaction between the radiation energy and the electronic system raising the ground level to an excited state, and the polarity of the excited state. The intensity of absorption may be expressed as transmittance according to the relationship:

$$\text{Transmittance (T)} = I/I_o,$$

where I_o is the intensity of the radiant energy striking the sample and I is the intensity of the radiation emerging from the sample. A more convenient expression of the absorption intensity is derived from the Beer-Lambert law as represented by the formula

$$A = \log_{10} I_o/I = kcb,$$

where k = a constant characteristic of the solute, c = concentration of solute, b = path length through the sample, and A = absorbance.

Depending on the units in which concentration c and path length b are expressed, the constant k can be ε, $E^{1\%}_{1\,cm}$, or D as defined in Table 5.1.1.

For lignin model compounds and lignin samples of known molecular weight, c is expressed in moles per liter. When the path length (b) is in centimeters, the Beer-Lambert equation becomes

$$A = \varepsilon cb.$$

The term ε (in $l\,mol^{-1}cm^{-1}$) is known as the molar absorptivity. The intensity of an absorption band in the ultraviolet spectrum is usually expressed as molar absorptivity at the maximum, ε_{max} or log ε_{max}. However, the molecular

Table 5.1.1. Terms and symbols used in lignin ultraviolet spectroscopy

Term	Symbol (k)[a]	Concentration (c)[a]
Molar absorptivity	ε	mol/l
Specific absorptivity	$E^{1\%}_{1\,cm}$	g/100 ml
Specific absorptivity	D (or a)	g/l

[a] Symbol used in Beer-Lambert equation (A = kb).

weight and constitution of most lignin samples are not known and the absorptivity is often expressed on different bases. For example, the absorptivity may be expressed as

$$E_{1\,cm}^{1\%} = A/cb,$$

where c = concentration in g/100 ml, b = path length through the sample in centimeters;

or as $$D(l\,g^{-1}cm^{-1}) = A/cb,$$

where c = concentration in g/l.

The term "absorptivity" is synonymous with extinction coefficient, absorbancy index, and specific extinction (Browning 1967).

A few terms frequently used in a discussion of electronic spectra are:

Chromophore: a covalently unsaturated group responsible for electronic absorption.
Bathochromic shift: the displacement of absorption to a longer wavelength due to substituent or solvent effects.
Hypsochromic shift: the displacement of absorption to a shorter wavelength due to substituent or solvent effects.
Hyperchromic effect: an increase in absorption intensity.
Hypochromic effect: a decrease in absorption intensity.

The characteristic electronic absorption spectra of lignin and lignin-related compounds may be interpreted, in principle, on the basis of the UV absorption properties of polysubstituted benzene. Using the designations of Braude (1955), the bands in the spectra are grouped into three types: B band, E_1 band and E_2 band. The details of these bands are discussed in the section on absorption characteristics.

5.1.3 Method

5.1.3.1 Solvent Selection

Proper selection of a solvent for preparing lignin solutions is of great importance in ultraviolet spectroscopy since only true solutions yield maximum absorptivity

Table 5.1.2. Solvents for use in the ultraviolet spectroscopic measurement of lignin and lignin model compounds

Sample	Solvent
Kraft, soda, organosolv lignins	Dimethylformamide, 2-methoxyethanol/water (8:2, v/v)
Lignosulfonates	Water, ethanol/water (2:8, v/v)
Milled wood lignins	Dimethylformamide, ethanol/water (8:2, v/v), 2-methoxyethanol/water (8:2, v/v)
Lignin model compounds	Water, cyclohexane, ethanol, 2-methoxyethanol/water (8:2, v/v)

values. Any insoluble material will not contribute to light absorption and consequently will result in an underestimation of molar absorptivity. Although it is not difficult to find good solvents for lignin model compounds and low-molecular weight lignin samples, high-molecular weight, water-insoluble lignins of industrial origin may contain a small quantity of colloidal material of very low solubility. Some recommended solvents for lignins and model compounds are listed in Table 5.1.2. The reader is encouraged to search for the solvent best suited for a particular lignin sample under study.

5.1.3.2 Preparation of Solutions for Spectral Measurements

Lignin and lignin-related model compounds are subject to air oxidation and photodegradation, particularly in a dilute, alkaline solution. It is strongly recommended that a fresh solution of the sample be prepared immediately before making spectral measurements. The following precautions should be taken in the preparation of solutions:

1. Dissolve the lignin and lignin model compounds in a good solvent at room temperature; do not heat the solution to dissolve the sample.
2. Avoid exposing the solution to strong daylight or ultraviolet light; flush the solution with nitrogen if possible.
3. In preparing an alkaline solution (for ionization difference spectra measurement), the alkali is added to a neutral solution immediately before recording the spectrum. Do not stir alkaline solutions in air for an extended period of time.

5.1.3.3 Procedure for Quantitative Determination (Neutral Spectra)

This procedure is followed for quantitative measurements of lignin in an unknown sample, such as for the determination of lignin concentration in water and organic solvents.

1. Accurately weigh 0.5 g (to 0.001 g) of the lignin sample into a 1000-ml volumetric flask and dilute to the mark with the solvent. For an aqueous solution, adjust the pH to 5 with acetic acid prior to dilution.

2. Using a 1-ml pipet (0.01 subdivisions), a suitably-sized aliquot (e.g., 0.3 ml) of the above solution is transferred to a 1-cm path quartz curvette (≧3-ml capacity) and diluted ten fold (i.e., to 3.0 ml) with the same solvent used to dissolve the sample (Note 1).

3. Determine the UV light absorption in the 210 to 400 nm range in a double-beam spectrophotometer against the solvent in a matching cuvet. The absorbance reading should be between 0.2 and 0.8 at the 280 nm maximum (Note 2).

4. Calculate the concentration of lignin (c) as follows:

$$c(\%) = (10A/DW)(100),$$

Table 5.1.3. Absorptivity values (D) of lignins at 280 nm

Lignin	D ($l\,g^{-1}cm^{-1}$)	Solvent	Reference
Spruce MW	16.7	2-methoxyethanol	Fengel et al. 1981
Spruce MW	20.7	Formamide	Brown et al. 1968
Spruce MW	19.5	Dioxane	Rezanowich et al. 1963
Pine MW	18.8	2-methoxyethanol/ethanol	Sarkanen et al. 1967
Beach MW	13.3	Formamide	Brown et al. 1968
Maple MW	12.9	2-methoxyethanol/ethanol	Sarkanen et al. 1967
Poplar dioxane	12.6	Dioxane	Rezanowich et al. 1963
Spruce lignosulfonate	11.9	Water	Fengel et al. 1981
Softwood lignosulfonate	11.9	Water	Lin and Detroit 1981
Beech lignosulfonate	10.4	Water	Fengel et al. 1981
Pine kraft	24.6	Water	Fengel et al. 1981
Pine kraft	26.4	2-methoxyethanol/water (8:2, v/v)	Lin and Detroit 1981

MW (milled wood); 2-methoxyethanol (methyl cellosolve)

where A = absorbance at 280 nm, D = absorptivity in $l\,g^{-1}cm^{-1}$ from Table 5.1.3 (Note 2), W = sample weight in grams.

Note 1. Adjust the concentration of the lignin solution as needed to ensure the absorbance reading being within the specified range. Too high a concentration may result in deviation from Beer's law (linearity between absorbance and concentration).

Note 2. The absorptivity values depend on the type of lignin and solvent. It is important to use an appropriate value for the lignin sample under study. The values listed in Table 5.1.3 are for general reference only.

5.1.3.4 Procedure for Measurement of Ionization Difference Spectra

Ionization of phenolic hydroxyl groups in lignin with alkali causes a bathochromic shift and a hyperchromic effect in the absorption spectrum. An alkaline ionization difference spectrum is obtained by subtracting the spectrum of the solute in a neutral solution from the corresponding spectrum measured in an alkaline medium. The difference spectrum may be measured directly with a spectrophotometer according to the following procedure:

1. Accurately weigh 0.5 g (to 0.001 g) of the lignin sample into a 1000-ml volumetric flask and dilute with a good solvent to the mark. For an aqueous solution, adjust the pH to 5 with acetic acid prior to diluting.

2. Using a 1-ml pipet (0.01 ml subdivision), a suitably-sized aliquot (e.g., 0.3 ml) of the above solution is transferred to a 1-cm path quartz (reference) cuvette (≧3-ml capacity) and diluted 10-fold (i.e., to 3.0 ml) with the same solvent used to dissolve the sample. A second aliquot, identical in volume to the first, is similarly transferred to a matching 1 cm path quartz (sample) cuvet and diluted

tenfold with a buffer solution or an aqueous sodium hydroxide solution of sufficient alkalinity to provide a final pH of 11.0–11.5 (Note 1).

3. Record the spectrum of the sample solution against the reference solution in the 210 to 400 nm range on a double-beam spectrophotometer to obtain a difference spectrum (Note 2).

Note 1. The solvent in the reference and sample cells should contain, respectively, 20% water and 20% buffer or aqueous sodium hydroxide solution.

Note 2. Lignin in dilute alkaline solution is readily oxidized in ultraviolet light. Therefore, it is recommended that the spectrum be scanned immediately after placing the sample cuvette in the spectrophotometer.

Other difference spectra (e.g., $NaBH_4$ reduction, ethanolysis, hydrogenation, and photo-reduction) are obtained similarly by subtracting from the spectrum of chemically treated lignin in the sample cuvet, the spectrum of the untreated lignin at an equal concentration.

5.1.3.5 Second Derivative Photometry

A second derivative UV absorption spectrum of lignin can be obtained on a double-beam spectrophotometer equipped with a microcomputer (e.g., Perkin-Elmer Model 552A). The built-in microcomputer performs the electronic processing of signals from the spectrophotometer and automatically prints out the absorbance derivative with respect to wavelength. The recommended keyboard control settings are: ordinate range, 0.25; response time, 0.5 s; slit, 2 nm; scan speed, 120 nm/min. Sample solution preparation and operational procedures are similar to those described in Section 5.1.3.3.

5.1.4 Ultraviolet Absorption Characteristics of Lignins and Lignin-Related Model Compounds

5.1.4.1 Solvent Effect

Solvents exert a significant effect on the absorption maxima in the ultraviolet spectrum of lignin. This explains, to a large extent, why the spectral data reported in the literature for a particular lignin vary considerably, depending on the solvent system employed by the investigator. Typical solvent effects are illustrated by the data in Table 5.1.4. The degree of solvent effect depends on the polarity of the solvent, and on the nature of the substituents on the aromatic ring of the lignin basic unit. Polar solvents, such as water, ethanol, and methyl cellosolve, tend to cause a hypsochromic shift in the spectra of lignin units that contain electron-withdrawing substituents, whereas the same solvents bring about a bathochromic shift in the spectra of structures having electron-donating substituents. A bathochromic shift is usually accompanied by a hyperchromic effect and a hypsochromic shift by a hypochromic effect.

Table 5.1.4. Effect of solvent on the absorption maxima (λ max) of guaiacyl model compounds. (Lin et al. 1974)

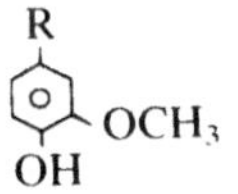

R	λ max (nm)			
	Water	Ethanol	2-Methoxyethanol	Cyclohexane
CH_2OH	278	279.5	280	–
H	275.5	277	278	282
COOH	289	292	293	293.5
CHO	314	312	310	307.5
$COCH_3$	306	305	304.5	–

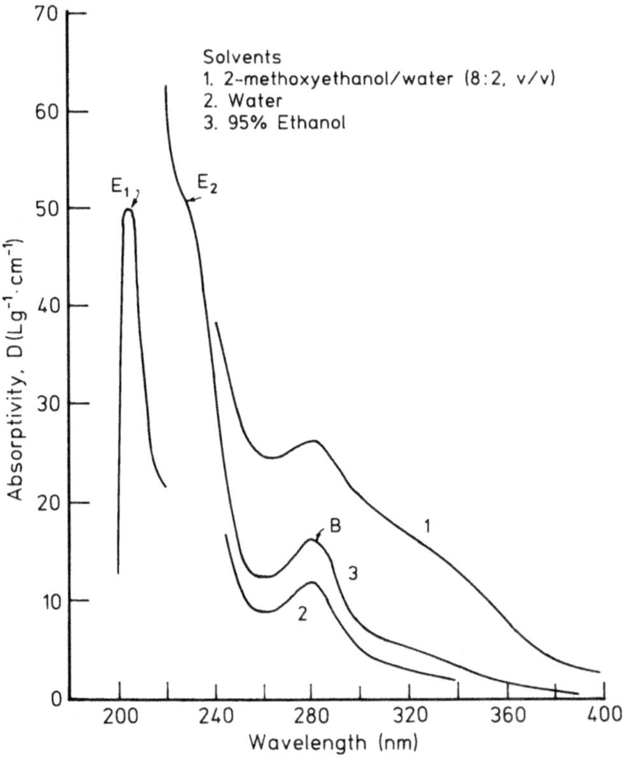

Fig. 5.1.1. Ultraviolet spectra of pine kraft lignin (*1*), spruce lignosulfonate (*2*), and milled-wood lignin (*3*)

5.1.4.2 Absorption Bands

Absorption bands in lignin spectra are typically broad and indistinct due to the interactions between electronic transition and other forms of energy transitions, namely vibrational and rotational, and the overlapping of electronic transition bands originating from numerous light-absorbing units in the polymeric lignin structure. The spectrum of softwood lignin consists of a maximum at 280 nm, a shoulder at 230 nm and a sharp peak at 200–210 nm. These bands are designated B, E_2, and E_1, respectively (see Fig. 5.1.1). The spectra of hardwood lignins, by contrast, exhibit B bands at somewhat shorter wavelengths, specifically in the 268–277 nm range. The B band absorptivity values for softwood lignins (18–21 $l\,g^{-1}cm^{-1}$) are much greater than those for hardwood lignins (12–14 $l\,g^{-1}cm^{-1}$). Chemical pulping processes cause significant changes in the light absorption properties of lignin. Thus, kraft lignin exhibits a greatly enhanced absorptivity due to stilbene, α-carbonyl and other conjugated groups formed in the pulping process. Sulfite pulping, on the other hand, has a bleaching effect on lignin, yielding lignosulfonates of lower light absorptivity. Both kraft lignin and lignosulfonate from the pulping of softwood exhibit a B band at 280 nm indicating that the two major pulping processes do not significantly alter the guaiacyl structural units which are predominant in softwood lignin.

5.1.4.3 Ionization Difference Spectra

Ionization of phenolic hydroxyl groups in lignin results in bathochromic and hyperchromic alterations of the absorption spectra maxima. These spectral changes are often used to determine the type and frequency of phenolic hydroxyl units in lignin samples.

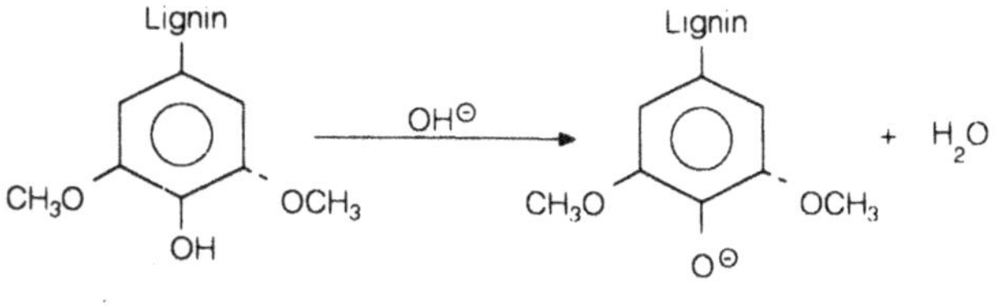

For simplicity, lignin structures are grouped into six types, each of which is characterized by a specific side chain substituent. Except for Types 5 and 6 (Fig. 5.1.2), there is a linear relationship between the difference molar absorptivity ($\Delta\varepsilon$) and the frequency (ν_m) of ultraviolet light at the maximum absorption, as expressed by the equation $\Delta\varepsilon = 1.435 \times 10^5 - 4.15\,\nu_m$ (Fig. 5.1.3). Since it is derived from extensive lignin model compound data (Lin et al. 1974), this equation is an accurate expression of the proportionality between the magnitude of the bathochromic shift and the hyperchromic effect. The few exceptions to the rule are observed in lignin structures where (1) two guaiacyl groups are conjugated via C=C double bonds as, for example, in p,p'-dihydroxystilbene units, where only one of the two phenolic hydroxyl groups is ionized, or (2) one of the phenolic hydroxyl groups in compounds

Type 1
Structures with saturated side chain

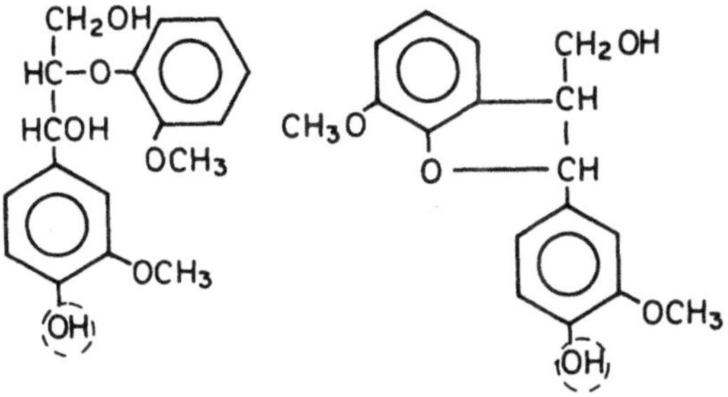

Type 2
Structures with conjugated double bond

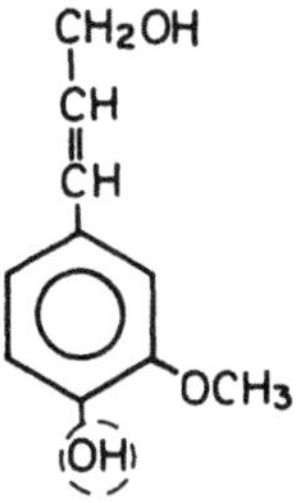

Type 3
Structures with α-carbonyl group

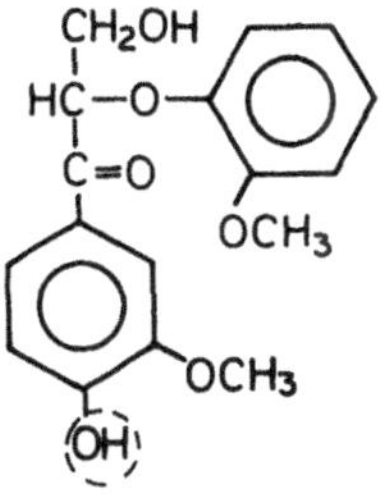

Type 4
p,p'-Stilbene

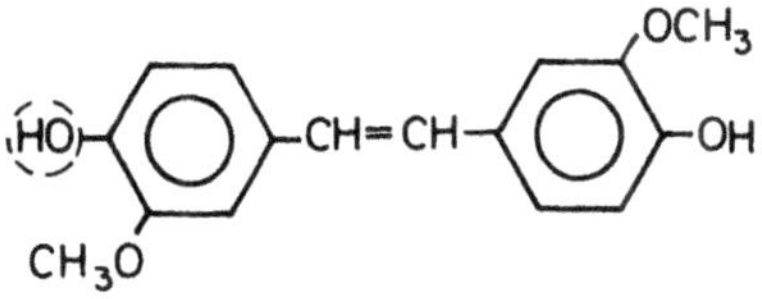

Type 5
o,p'-Stilbene

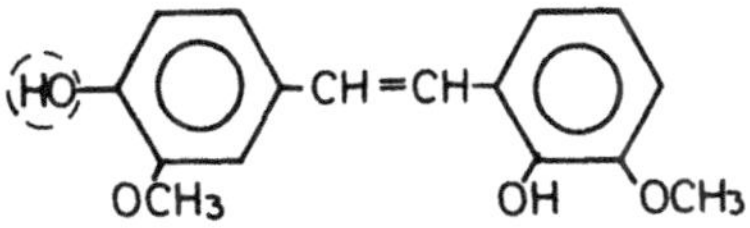

Type 6
Biphenyl structure

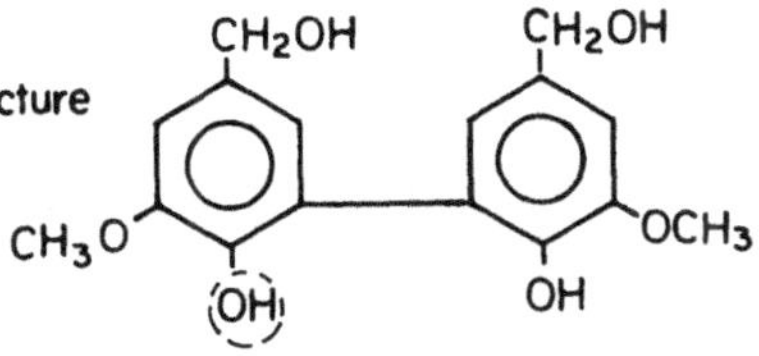

Fig. 5.1.2. Types of phenolic structures in lignin

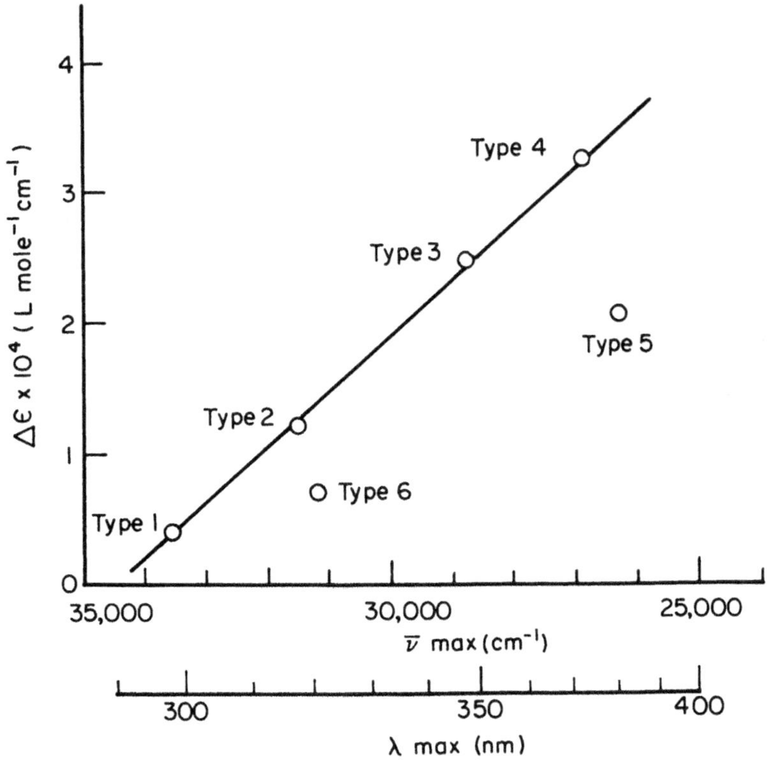

Fig. 5.1.3. The relationship between wave numbers and difference molar extinction of secondary bands in the alkaline ionization spectra of lignin-related model compounds. (Lin et al. 1974) Solvent: methyl cellosolve-water (80 : 20, v/v)

classified as Types 5 and 6 is resistant to ionization due to steric hindrance and electronic effects. In the first case, the observed difference in molar absorptivity is one half of the theoretical value, and in the second case, the observed absorptivity is less than, but more than one half of, the calculated value.

Determination of Phenolic Hydroxyl Content by Ultraviolet Difference Spectroscopy

The difference spectra in Fig. 5.1.4 have ΔD_{max} values of 12.8 and 5.3 at 250 nm and 298 nm, respectively, for spruce lignosulfonate, and 15.2, 5.7, and 8.1 at 250, 300, and 371 nm, respectively, for pine kraft lignin. Using the information available in Fig. 5.1.3, the types and quantities of phenolic hydroxyl groups are determined as follows:

Lignosulfonate

Type 1: Phenolic OH (at 298 nm) $= \Delta D/\Delta\varepsilon$

$= 5.3/4100$

$= 1.29$ mEq/g (2.19%)

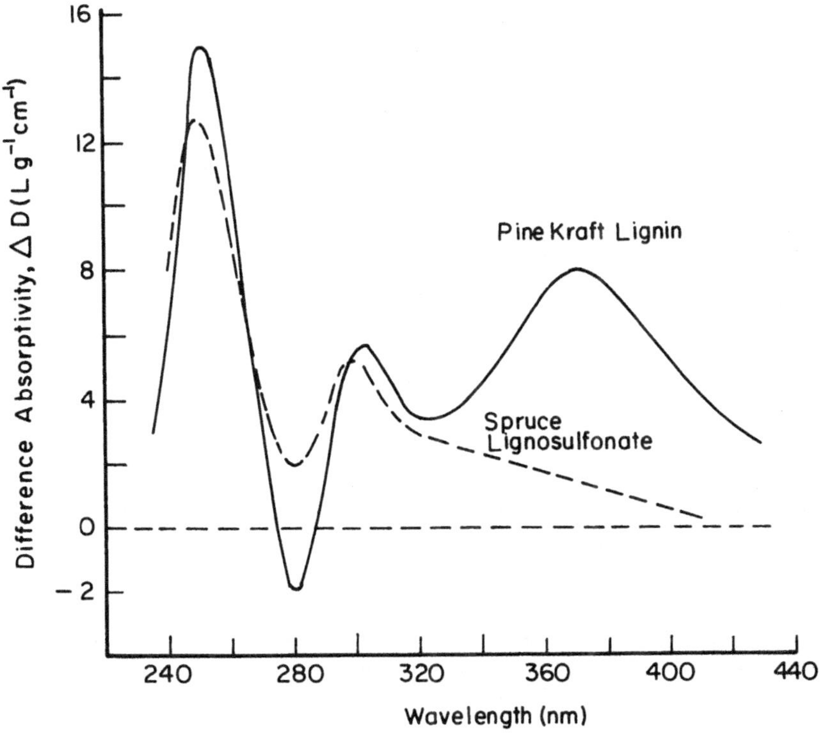

Fig. 5.1.4. Ionization difference spectra (ΔD_i-curves) of spruce lignosulfonate and pine kraft lignin. Solvents: water (lignosulfonate), methyl cellosolve-water (8:2, v/v) (kraft lignin)

Kraft lignin

Type 1: Phenolic OH (at 300 nm) = 5.7/4100
= 1.39 mEq/g (2.36%)

Type 4: Phenolic OH (at 371 nm) = 8.1/37250
= 0.22 mEq/g (0.37%).

5.1.4.4 $NaBH_4$ Reduction Difference Spectra

Lignin structures containing α-carbonyl groups are determined by $NaBH_4$ reduction difference spectra measurement. This approach is based on the spectral changes occurring on borohydride reduction of the α-keto groups, i.e., on hypsochromic and hypochromic changes in the spectrum resulting from transformations typified by the following reaction:

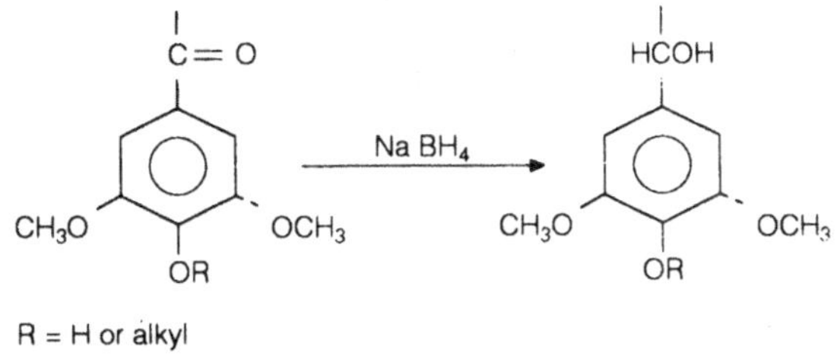

Table 5.1.5. Estimation of various conjugated carbonyl groups in spruce milled wood lignin (MWL) (Adler and Marton 1959) and kraft lignin (KL) (Marton 1964) by $NaBH_4$ reduction difference spectroscopy

Structure	Groups in 100 C_9 units	
	MWL	KL
G—CH=CHCHO	<1	0
R—O—4—G—CH=CHCHO	3	0
G—CO—R	<1	5
R—O—4—G—CO—R	5–6	0

G = guaiacyl unit
R = alkyl unit

By subtracting the absorption spectra of spruce milled wood lignin samples reduced with alkaline borohydride from those of the corresponding unreduced material, Adler and Marton (1959) obtained $NaBH_4$ reduction difference curves from which they estimated the amounts of four types of conjugated carbonyl groups (Table 5.1.5). In contrast to milled wood lignin, which contains few phenolic α-keto groups, kraft lignin examined by the same method is comprised of approximately 5% of these groups.

5.1.4.5 Second Derivative Photometry

Derivative spectrophotometry is a simple technique for emphasizing the fine structure of spectral curves. The two major advantages of the technique are improved resolution and an accentuation of spectral features. When applying the second derivative technique to lignin model compounds, milled wood lignins, and lignosulfonates, detailed spectra are obtained which show more structural features than do absorbance spectra (Lin 1982). In addition, the wavelength of absorption maxima in derivative spectra can be identified precisely. Guaiacyl model compounds and lignins display absorption bands in three regions, 200–208 nm (E_1 band), 227–233 nm (E_2 band), and 268–287 nm (B band). Syringyl structures absorb additionally in the region between the guaiacyl E_2 band and B band, specifically at 235.9, 239, 242.4, 262.6, and 267.6 nm. Conjugated systems such as those represented by stilbene, isoeugenol, α-carbonyl, and biphenyl compounds cause bathochromic shifts of the basic guaiacyl B bands by as much as 25, 28, 30, and 10 nm, respectively. Figure 5.1.5 compares the absorbance and second derivative spectra of two stilbene model compounds, 4-hydroxy-3-methoxystilbene and 4,4′-dihydroxy-3,3′-dimethoxystilbene.

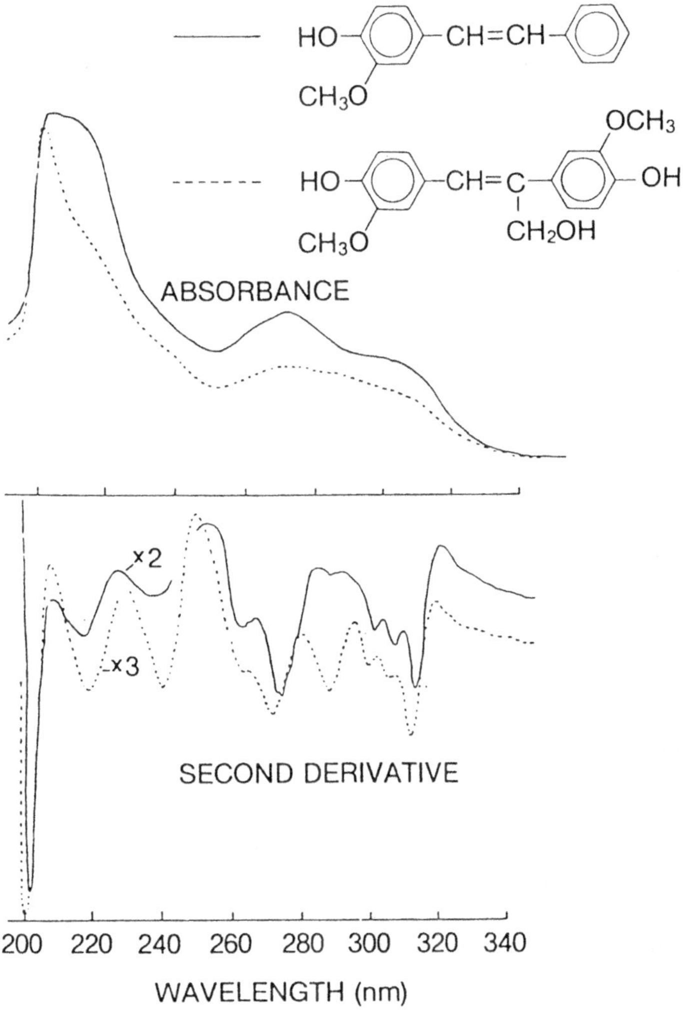

Fig. 5.1.5. Comparison of the absorbance and second derivative spectra of stilbene models. (Lin 1982)

Discrimination Effect

One general effect of plotting the second derivative of a spectrum is an accentuation of its spectral features, the so-called discrimation effect. This effect is particularly useful for the quantitative analyses of mixtures of lignin and certain low molecular weight organic compounds. As an illustration, the absorbance and second derivative spectra of mixtures of a softwood lignosulfonate and phenol at various ratios are shown in Fig. 5.1.6. In the absorbance spectra,

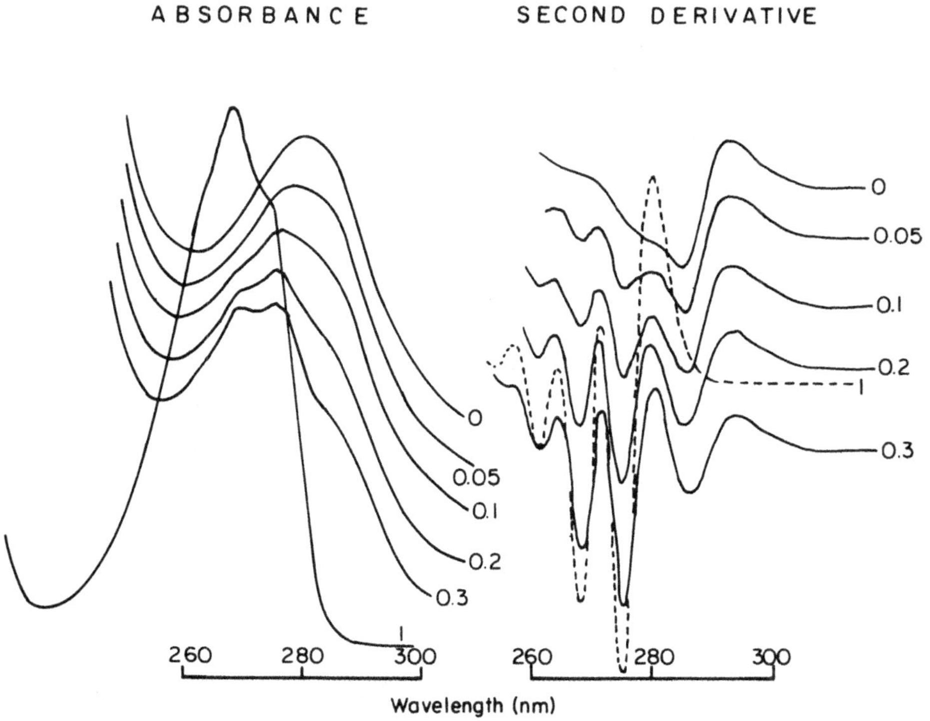

Fig. 5.1.6. Concentration of phenol in softwood lignosulfonate (weight fraction in the mixture) as revealed by second derivative spectra. (Lin 1982)

the sharp peaks of phenol are hidden in the broad absorption bands of the lignosulfonate. By contrast, the presence of phenol exerts a profound effect on the second derivative spectra of the mixture. The unique peaks for phenol can be used for identification and quantitative determination.

5.1.4.6 The Effect of Chemical Modification on Ultraviolet Absorption

Chemical modification of lignin affects its light absorption properties. As a general rule, modification reactions that introduce unsaturation into the side chain (e.g., formation of α-carbonyl groups and ring-conjugated double bonds), or that add substituent groups to the aromatic ring, cause a bathochromic shift of the absorption maxima. Conversely, reactions which result in the blocking of phenolic hydroxyl groups (e.g., reactions with ethylene oxide, epichlorohydrin and chloroacetic acid) and in the reduction of chromophoric structures (e.g., $NaBH_4$ reduction and hydrogenation), bring about hypsochromic and hypochromic changes in the spectrum. These spectral changes are useful for determining the extent to which lignin has been modified chemically.

5.1.5 Discussion

As an analytical method, ultraviolet spectrophometry is probably the most informative and convenient technique available for analyzing lignin. Sample preparation and spectral measurements are considerably simpler than those required for other spectrometric analyses (e.g., IR and NMR). Ultraviolet absorption spectra are valuable for identifying lignin source, determining lignin concentration, and for estimating the concentration of phenolic hydroxyl and other functional groups. Derivative ultraviolet spectrophotometry has many unique features that add to the benefits of ultraviolet spectrophotometry. Its fingerprinting capability is particularly useful to industrial chemists who deal with complex mixtures of lignin and synthetic organic compounds.

In spite of the advantages connected with UV spectroscopy, a few words of caution are in order. First, lignins isolated from wood (e.g., milled wood lignin, kraft lignin, and lignosulfonate) are complex, and their precise structures are not known. Therefore, the molar absorbtivities of lignin model compounds at best are only approximations of the true absorptivity of lignin. The structural variations and steric constraints typifying lignin that are not present in model compound structures undoubtedly influence the shape and intensity of absorption bands and the wavelength of maximum absorption. These factors introduce some degree of uncertainty in the use of absorption data based on model compounds for calculating lignin and functional group concentrations.

Although, the ultraviolet spectroscopy of softwood lignin and guaiacyl model compounds has been thoroughly studied, the same cannot be said about hardwood lignins. It has been shown that the absorptivity of hardwood milled wood lignins at 280 nm decreases linearly with the methoxyl/carbon ratio (Sarkanen et al. 1967), but the effects of side chain structures on the ultraviolet absorption properties of hardwood lignins have yet to be determined.

Difference spectroscopy utilizes certain spectral changes associated with modification of lignin structure to determine functional groups such as ring-conjugated carbon-carbon double bonds and α-carbonyl groups. The accuracy of this method relies heavily on whether the modification also causes structural changes other than those intended, and whether the modification of a specific functional group under study occurs quantitatively. In ionization difference spectroscopy, the overlapping of many absorption bands causes a serious error in estimating the phenolic hydroxyl content. One remedy is to eliminate all side chain unsaturation (α-carbonyl groups and conjugated carbon-carbon double bonds) before recording difference curves by converting all types of phenolic hydroxyl groups to those represented by Type 1 units.

One major drawback of second derivative spectroscopy is its inability to resolve the absorption of lignin tailing into the 300 to 400 nm region. The derivative technique is only capable of detecting major structural features of lignin and not minor functional groups. Nevertheless, its fingerprinting capability brings a unique advantage to ultraviolet spectroscopy as a qualitative analytical tool for lignin.

References

Adler E, Marton J (1959) Zur Kenntnis der Carbonyl-gruppen in Lignin. I. Acta Chem Scand 13:75–96

Braude EA (1955) Ultraviolet light absorption and the structure of organic compounds. Ann Rep Prog Chem, Chem Soc Lond 17:105–130

Brauns FE (1952) The chemistry of lignin. Academic Press New York, 217–230

Brauns FE, Brauns DA (1960) The chemistry of lignin, Suppl. Vol. Academic Press, New York, 199–219

Brown W, Cowling EB, Falkehag SI (1968) Molecular size distributions of lignins liberated enzymatically from wood. Sven Papperstidn 71:811–821

Browning BL (1967) Methods of wood chemistry, Vol. 1. Interscience, New York, 179–180

Fengel D, Wegener G (1984) Wood, chemistry, ultrastructure, reactions. Walter de Gruyter, Berlin, 157–160

Fengel D, Wegener G, Feckl J (1981) Contribution to characterization of analytical and technical lignins. (2) Physical-chemical and electron microscopical studies. Holzforschung 35:111–118

Goldschmid O (1971) Ultraviolet spectra. In: Sarkanen KV, Ludwig CH (eds) Lignin. Occurrence, formation, structure and reactions. Wiley-Interscience, New York, 241–266

Lin SY (1982) Derivative ultraviolet spectroscopy of lignin and lignin model compounds – a new analytical technique. Sven Papperstidn R162–171

Lin SY, Detroit WJ (1981) Chemical heterogeneity of technical lignins – its significance in lignin utilization. Ekman-Days 1981 (Stockholm). SPCI Rep 38, 4:44–52

Lin SY, Falkehag SI, Gierer J, Szabo-Lin I (1974) A contribution to lignin spectroscopy. 167th ACS National Meeting, Los Angeles, CA

Marton J (1964) On the structure of kraft lignin. Tappi 47:713–719

Rezanowich A, Yean WQ, Goring DAI (1963) The molecular properties of milled wood and dioxane lignins: sedimentation, diffusion, viscosity, refractive index increment and ultraviolet absorption. Sven Papperstidn 66:141–149

Sarkanen KV, Chang H-M, Allan GG (1967) Species variations in lignin. (2) Conifer lignins. (3) Hardwood lignins. Tappi 50:583–590

5.2 Fourier Transform Infrared Spectroscopy

O. FAIX

5.2.1 Introduction

Typically, lignins are characterized by IR spectroscopy in the solid state. The examination of lignins in solution has long been a domain of UV and NMR spectroscopy. However, with the advent of the FTIR technique, the IR spectra of liquid samples now can be readily obtained. The need to measure IR spectra in solution exists when there is not enough time to isolate lignin from solution as, for example, is the case when spent pulping liquors have to be analyzed rapidly during a pulping process for process control. This section describes briefly the common liquid sampling accessories and applications of solution state FTIR spectroscopy to lignins. For more details on FTIR spectroscopy, the reader is referred to Chapter 4.1.

5.2.2 Method

5.2.2.1 Transmission Cells

Sealed and semi-permanent cells are most common, but variable pathlength cells are also available. A simple version of a liquid sample cell is the demountable "Nujol" cell which is suitable for qualitative examination of pastes or viscous liquids. The use of spacers in these devices may or may not be necessary, depending on the viscosity of the sample under investigation. NaCl, KBr, CsI, and polyethylene window materials are convenient if the sample does not contain water. A drop of liquid or paste is placed on one of the windows, and the two windows are clamped with front and back metal plates. Semi-permanent cells are suitable for quantitative work and can be cleaned easily after use (Fig. 5.2.1).

The best-known window materials for liquid cells are listed below with the spectra accessibility range (in cm^{-1}) given in parenthesis: (1) NaCl (40000–590), (2) KBr (40000–340), (3) CsI (40000–200), (4) polyethylene (625–50), (5) ZnS, "Irtran-2" (17000–835), (6) ZnSe (20000–600), (7) Ge (5500–600), (8) Si (8300–660), and (9) sapphire (50000–1780). Materials (1)–(3) are not suitable for aqueous solutions because of their high hygroscopicity. ZnS and ZnSe are suitable for use with acidic and alkaline solutions. The path length spacers are made of lead and are available with 0.025, 0.05, 0.1, 0.2, 0.5, and 1.0 mm path lengths. Permanent cells have an amalgamated lead spacer between the

Springer Series in Wood Science
Methods in Lignin Chemistry
(Edited by S.Y. Lin and C.W. Dence)

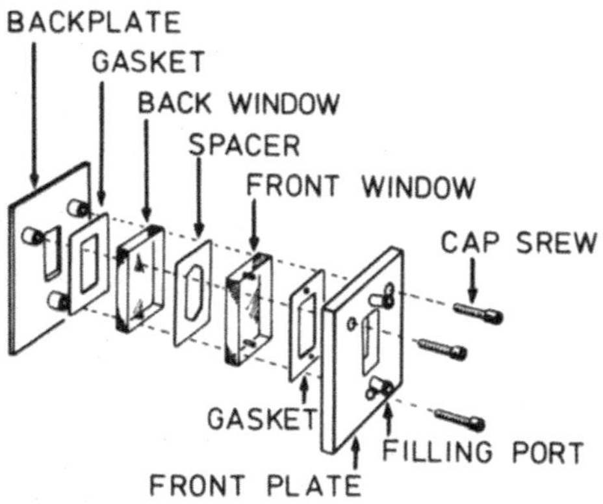

Fig. 5.2.1. Liquid sampling in the transmission mode. Construction of a semi-permanent liquid cell

windows. Large path length cells are used for solutions of low concentration. For aqueous solutions, the smallest available spacer (0.025 mm path length) should be used to avoid the excessive absorbance of water which has the effect of reducing spectral accessibility.

For demanding applications where different path lengths are required, variable path length cells are recommended. Cells with "nonrotating windows" have been designed to avoid accidental damage. The path length is adjusted to between a few μm and 6 mm micrometer screw device. Manufacturers often warrant perfect parallelism and reproducibility of path lengths.

Specialist suppliers offer variable temperature cells which operate in the range −185 to 250 °C. Micro cells with volume of 6 μl are also available. The use of microscope cells of low aperture size (3.5 × 0.5 mm) and a volume of less than 0.4 μl requires a beam condenser.

5.2.2.2 Attenuated Total Reflection (ATR)

ATR is a technique for analyzing films, powders, pastes, and liquids by the aid of an internal reflectance accessory. This technique is also referred to as *f*rustrated *m*ultiple *i*nternal *r*eflection (FMIR). The internal reflectance accessory consists essentially of a long crystal of high refractive index made of ZnSe, ZnS, Ge, KRD-5 (a mixed crystal of thallous bromide and thallous iodide), or AgCl. The crystal has a trapezoidal form or has the appearance of a parallelogram because the ends are beveled. The IR light is focused with a mirror system to one of the beveled edges with an incidence angle below the critical angle. The IR light beam is not reflected outside the crystal, and thus is forced to cross the crystal by repeated internal reflections on the inner surfaces. However, the reflection is such that the beam passes beyond the crystal's surface by up to a half wavelength. The light leaving the crystal on the other beveled end is directed toward the detector. The reflected IR beam interacts with the sample, liquid or paste, pressed against the crystal surface while slightly exceeding

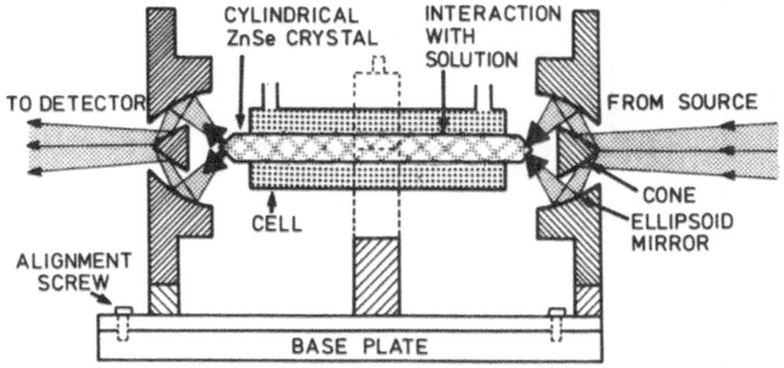

Fig. 5.2.2. Liquid sampling according to the principle of cylindrical internal reflection (cylindrical attenuated total reflection = CATR). The CIRCLE (Spectra-Tech Inc.) device

the border between the crystal and surrounding medium resulting in light absorption. The Contact Sampler (Spectra Tech, Inc.) is a horizontal ATR accessory which can be used conveniently for characterization of viscous technical lignin samples.

5.2.2.3 Circular Attenuated Total Reflection (CATR)

CATR (Spectra-Tech: CIRCLE) is a further refinement of the ATR technique for analyzing liquids and solutions of low viscosity. Here, the reflecting crystal is cylindrical with conical ends (Fig. 5.2.2). It has the optimal geometry required for the circular cross-section of the FTIR beam. The liquid material to be investigated is placed in a cell with circular cross-section surrounding the crystal. An optical design consisting of a cone and a concave mirror focuses the IR beam onto the end of the crystal, a design that provides the highest energy throughput. The cell is also suitable for measurement of dilute aqueous solutions. The working principle is the same as that described for ATR.

Several cell designs are available from the manufacturers. The open boat cell is very convenient for rapid sample changing and for cleaning. The flow-through cell is best adapted for investigations involving chemical kinetic and dynamic measurements. A standard CIRCLE equipped with ZnSe crystal may be used occasionally for analysis of spent pulping liquors. For frequent use or for use in stronger acid and alkali, commercially available Ge, Si, and AMTIR-1 (a germanium/arsenic/selenium glass crystals with accessibility between 11 000 and 625 cm^{-1}) are more suitable.

5.2.3 Liquid State FTIR Spectroscopy of Lignins

Good lignin solvents (e.g., acetone/water, dioxane/water, pyridine, ethylene chloride/ethanol, and acetic acid) exhibit strong IR bands and therefore are not commonly used for solution state IR investigations. An exception is chloroform.

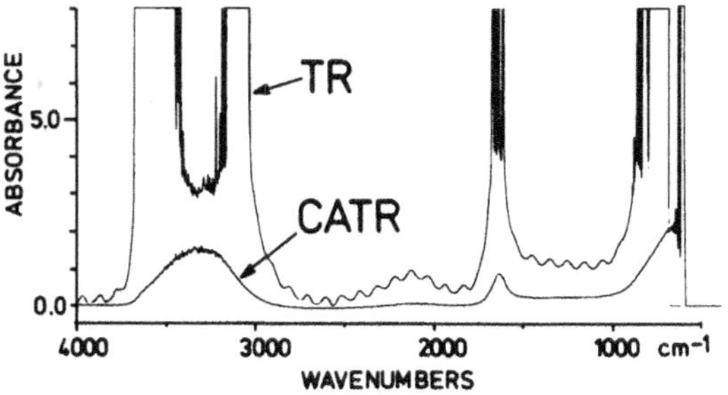

Fig. 5.2.3. Absorption spectra of water recorded in the transmission mode (TR, 0.0025 mm path length) and in CIRCLE cell (CATR, ZnSe macro rod crystals). Resolution $4\,cm^{-1}$, 32 scans

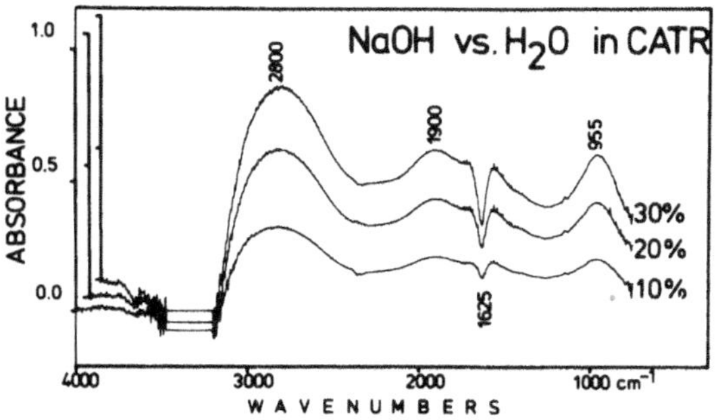

Fig. 5.2.4. Absorption spectra of alkaline solutions in a CIRCLE cell (resolution $4\,cm^{-1}$, 32 scans). The spectra are measured against water

It is a good solvent for acetylated lignins and exhibits only a small band around $1210\,cm^{-1}$ and a broad band between $780-720\,cm^{-1}$.

With grating-type instruments, it is difficult to record IR spectra in aqueous solutions. In contrast, FTIR instruments with their high energy throughput can cope well with this problem (Fig. 5.2.3) and can be used to measure IR spectra of lignosulfonates in aqueous solutions and nonsulfonated technical lignins in alkaline aqueous media. The advantage of CATR cells for aqueous solutions over TR cells can be seen in Fig. 5.2.3. The water bands between 3700–$2950\,cm^{-1}$, $1550-1750\,cm^{-1}$, and below $900\,cm^{-1}$ are much more intense for TR cells than for CATR cells. Fringes (regular waves), which are strong in the water spectrum measured in a TR cell, are absent when CATR cells are used. (Fringes are due to the reflection between the window plates which are close to each other and are unavoidable in TR cells. The distance of the waves is a measure of cell thickness.)

Sodium hydroxide solutions have a degree of accessibility for FTIR spectroscopy similar to that of water (Fig. 5.2.4). A negative band at $1625\,cm^{-1}$ (attributable to water) is typical when alkaline solutions are measured against water.

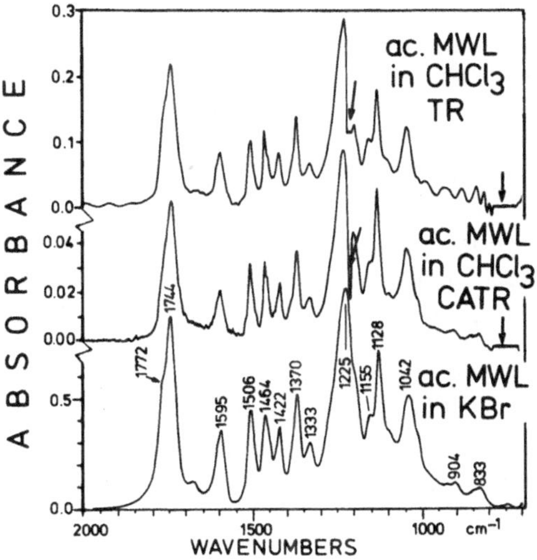

Fig. 5.2.5. Absorption spectra of acetylated olive tree (*Olea europea*) MWL in chloroform (2.5%, w/w) and in KBr pellet. The *arrows* indicate the spectral regions inaccessible because of the absorption bands of chloroform (resolution 4 cm^{-1}, 32 scans)

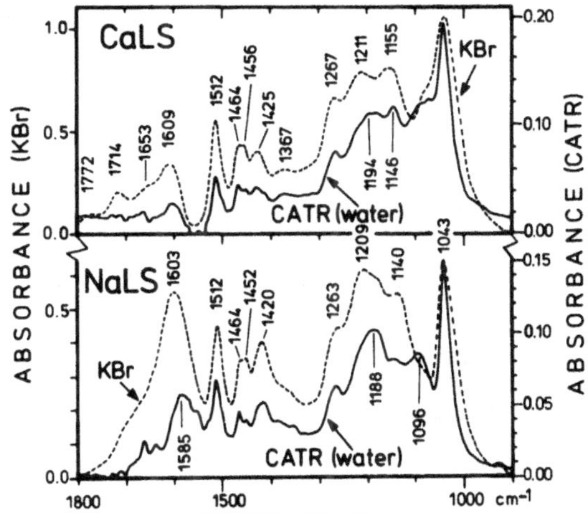

Fig. 5.2.6. Absorption spectra of calcium lignosulfonate (*CaLS*) and sodium lignosulfonate (*NaLS*) in water (10%, w/w) and in KBr pellets (resolution 4 cm^{-1}, 32 scans)

5.2.3.1 Acetylated Lignins in Chloroform

Figure 5.2.5 displays the FTIR spectra of an acetylated olive tree (*Olea europea*) MWL in chloroform in TR and CATR cells and its solid state spectrum in KBr. Nearly the whole fingerprinting region of interest is accessible in chloroform. The solution and solid state spectra are very similar. The spectra recorded in the

TR cell (0.025 mm spacer) and CATR cell (ZnSe, macro version) are also very similar. The arrows in the spectra indicate the wavelength regions inaccessible because of the absorption bands of chloroform. This range is smaller in the 1210 cm^{-1} region when the CATR cell is used instead of the TR cell. Interference fringes in the TR spectrum and the different absorbance scales are evident for the CATR cell (0–0.06), TR cell (0–0.25), and the KBr pellet (0–1). The low absorbance values in the CATR cell are caused by the interaction between solute and IR beam in a very thin layer around the internal reflection crystal. The effect is the same as using a TR cell of about one-fifth of the path length. To avoid or reduce the noise level in the CATR spectra of liquids in the 1500–1800 cm^{-1} range, the background spectrum of the solvent and the sample spectrum must be recorded under the same conditions. It is important to maintain the same purging time for the instrument before measurement of both the background and the sample, since otherwise the rotational vibrations of residual water vapor cause noise.

5.2.3.2 Lignosulfonates in Water

The spectra of calcium and sodium lignosulfonates (Fig. 5.2.6) show similar characteristics when recorded in aqueous solutions and in KBr pellets. However, the peak intensities in the CATR spectra in water (particularly in the carbonyl stretching region, including the 1600 cm^{-1} band) are greatly reduced. The prominent band for the lignosulfonates at 1040–1043 cm^{-1} ($R—SO_2—O^-$ sym.)

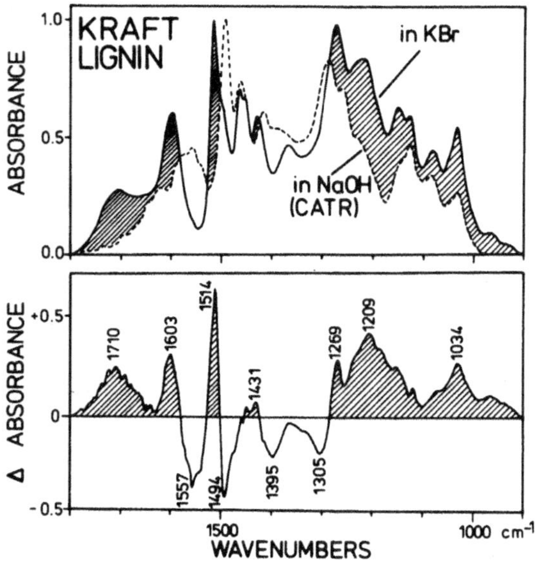

Fig. 5.2.7. Absorption spectrum of softwood kraft lignin in 2% NaOH solution (10%, w/w) and in KBr pellet. The difference spectrum below is obtained by subtraction of the alkaline solution spectrum from the KBr spectrum

exhibits a decreased bandwidth and the 1609–1603 cm^{-1} band is shifted 10–15 cm^{-1} to lower wavenumbers. The broad overlapping band at 1210 cm^{-1} in KBr now appears between 1188 and 1194 cm^{-1} in aqueous solution. These shifting bands in aqueous solutions are due to the ionic bonds between R—O^- or R—OO^- groups and cations. The good peak resolution in the FTIR spectra of the lignosulfonate samples in aqueous solutions and in KBr pellets indicates the suitability of the technique for qualitative comparisons of technical lignins.

5.2.3.3 Lignins in Alkaline Solutions

The spectra of kraft lignin in KBr pellets and in alkaline solutions show striking differences, as clearly illustrated by the difference spectrum (Fig. 5.2.7). The absorbances at 1560, 1494, 1395, and 1305 cm^{-1} are more intense, whereas the intensities in the spectral ranges 1710–1603 and 1270–900 cm^{-1}, and the aromatic skeletal vibration at 1514 cm^{-1}, are lower in NaOH solution. Similar qualitative spectral differences are observed when larch (*Larix decidua*) MWL is measured in KBr and in aqueous alkaline solutions (Fig. 5.2.8) although the magnitude of the differences is smaller with the hardwood MWL.

The spectral differences in the solid state (neutral) and solution state spectra (alkaline) are due to the presence of ionizable functional groups in lignin, e.g.,

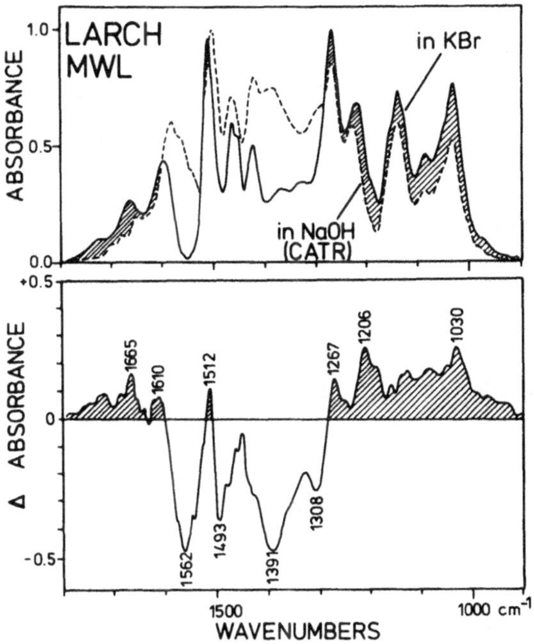

Fig. 5.2.8. Absorption spectrum of larch (*Larix decidua*) MWL in 2% NaOH solution (10%, w/w) and in a KBr pellet. The difference spectrum below is obtained by subtraction of the alkaline solution spectrum from the KBr spectrum

carboxyl and phenolic hydroxyl groups. This property can be used for determining functional groups in a manner paralleling the $\Delta\varepsilon_i$ technique used in UV spectroscopy (see Sect. 5.1).

5.2.3.4 FTIR Spectra of Pulping Liquors

A practical use of liquid state FTIR spectroscopy is in the analysis of pulping liquors. Using one of the advanced multivariate analysis techniques, the chemical composition of liquors (lignin, carbohydrate, and inorganic compounds) can be determined from the FTIR spectra. The short analysis time required by liquid state FTIR spectroscopy is a valuable asset for process control, e.g., kappa number determination (Patt et al. 1987, Welkener 1987, Faix et al. 1987, 1988, 1989, Vogel 1991). Figure 5.2.9 shows the changes in the FTIR spectra of a magnesium bisulfite liquor with cooking time. Some systematic changes in the absorption spectrum are clearly discernible.

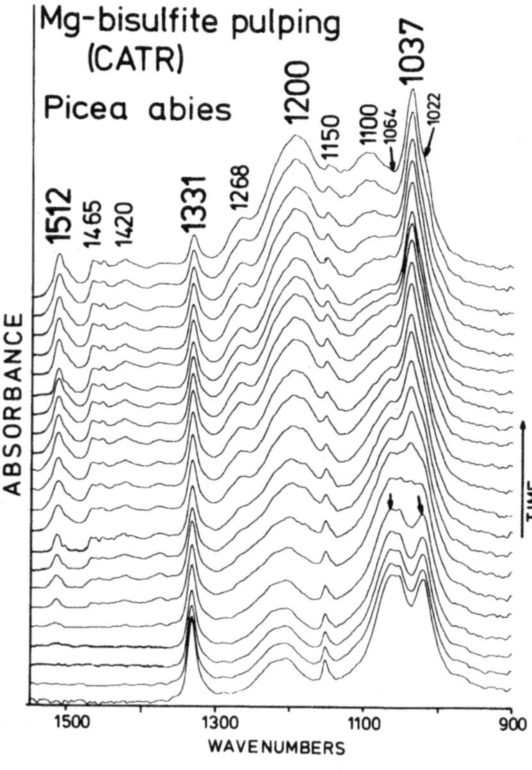

Fig. 5.2.9. Absorption spectra of spent liquors of a magnesium bisulfite pulping sequence for spruce. The spectra are measured against water (resolution 4 cm $^{-1}$, 64 scans). *1331*, *1150* free SO_2; *1200*, *1064*, *1022* HSO_3 $^{-1}$; *1037* lignosulfonates; *1512* aromatic skeletal vibrations of lignin; *other identified bands* indicate increases in the concentrations of organic constituents in the liquor

References

Faix O, Patt R, Beinhoff O (1987) Grundlagen und Anwendung von FTIR-Spektroskopie bei der Herstellung und Analyse von Zellstoffen Papier 41(12):657–667

Faix O, Schubert HL, Patt R (1989) Continuous process control of pulping by FTIR spectroscopy, In: Proc 5th Int Symp Wood Pulp Chem, Raleigh NC, 1–8 Tappi Press, Atlanta, GA

Faix O, Welkener U, Patt R (1988) Method for controlling the digestion of pulp by IR spectroscopy. US Patent 4,743,339; European patent 0,245,536 (1989)

Patt R, Faix O, Beinhoff O (1987) Kontinuierliche Aufschlusskontrolle beim Sulfitverfahren. Papier 41(10A):V7–V13

Vogel E (1991) FTIR-spektroskopische Untersuchungen in Wässriger Phase zur Koutrolle von Zellstoffgewinnungsverfahren. Dissertation, Universität Hamburg, Fachbereich Biologie

Welkener U (1987) Untersuchungen zur Einsatzmöglichkeit der FTIR-Spektroskopie für die Aufschlusskontrolle von Sulfitzellstoffkochungen. Dissertation, Universität Hamburg, Fachbereich Biologie

5.3 Proton (^{1}H) NMR Spectroscopy

K. Lundquist

5.3.1 Introduction

^{1}H NMR spectroscopy can be used for the characterization and classification of lignins and determination of lignin structure. Ludwig (1971) reviewed the literature dealing with the ^{1}H NMR spectroscopy of lignins through 1968, and a comprehensive ^{1}H NMR study of lignins and lignin model compounds was published in 1964 by Ludwig et al. (1964a,b). In this latter instance, the lignins were analyzed as acetate derivatives using deuteriochloroform as solvent. The methodology developed by Ludwig et al. (1964) has been applied in numerous studies (e.g., Šutý and Golis 1972, Košiková et al. 1973, Faix and Schweers 1974, Simionescu et al. 1981). During the 1970s and 1980s, important developments relating to the techniques used to record NMR spectra have taken place. For example, modern high-frequency instruments of the Fourier transform-type provide comparatively well-resolved ^{1}H NMR spectra. With such instruments, high signal/noise ratios and very accurate NMR data are obtained. Owing to these improvements, it is possible to enlarge regions of the spectra for closer examination and to determine trace amounts (~0.1%) of certain types of lignin units (e.g., benzaldehyde units).

Several ^{1}H NMR studies of lignins have been published in which conditions other than those used by Ludwig et al. (1964b) (acetate derivatives, chloroform solutions) have been employed. Acetylation or other derivatization procedures may cause unwanted chemical modifications of the lignin samples. This complication is avoided when underivatized lignins are examined (e.g., Bland and Menshun 1970, Gagnaire and Robert 1977, Lundquist 1981a,b, 1987, Marchessault et al. 1982, Hawkes et al. 1986). In the preferred method for ^{1}H NMR analysis of lignins (described below), the samples are examined as acetate derivatives using chloroform as a solvent. In these respects, the conditions used by Ludwig et al. (1964) have thus been retained. One advantage in doing this is that a vast amount of the reported ^{1}H NMR spectral data for lignins and lignin model compounds remains relevant for reference purposes. A modern high-frequency Fourier transform spectrometer is, however, specified in the preferred method which implies that the conditions presently used for recording the spectra differ greatly from those used by Ludwig et al. (1964).

Springer Series in Wood Science
Methods in Lignin Chemistry
(Edited by S.Y. Lin and C.W. Dence)

5.3.2 Description of Preferred Method

5.3.2.1 Acetylation

The lignin sample (~100 mg) is acetylated with 1–2 ml of acetic anhydride/pyridine (1:1, v/v) at room temperature overnight in a 50-ml flask. Ethanol (25 ml) is added, and, after 30 min, the solvents are removed by film evaporation. Repeated addition and removal (film evaporation) of ethanol (five to ten times) results in the removal of acetic acid and pyridine from the sample. The acetate is dried in vacuo over KOH and P_2O_5 or purified as described below prior to drying.

5.3.2.2 Purification of the Acetate Derivative

The acetate is dissolved in 1–2 ml of chloroform and the solution is dripped into 100 ml of magnetically stirred ether. The precipitated acetate is centrifuged off and washed with ether.

5.3.2.3 NMR Examination

Recording. The lignin acetate (50–100 mg, purified or unpurified) is dissolved in ~0.5 ml of deuteriochloroform and the solution examined by NMR spectroscopy using a spectrometer operated in the pulse Fourier mode (spectrometer frequency preferably >200 MHz). The internal reference is tetramethylsilane; (chloroform should not be used as an internal reference as the position of its signal is strongly dependent on sample concentration). Approximately 1000 scans should be made at a temperature of ~300 K. To facilitate phase correction compounds exhibiting signals at suitable positions may be added. A pulse interval of 4–5 s is required to ensure accuracy in quantitative determinations. The sweep width should cover at least 11 δ units (the spectra should be comprised of about 8 K or more data points).

Interpretation. Examples of spectra are given in Figs. 5.3.1 and 5.3.2. Peak positions, as well as interpretations, appear in Tables 5.3.1 and 5.3.2. Peak assignments in these instances are made primarily on the basis of model compound data. To obtain data with satisfactory accuracy, examination of lignin model compounds with modern high-frequency spectrometers is required. Lignin model compound data published through 1968 have been reviewed by Ludwig (1971). Among the numerous papers published later than 1968 which include ^1H NMR data for lignin model compounds in chloroform solution, the following have been selected to provide a basis for the interpretation of lignin spectra: Gellerstedt and Gierer 1971, Johansson and Miksche 1972, Glasser et al. 1973, Sarkanen and Wallis 1973, Wallis 1973, Nakatsubo and Higuchi 1975, Gellerstedt et al. 1976, Lundquist and Olsson 1977, Lundquist et al. 1979, 1987,

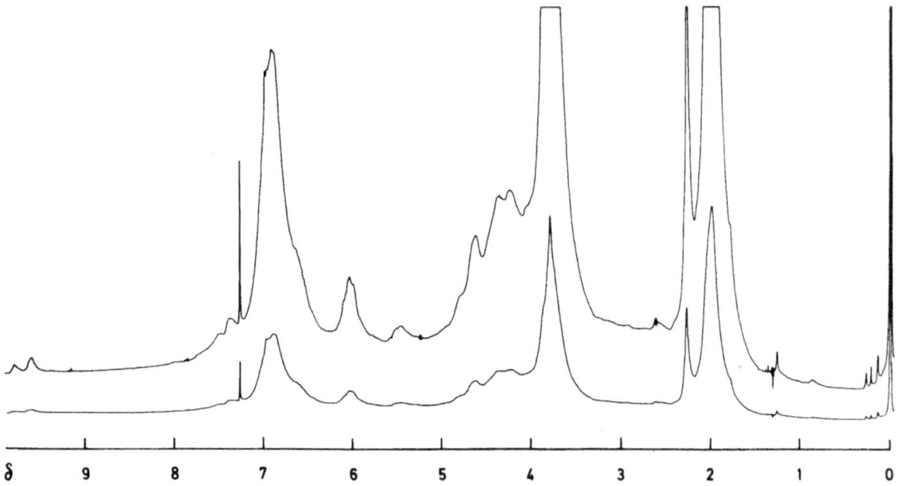

Fig. 5.3.1. ^{1}H NMR (270 MHz) spectrum of acetylated milled wood lignin from spruce. (Lundquist 1980). For peak positions and assignments of peaks, see Table 5.3.1

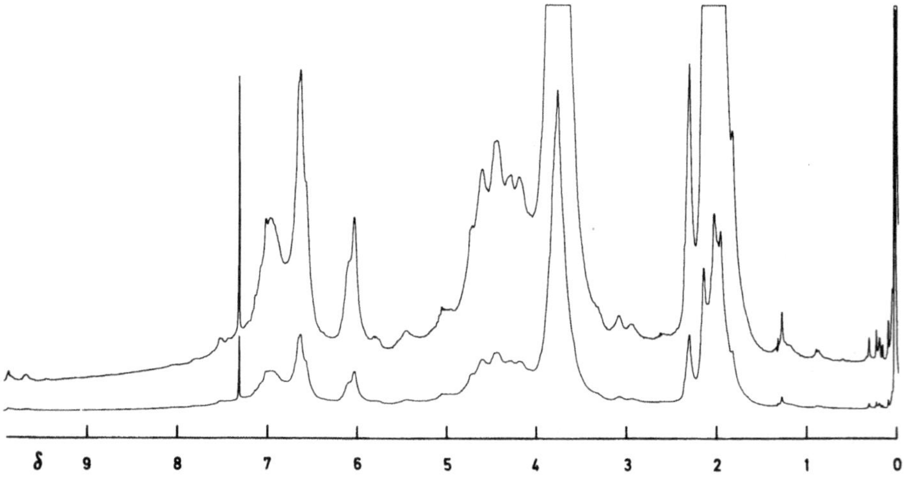

Fig. 5.3.2. ^{1}H NMR (270 MHz) spectrum of acetylated milled wood lignin from birch. (Lundquist 1979a). For peak positions and assignments of peaks, see Table 5.3.2

Lundquist 1979b, Ralph and Young 1983, Hauteville et al. 1986, Ralph et al. 1987, Brunow et al. 1989.

Integrals can be determined by the spectrometer or by measurements of peak areas on a recorded spectrum (Adler et al. 1987). When a peak of interest overlaps other peaks, it may be desirable to determine an upper and lower value for the peak area (Adler et al. 1987). The following methods have been employed to express the results of integrations in structural terms: calculations starting from the total integral of the lignin spectrum and the amount of hydrogen/

Table 5.3.1. Assignment of signals in the ^{1}H NMR spectrum of acetylated spruce lignin (Fig. 5.3.1)

δ[a]	Assignment
1.26	Hydrocarbon contaminant
2.01	Aliphatic acetate
2.28	Aromatic acetate
2.62	Benzylic protons in β-β structures of secoisolariciresinol type, benzylic protons in 3-aryl-1-propanol units
3.81	Protons in methoxyl groups
4.27	H_γ in several structures
4.39	H_γ in, primarily, β-O-4 structures (erythro forms) and β-5 structures
4.65	H_β in β-O-4 structures (methylene protons in cinnamyl alcohol units)
~4.80[b]	Inflection possibly due to H_α in pinoresinol units and H_β in noncyclic benzyl aryl ethers (threo forms).
5.49	H_α in β-5 structures (H_α in noncyclic benzyl aryl ethers, H_β in 2-aryloxypropiophenones).
6.06	H_α in β-O-4 structures (H_α in β-1 structures, H_β in cinnamyl alcohol units)
6.93	Aromatic protons (certain vinyl protons)
7.29	Chloroform (solvent)
7.41	Aromatic protons in benzaldehyde units and vinyl protons on the carbon atoms adjacent to aromatic rings in cinnamaldehyde units
7.53	Aromatic protons in benzaldehyde units
9.64	Formyl protons in cinnamaldehyde units
9.84	Formyl protons in benzaldehyde units

[a] Values refer to the highest point of the peak.
[b] Inflection includes contributions from unidentified signals (Ede et al. 1990).

phenylpropane unit (determined from analytical data) (Ludwig et al. 1964b), determinations involving the use of internal standards, e.g., 1,3,5-trinitrobenzene (Bland and Sternhell 1965) and docosane (Lundquist 1980), and calculations based on the determination of integrals attributable to aromatic protons and assumptions regarding the number of aromatic protons/phenylpropane unit (Lundquist 1979a, Adler et al. 1987).

Complementary Examinations. Peak intensity data (obtained from the spectrometer) may be of value in comparative studies (Lundquist et al. 1981). In some cases, peak assignments derived from model compound studies may be verified by decoupling experiments (Lundquist 1979a, 1980, Hauteville et al. 1986) or by the recording and examination of difference spectra (Lundquist and Olsson 1977, Brunow and Lundquist 1980, Lundquist et al. 1981, Adler et al. 1987). 2-D NMR spectroscopic methods may also be used for the same purpose (Lundquist and Stern 1989, Ede et al. 1990). Thus, it was concluded from 2-D COSY ^{1}H NMR experiments that the peak at δ 2.62 in Fig. 5.3.1 is partly due to benzylic protons in 3-aryl-1-propanol units (Lundquist and Stern 1989).

Table 5.3.2. Assignment of signals in the ^1H NMR spectrum of acetylated birch lignin (Fig. 5.3.2)

δ^a	Assignment
1.26	Hydrocarbon contaminant
1.95	Aliphatic acetate
2.02	Aliphatic acetate (including acetyl groups in xylan)
2.13	Benzylic acetate in erythro forms of β-O-4 structures of syringyl ether type
2.29	Aromatic acetate
2.94	Unknown; the peak is absent in the spectrum of $NaBH_4$-reduced and acetylated lignin
3.08	H_β in β-β structures of syringaresinol type
3.76	Protons in methoxyl groups
4.18	H_γ in several structures
4.28	H_γ in several structures
4.43	H_γ in, primarily, erythro forms of β-O-4 structures
4.60	H_β in β-O-4 structures
4.70	H_α in β-β structures (methylene protons in cinnamyl alcohol units, xylan)
5.04	Xylan
5.44	H_α in β-5 structures and noncyclic benzyl aryl ethers
5.76	Unknown
6.01	H_α in β-O-4 structures (H_α in β-1 structures, H_β in cinnamyl alcohol units)
6.60	Aromatic protons in syringyl units (certain vinyl protons)
6.94	Aromatic protons in guaiacyl units (a sharp peak at δ 6.99 has about the same intensity as the broad δ 6.94 peak) (certain vinyl protons)
7.29	Chloroform (solvent)
7.50	Aromatic protons located ortho to carbonyl groups
9.67	Formyl protons in cinnamaldehyde units
9.86	Formyl protons in benzaldehyde units

[a] Values refer to the highest point of the peak.

5.3.3 Discussion

5.3.3.1 Comments on the Preferred Method

Several methods describing the acetylation of lignin have been published (e.g., Bland and Sternhell 1965, Becker and Nimz 1974, Lundquist and Olsson 1977, Mörck and Kringstad 1985). Unwanted modifications of the lignin probably occur to some extent during acetylation. Thus, α-aryloxyacrolein structures in lignin acetates may be products of glyceraldehyde-2-aryl ether structures in the original lignin (Lundquist and Olsson 1977). The sample should be soluble in the acetylation mixture; if not, the addition of pyridine may help to achieve complete acetylation. It has been reported that the number of acetate groups introduced increases when 4-dimethylaminopyridine is present in the reaction mixture (Månsson 1983).

Chemical shifts in the ^1H NMR spectra of lignin acetates in chloroform solutions are relatively insensitive to sample concentration and are only slightly affected by moderate temperature changes. According to model compound studies, changes in chemical shifts are at most 0.01 δ unit when the temperature is varied in the range 295–305 K.

The peaks in the spectra shown in Figs. 5.3.1 and 5.3.2 are very broad, as is typically found for the ^{1}H NMR spectra of lignins. Whether this is merely due to chemical shift variation or is also related to the polymeric nature of the sample is a question which deserves further attention.

Examples of the use of ^{1}H NMR spectroscopy for the characterization of lignin acetates in chloroform solutions have been described by Becker and Nimz (1974), Kirk and Chang (1975), Lundquist et al. (1979, 1980), Hemmingson (1983), Lapierre et al. (1982), Bardet et al. (1985), Hagström-Näsi and Sjöström (1987). An example of the use of this technique for the elucidation of a specific structural problem is the analysis of the distribution of threo and erythro forms of β-O-4 structures in lignins (Lundquist 1980, Hauteville et al. 1986, Tollier et al. 1986). It is noteworthy that the ^{1}H NMR properties of lignin model compounds also have been the subject of theoretical treatments (Liptaj et al. 1980).

5.3.3.2 Alternative Conditions for the Recording of Spectra

^{1}H NMR studies of lignin acetates in acetone (Lundquist et al. 1983) and pyridine (Lundquist and von Unge 1986) have been described. Changes in chemical shifts due to solvent effects may provide structural information about the lignins which cannot be derived from examination of lignin acetates in chloroform solution alone (for examination of lignin acetates in chloroform solutions, see Lundquist 1979a, 1980). Further, it is possible to obtain a considerable amount of structural information by recording the spectra of lignin acetates in a series of different solvents (e.g., chloroform, acetone, and pyridine).

As pointed out above, derivatization may cause unwanted chemical modifications of a lignin sample. This risk constitutes an argument for the use of underivatized lignins in ^{1}H NMR studies. Hydroxyl groups are present in such samples, however, and it is well known that there are difficulties associated with the recording of signals from hydroxyl group protons. Moreover, almost all the signals from side chain protons attached to carbon atoms are found in the neighborhood of the large methoxyl peak (near δ 4) in the spectra of underivatized lignins which makes it difficult to interpret the spectra in structural terms. Examples which illustrate the potential of ^{1}H NMR examinations of underivatized lignins have been reported by Bland and Menshun (1970), Gagnaire and Robert (1977), Lundquist (1981a,b, 1987), Marchessault et al. (1982), and Hawkes et al. (1986).

References

Adler E, Brunow G, Lundquist K (1987) Investigation of the acid-catalyzed alkylation of lignins by means of NMR spectroscopic methods. Holzforschung 41:199–207

Bardet M, Robert DR, Lundquist K (1985) On the reactions and degradation of the lignin during steam hydrolysis of aspen wood. Sven Papperstidn 88:R61–67

Becker H, Nimz H (1974) Untersuchungen des Lignins der Mistel (Viscum album L.) in Abhängigkeit von der jeweiligen Wirtspflanze. Z Pflanzenphysiol 72:52–63

Bland DE, Menshun M (1970) The composition and reactive groups of the lignin of Eucalyptus regnans. Appita 23:427–433

Bland DE, Sternhell S (1965) Estimation of aromatic protons in methanol lignins of Pinus radiata and Eucalyptus regnans from proton magnetic resonance spectra. Aust J Chem 18:401–410

Brunow G, Lundquist K (1980) Comparison of a synthetic dehydrogenation polymer of coniferyl alcohol with milled wood lignin from spruce, using ^{1}H NMR spectroscopy. Pap Puu 62:669–671

Brunow G, Sipilä J, Mäkelä T (1989) On the mechanism of formation of non-cyclic benzyl ethers during lignin biosynthesis. Holzforschung 43:55–59

Ede RM, Brunow G, Simola LK, Lemmetyinen J (1990) Two-dimensional ^{1}H-^{1}H chemical shift correlation and J-resolved NMR studies on isolated and synthetic lignins. Holzforschung 44:95–101

Faix O, Schweers W (1974) Vergleichende Untersuchungen an Polymermodellen des Lignins (DHP's) verschiedener Zusammensetzungen. Holzforschung 28:179–185

Gagnaire D, Robert D (1977) A polymer model of lignin (DHP) ^{13}C selectively labelled at the benzylic positions: synthesis and NMR study. Makromol Chem 178:1477–1495

Gellerstedt G, Gierer J (1971) The reactions of lignin during acidic sulphite pulping. Sven Papperstidn 74:117–127

Gellerstedt G, Gierer J, Pettersson E-L, Szabo-Lin I (1976) Über die Reaktionen des Lignins während der Neutralsulfitkochung, VI Reaktionen von Zwischenprodukten des Stilbentyps. Liebigs Ann Chem 1976:2021–2036

Glasser WG, Gratzl JS, Collins JJ, Forss K, McCarthy JL (1973) Lignin. XVI. Synthesis, nuclear magnetic resonance and mass spectroscopy of several monomeric and dimeric lignin-like sulfonates. Macromolecules 6:114–128

Hagström-Näsi C, Sjöström E (1987) Oxidative pretreatment of pine wood prior to alkaline pulping. Part II. Characterization of lignin by spectroscopic methods. Holzforschung 41:39–45

Hauteville M, Lundquist K, von Unge S (1986) NMR studies of lignins. 7. ^{1}H NMR spectroscopic investigation of the distribution of erythro and threo forms of β-O-4 structures in lignins. Acta Chem Scand B40:31–35

Hawkes GE, Smith CZ, Utley JHP, Chum HL (1986) Key structural features of acetone-soluble phenol-pulping lignins by ^{1}H and ^{13}C NMR spectroscopy. Holzforschung Suppl 40:115–123

Hemmingson JA (1983) The structure of lignin from Pinus radiata exploded wood. J Wood Chem Technol 3:289–312

Johansson B, Miksche GE (1972) Über die Benzyl-arylätherbindung im Lignin II. Versuche an Modellen. Acta Chem Scand 26:289–308

Kirk TK, Chang H-m (1975) Decomposition of lignin by white-rot fungi II. Characterization of heavily degraded lignins from decayed spruce. Holzforschung 29:56–64

Košiková B, Polčin J, Joniak D (1973) NMR Studies of lignin-carbohydrate complexes. Holzforschung 27:59–64

Lapierre C, Lallemand JY, Monties B (1982) Evidence of poplar lignin heterogeneity by combination of ^{13}C and ^{1}H NMR spectroscopy. Holzforschung 36:275–282

Liptaj T, Remko M, Polčin J (1980) Analysis of ^{1}H-NMR spectra of cinnamaldehyde type model substances of lignin. Collect Czech Chem Commun 45:330–334

Ludwig CH (1971) Magnetic resonance spectra. In: Sarkanen KV, Ludwig CH (eds) Lignins – occurrence, formation, structure and reactions. Wiley, New York, 299–344

Ludwig CH, Nist BJ, McCarthy JL (1964a) Lignin. XII. The high resolution nuclear magnetic resonance spectroscopy of protons in compounds related to lignin. J Am Chem Soc 86:1186–1196

Ludwig CH, Nist BJ, McCarthy JL (1964b) Lignin. XIII. The high resolution nuclear magnetic resonance spectroscopy of protons in acetylated lignins. J Am Chem Soc 86:1196–1202

Lundquist K (1979a) NMR studies of lignins. 2. Interpretation of the ^{1}H NMR spectrum of acetylated birch lignin. Acta Chem Scand B33:27–30

Lundquist K (1979b) NMR studies of lignins. 3. ^{1}H NMR spectroscopic data for lignin model compounds. Acta Chem Scand B33:418–420

Lundquist K (1980) NMR studies of lignins. 4. Investigation of spruce lignin by ^{1}H NMR spectroscopy. Acta Chem Scand B34:21–26

Lundquist K (1981a) NMR studies of lignins. 5. Investigation of non-derivatized spruce and birch lignin by ^{1}H NMR spectroscopy. Acta Chem Scand B35:497–501

Lundquist K (1981b) Analysis of lignins by spectral methods. Proc. The Ekman Days Int Symp on Wood and Pulping Chem Stockholm 1:81–85

Lundquist K (1987) On the occurrence of β-1 structures in lignins. J Wood Chem Technol 7:179–185

Lundquist K, Olsson T (1977) NMR studies of lignins. 1. Signals due to protons in formyl groups. Acta Chem Scand B31:788–792

Lundquist K, Paterson A, Ramsey L (1983) NMR studies of lignins. 6. Interpretation of the ^{1}H NMR spectrum of acetylated spruce lignin in a deuterioacetone solution. Acta Chem Scand B37:734–736

Lundquist K, Simonson R, Tingsvik K (1979) On the occurrence of carbohydrates in milled wood lignin preparations. Sven Papperstidn 82:272–275

Lundquist K, Simonson R, Tingsvik K (1980) Studies on lignin carbohydrate linkages in milled wood lignin preparations. Sven Papperstidn 83:452–454

Lundquist K, Simonson R, Tingsvik K (1981) Investigation of lignins from soda and soda/anthraquinone cooking of spruce wood. Pap Puu 63:709–712, 734

Lundquist K, Stern K (1989) Analysis of lignins by ^{1}H NMR spectroscopy. Nord Pulp Pap Res J 4:210–213

Lundquist K, Stomberg R, von Unge S (1987) Stereochemical assignment of the threo and erythro forms of 2-(2,6-dimethoxyphenoxy)-1-(3,4-dimethoxyphenyl)-1,3-propanediol from X-ray analyses of the synthetic intermediates (Z)-2-(2,6-dimethoxyphenoxy)-3-(3,4-dimethoxyphenyl)-2-propenoic acid and threo-2-(2,6-dimethoxyphenoxy)-3-(3,4-dimethoxyphenyl)-3-hydroxypropanoic acid. Acta Chem Scand B41:499–510

Lundquist K, von Unge S (1986) NMR studies of lignins. 8. Examination of pyridine-d_5 solutions of acetylated lignins from birch and spruce by ^{1}H NMR spectroscopy. Acta Chem Scand B40:791–797

Månsson P (1983) Quantitative determination of phenolic and total hydroxyl groups in lignin. Holzforschung 37:143–146

Marchessault RH, Coulombe S, Morikawa H, Robert D (1982) Characterization of aspen exploded wood lignin. Can J Chem 60:2372–2382

Mörck R, Kringstad KP (1985) ^{13}C-NMR spectra of kraft lignins. II. Kraft lignin acetates. Holzforschung 39:109–119

Nakatsubo F, Higuchi T (1975) Synthesis of 1,2-diarylpropane-1,3-diols and determination of their configurations. Holzforschung 29:193–198

Ralph J, Ede RM, Robinson NP, Main L (1987) Reactions of β-aryl lignin model quinone methides with anthrahydroquinone and antranol. J Wood Chem Technol 7:133–160

Ralph J, Young RA (1983) Stereochemical aspects of lignin model β-aryl ether quinone methide reactions. J Wood Chem Technol 3:161–182

Sarkanen KV, Wallis AFA (1973) Oxidative dimerizations of (E)- and (Z)-isoeugenol (2-methoxy-4-propenylphenol) and (E)- and (Z)-2,6-dimethoxy-4-propenylphenol. JCS Perkin I 1973:1869–1878

Simionescu CI, Dragovova R, Kusmanova D (1981) NMR-Spektroskopische Untersuchungen von Dioxanligninfraktionen. Cellul Chem Technol 15:455–464

Šutý L, Golis E (1972) Chemical structure of beech lignin and its changes during water hydrolysis. Cellul Chem Technol 6:485–491

Tollier M-T, Monties B, Lapierre C, Hervé du Penhoat C, Rolando C (1986) Inhomogeneity of angiosperm lignin: Comparison of the monomeric composition of lignin fractions isolated from different wood species. Holzforschung Suppl 40:75–79

Wallis AFA (1973) Oxidation of (E) and (Z)-2,6-dimethoxy-4-propenylphenol with ferric chloride – a facile route to the 2-aryl ethers of 1-arylpropane-1,2-diols. Aust J Chem 26:585–594

5.4 Carbon-13 Nuclear Magnetic Resonance Spectrometry

D. ROBERT

5.4.1 Introduction

Nuclear magnetic resonance spectrometry (NMR) has progressed at a vigorous pace since the introduction of the first commercial spectrometers in the early 1960s, in part due to the introduction of superconducting magnets which can maintain stable higher fields than electromagnets. In parallel, microprocessor development for signal acquisition and data manipulation, coupled with a deeper and evolving understanding of magnetic resonance phenomena by NMR investigators, have led to invaluable methods which spurred structural determinations on numerous natural products and polymers (Snyder et al. 1989). The Fourier transform method applied to ^{13}C NMR, in combination with other spectrometric measurements, has increased our knowledge of the structure of organic molecules in general, and lignins in particular.

NMR parameters such as chemical shifts (δ), coupling constants (J), relaxation times, and signal intensities are closely related to electronic structure and chemical environments of nuclei involved in the resonance phenomenon. NMR is an appropriate tool for structural investigation: it reflects with high accuracy chemical structures, functionalities, and the nature of chemical bonding and environment in the lignin macromolecule. Lignin analyses using chemical methods often involve degradation of the polymer. These methods suffer from the drawback that both the lignin and resulting degradation products may undergo structural changes other than those generally associated with the procedure. Furthermore, the isolation of degradation products is time-consuming and product yield may be low. Thus, physical methods having as high a sensitivity as possible are preferred. Carbon-13 NMR spectrometry is one of the most powerful techniques available for lignin analyses, in particular for determining minor structures not amenable to analysis by other methods.

In the late 1970s, when ^{13}C NMR became available, about 40 signals could be distinguished in ^{13}C NMR lignin spectra using broad band proton decoupling which collapses ^{13}C multiplets and produces one sharp signal for every chemically unique carbon atom. Since carbon atoms from the backbone of organic molecules, ^{13}C NMR signal locations depend much more on structural environment and substitution patterns than those in ^{1}H NMR; the ^{13}C spectrum contains more structural information over a much wider chemical shift range. The ^{13}C NMR spectrum of lignin can be divided into three main segments: the first (200 to 165 ppm) contains signals assignable to carbonyl carbons; the second (165 to

Springer Series in Wood Science
Methods in Lignin Chemistry
(Edited by S.Y. Lin and C.W. Dence)

100 ppm) is ascribable to aromatic and olefinic carbons; and the third (100 to 10 ppm) is attributed to aliphatic carbon atoms (Nimz et al. 1981).

5.4.2 Principles of Pulsed Fourier Transform ^{13}C NMR

The principles of ^{13}C NMR spectrometry are described in many textbooks, some of which are listed in the references. A glossary of terms and symbols used in the following description is found below.

5.4.2.1 Glossary of Terms and Symbols

θ_1	Flip angle, radians or degrees
γ	Magnetogyric ratio, $T^{-1}s^{-1}$
δ	Chemical shift, ppm
μ	Magnetic moment
ν_0	Larmor frequency, MHz
ν_1	Frequency of alternating rf field (B_1), MHz
$\Delta\nu_{1/2}$	Linewidth at one-half peak amplitude, Hz
AT	Acquisition time, s
σ	Magnetic shielding constant
B_0	Static magnetic field of strength B_0, T
B_1	Radiofrequency field of strength, B_1, G
DW	Dwell time, μs
ΔE	Energy difference between two energy levels
FID	Free induction decay
h	Planck's constant
I	Nuclear spin quantum number
J	Spin-spin coupling constant, Hz
k	Boltzmann constant
M	Nuclear magnetization vector
m	Total magnetic quantum number for a spin system
NP	Number of data points
$N\alpha$, $N\beta$	Spin state populations
NOE	Nuclear Overhauser effect
p	Angular momentum
R	Digital resolution Hz/pt
S	Signal intensity
SW	Spectral width, Hz
T	Absolute temperature, K
T_1	Spin-lattice relaxation time constant, s
T_2	Spin-spin relaxation time constant, s
t_p	Pulse length, μs
t_1	Evolution time, s, in 2-D NMR
t_2	Acquisition time, s, in 2-D NMR

5.4.2.2 Basic Principles and Elements of an NMR Experiment

Nuclear spin, characterized by a spin quantum number I, is a quantum mechanical property possessed by some nuclei. This rotation of the nucleus around its spinning axis gives rise to an angular momentum, p, whose value is defined by

$$p = mh/2\pi \tag{1}$$

where h is the Planck's constant and m the total quantum magnetic number, with 2I + 1 allowed values, I, I-1, I-2, . . . -I. The value of I is characteristic of a particular isotope (I = 1/2 for ^{1}H, ^{13}C, ^{31}P, and ^{15}N; I = 1 for ^{2}H and ^{14}N, and I = 0 for ^{12}C and ^{16}O). Nuclei with I = 0 and p = 0 are not suitable for an NMR experiment. If $I \neq 0$, the spinning nucleus assumes a magnetic moment, μ, that is aligned with p such that

$$\mu = \gamma p, \tag{2}$$

where γ is the magnetogyric ratio, a characteristic constant of each nucleus. When exposed to a static magnetic field, B_0, along the z direction, the net result of the torque exerted by B_0 on the spinning nucleus is a precession movement of μ and p about B_0, known as Larmor precession, whose frequency,

$$\nu_0 = \gamma(B_0/2\pi). \tag{3}$$

(Fig. 5.4.1a).

The interaction energy between μ and B_0 is

$$E = -\mu B_0. \tag{4}$$

Following quantum mechanical rules, a nucleus with spin I may occupy (2I + 1) different energy levels when placed in a magnetic field. For ^{13}C nuclei, I = 1/2,

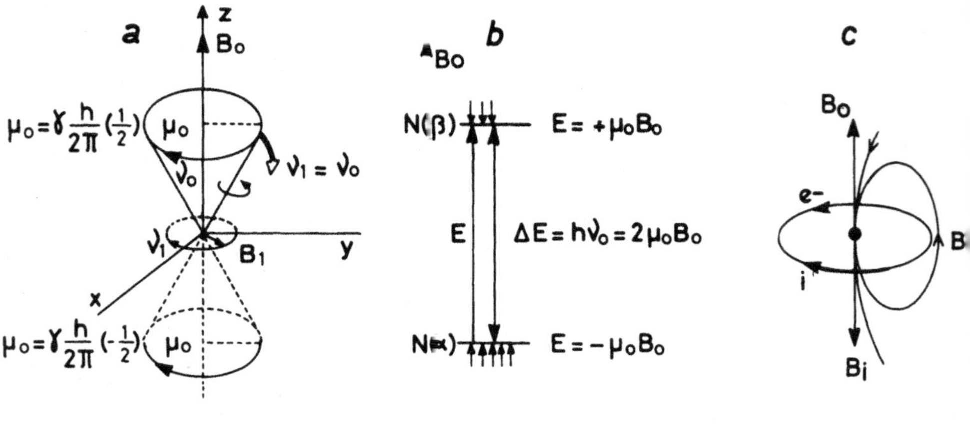

Fig. 5.4.1a–c. Basic NMR experiment. **a** Action of rf field B_1 (ν_1) on a nucleus (I = 1/2) precessing about B_0 with the Larmor frequency ν_0. When $\nu_1 = \nu_0$: change occurs in the orientation of μ (+1/2 to −1/2). **b** Characterization of the two possible spin states, +1/2 and −1/2, by their energy levels $-\mu_0 B_0$ and $\mu_0 B_0$, their energy difference, ΔE, and their spin populations, Nα and Nβ. **c** Electronic shielding of the nucleus: circulation of electrons under the influence of B_0, and the resulting secondary magnetic field B_i

and it follows from Eqs. (1), (2), (3), and (4) that the energy difference between two energy levels is

$$\Delta E = 2\mu_0 B_0 = \gamma(h/2\pi)B_0 \quad (5)$$

(Fig. 5.4.1b). If the nucleus is supplied with sufficient energy to fulfill the Bohr frequency condition or resonance condition, $\Delta E = h\nu_0$, a transition becomes possible. This condition can be achieved by applying an alternating radio frequency field (rf), B_1, perpendicular to B_0 and rotating in the xy plane with a frequency, ν_1 equal to ν_0. In this situation, $\nu_1 = \nu_0 = (\gamma/2\pi)B_0$ [Eq. (3)] defines the resonance condition. As the nucleus absorbs energy, its magnetic moment μ rotates away from B_0 toward the xy plane (Fig. 5.4.1a).

In collections of identical individual nuclei, there is a very slight excess of nuclei in the lower state, on the order of 1 in 10^5 at 25 °C, such that

$$N\beta/N\alpha = \exp(-\Delta E/kT), \quad (6)$$

where $N\beta$ and $N\alpha$ are the populations of the upper and lower spin energy states, respectively, k the Boltzmann constant, T the absolute temperature, and ΔE is about 10^{-2} cal/mol in a magnetic field of 1.4 to 6 Tesla. In practice, the macroscopic magnetization, M, is considered, which is the resultant sum of the individual μ, and arises from the slight excess of nuclei aligned along B_0 (Fig. 5.4.1b).

5.4.2.3 Pulsed NMR in the Rotating Frame of Reference

The interaction of the field B_1 with M is most easily seen in a system of coordinates x'y'z' (Fig. 5.4.2b) which rotates around B_0 at the Larmor fre-

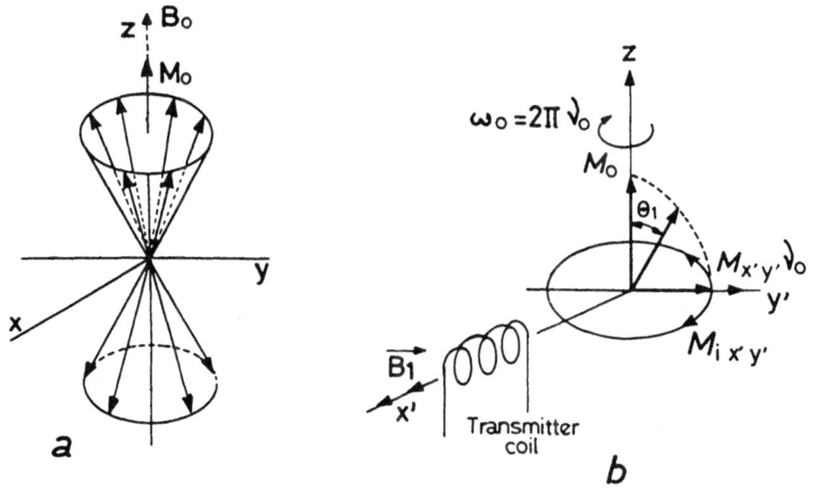

Fig. 5.4.2. a Macroscopic resultant magnetization M_0 at equilibrium. **b** At resonance the interaction of the rf pulse B_1 with M_0 produces, in the rotating frame, a tipping of M_0 through an angle θ about B_1

quency ν_0. For an observer in this system, in the absence of B_1, M is static and parallel to B_0. B_1 itself becomes static in the x'y' plane along the x' axis. If B_1 is switched on for a time t_p, the response of M is to rotate around B_1 in the y'z' plane, by an angle θ and with frequency ν_1 such that

$$\theta \text{ (in radians)} = 2\pi\nu_1 t_p = \gamma B_1 t_p \quad (7)$$

(Fig. 5.4.2b). Theta is designated the pulse or flip angle. The receiver coil can detect only the M components in the x'y' plane, along the y' axis.

Modern NMR spectrometers use pulsing techniques to overcome the drawbacks inherent in the low sensitivity and natural abundance (1.1%) of the ^{13}C nucleus. The B_1 field is applied as a brief (<50 μs) but intense rf pulse having sufficient power that during, time t_p, it simultaneously excites the frequency range $\nu_0 \pm 1/t_p$. The effective B_1 power is such that

$$\gamma B_1/2\pi \gg SW, \quad (8)$$

where SW is the spectral width, the frequency range for the nuclei under observation.

5.4.2.4 Relaxation Times

After pertubation, the system returns to its equilibrium state via two relaxation processes:

1. Spin-lattice or longitudinal relaxation: M_0 is restored along z in a process having the first-order time constant T_1 (the spin-lattice relaxation time). Due to motions of the nuclei, random fluctuations of local magnetic fields occur which allow transfer of the resonance energy from the nuclei to the lattice.

2. Spin-spin or transverse relaxation: because of dipole-dipole interactions between spins, each magnetic moment vector splits into components which dephase in the x'y' plane; the resultant vector fans out and decays exponentially with the time constant T_2. This phenomenon is responsible for the signal line width, which depends also on magnetic field inhomogeneities. Line width, $\Delta\nu_{1/2}$, is measured at one-half the peak height.

5.4.2.5 Acquisition of the Free Induction Decay (FID) and Fourier Transformation (FT) of the FID

The pulse length, t_p, can be chosen to give a 90° flip angle, which converts the longitudinal magnetization M_0 along z into a transverse magnetization $M_{x'y'}$ along y' (Fig. 5.4.2b). After the pulse, the magnetic moment vectors $M_{x'y'}$ are no longer in phase and split into components each having its own Larmor frequency, namely in the x'y' plane. As a consequence of this dephasing (Sect. 5.4.2.4), each $M^{(i)}{}_{x'y'}$ projection decays exponentially with a modulation corresponding to ν_i. The receiver coil along y will detect, for each different nucleus, a decaying signal voltage or induction current, termed free induction decay or FID, which is a function of time (Fig. 5.4.3a).

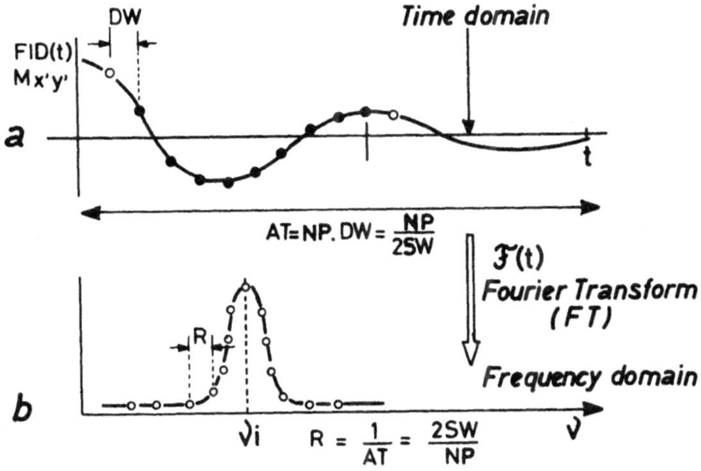

Fig. 5.4.3. a Free induction decay pattern FID(t). **b** NMR signal obtained after Fourier transformation of the FID, and the relation between acquisition parameters and digitization

The NMR information contained in the FID is a signal-time function that is not easily interpreted. Therefore, a mathematical operation (FT) is used to convert the time-domain FID(t) into the frequency-domain NMR spectrum, F(ν) (Fig. 5.4.3b).

The time-variant signal is sampled at equal time intervals, the dwell time (DW), with a digital converter and stored in the computer, usually adding several thousand acquisitions together. The digitization process influences the following spectral acquisition parameters (Fig. 5.4.3b):

1. Dwell time (DW) and spectral width (SW). The signal must be sampled at a rate at least twice the highest frequency (Nyquist theorem), such that DW $\leqslant 1/(2\,SW)$. Thus, for SW = 5 kHz, a minimum DW is 100 μs.
2. Acquisition time (AT) is obtained by multiplying the dwell time by the number of data points or $AT = NP \cdot DW = NP/(2\,SW)$.
3. Digital resolution (R) is the frequency distance from point to point, and is given by $R = 2\,SW/NP = 1/AT$.
4. Number of data points (NP) is a number adjusted to the memory structure of digital computers, expressed as a power of 2, such as $2^{12} = 4096$ or 4K.

The signal-to-noise ratio (S/N) can be improved by multiplying the FID by a decaying exponential function, $\exp(-\pi t \cdot LB)$, which attenuates the noise at the end of the FID but broadens peaks in the transformed spectrum by the amount LB (Hz).

5.4.2.6 Chemical Shifts

Chemical shifts arise from the shielding of the nuclei from B_0 by the surrounding electrons which, under the effect of this field, are induced to circulate around the nuclei (Fig. 5.4.1c). With this motion of charges is associated a magnetic field $B_i = \sigma_i B_0$, where σ_i is the magnetic shielding constant, a characteristic of each nucleus. According to Lenz's law, B_i opposes B_0 at the nucleus (Fig. 5.4.1c). The effective field experienced by the nucleus becomes $B_{eff} = B_0 (1 - \sigma_i)$ and the corresponding Larmor frequency $\nu_i = (\gamma/2\pi)B_0(1 - \sigma_i)$. The frequency difference is given by $\Delta\nu = \nu_{0s} - \nu_{0r}$ in Hz where ν_{0r} is the frequency of a reference sample (usually tetramethylsilane, TMS), and ν_{0s} is the sample frequency. To make the parameter independent of the field strength, a chemical shift scale is introduced as:

$$\delta = \frac{\nu_{0s} - \nu_{0r}}{\nu_1} \cdot 10^6 \, \text{ppm}. \tag{9}$$

Because $\Delta\nu$ between nuclei of the same type is very small (several Hz) compared with ν_1 (several MHz), the shifts on the δ scale are given in parts per million (ppm).

5.4.2.7 Spin-Spin Couplings: Scalar and Dipolar

The nuclear spin of atom A bonded to atom B is affected by the orientation of the nuclear spin of B, aligned or against the direction of B_0. These magnetic interactions, transmitted by the spins of the bonding electrons of bond A-B, are responsible for the scalar coupling. Consequently, the energy levels of A are split according to the different spin states of B and a new set of transitions arises with their corresponding frequencies. For a $^{13}C—^1H_n$ system, where n is the number of chemically and magnetically equivalent protons, there will be $(2nI + 1) = n + 1$ lines for the ^{13}C signal. The scalar coupling constant, J, is the frequency difference in Hz between adjacent lines in such a pattern.

Although the coupling constant, J, provides important structural information, in classical ^{13}C NMR spectrometry scalar coupling is usually suppressed by broad-band ^1H decoupling (BB) to collapse the ^{13}C signal to a single peak.

The local magnetic field created by a spinning nucleus affects the magnetic moment of its neighbours. This magnetic interaction or dipolar coupling is transmitted through space and depends on the relative geometry of the atoms. In the liquid state, the value of this dipolar coupling is averaged to zero because of rapid isotropic tumbling.

5.4.2.8 Signal Intensities

The signal intensities, S, are theoretically proportional to the number of spins promoted from the lower to the upper spin state at resonance, but in pulsed FT NMR these intensities are perturbed in a nonuniform way by several factors that need to be suppressed for quantitative analysis:

1. Spin-lattice relaxation

Spin-lattice relaxation time constants, T_1, of various ^{13}C nuclei are different. For instance, quaternary ^{13}C nuclei may have T_1 as long as several seconds.

During normal acquisition time of approximately 1 s, these nuclei cannot relax between pulses; the spin system becomes saturated and the signal-to-noise ratio decreases. A common solution for slow relaxation is a recovery delay after signal acquisition. The pulse delay should be of the order of four to five times the longest T_1 in the sample, allowing all nuclei to return to the Boltzmann equilibrium state before being flipped by the next pulse.

2. *Nuclear Overhauser Effect (NOE)*

NOE is observed upon proton decoupling of ^{13}C resonances and results from dipolar relaxation induced by the molecular motion in liquids. NOE causes an increase in spin populations in the ground energy levels, thereby enhancing the ^{13}C signal intensity. The maximum NOE is given by $(\gamma^{1}H/2\gamma^{13}C) = 1.988$, although actual NOE's vary among carbons in a sample.

3. *Rf Field Power Level*

This level must be kept constant over the whole range of spectral frequencies, a condition met when $\gamma(B_1/2\pi) \gg SW$ [cf. Eq. (8)].

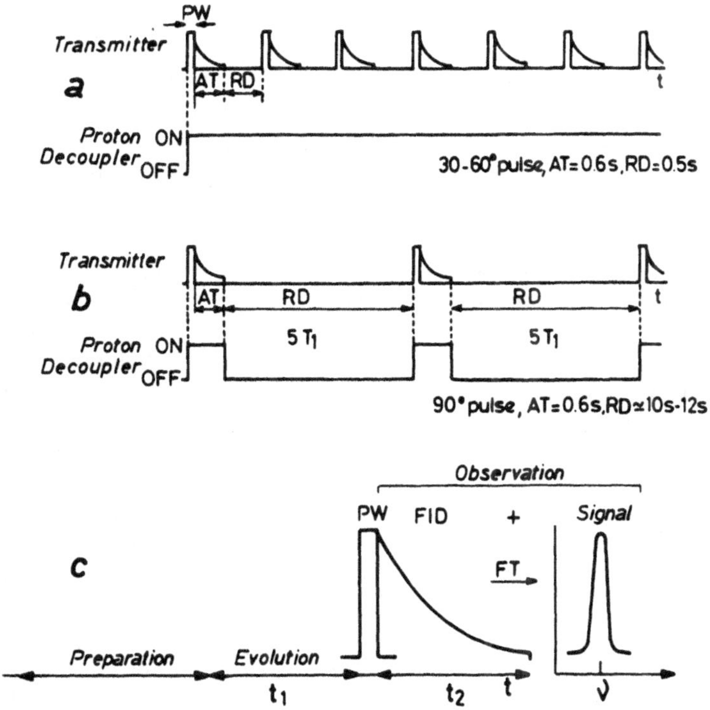

Fig. 5.4.4a–c. ^{13}C NMR sequences. **a** Routine sequence with broad band ^{1}H decoupling. **b** Inverse gated decoupling sequence (IGD) for quantitative analysis. **c** Time sequence for 1-D and 2-D pulse NMR experiments

5.4.2.9 One- and Two-Dimensional NMR Experiments

These experiments (Fig. 5.4.4) consist of a time sequence of pulses including a preparation time, an evolution time, t_1, and a detection or acquisition time, t_2 (Benn and Gunther 1983). During the preparation time, a series of suitable pulses on ^{13}C and/or ^1H nuclei are used and scalar spin-spin interactions are kept or canceled by gating the decoupler; magnetization transfers occur and dipolar coupling plays a role in the perturbation of the spin states. During the evolution time, t_1, the spin system evolves and, after the normal excitation pulse, data collection starts for a time t_2, as in a simple FT experiment. Various types of information are obtained based on the way the spins respond to the pertubations, which in turn is selective and reflects the chemical structure, spatial arrangement, and dynamic properties of the nuclei.

In a one-dimensional (1-D) NMR experiment, t_1 is kept constant, t_2 varies and the FT is taken with respect to t_2 only; in two-dimensional (2-D) NMR experiments, both t_1 and t_2 are varied and the FT is taken with respect to both time variables.

5.4.3 Experimental Procedures

5.4.3.1 Preparation of Samples

Solvents

Suitable solvents are those which dissolve lignin to the maximum extent possible without interfering significantly with the lignin NMR signals. Usually the solvent is deuteriated. The solvent deuterium NMR resonance is usually used for the field-frequency lock signal. The most common lignin solvents and their relevant properties are listed below:

Hexadeuteriated dimethylsulfoxide (DMSO-d_6) is currently the most widely used solvent for lignin samples. Its ^{13}C signal at δ = 39.6 ppm falls in a frequency range where no significant lignin signals are observed. To lower the solvent viscosity, the spectra are recorded at 323 K or higher, up to 355 K. One disadvantage connected with this solvent is the difficulty encountered in recovering the lignin dissolved in it; in this case, the lignin is best recovered by pouring the solution into twenty times its volume of water and freeze-drying the precipitate.

Hexadeuterated acetone: deuterium oxide ($CD_3COCD_3:D_2O$, 9:1, v/v) (δ ^{13}C = 29.2 ppm) and hexadeuterio dioxane: deuterium oxide (dioxane-d_6:D_2O 9:1, v/v) (δ = 69 ppm): the NMR signal of dioxane, although sharp, falls in the frequency range of carbon atoms in the lignin side chain and hence prevents accurate quantitative analyses of these carbons. Due to the low viscosity of the solvents, spectra are recorded at room temperature, and lignin samples are readily recovered by solvent evaporation. These solvents are not particularly well-suited for high molecular weight lignin fractions, but the solubility may be improved by adjusting the ratio of the two solvents in the range of 9:1 to 7:3 (v/v).

Deuterium oxide (D_2O) alone or mixed with DMSO-d_6 is suitable for dissolving lignosulfonates (Hassi 1984) that are not completely soluble in any other solvents.

Trideuterioacetonitrile/deuterium oxide ($CD_3C \equiv N/D_2O$ 4:1, v/v/) (δ $^{13}CD_3$ = 1.3 ppm, δ $^{13}C \equiv N$ = 117.7 ppm) is used when the 25–45 ppm (aliphatic) region has to be free of solvent signals (Gellerstedt and Robert 1987).

Hexadeuterioacetone (CD_3COCD_3) (δ $^{13}CD_3$ = 29.2 ppm/) and deuterio-chloroform ($CDCl_3$) (δ ^{13}C = 76.9 ppm) are good solvents for acetylated samples which are easily recovered from the solution. The $CDCl_3$ and sample signals may overlap.

Solution Preparation

Solutions of the lignin sample can be prepared by heating and stirring (mechanical or ultrasonic) to facilitate dissolution. However, this operation should not be conducted directly in the NMR tube. Insoluble fractions or impurities must be filtered through a syringe plugged with cotton wool to reduce broadening of the NMR signal and improve resolution. However, filtering usually removes cross-linked, insoluble lignin, and thus the spectrum may not be representative of that for the whole original sample. In such instances, reported spectra should be accompanied by an indication of the percent soluble lignin for simplifying cross-comparison with other samples.

Lignin Concentration

For highest signal/noise ratio, the concentration (C) has to be as high as possible ($S/N \alpha C$) as long as the solutions are not too viscous. A suitable concentration is between 20 and 30% (w/w) (400 to 600 mg in 1.8 ml DMSO-d_6 in a 10-mm tube). Nevertheless, all concentration ranges down to 100 mg in 1.8 ml solvent may be used provided the number of acquisitions, NA, is increased since S/N improves proportionately with $\sqrt{NA}$.

A 10-mm OD tube is filled to a depth of at least 4 cm to provide adequate sample height; to minimize any vortex formation, a Teflon vortex plug is used.

Care of NMR Tubes

NMR tubes are thoroughly cleaned by successive washing with acetone and distilled water and subsequently dried in an oven. All metal ion-containing cleaning solutions should be avoided.

5.4.3.2 Spectrometer Specifications

Commercially available FT NMR spectrometers (e.g., Bruker, Varian, GE, JEOL) are designated by their ^1H operating frequency ranges (Table 5.4.1). Satisfactory ^{13}C NMR spectra of lignin can be obtained with 200 and 250 MHz spectrometers. Higher field provides greater sensitivity and is used when enhanced sensitivity is a critical requirement. The following instrument specifications are important:

Table 5.4.1. ^1H and ^{13}C operating frequencies

B_O (T)	2.3	4.7	5.9	9.1	14.1
^1H (MHz)	100	200	250	400	600
^{13}C (MHz)	25.15	50.3	62.8	100.6	150.9

1. Stability of B_0.
2. Homogeneity of B_0 (obtained by adjusting gradient coils or shimming).
3. Intrinsic resolution, R_i, is defined by the minimum line width at half height, $\Delta\nu_{1/2}$, of the C_6H_6 signal in a $CDCl_3$ solution; 60 to 80 Hz are typical values observed in lignin spectra, much higher than instrument specifications.
4. Sensitivity is defined by the signal-to-noise ratio of one acquisition on an 8% solution of ethylbenzene in $CDCl_3$.

General operating conditions are described in the user's manual, including lists of commands, together with the program and microprogram sequences that are available. These sequences are periodically upgraded by the manufacturers. Once the NMR experiment and its sequence are selected, the operator has to choose the appropriate experimental parameters.

5.4.3.3 Recording of ^{13}C NMR Spectra: Summary of Acquisition Parameters

Spectra are usually recorded with complete decoupling of protons by broad-band noise decoupling. The general scheme of the sequence for a routine spectrum is illustrated in Fig. 5.4.5a and a set of recording values are listed in Table 5.4.2.

5.4.3.3.1 Routine Spectra

1. Carbon-13 operating frequency is determined by the spectrometer available (cf. Table 5.4.1).
2. The pulse length, tp, which produces a 90° flip angle is determined by finding the tp value which yields the highest signal intensity. However, for routine spectra, smaller flip angles (between 30° and 60°) can be used, reducing t_p proportionately from the 90° value.
3. Spectral width, SW, should exceed the range of expected frequencies, which is about 240 ppm for lignins.
4. Memory size, NP, is a power of 2 and is expressed as multiples of K (1K = 1024 memory). Usually, NP = 8K, 16K, or greater.
5. Acquisition time, AT, is automatically fixed when SW and NP are selected since AT = NP/2SW.
6. Relaxation delay, RD, between acquisitions can range from 0 to 1 s to improve observation of the quaternary carbons in carbonyl or carboxyl groups, which have longer relaxation times (T_1) than other carbons.

Table 5.4.2. Selected recording parameters for routine ^{13}C NMR spectra of lignins

^{13}C (MHz)	Spectral width (SW) (Hz)	NP (k)	Acquisition time (AT) (s)	Relaxation delay (RD) (s)	Digital resolution (R) (Hz/point)	Number of acquisitions (NA/1000)	Recording time (h)
50.3	11627[a]	16	0.705	0.5	1.41	15–20	~5–6
62.7	14285[b]	16	0.573	0.5	1.74	15–20	~5–6

[a] SW = 231 ppm.
[b] SW = 227 ppm.

7. The digital resolution, R, is expressed as R = 2SW/NP.
8. Number of acquisitions NA: Since the signal-to-noise ratio is directly proportional to both concentration and $\sqrt{NA}$, an appropriate NA value depends upon concentration (Sect. 5.4.3.2) and upon the spectrometer time available; the total time to acquire a spectrum is about NA(RD+AT).
9. Processing: Exponential multiplication with line broadening of 5–8 Hz gives acceptable line widths for lignin signals.
10. Temperature (see solvents, Sect. 5.4.3.1).

5.4.3.3.2 Spectra for Quantitative Analysis

A 90° pulse is chosen to give the highest transverse magnetization and, hence, the maximum signal intensity. The inverse gated decoupling (IGD) sequence, an available software package, is represented in Fig. 5.4.4b. In this sequence, the decoupler is switched on only during FID acquisition. During the recovery delay (RD) between the pulses, which is at least five times the longest T_1 (cf. Sect. 5.4.2.8), the proton decoupler is turned off so that a proton-decoupled ^{13}C NMR spectrum with suppressed NOE is obtained (see Breitmaier and Voelter, p. 50).

The total data acquisition time is usually long. It is recommended that the acquired data be stored in successive blocks to save the data in case of an accidental power interruption. Otherwise, except for the longer RD, acquisition parameters are the same as those considered in the recording of routine spectra (cf. Table 5.4.2).

T_1 relaxation times can be determined with the inversion recovery method (part of the spectrometer software package) using the 180°, τ, 90° sequence and varying τ.

5.4.3.3.3 DEPT (Distortionless Enhancement by Polarization Transfer) Spectra

The DEPT experiment is generated from available computer software with the following concomitant pulse sequence:

on ^{1}H: $90°_x$—τ—$180°_x$—τ— θ_y —τ—BB
on ^{13}C: $90°_x$—τ—$180°_x$—τ—FID-RD,

where x and y denote the axes along which pulses are directed and around which the magnetization vectors rotate by the flip angles specified.

There are three different θ_y pulses applied to ^{1}H with different values of pulse angle ($\theta_1 = \pi/4$; $\theta_2 = \pi/2$; $\theta_3 = 3\pi/4$) which generate three basic spectra $[\theta_1]$, $[\theta_2]$ and $[\theta_3]$.

The delay, $\tau = (2J)^{-1}$ with J = 150 Hz, is selected from the average J values (^{13}C—^{1}H coupling constants) found for lignin structures. The three pure subspectra CH, CH_2, and CH_3, are obtained from appropriate linear combinations of the three basic spectra:

subspectrum CH $= [\theta_2]$
subspectrum $CH_2 = [\theta_1] - [\theta_3]$
subspectrum $CH_3 = [\theta_1] + [\theta_3] + 0.707[\theta_2]$

It is recommended that the 90° pulse on ^{1}H be calibrated carefully using the lignin sample itself (Bardet 1987). One can use the DEPT sequence to look for the ^{1}H pulse length that gives the best nulling of the CH_2 signals. The average time for a routine DEPT experiment on a 20% solution is about 8 h with 2000 transients for $\theta_y = \pi/4$ and $3\pi/4$ and 4000 for $\theta_y = \pi/2$. For recovery delay RD, 2 s is chosen to be greater than the ^{1}H T_1. For quantitative analysis, the number of acquisitions can be doubled and RD increased to 3 s; in this case about 18 h are required for recording the spectra.

5.4.3.4 *Integration of NMR Signals*

The following general rules apply: before integration, the spectrum itself has to be phased carefully using the subroutine phasing program immediately after

Table 5.4.3. Functional group frequency ranges in ^{13}C NMR spectra of lignins

Structure or functional group	*Frequency range, ppm*
Aromatic C total	102–162
Aromatic-CH	102–127
Aliphatic-COR	58–95
Methoxyl	54.5–56.5
C-3/C-5 in β-O-4 S e	150.8–153.3
$C_\gamma H_2OH$[a]	58.4–64
C-2/C-6[b] in S	102.3–107.5
C_β in β-β and β-5	53.4–54
Acetyl carboxyl carbons	
Primary hydroxyl groups	170.4–171.6
Secondary hydroxyl groups	169.5–170.4
Phenolic groups	168.5–169.5

[a] Except C-γ in β-β units found at 71.2 ppm.
[b] Except C-2/C-6 in p-hydroxyphenyl units.

the FT. Minimum baseline correction is recommended as the integral is very sensitive to small changes in the baseline. For the integral itself, phasing is necessary to minimize slope and bias. Once phasing and baseline correction are completed, they must not be changed during analysis of the spectrum, as any comparison made between signals or clusters of signals phased differently is meaningless. Another important requirement is to choose accurate and useful frequency ranges for the signal integration limits and keep them identical when comparative studies are being performed (Table 5.4.3).

To measure the area under a resonance line, one can measure the integral itself or use the automatic printout of the integral values given by the spectrometer. Good agreement should exist between the two values. Carboxyl and methyl signals of acetoxyl groups and other signals that are very well-defined and separated from other signals may be recorded and integrated on an expanded scale (Fig. 5.4.10); alternatively, photocopies of the peaks are made, cut out, and weighed to determine the relative peak areas.

5.4.4 Results and Discussion

Most of the data accumulated to date reflect structural information at the molecular level. Very little effort has been invested in studies of the polymer properties of lignin by ^{13}C NMR techniques.

5.4.4.1 Qualitative Analysis

The assignment of ^{13}C signals in the NMR spectra of lignins (Table 5.4.4, Fig. 5.4.5a and b) is based on spectral information derived from studies of

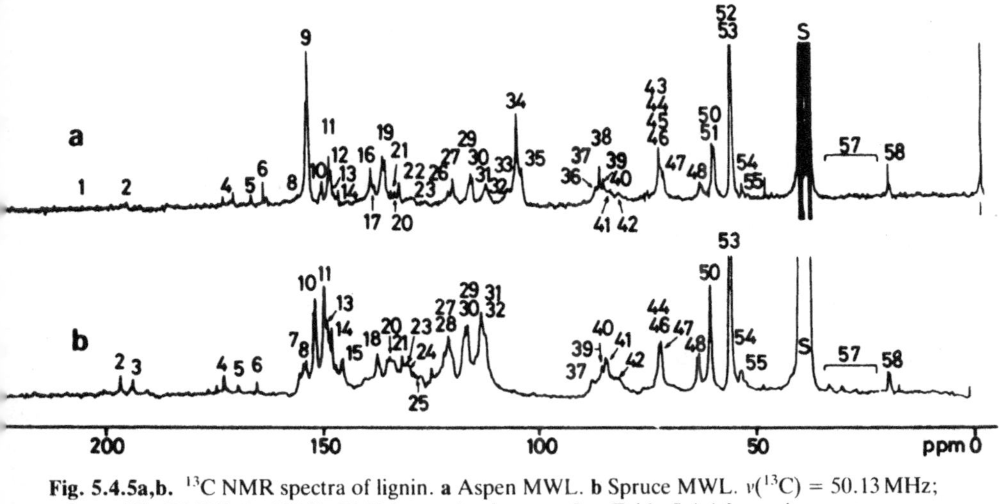

Fig. 5.4.5a,b. ^{13}C NMR spectra of lignin. **a** Aspen MWL. **b** Spruce MWL. $\nu(^{13}C)$ = 50.13 MHz; solvent (*S*), DMSO-d_6; 323 K; TMS internal reference. See Table 5.4.4 for assignments

Table 5.4.4. ^{13}C Chemical shifts and signal assignments for spruce and aspen milled wood lignins. Solvent: DMSO-d_6

Signal no.	δ in ppm/TMS	Assignments[a]
1	210–200	non-conjugated C=O in ketones
2	194	γ-CHO in cinnamaldehyde[b]
3	191.6	α-CHO in benzaldehyde
4	171–173	C=O in aliphatic RC(=O)OR
5	165–167	αC=0 in ArC(=O)OH
6	162	C-4 in ArC(=O)OH
7	152.9	C-3/C-3′ in etherified 5-5 units
8	152.6	C-α in cinnamaldehyde
9	152.1	C-3/C-5 in S β-O-4 e
10	149.1	C-3 in G e unit
11	147.4	C-4 in G e unit
12	147.1	C-3/C-5 in S β-O-4 ne
13	146.9	C-3 in G ne
14	145.3	C-4 in G β-O-4 ne
15	143.4	C-4/C-4′ in 5-5 units e
16	138	C-1 in S β-O-4 e
17	137.7	C-1 in S β-β
18	135.3	C-1 in G β-O-4 e
19	id	C-4 in S β-O-4 e and ne
20	133	C-1 in G β-O-4 ne
21	132.4	C-5/C-5′ in 5-5 units, C-1 in S β-O-4 ne
22	131	C-2/C-6 in H benzoate unit
23	128–130	–
24	128.3–129.3	C-α and C-β in ARCH=CH—CH_2OH
25	126.3	C-β in cinnamaldehyde
26	121.5	C-1 in H benzoate unit
27	120.3	C-6 in G units e and ne
28	119.5	"
29	115.2	C-5 in G units e and ne
30	114.6	C-3/C-5 in H benzoate units
31	111.4	C-2 in G units
32	110.9	C-2 in G-G stilbenes units[c]
33	106.8–107	C-2/C-6 in S with αC=O or αC=C
34	104.5–106.8	C-2/C-6 in S in general
35	103.6	C-2/C in β-β
36	87.2	C-β in S β-O-4 *threo*
37	87.2	C-α in β-5
38	86	C-β in S β-O-4 *erythro*
39	85.3	C-α in β-β
40	84.5	C-β in G β-O-4 *threo*
41	83.5	C-β in G β-O-4 *erythro*
42	80–82	C-β and C-α in β-O-4/α-O-4 units
43	72.5	C-α in β-O-4 G and S *erythro*
44	71.8	C-α in β-O-4 G and S *erythro*
45	71.8	C-α in β-O-4 G and S *threo*
46	71.4	C-α in β-O-4 G and S *threo*

Table 5.4.4. (continued)

Signal no.	δ in ppm/TMS	Assignments[a]
47	71.2	C-γ in β-β units
48	63.0	C-γ in β-5 and β-O-4 with α-C=O
49	61.7	C-γ in cinnamyl alcohol units
50	60.2	C-γ in β-O-4 S and G *threo* and *erythro*
51	59.8	C-γ in β-O-4 S and G *threo* and *erythro*
52	55.7–55.9	Aromatic OMe in G and S units
53	"	"
54	53.7	C-β in β-β
55	53.1	C-β in β-5
56		
57	20–40	—CH_3 and CH_2 in saturated aliphatic chain
58	20	—CH_3 in acetyl groups

G = guaiacyl; H = p-hydroxyphenyl; S = syringyl; e = etherified; ne = nonetherified
[a] For nomenclature, see Fig. 5.4.6.

[b] 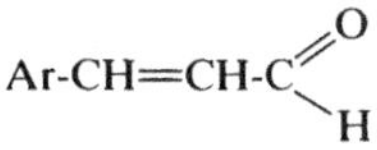

[c] G-CH=CH-G.

model compounds. This indirect technique was first used by Lüdemann and Nimz (1973, 1974a,b) who compared the chemical shifts of carbon atoms in monomeric and dimeric lignin model compounds with those in lignin spectra. The ^{13}C NMR spectra of partly derivatized trimers also have been recorded (Nimz and Lüdemann 1974). Tables of assignments for various types of lignins are available in the literature and correlations, mostly summative, exist for a variety of substituents (Chen et al. 1982, Chua et al. 1982, Lapierre et al. 1982, Hemmingson et al. 1983, Kringstad and Mörck 1983, Landucci 1984, Bardet 1987, Chen and Robert 1988). Chemical shifts for acetylated lignins have been similarly assigned (Nimz et al. 1981, Mörck and Kringstad 1985). The three most commonly studied lignin dimers are of the β-O-4, β-β and β-5 types (Fig. 5.4.6).

In ^{13}C NMR analyses, matching the chemical shift values of lignins and model compounds is done with spectra recorded under exactly the same conditions (solvent, temperature, and setting of the TMS δ value) to ensure valid assignments. Nevertheless, the matching technique is not completely satisfactory because a specific carbon atom in a dimeric model compound, for example, may not exhibit the same chemical shift when present in a polymer, as is often assumed in NMR spectrometry. Indeed, there is evidence that a slight downfield shift (to higher δ) occurs on proceeding from dimeric to polymeric structures (Kringstad and Mörck 1983).

The accuracy of assignments is also limited by the availability of relevant model compounds for spectral measurements and by spectrometer resolution. Because of the structural heterogeneity and complexity of lignin, numerous signals are hidden in shoulders and often overlooked. In time, this difficulty can be minimized through an expanding data bank as more oligomeric lignin model compounds become available, by applying labeling techniques to lignin syn-

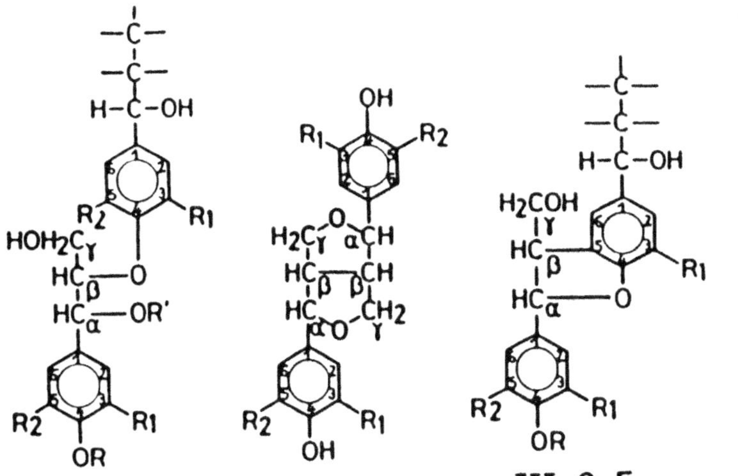

I β-O-4 II β-β III β-5

Fig. 5.4.6. Main dimer structures in ligin. R_1=R_2 = —OCH_3: syringylpropane unit, S; R_1 = —OCH_3, R_2=H guaiacylpropane unit, G; R_1=R_2=H: p-hydroxyphenylpropane unit, H; R=H: non-etherified unit, *ne*; R=C − *β*: etherified unit, *e*; R_2=C − 5: 5 − 5 unit; R=H; if R′=C − 4 in I: *β*-O-4/*α*-O-4 unit

thesized in vitro, and by polymerization of selectively ^{13}C-enriched coniferyl alcohol. Precise assignments of the labeled carbons in this lignin polymer model (DHP) are made from knowledge of the ^{13}C positions in the starting monomer (Gagnaire and Robert 1977, Haider et al. 1985).

An empirical substituent chemical shift (SCS) additivity rule has been established by which the shift of a particular aromatic carbon in the *β*-O-4, *β*-5 and 5-5 substructures of lignin can be estimated by summation of the different substituent effects (Hassi et al. 1987, Drumond et al. 1989). This method helps to distinguish among overlapping signals and improves assignment.

Recent advances in 2-D NMR are particularly applicable to lignin. Sophisticated sequences have been applied to lignin model compounds (Ralph 1988), revealing the long-range C—H correlations critical in proving structure. The 2-D INADEQUATE (Incredible Natural Abundance Double Quantum Transfer Experiment) sequence successfully mapped the carbon skeleton of a ^{13}C-enriched aspen MWL sample through observation of directly bonded C—C atoms (Guittet et al. 1985, Bardet et al. 1986). This sequence also provides stereochemical information about lignin structure. For example, INADEQUATE spectra support previous data obtained from gated decoupled spectra concerning the diasteroisomeric forms of *β*-O-4 units (Nimz et al. 1981, Ralph 1983) and allows new and more detailed assignments of signals 36, 38, 40, 41, and 43–46 (Table 5.4.4) (Robert et al. 1989, Robert and Lundquist, unpubl. results). The ability of the method to distinguish between diastereoisomers is a unique feature that offers a new opportunity to elucidate the stereochemistry of lignin and its effect on chemical reactivity.

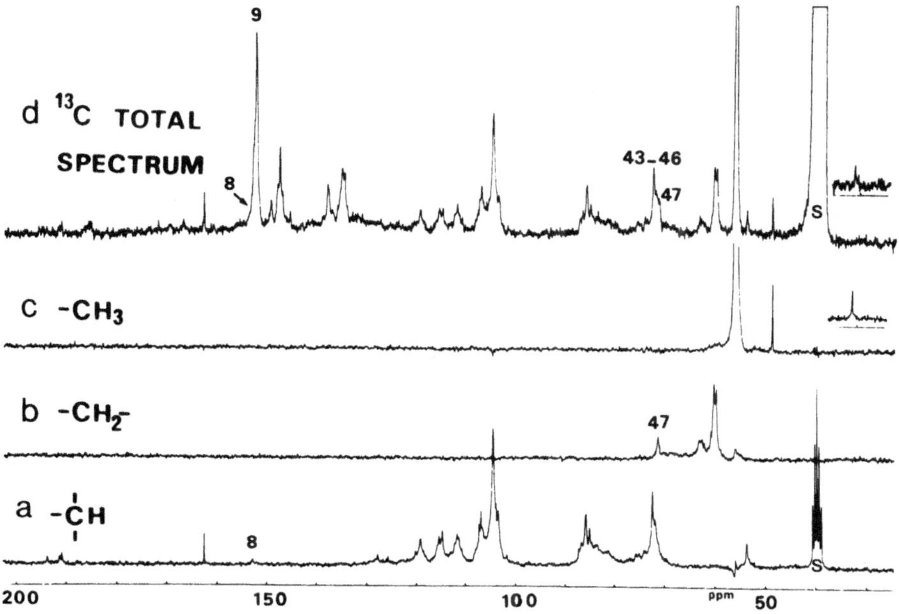

Fig. 5.4.7a–d. DEPT edited subspectra of birch lignin (MWL) at 50.3 MHz. Solvent (*S*): DMSO-d_6; reference TMS. **a** CH signals. **b** CH_2 signals. **c** CH_3 signals. **d** Total ^{13}C spectrum

5.4.4.2 ^{13}C NMR DEPT Spectra

This 1-D pulse sequence (Doddrell et al. 1982) has two main advantages: it edits spectra by proton multiplicities, giving separate observations of the CH, CH_2 and CH_3 signals (Fig. 5.4.7), and enhances signal intensities. The enhancement is due to transfer of polarization during the pulse sequence from the more sensitive ^{1}H nuclei to the less sensitive ^{13}C nuclei. When applied to lignins (Lapierre et al. 1984a,b, Bardet et al. 1985), it overcomes, to a large extent, signal overlap; cf. the $C_\gamma H_2$ signal 47 in the β-β unit and the $C_\alpha H$ signal 8 in cinnamaldehyde units in Fig. 5.4.7b and a, respectively, without overlap and signals 43, 44, 45, and 9 overlapping these peaks in the total spectrum (Fig. 5.4.7d). Equally important (cf. Sect. 5.4.4.3) is the clean break between quaternary and tertiary aromatic carbons at 127 ppm (Table 5.4.3). The advantage of the DEPT experiment lies in its clear presentation of structural information.

5.4.4.3 Quantitative Analysis

There are three major experimental requirements for obtaining reliable and useful quantitative ^{13}C NMR spectra:

1. Use of the inverse gated decoupling sequence (IGD).
2. Use of a solution with as high lignin concentration as possible to maximize the signal-to-noise ratio. This improves the spectrum baseline and facilitates phasing and recording of reliable integrals.
3. A lignin sample free of contaminants not chemically bonded to lignin such as carbohydrates, extractives, and monomeric, dimeric, and oligomeric lignin degradation products.

The longest T_1 is 2.1 s for C_3/C_5 in the S β-O-4 e unit in aspen MWL in DMSO solution (Bardet 1987), and 2.5 s for the carboxyl carbons in acetylated MWL samples (Nimz et al. 1982). Thus, unless exact T_1 values are needed and have to be determined, one can use RD values of 10 and 12 s, respectively, for nonacetylated and acetylated samples.

Once appropriate parameters for the IGD sequence are selected (Robert and Gagnaire 1981) and the integrals recorded, sound quantitation will rely both on their interpretation and precision. Since lignins have neither precise chemical formulas nor regular repeating units, quantitative data cannot be reported as absolute figures. Thus, quantitative structural data are reported on the basis of a structural element, usually the phenylpropane ($C_9(C_6$—$C_3)$) unit. Standard practice is to report a quantitative result as the ratio of the integral of a given carbon signal to one-sixth the integral of the aromatic carbons, whose signals appear between 102 and 162 ppm. A correction can be made (Table 5.4.5) for the minor amounts of vinyl carbons that are found in this region.

The results obtained from the analysis of the spectrum in Fig. 5.4.8a for spruce MWL are reported in Table 5.4.5 as the number of carbon atoms per aryl

Table 5.4.5. Number of carbon atoms in chemical shift ranges in the ^{13}C NMR spectrum of spruce milled wood lignin (MWL) (Fig. 5.4.8a)

Chemical Shift Range (δ in ppm)	Integral[a]	Carbons[b]/Aryl group
194–193	1.1	0.06
192–191	0.7	0.04
158–154	0.8	0.04
154–150.5	6.4	0.34
150.5–145	26.0	1.37
145–140	5.8	0.31
140–124	28.4	1.50
124–103	48.6	2.56
90–57	44.0	2.32
57–54.5	18.6	0.98
54.5–52.5	3.4	0.18

[a] Total integral for chemical shift range 160–100 ppm = 116. Thus, the integral for one aromatic carbon = 116/6.12 = 18.95, assuming that spruce MWL contains 3 Ar—CH=CH—CHO units and 3 Ar—CH=CH—CH_2OH units per 100 C_9 units (cf. Robert and Chen 1989).
[b] See Table 5.4.6 for assignments.

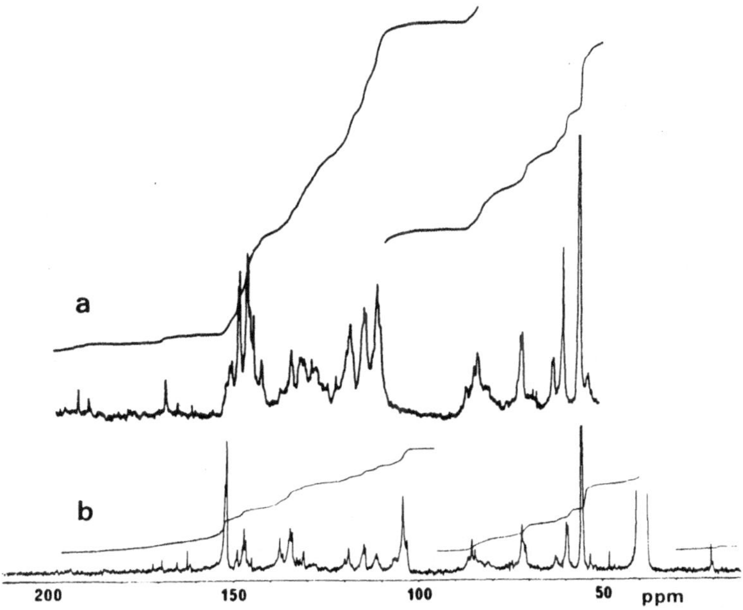

Fig. 5.4.8a,b. ^{13}C NMR spectra and integrals for quantitative analysis recorded with the inverse gated decoupling (IGD) sequence. **a** Spruce MWL. **b** Aspen MWL. $\nu(^{13}C) = 50.13$ MHz; solvent (S), DMSO-d_6

group (Robert and Chen 1989). The methoxyl content of 0.98 carbon per aryl group agrees quite well with the value of 0.94 per C_9 unit obtained from methoxyl analysis for the same spruce MWL sample. This indicates that ^{13}C NMR is a viable quantitative analytical technique for lignin. The amounts of various structural moieties in aspen and spruce MWLs, as estimated by ^{13}C NMR, are shown in Table 5.4.6 (Fig. 5.4.8b, Robert, unpubl.).

Quantitative ^{13}C NMR has been used to determine the methoxyl contents of lignin in many wood species (Obst and Landucci 1986). The syringyl/guaiacyl composition in aspen MWL also has been measured using routine NMR acquisition parameters, but this method is restricted to carbons having the same chemical substituents and T_1 values and the results should be very carefully interpreted (Lapierre et al. 1984b). Semi-quantitative analysis can also be performed (Nimz et al. 1982).

A combination of DEPT and IGD spectra allows more precise quantitative analysis of, for example, the number of quaternary aromatic carbons. From such results, the degree of carbon-carbon condensation on the aromatic ring can be defined as the number of quaternary carbon atoms that are neither C_1 nor C_4 nor are methoxyl-substituted C_3/C_5. This kind of information is particularly useful for following condensation reactions during such processes as steam explosion treatment of aspen wood (Robert et al. 1986, Bardet 1987) and kraft cooking (Robert et al. 1984, Gellerstedt and Robert 1987).

Estimation of the different types of hydroxyl groups can be performed readily on acetylated lignin samples (Fig. 5.4.9) (Robert and Brunow 1984,

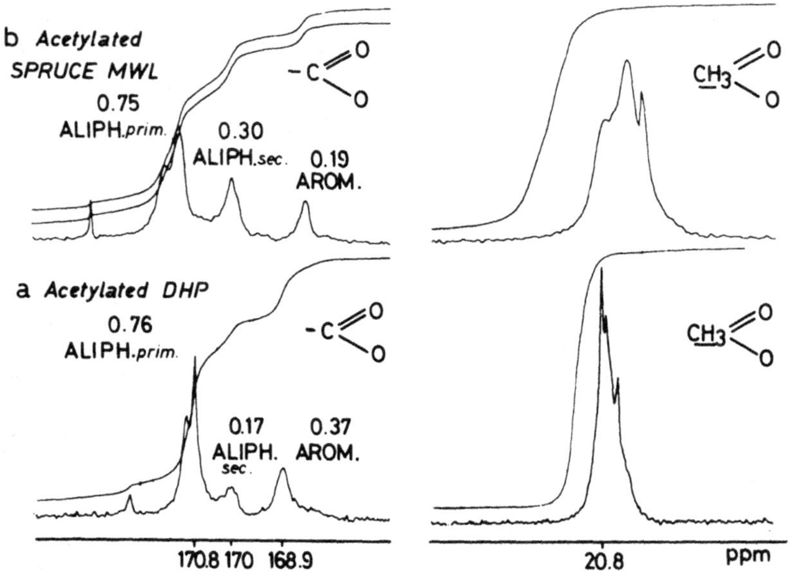

Fig. 5.4.9a,b. Expansion of the carboxyl and methyl signals of the acetyl groups in acetylated samples showing the peak areas measured for quantitative analysis. **a** DHP. **b** Spruce MWL

Table 5.4.6. Quantitative estimation of selected chemical structures in lignin (^{13}C NMR Spectra Fig. 5.4.8a and b)

	Aliphatic C—O—R	—$C_\gamma H_2$OR	Quaternary aromatic C	—OCH_3
	Number/Aryl group			
Aspen MWL	0.90	0.92	3.54	1.43
Spruce MWL	0.75	0.84	3.24	0.98

Landucci 1985). The equality of the acetyl methyl and carboxyl carbon integrals is a good test of the reliability of the IGD sequence, which compensates for the large differences in the values of T_1 and NOE for these two carbons.

Several recordings of the integrals give an average reproducibility of ±5% relative or less for the data shown in Tables 5.4.5, 5.4.6 and Fig. 5.4.9. An alternative method using optimized NMR parameters and several replicates affords 1–3% reproducibility of quantitative results (Sect. 5.4.2.8, item 2) (Landucci 1985). Such analytical precision in NMR results is impressive, considering that it may exceed the accuracy limits imposed by the chemical heterogeneity and ill-defined nature of lignin and the difficulties in isolating it reproducibly from wood. On the other hand, NMR spectrometry is not as sensitive a method as UV or IR spectroscopy for determining trace amounts of a material.

5.4.5 Concluding Remarks

^{13}C NMR is a sensitive method for the qualitative structural analysis of lignin. It provides detailed structural information on nearly all types of carbons including those in the minor structures. This is an important advantage compared to most other methods, which provide information only on specific groups. The capacity for producing reliable quantitative analyses of functional groups in lignin is another significant feature of ^{13}C NMR spectrometry. Methods using new pulse strategies in NMR spectrometry open new frontiers in the understanding of lignin structure and may promote NMR studies of lignin polymer properties.

References

Bardet M (1987) Analyse structurale des lignines par RNM: utilisation des nouvelles sequences impulsionnelles et étude des lignines extraites de bois traité par autohydrolyse rapide. Thèse de Doctorat d'Etat Université de Grenoble I, France 258 pp

Bardet M, Foray MF, Robert D (1985) Use of the DEPT pulse sequence to facilitate the ^{13}C NMR structural analysis of lignins. Makromol Chem 186:1495–1504

Bardet M, Gagnaire D, Nardin R, Robert D, Vincendon M (1986) Used ^{13}C enriched wood for structural NMR investigation of wood and wood components, cellulose and lignin. Holzforschung Suppl 40:17–24

Benn R, Gunther H (1983) Modern pulse methods in high resolution NMR spectroscopy. Angew Chem Int Ed Engl 22:350–380

Chen C-L, Chua MGS, Evans J, Chang H-m (1982) ^{13}C NMR spectroscopic study of spruce lignin degraded by Phanerochaete chrysoporium I Synthesis and chemical shifts of model compounds. Holzforschung 36:239–247

Chen C-L, Robert D (1988) ^{1}H and ^{13}C NMR spectroscopy of lignin. In: Wood WA, Kellogg ST (eds) Methods in enzymology biomass, Part B, Vol. 161. Academic Press, New York, 137–174

Chua MGS, Chen C-L, Chang H-m, Kirk TK (1982) ^{13}C NMR spectroscopic study of spruce lignin degraded by Phanerochaete chrysoporium II New structures. Holzforschung 36: 165–172

Doddrell DM, Pegg DT, Bendall MR (1982) Distorsionless enhancement of NMR signals by polarization transfer. J Mag Res 48:323–327

Drumond M, Aoyama M, Chen C-L, Robert D (1989) Substituent effects on C-13 chemical shifts to aromatic carbons in biphenyl type lignin model compounds. J Wood Chem Technol 9:421–441

Gagnaire G, Robert D (1977) A polymer model of lignin (DHP) ^{13}C selectively labeled at the benzylic positions: synthesis and NMR study. Makromol Chem 178:1477–1495

Gellerstedt G, Robert D (1987) Quantitative ^{13}NMR analysis of kraft lignins. Acta Chem Scand B 41:541–546

Guittet E, Lallemand JY, Lapierre C, Monties B (1985) Applicability of the ^{13}C NMR INADEQUATE experiment to lignin, a natural polymer. Tetrahedron Lett 26:2671–2674

Haider K, Kern HW, Ernst L (1985) Intermediate steps of microbial lignin degradation as elucidated by ^{13}C NMR spectroscopy of specifically ^{13}C enriched DHP lignins. Holzforschung 39:23–32

Hassi HY, Aoyama M, Tai D, Chen C-L, Gratzl JS (1987) Substituents effect on ^{13}C chemical shifts of aromatic carbons in β-O-4 and β-5 type lignin model compounds. J Wood Chem Technol 7:555–567

Hassi HY, Chen C-L, Gratzl JS (1984) Carbon 13 NMR characterization of lignosulfonates. Proc TAPPI R and D Conf, Appleton, WI, 249–260

Hemmingson J et al. (1983) The Structure of lignin from Pinus radiata exploded wood. J Wood Chem Technol 3:289–312

Kringstad KP, Mörck R (1983) ^{13}C NMR spectra of kraft lignins. Holzforschung 37:237–244
Landucci L (1984) ^{13}C NMR characterization of soda and soda/AQ spent liquor fractions. J Wood Chem Technol 4:171–186
Landucci L (1985) Quantitative C-13 NMR characterization of lignon. A methodology for high precision. Holzforschung 39:355–359
Lapierre C, Lallemand JY, Monties B (1982) Evidence of poplar lignin heterogeneity by a combination of ^{13}C and ^{1}H NMR spectroscopy. Holzforschung 36:275–282
Lapierre C, Monties B, Guittet E, Lallemand JY (1984a)Photosynthetically ^{13}C labeled poplar lignins: ^{13}C NMR experiments. Holzforschung 38:333–342
Lapierre C, Monties B, Guittet E, Lallemand JY (1984b) Quantitative measurements in hardwood lignin ^{13}C NMR spectra. Holzforschung 39:367–368
Lüdemann HD, Nimz HH (1973) C-13 Nuclear Magnetic Resonance spectra of lignins. Biochem Biophys Res Commun 52:1162–1169
Lüdemann HD, Nimz HH (1974a) ^{13}C Kernresonanz-Spektren von Ligninen. 1. Chemische Verschiebungen der monomeren und dimeren Modellsubstanzen. Makromol Chem 175: 2392–2407
Lüdemann HD, Nimz HH (1974b) ^{13}C Kernresonanz-Spektren von Ligninen. Buchen und Fichten Björkman Lignin. Makromol Chem 175:2409–2422
Mörck R, Kringstad KP (1985) ^{13}C NMR spectra of kraft lignins II. Holzforschung 39:109–119
Nimz H, Lüdemann HD (1974) ^{13}C Kernresonanzspektren von ligninen 5. Makromol Chem 175:2577–2583
Nimz H, Robert D, Faix O, Nemr M (1981) C-13 NMR spectra of lignins 8. Structural differences between lignins of hardwoods, softwoods and compression wood. Holzforschung 35:16–26
Nimz HH, Nemr M, Schmidt P, Margot C, Schaub B, Schlosser M (1982) Carbon 13 NMR spectra of lignins 9. Spin lattice relaxation times and determination of interunit linkages in three hardwood lignins (Alnus glutinosa, Corylus avellana and Acer pseudoplatanus). J Wood Chem Technol 2:371–382
Obst JR, Landucci LL (1986) Quantitative ^{13}C NMR of lignins. Methoxyl: aryl ratio. Holzforschung 40 Suppl: 87–92
Ralph J (1983) Stereochemical aspects of addition reactions involving lignin model quinonemethides. J Wood Chem Technol 3:161–181
Ralph J (1988) NMR of lignin model compounds: applications of long range C-H correlations through oxygen. Holzforschung 42:273–275
Robert D, Bardet M, Gellerstedt G, Lindfors EL (1984) Structural changes in lignin during kraft cooking. Part 3 On the structure of dissolved lignins. J Wood Chem Technol 4:239–263
Robert D, Bardet M, Lundquist K, von Unge S (1989) ^{13}C NMR spectroscopic studies of the distribution of erythro and threo forms of various types β-O-4 structures in lignins. TAPPI Proc Int Symp Wood Pulp Chem Raleigh, NC, TAPPI Press, Atlomta, 217–219
Robert D, Brunow G (1984) Quantitative estimation of hydroxyl groups in milled wood lignin from spruce and in a dehydrogenation polymer from coniferyl alcohol using ^{13}C NMR spectroscopy. Holzforschung 38:85–90
Robert D, Chen C-L (1989) Biodegradation in spruce wood by Phanerochaete chrysosporium: quantitative analysis of biodegraded spruce lignins by ^{13}C NMR spectroscopy. Holzforschung 43:325–334
Robert D, Gagnaire D (1981) Quantitative analysis of lignins by ^{13}C NMR spectroscopy. Proc The Ekman Days Int Symp Wood Pulp Chem, Stockholm, 1:86–88
Robert D, Gellerstedt G, Bardet M (1986) ^{13}C NMR analysis of lignins obtained after sulfonation of steam exploded aspen wood. Nord Pulp Pap Res J 3:18–25
Snyder J, Breuning R, Derguini F, Nakanishi K (1989) Fractionation and proof of structure of natural products, In: Lowe JW (ed) Natural products of woody plants I. Chemicals extraneous to the lignocellulosic cell wall. Springer Series in Wood Science, Springer, Berlin Heidelberg New York Tokyo, 60–84

NMR Textbooks

Breitmaier E, Voelter W (1987) Carbon-13 NMR spectroscopy, 3rd edn. VCH, Weinheim, 515 pp

Brevard C, Granger P (1981) Handbook of high resolution multinuclear NMR. Wiley-Interscience, New York, 229 pp

Derome AE (1987) Modern NMR techniques for chemistry Pergamon, Oxford, 280 pp

Freeman R (1987) A handbook of NMR. Longman Scientific and Technology, UK, 312 pp

Harris RK (1983) Nuclear Magnetic Resonance Spectroscopy. Pitman Books, London, 250 pp

Levy GC, Lichter RL, Nelson GL (1980) Carbon 13 Nuclear Magnetic Resonance Spectroscopy. 2nd edn. Wiley, New York, 338 pp

Martin ML, Delpuech JJ, Martin GL (1980) Practical NMR spectroscopy. Heyden, London, 460 pp

Sanders JKM, Hunter BK (1988) Modern NMR spectroscopy, a guide for chemists. Oxford University Press, Oxford, 308 pp

5.5 Electron Spin Resonance (ESR) Spectroscopy

D.N.-S. Hon

5.5.1 Introduction

Electron spin resonance (ESR) or electron paramagnetic resonance (EPR) spectroscopy has developed at an outstanding pace since its discovery in 1945 (Zavoiskii 1945), so that at present the technique is very well understood in its many aspects. In wood chemistry, ESR has become an essential tool for the study of the structure and dynamics of molecular systems containing one or more unpaired electrons, i.e., free radicals. ESR has found applications as a highly sensitive tool for the detection and identification of free radical species in lignin and lignin model compounds (Steelink 1966, Kringstad and Lin 1970). A recent literature review of free radicals in wood chemistry is available (Simkovic 1986).

5.5.2 Principle of ESR

The unique feature of a radical is its unpaired electron which give rise to a magnetic moment or paramagnetism, μ, given by

$$\mu = -g\beta S, \tag{1}$$

where g is the so-called spectroscopic splitting factor, or more simply, g-factor with a value close to 2.0023 for the free electron, β is the Bohr magneton (β = 9.2740×10^{-21} erg $gauss^{-1}$), and S the spin vector of the electron. When the radical is subjected to a magnetic field, it may exist in one of the two possible energy states due to the quantum mechanical restriction on the orientation of the electronic magnetic moment, either with or against the applied field. The separation of the energy states, ΔE, increases with the applied field, H_o, according to the relation,

$$\Delta E = g\beta H_o, \tag{2}$$

Just as in other spectroscopic experiments, there is a certain frequency of electromagnetic radiation that will induce transitions between these energy states. The frequency associated with this energy difference satisfies the equation,

$$h\nu = g\beta H_o, \tag{3}$$

where h is Planck's constant (6.626×10^{-27} erg s).

Springer Series in Wood Science
Methods in Lignin Chemistry
(Edited by S.Y. Lin and C.W. Dence)

ESR experiments are usually conducted at a fixed frequency and the resonance condition is found by varying the intensity of the static magnetic field. Most ESR spectrometers operate with microwave fields at 9.5 GHz (X band) or 35 GHz (Q band), the resonant fields in these two cases being about 3400 and 12 500 gauss, respectively. An ESR experiment consists of irradiating the sample held in the magnetic field with microwave radiation of a constant frequency while varying the magnitude of the field, H_o, until the conditions in Eq. (3) are satisfied; at this point energy is absorbed and detected electronically (Fig. 5.5.1a). Most modern instruments use a phase-sensitive method of detection, and hence produce the first derivative of the absorption spectrum (Fig. 5.5.1b). This is advantageous in several respects, one being that small inflections in the absorption spectrum are much more obvious in the derivative spectrum. In addition, the peak position in the first derivative is the baseline crossover; the width is conveniently measured between the points of maximum slope. If the spectral features are very poorly resolved, second derivative display may help make them more apparent (Fig. 5.5.1c).

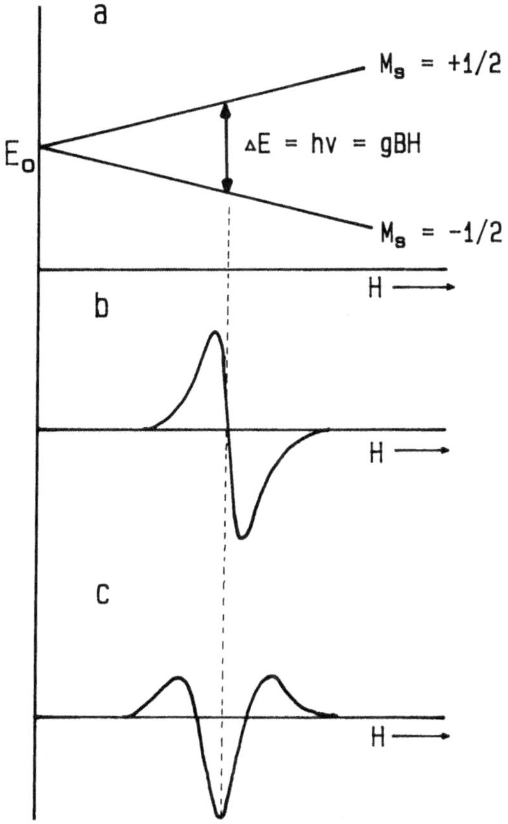

Fig. 5.5.1a–c. ESR experiment. **a** Splitting of the spin levels of an electron in an applied magnetic field of intensity H. **b** ESR first derivation signal. **c** ESR second derivative signal

5.5.3 Spectral Parameters

ESR is a sensitive tool for detecting free radicals. In addition, ESR represents further improvement over indirect methods normally employed in lignin investigations of free radicals by providing the possibility of specific free radical identification and characterization through spectroscopic detail. The characteristics of an ESR spectrum useful for deductions about physical interactions can be conveniently grouped into the following four parameters: (1) g-value, (2) intensity, (3) line shape, and (4) hyperfine structure.

5.5.3.1 The g-Value

The g-value is used to characterize the position of a resonance. It is a measure of the local magnetic field experienced by the electron. The g-value is a unique property of the molecule as a whole, and is independent of any electron-nuclear hyperfine interactions that may be present. The g-value is defined as

$$g = \frac{h\nu}{\beta H_o} = 0.7144 \times \frac{(GH_z)}{H_o(KG)}, \qquad (4)$$

where H_o is the applied (or external) magnetic field at resonance. The g-value for a molecule of unpaired electrons is characteristic of its electronic environment. It aids in the identification of an unknown signal. The g-value for a free electron is 2.0023 and for the phenoxy radicals in lignin, 2.0022. In transition metals, a g-value as high as 10 has been observed.

The absolute measurement of a g-value requires knowledge of the microwave frequency and the resonant magnetic field. However, standards with known g-value (Table 5.5.1) are often used to calculate the g-value of a lignin sample, according to the following formula:

$$g_x = \frac{g_s H_s}{H_x} = \frac{g_s H_s}{H_s - H} \approxeq g_s\left[1 + \frac{\Delta H}{H_s}\right], \qquad (5)$$

where g_s is the g-value of the standard and H_s and H_x are the resonant magnetic fields for standard and lignin, respectively, and $\Delta H = H_s - H_x$.

5.5.3.2 Intensity

The intensity of a spectral signal or, more accurately, the integrated area under the signal is proportional to the concentration of the unpaired spins giving rise to the spectrum. Comparison of signal intensities obtained in ESR studies are valid only if the spectra are recorded under identical experimental conditions. Alternatively, the intensity of the signal can be estimated using the following equation:

$$\text{Intensity} = \text{Derivative Height} \times (\text{Line Width})^2, \qquad (6)$$

where line width is the distance between peak to peak of a derivative signal.

Table 5.5.1. g-Values for standard samples

Standard sample	g-Value
DPPH (α, α'-diphenyl-β-picrylhydrazyl) powder	2.0037
Wurster's blue perchlorate in degassed ethanol	2.0031
Manganous ion (Mn^{2+}) in SrO powder	2.0012
Fremy's salt (Potassium peroxylamine disulfonate)	2.0055
Varian pitch	2.0028
Diphenyl nitroxide	2.0065

From Avogadro's number, it is known that one mole of any compound contains 6.0247×10^{23} molecules. Assuming that there is one electron spin in each molecule, then one mole of a compound contains approximately 6×10^{23} spins, and one gram of the compound contains

$$\frac{6 \times 10^{23}}{\text{Molecular Weight}} \text{ spins.} \tag{7}$$

The intensity (or area) of a spectrum is calibrated with standards such as α, α'-diphenyl-β-picrylhydrazyl (DPPH), potassium peroxylamine disulfonate [Fremy's salt $K_2NO(SO_3)_2$], or carbon. One gram of DPPH contains 1.5×10^{21} unpaired spins, and the "strong pitch" standard (0.1% powdered coal in KCl) has 3×10^{15} spins/cm of length. A computer is useful for integrating the spectral area. Alternatively, techniques have been developed in which the absorption line is cut out from the paper trace, weighed, and compared with that of a standard sample.

5.5.3.3 Line Shape

Theoretically, ESR lines should be infinitely narrow; experimentally, they are broadened by various mechanisms, both intrinsic (determined by sample properties and the physics of the resonance experiment) and extrinsic (dependent on spectrometer operating conditions). Line shapes of ESR signals of peroxy radicals in lignin are often asymmetric because of the "solid-state" or "power spectrum" effects of the various anisotropic interactions.

The line shape or line width of an ESR signal can be treated, rather simplistically, as comprising two kinds of contributions. The first contribution is related to the spin-lattice relaxation. Relaxation results from interactions between an electron in an excited energetic level and its surroundings. The relaxation process shortens the lifetime of the magnetic levels and causes broadening of the spectral lines because of the Heisenberg Uncertainty Principle. The second contribution arises because the local magnetic fields surrounding each molecule are slightly different. This difference results in a spread in the resonance frequency values and a corresponding broadening of the resonance, which is sometimes known as inhomogeneous broadening to distinguish it from

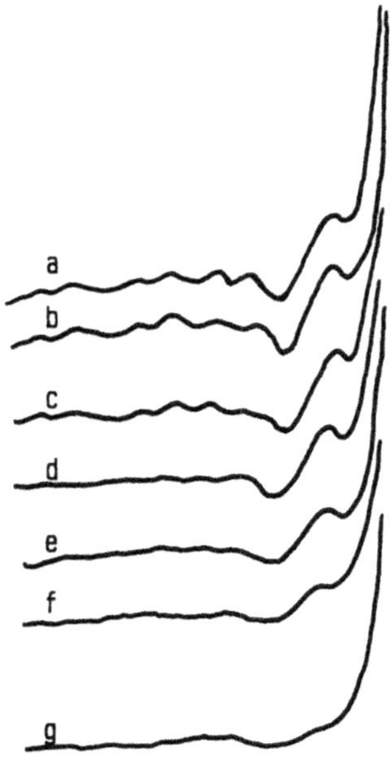

Fig. 5.5.2. Power saturation of ESR spectra of photo-irradiated milled wood lignin at 77 K. Microwave power (mW): *a* 5; *b* 10; *c* 15; *d* 20; *e* 30; *f* 35; *g* 40

pure relaxation broadening. This situation occurs when molecules are either static or tumbling relatively slowly. If the molecules tumble rapidly, static broadening is partially reduced by averaging the different local fields of the molecules.

Power saturation also leads to line broadening. Power saturation occurs when the incident radiation is sufficiently intense to lift the molecules from the lower to the upper energy level at a rate greater than the relaxation processes can restore thermal equilibrium. This causes a reduction in the peak intensity and an increase in the effective line width. Hence, ESR measurement should be performed at a power level as high as possible to obtain maximum peak intensity but before power saturation occurs. The most practical solution is to study a sample at different power levels to determine its particular power saturation properties and then to select the best operational power level for that sample. Nonetheless, power saturation is a valuable technique when two signals overlap. If one signal saturates readily and the other does not, the use of high microwave powers will favor the signal that does not saturate. This technique is illustrated in Fig. 5.5.2.

5.5.3.4 Hyperfine Structure

Hyperfine structure arises from the interaction of the electron spins (S) with the nuclear spins (I). In general, there are (2I + 1) orientations of the nuclear spin in a magnetic field, each of them corresponding to a different energy. The unpaired electron spin recognizes these different orientations with the result that

Table 5.5.2. Binomial coefficients for the relative intensities of hyperfine lines (spin = 1/2)

n	Relative intensity of ESR lines	Number of lines
0	1	1
1	1 1	2
2	1 2 1	3
3	1 3 3 1	4
4	1 4 6 4 1	5
5	1 5 10 10 5 1	6
6	1 6 15 20 15 6 1	7

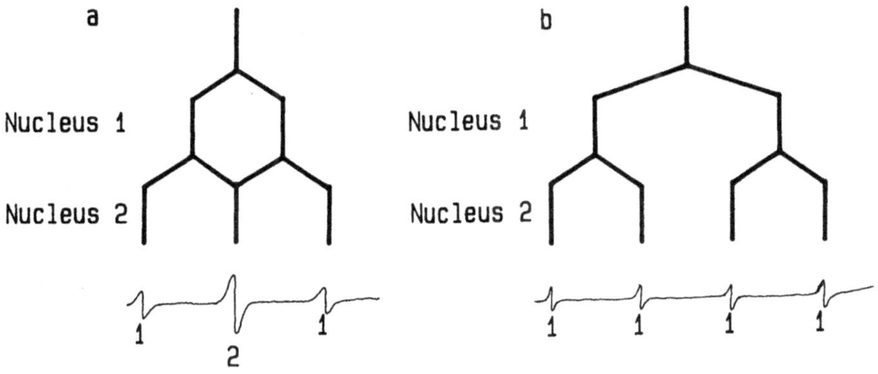

Fig. 5.5.3a,b. Evolution of ESR spectra for radicals containing two nuclei with I = 1/2 when they are (**a**) equivalent and (**b**) nonequivalent

the ESR spectrum is split into (2I + 1) lines of equal intensity. The magnitude of the splitting between the lines is termed the hyperfine splitting constant and is measured in gauss. For n equivalent nuclei, the ESR spectrum consists of (2nI + 1) lines whose relative intensities are given by the binomial coefficients obtained in the expansion of $(1 + x)^n$ (Table 5.5.2). For example, if n = 2, then the relative intensities are 1:2:1 (Fig. 5.5.3a). If there is more than one set of equivalent nuclei, each of the (2nI + 1) lines is further split according to the same principle. For equivalent nuclei, the splitting constant may be obtained by repetitive equal splitting of hyperfine levels. However, for nonequivalent nuclei, the splitting constant can be different, as shown in Fig. 5.5.3b. Obviously, the more nuclei that interact with the unpaired electron, the more complex the ESR spectrum.

Spectra that have overlapping peaks are difficult to interpret. Sometimes a computer is used to help interpret these spectra. The normal approach is to use measured or theoretical splitting constants to obtain a computer-calculated spectrum which is compared to the actual spectrum. If the spectra match, the sample is assumed to correspond to the single substance.

5.5.4 Methods

5.5.4.1 Sample Preparation

Lignin in either solid or solution form can be used in ESR investigations. Most ESR spectrometers operate at X-band frequencies where about 0.2 ml of a solution or 0.2 g of a solid sample is sufficient. Too large a sample may damp the cavity and lower the sensitivity. Lignin samples dissolved in a dioxane/water mixture are best contained in capillary tubes or as a relatively thin solution film in flat cells that are placed in a suitable region of the microwave cavity. A solid sample should be uniformly packed in a sample tube, which must be chosen with careful attention to the physical, chemical, and magnetic properties of each sample. Most glass and silica materials give strong ESR signals and are not suitable. Fused silica of high purity is, however, suitable. This material has a very low dielectric loss and hence is a good material for Dewar inserts. However, the irradiation of many materials, including fused silica by UV and other high energy radiation, often gives rise to substances having ESR signals. When polar solvents of high dielectric constant are used, even quartz tubes are generally not usable. In this case, capillaries or thin rectangular glass cells are used instead.

All samples are properly degassed before ESR measurements to eliminate oxygen that, as a paramagnetic substance, can contribute to the line broadening. In addition, oxygen may react readily with organic molecules having unpaired electrons to form peroxy radicals, as is frequently observed with lignin samples in the presence of air or oxygen. This, of course, reduces the radical concentration and, consequently, the ESR spectral intensity of the lignin sample.

Variable temperature controls are necessary for determining activation energies and associated thermodynamic parameters. Although sample temperature does not govern the line width at maximum sensitivity, one should work at as low a temperature as feasible because signal amplitude is inversely proportional to the absolute temperature (Curie's Law). Freezing lignin solutions removes the dielectric losses associated with liquid water, thus increasing sensitivity and allowing the use of larger samples. Moreover, unstable free radicals may decay at ambient temperature before they can be detected by ESR. For these materials, samples should be kept at the low temperature of liquid nitrogen or liquid helium. The presence of an alkali additive, usually potassium hydroxide or ammonia, is useful for maintaining the stability of some organic free radicals from hours to up to days. It is interesting to note that these free radicals are closely associated with color. A general rule is that clear individual color, such as green, blue, or yellow, denotes an abundance of free radicals and a change to colorless or brown connotes radical decay. Radicals formed by high energy X- or gamma-ray irradiation are usually quite stable and often survive for several months.

If the experiment is conducted in a Dewar flask at liquid nitrogen temperature, it is advisable to purge the ESR cavity with dry nitrogen gas to avoid ice formation around the Dewar.

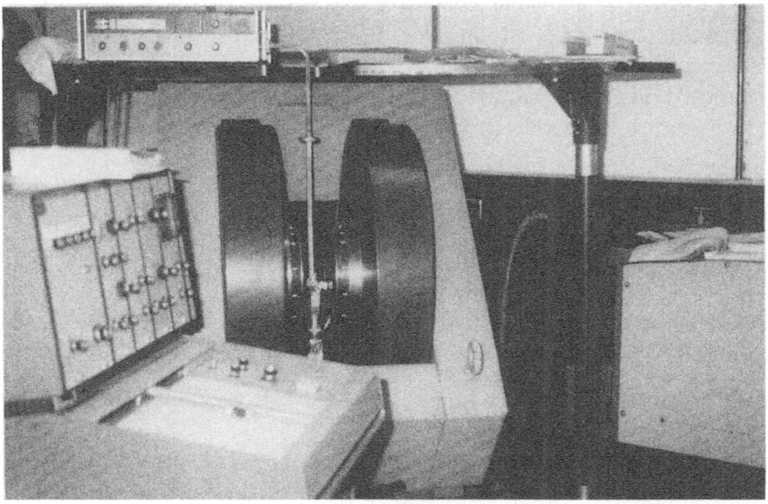

Fig. 5.5.4. A modern E-12 ESR spectrometer (*above*) and a typical ESR cavity situated between the poles of an electromagnetic field (*below*)

5.5.4.2 ESR Measurements

The simplest type of ESR spectrometer consists basically of an operator's console, microwave bridge, magnet, magnet power supply, sample cell or cavity, and a detector/recorder. Figure 5.5.4 shows a commercial ESR spectrometer.

On the left is a partial view of the console; on the right is the sample cavity between the poles of an electromagnet. Most instruments are designed to operate at constant frequency. Common frequencies are 9.5 (X band), 23 (K band), and 35 (Q band) GHz. Electromagnetic radiation, commonly provided by a reflex Klystron, that oscillates at a frequency of 9.5 GHz, is most often used.

5.5.4.3 ESR Operation

It is assumed that a well-tuned ESR instrument is being used. The following operating procedures are applicable to most commercial ESR instruments:

1. Open coolant water supply valve to ESR coolant system.
2. Turn the main power on.
3. Position a sheet of calibrated chart paper on the recorder platen with the center line of the chart aligned with the red arrow on the top edge of the platen.
4. Fill and assemble the pen for recording. Set recorder to OFF.
5. Set bridge mode switch to STANDBY and ATTENUATION to 0.2 mW.
6. Press console POWER switch ON.
7. Insert a strong pitch sample (or other standards) in cavity so that the active volume is centered in cavity.
8. Set ATTENUATION control to 0.2 mW, turn mode switch to TUNE, and POWER LEVELER switch to OFF (Klystron mode trade appears on oscilloscopy).
9. Adjust FREQUENCY control until cavity signal dip is centered on the Klystron mode trace.
10. Adjust cavity iris to the point where maximum dip is obtained.
11. Turn mode switch to OPERATE.
12. Slowly rotate ATTENUATION to 5 mW and simultaneously keep DETECTOR CURRENT indicator at 250 μA by adjusting iris screw on cavity as necessary.
13. Adjust FREQUENCY control bridge to minimize the AFC output.
14. Set FIELD SET controls to 3300 G.
15. Turn Recorder switch to ON.
16. Depending on samples, adjustment of RECEIVE GAIN and MODULATION AMPLITUDE may be necessary to display a well-defined signal.

5.5.5 ESR of Lignin

Lignin is a polyphenolic substance on which phenoxy radicals can be generated readily with many forms of energy. Carbon radicals are generated simultaneously. Thus, radicals have been detected in lignin after oxidation, mechanical treatments, and ultraviolet light and high energy irradiation.

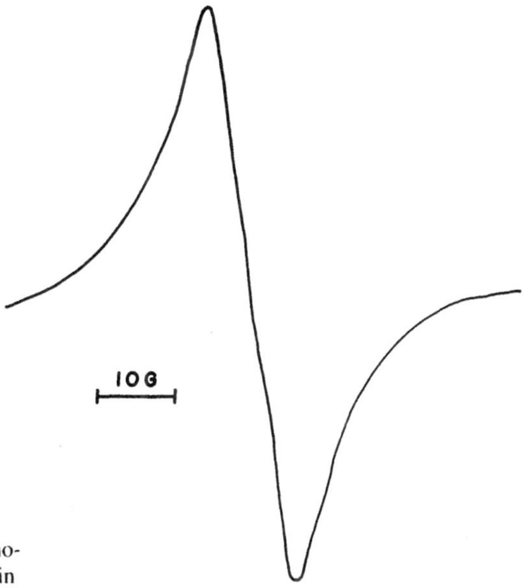

Fig. 5.5.5. ESR spectrum of mechanoradicals generated in milled wood lignin

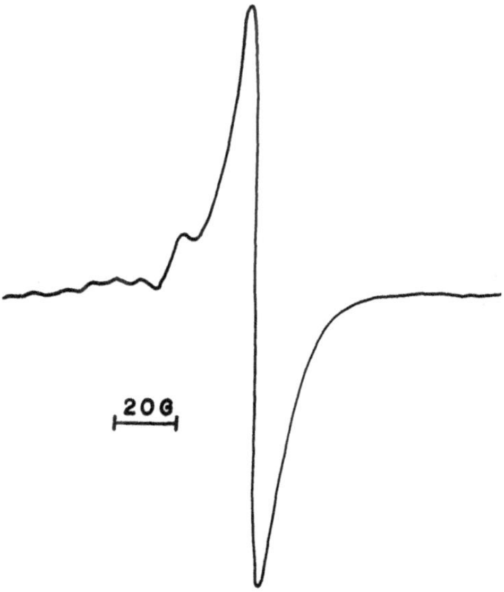

Fig. 5.5.6. ESR spectrum of milled wood lignin irradiated with an ultraviolet light of $\lambda > 260$ nm at 77 K

5.5.5.1 Detection of Mechanoradicals in Lignin

Milled wood lignin is carefully packed into a quartz sample tube and inserted into the ESR cavity. Turn on the instrument and by following the procedure described above, a singlet signal with a g-value of 2.0023 and a line width of 16 gauss will be detected (Fig. 5.5.5). The concentration of this particular lignin

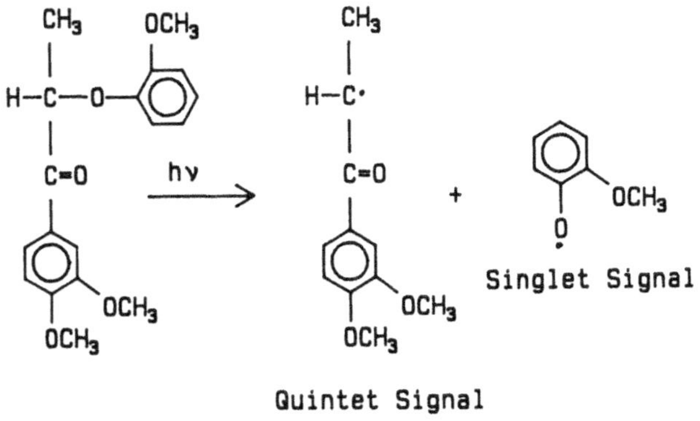

Fig. 5.5.7. Typical lignin compound containing an aryl-β-ether unit

free radical is 2.446×10^{17} spin/g. The singlet is a typical signal of phenoxy radicals generated in lignin samples.

5.5.5.2 Detection of Free Radicals in Photoirradiated Lignin

Milled wood lignin is carefully packed into a quartz sample tube and inserted into a Dewar flask filled with liquid nitrogen. The Dewar flask has a finger tip which can be inserted into the ESR cavity. An ultraviolet light source with a wavelength greater than 260 nm is used to irradiate the lignin sample in the ESR cavity. When lignin is irradiated for 60 min, an ill-defined multiplet signal can be detected (Fig. 5.5.6). A singlet signal with a line width of 14 gauss located at the center is predominant and several side peaks also appear in the lower magnetic field to the left of the principal line. The principal singlet signal has a g-value of 2.0023 and the overall concentration of free radicals is 1.296×10^{18} spins/gram.

5.5.5.3 Detection of Free Radicals in a Photoirradiated Lignin Model Compound

Since only an ill-defined signal can be detected from photoirradiated lignin, it is sometimes helpful to study free radicals generated in lignin model compounds. A typical lignin compound containing an aryl-β-ether unit is here selected for demonstration (see Fig. 5.5.7). The model compound is placed in a quartz sample cell and irradiated with ultraviolet light as discussed in the previous section. After irradiation for 60 min, a five-line spectrum is observed as shown in Fig. 5.5.8. When the irradiated sample is warmed to 298 K for 3 min and the spectrum is recorded again at 77 K, the five-line spectrum collapses to a singlet signal with a line width of 15 gauss and a g-value of 2.0024. It is apparent that the five-line signal consists of superimposed quintet and singlet signals. The free radicals giving rise to these ESR signals are formed by the cleavage of the β-O-4 linkage. The free radical at the β-carbon interacts equivalently with four protons

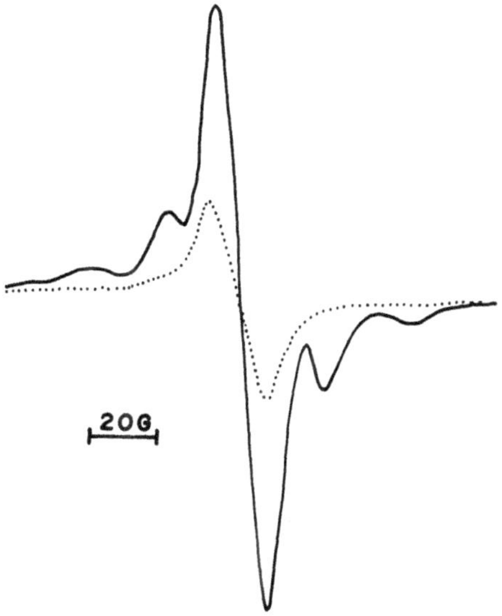

Fig. 5.5.8. ESR spectrum of a lignin model compound containing an aryl-β-ether unit irradiated with an ultraviolet light of $\lambda >$ 260 nm at 77 K. Heating of the irradiated sample at 298 K for 3 min reduces the spectrum to a singlet signal (*dotted line*)

at the β- and σ-positions to generate the quintet signal, while the phenoxy radical generates the singlet signal (Fig. 5.5.7).

Many free radicals generated in lignin have been studied. Because of the structural complexity of lignin, multiplet spectra are obtained. The analysis and interpretation of the spectrum can be performed by different techniques depending on the experimental conditions (Hon 1982, 1987, Zavoiskii 1945).

5.5.6 Discussion

As stated in the introduction, ESR is becoming an important tool for the study of lignin free radicals. However, the usefulness of this analytical technique, when applied to polymeric lignin, is limited to the quantitative measurement of spin (or free electron) content. Because of the unresolved nature of lignin ESR spectra, the detailed structure of lignin free radicals usually cannot be deduced. This difficulty is attributed to the complex polymeric structure of lignin which unavoidably gives rise to a variety of free radicals. The spectra of these radicals are overlapped and only a overall diffuse signal (singlet) can be detected.

ESR studies of lignin model compounds, on the other hand, have yielded information on the detailed structures of phenoxy radicals that are involved in some homolytic splitting reactions of lignin. Such information has proven useful in elucidating the mechanisms of these reactions. As with other spectrometric methods, ESR also suffers from its inability to shed light on the true nature of free radicals, phenoxy or otherwise, that exist in lignin macromolecules.

References

Alger RS (1968) Electron paramagnetic resonance: techniques and applications. Wiley-Interscience, New York, 580 pp

Hon DNS (1982) Photochemical degradation of lignocellulosic materials. In: Grassie N (ed) Development in polymer degradation-3. Applied Science Publishers, Essex, England, 229–281

Hon DNS (1987) Mechanochemistry of lignocellulosic materials. In: Grassie N (ed) Development in polymer degradation-7. Applied Science Publishers, Essex, England, 165–191

Kringstad KP, Lin SY (1970) Mechanism in the yellowing of high-yield pulps by light. Structure and reactivity of free radical intermediates in the photodegradation of lignin. TAPPI 53:2296–2301

Poole CP (1967) Electron Spin Resonance. Wiley-Interscience, New York, 922 pp

Simkovic I (1986) Free radicals in wood chemistry. J Macromol Sci – Rev Macromol Chem Phys C26:67–80

Steelink C (1966) Stable free radicals in lignin and lignin oxidation products. In: Marton J (ed) Lignin structure and reactions. Adv Chem Ser 59:51–64

Symons M (1978) Chemical and biochemical aspects of electron Spin Resonance Spectroscopy. Wiley, New York, 190 pp

Wertz JE, Bolton JR (1972) Electron Spin Resonance elementary theory and practical applications. McGraw-Hill, New York, 497 pp

Zavoiskii E (1945) Paramagnetic relaxation of liquid solution for perpendicular fields. J Phys (USSR) 9:211–216

Chapter 6

Characterization in Solution: Chemical Degradation Methods

6.1 Acidolysis

K. Lundquist

6.1.1 Introduction

In this section, the term acidolysis refers specifically to the refluxing (or heating at 100 °C) of lignin or lignocellulose with 0.2 M HCl in dioxane-water (9:1, v/v). This treatment was originally introduced in lignin chemistry as a procedure for isolating lignin from plant materials (Pepper et al. 1959) and as an analytical tool in lignin studies for determining the occurrence of β-O-4 structures and β-5 structures in spruce lignin (Adler et al. 1957, Adler 1977). It is noteworthy that, besides the acidolysis reagent, other mixtures containing dioxane and hydrogen chloride have been used for the extraction of lignin from plant materials (Stumpf and Freudenberg 1950, Stumpf et al. 1953, Pla et al. 1984).

Succeeding studies of acid lignin degradation have provided detailed knowledge about the reactions of lignin during acid treatments (Lundquist 1976, Hoo et al. 1983, Brunow and Lundquist 1984, Lapierre et al. 1985, Davies et al. 1987, Ede and Brunow 1989). Acidolysis studies have offered evidence for the occurrence of structural elements in lignins of the β-O-4, β-5, β-β, β-1, glyceraldehyde-2-aryl ether, 2-aryloxypropiophenone, cinnamaldehyde, cinnamic acid, benzaldehyde, benzoic acid, and quinonoid types. Such investigations also provide possibilities for making quantitative estimates of certain types of lignin structures. As a result of this work, most low-molecular weight lignin acidolysis products can be related to specific lignin structures with confidence. Acidolysis reactions in the presence of other catalysts than hydrogen chloride have recently been studied (Yasuda and Ota 1987 and preceding work, Karlsson et al. 1988).

Acidolysis causes the selective cleavage of arylglycerol-β-aryl ethers (4, Fig. 6.1.2) and some other types of labile ether linkages, and this has been utilized on several occasions, for example by Kirk and Adler (1970). Treatment with the acidolysis reagent at 50 °C (Adler et al. 1968) has been used in lignin chemistry for the selective hydrolysis of noncyclic benzyl aryl ethers.

Lignins can be detected and characterized by analysis of their low-molecular weight acidolysis products, and it is this application of acidolysis that is the subject of the current presentation. Routine analytical procedures have been developed only for monomeric acidolysis products. Therefore, analysis of such compounds is the primary concern in this section. Analysis of dimeric acidolysis products is, however, of great value for the characterization of lignins and is treated under a separate heading (Sect. 6.1.6).

Acidolysis has a close relationship to ethanolysis (i.e., heating of lignin with hydrogen chloride in ethanol). Both treatments result in the degradation of

Springer Series in Wood Science
Methods in Lignin Chemistry
(Edited by S.Y. Lin and C.W. Dence)

lignin with formation of substantial amounts of arylpropanones. Moreover, it has been shown (Karlsson et al. 1988) that the presence of chloride ion affects certain reactions involved in acid-catalyzed lignin degradation. Acidolysis will not be compared with ethanolysis in detail here, but it should be pointed out that dioxane-water (9:1, v/v) is a better solvent for lignins than is ethanol.

6.1.2 Characterization of Lignins by Analysis of Low-Molecular Weight Acidolysis Products

A method for the characterization of lignins based on the analysis of acidolysis monomers was developed in 1971 (Lundquist and Kirk 1971). Monomeric phenolic degradation products were analyzed as trimethylsilyl derivatives by gas chromatography (GC) and gas chromatography-mass spectroscopy (GC-MS). Alternatively, lignin acidolysis products may be analyzed by high performance liquid chromatography (HPLC) (Lapierre et al. 1983a).

Most of the acidolysis monomers detected are shown in Fig. 6.1.1 (Higuchi et al. 1972b, Nakatsubo et al. 1972, Lundquist 1976, Lapierre et al. 1983a). The origin of these products has been widely studied (Lundquist and Lundgren 1972, Yamasaki et al. 1972, Lundquist 1976, Lapierre et al. 1983a, Hoo et al. 1983, Brunow and Lundquist 1984, Gellerstedt and Lindfors 1987). The majority of the acidolysis monomers originate from arylglycerol β-aryl ether structures (4, Fig. 6.1.2). The formation of acidolysis products from such structures is outlined in Fig. 6.1.2; detailed mechanistic sequences illustrating the various acidolysis reactions are reported elsewhere (Lundquist 1976). The diketones (1e, 2e, 3e, Fig. 6.1.1) may also originate from 2-aryloxypropiophenone structures of type 5. Cinnamaldehydes (1i, 2i, 3i) are primarily liberated by cleavage of β-ether linkages.

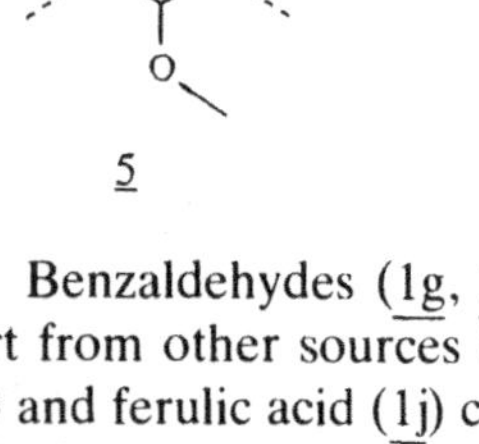

5

Benzaldehydes (1g, 2g, 3g) are liberated in the same way but originate in part from other sources (cf. Fig. 6.1.2). The presence of benzoic acids (1h, 2h, 3h) and ferulic acid (1j) can be explained by hydrolysis of ether or ester linkages. p-Hydroxybenzoic and p-coumaric acids (3h and 3j, respectively) are attached to certain types of lignins by ester linkages and their occurrence in the acidolysis mixtures is, in these cases, at least primarily due to the hydrolysis of these linkages.

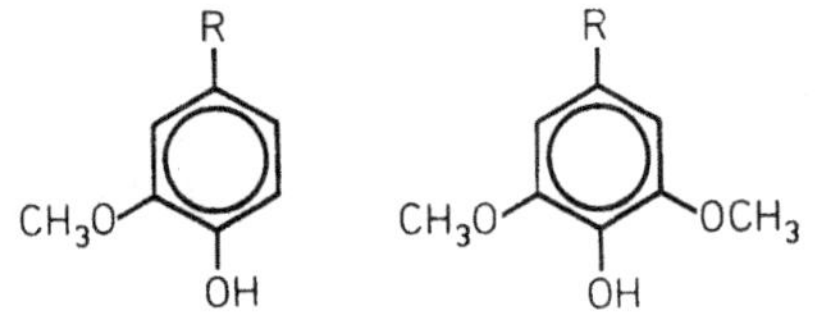

1	2	3	R
a	a	a	a: —CH_2COCH_2OH
b	b	b	b: —$CH(OH)COCH_3$
c	c	c	c: —$COCH(OH)CH_3$
d	d	d	d: —CH_2COCH_3
e	e	e	e: —$COCOCH_3$
f	f	–	f: —CH_2CHO
g	g	g	g: —CHO
h	h	h	h: —COOH
i	i	i	i: —CH=CHCHO
j	–	j	j: —CH=CHCOOH

Fig. 6.1.1. Monomeric phenols detected in lignin acidolysis reaction mixtures

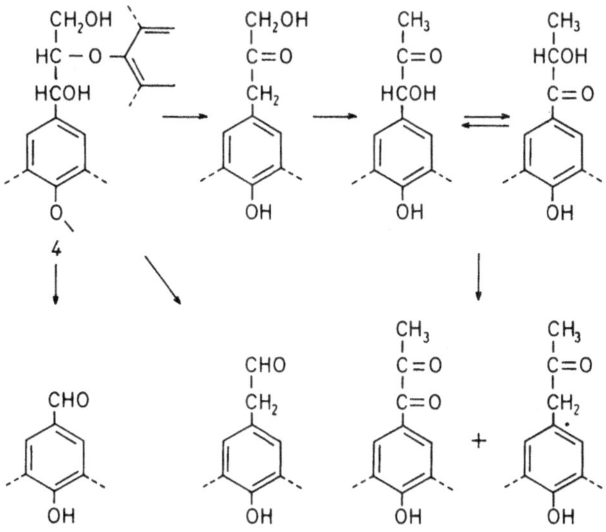

Fig. 6.1.2. Formation of monomeric phenols from *β*-O-4 structures in lignin during acidolysis

Analysis for low molecular weight acidolysis products constitutes a comparatively simple method for the detection of lignin in plant materials as well as for the recognition of the type of lignin present, since this information can be derived from product composition. The quantified results from such an analysis may also be used as an estimate of lignin content. Since the lignin structures that give rise to acidolysis monomers are rather sensitive to chemical changes, acidolysis results also provide information about the residual intact lignin in the sample. In some cases, compounds have been detected in acidolysis mixtures originating from modified lignins that reflect the types of changes the lignin has undergone (Kirk and Chang 1975, Crawford et al. 1982, Lapierre et al. 1983a, Gellerstedt et al. 1984).

6.1.3 Applications of the Acidolysis Method

Analysis of acidolysis monomers has been used as a tool for the characterization of a variety of isolated lignins and lignin components in plant materials (Lundquist and Kirk 1971, Higuchi et al. 1972a,b, Wolter et al. 1974, Kutsuki and Higuchi 1978, Lapierre et al. 1982, 1983a,b, 1984, 1985, 1986, Westermark 1985). Acidolysis has also been used for the characterization of dehydrogenation polymers of p-hydroxycinnamyl alcohols (Yamasaki et al. 1972, 1973, Kirk et al. 1975b).

The yield of acidolysis monomers reflects the amount of intact lignin in a sample. This fact constitutes the basis for the use of acidolysis in studies on the biodegradation of lignin by fungi (Konishi et al. 1972, Kirk 1975, Kirk and Chang 1975, Kirk et al. 1975a), by bacteria (Crawford et al. 1982, 1983, Borgmeyer and Crawford 1985, Deobald and Crawford 1987) and by termites (Esenther and Kirk 1974). Fungal (Kirk and Chang 1975) as well as bacterial (Crawford et al. 1982) degradation of lignin has been shown to lead to groups which release vanillic acid on acidolysis. Acidolysis has also been used in studies of the chemical changes in lignin occurring during kraft pulping (Gellerstedt et al. 1984, Gellerstedt and Lindfors 1987).

6.1.4 Preferred Acidolysis Procedure

6.1.4.1 Preparation of the Acidolysis Reagent

Reagent grade dioxane is refluxed with sodium until the sodium acquires a metallic luster (~12 h). The dioxane is distilled and stored frozen in bottles. The acidolysis reagent is prepared by dilution of 2 M hydrochloric acid with dioxane to a 0.2 M solution in a volumetric flask.

6.1.4.2 Acidolysis Reaction

A solution (or slurry) of the sample is refluxed for 4 h. A sample/reagent ratio of <0.01 is recommended. The boiling point of the dioxane-water azeotrope (81.6% dioxane) is 87.8 °C. Losses of solvent due to evaporation during the treatment should be avoided (Susarev and Guseva 1964, Froment and Robert 1970). (Note 1).

6.1.4.3 Work-Up of Acidolysis Mixture

The acidity of the cooled acidolysis mixture (V ml) is reduced by addition of 0.45 × V ml of 0.4 M aqueous $NaHCO_3$. The acidolysis products are extracted with chloroform (V ml and 3 × 0.5 × V ml). It is recommended that the completeness of the extraction be verified by UV spectroscopy or thin layer chromatography. If necessary, a complementary extraction with dioxane-chloroform (1:1, v/v) and chloroform may be performed. The extract is dried (Na_2SO_4) and solvents removed by film evaporation. The residue is dried in vacuo over KOH and P_2O_5. (Note 2).

6.1.4.4 Analysis of Acidolysis Products

A portion of the residue (5–10 mg) is treated with 60 μl of N,O-bis-trimethylsilylacetamide (BSA, Pierce) in 100 μl of pyridine for 30 min. The reaction mixture is analyzed by gas chromatography (and gas chromatography-mass spectrometry) according to Lapierre et al. (1983a). (Note 3).

In qualitative work, an acidolysis mixture from a previously investigated lignin, for example, spruce lignin (Lundquist and Kirk 1971) or aspen lignin (Lapierre et al. 1983a), may serve as a reference sample. In quantitative work, especially, there is a need for reference compounds. Several of the acidolysis monomers are commercially available. Syntheses of most of the arylpropanones present in lignin acidolysis mixtures have been described by Fischer and Hibbert (1947) and Hibbert and co-workers in preceding work, Kratzl and Schweers (1956), Gorecki and Pepper (1959), Lundquist and Hedlund (1967), Lundquist (1970), Lundquist and Lundgren (1972).

Note 1. An inert atmosphere, for example N_2, may be of value when special attention is paid to oxidative changes in the sample but does not seem to be a general requirement for acidolysis experiments.

Note 2. Drying in vacuo may cause loss of volatile compounds such as guaiacol.

Note 3. The conditions for the derivatization are selected to ensure complete derivatization. Alternatively, N,O-bis-trimethylsilyltrifluoroacetamide can be used as silylating agent. Docosane or tetracosane (in pyridine or dioxane solution) can be used as internal standard. The homogeneity of the reaction mixtures has to be considered in quantitative work.

6.1.5 Discussion of Procedure

Lundquist and Kirk (1971) treated spruce lignin with the acidolysis reagent at reflux temperature for 4, 8, and 24 h. Ketol 1a (Fig. 6.1.1) predominates in the 4-h acidolysis mixture (yield, 4.6% of the lignin) and prolonged treatment does not result in any appreciable increase in the yield of acidolysis monomers. The yields of the primary ketols 1a and 2a (Fig. 6.1.1) from birch lignin (a typical

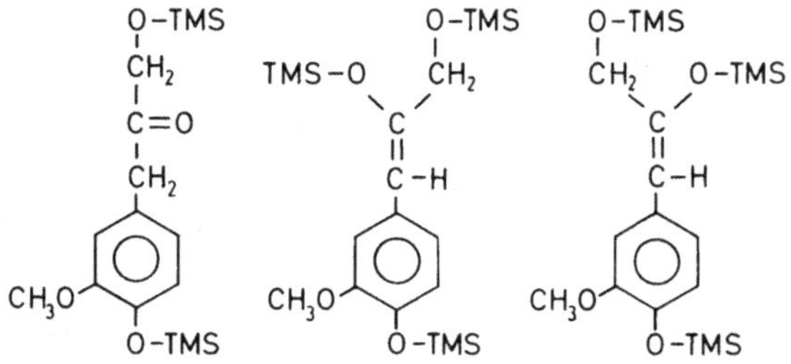

Fig. 6.1.3. Trimethylsilyl (TMS) derivatives of 1-hydroxy-3-(4-hydroxy-3-methoxyphenyl)-2-propanone (1a, Fig. 6.1.1)

hardwood lignin) were of the order of 3 and 5%, respectively. In most acidolysis studies, the sample is treated with the acidolysis reagent for 4 h at reflux temperature. Pometto and Crawford (1985) suggest heating the sample for 6 h at 87 °C in a sealed ampoule. Studies of the formation of acidolysis monomers at 100 °C (water bath, sealed ampules) showed that the yield increased only slightly when the reaction time was prolonged from 2 to 4 h (Lapierre et al. 1986). Acidolysis experiments have also been performed at 120 °C (Higuchi et al. 1972a).

A variety of work-up procedures have been described in the literature. Lundquist and Kirk (1971) applied gel permeation chromatography to separate the acidolysis monomers from polymeric materials in the initial step. This is, however, not a requirement when the acidolysis monomers are analyzed by gas chromatography as their trimethylsilyl derivatives.

Acidolysis monomers are generally analyzed as their trimethylsilyl derivatives by gas chromatography as described by Lundquist and Kirk (1971), Higuchi et al. (1972b), Lapierre et al. (1983a), Westermark (1985) and Pometto et al. (1985). Use of a capillary column appears to be advantageous (Lapierre et al. 1983a, Westermark 1985). Alternatively, acidolysis monomers may be analyzed by high performance liquid chromatography (Lapierre et al. 1983a).

The rate of derivatization of different compounds varies when trimethylsilyl derivatives are prepared (Lundquist and Kirk 1971). Another complication arises from the fact that enolization and subsequent formation of enol derivatives occur during the derivatization treatment (Lundquist and Kirk 1971). Ketol 1a (Fig. 6.1.1), for example, has been shown to give rise to three different derivatives (Fig. 6.1.3) (Lapierre et al. 1983a). One approach for overcoming these difficulties consists of derivatization under mild conditions in combination with a gas chromatographic study of the rate of derivatization (Lundquist and Kirk 1971, Pometto and Crawford 1985). A second solution to this problem consists of performing the silylation under conditions that secure complete derivatization and analysis of all types of derivatives formed (Lapierre et al. 1983a). The latter principle has been adopted in the preferred acidolysis procedure.

In an attempt to circumvent the analytical problems discussed above, a method for the analysis of acidolysis monomers has been developed which involves reduction of carbonyl and carboxyl groups with borane dimethyl sulfide

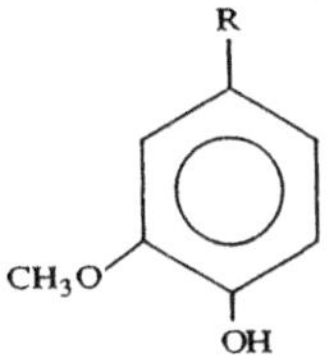

6

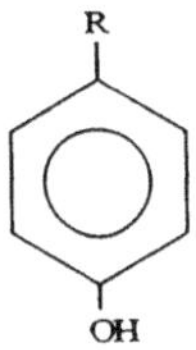

7

Compound	*R*
6a	$—CH(OH)CH(OH)CH_3$
6b,7b	$—CH_2OH$
6c,7a	$—CH_2CH(OH)CH_2OH$
6d	$—CH_2CH_2OH$
6f	$—CH_2CH(OH)CH_3$
6g	$—CH_2CH_2CH_2OH$

Fig. 6.1.4. Compounds identified after reduction of the acidolysis products with borane-dimethyl sulfide

Table 6.1.1. HPLC[a] elution volumes of lignin acidolysis products after reduction with borane-dimethyl sulfide and acetylation

Compounds[b]	Relative elution volume
Diacetate of 7b	0.39
Diacetate of 6f	0.51
Diacetate of 6g	0.56
Diacetate of 6b	0.56
Diacetate of 6d	0.60
Triacetate of 7a	0.65
Triacetate of 6a (*erythro*)	0.74
Triacetate of 6a (*threo*)	0.84
Triacetate of 6c	1.00

[a] Separation was performed with a normal phase column (Radial-PAK Si) using hexane-ethyl acetate (3:1 v/v) as eluent. Elution volumes are listed relative to that of the triacetate of 6c (elution volume = 20 ml); the void volume of the system was ~3.6 ml.
[b] See Fig. 6.1.4 for structures of compounds.

$[BH_3 \cdot S(CH_3)_2]$ and acetylation followed by gas chromatographic analysis (Kristersson et al. 1980). Alternatively, the reduction products possibly may be analyzed as trimethylsilyl derivatives by gas chromatography.

Reduction of carboxylic acids with borane-dimethyl sulfide generally proceeds rapidly. In connection with reduction experiments with p-hydroxybenzoic acid, it was found that this compound gave only low yields of p-hydroxybenzyl alcohol when treated with borane-dimethyl sulfide according to the procedure described by Kristersson et al. (1980). This is in accordance with studies of the

Table 6.1.2. HPLC[a] elution volumes of lignin acidolysis products after reduction with borane-dimethyl sulfide

Compounds[b]	Ralative elution volume	
	Radial-PAK C_{18}	Radial-PAK C_8
6a (*erythro*)	0.19	0.28
7a	0.27	0.35
6a (*threo*)	0.31	0.40
6b	0.33	0.42
6c	0.36	0.42
6d	0.65	0.68
6e	1.00	1.00
6f	1.41	1.21
6g	–	1.30
7b	–	0.35

[a] Separation was performed on reverse phase columns (Radial-PAK C_{18} and Radial-PAK C_8) using water-acetonitrile-acetic acid (90:10:1 v/v/v) as eluent. Elution volumes are listed relative to that of α-hydroxypropioguaiacone 6e; elution volume 25 ml (Radial-PAK C_8) and ~42 ml (Radial-PAK C_{18}); the void volumes of both systems were ~ 3.6 ml.
[b] See Fig. 6.1.4 for structures of compounds.

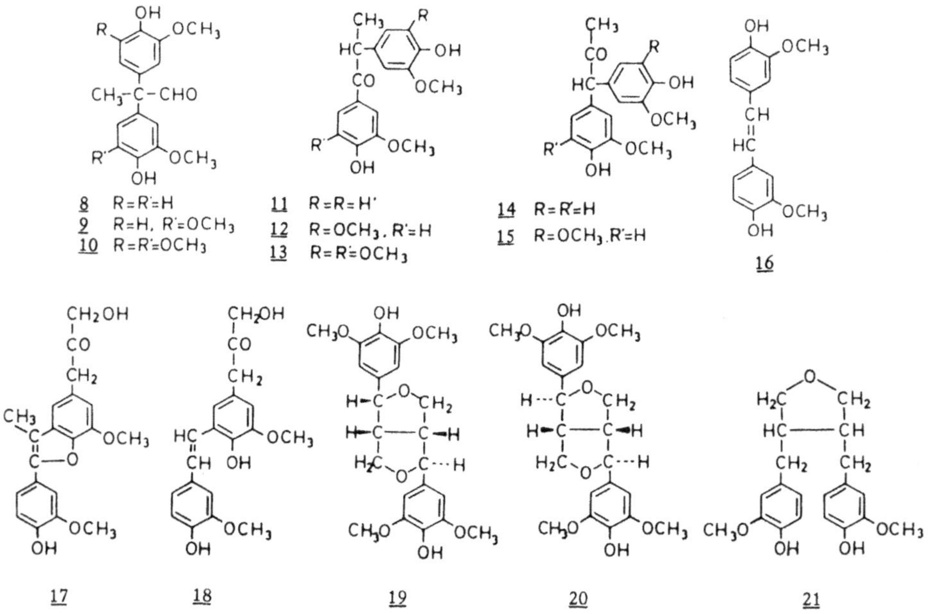

Fig. 6.1.5. Dimeric lignin acidolysis products

reduction of other aromatic carboxylic acids by borane-dimethyl sulfide (Lane et al. 1974). Thus, the analysis of aromatic acids in lignin acidolysis mixtures as their borane-dimethyl sulfide reduction products requires further investigation.

Several derivatives obtained by reduction/acetylation of acidolysis monomers have been analyzed by high performance liquid chromatography (HPLC) (Kristersson and Lundquist unpubl. work 1983). The relative elution volumes for these compounds are listed in Table 6.1.1. Nonacetylated reduction products of acidolysis monomers have also been examined by HPLC (Kristersson and Lundquist unpubl. work 1983). From the results shown in Table 6.1.2, it is evident that good separation was achieved but, unfortunately, compounds 6a (*threo* form), 6b and 6c have very similar elution volumes.

6.1.6 Dimeric Acidolysis Products

Acidolysis of spruce and/or birch lignins results in the formation of the dimeric products 8–21 (Fig. 6.1.5) (Lundquist 1976). Products 8–16 originate from structural elements of the β-1 type (22). Acidolysis of trimeric sequences of lignin units of type 23 gives rise to phenylcoumarone 17 and stilbene 18. Structures of β-β type are the sources of compounds 19–21.

Compounds 8–16 are of particular interest in the detection and characterization of lignins since they are comparatively volatile and therefore can be analyzed by gas chromatography. This has been done after reduction with sodium borohydride followed by acetylation (Lundquist 1970, 1973). The dimeric constituents are separated by gel permeation chromatography and column chromatography prior to derivatization and gas chromatographic analysis. However, it should be possible to derivatize the crude lignin acidolysis mixture using the method described by Kristersson et al. (1980) and analyze the product mixture for monomeric as well as dimeric constituents by gas chromatography.

Compounds 8–16 are prominent constituents of the dimer fractions from lignin acidolysis mixtures and this finding serves as evidence for the presence of substantial amounts of β-1 structures in the original lignins. However, other studies do not provide support for this conclusion (Lundquist 1987, Habu et al.

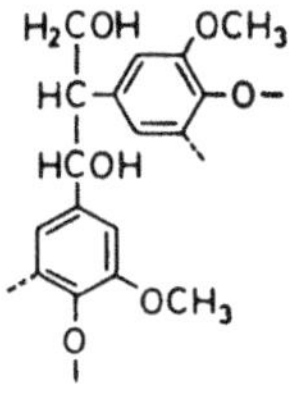

22

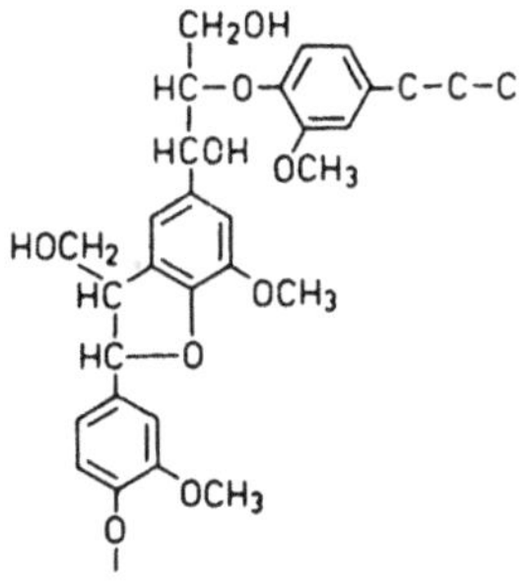

23

1990). Nevertheless, lignin degradation products assumed to be related to the β-1 type of structure are obtained from isolated lignins as well as lignocellulosic materials, thus reinforcing the opinion that they are true lignin degradation products.

References

Adler E (1977) Lignin chemistry – past, present and future. Wood Sci Technol 11:169–218

Adler E, Miksche GE, Johansson B (1968) Über die Benzyl-arylätherbindung im Lignin. I. Freilegung von phenolischem Hydroxyl in Ligninpräparaten durch Spaltung leicht hydrolysierbarer Alkyl-arylätherstrukturen. Holzforschung 22:171–174

Adler E, Pepper JM, Eriksoo E (1957) Action of mineral acid on lignin and model substances of guaiacylglycerol-beta-aryl ether type. Ind Eng Chem 49:1391–1392

Borgmeyer JR, Crawford DL (1985) Production and characterization of polymeric lignin degradation intermediates from two different Streptomyces spp. Appl Environ Microbiol 49:273–278

Brunow G, Lundquist K (1984) On the synthesis of certain lignin-related α-aryloxycinnamic acids and enol ethers. Acta Chem Scand B38:323–325

Crawford DL, Barder MJ, Pometto AL III, Crawford RL (1982) Chemistry of softwood lignin degradation by Streptomyces viridosporus. Arch Microbiol 131:140–145

Crawford DL, Pometto AL, Crawford RL (1983) Lignin degradation by Streptomyces viridosporus: Isolation and characterization of a new polymeric lignin degradation intermediate. Appl Environ Microbiol 45:898–904

Davies JL, Nakatsubo F, Murakami K, Umezawa T (1987) Organic acid pulping of wood. IV. Reactions of arylglycerol-β-guaiacyl ethers. Mokuzai Gakkaishi 33:478–486

Deobald LA, Crawford DL (1987) Activities of cellulase and other extracellular enzymes during lignin solubilization by Streptomyces viridosporus. Appl Microbiol Biotechnol 26:158–163

Ede RM, Brunow G (1989) Formic acid/peroxyformic acid pulping. II. Synthesis of 3-aryl-2,3-dihydro-7-methoxy-2-benzofuranmethanols – model compounds for lignin acidolysis products. Holzforschung 43:127–129

Esenther GR, Kirk TK (1974) Catabolism of aspen sapwood in Reticulitermes flavipes (Isoptera: Rhinotermitidae). Ann Entomol Soc Am 67:989–991

Fischer HE, Hibbert H (1947) Studies on lignin and related compounds. LXXXIII. Synthesis of 3-hydroxy-1-(4-hydroxy-3-methoxyphenyl)-2-propanone. J Am Chem Soc 69:1208–1210

Froment P, Robert A (1970) Étude des conditions d'extraction par des solution dioxane-eau-acide chlorhydrique. Rev ATIP 24:189–193

Gellerstedt G, Lindfors EL (1987) On the formation of enol ether structures in lignin during kraft cooking. Nord Pulp Pap Res J 2:71–75

Gellerstedt G, Lindfors EL, Lapierre C, Monties B (1984) Structural changes in lignin during kraft cooking. 2. Characterization by acidolysis. Sven Papperstidn 87:R61–R67

Gorecki EW, Pepper JM (1959) The synthesis of a lignin model substance: 3-hydroxy-1-(4-hydroxy-3,5-dimethoxyphenyl)-2-propanone. Can J Chem 37:2089–2092

Habu N, Matsumoto Y, Ishizu A, Nakano J (1990) The role of the diarylpropane structure as a minor constituent in spruce lignin. Holzforschung 44:67–71

Higuchi T, Tanahashi M, Nakatsubo F (1972a) Acidolysis of bamboo lignin. III. Estimation of arylglycerol-β-aryl ether groups in lignins. Wood Res (Kyoto) 54:9–18

Higuchi T, Tanahashi M, Sato A (1972b) Acidolysis of bamboo lignin. I. Gas-liquid chromatography and mass spectrometry of acidolysis monomers. Mokuzai Gakkaishi 18:183–189

Hoo LH, Sarkanen KV, Anderson CD (1983) Formation of C_6C_2-enol ethers in the acid-catalyzed hydrolysis of erythro-veratrylglycerol-β-(2-methoxyphenyl) ether. J Wood Chem Technol 3:223–243

Karlsson O, Lundquist K, Meuller S, Westlid K (1988) On the acidolytic cleavage of arylglycerol β-aryl ethers. Acta Chem Scand B42:48–51
Kirk TK (1975) Effects of a brown-rot fungus, Lenzites trabea, on lignin in spruce wood. Holzforschung 29:99–107
Kirk TK, Adler E (1970) Methoxyl-deficient structural elements in lignin of sweetgum decayed by a brown-rot fungus. Acta Chem Scand 24:3379–3390
Kirk TK, Chang H-m (1975) Decomposition of lignin by white-rot fungi. II. Characterization of heavily degraded lignins from decayed spruce. Holzforschung 29:56–64
Kirk TK, Chang H-m, Lorenz LF (1975a) Topochemistry of the fungal degradation of lignin in birch wood as related to the distribution of guaiacyl and syringyl lignins. Wood Sci Technol 9:81–86
Kirk TK, Connors WJ, Bleam RD, Hackett WF, Zeikus JG (1975b) Preparation and microbial decomposition of synthetic [^{14}C] lignins. Proc Natl Acad Sci 72:2515–2519
Konishi K, Inoue Y, Higuchi T (1972) Decomposition of lignin by Coriolus versicolor. IV. Effect of laccase type enzyme on the interphenylpropane linkage of lignin. Mokuzai Gakkaishi 18:571–576
Kratzl K, Schweers W (1956) Über die papierchromatographische Trennung der monomeren Äthanolyseprodukte des Lignins. III. Chem Ber 89:186–192
Kristersson P, Lundquist K, Strand A (1980) Derivatization and analysis of low molecular weight lignin acidolysis products. Wood Sci Technol 14:297–300
Kutsuki H, Higuchi T (1978) The formation of lignin of Erythrina crista-salli. Mokuzai Gakkaishi 24:625–631
Lane CF, Myatt HL, Daniels J, Hopps HB (1974) Organic synthesis using borane-methyl sulfide. II. Reduction of aromatic carboxylic acids in the presence of trimethyl borate. J Org Chem 39:3052–3054
Lapierre C, Rolando C, Monties B (1982) Chromatographie en phase gazeuse sur colonne couplée à la spectrométrie de masse (CGCC-SM) et chromatographie en phase liquide à haute performance (CLHP) des monomères d'acidolyse des lignines. Bull Liaison Groupe Polyphenols 11:381–387
Lapierre C, Rolando C, Monties B (1983a) Characterization of poplar lignins acidolysis products: capillary gas-liquid chromatography of monomeric compounds. Holzforschung 37:189–198
Lapierre C, Gaudillere JP, Monties B, Guittet E, Rolando C, Lallemand JY (1983b) Enrichissement photosynthetique en carbone 13 de lignines de peuplier: caractérisation preliminaire par acidolyse et RMN^{13}C. Holzforschung 37:217–224
Lapierre C, Monties B, Rolando C (1984) Structure des lignines: évaluation de liaisons arylglycérolarylethers par thioacidolyse. C R Acad Sci Ser 3 299:441–444
Lapierre C, Monties B, Rolando C (1985) Thioacidolysis of lignin: comparison with acidolysis. J Wood Chem Technol 5:277–292
Lapierre C, Monties B, Rolando C (1986) Thioacidolysis of poplar lignins: identification of monomeric syringyl products and characterization of guaiacylsyringyl lignin fractions. Holzforschung 40:113–118
Lundquist K (1970) Acid degradation of lignin. II. Separation and identification of low molecular weight phenols. Acta Chem Scand 24:889–907
Lundquist K (1973) Acid degradation of lignin. Part VIII. Low molecular weight phenols from acidolysis of birch lignin. Acta Chem Scand 27:2597–2606
Lundquist K (1976) Low-molecular weight lignin hydrolysis products. Appl Polymer Symp 28:1393–1407
Lundquist K (1987) On the occurrence of β-1 structures in lignins. J Wood Chem Technol 7:179–185
Lundquist K, Ericsson L (1970) Acid degradation of lignin. III. Formation of formaldehyde. Acta Chem Scand 24:3681–3686
Lundquist K, Hedlund K (1967) Acid degradation of lignin. I. The formation of ketones of the guaiacylpropane series. Acta Chem Scand 21:1750–1754
Lundquist K, Kirk TK (1971) Acid degradation of lignin. IV. Analysis of lignin acidolysis products by gas chromatography, using trimethylsilyl derivatives. Acta Chem Scand 25: 889–894

Lundquist K, Lundgren R (1972) Acid degradation of lignin. VII. The cleavage of ether bonds. Acta Chem Scand 26:2005–2023

Nakatsubo F, Tanahashi M, Higuchi T (1972) Acidolysis of bamboo lignin. II. Isolation and identification of acidolysis products. Wood Res 53:9–18

Pepper JM, Baylis PET, Adler E (1959) The isolation and properties of lignins obtained by the acidolysis of spruce and aspen woods in dioxane-water medium. Can J Chem 37: 1241–1248

Pla F, Froment P, Mouttet B, Robert A (1984) Étude de la délignification des végétaux par acidolyse. Holzforschung 38:127–132

Pometto AL III, Crawford DL (1985) Simplified procedure for recovery of lignin acidolysis products for determining the lignin-degrading abilities of microorganisms. Appl Environ Microbiol 49:879–881

Stumpf W, Freudenberg K (1950) Lösliches Lignin aus Fichten- und Buchenholz. Angew Chem 62:537

Stumpf W, Weygand F, Grosskinsky O-A (1953) Synthese von Dioxan-[^{14}C] und Anwendung als Extraktionsmittel für lösliches Lignin. Chem Ber 86:1391–1401

Susarev MP, Guseva VV (1964) Study of the phase equilibria in the system water-hydrogen-chloride-dioxane at 50 °C. J Appl Chem USSR (Engl Trans) 37:1067–1072

Westermark U (1985) The occurrence of p-hydroxyphenylpropane units in the middle-lamella lignin of spruce (Picea abies). Wood Sci Technol 19:223–232

Wolter KE, Harkin JM, Kirk TK (1974) Guaiacyl lignin associated with vessels in aspen callus cultures. Physiol Plant 31:140–143

Yamasaki T, Hata K, Higuchi T (1972) Chemical properties of dehydrogenation polymer from p-coumaryl alcohol. Mokuzai Gakkaishi 18:361–366

Yamasaki T, Hata K, Higuchi T (1973) Dehydrogenation polymer of sinapyl alcohol. Mokuzai Gakkaishi 19:299–300

Yasuda S, Ota K (1987) Chemical structures of sulfuric acid lignin. X. Reaction of syringylglycerol-β-syringyl ether and condensation of syringyl nucleus with guaiacyl lignin model compounds in sulfuric acid. Holzforschung 41:59–65

6.2 Nitrobenzene and Cupric Oxide Oxidations

C.-L. Chen

6.2.1 Introduction

6.2.1.1 Nitrobenzene Oxidation

Nitrobenzene oxidation of lignins in alkaline solution was introduced in 1939 by Freudenberg and co-workers (Freudenberg 1939, Freudenberg and Lautsch 1939, Freudenberg et al. 1940) to obtain further evidence for the aromatic nature of lignins, which had been established earlier by potassium permanganate oxidation (Freudenberg et al. 1936, 1937, Freudenberg and Müller 1938) (Sect. 6.3). Vanillin (2) (Fig. 6.2.1) was produced in a yield of about 25% (based on Klason lignin) on oxidation of extractive-free spruce (*Picea abies*) wood meal with nitrobenzene in 2 M sodium hydroxide solution at 160 °C for 3 h. When the wood residue from the oxidation was treated again under the same conditions, about 2% additional vanillin was obtained, in addition to insignificant amounts of guaiacol (6), vanillic acid (10), 5-carboxyvanillin (13) (Fig. 6.2.1), and some unidentified phenols. The method applied was a modified Bischler procedure for conversion of isoeugenol into vanillin. Earlier, Bischler produced vanillin (2) in a yield of about 90 mol % by oxidation of isoeugenol with nitrobenzene in alkaline solution at high temperature (Freudenberg et al. 1940).

On nitrobenzene oxidation, normal softwoods (Gymnospermae) and their lignins give rise to vanillin (2) as the major product in a yield range of 24–28% based on the Klason lignin or lignin, respectively. In addition, insignificant amounts of p-hydroxybenzaldehyde (1), syringaldehyde (3), and other minor oxidation products (Leopold and Malmström 1952) are obtained except for lignins from exceptional conifers such as members of the genera *Welwitschia*, *Ephedra*, *Gnetum*, *Tetraclinis*, and *Podocarpus*. Woods from trees in these genera yield vanillin (2) and syringaldehyde (3) (Fig. 6.2.1) as the major products of nitrobenzene oxidation, a characteristic of hardwood lignins. Table 6.2.1 shows the compounds identified in the nitrobenzene oxidation mixture of Norway spruce (*Picea abies*) (Leopold 1952a, Pew 1955). In contrast, normal hardwoods (dicotyledoneous Angiospermae) and their lignins give vanillin (2) and syringaldehyde (3) as the major products on nitrobenzene oxidation (Creighton et al. 1944, Creighton and Hibbert 1944), while grasses (monocotyledoneous Angiospermae) and their lignins afford p-hydroxybenzaldehyde (1), vanillin (2), and syringaldehyde (3) as the major products under the same reaction conditions (Higuchi et al. 1967).

Springer Series in Wood Science
Methods in Lignin Chemistry
(Edited by S.Y. Lin and C.W. Dence)

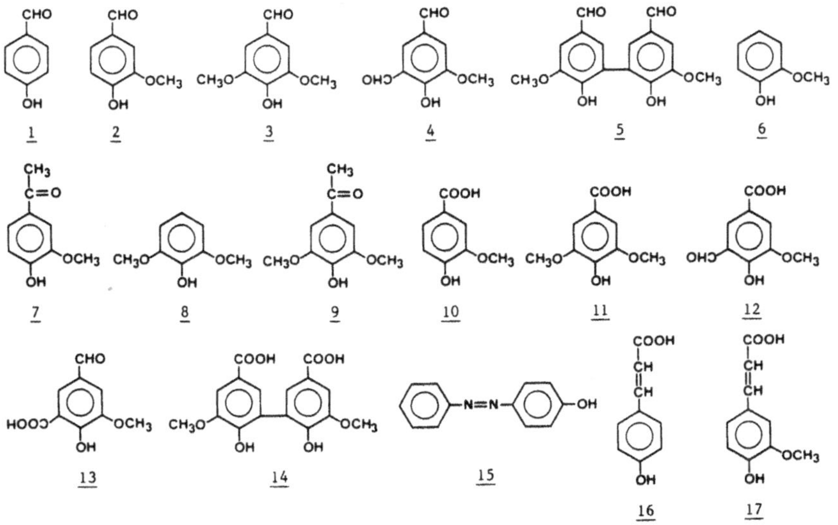

Fig. 6.2.1. Nitrobenzene and cupric oxide oxidation products of lignins

In general, the yield and molar ratio of the phenolic aldehydes, (1)–(3), depend on the plant species being investigated. Thus, the nitrobenzene oxidation of lignin is not only important in terms of the characterization of lignins, by providing information on the *minimal* quantities and the relative amounts of the uncondensed p-hydroxyphenyl-, guaiacyl-, and syringylpropane units present in a lignin, but also in terms of the taxonomy of vascular plants. Moreover, the method is relatively simple compared with the potassium permanganate oxidation of lignins. These advantages also hold true for the cupric oxide oxidation of lignins which is discussed later.

Studies on the nitrobenzene oxidation of lignin model compounds (Leopold and Malmström 1951, Leopold 1951, 1952a,b, Pew 1955) indicate that phenolic aldehydes (1)–(3) are derived from oxidative degradation of the corresponding 4-hydroxyphenylpropane units and their ethers, in particular the corresponding 4-O-alkylated and α-O-4 and β-O-4 lignin substructures, under the imposed reaction conditions. Moreover, aromatic acids related to phenolic aldehydes (1)–(3) are produced only in negligible amounts from either lignin model compounds or lignins. It is well established that benzaldehyde and a 4-O-alkylated vanillin, such as veratraldehyde, readily undergo the Canizzaro reaction under the conditions of nitrobenzene oxidation to produce benzoic acid and benzyl alcohol, and 4-O-alkylated vanillic acid and 4-O-alkylated vanillyl alcohol, respectively. Collectively, these findings indicate that the Canizzaro reaction does not play an important role during nitrobenzene oxidation of lignins. It follows that base-catalyzed cleavage of α- and β-aryl ether bonds producing the corresponding phenolate anions precedes the oxidative cleavage of C_α—C_β bonds. Because of their stability toward alkaline hydrolysis under the test conditions, aryl-O-4-linked lignin units are resistant to nitrobenzene oxidation.

Table 6.2.1. Yields and oxidation products from Norway spruce (*Picea abies*) wood on nitrobenzene oxidation

Oxidation products	Yield (% of Klason lignin) *A*[a]	*B*[b]
p-Hydroxybenzaldehyde (1)	0.25	–
Vanillin (2)	27.5	25.8
Syringaldehyde (3)	0.06	–
5-Formylvanillin (4)	0.23	–
Dehydrodivanillin (5)	0.80	2.2
Guaiacol (6)	–	–
Acetoguaiacone (7)	0.05	–
Syringol (8)	–	–
Acetosyringone (9)	–	–
Vanillic acid (10)	4.8	1.3
Syringic acid (11)	0.02	–
5-Formylvanillic acid (12)	0.1	–
5-Carboxyvanillin (13)	1.2	0.6
Dehydrodiveratric acid (14)	0.03	–

[a] Leopold (1952a).
[b] Pew (1955).

The reaction mechanism for nitrobenzene oxidation of lignin is still not well understood. Thus, although nitrobenzene in an alkaline solution has been usually regarded as two-electron-transfer oxidant (Chang and Allan 1971), recent investigations on the nitrobenzene oxidation of lignins and related compounds indicate that nitrobenzene in alkaline solution acts as an one-electron-transfer oxidant (Schultz and Templeton 1986).

The optimum reaction conditions for nitrobenzene oxidation of lignins have been elucidated by Leopold (1950) from model compounds experiments. The highest yields of phenolic aldehydes were obtained when lignin-containing plant tissues and lignin preparations were oxidized in 2 M sodium hydroxide solution with nitrobenzene at 170–180 °C for 2–3 h. These results were confirmed in a later investigation (Kavanaugh and Pepper 1955). In general, nitrobenzene oxidation of lignins is conducted in the laboratory according to the procedure of Leopold (1950). However, extensive modifications have been made in the qualitative and quantitative determination of the oxidation products because of the recent advances in analytical instrumentation. Thus, constituents of the oxidation mixture are now qualitatively and quantitatively determined by means of high performance liquid chromatography (HPLC), gas chromatography (GC), and gas chromatography-mass spectrometry (GC-MS).

6.2.1.2 Cupric Oxide Oxidation

The successful conversion of lignins into vanillin (2) and its analogs by nitrobenzene oxidation stimulated a search for economically viable processes for the manufacture of vanillin (2) from technical lignins. In 1942, Pearl found that vanillin (2) could be obtained in a yield of about 20% of the lignin when

Table 6.2.2. Yields on nitrobenzene and cupric oxide oxidation products from Norway spruce (*Picea abies*) and aspen (*Populus tremuloides*) woods

Oxidation Products	Yield (Percent of Klason lignin) Nitrobenzene	Cupric oxide
Norway spruce		
Guaiacol (6)	–	1.4
Vanillin (2)	27.5[a]	15.9
Acetoguaiacone (7)	–	3.9
Total yield	27.5[a]	21.9
Aspen		
p-Hydroxybenzaldehyde (1)	–	1.8
Vanillin (2)	12.4	7.8
Acetoguaiacone (7)	–	2.0
Syringaldehyde (3)	30.0	20.0
Acetosyringone (9)	–	5.3
Total yield	42.4	36.9

[a] Leopold (1952a); all other values from Pepper et al. (1967).

hemlock sulfite liquor solids containing about 40% of lignosulfonate were heated with 4 M sodium hydroxide solution in the presence of cupric sulfate in an autoclave at 160 °C for several hours. Cupric sulfate was later replaced by cupric hydroxide when the latter became commercially available (Pearl and Beyer 1950). Cupric hydroxide reverts to cupric oxide with loss of one mole water upon heating.

The cupric oxidation of lignins is optimally performed by heating a mixture of lignin preparation, cupric oxide, and sodium hydroxide in an approximate weight ratio of 1:6:2.9 in an autoclave at 170 °C for 2–5 h, the concentration of sodium hydroxide being 1.5–2 M (Pearl and Beyer 1950, Pepper et al. 1967). In the laboratory, cupric oxide oxidation of lignins is usually conducted according to the procedure of Pepper and co-workers (1967). The constituents of the oxidation mixture are qualitatively and quantitatively determined as described for the compounds in nitrobenzene oxidation mixtures.

As shown in Table 6.2.2, cupric oxide oxidation of a lignin generally provides results comparable to those obtained from the nitrobenzene oxidation of the same lignin under optimum conditions. However, on cupric oxide oxidation, softwood or hardwood lignins yield, respectively, an appreciably smaller amount of vanillin (2) or vanillin and syringaldehyde (3), and a corresponding larger amount of acetoguaiacone (7) or acetoguaiacone and acetosyringone (9) (Fig. 6.2.1) than nitrobenzene oxidation of the corresponding lignins. Pepper and co-workers (1967) have shown that compounds (7) and (9) in an alkaline solution are stable toward oxidation by cupric oxide but susceptible to oxidation by nitrobenzene. Thus, the higher yields of compounds (7) and (9) are gained in part at the expense of the yields of compounds (2) and (3). The reaction mechanism for cupric oxide oxidation of lignins is not well established, although cupric oxide is known to be an one-electron-transfer oxidant.

6.2.1.3 Nitrobenzene and Cupric Oxide Oxidations of Lignins in Grass Tissues

Lignins in grasses (Gramineae) are composed of a lignin core and peripheral groups (Higuchi and Kawamura 1966, Higuchi et al. 1967, Shimada et al. 1971, Nakamura and Higuchi 1976, Nimz et al. 1981, Scalbert et al. 1985). The lignin core consists mainly of guaiacyl- and syringylpropane units, the quantity of p-hydroxyphenylpropane units being negligible. Thus, the lignin core is a guaiacyl-syringyl-type lignin. The peripheral groups include p-hydroxycinnamic acid (16) and ferulic acid (17) units, the former being the major component.

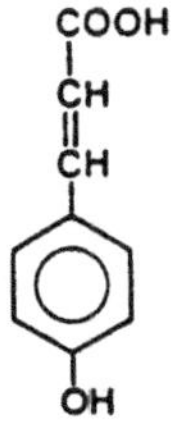

16

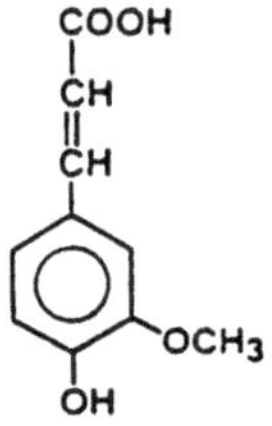

17

These units are chemically bonded to the lignin core, mostly in the form of ester linkages with hydroxyl groups at C-α and C-γ of the side chain and to a lesser extent in the form of aryl ether linkages at C-α. It is desirable, therefore, to conduct nitrobenzene or cupric oxide oxidation of lignins in grass tissues or milled grass lignins (MGLs) both before and after saponification of the specimens under mild reaction conditions according to the procedure of Scalbert and co-workers (1985). The quantities of p-hydroxycinnamic and ferulic acid ester groups in grass lignins may be estimated by quantitative determination of p-hydroxycinnamic and ferulic acids (16 and 17) in the alkaline hydrolyzate.

6.2.2 Experimental Procedures

6.2.2.1 Sample Preparation

For the characterization of lignins in woody plant tissues and pulps, an air-dried sample is ground in a Wiley mill to pass a 40–60-mesh screen. Unnecessarily vigorous milling of the sample for a prolonged period should be avoided, as this results in changes of lignin structure, such as the introduction of α-carbonyl groups. The resulting wood or pulp meal is extracted continuously with benzene-ethanol (2:1, v/v) or acetone-water (10:1, v/v) in a Soxhlet extractor for 48 h to remove the extractives and any low-molecular weight contaminants. The extracted sample is air-dried, then dried over P_2O_5 in a vacuum desiccator for several days. Before conducting either nitrobenzene or cupric oxide oxidation of a plant tissue, the following analyses of the sample should be conducted: (a) moisture content of the dried, extracted sample expressed as weight percentage, (b) extractives content in the original sample in weight percentage based on the oven-dried original sample, and (c) total lignin content of the sample, i.e., Klason lignin and any acid-soluble lignin (Chap. 2.2.2.1). The total lignin content based on oven-dried original plant tissue is calculated from Eq. (1):

$$\text{Total lignin content as \% of oven-dried original plant tissue} = \frac{100L}{W(1 - M/100)(1 - E/100)}, \qquad (1)$$

where L is the weight of total lignin (g), W is the weight of the extracted sample (g), M is the moisture content of the extracted specimen (%), and E is the extractives content of oven-dried original plant tissue (%).

For the characterization of lignin preparations, the lignin sample must be purified by an appropriate method, e.g., milled wood lignins (MWLs) and kraft lignins according to the procedures of Björkman (1956) and Lundquist (Lundquist et al. 1977, Lundquist and Kirk 1980). Before conducting either nitrobenzene or cupric oxide oxidation of a lignin preparation, the following determinations need to be performed: (a) elemental composition and methoxyl content, (b) moisture content, (c) total lignin content, (d) carbohydrate content by the procedure described by Borchardet and Piper (1970) or Fengel et al. (1977). If the carbohydrate content of a lignin is appreciably greater than 5% or if a carbohydrate-free lignin preparation is required, the lignin should be purified again according to the procedure of Lundquist. Table 6.2.3 shows the carbohydrate content, elemental composition with carbohydrate correction, and C_9 formula of several lignin preparations. In this table, the methoxyl content/C_9 unit provides an indication of the syringyl/guaiacyl (S/V) ratio in the hardwood lignins. Thus, for birch lignin, the calculated percentages of syringyl and guaiacyl units are about equal.

6.2.2.2 Nitrobenzene Oxidation

Weigh accurately about 200 mg of a dried, extracted wood meal (40–60 mesh) with known moisture and extractives contents or 50 mg of a dried, purified lignin preparation. The specimen, 7 ml of 2 M NaOH, and 0.4 ml of freshly distilled nitrobenzene are placed in a 10-ml stainless steel bomb (Note 1). The bomb is sealed tightly with a screw cap fitted with a Teflon gasket and heated at 170 °C for 2.5 h in an electrically preheated (170 °C), thermostated aluminum block with six holes to accommodate the cylindrically shaped bombs or in a thermostated oil bath. The bomb is shaken occasionally and, after the heating period, is removed from the heater and cooled with ice water. The oxidation mixture is transferred to a liquid-liquid extractor, and extracted continuously with $CHCl_3$, for 4 h to remove nitrobenzene reduction products and any excess nitrobenzene. The oxidation mixture is acidified to pH 3–4 with conc. HCl, and further extracted continuously with $CHCl_3$ for 48 h. The solvent from the second $CHCl_3$ solution is removed at 40 °C under reduced pressure to obtain the nitrobenzene oxidation mixture. The mixture is then dissolved in 5 ml of CH_2Cl_2 and transferred to a 10-ml volumetric flask. The solution is made up to 10 ml

Table 6.2.3. Carbohydrate content, elemental composition, and C_9 unit formula of some lignin preparations

Lignin preparation	Carbohydrate content Percent of lignin	Elemental composition[a], % C	H	O	S	OCH_3	C_9 Unit Formula
Milled wood or bamboo lignins							
Spruce (*Picea glauca*)	2.2	64.30	6.21	29.48	–	15.72	$C_9H_{7.66}O_2(H_2O)_{0.48}(OCH_3)_{0.94}$
Birch (*Betula papyrifera*)	4.8	61.88	6.32	33.44	–	22.53	$C_9H_{6.87}O_2(H_2O)_{0.86}(OCH_3)_{1.52}$
Sweetgum (*Liquidambar stryraciflua*)	3.6	58.74	5.67	35.59	–	22.16	$C_9H_{5.00}O_2(H_2O)_{1.28}(OCH_3)_{1.55}$
Zhong-yang Mu (*Bischofia polycarpa*)	6.2	61.36	5.96	32.68	–	17.75	$C_9H_{6.49}O_2(H_2O)_{0.92}(OCH_3)_{1.13}$
Bamboo (*Phyllostachys pubescence*)	5.0	61.66	5.53	32.81	–	19.42	$C_9H_{5.52}O_2(H_2O)_{0.84}(OCH_3)_{1.25}$
Technical lignins							
Sweetgum HCl lignin	–	59.21	5.67	35.12	–	22.95	$C_9H_{5.06}O_2(H_2O)_{1.15}(OCH_3)_{1.60}$
Sweetgum kraft lignin	–	60.04	5.68	34.28	1.39	19.06	$C_9H_{5.99}O_2(H_2O)_{0.93}S_{0.09}(OCH_3)_{1.27}$
Bamboo kraft lignin	–	61.81	5.25	31.26	1.68	15.99	$C_9H_{5.52}O_2(H_2O)_{0.80}S_{0.08}(OCH_3)_{1.00}$

[a] Corrected for carbohydrate content; based on average of two elemental analyses.

with the solvent. This solution is the stock solution for further analysis of the oxidation mixture.

Note 1. In case a stainless steel bomb is not available, use a round-bottom Pyrex ampule of appropriate volume and strength. Place the specimen and reagents in the ampule and seal it with a flame-sealing torch. After the reaction, the ampule is cooled with air to a temperature below 100 °C and then with cold water.

6.2.2.3 Cupric Oxide Oxidation

Weigh accurately about 200 mg of a dried, extracted wood meal (40–60 mesh) with known moisture and extractives contents or 50 mg of a dried and purified lignin preparation. The specimen, 1 g of CuO, and 7 ml of 2 M NaOH are placed in a 10-ml stainless steel bomb (Note 1). The bomb is flushed with nitrogen, sealed tightly with a screw cap fitted with a Teflon gasket, then heated at 170 °C for 2 h with occasional shaking in an electrically preheated (170 °C) aluminum block (see Sect. 6.2.2.3). Allow a warm-up period of about 10 min for the bomb. After the bomb is removed and cooled with ice water, its contents are filtered through a glass filter to remove any insoluble materials. The insoluble material is washed twice with 5 ml of 2 M NaOH and then with 5 ml of water. The filtrate and washings are combined, placed in a liquid-liquid extractor and extracted continuously with $CHCl_3$ for 4 h to remove any neutral substances. Subsequent acidification, solvent extraction, concentration and volumetric dilution steps are identical with those described for the nitrobenzene oxidation product mixture (Sect. 6.2.2.2).

6.2.2.4 Qualitative Determination of Oxidation Products

6.2.2.4.1 Gas Chromatography (GC)

Place 1 ml of the stock solution in a 2-ml GC vial fitted with a Teflon gasket and screw cap, and make up the solution of 2 ml with CH_2Cl_2. The resulting solution is filtered through a Millipore HVHP membrane (pore size 0.45 μm) to remove any high-molecular weight contaminants. The sample solution is stored in a GC vial fitted with a Teflon gasket and screw cap (Note 2). Inject about 0.5 μl of the sample solution into a suitable GC system (e.g., Hewlett-Packard Model 5880A) equipped with a DB5-bonded phase, fused silica capillary column [30 m × 0.32 mm i.d., film thickness, 0.25 μm (J & W Scientific)] and a flame ionization detector (temperature, 270 °C). (The temperature should be at least 10 °C higher than the post value to prevent condensation of decomposition products on the detector). The carrier gas is helium (column flow rate, 2 ml/min). The injection port (temperature, 200 °C) is fitted with a split liner (split ratio, 10:1) and a septum purge system. The temperature profile for GC analysis is as follows:

Table 6.2.4. Conversion factors (C) and GC retention times for the major nitrobenzene and cupric oxide oxidation products of lignins

Compounds	Conversion factor[a]	GC Retention time, min
p-Hydroxybenzaldehyde (1)	1.2886	9.0
Vanillin (2)	1.3158	9.7
Acetoguaiacone (7)	1.3832	11.1
Syringaldehyde (3)	1.4302	14.2
Acetosyringone (9)	1.4674	15.6
m-Meconin	–	

[a] Calculated from appropriate calibration curves and relative to m-meconin, the internal standard; see Section 6.2.2.5 for GC conditions.

hold 1 min at 60 °C, 60–140 °C (5 °C/min), 140–250 °C (5 °C/min), final temperature, 255 °C (10 min), and post value, 260 °C. The GC temperature profile is determined by the nature of the sample being analyzed.

The GC retention times of peaks in the gas chromatogram are compared with those of the authentic compounds. The identity of compounds giving rise to these peaks is verified by co-injecting a solution of the suspected compound and the sample (Note 3). The possible presence of a particular compound in the oxidation mixture is indicated by an increase in intensity or area ratio of the peak having a GC retention time corresponding to the authentic compound relative to those of other peaks in the chromatogram obtained after co-injection. Table 6.2.4 shows the GC retention times for the major oxidation products from nitrobenzene and cupric oxide oxidations of lignins obtained under the aforementioned GC operating conditions. Further verification of the identity of the oxidation products may be obtained by GC-MS analysis of the oxidation products.

Note 2. A vial with a plastic cap other than one made of Teflon should not be used. This is of primary importance in obtaining a clean gas chromatogram, i.e., a gas chromatogram free of GC peaks for artifacts. Otherwise, plasticizers such as bis(2-ethylhexyl)phthalate in the plastic cap will dissolve in the sample solution. The presence of such artifacts in the sample solution also affects the resolution of GC peaks in the gas chromatograms and the mass spectra obtained from GC-MS analysis as well.

Note 3. The concentration of the authentic compound should be about 1 mg/ml of solvent. Draw 0.5 μl of the sample solution, a bubble of air, and 0.1–0.2 μl of the solution of the compound to be co-injected into a microsyringe and inject the mixture into the GC system.

6.2.2.4.2 Gas Chromatography-Mass Spectrometry (GC-MS)

It is of primary importance to find the optimum operating conditions for GC analysis of the oxidation products before conducting a GC-MS analysis of the

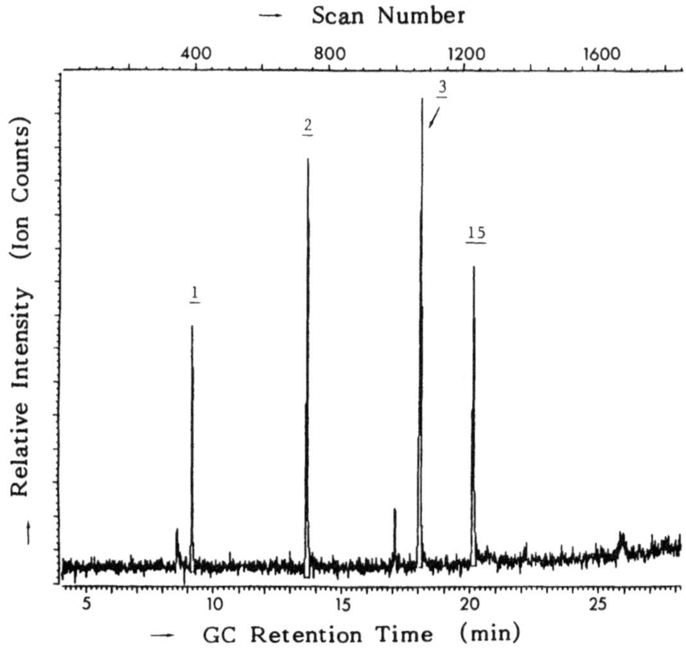

Fig. 6.2.2. Total ion chromatogram of nitrobenzene oxidation mixture from milled bamboo lignin (MBL) prepared from *Phyllostachys pubescence*. See Section 6.2.2.4.2 for GC conditions

oxidation products. In general, the optimum conditions for GC-MS analysis are determined by the optimum conditions previously found for GC analysis of the sample.

Inject about 1–3 μl of the sample solution, prepared as described in the previous section, into a GC-MS system (e.g., Hewlett-Packard Model 5985B) with a DB5-bonded phase, fused silica capillary column [30 m × 0.32 mm i.d., film thickness 0.25 μm (J & W Scientific)]. The temperature profile for GC operation is as follows: hold 1 min at 60 °C, 60–250 °C (10 °C/min). The MS conditions are electron impact (EI) at 70 eV and a scanning ratio in the range of 50–60 scans/min. Figure 6.2.2 shows a total ion chromatogram of the nitrobenzene oxidation mixture of milled bamboo lignin (MBL) from *Phyllostachys pubescence* (Tai et al. 1990).

6.2.2.5 Quantitative Determination of Oxidation Products

6.2.2.5.1 Gas Chromatography (GC)

Place 2.5 ml of the stock solution (Sects. 6.2.2.2 and 6.2.2.3) and 1 or 2 ml of a solution of 97.1 mg of m-meconin (Edwards et al. 1925) in 50 ml of CH_2Cl_2 (1 ml = 0.01 mmol) in a 10-ml volumetric flask and make up to 10 ml with the solvent. The internal standard for the quantitative determination of the oxida-

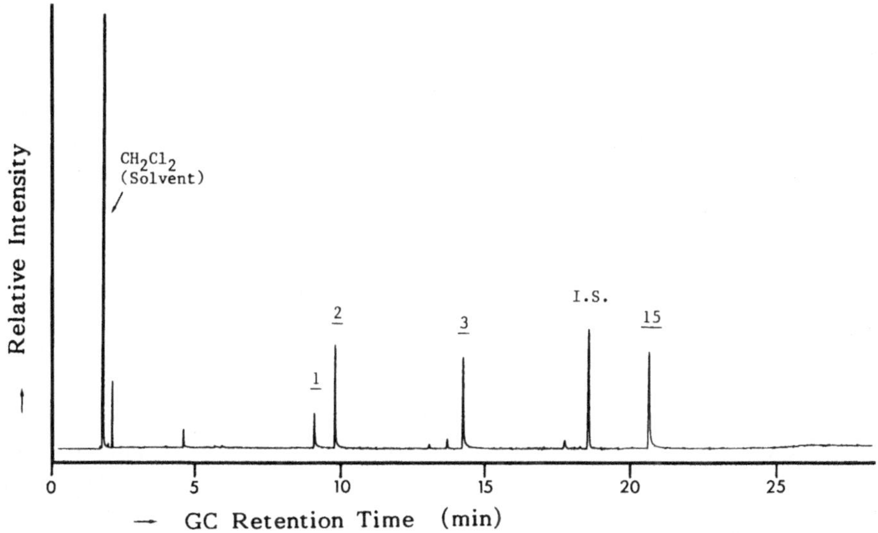

Fig. 6.2.3. Gas chromatogram of nitrobenzene oxidation mixture from milled bamboo lignin (MBL) prepared from *Phyllostachys pubescence* with m-meconin as internal standard (*I.S.*). See Section 6.2.2.5.1 for GC conditions

tion products is m-meconin (Note 4). The resulting sample solution is filtered through a Millipore HVHP membrane (pore size 0.45 μm) to remove any high-molecular weight contaminants. The sample solution is then stored in a GC vial fitted with a Teflon gasket and screw cap. Inject 0.5 μl of the sample solution into a GC system (e.g., Hewlett-Packard Model 5880A) as described in Sect. 6.2.2.4.1) under the same optimum operating conditions used for GC analysis of the oxidation mixture, including the same temperature profile. Figure 6.2.3 shows a gas chromatogram of the nitrobenzene oxidation mixture of milled bamboo lignin from *Phyllostachys pubescence* (Tai et al. 1990).

The yield of the nitrobenzene oxidation products is calculated in mmol/100 mg of lignin (or Klason lignin in lignocellulose samples) from Eq. (2) and in mol % from Eq. (3).

$$Y_1(\text{mmol/100 mg of lignin}) = \frac{400\ CA_pI}{A_iL}, \tag{2}$$

where Y_1 is the yield of a particular product, A_p is the peak area of that product, A_i is the peak area of an internal standard (e.g., m-meconin), I is the quantity of an internal standard added (mmol), L is the weight of lignin used (mg), and C is the conversion factor including extraction efficiently and response factor of the particular product toward the detector (see Table 6.2.1).

$$Y_2(\text{mol \%/}C_9\text{ unit}) = Y_1M, \tag{3}$$

where Y_1 is the yield of a particular product in mmol/100 mg of lignin or in mol/100 g of lignin [see Eqs. (2) or (4)], Y_2 is the yield of that product in mol %/C_9 unit, and M is the weight of a C_9 lignin unit.

Note 4. Compounds other than m-meconin can be used as internal standards provided that the GC peak of the compound does not overlap with the GC peaks of the oxidation products. In addition, the quantity of the internal standard added to the sample solution should be selected so that the intensity of the GC peak for the internal standard is about the average of the most and least intensive GC peaks to be determined quantitatively.

6.2.2.5.2 High Performance Liquid Chromatography (HPLC)

Place 2 ml of the stock solution (see Sects. 6.2.2.2 and 6.2.2.3) and 1 or 2 ml of a solution of 97.1 mg of m-meconin in 50 ml of CH_3CN (1 ml = 0.01 mmol) (Note 4) in a 25-ml volumetric flask and make up to 25 ml with CH_3CN-H_2O (1:2, v/v). The resulting sample solution is filtered through a Millipore HVLP membrane (pore size 0.45 μm) to remove high-molecular weight contaminants. Inject about 10 μl of the sample solution into a HPLC system (e.g., Waters Associates HPLC inject port with pump Model M-6000 A) equipped with a Chromega bond C_{18} column (particle size 10 μ, 30 cm × 4.6 mm i.d., ES Industries, Marlton, New Jersey) to determine quantitatively the oxidation products. A CH_3CN-H_2O mixture (1:8, v/v) containing 1% acetic acid is used as eluent (flow rate, 2 ml/min). The eluent is monitored with a UV detector (e.g., Waters Associates Model 440) at 280 nm. Figure 6.2.4 shows a high perform-

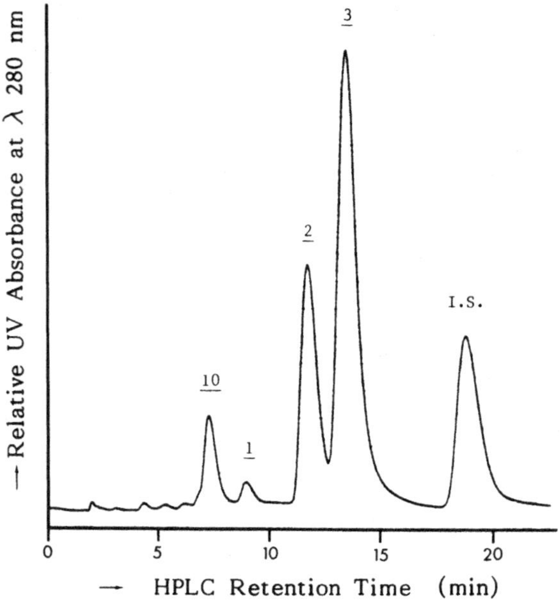

Fig. 6.2.4. High performance liquid chromatogram of nitrobenzene oxidation mixture from lignin in sweetgum wood (*Liquidambar styraciflua*) with m-meconin as internal standard. See Section 6.2.2.5.2 for HPLC conditions

Table 6.2.5. Conversion factors (C) and HPLC retention times for the major saponification, nitrobenzene, and cupric oxide oxidation products of lignins

Compounds	Conversion factor[a]	HPLC Retention time, min
p-Hydroxybenzoic acid	–	6.1
Vanillic acid (10)	0.4407	7.0
Syringic acid (11)	0.7134	7.5
p-Hydroxybenzaldehyde (1)	0.4261	8.8
Vanillin (2)	0.5507	11.6
Syringaldehyde (3)	0.7706	13.4
p-Hydroxycinnamic acid (16)	0.6456[b]	14.4
Acetoguaiacone (7)	0.5757	14.8
Acetosyringone (9)	0.6621	16.9
Ferulic acid (17)	0.9332[b]	18.4
m-Meconin	–	18.8

[a] Calculated from appropriate calibration curves and relative to m-meconin, the internal standard except for compounds 16 and 17; see Section 6.2.2.5.2 for HPLC conditions.
[b] Conversion factors for compounds 16 and 17 are relative to vanillin (2), the internal standard (see Sect. 6.2.2.6).

ance liquid chromatogram of a nitrobenzene oxidation mixture from lignins in sweetgum (*Liquidambar styraciflua*) wood (Winston et al. 1986). Table 6.2.5 shows the HPLC retention times for the major oxidation products from the nitrobenzene oxidation of lignins under the aforementioned HPLC operating conditions.

The yield of the nitrobenzene oxidation products in mmol/100 mg of lignin (or Klason lignin in lignocellulose samples) is calculated from Eq. (4), and in mol % from Eq. (3) in Sect. 6.2.2.5.1.

$$Y_I(\text{mmol}/100\,\text{mg of lignin}) = \frac{500\ CA_pI}{A_iL}, \tag{4}$$

where Y_I is the yield of a particular product, A_p is the peak area of that product, A_i is the peak area of an internal standard (e.g., m-meconin), I is the quantity of an internal standard added (mmol), L is the weight of lignin oxidized (mg), and C is the conversion factor including extraction efficiency and the detector response factor of the particular product (see Table 6.2.5).

6.2.2.6 Determination of p-Hydroxycinnamic and Ferulic Acid Ester Units in Grass Lignins

Weigh accurately about 200 mg of a dried, extracted grass meal (40–60 mesh) or 50 mg of a dried, purified milled grass lignin (MGL). The specimen, in 12 ml of 1 M NaOH, is placed in an Erlenmeyer flask and the resulting mixture or solution is kept at room temperature for 48 h under a nitrogen atmosphere while being stirred mechanically. The reaction mixture is acidified to pH 2–3 with 1 M HCl and the resulting precipitate is centrifuged off and washed twice with 15 ml of distilled water. The precipitate serves as the specimen for nitrobenzene or cupric oxide oxidation described in Sections 6.2.2.2 or 6.2.2.3, respectively. The supernate and washings are combined and continuously extracted in a liquid-liquid extractor with Et_2O for 48 h. After removal of the solvent from the ether extract, the p-hydroxycinnamic and ferulic acids are determined quantitatively by HPLC as described in Section 6.2.2.5.2 for the determination of vanillin and its homologs by HPLC, but using vanillin (2) as the internal standard instead of m-meconin.

6.2.2.7 Interpretation of Mass Spectra

As mentioned previously (Chap. 2.2.2.4.1), the major purpose of the GC-MS analysis of a nitrobenzene or cupric oxide oxidation mixture is to verify the identity of the oxidation products established previously by GC or HPLC analysis, and to elucidate the structure of unknown constituents. For example, GC-MS analysis of the nitrobenzene oxidation mixture of milled bamboo lignin from *Phyllostachys pubescence* showed unequivocally that compounds (1)–(3) in the total ion chromatogram of the oxidation mixture (Fig. 6.2.2) are indeed p-hydroxybenzaldehyde, vanillin, and syringaldehyde, respectively (Tai et al. 1990) (see Chap. 9.1 for a discussion of the GC-MS technique). In addition, the unknown compound in the chromatogram was identified as p-hydroxyazobenzene (15) (Fig. 6.2.1), one of the phenolic reduction products of nitrobenzene.

In general, mass spectra of the major oxidation products are unambiguous. As shown in Table 6.2.6, the major mass spectral fragmentation pattern of p-hydroxybenzaldehyde (1), vanillin (2), and syringaldehyde (3) is loss of a hydrogen atom from the aldehyde group of the corresponding molecular ions, (M^+), to give the (M-1) ions. A subsequent elimination of CO from the (M-1) ions produces the (M-29) ions. The M and (M-1) ions are the most abundant ions in the mass spectra of compounds (1)–(3) with intensities in the 60–100% range relative to that of the base peak. The relative intensities of the (M-29) ions are usually in the 20–30% range.

The major mass spectral fragmentation pattern of acetoguaiacone (7) and acetosyringone (9) (Fig. 6.2.1) is loss of $\cdot CH_3$ from the corresponding molecular ion, (M^+), to afford stable (M-15) ions which are the base peaks in the spectra and correspond to the (M-1) ions of compounds (2) and (3), respectively. Subsequent loss of CO from the (M-15) ions provides (M-43) ions. In general, the (M-15) ion is the base peak in the mass spectra of compounds (7) and (9), and their homologs. The relative intensity of the M^+ ions is in the 40–50% range, while that of the (M-43) ions is in the 20–30% range.

6.2.3 Discussion of the Methods

Comparison of Nitrobenzene and Cupric Oxide Oxidations

For the characterization of lignins, nitrobenzene oxidation has the following advantages over cupric oxide oxidation: (a) the constituents of the oxida-

Table 6.2.6. Major mass spectrometric fragmentation patterns for nitrobenzene and cupric oxide oxidation products of lignins

Compounds	Relative intensity of major ion species (percent of base peak)				
	M^+	(M − 1)	(M − 15)	(M − 29)	(M − 43)
p-Hydroxybenzaldehyde (1)	75–85	100 (BP)[a]	+	20–30	–
Vanillin (2)	85–95	100 (BP)[a]	+	15–25	–
Syringaldehyde (3)	100 (BP)[a]	60–70%	+	5–10	–
Acetoguaiacone (7)	35–45	–	100 (BP)[a]	–	20–30
Acetosyringone (9)	45–55	–	100 (BP)[a]	–	15–15

[a] BP = base peak; + = present in insignificant intensity.

tion mixture are less complex, and (b) the total yield of degradation products detectable by GC and HPLC is higher. However, the nitrobenzene oxidation mixture contains numerous nitrobenzene reduction products which must be removed before qualitative and quantitative determinations of the oxidative degradation products are undertaken. In general, the accuracy of the analysis is in the range of 3–5% for the quantitative determination of both nitrobenzene and cupric oxide oxidation products by GC using a capillary column and by HPLC.

Significance of the Results from Nitrobenzene and Cupric Oxide Oxidations

In addition to its utility in serving as a means of confirming the aromatic nature of lignin, alkaline nitrobenzene oxidation has found a host of applications, several of which have been documented by Brauns and Brauns (1960) and by Pearl (1967). Representative uses to which the procedure has been put include the botanical classification of plants (Creighton et al. 1944, Creighton and Hibbert 1944), the characterization of lignin in various morphological regions of wood tissue (e.g., Meshitsuka and Nakano 1985, Obst et al. 1988), estimation of the proportion of uncondensed guaiacyl nuclei in softwood lignin (Leopold 1952c), the monitoring of changes in lignin during plant growth (Abbot et al. 1986), and the assessment of the effect on lignin of various chemicals (e.g., Kratzl and Silbernagel 1952, Jayne 1953, Cymbaluk and Neudorffer 1970, Van Buren and Dence 1970) and biological (e.g., Blanchette et al. 1988, Nilsson et al. 1989) treatments. The scope and versatility of the nitrobenzene and cupric oxide oxidation methods are illustrated in the examples described below.

Table 6.2.7 shows the yields of the phenolic aldehydes obtained from lignins in several representative plant species by nitrobenzene oxidation. The results indicate both the types of lignin and the minimum quantity of uncondensed phenylpropane units present, i.e., units whose aromatic rings are not substituted at C-5 and/or C-6 with aryl, aroxyl, or alkyl groups. Thus, lignin in spruce wood (*Picea glauca*) is a guaiacyl lignin and contains a minimum of about 33 uncondensed guaiacylpropane units per 100 C_9 units. In contrast, lignin in Zhong-Yang Mu wood (*Bischofia polycarpa*) is a guaiacyl-syringyl lignin and is

Table 6.2.7. Yields of phenolic aldehydes from several plant species and their lignin preparations on nitrobenzene oxidation (yield in mol %/C_9 unit)[a]

Plant species	Specimen	Total lignin Content (%)	C_9 Unit weight	p-Hydroxy- -benzaldehyde	Vanillin	Syring- aldehyde	Total aldehyde	Molar Ratio (H:V:S)[b]
Spruce	Wood	27.6		+	33.4	+	33.4	
(*Picea glauca*)	MWL	97.8	185.6	+	33.9	+	33.9	
Birch	Wood	22.6		–	14.5	36.2	50.7	0:1:2.5
(*Betula papyrifera*)	MWL	95.6	209.7	–	14.1	34.2	49.5	0:1:2.4
Sweetgum	Wood	25.8		–	11.2	32.1	43.3	0:1:2.9
(*Liquidambar stryraciflua*)	MWL	96.4	216.3	–	10.3	23.4	33.7	0:1:2.3
	HCl lignin	–	215.6	–	7.6	23.8	31.4	0:1:3.1
	Kraft lignin	–	205.2	–	5.4	11.0	16.4	0:1:2.0
Zhong-yang Mu	Wood	33.3		–	27.0	10.6	37.6	0:1:0.39
(*Bischofia polycarpa*)	MWL	93.8	197.3	–	27.1	12.0	39.1	0:1:0.44
Bamboo	MBL	88.1[c]	199.6	7.9	19.0	25.7	52.6	0.4:1:1.4
(*Phyllostachys pubescence*)	MBL-S[d]	–		0.8	17.8	24.3	42.9	0.05:1:1.4
	Kraft lignin	–	194.3	+	7.2	4.5	11.7	0:1:0.6

[a] + = trace amount; – = not detected; C_9 unit formula, see Table 6.2.3.
[b] H = p-hydroxygenzaldehyde (1); V = vanillin (2); S = syringaldehyde (3).
[c] The milled bamboo lignin (MBL) from *P. pubescence* contains approximately 5% carbohydrates and 6.9% proteins (Tai et al. 1990).
[d] MBL-S = MBL after saponification with alkaline solution.

comprised of at least 27 and 10 uncondensed guaiacyl- and syringylpropane units per 100 C_9 units, respectively (Pan et al. 1990). Since the oxidation efficiencies of lignin substructures differ considerably from one another or are undetermined, the results cannot be interpreted to mean that spruce lignin necessarily contains approximately 67 condensed phenylpropane units per 100 C_9 units, or that the Zhong-Yang Mu lignin of necessity contains a total of about 37 uncondensed phenylpropane units per 100 C_9 units.

In addition, the syringaldehyde/vanillin (S/V) molar ratio obtained from nitrobenzene oxidation of a lignin does not imply that uncondensed phenylpropane units in the lignin consist of guaiacyl- and syringylpropane units in a molar ratio corresponding to the S/V molar ratio. Thus, the S/V and H/V/S molar ratios (H = p-hydroxybenzaldehyde) are not absolute measures of the amounts of uncondensed phenylpropane units in lignins. However, the results do provide a relative measure of the extent of condensation in the aromatic moieties of lignins. In nitrobenzene and cupric oxide oxidations, the total yield of vanillin (2) and acetoguaiacone (7) and that of syringaldehyde (3) and acetosyringone (9) in mol % per C_9 unit correspond to the minimum quantities of uncondensed guaiacyl- and syringylpropane units, respectively, per 100 C_9 lignin units. A decrease in the total yield of nitrobenzene or cupric oxide oxidation products resulting from a particular treatment of lignin constitutes a sensitive measure of secondary ring condensation reactions.

The S/V molar ratio of the products in the nitrobenzene oxidation mixture from sweetgum HCl-lignin (Table 6.2.7) is about 3.1 compared with 2.9 for sweetgum wood. This indicates that the uncondensed guaiacylpropane units of the lignins in the wood are more susceptible to acid-catalyzed condensation than are uncondensed syringylpropane units. Thus, the results both reflect possible structural changes occurring in the sweetgum lignins and provide an indication of the type of reactions involved in the structural changes occurring during the preparation of sweetgum HCl-lignin from the wood (Winston et al. 1986). In contrast, the S/V molar ratio for compounds in the nitrobenzene oxidation mixture from sweetgum kraft lignin is approximately 2.0, indicating that, during kraft pulping, the uncondensed syringylpropane units in the sweetgum lignins in the wood are more susceptible to base-catalyzed cleavage of α- and β-aryl ether bonds and concomitant condensation than are the uncondensed guaiacylpropane units.

Except for saponification of p-hydroxycinnamic acid ester bonds in bamboo lignins, the same conclusion is reached from the results observed on nitrobenzene oxidation of the milled bamboo lignin (MBL) from *Phyllostachys pubescence* and bamboo kraft lignin (Tai et al. 1990). The S/V molar ratio of products in the nitrobenzene oxidation mixture from the MBL after saponification is practically the same as that before saponification. However, the H/V molar ratio decreases drastically from about 0.4 to less than 0.1. The quantitative analysis of the alkaline hydroyzate by HPLC shows that saponification led to the formation of approximately 9 mol % p-hydroxycinnamic acid (16/C_9 unit and only a trace amount of ferulic acid (17). This is comparable to about a 7 mol %/C_9 unit decrease in the formation of p-hydroxybenzaldehyde (1) on nitrobenzene oxidation of the MBL after saponification. Thus, the results indicate that the lignin

Table 6.2.8. Alkaline nitrobenzene oxidation products of the fiber fraction from birch wood soft xylem[a]

Products	Fiber fraction	Whole wood
Total aldehydes	9.5	42.7[b]
Vanillin	7.3	8.0
Syringaldehyde	2.2	34.7
S/V Molar ratio	0.25	3.62

[a] Data taken from Meshitsuka and Nakano (1985).
[b] Yield (%) based on lignin content determined by the acetyl bromide method.

core is composed mostly of guaiacyl- and syringylpropane units, while the peripheral groups consist predominantly of p-hydroxycinnamic acid ester groups bonded to the lignin core, probably at C-α and C-γ, whereas the quantity of ferulic acid ester groups in the MBL is negligible.

Nitrobenzene oxidation has been used as a means of characterizing lignin in different morphological regions of wood cell walls. As an example, Meshitsuka and Nakano (1985) compared the nitrobenzene oxidation products of birch whole wood with those of a vessel-free fiber fraction separated from soft xylem of the same wood. Their results, summarized in Table 6.2.8, indicated that, based on individual lignin content, the total yield of aldehydes from the fiber fraction was approximately only 20% that from the whole wood. Further, the S/V ratio of the soft xylem fiber fraction was much lower than the corresponding ratio for the lignin in the whole wood. Collectively, these results indicated that the lignin in the fiber fraction of the soft xylem (i.e., lignin in a compound middle lamella-rich region) was guaiacyl-rich and highly condensed.

As a further illustration of the kind of information derivable from alkaline nitrobenzene oxidation, the effect of several white-rot fungi on the lignin in birch wood (*Betula verrucosa* Ehrh) has been reported using this technique (Nilsson et al. 1989). In this case, the decrease in S/V ratios shown in Table 6.2.9 reveal that the syringylpropane lignin units were preferentially degraded and that the extent of decrease in the ratios mirrors the weight loss of lignin.

In a related application, the cupric oxide oxidation method has been used to monitor structural changes taking place during the chlorination of softwood kraft and bisulfite pulps (Van Buren and Dence 1970). In this case the oxidation products identified and measured were, in addition to vanillin, acetoguaiacone, 5- and 6-chlorovanillin, and 5,6-dichlorovanillin. The findings, summarized in Table 6.2.10, provided an indication of the effect of increasing chlorine charge on the amount and rate of aromatic ring chlorination and on the overall disappearance of uncondensed aromatic nuclei via oxidative degradation and other processes.

As stated above, nitrobenzene and cupric oxide oxidations of lignins are simple but by no means perfect methods for the characterization of lignins. However, the methods are effective in the characterization of lignins if they are

Table 6.2.9. Nitrobenzene oxidation of birch wood decayed by four ascomycetes[a]

	Lignin Loss, %	Syringaldehyde[b] (S), mg	Vanillin[b] (V), mg	S/V Molar ratio
Sound birch	–	3.17	0.72	3.68
Decayed by:				
H. fuscum	24.0	2.74	0.88	2.60
L. betulina	28.8	2.96	1.00	2.47
H. multiforme	31.6	3.24	1.05	2.58
D. concentrica	44.0	3.24	1.74	1.55

[a] Data taken from Nilsson et al. (1989).
[b] Yield, mg/ca. 40 mg wood sample.

Table 6.2.10. Aromatic products from the alkaline cupric oxide oxidation of chlorinated pine kraft pulp[a]

Applied chlorine, % o.d. pulp	Product yield (%)[b]					Total
	Vanillin	Acetoguaiacone	5-Chloro-vanillin	6-Chloro-vanillin	5,6-Dichloro-vanillin	
0	9.5	1.5	–	–	–	11.0
1	4.8	0.91	0.08	0.10	0.00	5.9
3	1.4	0.59	0.23	1.5	0.00	3.7
5	0.38	0.16	0.05	0.48	Trace	1.0

[a] Data taken from Van Buren and Dence (1970).
[b] Based on Klason lignin content of original pulp.

applied imaginatively. For example, nitrobenzene or cupric oxide oxidation, combined with ^{13}C NMR spectroscopy (Tai et al. 1990, Pan et al. 1990) or especially with nucleus exchange (Chap. 6.6) (Chiang and Funaoka 1990a,b) is a very useful approach for characterizing lignins.

References

Abbot TP, James C, Bayby MO (1986) Oxidation and quantification of ^{14}C-lignin at different ages in wheat, pine, oak and kenaf. J Wood Chem Technol 6:473–486

Björkman AJ (1956) Studies on finely divided wood. Part I. Extraction of lignin with neutral solvents. Sven Papperstidn 59:477–485

Blanchette RA, Obst JR, Hedges JI, Weliky K (1988) Resistance of hardwood vessels to degradation by white rot Basidiomycetes. Can J Bot 66:1841–1847

Borchardet LG, Piper CV (1970) A gas chromatographic method for carbohydrates as alditol-acetates. Tappi 53:257–260

Brauns FE, Brauns DA (1960) The chemistry of lignin: supplement volume. Academic Press, New York, 531–546

Chang H-m, Allan GG (1971) Oxidation. In: Sarkanen KV, Ludwig CH (eds) Lignins: occurrence, formation, structure and reactions. Wiley-Interscience, New York, 433–483

Chiang VL, Funaoka M (1990a) The dissolution and condensation reactions of guaiacyl and syringyl units in residual lignin during kraft delignification of sweetgum. Holzforschung 44:147–155
Chiang VL, Funaoka M (1990b) The difference between guaiacyl and guaiacyl-syringyl lignins in their responses to kraft delignification. Holzforschung 44:309–313
Creighton RHJ, Gibbs RD, Hibbert H (1944) Studies on lignin and related compounds. LXXV. Alkaline nitrobenzene oxidation of plant materials and application to taxonomic classification. J Am Chem Soc 66:32–37
Creighton RHJ, Hibbert H (1944) Studies on lignin and related compounds. LXXVI. Alkaline nitrobenzene oxidation of corn stalks. Isolation of p-hydroxybenzaldehyde. J Am Chem Soc 66:37–38
Cymbaluk NF, Neudorffer TS (1970) Quantitative gas-liquid chromatographic determination of aromatic aldehydes and acids from nitrobenzene oxidation of lignins. J Chromatogr 51: 167–174
Edwards GA, Perkin WH, Stoyle FW (1925) New synthesis of the meconines. J Chem Soc 195–199
Fengel D, Wegener G, Heizmann A, Przklent M (1977) Trifluoressigsäure zur schnellen and schonenden Hydrolyse von Cellulose und anderen Polysacchariden. Holzforschung 31: 65–71
Freudenberg K (1939) Über Lignin. Angew Chem 52:362–363
Freudenberg K, Janson A, Knopf E, Haag A (1936) Zur Kenntnis des Lignins (15.Mitteil.). Ber Dtsch Chem Ges 69:1415–1425
Freudenberg K, Lautsch W (1939) Zur Konstitution des Fichtenlignins. Naturwissenschaften 27:227–228
Freudenberg K, Lautsch W, Engler K (1940) Die Bildung von Vanillin aus Fichtenlignin. Ber Dtsch Chem Ges 73:167–171
Freudenberg K, Meister M, Flickinger E (1937) Fichtenlignin (XVI.Mitteil. Über Lignin). Ber Dtsch Chem Ges 70:500–514
Freudenberg K, Müller H (1938) Zur Kenntnis des Buchenlignins (XVIII.Mitteil. Über Lignin). Ber Dtsch Chem Ges 71:1821–1825
Higuchi T, Ito Y, Shimada M, Kawamura I (1967) Chemical properties of milled wood lignin of grasses. Phytochemistry 6:1551–1556
Higuchi T, Kawamura I (1966) Occurrence of p-hydroxyphenylglycerol-β-aryl ether structure in lignins. Holzforschung 20:16–21
Jayne JE (1953) A study of the alkaline nitrobenzene oxidation of chlorite lignin. Tappi 36:571–576
Kavanaugh KR, Pepper JM (1955) The alkaline nitrobenzene oxidation of aspen wood and lignin model substances. Can J Chem 33:24–30
Kratzl K, Silbernagel H (1952) The thermal treatment of spruce and lignin with water. Monatsh Chem 82:1022–1037
Leopold B (1950) Aromatic keto- and hydroxy-polyethers as lignin models. III. Acta Chem Scand 4:1523–1537
Leopold B (1951) Nitrobenzene oxidation of the products formed by the condensation of resorcinol with lignin models. Acta Chem Scand 5:1393–1421
Leopold B (1952a) Studies on lignin. III. Oxidation of wood from Picea abies (L.) Karst. (Norway spruce) with nitrobenzene and alkali. Acta Chem Scand 6:38–39
Leopold B (1952b) Studies on lignin. VI. Nitrobenzene oxidation of lignin sulfonic acids. Acta Chem Scand 6:55–63
Leopold B (1952c) Lignin. XI. The alkaline nitrobenzene oxidation of lignin and lignin models. Sven Kem Tidskr 64:18–26
Leopold B, Malmström IL (1951) Nitrobenzene oxidation of compounds of the lignin type. Acta Chem Scand 5:936–940
Leopold B, Malmström IL (1952) Studies on lignin. IV. Investigation on the nitrobenzene oxidation products of lignin from different woods by paper partition chromatography. Acta Chem Scand 6:49–54
Lundquist K, Kirk TK (1980) Fractionation-purification of an industrial kraft lignin. Tappi 63:80–82

Lundquist K, Ohlsson B, Simonson R (1977) Isolation of lignin by means of liquid-liquid extraction. Sven Papperstidn 80:143–144

Meshitsuka G, Nakano J (1985) Structural characteristics of compound middle lamella lignin. J Wood Chem Technol 5:391–404

Nakamura Y, Higuchi T (1976) Ester linkage of p-coumaric acid in bamboo lignin. Holzforschung 30:187–191

Nilsson T, Daniel G, Kirk TK, Obst JR (1989) Chemistry and microscopy of wood decay by some higher Ascomycetes. Holzforschung 43:11–18

Nimz HH, Robert D, Faix O, Nemer M (1981) Carbon-13 NMR spectra of lignin 8. Holzforschung 35:16–26

Obst JR, Sachs IB, Kuster TA (1988) The quantity and type of lignin in tyloses of bur oak (Quercus macroscarpa). Holzforschung 42:229–231

Pan D, Tai D, Chen C-L, Robert D (1990) Comparative studies on chemical composition of wood components in recent and ancient woods of Bischofia polycarpa. Holzforschung 44:7–16

Pearl IA (1942) Vanillin from lignin materials. J Am Chem Soc 64:1429–1431

Pearl IA (1967) The chemistry of lignin. Marcel Dekker, New York, 202–205

Pearl IA, Beyer DL (1950) Studies on lignin and related products. Tappi 33:544–548

Pepper JM, Casselman BW, Karapally JC (1967) Lignin oxidation. Preferential use of cupric oxide. Can J Chem 45:3009–3012

Pew JC (1955) Nitrobenzene oxidation of lignin model compounds, spruce wood and spruce "native lignin". J Am Chem Soc 77:2831–2833

Scalbert A, Monties B, Lallemand J-Y, Guittet E, Roland C (1985) Ether linkages between phenolic acids and lignin fractions from wheat straw. Phytochemistry 24:1359–1362

Schultz TP, Templeton MC (1986) Proposed mechanism for the nitrobenzene oxidation of lignin. Holzforschung 40:93–97

Shimada M, Fukuzuka T, Higuchi T (1971) Ester linkages of p-coumaric acid in bamboo and grass lignins. Tappi 54:72–78

Tai D, Chen C-L, Gratzl JS (1990) Chemistry of delignification during kraft pulping of bamboos. J Wood Chem Technol 10:75–99

Van Buren JB, Dence CW (1970) Chlorination behavior of pine kraft lignin. Tappi 53: 2246–2253

Winston MH, Chen C-L, Gratzl JS, Goldstein IS (1986) Characterization of the lignin residue from hydrolysis of sweetgum wood with superconcentrated hydrochloric acid. Holzforschung 40:Suppl. 45–50

6.3 Chemical Degradation Methods: Permanganate Oxidation

G. GELLERSTEDT

6.3.1 Introduction

Much of our current knowledge about the structure of lignin in wood and pulp is based on results obtained from permanganate oxidation. This method involves the selective degradation of all aliphatic side chains attached to aromatic moities in lignin, resulting in the formation of a mixture of aromatic carboxylic acid structures henceforth referred to as "acids". The identity of these as well as the amount of each individual acid provides information about the substitution pattern in a particular lignin and the frequency with which individual linkages between phenylpropane units occur.

The major drawback of the method is that only structures in lignin carrying free phenolic hydroxyl groups can be analyzed. Therefore, the information obtained accounts for only a fraction of the total lignin structure. However, in many cases, this information is sufficient to allow conclusions about the structure of a particular lignin. Where such is not the case, alternative approaches are available to increase the number of free phenolic hydroxyl groups prior to analysis.

In the original analytical method developed by Freudenberg and co-workers, wood was first subjected to alkaline hydrolysis at high temperature followed by a methylation with dimethyl sulfate. The free phenolic hydroxyl groups were converted by this treatment to methoxyl groups and the phenolic units were thus stabilized toward oxidative breakdown. The oxidation step was carried out by addition of portions of solid potassium permanganate at a pH at or near 7 until the purple color of the solution remained (Freudenberg et al. 1936).

Later, this procedure was modified such that the hydrolysis was preceded by a primary methylation step performed under mild conditions, thereby increasing the yield of acids. Altogether, the Freudenberg group was able to identify nine different acids in the veratryl series by this technique, that is, acids originating from guaiacyl structures in lignin. Furthermore, five acids in the anisyl series and three in the trimethoxybenzene series were found, corresponding to p-hydroxyphenyl and syringyl structures, respectively. All results were obtained on lignin from spruce wood (Freudenberg 1968).

The method of oxidative degradation using permanganate was later adopted by Miksche and co-workers. In their comprehensive work, covering approximately a 10-year period, the methodology was greatly improved and modern spectroscopic methods were introduced. Thus, it was found that the yield of

Springer Series in Wood Science
Methods in Lignin Chemistry
(Edited by S.Y. Lin and C.W. Dence)

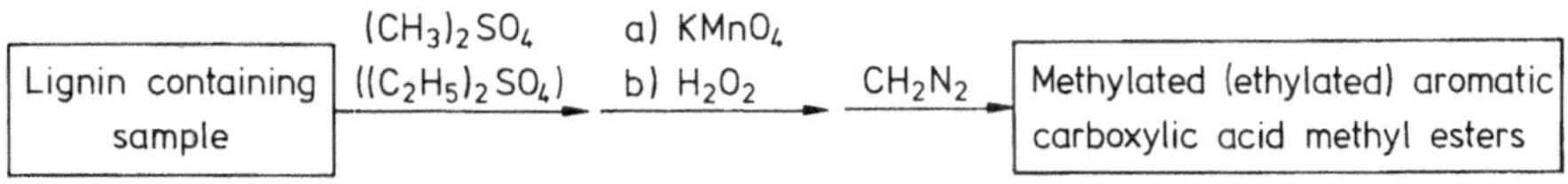

Fig.6.3.1. Reaction sequence for the oxidative degradation of lignin with potassium permanganate

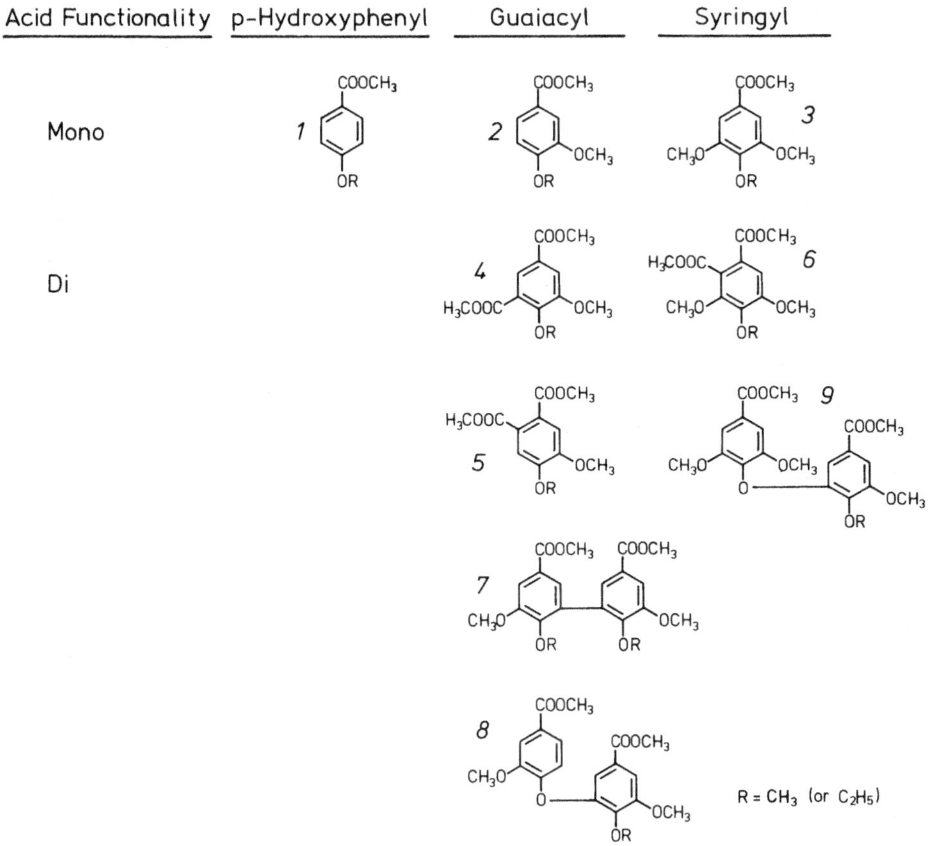

Fig.6.3.2. Major carboxylic acid methyl esters formed in the oxidation of lignin with potassium permanganate. (Compounds 3, 6 and 9 are from hardwood lignin)

degradation products was greatly improved when the permanganate oxidation was conducted under alkaline instead of neutral conditions. However, under such conditions the reaction mixture contained appreciable amounts of phenylglyoxylic acids in addition to the aromatic acids. Therefore, a second oxidation step was introduced, an oxidation under mild conditions with alkaline hydrogen peroxide. This oxidation converted all intermediate degradation products to aromatic acids. Finally, the mixture of acids was methylated with diazomethane to obtain derivatives suitable for gas chromatographic separation (Larsson and Miksche 1967). The complete reaction sequence is outlined in

Fig. 6.3.1. The predominant degradation products obtained from softwood and hardwood lignins are listed in Fig. 6.3.2. For softwood lignin samples it has been shown, however, that a total of approximately 40 different degradation products are formed (Larsson and Miksche 1969b, 1971b, 1972).

Following their studies dealing with the structure of lignin in spruce wood, Miksche and co-workers applied the method to birch and beech lignins, to compression wood from pine, to bark and to a large number of nonwoody plant materials (Andersson et al. 1973, Erickson and Miksche 1974a,b,c,d, Erickson et al. 1973a,b,c,d,e, Larsson and Miksche 1967, 1969a,b, 1971a,b, 1972, Miksche and Yasuda 1976).

More recently, oxidative degradation with permanganate has also been used for elucidating the structure of lignins, modified in various technical processes. Thus, comprehensive studies on the structure of residual lignins in both neutral sulfite (Glasser and Morohoshi 1979, Morohoshi and Glasser 1979a,b) and in kraft pulps (Gellerstedt and Lindfors 1984a, Gellerstedt and Gustafsson 1987a, Gellerstedt et al. 1984, 1986, 1988) have been conducted. The effect on lignin of bleaching kraft pulp with chlorine and chlorine dioxide has also been investigated (Erickson and Dence 1976, Lindström and Österberg 1984, Österberg and Lindström 1985a,b). The experimental procedure followed, however, remained essentially that described by Erickson et al. (1973a).

6.3.2 Experimental Procedure

6.3.2.1 Preparation of Samples

In the analysis of wood, pulp and other plant materials, a grinding procedure; e.g., milling in a Wiley mill, is used to reduce the particle size of the sample to facilitate penetration of the chemicals. Vigorous milling for prolonged periods should be avoided since such action may lead to irreversible chemical changes in the lignin structure (see below). After milling, the material is Soxhlet-extracted with a neutral solvent such as acetone or benzene-ethanol (2:1, v/v) to remove the extractives and any other low-molecular weight compounds present.

In the analysis of isolated lignins and lignin model compounds, the experimental procedure starts directly with the alkylation step.

6.3.2.2 Alkylation

Irrespective of whether the starting material is a wood, plant, or pulp sample, an isolated lignin sample, or a lignin model compound, it is essential that all free phenolic hydroxyl groups be protected by alkylation prior to oxidation to prevent degradation of the phenolic units. This is usually accomplished by methylation or ethylation in an alkaline solution with dimethyl sulfate or diethyl sulfate, respectively. The reaction step should be performed in a well-ventilated hood.

The sample, preferentially in the order of 10–100 mg calculated as lignin, is dissolved or suspended in 10 ml of 1,2-dimethoxyethane-methanol(ethanol)-water (35:35:30, v/v) in a 25-ml flat-bottomed three-necked flask placed on a magnetic stirrer. Through the middle neck a pH electrode is loosely inserted so that the tip extends below the surface of the solution. In the two other necks are placed glass tubes for the introduction of alkali solution and nitrogen gas (free from oxygen). The reaction is started by the addition of 2 ml of dimethyl sulfate (diethyl sulfate) to the solution. Simultaneously, alkali, in the form of an approximately 15% solution of KOH in the solvent mixture referred to above, is added to the reaction mixture via an automatic titration assembly consisting of an autoburet, titrator, and pH meter. During the reaction, the pH is maintained at 11.

The reaction is continued for approximately 24 h with stirring at room temperature or until the consumption of alkali ceases, after which the addition of alkali is interrupted and the reaction mixture is acidified to pH 3 by the addition of 0.5 M phosphoric acid. After 30 min, the pH is again adjusted to 6.5 by the addition of potassium hydroxide solution and the entire sample is transferred to a 500-ml round-bottomed flask. After evaporating the mixture to dryness, a small volume (20–30 ml) of t-butyl alcohol-water (3:1, v/v) is added and the solution is again evaporated to dryness. (The t-butyl alcohol should be free of sec-butyl alcohol and other impurities.)

6.3.2.3 Oxidation with Potassium Permanganate

The residue from the above evaporation is suspended in 40 ml of t-butyl alcohol-water (3:1, v/v), and the flask is placed in a constant temperature bath. After 40 ml of 0.5 M sodium hydroxide, 100 ml of 0.06 M sodium periodate, and 20 ml of 0.03 M potassium permanganate are added, in this order, to the stirred suspension, the flask is heated to 82 °C and kept at this temperature for 6 h. The periodate continuously reoxidizes any manganate (V or VI) to permanganate thereby keeping the permanganate concentration at a low level which minimizes the risk of any over-oxidation of the lignin degradation products. If the periodate is consumed in other reactions, however, a shortage of permanganate will result, thus leading to incomplete lignin oxidation. Since it is essential that permanganate be constantly present throughout the entire reaction, portions of solid sodium periodate (1.28 g) and potassium permanganate (0.095 g) should be added when necessary, that is, when the purple permanganate color fades.

The oxidation reaction is interrupted by the addition of 10 ml of ethanol. After 10 min, the manganese dioxide is filtered off through a layer of water-washed fumed silica ("Kieselguhr", Fluka AG Switzerland) and washed with a small amount of sodium bicarbonate (1% solution). The aqueous phase is extracted with 2 × 50 ml of diethyl ether, the ether in turn is extracted with 15 ml of sodium bicarbonate (1% solution) and, after adjustment of the pH to 6.5 with 9 M sulfuric acid, the combined aqueous extracts are evaporated to a small volume (30 ml).

6.3.2.4 *Oxidation with Hydrogen Peroxide*

To the above solution is added 0.25 ml of aqueous DTPA (diethylentriamine-pentaacetic acid) (0.3% solution, pH adjusted to 7) and 20 ml of t-butyl alcohol-water (1:1, v/v). The solution is made alkaline by the addition of 0.9 g of dry sodium carbonate, and the oxidation is started by adding 5 ml of hydrogen peroxide (30% solution). After 10 min at 50 °C, the oxidation is terminated by the addition of 100 mg of manganese dioxide. The mixture is allowed to stand at room temperature for 2 h to decompose all of the hydrogen peroxide and is then filtered. After washing of the precipitate with a small amount of water, the aqueous solution is acidified to pH 2 with 9 M sulfuric acid and extracted with acetone-methylene chloride (1:1, v/v) three times, each time using a volume of organic solvent approximately equal to the volume of water. Finally, the combined organic extracts are dried over anhydrous sodium sulfate and evaporated.

6.3.2.5 *Esterification*

The residue after evaporation is dissolved in 5 ml of methanol and 15 ml of diethyl ether (free of stabilizer) is added. Diazomethane is added to the resulting solution under a stream of nitrogen. (The reaction should be conducted in a well-ventilated hood.) The diazomethane is conveniently prepared from N-methyl-N-nitroso-p-toluenesulfonamide (Merck-Schuchardt, Aldrich, Fisher, Fluka, and others) in a special reaction flask as described by Schlenk and Gellerman (1960).

When the yellow color of the reaction mixture persists, the introduction of diazomethane is interrupted and the solution is left overnight. After evaporation, the residue is dissolved in diethyl ether and the solution filtered through a small column of silica gel (~150–200 mg). The resulting colorless solution, containing the aromatic carboxylic acid esters, is again evaporated, and the residue, after dissolution in a small amount of dichloromethane, is transferred to a vial equipped with a Teflon-lined screwcap.

6.3.2.6 *Gas Chromatographic Analysis*

Pyromellitic acid tetramethyl ester, mp 142–143 °C, prepared by diazomethane methylation of pyromellitic acid, is used as the internal standard and is stored as a dichloromethane solution (~1 mg/ml) in a freezer. (Any loss of solvent from the solution is conveniently verified by weighing.) From this solution, the exact volume corresponding to from 200 to 1000 μg of the internal standard is withdrawn and added to the ester mixture to be analyzed. The amount of internal standard chosen is dependent on the amount and type of sample. For samples where a high yield of esters is expected, the upper level of internal standard is used.

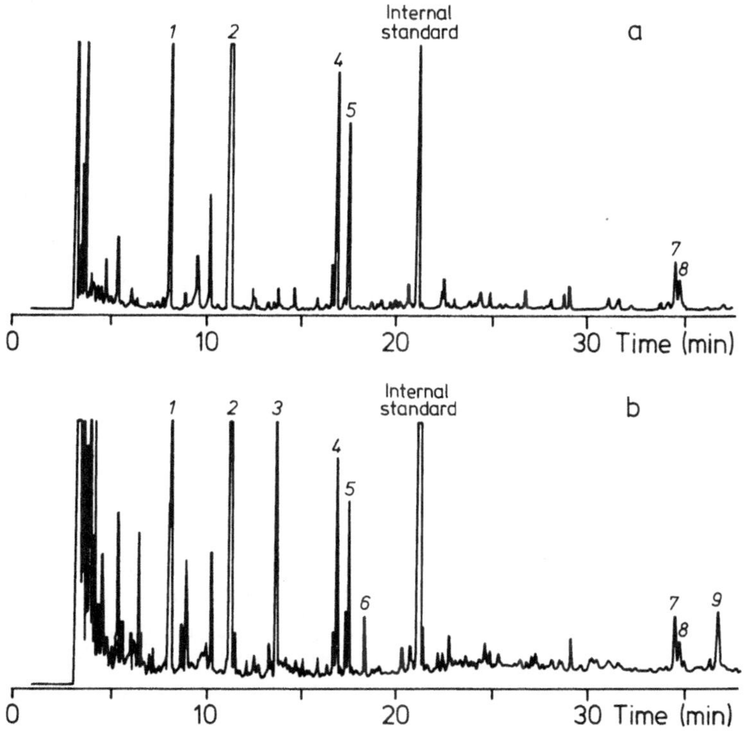

Fig.6.3.3. Typical chromatograms of products (analyzed as ethylated carboxylic acid methyl esters) obtained in the permanganate oxidation of spruce (**a**) and birch (**b**) wood. (*Numbered peaks* refer to structures listed in Fig. 6.3.2)

For quantification purposes it is convenient to use a synthetic mixture containing the methyl esters of the major acids together with the internal standard. At the Swedish Pulp and Paper Research Institute, major methyl esters 1, 2, 3, 4, 7 (Fig. 6.3.2) have been used with the amounts of each individual component chosen so that the detector response is of the same order of magnitude as that obtained from the lignin samples. (One ml of the mixture typically contains approximately 600, 9400, 2400, 2000, and 1900 μg of 1, 2, 3, 4, and 7, respectively, and 2400 μg of internal standard.) For other major acids present in the product mixture, e.g., those corresponding to 5 and 8 and to 6 and 9 (Fig. 6.3.2.), which are from hardwood lignin, it is assumed that the response factors in both cases are similar to those of 4 and 7.

All ester mixtures are separated by gas chromatography on a 30 m DB-1 fused silica capillary column using a temperature program of 5 °C/min from 150 to 270 °C and, thereafter 270 °C for 7 min. Prior to analysis of the ester mixture from a lignin sample, the synthetic mixture is analyzed. Quantification is conveniently performed by electronic integration of the chromatogram peaks. Typical gas chromatograms obtained from the analysis of spruce and birch wood are reproduced in Fig. 6.3.3.

6.3.3 Discussion of Method

6.3.3.1 Reaction Selectivity

The yield of products in the four-step reaction sequence leading to the formation of aromatic carboxylic acid esters has been thoroughly investigated by Erickson et al. (1973a). By subjecting a large number of lignin model compounds to oxidative degradation, information was obtained concerning the reactivity of some of the major substructures in lignin: β-aryl ether, biphenyl, and phenylcoumaran units. Furthermore, the influence of side chain structure on the oxidation was revealed. Thus, for similar model compounds having either a methylene, carbinol, or keto group at the position alpha to the aromatic ring, a pronounced difference in the yield of the corresponding degradation acid was found, the methylene structure being the least reactive.

For native lignins this latter observation is of little importance since the overwhelming majority of lignin substructures have an oxygen function at the α-carbon atom. For a variety of other lignin containing materials, e.g., kraft pulps and kraft lignins, this difference in reactivity may play a role, however, since such lignins contain significant amounts of methylene and methine groups in their side chains (Gellerstedt and Robert 1987 and unpublished).

For lignin substructures containing a carbon substituent in the aromatic C-5 position, or for stilbenes originating from phenylcoumaran structures, the oxidative degradation leads to the formation of isohemipinic acid methyl ester (4) (Fig. 6.3.2, R=CH_3). However, Gellerstedt and co-workers have shown that model compounds of the biphenyl type also give rise to isohemipinic acid in addition to the expected bis-vanillic acid structure (7). The reason seems to be incomplete alkylation (pH 11, 24 h) with the result that the substrate is alkylated at only one of the available positions. To achieve a more complete alkylation, it is necessary to increase the pH to 13 or alternatively, to extend the alkylation time to 84 h. In both cases, the yield of the bis-vanillic acid structure is substantially increased as shown in Table 6.3.1; nevertheless, the concomitant formation of isohemipinic acid cannot be completely avoided.

Oxidative degradation of native lignin samples (MWL) from spruce and from aspen at varying pH and reaction times during the alkylation step has shown, however, that an optimum yield of products (including dimethyl ester 7) is obtained under the conditions stated above, i.e., reaction at pH 11 for 24 h (Gellerstedt, unpubl.). The reason for this discrepancy in the results from model compounds and lignins is not known.

6.3.3.2 Identification and Estimation of Structural Unit Types

The major degradation products obtained by permanganate oxidation (Fig. 6.3.2) reflect the distribution of the various types of phenylpropane units in a lignin sample. Thus, structures (1), (2), and (3) originate from uncondensed lignin end groups in which the original propanoid side chain has been degraded to a carboxyl group in the oxidation procedure. From lignin structures contain-

Table 6.3.1. Product yields in the oxidative degradation of 5,5′-bis-n-propylguaiacol using different reaction conditions in the alkylation (ethylation) step

Model compound	Reaction time, h	pH	Product yield, % 4	7
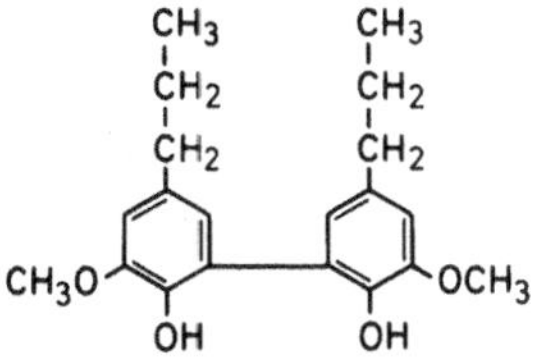	24	11	35	17
	84	11	8	50
	24	13	4	35

ing, in addition to the normal C-1 side chain, a carbon-carbon linkage between either the C-5 or the C-6 position of the aromatic ring and an adjacent lignin unit, degradation products (4), (5), and (6) are formed. Finally, products (7), (8), and (9) demonstrate the presence of diaryl and diaryl ether linkages in the lignin sample.

Besides providing information about the absolute content of various phenolic groups present in lignin, the results from the permanganate degradation analysis can be used to calculate the relative abundance on a molar basis of individual lignin substructures, thereby furnishing information about the structural composition of a particular lignin.

From the absolute yields of individual acids obtained in the gas chromatographic analysis, the corresponding theoretical yields can be conveniently calculated by assuming an overall yield in the oxidative degradation procedure. Based on a large number of experiments performed on lignin model compounds (Erickson et al. 1973b), this overall yield is assumed to be 60% per carbon substituent attached to the same aromatic ring.

It is obvious that this figure is only an estimated average value since individual lignin substructures react at different rates depending on such factors as accessibility and side chain structure. The absolute yield of degradation acids containing two carboxylic acid groups consequently is assumed to be only 36%. Therefore, in the calculation of the corresponding theoretical yields, any discrepancy will increase the error considerably. Thus, based on the work of Gellerstedt et al. (1986), it has been found that, for lignins containing relatively large amounts of "condensed" lignin structures, such as are found in oxygen-bleached pulps and lignins, oxidative degradation seems to underestimate the amounts of the corresponding degradation acids.

Once the theoretical yield of individual degradation acids have been calculated, their corresponding relative frequencies of occurrence can be obtained. The figures are usually expressed in percent on a molar basis and are related to

Table 6.3.2. Relative frequency of aromatic carboxylic acids formed in the permanganate oxidation of selected woods and pulps

Sample	Relative frequency, mol %					
	1[a]	2	4	5	7	8
Wood (spruce)	6	67	10	8	6	3
Wood (pine)	3	69	10	7	7	4
MWL (spruce)	4	54	12	8	15	7
CTMP (spruce)	6	60	10	15	6	3
Sulfite pulp (spruce, kappa number 15)	4	40	19	19	7	11
Kraft pulp (pine, kappa number 36)	3	42	16	6	21	12
O_2-Bleached kraft pulp (pine, kappa number 17)	1	38	19	6	24	12

[a] See Fig. 6.3.3 for structures of compounds.

the basic C_9 (phenylpropane) unit in lignin. These values are not dependent on the lignin content in a particular sample but indicate the relative distribution of individual types of structural units in lignin. The relative abundance of carboxylic acids resulting from the permanganate oxidation of several representative lignin-containing materials is indicated by the values presented in Table 6.3.2.

To obtain quantitative information about the most abundant interunit linkage in lignin, the phenylpropane-β-aryl ether structure, it is necessary to perform the oxidative degradation on a sample before and after a selective β-aryl ether cleavage reaction, as has been done by Larsson and Miksche (1971a) and by Morohoshi and Glasser (1979a). The method used involves treatment of the lignin with alkali and cupric oxide at an elevated temperature. Whether this method in fact results in a quantitative cleavage of all β-aryl ether (and α-aryl ether) structures in a lignin is, however, not known with certainty. Furthermore, the reaction can only be applied to isolated lignins. As an alternative approach for liberating phenolic structures in β-aryl ether units, kraft cooking has been suggested (Larsson and Miksche 1971a). However, it has been shown that kraft lignins still contain appreciable amounts of β-aryl ether structures (Gellerstedt et al. 1984).

Even without quantification of β-aryl ether structures, calculation of the relative frequencies of individual lignin substructures is informative (cf. Table 6.3.2). From the values obtained, a judgment can be made about the degree of branching (and/or cross-linking) in a particular lignin. Furthermore, by comparing samples containing modified lignins, for example, chemical pulps, with each other and with reference samples having "known" lignin structures, it is possible to draw conclusions about the structural changes that may have occurred. It is also possible to identify, for example, condensation reactions involving aromatic rings, such as those taking place in various sulfite pulping processes. In technical processes involving oxidation and substitution of lignin as in pulp bleaching, the degree and type of aromatic substitution can be evaluated. Thus, a large number of chlorinated aromatic acids have been identified after oxidative degradation of chlorinated pulps with permanganate (Österberg and Lindström 1985b).

6.3.3.3 Estimation of Phenolic Hydroxyl Groups

The theoretical yield of aromatic carboxylic acids obtained in the analytical method can be used to calculate the approximate total number of free phenolic hydroxyl groups present in lignin. This number is often of great importance since it affords an indication of the reactivity and the degree of polymerization of a particular lignin. For lignins which have been subjected to hydrolytic reactions such as occur in kraft or sulfite pulping, an increase in the yield of acids (or phenolic group content) compared with that in the original material is normally encountered. The opposite is true for lignins which have been subjected to oxidative conditions. In the latter case, the amount and types of remaining phenolic structures provide information about the efficiency and selectivity of the oxidation.

For a series of kraft pulps, the number of phenolic structures present in the residual lignin has been calculated from results obtained using permanganate degradation analysis (Gellerstedt and Lindfors 1984a). Analysis of the phenolic hydroxyl group content of the same pulps by other means has shown that the figures are in fair agreement with each other (Gellerstedt and Lindfors 1984b).

For wood samples, however, indirect determination of phenolic hydroxyl groups using the results from permanganate oxidation tends to give values which are abnormally high. This is because the milling of wood, which is normally used to decrease the particle size prior to analysis, is accompanied by chemical changes in the lignin. Thus, new phenolic structures are created by the mechanical action on the wood. Such reactions are particularly harmful in lignin samples containing β-aryl ether structures of the syringyl type found in hardwoods. Thus, it has been shown that the yield of acids from the oxidative degradation of birch wood may increase by as much as 35% if the milling is done in a ball mill instead of a Wiley mill (Gellerstedt and Northey 1989).

Another factor which may influence the yield of acids obtained from lignin in wood is the presence of easily hydrolyzable aryl ether linkages such as those represented by phenylpropane-α-aryl ether structures. In phenolic units, such structures are easily cleaved even under very mild alkaline conditions (Adler et al. 1968). Therefore, it can be assumed that, in the alkylation step of the oxidative degradation procedure, a partial hydrolysis of such structures may occur, thereby creating new phenolic end groups that may affect the analysis.

Native, as well as technical lignins, whether present in pulp (wood) or in the form of isolated samples, usually contain trace amounts of catechol structures. These can be identified and quantified if the alkylation step in the oxidative degradation procedure is performed with diethyl sulfate since, in this case, various types of catechol units in the form of 1,2-diethoxybenzene derivatives are distinguished from guaiacyl and syringyl structures (Gellerstedt and Lindfors

1984a). Catechol structures, which constitute an important leucochromophoric system, can easily be oxidized, for example, in the presence of air (oxygen) to o-benzoquinones. The amount of such structures is therefore of particular importance when analyzing lignin in different types of high-yield pulp (Gellerstedt and Gustafsson 1987b).

References

Adler E, Miksche GE, Johansson B (1968) Über die benzyl-arylätherbindung in Lignin. I. Freilegung von phenolischen Hydroxyl in Ligninpräparaten durch Spaltung leicht hydrolysierbarer Alkylarylätherstrukturen. Holzforschung 22:171–174

Andersson A, Erickson M, Fridh H, Miksche GE (1973) Zur Struktur des Lignins der Rinde von Laub- und Nadelhölzern. Holzforschung 27:189–193

Erickson M, Dence CW (1976) Phenolic and chlorophenolic oligomers in chlorinated pine kraft pulp and in bleach plant effluents. Sven Papperstidn 79:316–322

Erickson M, Miksche GE (1974a) Two dibenzofurans obtained on oxidative degradation of the moss Polytrichum commune Hedw. Acta Chem Scand B28:109–113

Erickson M, Miksche GE (1974b) Charakterisierung der Lignine von Gymnospermen durch oxidativen Abbau. Holzforschung 28:135–138

Erickson M, Miksche GE (1974c) Charakterisierung der Lignine von Pteridophyten durch oxidativen Abbau. Holzforschung 28:157–159

Erickson M, Miksche GE (1974d) On the occurrence of lignin or polyphenols in some mosses and liverworts. Phytochemistry 13:2295–2299

Erickson M, Larsson S, Miksche GE (1973a) Gaschromatografische Analyse von Ligninoxydationsprodukten. VII. Ein verbessertes Verfahren zur Charakterisierung von Ligninen durch Methylierung und oxydativen Abbau. Acta Chem Scand 27:127–140

Erickson M, Larsson S, Miksche GE (1973b) Gaschromatografische Analyse von Ligninoxydationsprodukten. VIII. Zur Struktur des Lignins der Fichte. Acta Chem Scand 27:903–914

Erickson M, Larsson S, Miksche GE (1973c) Zur Struktur des Lignins des Druckholzes von Pinus mugo. Acta Chem Scan 27:1673–1678

Erickson M, Miksche GE, Somfai I (1973d) Charakterisierung der Lignine von Angiospermen durch oxydativen Abbau. I. Dikotylen. Holzforschung 27:113–117

Erickson M, Miksche GE, Somfai I (1973e) Charakterisierung der Lignine von Angiospermen durch oxydativen Abbau. II. Monokotylen. Holzforschung 27:147–150

Freudenberg K (1968) In: Freudenberg K, Neish AC (eds) Constitution and biosynthesis of lignin. Springer, Berlin Heidelberg New York 78–82

Freudenberg K, Janson A, Knopf E, Haag A (1936) Zur Kenntnis des Lignins. Chem Ber 69:1415–1425

Gellerstedt G, Gustafsson K (1987a) Structural changes in lignin during kraft cooking. Part 5. Analysis of dissolved lignin by oxidative degradation. J Wood Chem Technol 7:65–80

Gellerstedt G, Gustafsson K (1987b) Reaction of lignin in chemimechanical pulping. Proc Fourth Int Symp Wood Pulp Chem Paris 1:15–19

Gellerstedt G, Gustafsson K, Lindfors E (1986) Structural changes in lignin during oxygen bleaching. Nord Pulp Pap Res J 3:14–17

Gellerstedt G, Gustafsson K, Northey RA (1988) Structural changes in lignin during kraft cooking. Part 8. Birch lignins. Nord Pulp Pap Res J 3:87–94

Gellerstedt G, Lindfors E (1984a) Structural changes in lignin during kraft pulping. Holzforschung 38:151–158

Gellerstedt G, Lindfors E (1984b) Structural changes in lignin during kraft cooking. Part 4. Phenolic hydroxyl groups in wood and kraft pulps. Sven Papperstidn 87:R115–R118

Gellerstedt G, Lindfors E, Lapierre C, Monties B (1984) Structural changes in lignin during kraft cooking. Part 2. Characterization by acidolysis. Sven Papperstidn 87:R61–R67

Gellerstedt G, Northey RA (1989) Analysis of birch wood by oxidative degradation. Wood Sci Technol 23:75–83
Gellerstedt G, Robert D (1987) Quantitative ^{13}C NMR analysis of kraft lignins. Acta Chem Scand B41:541–546
Glasser WG, Morohoshi N (1979) The structure of lignins in pulps. The distribution of interunit linkages. Tappi 62(12):101–105
Larsson S, Miksche GE (1967) Gas chromatographic analysis of lignin oxidation products. The diphenyl ether linkage in lignin. Acta Chem Scand 21:1970–1971
Larsson S, Miksche GE (1969a) Gaschromatografische Analyse von Ligninoxydationsprodukten. II. Nachweis eines neuen Verknüpfungsprinzips von Phenylpropaneinheiten. Acta Chem Scand 23:917–923
Larsson S, Miksche GE (1969b) Gaschromatografische Analyse von Ligninoxydationsprodukten. III. Oxydativer Abbau von methyliertem Björkman-Lignin (Fichte). Acta Chem Scand 23:3337–3351
Larsson S, Miksche GE (1971a) Gaschromatografische Analyse von Ligninoxydationsprodukten. IV. Zur Struktur des Lignins der Birke. Acta Chem Scand 25:647–662
Larsson S, Miksche GE (1971b) Gaschromatografische Analyse von Ligninoxydationsprodukten. V. Zwei trimere Abbauprodukte aus Fichtenlignin. Acta Chem Scand 25:673–679
Larsson S, Miksche GE (1972) Gaschromatografische Analyse von Ligninoxydationsprodukten. VI. 4,4-Bis-aryloxy-cyclohexa-2,5-dienonstrukturen im Lignin. Acta Chem Scand 26: 2031–2038
Lindström K, Österberg F (1984) Characterization of the high molecular mass chlorinated matter in spent bleach liquors (SBL). Part 1. Alkaline SBL. Holzforschung 38:201–212
Miksche GE, Yasuda S (1976) Zur Struktur des Lignins der Buche (Fagus silvatica L.). Liebigs Ann Chem 1323–1332
Morohoshi N, Glasser WG (1979a) The structure of lignins in pulps. Part 4. Comparative evaluation of five lignin depolymerization techniques. Wood Sci Technol 13:165–178
Morohoshi N, Glasser WG (1979b) The structure of lignins in pulps. Part 5. Gas and gel permeation chromatography of permanganate oxidation products. Wood Sci Technol 13:249–264
Österberg F, Lindström K (1985a) Characterization of the high molecular mass chlorinated matter in spent bleach liquors (SBL). Part 2. Acidic SBL. Holzforschung 39:149–158
Österberg F, Lindström K (1985b) Characterization of the high molecular mass chlorinated matter in spent bleach liquors (SBL). Part 3. Mass spectrometric interpretation of aromatic degradation products in SBL. Org Mass Spectrom 20:515–524
Schlenk H, Gellerman JL (1960) Esterification of fatty acids with diazomethane on a small scale. Anal Chem 32:1412–1414

6.4 Thioacidolysis

C. ROLANDO, B. MONTIES, and C. LAPIERRE

6.4.1 Introduction

Thioacidolysis, that is, solvolysis in dioxane-ethanethiol with boron trifluoride etherate, is an acid-catalyzed reaction which results in the depolymerization of lignins. The method may be used to estimate the amount and compositon of uncondensed alkyl aryl ether structures, with the exception of methyl aryl ethers (methoxyl substituents in lignin), and of several minor structural types in both isolated and in situ lignins. The detection of C_6C_3 trithioethyl phenylpropane compounds in the thioacidolysis of a plant material provides unambiguous evidence for the occurrence of arylglycerol aryl ether structures, the most characteristic ones in lignin. On this basis, thioacidolysis may be viewed as a diagnostic test for the presence of lignin where other chemical or spectroscopic methods fail for lack of specificity or sensitivity, respectively.

6.4.2 Thioacidolysis Reactions of Lignin

As in acidolysis (Chap. 6.1), thioacidolysis proceeds mainly by cleavage of arylglycerol-β-aryl ether linkages. In contrast to acidolysis, however, thioacidolysis is performed in anhydrous media. The ether-cleaving reagent combines a hard Lewis acid, boron trifluoride etherate, and a soft nucleophile, ethanethiol (Node et al. 1976, Fuji et al. 1979).

The mechanism by which various lignin structures are degraded has been investigated by subjecting appropriate lignin model compounds to thioacidolysis (Lapierre 1986, Lapierre et al. 1987). The prominent β-aryl ether structures in lignins are degraded by the mechanism depicted in Fig. 6.4.1. The first step of the reaction takes place at the benzyl alcohol or benzyl ether group. Acetylation of the benzyl alcohol group completely represses thioacidolysis when the reagent mixture is composed of BF_3 etherate and EtSH, but not when a mixture of HBF_4 and EtSH are used. Practically, this limitation in the case of BF_3 etherate-EsSH mixture has to be kept in mind when lignins modified by acetylation or sulfonation are analyzed. Reaction of C_β with accompanying ether bond cleavage takes place with participation of the thioethyl group previously introduced at C_α. This mechanism affords a pair of erythro and threo isomers regardless of the stereochemistry of the initial structure. Substitution of the oxygenated function at C_γ proceeds in a similar way.

Springer Series in Wood Science
Methods in Lignin Chemistry
(Edited by S.Y. Lin and C.W. Dence)

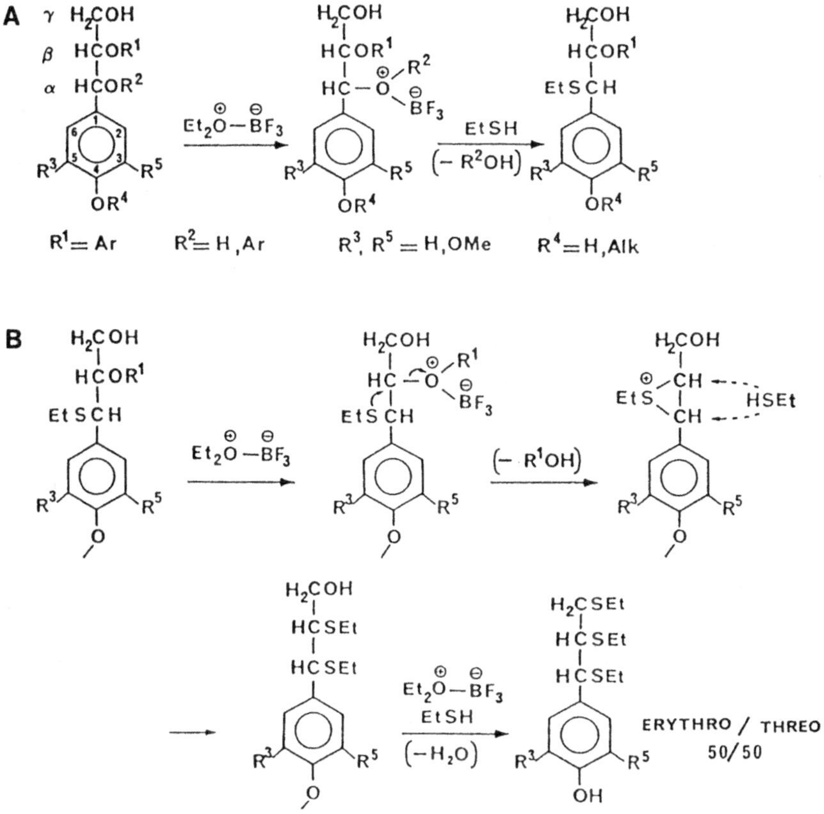

Fig. 6.4.1A,B. Proposed mechanism of lignin thioacidolysis. **A** Substitution at C_α. **B** Substitution at C_β, then at $C\gamma$ via a neighboring group participation reaction

Phenylpropane β-aryl ether structures containing an α-carbonyl substituent give rise to C_6C_2 products in low yield (Lapierre et al. 1987). Therefore, for lignin samples having a high carbonyl content, it is advisable to preface thioacidolysis with a sodium borohydride reduction step. Lignin structures containing a conjugated carbon-carbon double bond generally afford, either partially or totally, the Michael addition product. The extent of the addition varies according to the activating effect of the side chain terminal group.

Particular attention has been devoted to the fate of C_6C_2 enol ether structures present in kraft or soda lignins (Gellerstedt et al. 1984). It has been shown that such structures are essentially converted to a C_6C_2 product with two thioethyl groups at C_β. This dithioacetal compound is formed only in small amounts on thioacidolysis of C_6C_3 β-aryl ether structures. Therefore, when present in significant amounts in the thioacidolyzate of a sample, the dithioacetal compound is indicative of C_6C_2 enol ether structures preexisting in the sample.

Whatever the mechanism of their formation, the various monomeric products recovered from the thioacidolysis of a lignin sample reflect the structure of the original C_6C_3 or C_6C_2 unit, as illustrated in Table 6.4.1. Most of the initial

Table 6.4.1. Main products obtained by thioacidolysis of various phenolic structures

C_6C_3 Arylglycerol-β-aryl ether	R^a—CHSEt—CHSEt—CH_2SEt erythro/threo 50/50
C_6C_3 β-Aryl ether with α-CO substituent	R—CHSEt=CHSEt (Z and E)
p-Hydroxycinnamaldehyde end group	R—CHSEt—CH_2—CH(SEt)$_2$
p-Hydroxycinnamyl alcohol end group	R—CH=CH—CH_2SEt
	R—CHSEt—CH_2—CH_2SEt
p-Hydroxycinnamic acid unit[b]	R—CH=CH—COOH
	R—CHSEt—CH_2—COOH
p-Hydroxybenzaldehyde end group	R—CH(SEt)$_2$
C_6C_2 enol ether[c]	R—CH_2—CH(SEt)$_2$

[a] R=p—Hydroxyphenyl, guaiacyl and/or syringyl rings.
[b] p—Coumaric and ferulic acid units in grass samples.
[c] Present in kraft lignins (Gellerstedt et al. 1984).

structures, except C_6C_3 units containing an α-carbonyl substituent, are degraded in high yield and without secondary reactions. The quantitative determination of the recovered products may therefore reflect the amount of the starting structure.

6.4.3 Description of Method

6.4.3.1 Reagents

Dioxane is freshly distilled from a benzophenone-sodium mixture, or, as is more convenient for routine analysis, is used without further purification when obtained as an analytical grade reagent.

Boron trifluoride etherate is freshly distilled over CaH_2 or used directly when obtained as an analytical grade chemical. This reagent is kept at 4 °C under nitrogen as it is a moisture-sensitive reagent.

Ethanethiol and HBF_4 etherate, which replaces BF_3 etherate in the thioacidolysis of acetylated samples, are of analytical grade quality.

The thioacidolysis reagent is prepared immediately before use; 2.5 ml of BF_3 etherate and 10 ml of ethanethiol are successively introduced into a 100-ml volumetric flask containing 20 ml of dioxane and the final volume is adjusted to 100 ml with dioxane. This reagent is therefore a solution of 0.2 M BF_3 etherate (Note 1) in a 8.75:1 (v/v) dioxane/ethanethiol mixture. It is a colorless reagent immediately after preparation. When an inferior grade of dioxane is used, a yellow color is observed; this reagent is not suitable for analytical use.

6.4.3.2 Treatment of Lignin with Thioacidolysis Reagent

Extractive-free wood and straw samples are prepared by successive 24-h extraction in Soxhlet apparatus with ethanol-toluene (1:2, v/v), ethanol and water. Pulps are similarly extracted for 24 h with acetone.

Soluble lignins (5 mg) (Note 2) or insoluble lignocellulose residues (10 mg) (Note 3) are added to 5 or 10 ml, respectively, of thioacidolysis reagent in a tube fitted with a Teflon-lined screwcap (Note 4) under an atmosphere of nitrogen or argon. The thioacidolysis is allowed to proceed at 100 °C (oil bath) for 4 h (Note 5) with occasional shaking. The reaction tube is cooled in ice water and the reaction mixture, together with a few ml of water to rinse the tube (3 × 5 ml) are poured over CH_2Cl_2 (analytical grade) to which the gas chromatography (GC) internal standard (see below) has been added. The pH of the aqueous (upper) phase is adjusted to 3–4 as indicated by pH-indicator paper by the addition of aqueous 0.4 M $NaHCO_3$ and the two phases are together extracted with CH_2Cl_2 (3 × 50 ml). The combined organic extracts are dried over anhydrous Na_2SO_4, and then evaporated under reduced pressure at 40 °C. The oily residue is carefully redissolved in 0.5–1 ml of CH_2Cl_2 to recover a solution appropriately diluted for the following silylation and GC steps.

The entire extraction procedure preferably is performed without interruption; when kept in aqueous media overnight, the main products are slightly degraded as shown by decreased recovery. In contrast, the stability of thioacidolysis products in organic media (for example CH_2Cl_2) is excellent, at least at 4 °C, where standard solutions remained unchanged for more than 1 year.

6.4.3.3 GC Analysis of Monomeric Products

The GC internal standard is usually hexacosane (C_{26}) and/or tetracosane (C_{24}). For complex samples, such as those represented by lignocellulose residues, it is advisable to add both of these internal standards. The amount of each internal standard (C_{26} and/or C_{24}) to be added is 0.5 mg/20 mg of extractive-free wood, 0.25 mg/20 mg of extractive-free straw, and 1 mg/5 mg of milled wood lignin (Note 6).

Silylation: the final CH_2Cl_2 solution of the thioacidolysis sample is again carefully dried over Na_2SO_4; 5–10 μl of this dried organic solution is trimethylsilylated (TMS) at room temperature with 50 μl of N,O bis (trimethylsilyl) trifluoroacetamide (BSTFA) and 5 μl of GC-grade pyridine in a 200-μl reaction vial fitted with a Teflon-lined screwcap. For soluble lignin samples, the silylation is complete within a few minutes; for lignocellulose residues, silylation is complete within 2 to 4 h. In both cases and with well-dried samples, silylations performed in the aforementioned vials are stable for several days (at least 48 h).

Gas chromatography (GC) and gas chromatography-mass spectrometry (GC-MS) analyses: GC analyses are performed on a WCOT (wall-coated open tubular) polydimethylsiloxane fused silica capillary column (1 μm film thickness, 50 m × 0.32 mm I.D.). Shorter columns, for example 25 m, or columns with thinner films, for example 0.2 μm, may also be used successfully. Injection of the silylated sample (2 μl) is performed with a moving needle-type injector; other types of capillary GC injection systems may be used after optimization of the concentration and amount of sample to be injected. With a 50-m capillary

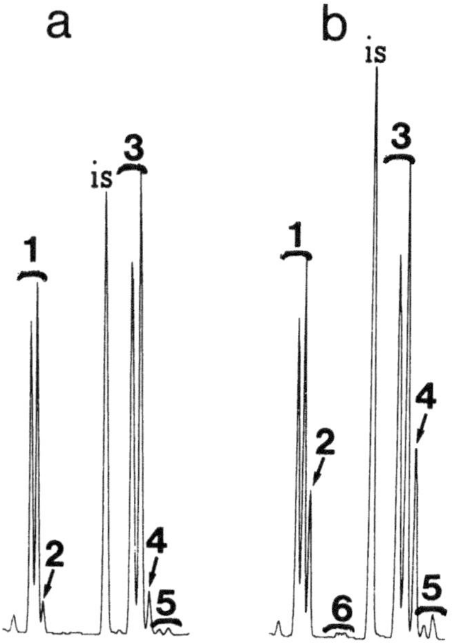

Fig. 6.4.2a,b. Partial gas chromatograms of products recovered from the thioacidolysis of poplar milled wood lignin with 0.2 N (**a**) and 1 N (**b**) BF_3 etherate. *Numbered peaks* correspond to the TMS derivatives of 4-hydroxyphenyl compounds with R^1, R^2 and R^3 substituents at C1, C3 and C5 carbons, respectively: 1 R^1=CHSEt-CHSEt-CH_2SEt, R^2=OMe, R^3=H; 2 R^1=CH_2-CHSEt-CH$(SEt)_2$, R^2=OMe, R^3=H; 3 R^1=CHSEt-CHSEt-CH_2SEt, R^2=R^3=OMe; 4 R^1=CH_2-CHSEt-CH$(SEt)_2$, R^2=R^3=OMe; 5 R^1=CHSEt-CHSEt-CH_2SEt, R^2=OH, R^3=OMe; 6 R^1=CHSEt-CHSEt-CH_2SEt, R^2=OH, R^3=H; *IS* internal standard

column, the carrier gas is helium at an inlet pressure of 1.5 bar and the temperature is programmed from 180 to 280 °C (+5 °C/min); for GC analysis, detection is achieved by a flame ionization detector. CG-MS analyses are performed with a similar injector and column but with a lower helium inlet pressure (0.7 to 1 bar) and a lower temperature program (130 to 260 °C at +5 °C/min). Detection is accomplished with a quadrupole spectrometer operating in the electron impact mode (70 eV) and with a source at 100 °C. GC-MS analyses are not performed routinely but are conducted periodically to confirm the identity of GC peaks, particularly when new sample types are being investigated.

6.4.3.4 Quantitative Determination of the Main Monomeric Products

The main syringyl (S) and guaiacyl (G) monomeric products from the thioacidolysis of lignins are isolated by preparative-scale experiments (Lapierre et al. 1986a,b). The only difference between these and the previously described analytical procedure is that the reaction mixture (500 ml of reagent and 3–6 g of lignin sample) is refluxed. The product yields are identical for both analytical and preparative procedures.

The response factors of the main syringyl, guaiacyl, and p-hydroxyphenyl (H) trithioethyl C_6C_3 diastereoisomers, defined as the ratio of relative concentration (underivatized compound) to relative area, are identical and equal to 1.50 using either tetracosane or hexacosane as an internal standard. For minor products, which are not isolated in preparative experiments, use of the same response factor is justified on the grounds that their structures are similar to those of the main products.

As the derivatized (TMS) products are very sensitive to active surfaces, it is critical that the GC column be perfectly deactivated and that any condition which might reactivate it be avoided (Note 7).

Note 1. When the concentration of BF_3 etherate is increased from 0.2 to 1 M or when this reagent is replaced by HBF_4 etherate, as is required for the analysis of acetylated samples, the side products corresponding to peaks 2 and 4 in the gas chromatograms shown in Fig. 6.4.2 increase in relation to those of the main thioacidolysis products, 1 and 3; the total yield of products corresponding to all four peaks decreases when the concentration of BF_3 etherate is 1.0 M. Increases in peaks 2 and 4 relative to peaks 1 and 3 is indicative of problems related to the quality of the reagent or of a leak in the screwcap of the reaction tube. Demethylation reactions, revealed by peaks 5 and 6, occur to a negligible extent with 0.2 N BF_3 etherate but to a significant degree with 1 N BF_3 etherate (Fig. 6.4.2).

Note 2. For soluble lignins the ratio of reagent volume to samples weighed is not critical and the same yields are observed when small (5 mg) or large (50 mg) amounts of lignin are reacted with 5 ml of the thioacidolysis reagent.

Note 3. For insoluble lignocellulose residues, it is necessary to add the thioacidolysis reagent in large excess. As the presence of glycosidic units causes a large consumption of the reagent, it is preferable to use a sample-to-reagent ratio lower than 2 mg/ml. In the case of wheat straw, the yields of the main lignin products decreased with ratios higher than 5 mg/ml. Although the standard procedure specifies the addition of 20 mg of lignocellulose residue to 10 ml of reagent, it is advisable, when analyzing a new type of insoluble residue, to test several different sample to reagent ratios in order to find the most suitable one.

Note 4. It is critical to avoid any leak through the screwcap to insure a maximum yield of degradation products.

Note 5. A kinetic study of the recovery of main monomeric products in the thioacidolysis of soluble lignin and unsoluble wood or straw samples has revealed that, between 2 and 6 h, the yields are remarkably constant when compared to those reported for acidolysis (Gellerstedt et al. 1984, Lapierre 1986). With prolonged reaction time (24 h), product yields decrease slightly.

Note 6. These amounts should be adjusted slightly according to the lignin content of the sample. The nature of the lignin under investigation also has to be considered. With heavily degraded lignins, such as those recovered at the end of kraft pulping or those obtained from drastic acid hydrolyses, the sample to internal standard ratio should be increased.

Note 7. To determine if the column is performing satisfactorily a standard sample, wood or a β-O-4 dimer, should be subjected perioidically to thioacidolysis and the results compared to those reported in the literature. Abnormally low yields are frequently indicative of an incompletely deactivated or reactivated column. It is specially important to perform this test when replacing an old column with a new one.

6.4.4 Discussion of Method

6.4.4.1 Sensitivity and Reproducibility

Tests performed on lightly lignified materials indicate that it is possible to characterize unambiguously as little as 15 nanograms of lignin. In such samples, the quantitative determination of the main compounds derived from alkyl aryl ether structures is impeded more by the overloading of the capillary column with carbohydrate degradation products of the dithioacetal type than by the detection limit of lignin degradation products. The high sensitivity of the method is traceable to the relatively high yield and small number of lignin degradation products.

For a soluble lignin sample, the standard deviation for the yield of main C_6C_3 monomeric products is less than 2%. For lignocellulosic materials, it is higher (5–10%) owing to the greater complexity of the chromatogram.

6.4.4.2 Identification of the Major Thioacidolysis Products

Thioacidolysis has been successfully applied to a wide array of lignocellulosic materials and isolated lignins, including extractive-free softwood and hardwood meals and grass straws, kraft and soda lignins and the corresponding hardwood and softwood pulps, milled wood lignins, enzymatically liberated lignin, and samples exhaustively methylated with diazomethane (Lapierre et al. 1988, Lapierre and Rolando 1988); the thioacidolysis behavior of sulfonated lignin has not been studied sufficiently to confirm the validity of the procedure in such a case. The following subsections describe the identification and structural significance of the major products obtained on thioacidolysis of softwood, hardwood, and straw lignins. Where warranted, the significance of some of the minor products is also discussed.

Hardwood Lignins

The gas chromatograms and mass spectral data which provide the basis for the identification of thioacidolysis products from extractive-free poplar and the

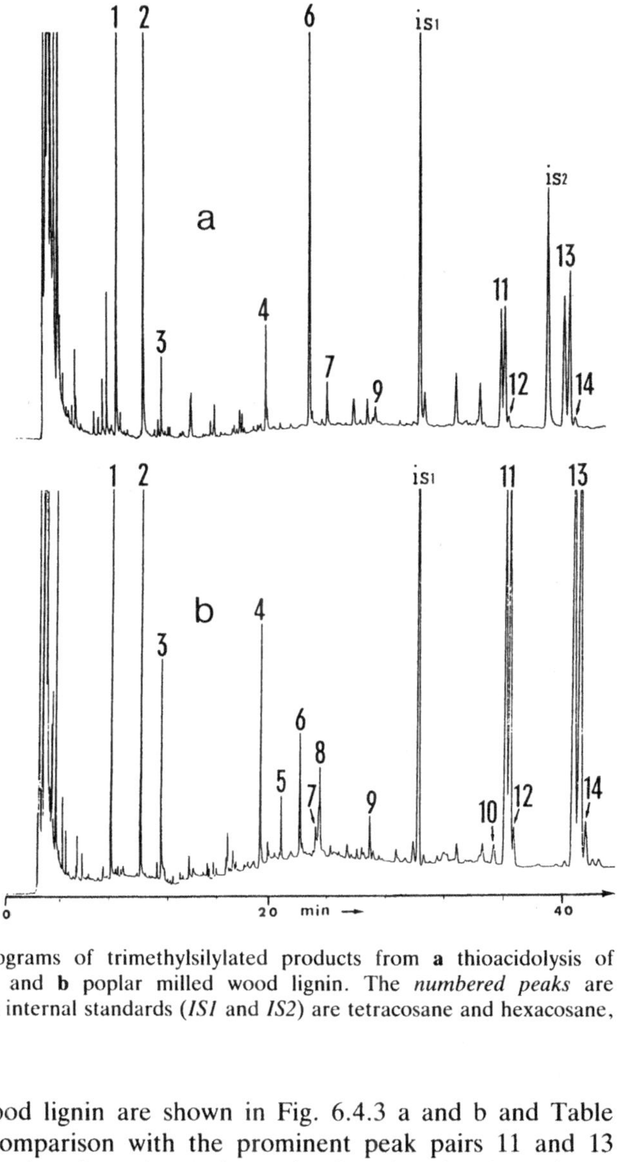

Fig. 6.4.3a,b. Gas chromatograms of trimethylsilylated products from **a** thioacidolysis of extractive-free poplar wood and **b** poplar milled wood lignin. The *numbered peaks* are identified in Table 6.4.2. The internal standards (*IS1* and *IS2*) are tetracosane and hexacosane, respectively

corresponding milled wood lignin are shown in Fig. 6.4.3 a and b and Table 6.4.2, respectively. In comparison with the prominent peak pairs 11 and 13 (Fig. 6.4.3) which are assigned to the main guaiacyl and syringyl products, respectively, the other assigned peaks originating from lignin are of minor importance. The amounts of uncondensed alkyl aryl ether-linked guaiacyl and syringyl units are estimated from the guaiacyl and syringyl C_6C_3 adducts corresponding to peaks 11 and 12 and to peaks 13 and 14, respectively. Peak 10 (Fig. 6.4.3b) is assignable to the degradation product from coniferyl aldehyde end groups.

C_6C_2 monomeric compounds, corresponding to peaks 7 and 9, originate from a minor reaction pathway via elimination of the γ-hydroxymethyl group similar to the minor degradation pathway identified in acidolysis (Lundquist

Table 6.4.2. Structure and mass spectral characteristics of thioacidolysis products from various isolated lignins and lignocelluloses

GC Peak[a]	Product (TMS derivative)	Molecular ion and prominent fragments m/z (rel. intensity)
1	CHR_3[b]	196(2),135(100),117(20)
2	CHR_2—CH_2R[b]	210(20),149(25),135(100),89(50),75(25)
3	H—COOH	282(25),267(75),223(60),193(50), 73(100)
4	CHR_2—CHR—CH_2R[b]	284(5),161(20),135(100),75(75)
5	G—CHR_2	330(1),315(3),269(100),239(8),73(30)
6	CHR_2—CH=COH—CHOH—CH_2OH	454(2),319(50),249(75),217(25), 205(50),189(20),176(20),163(15), 147(50),135(50),117(10),103(50), 75(25),73(100)
7	G—CH_2—CHR_2	344(10),283(3),222(5),209(15),192(8), 179(6),135(100),73(40)
8	S—CHR_2	360(1),345(2),299(100),269(10),73(20)
9	S—CH_2—CHR_2	374(10),313(2),252(5),239(20),222(10), 209(10),179(10),135(100),73(40)
10	G—CHR—CH_2—CHR_2	418(10),403(3),356(3),327(3),295(10), 269(100),239(5),209(5),192(5), 135(60),73(50)
11	G—CHR—CHR—CH_2R erythro and threo	418(3),403(2),295(10),269(100),235(8), 75(15),73(20)
12	G—CH_2—CHR—CHR_2	418(4)403(3),283(40),222(32),192(10), 135(100),75(20),73(10)
13	S—CHR—CHR—CH_2R erythro and threo	448(2),403(1),325(5),299(100),269(5), 265(10),239(15),75(30),73(40)
14	S—CH_2—CHR—CHR_2	448(4),433(3),313(40),252(30),222(10), 135(100),75(70),73(60)
15	H—CH=CH—COOH(E)	308(40),293(50),249(25),219(50), 179(10),139(10),73(100)
16	H—CHR—CH_2—COOH	370(2),309(20),265(15),239(15), 177(20),75(25),73(100)

1976). With milled wood lignins or arylglycerol-β-aryl ether dimers, the contribution of such C_6C_2 compounds amounts to less than 10% of that of the major C_6C_3 compounds.

C_6C_1 compounds, corresponding to peaks 5 and 8 (Fig. 6.4.3b), originate in benzaldehyde end groups. These end groups are identified in milled wood lignin (MWL) and their presence is ascribed to the effect of milling. These peaks disappear after $NaBH_4$ reduction of the MWL, while peaks corresponding to products of the corresponding benzyl alcohol groups, Ar-CH_2SEt, appear (Lapierre 1988, unpubl. results).

Peak 3 (Fig. 6.4.3) corresponds to p-hydroxybenzoic acid which is mainly ester-linked in poplar lignins. Although no study using appropriate model compounds has been made, it may be assumed that the thioacidolysis reagent is able to cleave, to a minor extent, benzoic and p-hydroxycinnamic ester linkages, as manifested by the recovery of benzoic and p-hydroxycinnamic acids from

Table 6.4.2. (continued)

GC Peak[a]	Product (TMS derivative)	Molecular ion and prominent fragments m/z (rel. intensity)
17	G—CH=CH—COOH(E)	338(90),323(50),308(50),293(30), 279(10),249(30),219(20),73(100)
18	G—CHR—CH_2—COOH	400(5),385(2),339(20),295(10),269(15), 207(20),192(25),73(100)
19	H—CHR—CHR—CH_2R erythro and threo	388(2),265(5),239(100),205(20),75(30), 73(50)
20	G—CH_2—CH_2—CH_2OH	326(40),311(20),236(20),221(15), 209(20),206(100),179(35),73(60)
21	G—CH=CHR	282(100),252(40),222(20),207(5), 192(5),179(10),149(40),73(50)
22	G—CH=CH—CH_2R	296(25),281(2),235(75),204(40), 131(60),75(10),73(100)
23	G—CHR—CH_2—CH_2R	358(10),343(2),297(30),235(30), 209(15),205(5),75(100),73(60)
24	G—CR=CHR	342(25),312(2),252(10),222(30),209(5), 207(10),190(20),160(10),89(30), 73(100)
25	C—CHR—CHR—CH_2R	476(3),461(1),353(15),327(100), 293(12),265(5),75(25),73(60)
26	G—CHR—CH_2—G	478(2),463(1),417(1),269(100),73(25)

H: p-Hydroxyphenyl (4-hydroxyphenyl)
G: guaiacyl (4-hydroxy-3-methoxyphenyl)
S: syringyl (4-hydroxy-3,5-dimethoxyphenyl)
C: catechol (3,4-dihydroxyphenyl)
R: SEt

[a] See Fig. 6.4.3–6.4.5.
[b] Compound is in underivatized form.
R=SEt

poplar and grass samples. Traces of vanillic and syringic acids are also recovered from poplar milled wood lignins.

Thioacidolysis performed with xylose or larch xylan has revealed that peaks 2, 4, and 6 (Fig. 6.4.3) essentially originate in xylose units from the xylan in wood meal or xylan impurities in the milled wood lignin.

Grass Lignins

Figure 6.4.4 and Table 6.4.2 provide gas chromatographic and gas chromatographic-mass spectral information relating to the thioacidolysis products derived from an extractive-free wheat straw and the corresponding lignin isolated by alkali (Scalbert 1984).

Grass lignins have been described as p-hydroxyphenyl-guaiacyl-syringyl lignins. However, as is apparent from Fig. 6.4.4, the p-hydroxyphenyl C_6C_3 compounds, represented by peak pair 19 (Fig. 6.4.4) are less than 5% (molar basis) of the corresponding guaiacyl and syringyl derivatives (peak pairs 11 and

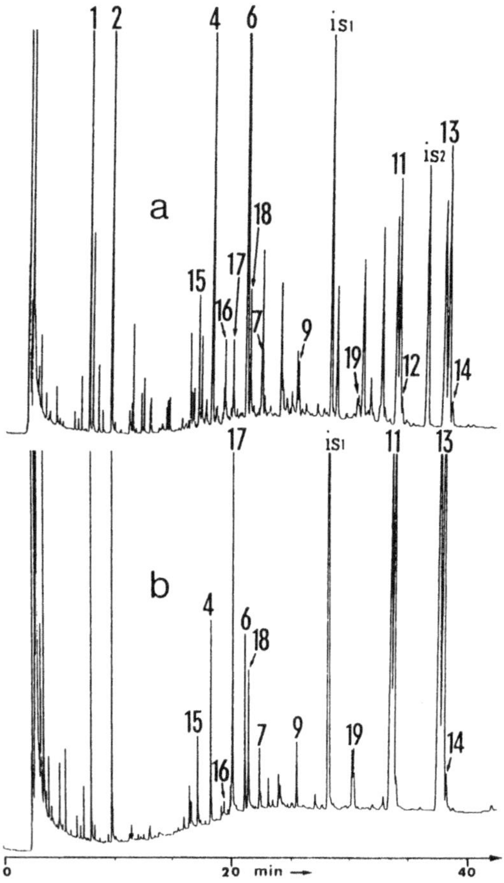

Fig. 6.4.4a,b. Gas chromatograms of trimethylsilylated products from the thioacidolysis of **a** extractive-free wheat straw and **b** alkali soluble wheat lignin (see Fig. 6.4.3 for description of notations and symbols)

13, respectively). This indicates that p-hydroxyphenyl units in grass lignin are very rare or are mainly linked by carbon-carbon bonds.

As shown by the thioacidolysis of p-coumaric and ferulic acids, peak combinations 15/16 and 17/18 (Fig. 6.4.4) are derived, respectively, from p-coumaric and ferulic acid units characteristic of grass samples. Peaks 16 and 18 originate from the Michael addition of EtSH to the olefinic double bond. It is noteworthy that the alkali-soluble wheat lignin is considerably enriched in ferulic acid units compared to the extractive-free wheat meal. The selective removal of ferulic acid units coincidentally with the alkali-soluble wheat lignin is apparent from the heights of peaks 17 and 18 relative to those of peaks 11 and 13 which are higher for the alkali-soluble lignins (Fig. 6.4.4b) than for the extractive-free wheat meal (Fig. 6.4.4a). This observation is in agreement with previously reported results (Scalbert 1984, Scalbert et al. 1985).

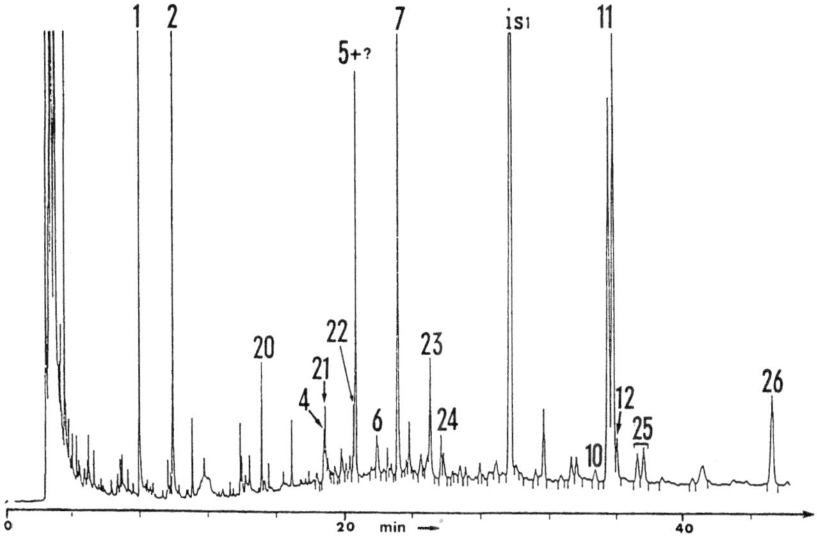

Fig. 6.4.5. Gas chromatrogram of trimethylsilylated products from the thioacidolysis of a pine kraft lignin corresponding to 51% pulp yield (see Fig. 6.4.3 for description of notations and symbols)

The origin of other peaks in Fig. 6.4.4 has already been indicated in the previous section on hardwood lignins.

Softwood Lignins

Thioacidolysis of extractive-free softwood meal or softwood MWL essentially leads to the guaiacyl structures listed in Table 6.4.2 and referred to previously in the section on thioacidolysis of hardwood lignins. When subjected to thioacidolysis, compression wood additionally gives rise to significant amounts of p-hydroxyphenyl C_6C_3 monomeric compounds (Lapierre et al. 1987).

Particular attention has been focused on the thioacidolysis of kraft lignin as this material has been thoroughly characterized previously by acidolysis (Gellerstedt et al. 1984). The GC separation of trimethylsilylated products resulting from thioacidolysis of kraft lignin and the structural assignment of the various peaks based on GC/MS data are shown in Fig. 6.4.5 and Table 6.4.2, respectively.

The high concentration of compound 7 relative to that of the guaiacyl products (11 in Fig. 6.4.5) is noteworthy, especially since this compound was detected only in comparatively small amounts in other thioacidolysis product mixtures. As shown by model compound experiments (Gellerstedt, Lindfors and Lapierre 1988, unpubl. results), the main source of compound 7 is the enol ether structures formed during kraft pulping. This finding is in agreement with acidolysis results (Gellerstedt et al. 1984, Gellerstedt and Lindfors 1987). Therefore, similar to acidolysis, thioacidolysis should provide information about the formation of enol ether structures during kraft or soda pulping.

Compound 20 is consistently detected in the softwood thioacidolysis mixtures. This compound has also been detected in acidolysis product mixtures (Gellerstedt et al. 1984). Its structure confirms the principle that in saturated side chains, substitution of the oxygenated group at C_γ by EtSH occurs only if the neighboring group at C_β has been previously substituted.

As shown by thioacidolysis of coniferyl alcohol, compounds 22 and 23 are formed from coniferyl alcohol end groups. They are recovered in larger amounts, relative to compound pair 11, from kraft lignin than from than from milled wood lignin. Compound 24 originates from C_6C_3 alkyl aryl ether structures with α-carbonyl substituents via elimination of the C_γ hydroxymethyl group (Lapierre 1986, Lapierre et al. 1987).

Compounds comprising peak pair 25 (Fig. 6.4.5) are most likely catechol adducts whose occurrence reflects demethylation reactions occurring at the end of kraft pulping. The structure of these compounds has been assigned on the basis of the 327 m/z mass spectral fragment which corresponds to the Ar-$CHSEt^+$ ion comprised of a trimethylsilylated catechol ring. Mass spectral evidence has also been obtained indicating that catechol groups on syringyl units are produced when poplar milled wood lignin is thioacidolyzed with 1 M BF_3 etherate (peak pair 5 in Fig. 6.4.2).

The determination of catechol groups in various lignins is of practical and fundamental importance (Chap. 5.2.3.3). That such compounds have not been detected in acidolysis studies (Gellerstedt et al. 1984) may be attributed to their unstability under the acidolysis conditions. The indicated detection of catechol groups implies that structures such as p-hydroxycinnamyl alcohol end groups and catechol units, which are unstable in acidic aqueous media, may be determined by thioacidolysis.

Based on a model compound study (Lapierre 1988, unpubl. results), compound 26 possibly may be attributed to stilbene structures in which EtSH has been added to the ethylenic groups.

Table 6.4.3. Yields (in μmol g^{-1} lignin) of monomeric guaiacyl (G) and syringyl (S) products recovered from poplar lignin fractions after degradation by various procedures. (Lapierre 1986)

		Thioacidolysis	Acidolysis	Nitrobenzene oxidation	Permangate oxidation
Poplar MWL[a]	S+G	1024	583	1660	274
	S/G	1.20	0.70	1.40	0.39
Poplar E1[a]	S+G	1243	630	1770	106
	S/G	1.67	0.91	1.87	0.48

[a] Milled wood lignin and enzymatically isolated lignin fractions successively isolated from extractive-free poplar wood.

Table 6.4.4. Yields of main p-hydroxyphenyl (H), guaiacyl (G) and syringyl (S) C_6C_3 monomeric thioacidolysis products recovered from insoluble extractive-free lignocelluloses and soluble lignin fractions

	H	G	S	Total	H/G	S/G
a) Lignocellulosic samples: yield, μmol g^{-1} of Klason lignin						
Spruce wood						
(*Picea abies*)	n.d.[a]	1006	n.d.	1006	–	–
Pine wood						
(*Pinus pinaster*)						
– compression	207	934	n.d.	1141	0.22	–
– opposite	25	992	n.d.	1017	0.03	–
Poplar wood						
(*Populus euramericana*)						
– tension	–	732	1080	1812	–	1.48
– opposite	–	763	1186	1949	–	1.55
Birch wood						
(*Betula verrucosa*)	–	525	1341	1866	–	2.55
Wheat straw						
(*Triticum aestivum*)						
– cv. Champlein	43	457	636	1136	0.08	1.07
– cv. Capitole	53	535	650	1238	0.10	1.21
Rice straw						
(*Oryza sativa*),						
– normal rice	95	287	251	632	0.33	0.87
– brittle-stem mutant	38	228	235	501	0.17	1.03
b) Lignin samples: yield, μmol g^{-1} of lignin						
Pine (*Pinus pinaster*) MWL[b]						
– compression wood	146	627	21	794	0.23	0.03
– opposite wood	15	803	8	800	0.02	0.01
Poplar (*Populus euramericana*) MWL[b]						
– tension wood	–	497	513	1010	–	1.03
– opposite wood	–	520	601	1021	–	1.16
Aspen (*Populus tremuloides*)						
steam explosion lignin[b,c]						
– 25 bars, 4.5 s	–	254	415	669	–	1.63
– 35 bars, 125 s	–	123	232	355	–	1.89
Wheat (*Triticum aestivum*)						
MWL[b]	27	410	343	780	0.06	0.84

[a] n.d.: Not determined owing to very low relative amounts.
[b] Values are not corrected for carbohydrate components.
[c] Reported values correspond to two different steam treatment conditions (pressure and treatment time).

6.4.4.3 Quantitative Aspects of the Thioacidolysis Method

In Table 6.4.3 are shown the yields of monomeric products recovered from milled wood lignin (MWL) and enzymatically liberated lignin (EL) fractions successively isolated from poplar wood and subjected to various degradation procedures (Lapierre 1986). These values reveal that acidolysis, which should allow the characterization of the same lignin structures as thioacidolysis, suffers a serious drawback in that syringyl units participate extensively in side reactions, as shown by the low product yields and low S/G ratios. Compared with thioacidolysis, nitrobenzene oxidation leads to higher product yields, which indicates that this oxidant is able to degrade structures other than alkyl aryl ethers, for example, diarylpropane units (Chang and Allan 1971), to monomeric compounds. However, nitrobenzene oxidation products may arise from phenolic components other than lignin, as is especially the case with grass lignins which cannot be unambiguously characterized by this technique (Sarkanen and Ludwig 1971) (Chap. 6.1.1). Permanganate oxidation, performed as described by Gellerstedt and Lindfors (1984), is applicable only to those phenylpropane units originally carrying a free phenolic hydroxyl group (Chap. 5.2.1).

The low product yield and S/G ratio indicates that these units are not abundant and that they are mainly of the guaiacyl type.

Based on the data in Table 6.4.3, it is apparent that none of the degradation methods currently in use is capable of providing a completely unequivocal characterization of lignin and hence use of a combination of methods is required to attain this objective. However, as a single method, thioacidolysis has a definite advantage in that it may be used to characterize unambiguously typical and prominent lignin structures.

Table 6.4.4 lists the thioacidolysis product yields for insoluble, extractive-free lignocelluloses and soluble, isolated lignins. The wide range of values for total yield and structural unit molar ratio emphasizes the large variation in lignin structure that can be associated with genetic origin and with the procedure followed to isolate soluble lignins. The effect of growth environment can be seen in the examples shown for reaction and opposite woods in pine and poplar woods.

Although a detailed description of all the examples presented in Table 6.4.4 is outside of the scope of this book, it is noteworthy that for in situ lignins, low yields of thioacidolysis products are observed for straw samples, indicating that grass lignins have more carbon-carbon linkages than hardwood lignins. Further, the p-hydroxyphenyl (H) units are relatively unimportant in the uncondensed moieties of grass lignins as shown by the low H/G ratio. Such findings cannot be obtained with oxidative procedures involving nitrobenzene or permanganate oxidation: for a grass sample, oxidation of the p-coumaric and ferulic acid ester substituents leads to the same compounds as do p-hydroxyphenyl and guaiacyl lignin-building units, respectively.

References

Chang H-m, Allan GG (1971) Oxidation. In: Sarkanen KV, Ludwig CH (eds) Lignins. Occurrence, formation, structure and reactions. Wiley-Interscience, New York, 433–485

Fuji K, Ichikawa K, Node M, Fujita E (1979) Hard acid and soft nucleophile system. New efficient method for removal of benzyl protecting group. J Org Chem 44:1661–1664

Gellerstedt G, Lindfors EL (1984) Structural changes in lignin during kraft cooking. Holzforschung 38:151–158

Gellerstedt G, Lindfors EL (1987) On the formation of enol ether structures in lignin during kraft cooking. Nord Pulp Pap Res J 2:71–75

Gellerstedt G, Lindfors EL, Lapierre C, Monties B (1984) Structural changes in lignins during kraft cooking. Part 2 – Characterization by acidolysis. Sven Papperstidn 9:R61–R67

Lapierre C (1986) Hétérogénéité des lignines de peuplier: mise an évidence systématique. Thèse de Doctorat d'Etat. Université Paris-Sud 305 pp

Lapierre C, Monties B, Rolando C (1986a) Thioacidolysis of poplar lignins: identification of monomeric syringyl products and characterization of guaiacyl-syringyl lignin fractions. Holzforschung 40:113–118

Lapierre C, Monties B, Rolando C (1986b) Preparative thioacidolysis of spruce lignin: isolation and identification of main monomeric products. Holzforschung 40:47–50

Lapierre C, Monties B, Rolando C (1987) Degradation of various lignins and lignin model compounds by thioacidolysis. 4th Int Symp Wood Pulp Chem, Paris, Proc 2:431–435

Lapierre C, Monties B, Rolando C (1988) Thioacidolysis of diazomethane-methylated pine compression wood and wheat straw in situ lignins. Holzforschung 42:409–411

Lapierre C, Rolando C (1988) Thioacidolysis of pre-methylated lignin samples from pine compression and poplar woods. Holzforschung 42:1–4

Lundquist K (1976) Low-molecular weight lignin hydrolysis products. Appl Polymer Symp 28:1393–1407

Node M, Hori H, Fujita E (1976) Demethylation of aliphatic methyl ethers with a thiol and boron trifluoride. J Chem Soc Perkin Trans I:2237–2240

Sarkanen KV, Ludwig CH (1971) Definition and nomenclature. In: Sarkanen KV, Ludwig CH (eds) Lignins. Occurrence, formation, structure and reactions. Wiley-Interscience, New York, 1–17

Scalbert A (1984) Caracterisation des lignines de paille de blé: fractionnement, association avec les oses et les acides phénoliques. Thèse de Docteur-Ingénieur. Institut National Agronomique de Paris-Grignon, 58 pp

Scalbert A, Monties B, Lallemand JY, Guittet E, Rolando C (1985) Ether linkage between phenolic acids and lignins fractions from wheat straw. Phytochemistry 24:1359–1362

6.5 Hydrogenolysis

A. Sakakibara

6.5.1 Introduction

The term "hydrogenation" is often used indiscriminately and imprecisely to describe a number of processes involving reactions of molecular hydrogen. In characterizing the reactions of hydrogen with lignin, the pertinent terms may be defined as follows:

Hydrogenation: the addition of hydrogen (H_2) to a multiple bond
Hydrogenolysis: a process in which cleavage of a carbon-carbon or carbon-hetero atom single bond is accomplished by reaction with hydrogen

Depending on the reaction conditions and catalyst, one or combinations of the foregoing processes may occur during the reaction of hydrogen with lignin.

Hydrogenolysis and hydrogenation may occur in the absence of a catalyst but for rapid stabilization of, or effective addition of hydrogen to, the primarily formed lignin fragments, the application of a suitable catalyst is necessary. The only exception to this principle are hydrogen-generating reactions which occur when primary alcohols and alkali are heated to temperatures in the vicinity of 350 °C.

Catalytic hydrogenolysis is one of the most effective methods of obtaining valuable information about the chemical structure of lignin since high yields of mono-, di-, tri-, and oligolignols are formed without the occurrence of secondary condensation or polymerization reactions. Subsequent characterization of these comparatively simple compounds reveals the nature of the structural groups comprising lignin and the manner in which they are linked. Hydrogenolysis also has been studied extensively, although with somewhat less success, in connection with the conversion of lignin to commercially valuable products.

The hydrogenolysis of lignin has been reviewed in detail by several authors (Brauns 1952, Brauns and Brauns 1960, Pearl 1967, Goheen 1971, Hrutfiord 1971) and will be dealt with only briefly here. Lindblad (1931) was among the first to subject wood to catalytic hydrogenolysis to obtain products of commercial importance; efforts to obtain significant amounts of useful low molecular weight phenols have since been continued by a number of investigators (e.g., Kashima and Osada 1961, Sakakibara et al. 1964, Parkhurst et al. 1980). As the first to examine the reactions of an isolated lignin with hydrogen, Harris et al. (1938) isolated several n-propylcyclohexane derivatives. Shortly thereafter, Godard et al. (1941) isolated similar products after

Springer Series in Wood Science
Methods in Lignin Chemistry
(Edited by S.Y. Lin and C.W. Dence)

treating maple wood with hydrogen. Identification of these phenylpropane hydrogenation products was particularly noteworthy since it demonstrated the predominantly aromatic character of lignin.

Hydrogenolysis has also found use in a lignin biosynthesis study in which ^{14}C-labeled lignin precursors were administered to wheat plants and maple twigs and ultimately recovered in the form of the typical lignin hydrogenolysis products, dihydroconiferyl and dihydrosinapyl alcohols (Brown and Neish 1959).

Lautsch (1941) and Freudenberg et al. (1941) thoroughly studied the high pressure hydrogenolysis of commercial lignin with various catalysts and obtained cyclohexanol derivatives in high yield. When hydrochloric acid lignin was reacted with ethanol and sodium hydroxide at 340–350 °C, 97% of the lignin was dissolved (Lautsch and Piazolo 1943). Under similar conditions, lignosulfonates were recovered as low-molecular weight neutral oils (Sakakibara et al. 1957).

Hydrogenolysis was applied to the pulping of wood by Bhattacharya and Schuerch (1960). In this study, arylmethanes, arylpropanols, and arylethanols were identified among the reaction products. Schuerch and co-workers detected several dilignols among the alkaline hydrogenolysis products of maple wood meal (Granath and Schuerch 1953) and obtained evidence for the presence of a mixed trilignol (Parker et al. 1966, Bhaskaran and Schuerch 1969).

Although heterogeneous catalysts have been used almost exclusively in hydrogenolysis studies, a few examples may be cited where homogeneous catalyst systems have been employed in hydrogenolysis studies of lignin. For example, Sakakibara et al. (1964) used the metal carbonyls, $Ni(CO)_4$, $Fe(CO)_5$ and $Co(CO)_3$, as catalysts in the hydrogenolysis of sulfuric acid (Klason) lignin and obtained a 37% yield of low boiling phenols. Coincidentally, Nahum (1965) employed a cobalt carbonyl catalyst in a study in which several lignin monomers and a dimer were isolated after hydrogenolysis of red spruce wood. Eachus and Dence (1975) employed hydrogen and a homogeneous catalyst system consisting of a nickel hydride complex to convert the phenolic nuclei of lignin model monomers and dimers to cyclohexyl derivatives.

Numerous studies on the hydrogenolysis of lignin to monomeric phenylpropanoids have been reported (Coscia et al. 1961, Olcay 1963, Pepper and Steck 1963, Pepper and Lee 1969, 1970, Pepper and Fleming 1978, Pepper and Supathna 1978, Pepper and Rahman 1987, Rahman and Pepper 1988). Using shorter reaction times and moderately active catalysts, lignin may be hydrogenolyzed to di- and trilignols (Hwang et al. 1981, Hwang and Sakakibara 1979, 1981, Matsukura and Sakakibara 1969, 1973a,b,c, Ohta and Sakakibara 1969, Sakakibara et al. 1980, Sudo et al. 1978, Sudo and Sakakibara 1973, 1974a,b,c, 1975, 1977, 1979, Wada and Sakakibara 1969, Yasuda S and Sakakibara 1975, 1976, 1977a,b, Yasuda T and Sakakibara 1981). The intent in this latter series of studies was to establish the manner in which the phenylpropane units in lignin are joined. The foregoing contributions constitute the basis for the hydrogenolysis procedures detailed below.

6.5.2 Proposed Sequence for Hydrogenolysis

It should be stressed from the outset that although it is convenient to use the term "hydrogenolysis" when referring to the overall sequence of reactions of lignin with hydrogen in the presence of a catalyst, a variety of individual processes may be represented other than those involving hydrogen. The exact characteristics of these latter reactions and their sequential relationships with transformations where hydrogen has an actual role await definition in many instances.

The steps postulated as occurring in the hydrogenolysis of lignin are shown in Fig. 6.5.1. This sequence is a modified version of the scheme proposed by Hrutfiord (1971). As shown in the figure, the initial phase of hydrogenolysis consists of hydrolysis and thermolysis reactions, which result in a partial breakdown and solubilization of the lignin. The fragments initially formed diffuse into the liquid phase where they are rapidly stabilized against condensation reactions through the addition of hydrogen atoms generated at the catalyst surface. Although sequence 1 to 2a (Fig. 6.5.1) appears reasonable, step 2b is difficult to explain since solid catalysts cannot function directly in situations where the reactant consists of suspended insoluble particles, as is the case for the unsolubilized lignin residues in the wood tissue.

Step 2b speculatively may be rationalized based on the results of a study reported by McMillen et al. (1985), who proposed a mechanism to explain catalyzed coal liquefaction. According to these investigators, a significant part of the structural fragmentation occurring during coal liquefaction results from a fundamental reaction involving a hydrogen radical transfer to aromatic portions of the coal structure bearing aliphatic alcohol (or ether) linkages. As applied to the catalytic hydrogenolysis of lignin, particularly native lignin in wood meal, this proposed mechanism suggests the possibility that the relatively small lignin fragments dissolved in the initial phase of the hydrogenolysis transfer free hydrogen atoms to residual lignin in the wood. The radical intermediates thereby produced are subsequently stabilized through intramolecular displacement-type reactions in which relatively large, soluble lignin fragments are liberated. The latter would then undergo further hydrogenolysis and hydrolysis to mono-, di-, tri-, and oligolignols.

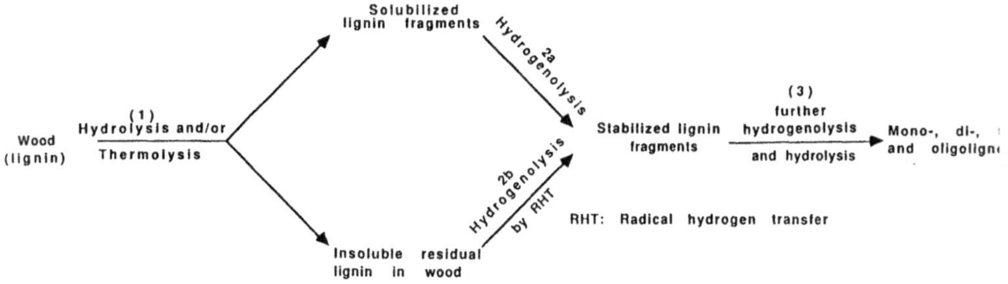

Fig. 6.5.1. Proposed sequence for hydrogenolysis of lignin

6.5.3 Hydrogenolysis and Hydrogenation Reactions

6.5.3.1 Cleavage of Interunitary Linkages

The principal reaction in the hydrogenolysis of lignins is the cleavage of ether linkages connecting the α, β, and γ carbon atoms of a side chain and the 4-position of a phenolic ring in an adjacent unit (Fig. 6.5.2). In general, the alkyl-O-aryl linkage is more readily hydrogenolyzed than is the alkyl-O-alkyl linkage. The α-C-O bond, especially if present as part of a benzyl aryl ether unit [(1) in Fig. 6.5.2], is so labile that hydrogenolysis products having this linkage have yet to be isolated. The cyclic dialkyl ether comprising the bis-dihydrofuran ring present in pinoresinol and syringaresinol units [(3) in Fig. 6.5.2] is moderately stable under hydrogenolysis conditions; on the other hand, the cyclic alkyl aryl ether constituting the phenylcoumaran ring in (2) (Fig. 6.5.2) is comparatively labile during hydrogenolysis giving rise only to β-5 units in which the furane ring

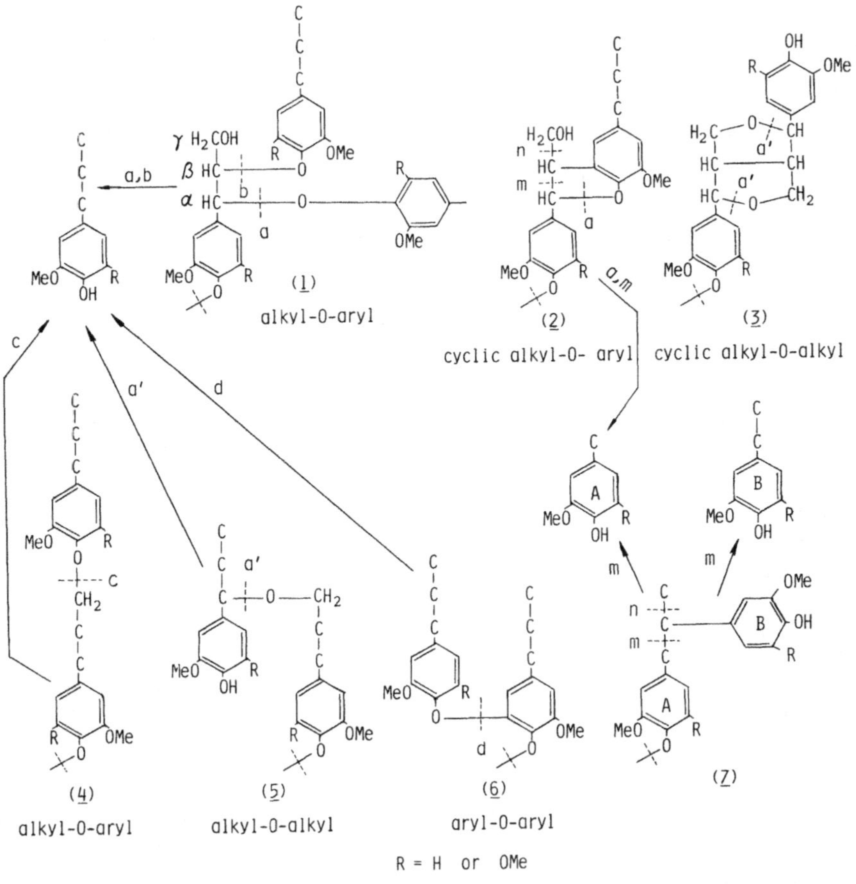

Fig. 6.5.2. Cleavage patterns in the hydrogenolysis of interunitary lignin linkages

is cleaved. The γ-O-4, α-O-γ and diphenyl ether linkages in (4), (5), and (6) (Fig. 6.5.2) are moderately stable to hydrogenolysis.

The foregoing generalizations notwithstanding, ether cleavage is in practice dependent on catalyst activity. Thus, for example, the α-O-γ linkage in pinoresinol [a′ in (3), Fig. 6.5.2] is not cleaved during hydrogenolysis in the presence of copper chromium oxide catalyst (Matsukura and Sakakibara 1973c), but instead requires more active catalysts (ruthenium on charcoal or Raney nickel) such as were employed by Pepper and Lee (1970).

Depending on the reaction conditions, carbon-carbon linkages may also be ruptured during hydrogenolysis especially at sites designated m and n in structures (2) and (7) (Fig. 6.5.2). Whether the rupture of carbon-carbon linkages is exclusively the result of hydrocracking reactions or in part stems from acid- and base-catalyzed retrograde aldol condensation reactions remains to be established.

6.5.3.2 Hydrogenation of Side Chain and Aromatic Ring

Hydrogenolysis is accompanied by hydrogenation, that is, reduction of the side chain and aromatic ring. Thus, ethylenic and carbonyl groups in the side chains of coniferyl alcohol and coniferaldehyde, respectively, are readily reduced to dihydroconiferyl alcohol, (9). Benzylic alcohol groups in lignin are readily reduced to α-methylene groups but the hydroxyl on the γ-carbon is very stable to reduction. Although a carbonyl group on the side chain is usually reduced to an alcohol or hydrocarbon, the carbonyl group in 4-hydroxyphenacyl units is moderately stable, as evidenced by the isolation of propiosyringone after hydrogenolysis of Yachidamo wood (Hwang and Sakakibara 1979).

If the catalyst is sufficiently active and/or the reaction conditions are severe enough, the aromatic ring may be hydrogenated to a cyclohexyl ring (Harris et al. 1938, Brewer et al. 1948, Pepper and Lee 1969). However, for the purpose of gaining information about lignin structure, it is desirable to retain the ring as a phenyl group.

6.5.4 Factors Affecting Hydrogenolysis

6.5.4.1 Catalyst Activity

A variety of catalysts have been used in the hydrogenolysis of lignin: Raney nickel, stabilized and unstabilized copper chromite (copper chromium oxide), cobalt octacarbonyl, and palladium, rhodium, and ruthenium on various supporting surfaces. Catalysts which have high hydrogenation activity, for example, stabilized copper chromite (Olcay 1963), ruthenium-charcoal, and ruthenium-alumina, or catalysts applied in large amounts (Pepper and Lee 1969) convert lignin mainly to cyclohexyl derivatives, while catalysts having medium activity give rise to aromatic derivatives such as (8)–(11) (e.g., Brewer et al.

1948, Coscia et al. 1961, Olcay 1963, Pepper and Fleming 1978, Rahman and Pepper 1988).

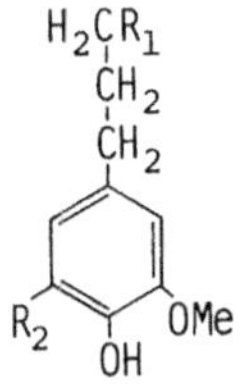

(8): R_1=R_2=H
(9): R_1=OH, R_2=H
(10): R_1=H, R_2=OMe
(11): R_1=OH, R_2=OMe

In general, hydrogenation and hydrogenolysis reactions compete with one another. If hydrogenation occurs first, the ensuing hydrogenolysis is extremely restricted. However, when this sequence is reversed, both cleavage and reduction are observed (Coscia et al. 1961, Pepper and Lee 1969). Therefore, when high yields of monomeric phenylpropanoids or cyclohexylpropanoids are desired, highly active catalysts should be selected but if dimeric, trimeric, or oligomeric phenylpropanoids are sought, the catalysts and reaction conditions should be milder even if the resulting total yield of products is lower.

6.5.4.2 Catalyst Amount

Pepper and Supathna (1978) studies the effect on product yields of Rh-C applications ranging in amount from 1 to 30% of the original wood in the hydrogenolysis of spruce wood meal. The yield of dihydroconiferyl alcohol, (9), was found to increase while the yield of 4-n-propylguaiacol, (8), decreased. When the catalyst application exceeded 10%, traces of cyclohexyl derivatives were detected. The maximum total yield of (8) and (9) was obtained with a 10–20% catalyst application.

With the addition of increasing amounts of Rh-C catalyst, the total yield of monomeric products increased, while the yield of chloroform extractables simultaneously decreased. This finding suggests that liberation of lignin from a lignin-carbohydrate complex is facilitated by an increase in the amount of catalyst.

6.5.4.3 Hydrogen Pressure

In the presence of Rh-C catalyst, higher initial hydrogen pressure has the effect of decreasing slightly the yield of 4-n-propylguaiacol (8), and increasing the yield of dihydroconiferyl alcohol (9) in the hydrogenolysis of spruce wood meal (Pepper and Supathna 1978). Increasing initial pressure in the hydrogenolysis of 4-n-propylguaiacol, (8), in the presence of sulfided $CoO\text{-}MoO_3/\gamma\text{-}Al_2O_3$ has the effect of altering the product distribution such that the formation of 4-n-propylcatechol is retarded while that of n-propylphenols and n-propylcresols changes very little (Vuori and Bredenberg 1984). However, the formation of n-propylcyclohexenes and n-propylcyclohexane increases.

6.5.4.4 Solvent

The choice of a solvent for hydrogenolysis is an important consideration; the solvent should promote dissolution of residual unsolubilized lignin and its hydrogenolysis products while being stable under the imposed reaction conditions. The swelling of wood in organic solvents is enhanced by the addition of water; in the hydrogenolysis of lignin, this effect is manifested by increased hydrolytic cleavage of lignin-carbohydrate bonds which in turn accelerates lignin dissolution (Wada and Sakakibara 1969).

Pepper and Supathna (1978) studied the effect of pH (0.1 to 12.7) in a dioxane-water (1:1, v/v) system on the hydrogenolysis of spruce wood meal and found that pH variation in the 3–10 range had little effect on the composition of the product mixture. However, at a very low initial pH (0.1 to 1.1), the total yield of monomeric products was considerably lower. It seems likely that, in the presence of strong acid, small lignin fragments formed initially by hydrolysis undergo secondary (possibly condensation or polymerization) reactions (Lai and Sarkanen 1971). However, such secondary reactions have not been observed under ordinary catalytic hydrogenolysis conditions over a broad pH range (3–10) at temperatures ranging from 180 to 240 °C.

6.5.5 Hydrogenolysis to Monomeric Products

Pepper and co-workers have extensively studied the effect of catalysts such as Raney nickel, palladium-charcoal, rhodium-charcoal, rhodium-alumina, ruthenium-charcoal, and ruthenium-alumina on the yields of monomeric hydrogenolysis products (Pepper and Lee 1969, Pepper and Fleming 1978). These and other studies indicate that the highest yields of monomeric hydrogenolysis products are obtained when aspen wood is reacted with hydrogen in the presence of Raney nickel (Pepper and Steck 1963) or Rh-C (Pepper and Fleming 1978). The combined yield of eight monomeric products in the first case was 52.2% based on the original Klason lignin while the total yield of compounds identified when Rh-C was the catalyst was 40%. The procedure followed when Raney nickel was used as catalyst is described below.

Preparation of Materials

Wood is ground to 2-mesh in a Wiley mill, extracted 48 h in a Soxhlet extractor with ethanol-benzene (1:1, v/v), and air-dried.

Raney nickel catalyst is prepared as described by Pavlic and Adkins (1946) with the exception that the washing under a hydrogen atmosphere is omitted. The catalyst is stored under cold dioxane previously purified by distillation over metallic sodium (Fieser 1955) or lithium aluminum hydride (Augustine and Caputo 1965).

Procedure

Pre-extracted, air-dried wood meal (50 g), Raney nickel (10 g), and 50% aqueous dioxane (400 ml) are stirred together in a 1410-ml capacity rocking autoclave (Aminco or equivalent). The sealed bomb is flushed twice with hydrogen and then filled with hydrogen to a pressure of 34.5×10^2 kPa (~500 psig). The temperature is raised (about 120 °C/h) and controlled automatically at the desired value for the required period. During this time, the bomb is rocked continuously but allowed to cool overnight without rocking.

The bomb contents are suction-filtered and the residual wood meal and nickel are washed with 90% aqueous dioxane until the washings are colorless. The washings and the original filtrate are combined, gravity-filtered (if cloudy), concentrated under reduced pressure to the point at which sludge begins to form, and extracted with chloroform (3 × 80 ml) or until the extract is no longer colored. The chloroform solution is back-extracted with 100 ml of water, dried over anhydrous sodium sulfate, filtered, and concentrated free of chloroform below 50 °C under reduced pressure. The resulting reddish oil (termed chloroform extract) contains residual dioxane besides lignin degradation products.

Gas Chromatographic Analysis of Chloroform Extract

The components in the chloroform extract are separated by gas chromatography using a 1.82 m × 0.635 cm (O.D.) copper column packed with 10% Apiezon N grease on Fluoropak at 205 °C (injector and detector at 232 °C); the helium flow is 36 ml/min at 50 psig and the filament current, 200 milliamperes (Note 1). When estimating the quantities of degradation products in the chloroform extract, corrections for the residual dioxane in the extract and the fraction of injected lignin passing through the column are made as described by Pepper and Steck (1963).

Identification of Compounds

Compounds represented by the four major peaks are collected and identified by comparison of the IR spectra, refractive indices, and retention times with those of authentic samples of probable hydrogenolysis monomers. Other components of the mixture are identified by matching retention times and relative retention times with those of authentic reference compounds (Note 2).

Note 1. As a consequence of the continuing development of separation techniques and equipment, the resolution of lignin hydrogenolysis product mixtures is not restricted to the foregoing set of conditions but may be achieved equally well with other equipment and combinations of conditions.

Table 6.5.1. Monomeric aromatic products formed in the hydrogenolysis of lignin

Compound	Aspen wood meal Raney nickel 195 °C, 5 h Dioxane-water (1:1) (Pepper and Steck 1963)	Spruce wood meal Rhodium-C 195 °C, 5 h Dioxane-water (1:1) (Pepper and Lee 1969)	*Birch MWL* Copper chromite 240–260 °C, 48 h Dioxane (Coscia et al. 1961)	*Oak MWL*	*Spruce MWL* Copper chromite 240 °C, 47 h Dioxane (Olcay 1963)
	Yield, % of original lignin				
4-n-Propylguaiacol (8)	0.3[a]	5.6[a]	2.3	2.6	5.9
Dihydroconiferyl alcohol (9)	10.3	10.6	2.0	0.6	8.1
4-n-Propylsyringol (10)	2.9	–	3.9	7.3	–
Dihydrosinapyl alcohol (11)	31.0	–	7.9	0.8	–
4-Methylguaiacol (12)	–	0.1	1.1	1.0	3.5
4-Methylsyringol (13)	1.7	–	2.0	3.1	–
4-Ethylguaiacol (14)	1.0	0.2	0.9	0.6	2.1
4-Ethylsyringol (15)	3.6	–	1.1	0.7	–
Total	50.7 (44.5)[b]	16.5 (16.2)[b]	21.2 (16.1)[b]	16.7 (11.3)[b]	19.6 (14.0)[b]

[a] Based on Klason lignin content.
[b] Yield of phenylpropane derivatives.

Note 2. The identification techniques which comprise part of the original procedure may be supplemented or supplanted with NMR and mass spectrometric methods of analysis (Faix and Meier 1987).

Discussion of Method

Aromatic monomeric products identified by Pepper and Steck (1963), Pepper and Lee (1969), Coscia et al. (1961), and Olcay (1963) are listed in Table 6.5.1. The occurrence of hydrogenolysis is evidenced by the detection of dihydroconiferyl and dihydrosinapyl alcohols, (9) and (11), respectively, and the corresponding n-propyl derivatives, (8) and (10).

The rupture of carbon-carbon single bonds is demonstrated by the identification of methyl- and ethylguaiacol (12) and (14), respectively, and the corresponding alkylsyringols, (13) and (15). The methyl derivatives in all likelihood are formed by cleavage of the lignin side chain at the m position [Fig. 6.5.2, structures (2) and (7)], while the ethyl homologs seemingly may arise from the B ring of structure (7) as shown in the figure. Undoubtedly, hydrogenation of the ethylenic groups in coniferyl and sinapyl alcohol end units also occurs, but the resulting products are indistinguishable from those produced by hydrogenolysis.

As illustrated by the values in the first two columns of Table 6.5.1, the total yield of monomeric products from hardwood lignin generally exceeds that from softwood lignin, possibly because of the greater abundance of β-O-4 units in the former. When aspen wood meal was hydrogenolyzed over Raney nickel, approximately 50% of the lignin was recovered as chloroform-soluble monomers, all of which were identified (Pepper and Steck 1963). Thus, under the imposed conditions, the ether linkages joining lignin units appear to be selectively and extensively cleaved.

The values recorded in Table 6.5.1 for hardwood lignins indicate a substantially greater abundance of syringyl-type monomers than guaiacyl-type monomers. This difference may be a manifestation of the amounts of each structural type in the lignin or a reflection of a difference in the manner in which these types of units are hydrogenolyzed. The significance of the relative abundance of syringyl and guaiacyl derivatives in the hydrogenolyzate has to be judged in the context of a lignin model compound study in which some conversion of syringyl to guaiacyl derivatives was found to occur during hydrogenolysis (Pepper and Steck 1963).

The lower total yields of monomeric products obtained after lignin is hydrogenolyzed in the presence of copper chromite (Table 6.5.1, last the three columns) is probably attributable to secondary reactions promoted by the more severe conditions used in these treatments (higher temperatures and longer reaction times). Differences in the total yields of monomeric products obtained by hydrogenolyzing aspen and oak milled wood lignins (Table 6.5.1, 3rd and 4th columns) under identical conditions suggest that this procedure is capable of demonstrating structural differences in lignins from different sources.

6.5.6 Hydrogenolysis to Dimeric and Trimeric Products

To obtain direct information about the interunitary linkage patterns in lignin, it is necessary to isolate and characterize dimeric and trimeric fragments from the hydrogenolysis mixture. For this purpose, the hydrogenolysis conditions need to be mild enough to suppress cleavage of carbon-carbon single bonds (hydrocracking) and to retain some of the carbon-oxygen bonds linking phenylpropane units. Thus, hydrogenolysis conditions are required which are somewhat milder than those employed to cleave extensively ether linkages and form predominantly monomeric fragments.

In a series of investigations, Sakakibara and co-workers employed copper chromite as a catalyst for hydrogenolysis performed under conditions selected so as to yield dimeric and trimeric structures. The procedure described by Hwang and Sakakibara (1979) serves as a typical example.

Preparation of Materials

Wood meal (<60-mesh) is exhaustively extracted successively with refluxing acetone-water (9:1, v/v), ethanol-benzene (1:2, v/v), and 95% ethanol in a Soxhlet apparatus for 48 h.

Copper chromite is prepared according to the method described by Adkins (1954).

Hydrogenolysis Procedure

Extracted, air-dried wood meal (50 g) is suspended in 400 ml of dioxane-water (9:1, v/v) and reacted with hydrogen (initial pressure 78.4×10^2 kPa) in the presence of copper chromite catalyst (10 g) for 1 h at 200–210 °C in a rocking autoclave. The contents of the bomb are cooled, filtered, and washed successively with ethanol and acetone. The combined filtrate and washings are concentrated at 40 °C to half of their original volume in a stream of nitrogen and fractionateed according to the scheme outlined in Fig. 6.5.3.

Separation of the Ether-Soluble (ES) Fraction by Column Chromatography

As most of the dimeric and trimeric lignin hydrogenolysis products are found in the ES-fraction, the procedure for this fractionation is described in detail; procedures for the other separations shown in Fig. 6.5.3 may be obtained by consulting the original literature source (see Table 6.5.2).

The ether-soluble (ES) fraction (64.4 g) is separated into six fractions (ES-A to ES-F) on a silica gel column (7 × 85 cm) using a solvent mixture of n-hexane-ethyl acetate (4:1, v/v) as eluent. The F-fraction is subsequently separated into five fractions (ESF-1 to ESF-5) on a 5 × 138 cm column of Sephadex LH-20 using dioxane-water (1:1, v/v) as an eluent.

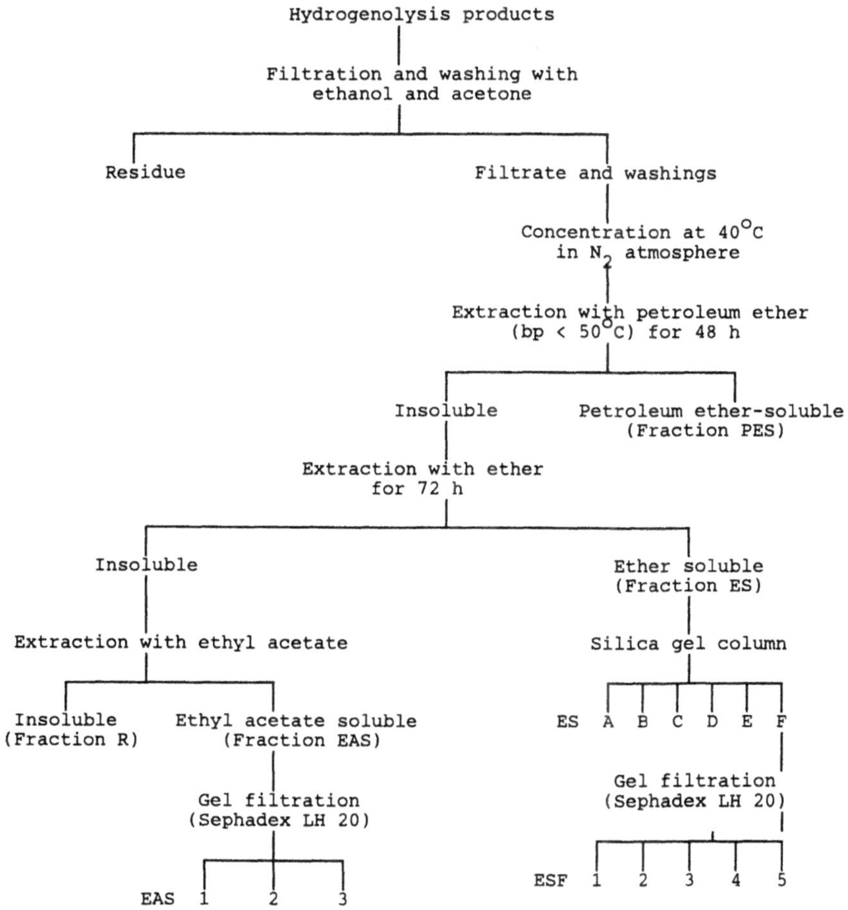

Fig. 6.5.3. Fractionation scheme for lignin hydrogenolysis products

Identification of Dimeric and Trimeric Hydrogenolysis Products

The hydrogenolysis products in the various fractions are identified by combinations of IR, UV, MS and ^{1}H-NMR measurements (see Chaps. 5.1, 5.3, 5.4, and 7.1). The spectral characteristics of the di- and trilignols are found in the original publications (see Table 6.5.2).

Discussion of Method

More than 50 di- and trilignols have been isolated from the hydrogenolysis product mixtures of softwoods and hardwoods. The structures of the phenylpropanoid members of this group are shown in Fig. 6.5.4 (dilignols) and Fig. 6.5.5 (trilignols). Reference citations relating to the isolation of individual compounds are compiled in Table 6.5.2.

Table 6.5.2. Fraction source for dimeric and trimeric lignin hydrogenolysis products

Compound number[a]	Fraction source[b]	Reference	
31	Chloroform solubles	Ohta and Sakakibara	1969
43, 44, 45	PES	Wada et al.	1969
16	PES	Matsukura and Sakakibara	1969
17, 18	ES	Ibid.	1973a
48	ES	Ibid.	1973b
36	ES	Ibid.	1973c
29	ES-D	Sudo and Sakakibara	1973
32	ES-E	Ibid.	1973
21, 22	ES-F	Ibid.	1974b
38, 47	ES-F	Ibid.	1974c
19, 20	ES-A	Yasuda S and Sakakibara	1975
23, 40	ES-A	Ibid.	1976
46	ES-B	Ibid.	1977a
37	ES-C	Ibid.	1977a
49	ES-F	Sudo and Sakakibara	1977
42	ES-C	Yasuda S and Sakakibara	1977b
35	ES-D	Ibid.	Ibid.
39	ES-D	Sudo and Sakakibara	1979
28	ES-F-3	Hwang and Sakakibara	1979
52, 53	ES-F-3	Ibid.	1979
25, 30	ES	Sakakibara et al.	1980
50, 51	ES-E	Yasuda T and Sakakibara	1981
26	EAS-3	Hwang et al.	1981
27	ES-F-3	Hwang and Sakakibara	1981
33	EAS-5	Ibid.	1981
24	ES-F-3	Ibid.	1981

[a] See Figs. 6.5.4 and 6.5.5 for structures of compounds.
[b] See Fig. 6.5.3 for fractionation scheme.

Among the compounds isolated but not included in Figs. 6.5.4 and 6.5.5 are several whose side chain was cleaved during hydrogenolysis. These include the dilignols, (17), (31) (Yasuda S and Sakakibara 1976), and (35) (Yasuda S and Sakakibara 1977b) and the trilignols, (48) (Sudo and Sakakibara 1976) and (49) (Sudo and Sakakibara 1974a), where C_β-C_γ cleavage of the side chain occurred and the trilignols, (47) (Sudo and Sakakibara 1975) and (49) (Matsukura and Sakakibara 1973c) where c_α-c_β scission took place.

Identification of the di- and trilignols listed in Figs. 6.5.4 and 6.5.5 has proven useful by providing evidence for the existence of previously unrecognized interunit linkages in lignin as well as by corroborating the presence of linkages revealed by other degradative methods. Additionally, identification of dimeric and trimeric hydrogenolysis products comprised of guaiacyl and syringyl units has shown that hardwood lignin is a copolymer rather than a mixture of homopolymers.

The major units revealed by the di- and trilignols appearing in Figs. 6.5.4 and 6.5.5 are summarized in Table 6.5.3. Dimers (31), (32), and (33), containing a β-1 linkage, were detected following hydrogenolysis of Ezomatsu and Yachidamo wood meals while the β-1-linked dimer (34) was isolated after aspen

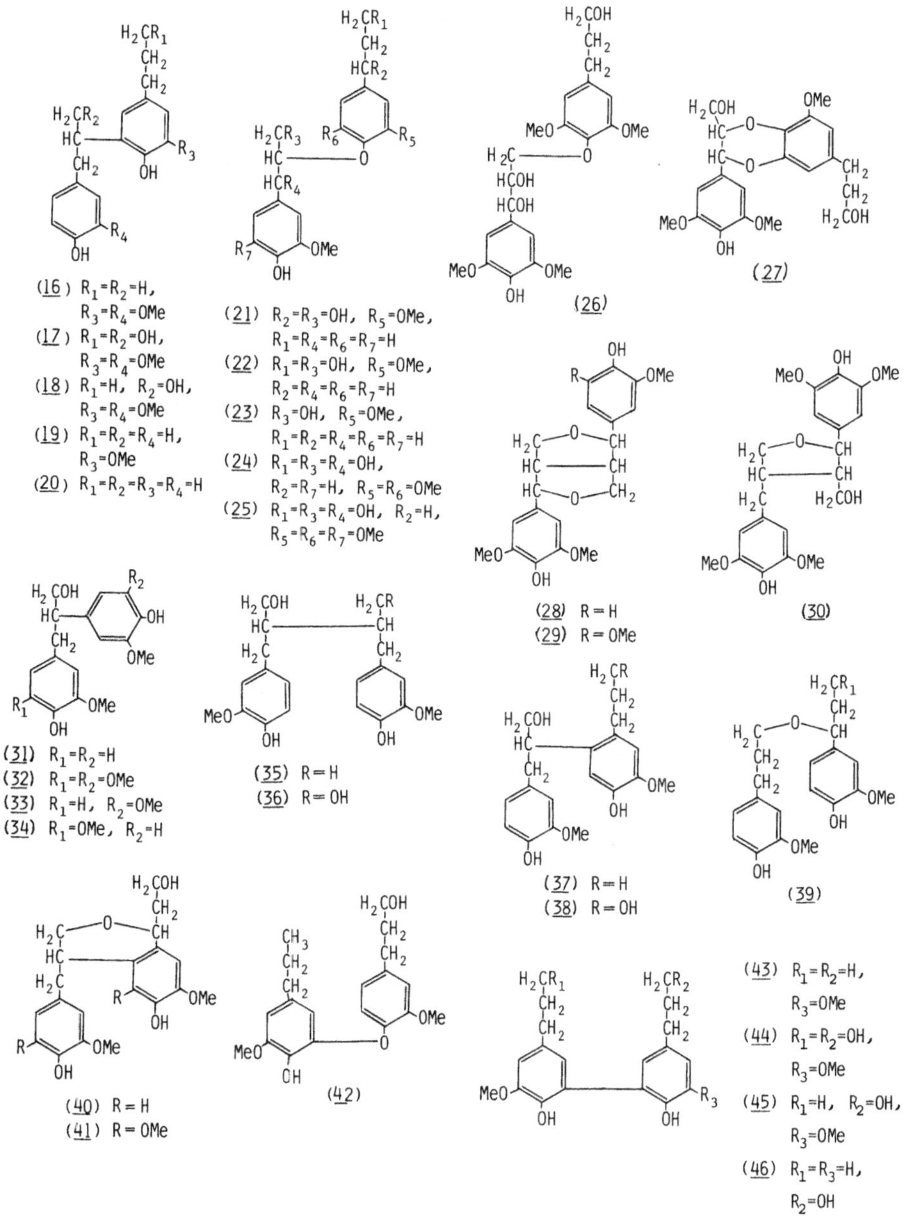

Fig. 6.5.4. Dilignols isolated from the hydrogenolysis of protolignins

wood meal was hydrogenolyzed using a Rh-C catalyst (Sudo et al. 1981). The biphenyl dilignol, (43), was first isolated by Nahum (1965) following hydrogenolysis of red spruce wood meal in the presence of cobalt octacarbonyl. Structures (30), (35), (36), (50), and (51) represent, in this particular instance, structural types initially present in the lignin rather than hydrogenolysis pro-

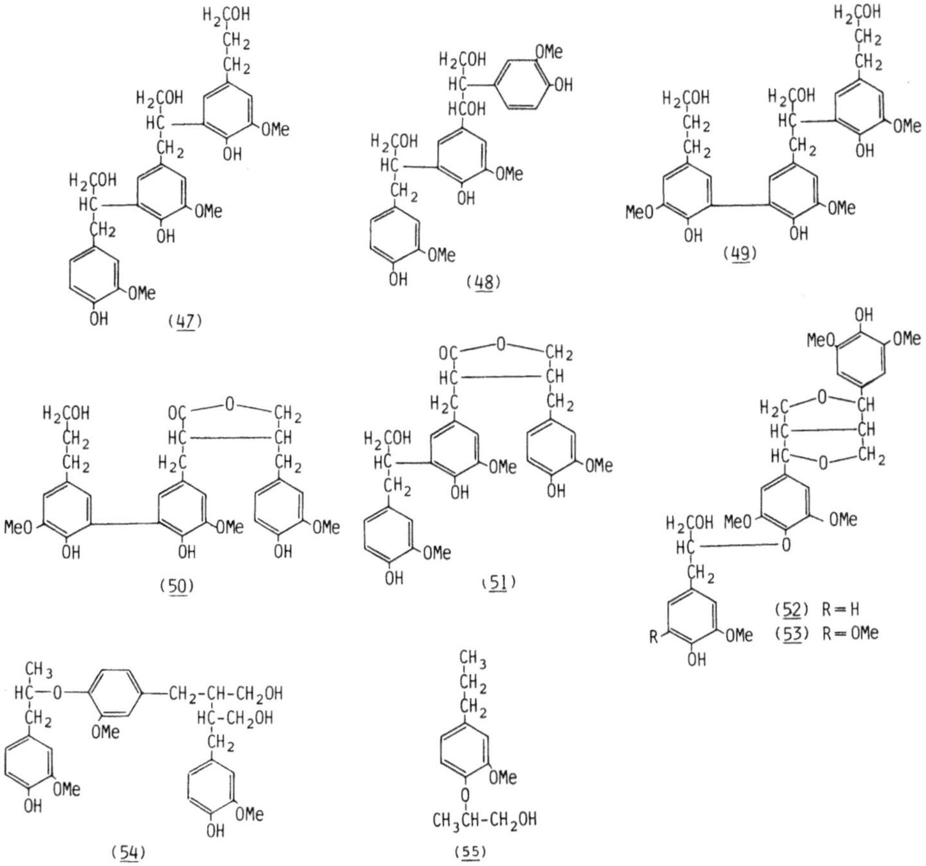

Fig. 6.5.5. Trilignols isolated from the hydrogenolysis of protolignins

ducts. On the other hand, trilignol, (54), (Fig. 6.5.5) identified by Lee and Pepper (1978), evidently is a cleavage product of the guaiacyl analog of (53) formed specifically when a highly active rhodium catalyst was used. The isolation of (50) and (51) confirms the existence of a γ lactone structure in lignin in agreement with the findings of Freudenberg and Geiger (1963), who identified a dimer containing a γ lactone group on a DHP lignin mixture.

In addition to the major lignin linkages shown in Table 6.5.3, the existence of several minor lignin linkages is manifested by the identification of dimeric hydrogenolysis products. These linkages and the structures wherein they appear are as follows: β-6 (37), (38), α-O-5 (27), α-O-γ (39), (40), (41), and γ-O-4 (26). Dilignol (41), the syringyl homolog of (40), was isolated from aspen lignin by Sudo and Pepper (1982) thereby indicating that the phenylisochroman linkage is not unique to compression wood lignin, the source of (40) (Yasuda S and Sakakibara 1976). The β-6 linkage is most likely formed as a consequence of the dienone-phenol rearrangement of an initial β-1 adduct.

Table 6.5.3. Major interunitary linkages found in lignin hydrogenolysis products[a]

Interunitary linkages	Hydrogenolysis products
β-O-4	20–25, 52, 53
β-5	16–20, 47–49, 51
β-1	31–34, 48
β-β′	28–30, 35, 36, 49, 50–53
4-0-5	42
5-5′ (biphenyl)	43–46, 49, 50

[a] See Figs. 6.5.4 and 6.5.5 for structures.

The occurrence of γ-O-4 dilignol (26) (Fig. 6.5.4) is difficult to reconcile in terms of the dehydrogenative polymerization of ρ-hydroxycinnamyl alcohols in lignin biogenesis, thus raising the possibility that (26) may be an artifact of the hydrogenation treatment. However, studies performed by Hwang et al. (1981) and Coscia et al. (1961) afforded evidence leading to the conclusion that the γ-O-4 linkage is indigenous in lignin and is not formed during hydrogenolysis. Later work conducted by Miki et al. (1986) in which β-O-4 lignin model compounds were oxidized with a peroxidase from *Phanerochaete chrysosporium* supplied evidence supporting the concept that structures with the foregoing linkage are rearranged to (26) or (27) by the peroxidase during the formation of lignin in wood tissue.

Klemola and Pepper (1969) isolated 2-(2-methoxyl-4-n-propylphenoxy)-propan-1-ol, (55), after hydrogenolysis of spruce wood in the presence of a Rh-C catalyst.

This product may be derived from a glyceraldehyde-2-aryl ether unit formed via a displacement reaction in the dehydrogenative polymerization of coniferyl alcohol (Lundquist et al. 1967).

References

Adkins H (1954) Catalytic hydrogenation of esters to alcohols. In: Adams R (ed) Organic reactions, Vol. VIII, Wiley, New York, 9

Augustine RL, Caputa AV (1965) $\Delta^{1(9)}$-Octalone-2 Org Syn 45:80–82

Bhaskaran TA, Schuerch C (1969) A study of non-volatile hydrogenated maple wood lignin. TAPPI 52:1948–1952

Bhattacharya A, Schuerch C (1960) Lignin from the alkaline hydrogenation of sugar maple-wood III. Chromatographic separation of higher molecular weight fractions and a structural study. TAPPI 43:840–844

Brauns FE (1952) The chemistry of lignin. Academic Press, New York, 511–535

Brauns FE, Brauns DA (1960) The chemistry of lignin: supplement volume. Academic Press, New York, 486–497

Brewer C, Cooke LM, Hibbert H (1948) Studies on lignin and related compounds LXXXIV. The high pressure hydrogenation of maple wood. J Am Chem Soc 70:57–59

Brown SA, Neish AC (1959) Studies of lignin biosynthesis using isotopic carbon. VIII. Isolation of radioactive hydrogenolysis products of lignin. J Am Chem Soc 81:2419–2424

Coscia CJ, Schubert WJ, Nord FF (1961) Investigations on lignins and lignification XXIV. The application of hydrogenation, hydrogenolysis, and vapor phase chromatography in the study of lignin and structure. J Org Chem 26:5085–5091

Eachus SW, Dence CW (1975) Hydrogenation of lignin model compounds in the presence of a homogeneous catalyst. Holzforschung 29:41–48

Faix O, Meier D (1987) Pyrolytic and hydrogenolytic studies on lignocelluloses, pulps, and lignins. Int Symp Wood Pulp Chem (Paris) 4th, Oral Presentations Imprimeric Jacques Poncet, 1:253–257

Fieser LF (1955) Experiments in organic chemistry. 3rd edn. D.C. Heath, Boston, 284–285

Freudenberg K, Geiger H (1963) Pinoresinolid und andere Zwischenprodukte der Ligninbildung. Chem Ber 96:1265–1270

Freudenberg K, Lautsch W, Piazolo G, Scheffer A (1941) Die Druckhydrierung des Lignins und Ligninhaltigen Ablaugen der Fichte. Chem Ber 74:171–183

Godard HP, McCarthy JL, Hibbert H (1941) Studies on lignin and related compounds LXII. High pressure hydrogenation of wood using copper chromite catalyst (Part 1). J Am Chem Soc 63:3061–3065

Goheen DW (1971) Low molecular weight chemicals. In: Sarkanen KV, Ludwig CH (eds) Lignins. Occurrence, formation, structure and reactions. Wiley-Interscience, New York, 818–821

Granath M, Schuerch C (1953) Preliminary separation of maple hydrol lignin. J Am Chem Soc 75:707–709

Harris EE, D'Ianni J, Adkins H (1938) Reaction of hardwood lignin with hydrogen. J Am Chem Soc 60:1467–1470

Hrutfiord BF (1971) Reduction and Hydrogenolysis. In: Sarkanen KV, Ludwig CH (eds) Lignins. Occurrence, formation, structure and reactions. Wiley-Interscience, New York, 487–509

Hwang BH, Sakakibara A (1979) Hydrogenolysis of protolignin XV. Further isolation of some compounds from hardwood lignin. Mokuzai Gakkaishi 25:647–652

Hwang BH, Sakakibara A (1981) Hydrogenolysis of protolignin XVIII. Isolation of a new dimeric compound with a heterocycle involving α,β-diether. Holzforschung 35:297–300

Hwang BH, Sakakibara A, Miki K (1981) Hydrogenolysis of protolignin XVII. Isolation of three dimeric compounds with γ-O-4, β-1, and β-O-4 linkages from hardwood lignin. Holzforschung 35:229–232

Kashima H, Osada T (1961) High pressure hydrogenolysis of lignin. J Soc Chem Ind Jpn 64:916–919

Klemola A, Pepper JM (1969) 2-(2-Methoxy-4-n-propylphenoxy)-propan-1-ol as a new spruce lignin hydrogenolysis product. Chem Commun 1048

Lai YZ, Sarkanen KV (1971) Isolation and structural studies. In: Sarkanen KV, Ludwig CH (eds) Lignins. Occurrence, formation, structure and reactions. Wiley-Interscience, New York, 182–185

Lautsch W (1941) Über die oxydative und hydrierende Destruktion des Holzes, des Lignins und der schwefelhaltige Ablaugen der Fichte. Cellul Chem 19:69–87

Lautsch W, Piazolo G (1943) Über die Hydrierung von Lignin und ligninhaltigen Stoffen mit Wasserstoff abgebenden Mitteln, insbesondere Alkoholen. Chem Ber 76:486–498

Lee YW, Pepper JM (1978) Lignin and related compounds VII. The isolation of a trimeric lignin compound by the hydrogenolysis of spruce wood. Tetrahedron Lett 5061–5062

Lindblad A (1931) Preparation of oils from wood by hydrogenation. Ing Vetenskaps Akad Handl 107:7–59

Lundquist K, Miksche GE, Ericsson L, Berndtson L (1967) Über das Vorkommen von Glyceraldehyde-2-Aryl-Ätherstrukturen. Tetrahedron Lett: 4587–4591

Matsukura M, Sakakibara A (1969) Hydrogenolysis of protolignin IV. Isolation of a dimeric material of "condensed type". Mokuzai Gakkaishi 15:297–302

Matsukura M, Sakakibara A (1973a) Hydrogenolysis of protolignin V. Isolation of some dimeric compounds with carbon to carbon linkage. Mokuzai Gakkaishi 19:131–135

Matsukura M, Sakakibara A (1973b) Hydrogenolysis of protolignin VI. Isolation of a new trimeric compound with carbon to carbon linkage. Mokuzai Gakkaishi 19:137–140

Matsukura M, Sakakibara A (1973c) Hydrogenolysis of protolignin VIII. Isolation of a dimer with C_{β}—$_{\beta}'$ linkage and a biphenyl. Mokuzai Gakkaishi 19:171–176

McMillen DF, Malhotra R, Hum GP, Chang S-J (1985) Hydrogen-transfer-promoted bond scission initiated by coal fragments. J Energy Fuels 1:193–198

Miki K, Renganathan V, Gold MH (1986) Mechanism of β-aryl ether dimeric lignin model compound oxidation by lignin peroxidase of *Phanerochaete chrysosporium*. Biochemistry 25:4790–4796

Nahum L (1965) Delignification of wood by hydrogenation in presence of dicobalt octacarbonyl. Ind Eng Chem 4:71–74

Ohta M, Sakakibara A (1969) Hydrogenolysis of protolignin III. Isolation of diguaiacylpropanol and chromatographic detection of some monomers. Mokuzai Gakkaishi 15:247–250

Olcay A (1963) Investigations on lignins and lignification XXIX. Hydrogenation products of spruce milled-wood lignin and of related model compounds. Holzforschung 17:105–109

Parker PE, Coalson RL, Schuerch C (1966) The structure of dimers from the alkaline hydrogenation of lignin. Adv Chem Ser 59:249–260

Parkhust HJ, Huibers DTA, Jones MWJ (1980) Production of phenol from lignin. Am Chem Soc Div Petrol Chem 25:657–667

Pavlic AA, Adkins H (1946) Preparation of a Raney nickel catalyst. J Am Chem Soc 68:141

Pearl IA (1967) The chemistry of lignin. Marcel Dekker, New York, 210–220

Pepper JM, Fleming RW (1978 Lignin and related compounds V. The hydrogenolysis of aspen wood lignin using rhodium-on-charcoal as catalyst. Can J Chem 56:896–898

Pepper JM, Lee YM (1969) Lignin and related compounds I. A comparative study of catalysts for lignin hydrogenolysis. Can J Chem 47:723–727

Pepper JM, Lee YM (1970) Lignin and related compounds II. Studies using ruthenium and Raney nickel as catalysts for lignin hydrogenolysis. Can J Chem 48:477–479

Pepper JM, Steck W (1963) The effect of time and temperature on the hydrogenation of aspen lignin. Can J Chem 41:2867–2875

Pepper JM, Supathna P (1978) Lignin and related compounds VII. A study of variables affecting the hydrogenolysis of spruce wood lignin using a rhodium-on-charcoal catalyst. Can J Chem 56:899–902

Pepper JM, Rahman MD (1987) Lignin and related compounds. XI. Selected degradation of aspen poplar lignin by catalytic hydrogenolysis. Cellul Chem Technol 21:233–239

Rahman RD, Pepper JM (1988) Lignin and related compounds. XII. Catalytic degradation of proto and isolated aspen lignins under initially alkaline conditions. J Wood Chem Technol 8:313–322

Sakakibara A, Araki T, Takahashi S (1957) Studies on pressure hydrogenolysis of lignin I. Heating of sulfite liquor with ethanol and sodium hydroxide under pressure. Bull Gov For Exp Stn 93:113–117

Sakakibara A, Oda K, Kubota M (1964) Hydrogenolysis of lignin V. Hydrogenolysis with the catalysts metal carbonyl and nickel carbonate. Bull Gov For Exp Stn 166:159–171

Sakakibara A, Sudo K, Kishi M, Aoyama M, Hwang (1980) Hydrogenolysis of protolignin XVI. Isolation of β-O-4 and β-β type compounds. Mokuzai Gakkaishi 26:628–632

Sudo K, Hwang BH, Sakakibara A (1978) Isolation of α-O-γ compound from hydrogenolysis products of lignin. Mokuzai Gakkaishi 24:424–425

Sudo K, Mullard DJ, Pepper JM (1981) Lignin and related compounds VIII. Lignin monomers and dimers from hydrogenolysis of aspen poplar wood using rhodium-on-charcoal catalyst. Can J Chem 59:1028–1031

Sudo K, Pepper JM (1982) Lignin and related compounds. IX. The isolation of a dimer with a benzylisochroman structure from the hydrogenolysis products of aspen lignin. Can J Chem 60:229–230

Sudo K, Sakakibara A (1973) Hydrogenolysis of protolignin VII. Isolation of D,L-syringaresinol, biphenyl, and diarylpropane from hardwood lignin. Mokuzai Gakkaishi 19:165–169

Sudo K, Sakakibara A (1974a) Hydrogenolysis of protolignin IX. Isolation of a trimeric compound linked with biphenyl- and C_{β}-C_5-linkages. Mokuzai Gakkaishi 20:327–330

Sudo K, Sakakibara A (1974b) Hydrogenolysis of protolignin X. Isolation of two dimeric compounds with β-aryl ether linkage. Mokuzai Gakkaishi 20:331–335

Sudo K, Sakakibara A (1974c) Hydrogenolysis of protolignin XI. Isolation of a dimer with C_β-C_6 and a trimer with two C_β-C_5 linkages. Mokuzai Gakkaishi 20:396–401

Sudo K, Sakakibara A (1975) Hydrogenolysis of protolignin XII. Isolation of dimeric and trimeric compounds with carbon to carbon linkages. Mokuzai Gakkaishi 21:165–169

Sudo K, Sakakibara A (1977) Hydrogenolysis of protolignin XIII. Isolation of trimeric biphenyl and lignin type compounds. Mokuzai Gakkaishi 23:151–155

Sudo K, Sakakibara A (1979) Hydrogenolysis of protolignin XIV. Isolation of a dimeric compound with α-O-γ linkage. Mokuzai Gakkaishi 25:61–66

Vuori A, Bredenberg JB-son (1984) Hydrogenolysis and hydrocracking of the carbon-oxygen bond 5. Hydrogenolysis of 4-propylguaiacol by sulfided CoO-MoO_3/γ-Al_2O_3. Holzforschung 38:253–262

Wada I, Ohta M, Sakakibara A (1969) Hydrogenolysis of protolignin II. Isolation of some hydrogenolysis products. Mokuzai Gakkaishi 15:218–221

Wada I, Sakakibara A (1969) Hydrogenolysis of protolignin I. Hydrogenolysis products under various reaction conditions. Mokuzai Gakkaishi 15:214–218

Yasuda S, Sakakibara A (1975) Hydrogenolysis of protolignin in compression wood I. Isolation of two dimers with C_β-C_5 and C_β-C_3 composed of p-hydroxyphenyl and guaiacyl nuclei and two p-hydroxyphenyl nuclei, respectively. Mokuzai Gakkaishi 21:370–375

Yasuda S, Sakakibara A (1976) Hydrogenolysis of protolignin in compression wood II. Isolation of two dimers with β-aryl ether linkage and phenylisochroman structure. Mokuzai Gakkaishi 22:606–612

Yasuda S, Sakakibara A (1977a) Hydrogenolysis of protolignin in compression wood III. Isolation of four dimeric compounds with carbon to carbon linkage. Mokuzai Gakkaishi 23:114–119

Yasuda S, Sakakibara A (1977b) Hydrogenolysis of protolignin in compression wood IV. Isolation of a diphenyl ether and three dimeric compounds with carbon to carbon linkage. Mokuzai Gakkaishi 23:383–387

Yasuda T, Sakakibara A (1981) Hydrogenolysis of protolignin in compression wood V. Isolation of two trimeric compounds with γ-lactone ring. Holzforschung 35:183–187

6.6 Nucleus Exchange Reaction

M. Funaoka, I. Abe, and V.L. Chiang

6.6.1 Introduction

A chemical method to degrade lignin in a medium consisting of boron trifluoride and excess phenol was originally developed by Funaoka and Abe (1978a,b, 1980a,b, 1982a,b,c,d,e,f,g). In this method, Cα-aryl carbon-carbon linkages in phenylpropane units are selectively and quantitatively cleaved, following phenolation at the Cα position, releasing phenyl nuclei typified by guaiacol (1) (see Fig. 6.6.1 for structures of compounds) from guaiacyl (softwood) lignins. The guaiacol subsequently is partially demethylated to catechol (2). These transformations are illustrated in Fig. 6.6.2. The key step in this sequence, the

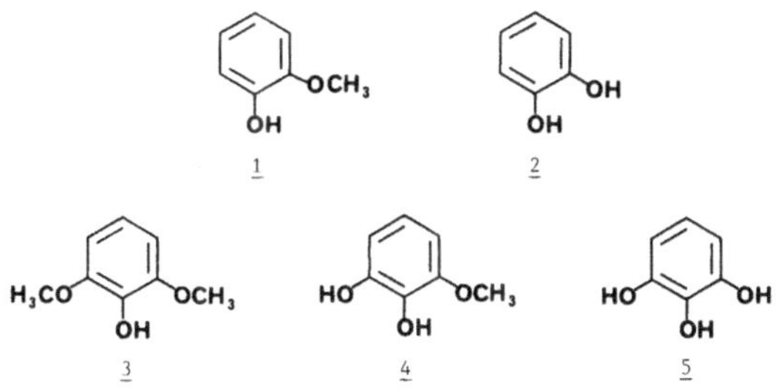

Fig.6.6.1. Main polyhydric phenols generated from lignins in the nucleus exchange method

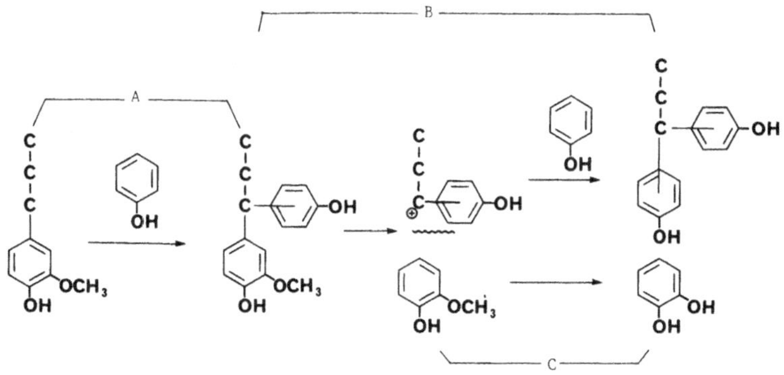

Fig.6.6.2. Formation of catechol from softwood lignin in the NE method

Springer Series in Wood Science
Methods in Lignin Chemistry
(Edited by S.Y. Lin and C.W. Dence)

release of a phenyl nucleus on reaction with phenol, was termed nucleus exchange (NE) by the original authors.

As applied to protolignin in softwoods, the NE reaction is restricted to the quantitative conversion of noncondensed phenyl nuclei (type Ga, Fig. 6.6.3) to guaiacol and catechol. For hardwood protolignin, NE results in the quantitative transformation of type Sa phenyl nuclei (Fig. 6.6.3) to pyrogallol-1,3-dimethyl ether (3), pyrogallol-1-methyl ether (4), and pyrogallol (5). The so-called "condensed" lignin structures, i.e., units consisting of phenyl (guaiacyl or syringyl) nuclei C-substituted at positions other than the side chain site, are excluded from the determination.

When protolignins are subjected to reactions under either acidic or alkaline conditions, such as occur during pulping, they are converted to subunits which are otherwise absent in protolignins. These modified subunits are known to be mainly of the diphenylmethane (DPM) type (Gierer 1982). Noncondensed units in DPM-containing structures give rise to the same simple phenols as are obtained when noncondensed structures in softwood and hardwood protolignin are similarly subjected to the NE reaction. Using the NE method and a com-

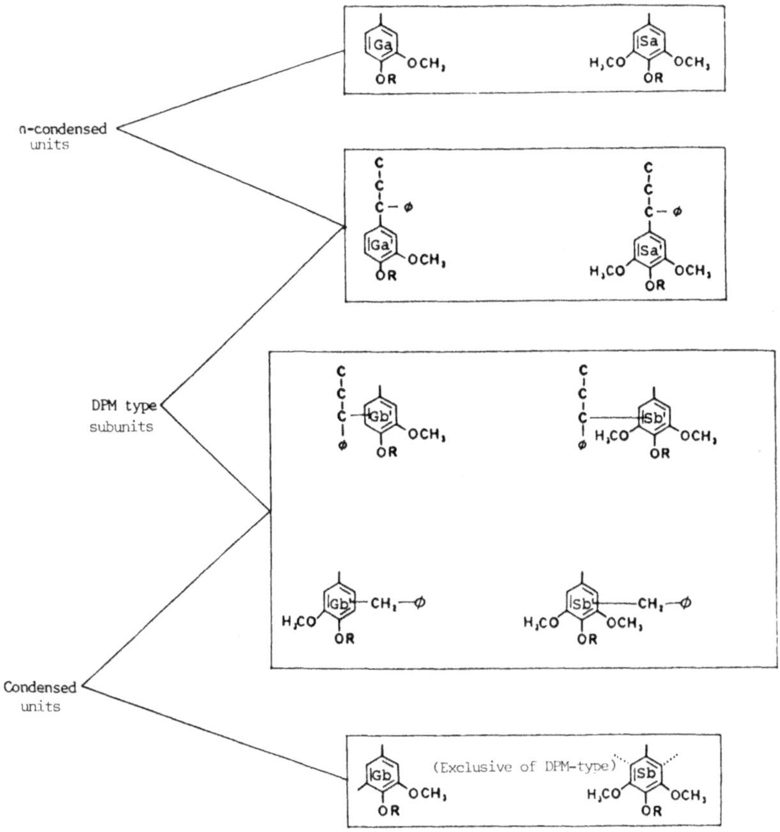

Fig.6.6.3. Types of lignin phenyl nuclei (R: H or alkyl)

bination of NE and alkaline nitrobenzene (Chap. 6.2) methods, the quantities of phenyl nuclei arising from Ga, Ga′, Gb′, Sa, Sa′ and Sb′ units (Fig. 6.6.3) can be individually determined. Based on the quantities of these phenyl nuclei, the amounts of Gb and Sb (condensed) nuclei can be calculated as described below.

As stated above, NE applied to softwood protolignin leads to the quantitative conversion of noncondensed guaiacyl nuclei (type Ga) having side chain structures *a*, *b*, *c*, *d*, and *e* (Fig. 6.6.4) to guaiacol (1) and catechol (2), whereas quantitative yields of pyrogallol-1,3-dimethyl ether (3), pyrogallol-1-methyl ether (4), and pyrogallol (5) can be obtained from noncondensed syringyl nuclei (type Sa). Therefore, the sum of the guaiacol and catechol (mol%) represents the quantity of noncondensed phenyl nuclei in softwood protolignins. In addition to guaiacol and catechol, the syringyl structures, 3, 4, and 5 are formed from the release and subsequent demethylation of syringyl nuclei in hardwood lignins.

Thus, the sum of (1) and (2) and the sum of (3), (4), and (5) represents the quantities of noncondensed phenyl nuclei in the guaiacyl and syringyl units of hardwood protolignins, respectively. Consequently, unlike other lignin degradation reactions, the NE reaction alone provides a rapid and convenient means to determine quantitatively the extent of phenyl nuclei condensation in protolignins. Furthermore, by applying the NE reaction to hardwood lignins, true syringyl/guaiacyl (S/G) ratios can be obtained. Thus far, this is the only technique by which these S/G ratios can be obtained for hardwood protolignins using a single chemical reaction.

In modified softwood lignins, NE quantitatively converts not only noncondensed guaiacyl nuclei (type Ga) having side chain structures *a*, *b*, *c*, *d*, and *e* (Fig. 6.6.4) into guaiacol and catechol, but also guaiacyl nuclei Ga′ and Gb′ in DPM moieties having these side chain structures into the same products. To

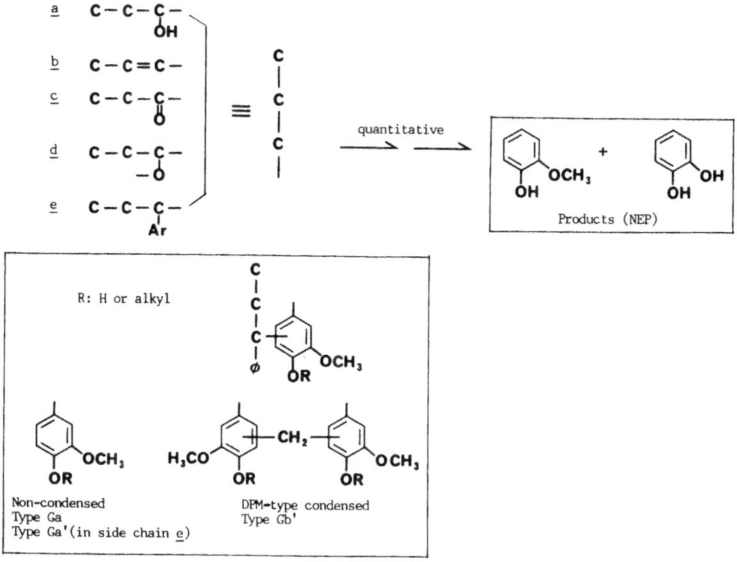

Fig.6.6.4. Lignin structural units giving rise to monomeric products (NEP) by NE method

differentiate between the quantities of Ga guaiacyl nuclei originating in Ga-type structures from those derived from Ga′ and Gb′ based on the total quantity of released guaiacol and catechol, nitrobenzene oxidation of the same lignin preparation also has to be applied. The quantities of these guaiacyl nuclei and syringyl nuclei (types Sa, Sa′, and Sb′) can similarly be obtained for chemically modified hardwood lignins. Consequently, a combination of the NE method and nitrobenzene oxidation can be extremely useful for determining structural changes occurring in the phenyl nuclei of lignin which lead to the formation of DPM moieties during various modification processes (Funaoka and Abe 1983a,b, 1984, Chiang and Funaoka 1988, 1990a,b, Chiang et al. 1988, 1989a,b). However, in this chapter only the use of the NE method to determine the proportion of condensed and noncondensed phenyl nuclei in protolignins will be described as nitrobenzene oxidation is covered in Chapter 6.2. The application of NE method and nitrobenzene oxidation to reveal the formation of DPM moieties during kraft pulping has been described in detail in previous publications (Chiang and Funaoka 1988, 1990a,b).

6.6.2 Reaction Mechanism

In the NE reaction, lignin is degraded by boron trifluoride in the presence of excess phenol. The key step in this process is the dealkylation of DPM-type structural units formed by phenolation the Cα side chain position of the phenylpropane units. To bring this reaction sequence into clear focus, it is appropriate to discuss the alkylation and dealkylation of simple aromatic compounds.

Under acidic conditions, the alkylation and dealkylation of aromatic compounds are reversible reactions involving several steps in which π- and σ-complexes are formed. However, dealkylation proceeds only under more drastic conditions compared with alkylation. Nevertheless, this is not always the case. For example, if the aromatic compound is of the DPM type, the dealkylation may proceed under mild conditions since the cations formed (Fig. 6.6.5) are resonance-stabilized. This statement is supported by the fact that DPM derivatives may be degraded even at room temperature by aluminum chloride to yield benzene, alkylbenzene, and alkyldiphenylmethane, together with some resinous substances (Tsuge and Tashiro 1962, 1965).

Based on the mechanism shown in Fig. 6.6.5, it is also clear that the cations formed after dealkylation may recombine with released phenyl nuclei as well as

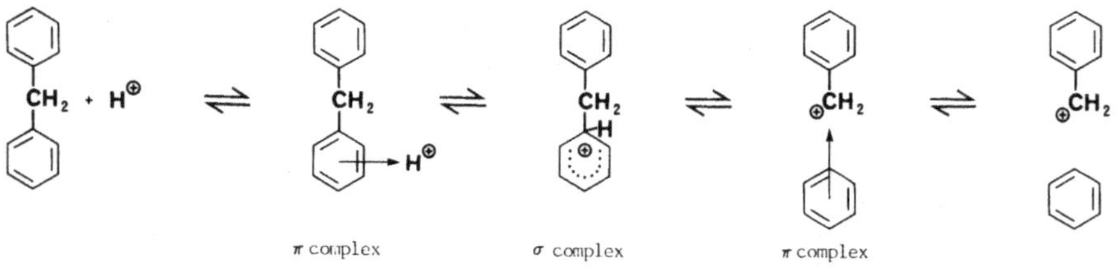

Fig.6.6.5. Dealkylation of phenyl nuclei in diphenylmethane-type structures

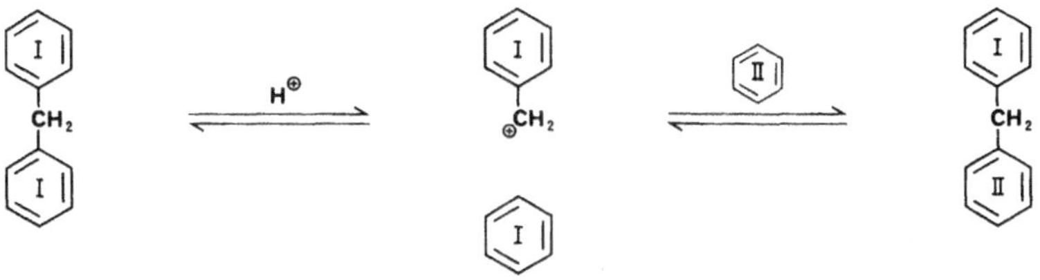

Fig.6.6.6. Dealkylation of phenyl nuclei in the presence of excess aromatic compound (II)

with the starting DPM compounds to form resin-like polymers (Radziewanowski 1894, Oda and Ogata 1941, Tsuge and Tashiro 1962, 1965). However, if to the reaction system shown in Fig. 6.6.5, another aromatic compound is added in excess, the newly formed cations will preferably react with this aromatic compound (Fig. 6.6.6). Consequently, the dealkylated phenyl nuclei (e.g., compound I in Fig. 6.6.6) remain as monomer units in the reaction system. The reaction scheme in this figure illustrates the basic principle of the NE reaction.

In the NE reaction system, lignin phenyl nuclei are converted into polyhydric phenols, such as catechol from guaiacyl nuclei and pyrogallol from syringyl nuclei, according to the three reaction steps shown in Fig. 6.6.2. Step A is the phenolation of phenylpropane side chain to form DPM-type structural units, step B is the exchange of the phenyl nuclei in lignin for phenol, and step C is the cleavage of methoxyl groups. The quantitative character of these reactions is described below.

6.6.2.1 Phenolation

The reaction of lignin with phenol under acidic conditions has been demonstrated to occur predominantly at the $C\alpha$ position of the phenylpropane side chain (Ishikawa 1958, Kratzl et al. 1962, Tai et al. 1968, Funaoka et al. 1977, Kratzl and Oburger 1980a,b). However, the phenolation reaction proceeds rapidly under even milder conditions if boron trifluoride is used as a catalyst. Using a series of monomeric guaiacyl and syringyl lignin model compounds having hydroxyl, carboxyl, carbonyl, ethylenic, and ether functionalities at the $C\alpha$ position of the side chain, Funaoka and Abe (1980a) demonstrated that almost all of the model compounds were quantitatively phenolated at the $C\alpha$ position using boron trifluoride as the acid after a 10-min reaction at 40 °C. Moreover, at higher temperatures, all the model compounds listed above were quantitatively phenolated at the $C\alpha$ position to afford DPM-type structures. Furthermore, quantitative phenolation occurred with both phenolic and nonphenolic model compounds (Funaoka and Abe 1980a, Funaoka 1984). Confirmation of this quantitative phenolation reaction was demonstrated by titrating the water formed during the reaction with phenol using a Karl-Fischer water titrator (Funaoka and Abe 1980a). For instance, for every mole of phenol reacted with the benzylic hydroxyl group in 4-hydroxy-3-methoxy benzyl alcohol, one mole of water was released.

6.6.2.2 Nucleus Exchange

The formation of DPM-type moieties in lignin through phenolation at the Cα position of the phenylpropane side chain is a key step for facilitating the release of phenyl nuclei, such as guaiacol, in the NE reaction. As was stated previously, the dealkylation of DPM derivatives is facilitated because of the formation of resonance-stabilized intermediate carbonium ions. Similarly, using the lignin model compounds mentioned previously, Funaoka and Abe (1980a,b, 1982e, 1987a,b) and Funaoka (1984) have shown that after the formation of DPM- type moieties, guaiacol and pyrogallol-1,3-dimethyl ether were liberated quantitatively from noncondensed guaiacyl, and noncondensed syringyl models, respectively, under mild conditions (40 to 60 °C). The sites from which guaiacol or pyrogallol-1,3-dimethyl ether were released were attacked immediately by the phenol, thus insuring that the liberated guaiacol or pyrogallol-1,3-dimethyl ether was preserved in monomeric form.

The liberation of phenyl nuclei is a second-order reaction (Funaoka and Abe 1980a,b, Funaoka 1984). The reaction rate increases in the presence of a group capable of stabilizing the intermediate carbonium ions after the liberation of the phenyl nuclei. The rate is decreased by etherification of the phenolic hydroxyl group. This indicates that the phenolic hydroxyl group contributes to an increase in the electron density of the aromatic ring, thus facilitating coordination of boron trifluoride with the aromatic ring. Because the coordination of boron trifluoride with the phenolic hydroxyl group is hindered by the methoxyl groups in syringyl nuclei, the rate at which pyrogallol-1,3-dimethyl ether is liberated is lower than that for the release of guaiacol from guaiacyl units. Therefore, to overcome these obstacles for obtaining quantitative yields of guaiacol or pyrogallol-1,3-dimethyl ether, the NE reaction should be conducted at elevated temperatures such as 100 °C and above. At these temperatures, partial demethylation of guaiacol to catechol, and of pyrogallol-1,3-dimethyl ether to pyrogallol-1-methyl ether, occurs.

6.6.2.3 Demethylation

Unlike the two previous reaction steps (A and B, Fig. 6.6.2), demethylation proceeds slowly in the NE reaction sequence. Demethylation is second-order and becomes significant only at temperatures above 100 °C (Funaoka and Abe 1978b). When guaiacol was used as a model compound to study demethylation in the NE method, it was found that extensive demethylation occurred in the temperature range of 160 to 180 °C. However, at any time and temperature investigated (160 to 180 °C), the sum of guaiacol and catechol concentrations in the reaction mixture was always equivalent to the starting concentration of guaiacol. This indicated that guaiacol was effectively demethylated to form only catechol and that catechol was completely stable in the NE reaction system.

Pyrogallol-1,3-dimethyl ether (3 in Fig. 6.6.1) is demethylated to pyrogallol-1-methyl ether (4) which in turn is demethylated to pyrogallol (5) by the NE treatment. The rates for demethylating 3 to 4 and 4 to 5 (Fig. 6.6.1) were found to be greater than the rate for demethylating guaiacol to catechol, as shown

Table 6.6.1. Rate constants and activation energies for the demethylation of the methoxyl group in the nucleus exchange reaction. (Funaoka and Abe 1982e)

	Rate constant at 130 °C (mol · mol^{-1} min^{-1})	Activation energy (kcal)
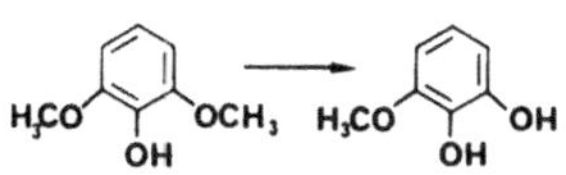	8.58×10^{-2}	24.09
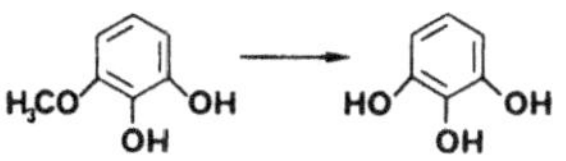	2.33×10^{-2}	28.45
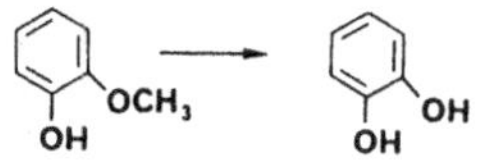	8.52×10^{-3}	30.55

in Table 6.6.1 (Funaoka and Abe 1982e). This observation was explained mechanistically in detail in previous publications (Funaoka and Abe 1982e, Funaoka 1984). As was demonstrated, the demethylation of syringyl nuclei occurred in the 90 to 140 °C temperature range. However, only at temperatures lower than 110 °C was the sum of 3, 4, and 5 equivalent to the starting concentration of 3. At temperatures higher than 110 °C, secondary modification of pyrogallol (5) occurred, lowering the combined yield of 3, 4, and 5 (Funaoka and Abe 1982e, Funaoka 1984). Consequently, when dealing with syringyl (hardwood) lignins, a low temperature should be used to obtain reliable quantitative information.

6.6.2.4 The Response of Various Structural Units in Lignins to the NE Treatment

The above discussion has stressed that lignin structural units, such as noncondensed guaiacyl and noncondensed syringyl nuclei having hydroxyl, carboxyl, carbonyl, carbon-carbon double bond, and ether functionalities at the $C\alpha$ position of the side chain, give quantitative yields of catechol derivatives from guaiacyl nuclei and pyrogallol derivatives from syringyl nuclei in the NE reaction. The ether linkages in the predominant β-aryl ether structural units of lignin are easily cleaved by the NE reagents under mild conditions. This was demonstrated by the rapid liberation of guaiacol at 40 °C from β-aryl ether model compounds, such as ω-(2-methoxyphenoxy) acetoveratrone (6) in Fig. 6.6.7) and 1-(3,4-dimethoxyphenyl)-2-(2-methoxyphenoxy) ethanol (Funaoka and Abe 1980a, Funaoka 1984). At this temperature (40 °C), demethylation does not occur. Consequently, the formation of guaiacol shown in Fig. 6.6.7 is due entirely to the cleavage of the β-aryl ether linkage, releasing the guaiacol moiety from these model compounds. Cleavage of the β-aryl ether linkage was

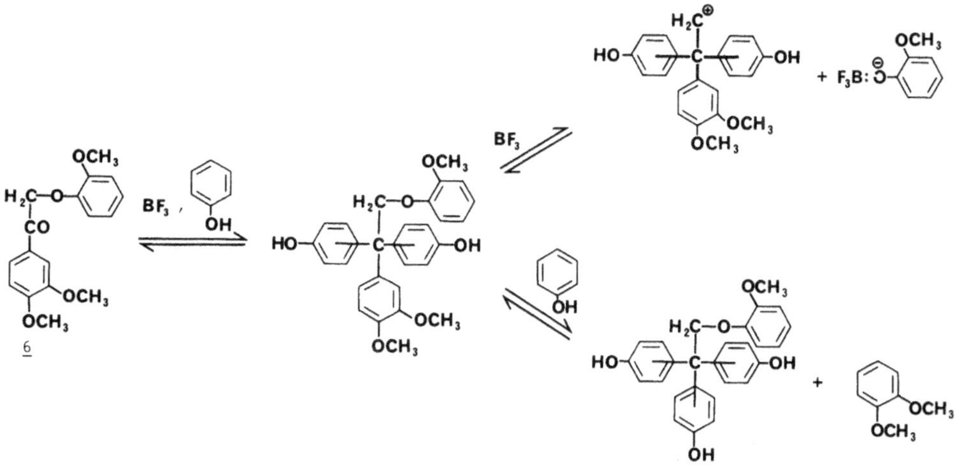

Fig. 6.6.7. The reaction of ω-(2-methoxy phenoxy)acetoveratrone in the NE reaction system

found to be a second-order reaction in which the rate-determining step was phenolation at the Cα position. In other words, the rate at which β-aryl ether linkages are cleaved is influenced by the reactivity of Cα functional groups on the side chain towards phenolation. After phenolation, cleavage of the β-aryl ether linkage and the NE reaction occur simultaneously (Fig. 6.6.7) (Funaoka and Abe 1980a, Funaoka 1984).

α-Aryl ether linkages in lignin structural units are also efficiently cleaved in the NE reaction system. Thus, guaiacol is released quantitatively from dehydrodiisoeugenol (Fig. 6.6.8) (Funaoka and Abe 1980a). Conversely, the 5-5 linkage in biphenyl units is stable in the NE reaction system, while the hydroxyl groups in 2,2′-dihydroxydiphenyl are dehydrated, forming a furan ring (Funaoka and Abe 1982b).

6.6.3 Description of the NE Method

6.6.3.1 Sample Preparation

An air-dried wood sample is ground successively in a Wiley mill and a vibrational mill to pass an 80-mesh screen. (Note 1) The wood meal is extracted continuously with benzene-ethanol (2:1, v/v) in a Soxhlet extractor for 48 h. The extracted sample is first air-dried and then dried over P_2O_5 under vacuum for at least one week to completely remove the moisture. A moisture-free sample is required for the NE reaction since boron trifluoride will be deactivated to some extent by any residual moisture in the sample.

Note 1. To obtain quantitative yields of NE products, the particle size of the wood meal must be small enough to pass an 80-mesh screen (Funaoka and Abe 1985).

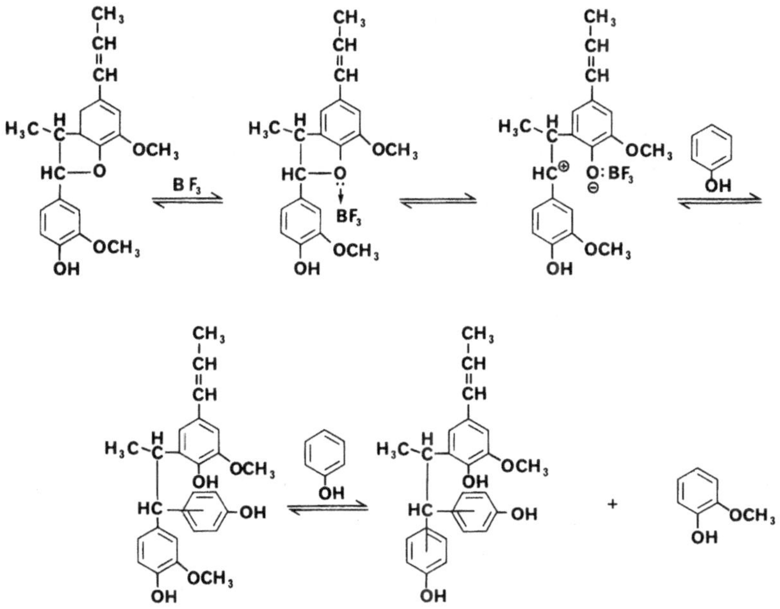

Fig.6.6.8. The formation of guaiacol from dehydrodiisoeugenol

6.6.3.2 Preparation of Reagent

The reagent is a mixture of phenol, boron trifluoride-phenol complex, and xylene. Xylene is used as an inert reaction medium. The standard composition of the mixture (by volume) is listed below.

Phenol (reagent grade)	19 parts
Xylene (reagent grade)	10 parts
Boron trifluoride-phenol complex	4 parts

(Nadarai Chemical Ltd., Kyoto, Japan. Lot #: M6R6604 (preferably), Sharpe Chem. Co., Burbank, CA)

This composition is used for the analysis of lignin in wood or pulp. For isolated lignins, such as milled wood lignin and kraft lignin, three parts of boron trifluoride-phenol complex are used with the amounts of phenol and xylene shown above. Prior to mixing these three reagents, the xylene is dried over anhydrous Na_2SO_4. The phenol is liquified at 40 to 50 °C just before mixing. All glassware should be moisture-free, and the preparation of the mixture should be done as quickly as possible.

The required volume of dry xylene is measured into a graduated cylinder and a portion is transferred to a reagent bottle. The appropriate amounts of liquified phenol and boron trifluoride-phenol complex are similarly measured out and quickly transferred to the same reagent bottle. After each of the additions, the cylinder is washed well with the remaining portion of the xylene which is also transferred to the reagent bottle. The bottle is sealed and stored in a warm, dark, moisture-free chamber.

The reagent should be warmed and mixed well prior to use and the phenol should be completely dissolved. If the reagent is used daily or, in other words, if the reagent is contacted with air moisture frequently, it is advisable to prepare a fresh quantity after a three-month period. However, the reagent should be checked periodically to confirm its reactivity.

6.6.3.3 Description of Procedure

Depending upon the sample size, the NE treatment may be conducted in a stainless steel autoclave which is conveniently made from a stainless steel tube of appropriate size fitted with screw caps (stainless steel) at both ends. For the sample size (~28 mg) described later, the autoclave is made from a 0.6 cm (diam.) × 7.5 cm (length) stainless steel tube. The autoclave is heated in an oil bath or comparable device in which the temperature can be controlled to ± 1~2 °C and which is provided with a means of shaking the autoclave continuously during the reaction. The detailed procedure is described below.

Accurately weigh approximately 28 mg of extractive- and moisture-free wood meal into the autoclave and add two small stainless steel balls. For each 28 mg of wood meal, 1 ml of reagent is added. The autoclave is sealed, shaken manually for about 5 min, then heated at 180 ± 1~2 °C for 4 h. The autoclave is shaken continuously throughout the heating operation.

At the end of the reaction, the autoclave is removed from the heat source and cooled in a cold water bath. One of the caps is removed and the autoclave contents are carefully poured into an 80-ml beaker. A small amount of ethyl ether is added to the autoclave and the residual material on the wall is scraped with a thin spatula. This solution is also added to the 80-ml beaker. This procedure is repeated until the washings become clear. The cap at the other end of the autoclave is removed and both caps are washed with ethyl ether which is combined with the other ether washings.

A 0.5-ml portion of an internal solution (5 mg of bibenzyl/ml benzene) is added to the combined reaction solution and washings in the 80-ml beaker and filtered through a glass filter paper which subsequently is washed several times with ethyl ether. The combined filtrate and washings from this operation are transferred to a separatory funnel and a NaCl-saturated aqueous solution is added to deactivate the boron trifluoride. The funnel is shaken vigorously and the ether layer is removed and concentrated to about 10 ml by film evaporation. The concentrate is transferred to a glass vial fitted with a Teflon-lined cap and dried overnight over anhydrous sodium sulfate.

A 0.05-ml aliquot of the dried ether solution is pipetted into a small glass vial containing a few drops of pyridine and 0.1 ml of N,O-bis(trimethylsilyl)acetamide (Pierce Chemicals). The vial is stoppered with a Teflon-lined cap and the TMS derivatization is performed at room temperature for 1 h. The TMS derivatives are separated and determined by gas chromatography.

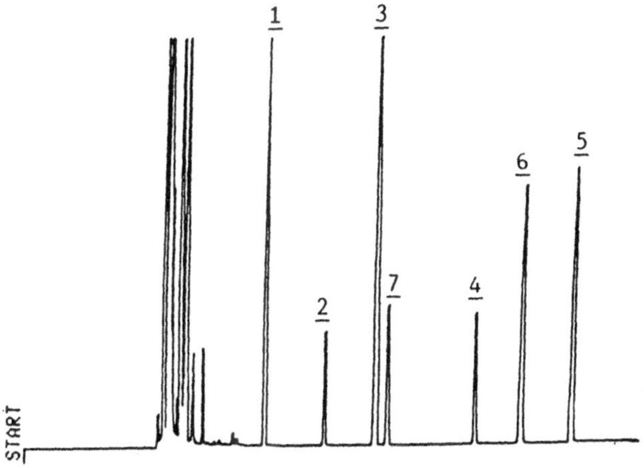

Fig.6.6.9. Gas chromatogram of NE products. 1 Guaiacol; 2 catechol; 3 pyrogallol-1,3-dimethyl ether; 4 pyrogallol-1-methyl ether; 5 pyrogallol; 6 internal standard (bibenzyl); 7 diphenyl ether

6.6.3.4 Quantitative Analysis of Products by Gas Chromatography

A Hewlett Packard 5890A Gas Chromatograph equipped with a computerized integrator (HP 9122 + HP 3393A) or equivalent and a FID detector are used. Approximately 0.5 μl is injected into the GC system. The separation is carried out on a 50-m crosslinked methyl silicone capillary column (Hewlett Packard HP-1, 50 m × 0.2 mm × 0.33 μm). The column temperature is held at 170 °C for 14 min and then is raised to 180 °C at a rate of 5 °C/min. After 8 min at 180 °C, the column is heated to a final temperature of 270 °C at a rate of 30 °C/min. The final temperature is maintained for 15 min. The injector and detector temperatures are 250 and 275 °C, respectively. Helium is used as the carrier gas with a flow rate of 30 ml/min. A typical chromatogram is shown in Fig. 6.6.9.

Applying the same GC separation conditions described above, calibration curves are made for guaiacol, catechol, pyrogallol-1,3-dimethyl ether, pyrogallol-1-methyl ether, and pyrogallol using the authentic compounds. For each compound, a linear relationship is established between (W_p/W_i) and (A_p/A_i), as shown in Eq. (1).

$$(W_p/W_i) = a + b \times (A_p/A_i), \tag{1}$$

where W_p and W_i are the weights of the product and the internal standard, respectively, A_p and A_i are the peak areas of the product and the internal standard, respectively, and b is slope of the straight line which intersects the X axis at a. From the calibration curves, the yields of NE products can be calculated and expressed as weight % (Y_w) based on lignin or mol % (Y_m) based on lignin according to Eqs. (2) and (3), respectively.

$$Y_w \text{ \% (weight \%, based on lignin)} = W_p/W_l \tag{2}$$

$$Y_m \text{ \% (mol \%, based on lignin)} = (W_p/MW_p) \times (MW\text{-}C_9/W_l), \tag{3}$$

where W_l is the weight of lignin, MW_p is the molecular weight of the product, and $MW\text{-}C_9$ is the molecular weight of a C_9 lignin unit. W_l can be determined from the Klason + acid soluble lignin contents (see Chap. 2.2). To obtain the molecular weight of the C_9 unit, milled wood lignin is isolated from the corresponding wood sample as described by Chiang and Funaoka (1988). From the elemental and methoxyl content analyses, a C_9 empirical formula and, thus, the molecular weight can be obtained. When a sample such as a kraft lignin is used, a C_9 molecular weight for this lignin should be determined and used in Eq. (3).

6.6.4 Quantitative Determination of Noncondensed and Condensed Guaiacyl Nuclei in Softwood Protolignins

For softwood protolignins, the yields of the NE products (NEP_g, guaiacol + catechol) represents the amount of noncondensed guaiacyl nuclei.

$$\text{mol \% of noncondensed guaiacyl nuclei} = NEP_g = Y_{mg}\,\% + Y_{mc}\,\%, \qquad (4)$$

whereY_{mg} and Y_{mc} % are mol % (based on lignin) yields of guaiacol and catechol, respectively.

Since the numbers of syringyl and p-coumaryl units are usually insignificant for most of the common normal softwood species, then:

$$\text{mol \% of condensed guaiacyl nuclei} = 100\text{-}NEP_g. \qquad (5)$$

6.6.5 Quantitative Determination of Noncondensed and Condensed Phenyl Nuclei in Hardwood Protolignins

There are two principal types of phenyl nuclei, guaiacyl and syringyl, in most of the common hardwood protolignins. There being no C-bonded substituents on ring positions 2 and 6, syringyl nuclei in hardwood protolignins are of the noncondensed type. Consequently, there are noncondensed and condensed guaiacyl and noncondensed syringyl nuclei in hardwood protolignins. The quantity of noncondensed guaiacyl nuclei can be obtained in the same way as for softwood protolignins described above. Thus, after the standard NE treatment (at 180 °C for 4 h), the mol % of NEP_g represents the noncondensed guaiacyl nuclei and can be calculated according to Eq. (4). However, to obtain the quantity of condensed guaiacyl nuclei, the quantity of noncondensed syringyl nuclei must be determined.

To obtain quantitative information for syringyl nuclei, the NE treatment is conducted first at 110 °C for 4 h to obtain the NE products (NEP_s^{110}), pyrogallol-1,3-dimethyl ether (3) + pyrogallol-1-methyl ether (4) + pyrogallol (5), from syringyl nuclei. At temperatures higher than 130 °C, pyrogallol tends to undergo secondary modification reactions which lower the total yield of products (see Sect. 6.6.2.3). This problem is prevented when the reaction temperature is reduced to 110 °C. However, the total of 3, 4, and 5 at 110 °C (NEP_s^{110}) is only a fraction of the maximum yield of NEP_s from syringyl nuclei. Consequently, the

NEP_s^{110} has to be multiplied by a factor, "f", to obtain the quantitative yield of NEP_s. Because of the similarity between guaiacyl and syringyl nuclei in their respones to the NE reaction (Funaoka and Abe 1982e), f can be conveniently obtained from the ratio of quantitative NEP_g yield obtained at 180 °C to NEP_g yield at 110 °C (NEP_g^{110}):

$$(NEP_g/NEP_g^{110}) = f \tag{6}$$

and NEP_s (quantitative mol % yield of 3 + 4 + 5 from syringyl nuclei)

$$= NEP_s^{110} \times f = NEP_s^{110} \times (NEP_g/NEP_g^{110}). \tag{7}$$

From the known quantity of noncondensed syringyl nuclei (NEP_s), the quantity of total guaiacyl nuclei is calculated according to Eq. (8). The quantity of condensed guaiacyl nuclei is then obtained from Eq. (9).

$$\text{mol \% of total guaiacyl nuclei} = 100\text{-}NEP_s \tag{8}$$

$$\text{mol \% of condensed guaiacyl nuclei} = \text{total phenyl nuclei-syringyl nuclei-noncondensed guaiacyl nuclei} = (100\text{-}NEP_s)\text{-}NEP_g. \tag{9}$$

When the sample under investigation is an isolated protolignin, e.g., milled wood lignin, the syringyl nuclei content can be directly calculated from the analytical composition expressed on a $C_9(OCH_3)_x$ basis.

6.6.6 Discussion of Method

6.6.6.1 Confirmation of the Quantitative Nature of the NE Reaction

When model compounds, representing guaiacyl and syringyl nuclei having hydroxyl, carboxyl, carbonyl, double bond, and ether functionalities at the $C\alpha$ position of the side chain, were subjected to the NE reaction, quantitative formation of catechol derivatives from guaiacyl nuclei and pyrogallol derivatives from syringyl nuclei was obtained only from those model compounds comprised of noncondensed phenyl nuclei (Funaoka and Abe 1980a,b, 1982a,b,c, Funaoka 1984). Thus, although all the above-mentioned functional groups are the reactive sites for phenolation (Sect. 6.6.2.1) and the ensuing cleavage of $C\alpha$-aryl carbon-carbon linkages, noncondensed guaiacyl nuclei are the sole source of guaiacol and catechol generated quantitatively in the NE system. Therefore, the guaiacol and catechol detected in the reaction mixture reflect directly the quantity of noncondensed guaiacyl nuclei. Since the above-mentioned side chain structures represent most, if not all, of the known side chain structures in protolignins (Marton 1964, Lai and Sarkanen 1971), the release of guaiacol and catechol from softwood protolignins in the NE method after phenolation at $C\alpha$ position of these side chains is from noncondensed guaiacyl nuclei.

In general, the fraction of condensed guaiacyl nuclei in softwood protolignins is in the range of 43 to 55%. The extent of condensation in guaiacyl nuclei has been estimated mainly using milled wood lignins isolated from various softwood species (Adler and Lundquist 1961, Ludwig et al. 1964, Lenz 1968, Morohoshi

Table 6.6.2. Fractions of condensed and noncondensed guaiacyl nuclei in softwood protolignins

Species	Mol % based on lignin	
	Condensed	Noncondensed
Douglas-fir[a] (*Pseudotsuga menziesii*)	50	50
Western hemlock[b] (*Tsuga heterophylla*)	55	45
Spruce (*Picea abies*)	48	52
Yezo spruce (*Picea jezoensis*)	47	53
Glehn's spruce (*Picea glehnii*)	45	55
Slash pine (*Pinus elliottii*)	50	50
Japanese red pine (*Pinus densiflora*)	43	57
Japanese black pine (*Pinus thunbergii*)	48	52
Japanese larch (*Larix leptolepis*)	55	45
Japanese fir (*Abies firma*)	49	51
Japanese hemlock (*Tsuga sieboldii*)	48	52
Sugi (*Cryptomeria japonica*)	45	55
Hinoki (*Chamaecyparis obtusa*)	45	55
Japanese torreya (*Torreya nucifera*)	53	47

[a] Chiang and Funaoka (1988).
[b] Chiang et al. (1989a).
The rest of the data are from Funaoka and Abe (1987a,b).

and Sakakibara 1971). For protolignins, Terashima et al. (1979), Tomimura et al. (1979), and Terashima and Seguchi (1987) have demonstrated, using radiotracer methods, that about 50% of the guaiacyl nuclei are condensed. Their finding is in excellent agreement with the results of the NE method applied to protolignins (Table 6.6.2). However, unlike the NE reaction, their methods are usually laborious and sometimes require sophisticated instrumental techniques.

6.6.6.2 Use of the NE Method

The NE method provides a rapid and relatively simple means of determining the degree of condensation in protolignins. As shown in Table 6.6.2, the quantities of noncondensed (type Ga, in Fig. 6.6.3) and condensed (type Gb) guaiacyl nuclei have been determined for 14 softwood protolignins using the NE reaction. The standard NE method, which involves the addition of an excess amount of phenol, is suitable for the analysis of protolignins in normal woody plants. However, it is not suitable for grass lignins, which contain a significant amount of p-hydroxyphenyl units which yield phenol in the NE reaction (Wang 1987, Chiang et al. 1988). However, the structure of p-hydroxyphenyl units can be determined by replacing phenol in the standard reagent with other phenol derivatives such as cresol.

Qualitatively, it has been demonstrated by Meshitsuka and Nakano (1985) that guaiacyl lignin in the middle lamella region is more condensed than the lignin in the secondary wall region. Quantitatively, this has been confirmed by applying the NE method to secondary wall- and middle lamella-enriched

Table 6.6.3. Characteristics of Douglas-fir lignin. (Chiang et al. 1989b)

	Whole wood	Secondary wall	Middle lamella
Lignin content	29.8 (% on wood)	25.9 (% on SW fraction)	46.2 (% on ML fraction)
Condensed / Noncondensed	50/50	48/52	71/29

Table 6.6.4. Characterization of hardwood protolignin by the NE method

Sample	Guaiacyl, mol % on lignin			Syringyl, mol % on lignin		S/G[b]	S/V[c]
	Noncondensed	Condensed	Total	Noncondensed	Total		
Sweetgum[a]	22	17	39	61	61	1.56	2.82

[a] Data from Chiang and Funaoka (1990a).
[b] S/G = Total syringyl nuclei/total guaiacyl nuclei.
[c] S/V = Syringaldehyde + syringic acid/vanillin + vanillic acid, from nitrobenzene oxidation.

fractions isolated from Douglas-fir wood meal (Chiang et al. 1989b). As shown in Table 6.6.3, the ratio of condensed guaiacyl units/noncondensed guaiacyl units is about unity for both whole wood and the secondary wall-enriched fraction, while a ratio of about 2.5/1 is found for the middle lamella-enriched fraction. This suggests that the frequencies of β-5, 5-5, and 4-O-5 linkages are significantly higher in middle lamella lignin than in the secondary wall lignin. This further indicates that the polymerization of lignin in the middle lamella region is mainly of the "bulk type", generating a highly condensed lignin macromolecule.

The quantities of noncondensed and condensed phenyl nuclei in hardwood protolignins have been reported infrequently due mainly to the difficulty of analyzing these heterogeneous materials by conventional methods. However, the NE method again services as a facile means for analyzing the heterogeneity of hardwood lignins. The concentrations of various phenyl nuclei in sweetgum protolignin, analyzed by the NE method, are shown in Table 6.6.4. Sweetgum protolignin has a syringyl/guaiacyl (S/G) ration of 1.56 which is in agreement with the estimated S/G ratios obtained in a solid state ^{13}C NMR study (Manders 1987). This value is the true ratio of syringyl to guaiacyl contents and differs from the conventional syringaldehyde + syringic acid/vanillin + vanillic acid (S/V) ratio indicated by alkaline nitrobenzene oxidation (Chap. 6.2). Nitrobenzene oxidation converts only a part of the noncondensed guaiacyl units to vanillin and vanillic acid, whereas the NE method quantitatively converts noncondensed guaiacyl nuclei into guaiacol and catechol. For this reason, the S/V ratio obtained from nitrobenzene oxidation is always much higher than the true S/G ratio. Consequently, it should be stressed that the S/V ratios obtained by nitrobenzene oxidation of lignin preparation may be misleading and deviate significantly from the true relationship between the quantities of syringyl and guaiacyl units.

The NE method is especially useful as an analytical tool when used to determine the content of total DPM moieties in lignins modified in various chemical processes such as pulping. The formation of DPM-type structural units in lignin is an important reaction and occurs readily under both acidic and alkaline conditions. For instance, the α-position on the lignin side chain may be attacked readily by adjacent phenyl nuclei under both acidic and alkaline conditions or merely by heating, forming $C\alpha$-aryl linkages (DPM-type of structures) (Nimz 1969, Funaoka and Abe 1978a, Gierer 1982, Funaoka et al. 1988, 1990a,b). It is also known that condensation of phenyl nuclei with formaldehyde derived from the γ-carbon yields DPM-type structures (Gierer 1982, Yasuda and Terashima 1982). DPM-type structural units greatly influence the color of lignin and other properties such as reactivity and solubility. Thus far, the quantity of these units secondarily formed within the lignin molecule can only be determined without isolation of the lignin by a technique combining the NE method and nitrobenzene oxidation. A detailed description of this new technique may be found in previous publications (Funaoka and Abe 1983a,b,c, 1984, 1987b, Chiang and Fuanoka 1988, Chiang et al. 1988, 1989a,b, Chiang and Funaoka 1990a,b).

In addition to its use as a quantitative analytical technique in the analysis of lignin structure, the NE method is also useful for the production of valuable chemicals such as resorcinol, phloroglucinol, catechol, and pyrogallol, in high yields from biomass such as bark and foliage, and catechol and pyrogallol derivatives from industrial lignins (Abe et al. 1987, Funaoka and Abe 1987a,b).

References

Abe I, Funaoka M, Kodama M (1987) Phenolic nuclei of condensed tannins. Approaches by the phenyl nucleus-exchange method. Mokuzai Gakkaishi 33:582–588

Adler E, Lundquist K (1961) Estimation of uncondensed phenolic units in spruce lignin. Acta Chem Scand 15:223–228

Chiang VL, Funaoka M (1988) The formation and quantity of diphenylmethane-type structures in residual lignin during kraft delignification of Douglas-fir. Holzforschung 42:385–391

Chiang VL, Funaoka M (1990a) The dissolution and condensation reactions of guaiacyl and syringyl units in residual lignin during kraft delignification of sweetgum. Holzforschung 44:147–155

Chiang VL, Funaoka M (1990b) The difference between guaiacyl and guaiacyl-syringyl lignins in their responses to kraft delignification. Holzforschung 44:309–313

Chiang VL, Funaoka M, Wang X (1988) Structural changes of lignin is soda and soda-AQ delignification of wheat straw. Nonwood plant fiber pulping series No. 18, Tappi Press, Atlanta, GA, 51–55

Chiang VL, Kolppo K, Stokke DD, Funaoka M (1989a) Structure changes of lignin in kraft and acid sulphite delignification of western hemlock. Proc Int Symp Wood Pulp Chem, Raleigh, NC. Tappi Press, Atlanta, GA, pp. 593–597

Chiang VL, Stokke DD, Funaoka M (1989b) Lignin fragmentation and condensation reactions in middle lamella and secondary wall regions during kraft pulping of Douglas-fir. J Wood Chem Technol 9:61–83

Funaoka M (1984) Studies on the reaction of lignin in the presence of phenol and boron trifluoride. Bull Mie Univ For Tsu, Japan 13:1–154

Funaoka M, Abe I (1978a) The reaction of lignin under the presence of phenol and boron trifluoride. I. On the formation of catechol from MWL, dioxane lignin and kraft lignin Mokuzai Gakkaishi 24:256–261
Funaoka M, Abe I (1978b) The reaction of lignin under the presence of phenol and boron trifluoride. II. The effect of methoxyl group and ethylene double bond on the formation of catechol. Mokuzai Gakkaishi 24:892–897
Funaoka M, Abe I (1980a) The reaction of lignin under the presence of phenol and boron trifluoride. III. The chemical structure of lignin required for the liberation of phenyl nucleus. Mokuzai Gakkaishi 26:334–341
Funaoka M, Abe I (1980b) The reaction of lignin under the presence of phenol and boron trifluoride. IV. The structure formed by the nuclear exchange for phenol. Mokuzai Gakkaishi 26:342–346
Funaoka M, Abe I (1982a) The reaction of lignin under the presence of phenol and boron trifluoride. V. The role of boron trifluoride and phenol on the formation of catechol. Mokuzai Gakkaishi 28:522–528
Funaoka M, Abe I (1982b) The reaction of lignin under the presence of phenol and boron trifluoride. VI. The behavior of lignin units unable to form the diphenylmethane structure. Mokuzai Gakkaishi 28:529–534
Funaoka M, Abe I (1982c) The reaction of lignin under the presence of phenol and boron trifluoride. VII. The cleavage of C-C linkage by disproportionation. Mokuzai Gakkaishi 28:563–569
Funaoka M, Abe I (1982d) The reaction of lignin under the presence of phenol and boron trifluoride. VIII. The decomposition of softwood lignin. Mokuzai Gakkaishi 28:627–634
Funaoka M, Abe I (1982e) The reaction of lignin under the presence of phenol and boron trifluoride. IX. The decomposition of hardwood lignin. Mokuzai Gakkaishi 28:635–643
Funaoka M, Abe I (1982f) The reaction of lignin under the presence of phenol and boron trifluoride. X. The properties of residual lignin and the degradation mechanism of lignin. Mokuzai Gakkaishi 28:705–717
Funaoka M, Abe I (1982g) The reaction of lignin under the presence of phenol and boron trifluoride. XI. The decomposition of industrial lignin. Mokuzai Gakkaishi 28:718–726
Funaoka M, Abe I (1983a) The nuclear exchange method for estimation of lignin structure. Proc Int Symp Wood Pulp Chem, Tsukuba City, Japan, Vol. 4. Tappi Press, Tokyo, pp. 79–84
Funaoka M, Abe I (1983b) Structures of softwood lignins dissolved by successive acidolysis. Mokuzai Gakkaishi 29:781–788
Funaoka M, Abe I (1983c) On the distribution of condensed structure in kraft lignin. Bull Mie Univ For, Tsu, Japan 12:27–39
Funaoka M, Abe I (1984) Analysis of chemical structures of hardwood lignins by nuclear-exchange method. Mokuzai Gakkaishi 30:68–73
Funaoka M, Abe I (1985) Degradation of protolignin by the nuclear exchange method. Mokuzai Gakkaishi 31:671–676
Funaoka M, Abe I (1986) Condensation degree of softwood protolignin. Bull Fac Agric, Mie Univ, Tsu, Japan 73:67–74
Funaoka M, Abe I (1987a) Phenyl nucleus-exchange method for the degradation of lignin. Wood Sci Technol 21:261–279
Funaoka M, Abe I (1987b) Selective cleavage of $C\alpha$-aryl bonds in lignin. Bull Mie Univ, For, Tsu, Japan 15:97–109
Funaoka M, Abe I, Yoshimura M (1977) On the reaction of hydrotropic lignin with phenol and formaldehyde. Mokuzai Gakkaishi 23:571–578
Funaoka M, Kako T, Abe I (1990a) Condensation of lignin during heating of wood. Wood Sci Technol 24:277–288
Funaoka M, Shibata M, Abe I (1990b) Structure and depolymerization of acid-condensed lignin. Holzforschung 44:357–366
Funaoka M, Shibata M, Kako T, Abe I (1988) Structure of highly condensed lignins and their depolymerization by phenolation. Proc Cellucon 88 Japan, Nov.28–Dec.1. Kyoto, Japan, The Society of Fiber Science and Technology, Kyoto, Japan, p 160
Gierer J (1982) The chemistry of delignification. A general concept. Holzforschung 36:43–51

Ishikawa H (1958) Phenolation of lignin. Memoirs of Ehime Univ, Sect 6 (Agric) 4:1–94

Kratzl K, Buchtela K, Gratzl J, Zauner J, Ettingshausen O, (1962) Lignin and plastics: the reactions of lignin with phenol and isocyanates. Tappi 45:113–119

Kratzl K, Oburger M (1980a) Kondensationsreaktionen von Lignin-Modellsubstanzen mit Phenol unter saurer Katalyse. Holzforschung 34:11–16

Kratzl K, Oburger M. (1980b) Kondensationsreaktionen von Lignin-Modellsubstanzen mit Phenol unter saurer Katalyse. 2. Mitt. Holzforschung 34:191–196

Lai YZ, Sarkanen KV (1971) Isolation and structural studies. In: Sarkanen KV, Ludwig CH (eds) Lignin. Occurrence, formation, structure and reaction. Wiley-Interscience, New York, 165–240

Lenz BL (1968) Application of nuclear magnetic resonance spectroscopy to characterization of lignin. Tappi 51:511–519

Ludwig CH, Nist BJ, McCarthy JL (1964) Lignin XIII. The high resolution nuclear magnetic resonance spectroscopy of protons in acetylated lignins. J Am Chem Soc 86:1196–1201

Manders WF (1987) Solid-state C-13 NMR determination of the syringyl/guaiacyl ratio in hardwoods. Holzforschung 41:83–88

Marton J (1964) On the structure of kraft lignin. TAPPI 47:713–719

Meshitsuka G, Nakano J (1985) Structural characteristics of compound middle lamella lignin. J Wood Chem Technol 5:391–404

Morohoshi N, Sakakibara A (1971) The chemical composition of reaction wood. Mokuzai Gakkaishi 17:393–399

Nimz H (1969) Kondensationsreaktionen des Lignins, Umsetzung von Lignin-Modellsubstanzen mit Resorcin und Salzaure. Holzforschung 23:84–88

Oda R, Ogata Y (1941) Reaction of diphenylmethane and its derivatives with aluminium chloride. Kogyo Kagaku Zasshi 44:961–963

Radziewanowski C (1894) Beiträge zur Kenntnis der Wirkungsweise des Aluminiumchlorids. Berichte 27:3235–3238

Tai S, Nakano J, Migita N (1968) Studies on utilization of lignin VI. Activation of thiolignin with phenol. Mokuzai Gakkaishi 14:40–45

Terashima N, Seguchi Y (1987) Factors affecting the formation of condensed structures in lignin. Proc Int Symp Wood Pulp Chem, Paris, France, Vol. 2. Imprimerie Jacques Poncet, Bresson, France, pp.1–4

Terashima N, Tomimura Y, Araki H (1979) Heterogeneity in formation of lignin III. Formation of condensed type structure with bond at position 5 of guaiacyl nucleus. Mokuzai Gakkaishi 25:595–599

Tomimura Y, Yokoi T, Terashima (1979) Heterogeneity in formation of lignin. IV. Various factors which influence the degree of condensation at position 5 of guaiacyl ring. Mokuzai Gakkaishi 25:743–748

Tsuge O, Tashiro M (1962) Studies on diarylalkanes (II). Reaction of diaryl-methanes and -ethanes with aluminium chloride (1). The Coal Tar 14:513–519

Tsuge O, Tashiro M (1965) Studies of diarylalkanes (III). The cleavage of diarylmethanes under the influence of aluminium chloride. Bull Chem Soc Jpn 38:184–188

Wang X (1987) Structure changes of lignin in soda and soda-anthraquinone delignification of wheat straw. M.S. Thesis, Michigan Technological University, Houghton, Michigan

Yasuda S, Terashima N (1982) Chemical structures of sulfuric acid lignin V. Reaction of three arylglycerol-β-aryl ether [α-, β-, γ-^{13}C] with seventy two percent sulfuric acid. Mokuzai Gakkaishi 28:383–387

6.7 Ozonation

K.V. Sarkanen, A. Islam, and C.D. Anderson

6.7.1 Introduction

Ozone has long been known to cleave carbon-carbon double and triple bonds and the products are often useful in elucidating the structure of the parent compound. The first ozonations of wood and lignins were carried out by Doree and Cunningham (1913). It was found that lignins were highly reactive toward ozone even at room temperature and became rapidly degraded to low-molecular weight compounds, while polysaccharides were much more resistant. In early work, the isolated reaction products, consisting largely of otherwise obtainable or excessively degraded acids, appeared to have little structural significance. It was not until comparatively recently, mainly thanks to Nakano and coworkers (but presaged by Aulin-Erdtman and Tomita in 1963), that it was discovered that extensive ozonation of lignin destroyed the aromatic moieties while leaving largely intact the side chains, which could be recovered in the form of identifiable mono- and dicarboxylic acids (Matsumoto et al. 1986 and earlier references therein). Thus, the side chain of the dominant structure in all native lignins, the arylglycerol-β-aryl ether moiety, can be obtained in the form of erythronic and threonic acids. Often, the only surviving aromatic carbons in the ozonation products are those directly attached to side chains. These carbons are recovered in the form of carboxylic acid groups.

It is of interest to note that in other oxidative degradation methods of structural significance, such as nitrobenzene and permanganate oxidations, the mode of oxidation is precisely the opposite. Here, it is the aromatic moiety that remains largely intact, while the side chain is shortened to an aldehyde or carboxylic acid group. Consequently, these oxidation methods provide virtually no information on the relative frequencies of the various side chain structures in lignins. Moreover, no alternative methods are available for determination of the frequency of these different side chains or for characterization of their stereochemistry. The relative incidence of different types of side chains proposed by Adler (1977) for softwoods and hardwoods are only estimates, derived mainly from products obtained on acidolysis, although they are still widely quoted for lack of better analytical methods. It would accordingly seem that ozonation provides the best opportunity for obtaining more reliable values. For this goal to be accomplished, experimental data are required for a large variety of native lignins.

Springer Series in Wood Science
Methods in Lignin Chemistry
(Edited by S.Y. Lin and C.W. Dence)

6.7.2 Mechanism of Ozonation

Ozone is highly reactive toward aliphatic multiple bonds and reacts only somewhat less readily with aromatic rings, degrading them eventually to low-molecular weight products such as glycolic, oxalic, and formic acids. The accepted reaction mechanism is that suggested by Criegee (1975) and Bailey (1978) and is illustrated in Fig. 6.7.1. The reaction between ozone and the ethylenic bond first produces the relatively unstable primary ozonide (1) which cleaves to an aldehyde or ketone and a carbonyl oxide (2) often called the Criegee intermediate or zwitterion. The latter may either recombine with the carbonyl compound, forming the final ozonide (3), or react irreversibly with a hydroxyl compound, generating a second carbonyl compound and a peroxide.

The rate of the reaction depends largely on the nature of substituents R and R[1]. In accordance with the electrophilic nature of ozone, electron-releasing groups, such as alkyl, aryl, or ether substituents, accelerate the reaction. In contrast, electron-attracting substituents, such as halogen, carbonyl, and nitro groups, have a retarding influence. In general, aromatic rings react more slowly with ozone than do alkenes. The effect of ring substituents is the same as in alkenes, that is, electron-releasing substituents increase and electron-withdrawing groups decrease the rate of ozonation. For example, both phenols and phenol ethers react fairly rapidly with ozone. The reaction rates observed for free phenols are, however, always higher than those of the corresponding ethers.

In accordance with these generalizations, studies with model compounds (Kaneko et al. 1983, Eriksson and Gierer 1985) indicate that the rates of ozonation decrease in the order stilbenes > styrenes > phenolic structures > muconic acid intermediates > nonphenolic structures > aroyl structures >

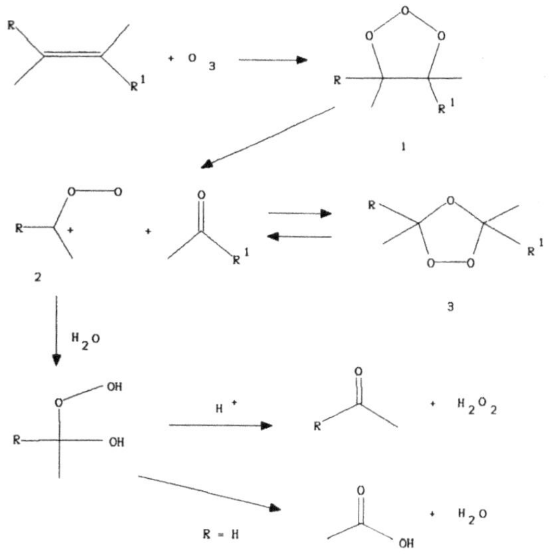

Fig. 6.7.1. Mechanism of ozonation. (Criegee 1975)

carbohydrates. Interestingly, it has been shown that methylation of phenolic hydroxyl groups in biaryl-type lignin model compounds decreases the rate much more than does methylation of phenolic hydroxyl groups in other types of lignin compounds. Also, in keeping with the electrophilic nature of ozone, the introduction of methoxyl groups into phenolic rings effects the reactivity in the following order: syringyl > guaiacyl > 4-hydroxyphenyl (Tanahashi et al. 1975, Haluk and Metche 1986). Introduction of electron-withdrawing substituents causes corresponding decreases in the reactivity. In veratryl and guaiacyl lignin units, ozone typically cleaves the ring between C3 and C4 first, although this selectivity can vary with the solvent and is diminished as a substituent at C1 becomes more electron-withdrawing (Kratzl et al. 1976).

6.7.3 Factors Involved in the Design of an Ozonation Procedure

This section will review the range of conditions and techniques that have been used for ozonation studies of lignins, focusing on those most related to determinations of side chain structures and their incidence.

6.7.3.1 Conditions Used for the Reaction with Ozone

The Ratio of Ozone to Substrate

Three ranges of this ratio will be considered, namely <10, 10–30, and >30 moles of ozone per mole of substrate, usually meaning per mole of lignin or lignin model monomer unit. In view of the relative rates of ozonation for structural units present in or formed from lignins (Sect. 6.7.2), the lowest ratios result in cleavage of alkene C=C bonds in stilbene and styrene units or in allyl units in allylated lignins (Kondo et al. 1987). The rate of consumption is also dependent on other factors, such as the design of the ozonation apparatus, the accessibility of the reacting structural units, and their stereochemistry (Deslongchamps et al. 1974). The lowest ratios typically have been used when the objective has been to modify the physical or chemical properties of the substrate at low cost or with a minimal reduction in its degree of polymerization (Soteland 1971, Godsay and Pearce 1984, Eriksson and Gierer 1987, Kondo et al. 1987, Bes et al. 1989, Bonnet et al. 1989, Hayashi et al. 1989, Sawada et al. 1989, Tomita et al. 1989), or when the intention has been to determine the relative rates and/or early sites of attack by ozone (Balousek et al. 1981, Kaneko et al. 1983, Eriksson and Gierer 1985).

The products obtained using the intermediate ozone-to-substrate ratios include, in addition, cleavage products of the more reactive aromatic rings, namely those containing a free phenolic hydroxyl. The initial products often contain oxo groups, usually ester links and/or free carboxyl groups, and the less ozone-reactive aromatic rings (Kratzl et al. 1976, Eriksson and Gierer 1985, Haluk and Metche 1986, Bonnet et al. 1989). When the highest ozone-to-substrate ratios are used, essentially all aromatic rings are completely destroyed

(Aulin-Erdtman and Tomita 1963, Tanahashi et al. 1975, Matsumoto et al. 1984a,b, 1986, Habu et al. 1990, Tsutsumi et al. 1990).

The aliphatic products from high ozone exposure are typically much less reactive than their precursors, but their further oxidation cannot be totally disregarded. From studies of the ozonation of malonic acid (Dobinson 1959), saccharides (Godsay and Pearce 1984, Matsumoto et al. 1984b, Bonnet et al. 1989), and erythronic and threonic acids (Tsutsumi et al. 1990), it is clear that aliphatic oxy, oxo, and carboxylic compounds all undergo further, usually slow, degradation by ozone.

Solvent/Dispersing Medium

The choice of liquid phase can affect the course of ozonation. This has been shown with phenylindenes by Miura et al. (1985) and with lignin models by Kratzl et al. (1976), who found that 4-methylveratrole gave a higher yield of dimethyl β-methyl-cis,cis-muconate in 45% aqueous acetic acid than in dichloromethane and that the yield of dimethyl oxalate from acetoveratrone was markedly dependent on the solvent. Eriksson and Gierer (1985) found that in methanol the benzene ring of vanillin is cleaved (albeit only very slowly, after acetalization has occurred), but that in nonprotic solvents, such as acetone, the initial reaction is oxidation of the aldehyde function to a carboxyl group.

Deslongchamps et al. (1974) found that the course of ozonations can be altered in a sometimes advantageous way by using as medium acetic anhydride containing sodium acetate. This mixture acetylates pre-existing or newly formed primary and secondary hydroxyl functions, stabilizing them against oxidation during long exposure to ozone; in at least one case, acetylation also protected a γ-hydroxy acid against lactonization.

In an early study on the ozonation of lignin, Freudenberg et al. (1935) found that both lignin and methylated lignin were more quickly oxidized in glacial acetic acid than in water. Nakano and his coworkers (Matsumoto et al. 1984a,b, 1986, Habu et al. 1990) used acetic acid/water/methanol (16:3:1, v/v/v) as the reaction medium. This mixture was also used by Sarkanen and his coworkers (Tsutsumi et al. 1990 and unpubl. work).

Dorland et al. (1939) used 95% formic acid in water and Miura et al. (1985) 50% formic acid in dichloromethane as reaction media. In the latter study, the results were not strikingly different from those obtained using acetic or propanoic acids in parallel experiments, even though formic acid is known to be consumed by ozone in dilute aqueous perchloric acid. This finding is also consistent with the appearance of formic acid as a lignin ozonation product. Bonnet et al. (1989) found that up to 36% of the total acidity that developed on ozonation of alkali lignin in water was due to the presence of formic acid.

When the solvent is or contains an alcohol, it is often incorporated in the ozonation products. For example, Kratzl et al. (1976) obtained about one mole of ethyl formate per mole of veratrole ozonized when the solvent was 10% ethanol in chloroform. Tanahashi et al. (1975) obtained methyl esters and, on hydrogenolysis of some ozonide/methanol adducts, methyl benzyl ethers using

Table 6.7.1. Media used in ozonation of lignins

Solvent/dispersing medium	Reference
A. Carboxylic acid media	
Acetic acid	Phillips and Goss (1933)
	Aulin-Erdtman and Tomita (1963)
B. Alcohol media	
Methanol	Kolsaker and Bailey (1967)
Methanol/CH_2ClCH_2Cl	Tanahashi et al. (1975)
	Haluk and Metche (1986)
Dioxane/methanol (1:2, v/v)	Tomita et al. (1989)
CH_2Cl_2/methanol (5:1, v/v)	Kondo et al. (1987)
Methyl cellosolve	Kratzl et al. (1976)
C. Aqueous media	
Water	Dobinson (1959)
Acetone/water (9:1, v/v)	Soteland (1971)
(4:1, v/v)	Balousek et al. (1981)
D. Aprotic media	
Acetic anhydride[a]	Tishchenko (1959)
Ethyl acetate[a]	Tishchenko (1959)
Chloroform	Kratzl et al. (1976)
Carbon tetrachloride	Kratzl et al. (1976)

[a] cf. Deslongchamps et al. (1974).

methanol-dichloroethane (4:1, v/v) as solvent during ozonation of styrene and/or phenylcoumarone derivatives.

It appears that ozone attacks lignin more selectively relative to carbohydrates if the concentration of byproduct hydroxyl and hydroperoxyl radicals is diminished (Eriksson and Gierer 1985). This situation is favored by the radical quenching action of methanol (Eriksson and Gierer 1987) or, in aqueous media, by keeping the pH low (Godsay and Pearce 1984, Bonnet et al. 1989). Other media used in the ozonation of lignin and lignin model compounds are listed in Table 6.7.1.

Ozonation Temperature

The use of lower temperatures and lower pH during ozonation of pulps helps to maintain viscosity (Godsay and Pearce 1984), apparently by reducing the attack of the ozone on the cellulose.

In the ozonation of indenes in dichloromethane, a decrease in temperature led to increased proportions of the more thermodynamically stable ozonide stereoisomer but had no effect on total ozonide yield. In methanol-rich solvent mixtures, a temperature decrease had the same effect on the ozonide stereoisomer ratio but markedly increased competition from solvent incorporation into ozonides (Miura et al. 1985).

Temperature can evidently influence the course of ozonation per se, but the literature does not indicate the nature or yield of the final ozone-cleavage products are temperature-dependent. Therefore, the fact that most ozonations of lignin have been carried out within the convenient temperature range of 0–25 °C would seem justified.

6.7.3.2 Work-Up of Ozonation Product Mixtures

In most lignin ozonation studies, residual ozone is purged from the system by a stream of nitrogen. Ozonides and ozone-generated peroxides can survive in the crude product mixture for several days (Kolsaker and Bailey 1967, Soteland 1971). Conversion of the ozonation products to stable, more easily identifiable products can be achieved by further oxidation, reduction, acid- or base-induced decomposition, or some combination of such methods.

To ensure complete oxidation of aldehydes to carboxylic acids, Aulin-Erdtman and Tomita (1963) treated the ozonation products of lignin model compounds with hydrogen peroxide. Soteland (1971) applied hydrogen peroxide to a groundwood ozonation product and found that it degraded the components to unidentified volatile substances. To simplify the product mixture, reduction is sometimes used to stop further oxidation. Tanahashi et al. (1975) applied palladium-catalyzed hydrogenation, while Kondo et al. (1987) added dimethyl sulfide.

A discussion of the acid- or base-induced decomposition of ozonides and derived peroxides has been provided by Kolsaker and Bailey (1967), who studied the effects of adding a strong acid or base to methanolic solutions of ozonation products obtained from substrates of the styrene type. Under neutral or acidic conditions, when the aryl group carries electron-donating substituents such as a 4-methoxy group, high yields of the phenols formed by rearrangement and solvolysis are obtained (see Fig. 6.7.2). In contrast, basic conditions give high yields of the normal ozonation product. Thus, base-promoted decomposition can be expected to give simpler cleavage results. For this and another reason noted below, basic conditions are usually used.

Lignin ozonation products contain some diacids that occur in nonesterified, partially esterified, and fully esterified forms (typically methyl esters derived from cleavage of the methoxylated aromatic rings). Bonnet et al. (1989) chose to esterify the free carboxyl groups by use of diazomethane to simplify the

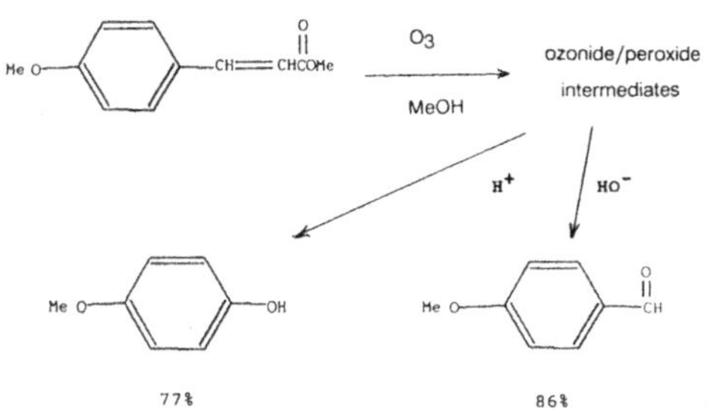

Fig. 6.7.2. Effect of acidity/basicity on decomposition of a styrene-type ozonide. (Kolsaker and Bailey 1967)

separation and identification of the diacids. However, since the ozonation products also contain ester (including lactone) groups other than methyl esters, the common approach is to simplify the mixture by hydrolyzing all ester linkages, usually by saponification. The resulting carboxylic acids, along with those that were present before hydrolysis, can then be suitably derivatized.

If so desired, the hydrolyzed ozonation products can first be separated on an anion exchange resin into what have been described as neutral, monocarboxylic and dicarboxylic fractions (Matsumoto et al. 1986, Taneda et al. 1989). In our experience, however, acids do not always appear in the expected fraction, and there is a complicating contamination by succinic acid, apparently originating in the resin. These difficulties together with the greatly increased resolution of capillary gas chromatography (GC) have combined to render this preliminary separation unnecessary.

For GC analysis, the salts of the lowest molecular weight acids present in ozonation products subjected to base-promoted hydrolysis have been converted to their benzyl esters by reaction with benzyl bromide (Bonnet et al. 1989). The salts of all acids produced have commonly been converted to the free acids, usually with the aid of a cation exchange resin. The acids have then been converted to methyl esters by reaction with diazomethane (Bonnet et al. 1989) or, more often, have been converted to trimethylsilyl (TMS) esters (Matsumoto et al. 1986, Taneda et al. 1989, Habu et al. 1990). Trimethylsilylation has the major advantage that alcoholic and phenolic hydroxyl groups are simultaneously converted to TMS ethers, thus greatly facilitating GC analysis.

Alternatively, the saponified ozonation products are converted from sodium to ammonium salts by ion exchange followed by direct trimethylsilylation, a procedure that avoids the complication of partial to complete lactonization of hydroxy acids on acidification (Hyppanen et al. 1983, Tsutsumi et al. 1990). A limition on the use of ammonium salts, however, is that salts of volatile acids may revert to the parent acid and ammonia and be lost.

6.7.3.3 Procedures for Separation and Identification of Ozonation Products

The most frequently used separation technique is GC, usually applied to trimethylsilylated samples. In recent investigations, advantage has been taken of the greater resolution inherent in capillary columns (Matsumoto et al. 1989, Habu et al. 1990, Tsutsumi et al. 1990), typically coated with a dimethylsilicone stationary phase. Trimethylsilylation has usually been performed using N,O-bis(trimethylsilyl)trifluoroacetamide (BSTFA), usually in pyridine (Habu et al. 1990, Tsutsumi et al. 1990), although acetonitrile was used by Bonnet et al. (1989). The principal alternative separation technique is high pressure liquid chromatography (HPLC) (Kaneko et al. 1983, Eriksson and Gierer 1985, 1987, Taneda et al. 1989). This method more readily permits collection of individual peak components for NMR and other studies.

Because of the power and convenience of combined gas chromatography-mass spectrometry (GC-MS), product identification has usually been attempted first through their mass spectra. MS data for some identified or likely components are given by Goodman and Markey (1981), Alen et al. (1984), Eriksson and Gierer (1985), and Tsutsumi et al. (1990). Because some of the components are identical, the GC-MS data for compounds in alkaline pulping mixtures are valuable (Niemelä and Sjöström 1986, Niemelä 1988). IR spectral data for ozonation products have been provided by Soteland (1971), Haluk and Metche (1986), and Kondo et al. (1987). Proton and carbon-13 NMR data for ozonation products have been reported by Tanahashi et al. (1975), Eriksson and Gierer (1985), Haluk and Metche (1986), and Matsumoto et al. (1986, 1989).

6.7.3.4 Procedures for Quantitative Determination of Ozonation Products

Aside from a preliminary application of carbon-13 NMR by Matsumoto et al. (1986), which employed visual peak height comparisons, efforts to determine the yields, relative or absolute, of ozonation products, have employed GC with flame ionization detection. Matsumoto et al. (1986) used the same visual comparison approach for GC data as for NMR data, and Eriksson and Gierer (1985) tested relative rates of ozonation of various lignin model compounds by determining which disappeared faster. Use of internal standards to measure accurately the consumption of substrate and/or yields of products has been made by Balousek et al. (1981); internal standards were 1-(4-benzyloxy-3-methoxyphenyl)-1-propanone or the combination of 2-[3,4-dimethoxyphenyl]ethanol and hexanedioic acid. Matsumoto et al. (1984a) used erythritol as a standard. Bonnet et al. (1989) used pentanoic and hexanoic acids for determining yields of formic and acetic acids. Pentaerythritol was used by Tsutsumi et al. (1990). Habu et al. (1990) used the relative yield of two ozonization products to determine the incidence of two types of interunit linkages in lignin, using a calibration curve based on the yields of these products from mixtures of appropriate model compounds of known composition.

In most of the above cases, an internal standard was added to the ozonation product immediately before trimethylsilylation. Prior to using pentaerythritol as an internal standard, Tsutsumi et al. (1990) used D-glucitol for this purpose. In some analyses, however, the area of the glucitol peak was abnormally small. Microbiological degradation of the stock glucitol solution used was ruled out, since a fresh solution occasionally also gave a small peak area at the standard dosage level. More rarely, pentaerythritol has also presented the same difficulty. It is known that some trimethylsilylating agents are useful for making acetals from mixtures of carbonyl compounds and alcohols (Chan et al. 1983). This may make it necessary to reduce ozonation products that contain significant amounts of aldehydes or ketones prior to trimethylsilylation if this interference is to be prevented.

6.7.4 Ozonation Procedure

The following procedure is a modified version of that used by Nakano and his coworkers (Habu et al. 1990 and references therein). It was turned to here after acidic permanganate oxidation had been found to give lower than expected yields of products derived from lignin side chains (Tsutsumi et al. 1990).

6.7.4.1 Ozonation, Saponification, and Derivatization

Substrate, 8 mg of model compound or lignin, or 50 mg of wood meal (Note 1), is dissolved or suspended in 40 ml of a mixture of acetic acid, water and methanol (16:3:1, v/v/v) (Note 2) in a 125-ml gas-washing bottle, which serves as the reactor. The latter is equipped with a glass-covered magnetic stirring bar and immersed in an ice-water bath. Oxygen with an ozone concentration of ~3% (Note 3) is bubbled into this mixture at the rate of 500 ml/min for the desired time period (Note 4) with continuous stirring. The system is immediately purged with nitrogen to remove residual ozone. Any insoluble material is removed by filtration (Note 5).

The ozonation mixture is spin-evaporated to dryness (40 °C) at water pump pressure. Traces of acetic acid are azeotropically removed by addition and spin evaporation of two separate 10-ml portions of water. The residue is re-dissolved in 15 ml of water and saponified by adding enough 2 M NaOH to adjust the pH to 12 and leaving the mixture at room temperature under nitrogen in the dark for 20 h. The mixture is then passed through a column (1-cm diameter) filled with 15 ml (50 ml for wood meal) of Amberlite IR-120 (NH_4^+ form) cation exchange rein (Hyppanen et al. 1983) (Note 6). The column is eluted with water until the total eluate volume is 140 ml (200 ml for wood meal). To the eluate, 5.00 ml of a solution containing 0.544 mg (4.00 μmol) of pentaerythritol in 80% aqueous methanol are added, and the mixture is evaporated as above. The residue is transferred in aqueous methanol to a 5-ml conical vial, from which the solvents are spin-evaporated as before, and the residue is dried in vacuo over P_2O_5 overnight. To the residue is added 1 ml of pyridine and 1 ml of N,O-bis(trimethylsilyl)trifluoracetamide (BSTFA) containing 1% chlorotrimethylsilane. The vial is septum-capped and heated for 1 h at 70 °C in a thermostated heating block and subsequently cooled to room temperature. The vial contents are removed with a syringe for GC and GC-MS analyses.

6.7.4.2 Gas Chromatography and Gas Chromatography-Mass Spectrometry

After trimethylsilylation, GC analysis is conducted using a low-polarity capillary column and a flame ionization detector (Note 7). The recommended temperature program is 4 min at 100 °C, then 4 °C/min to 240 °C, with the temperature of both injector port and detector at 260 °C. Programs of 2 min at 100 °C and subsequently 5 or 10 °C/min to 240 °C are used for GC-MS analysis. Sample injection volumes are 0.2–2.0 μl.

For quantitative analysis, GC peak area ratios (Grob 1985), using the added pentaerythritol or some other suitable compound as internal standard, or GC-MS selected ion monitoring (Chapman 1985) may be used when authentic samples of the compounds of interest are available.

Note 1. In contrast to the procedure used by Matsumoto et al. (1986), reduction with sodium borohydride prior to ozonation was not performed. Before ozonation, extracted wood meal was milled in a planetary ball mill in the dry state for a total of 36 h (9–12 h/day over 3 or 4 successive days to prevent excessive heat build-up).

Note 2. The glacial acetic acid used was reagent grade and methanol and water were HPLC grade. The use of well-cleaned glassware with inert caps and seals is necessary to eliminate artifacts.

Note 3. The ozone generator was a Welsbach model T-408 operated at 95 V. Measurement of ozone concentration at its source, or of unconsumed ozone during sample treatment, may be made by titration of the iodine generated from aqueous potassium iodide solutions placed in downstream bubblers.

Note 4. Ozonation times have varied from as short as 4 min for model compounds to as long as 60 min for wood meal. With wood meals, the yields of erythronic and threonic acids were highest at ~40 min.

Note 5. Insolubles are present when Klason lignin or wood meal is used. After a 40-min ozonation of wood meal, examination of the insoluble residue (in KBr) by FTIR showed the absence of characteristic lignin absorption bands.

Note 6. A commercial resin was washed successively with 2 M hydrochloric acid, 1 M acetic acid, and water. It was treated twice for 30-min periods with twice its volume of 7.5% aqueous ammonia and finally washed, first with deionized water, and subsequently with HPLC-grade water.

Note 7. Both an SE-30 fused silica capillary column (30 m, 0.25 mm inner diameter, 0.25-μ film thickness) and a similar, bonded dimethylsilicone column with helium as carrier gas (35–40 cm s^{-1} linear velocity) give good resolution.

6.7.5 Information on the Structure of the Lignin Side Chain

In this section, examples of information available from ozonation of lignin model compounds, related compounds, and lignins will be presented. Table 6.7.2 lists the monomeric and dimeric lignin model compounds that have been ozonated.

The earliest example of ozonation of a model compound with an alkene side chain is that of isoeugenol (Wood 1923). The anticipated product, vanillin, may be obtained in yields as high as 90% because of the retardation of further oxidation by the deactivating aldehyde group. Systematic studies on models of this type have been reported by Eriksson and Gierer (1985). The investigation by Kolsaker and Bailey (1967) is also relevant.

Balousek et al. (1981) used the γ-deoxy β-O-4 lignin model compound (4) (Fig. 6.7.3) and an intentionally very low ozone/substrate ratio ($\leqslant 0.4:1.0$). They found that, of the 30% of (4) consumed, the most abundant product identified was γ-deoxyveratrylglycerol (6), which accounted for 10.8% of the starting material (4) consumed. This compound and about 3% of other ident-

Table 6.7.2. Lignin model compounds subjected to ozonation

Substrate	Mol. form	Reference
Monomeric lignin models		
Vanillin	$C_8H_8O_3$	Eriksson and Gierer 1985
Veratrole	$C_8H_{10}O_2$	Kratzl et al. 1976
4-Methylguaiacol	$C_8H_{10}O_2$	Eriksson and Gierer 1985
Cinnamic acid (and esters)	$C_9H_8O_2$	Haluk and Metche 1986
p-Coumaric acid (and esters)	$C_9H_8O_3$	Haluk and Metche 1986
Guaiacol, acetate ester	$C_9H_{10}O_3$	Kratzl et al. 1976
Veratraldehyde	$C_9H_{10}O_3$	Kratzl et al. 1976, Eriksson and Gierer 1985
Acetoguaiacone	$C_9H_{10}O_3$	Kaneko et al. 1983
4-Methylveratrole	$C_9H_{12}O_2$	Kratzl et al. 1976
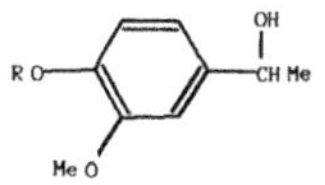		
R=H	$C_9H_{12}O_3$	Kaneko et al. 1983
R=Me	$C_{10}H_{14}O_3$	Kaneko et al. 1983
Ferulic acid (and esters)	$C_{10}H_{10}O_4$	Haluk and Metche 1986
Isoeugenol	$C_{10}H_{12}O_2$	Wood 1923
Coniferyl alcohol	$C_{10}H_{12}O_3$	Eriksson and Gierer 1985
Acetoveratrone	$C_{10}H_{12}O_3$	Kratzl et al. 1976, Kaneko et al. 1983
Veratric acid, methyl ester	$C_{10}H_{12}O_4$	Kratzl et al. 1976
Sinapic acid (and esters)	$C_{11}H_{12}O_5$	Haluk and Metche 1986
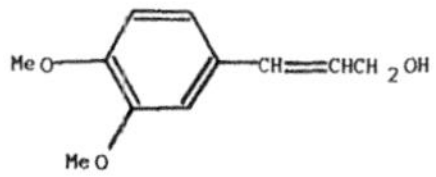	$C_{11}H_{14}O_3$	Eriksson and Gierer 1985
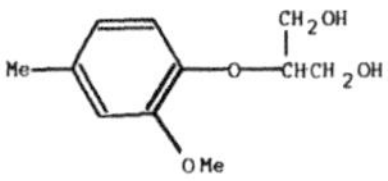	$C_{11}H_{16}O_4$	Matsumoto et al. 1984a
Veratrylglycerol	$C_{11}H_{16}O_5$	Tsutsumi et al. 1990
Dimeric Lignin Models		
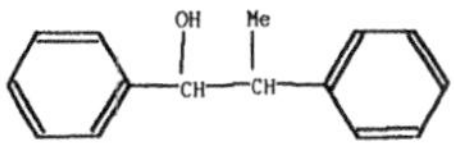	$C_{15}H_{16}O$	Aulin-Erdtman and Tomita 1963
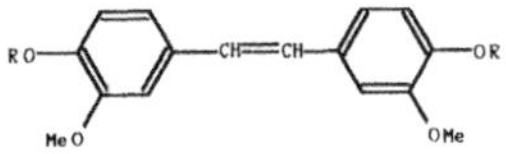		
R=H	$C_{16}H_{16}O_4$	Eriksson and Gierer 1985
R=Me	$C_{18}H_{20}O_4$	Eriksson and Gierer 1985

Table 6.7.2. (continued)

Substrate	Mol. form	Reference
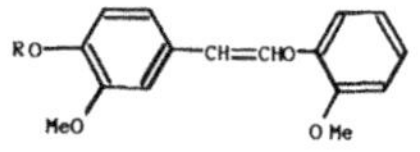		
R=H	$C_{16}H_{16}O_4$	Eriksson and Gierer 1985
R=Me	$C_{17}H_{18}O_4$	Eriksson and Gierer 1985
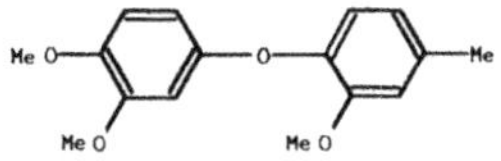	$C_{16}H_{18}O_4$	Eriksson and Gierer 1985
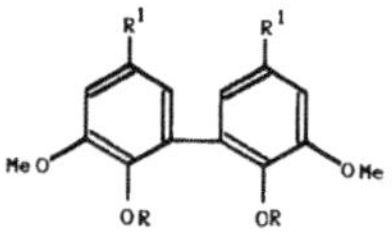		
R=H, R'=Me	$C_{16}H_{18}O_4$	Eriksson and Gierer 1985
R=H, R'=CH_2OH	$C_{16}H_{18}O_6$	Kaneko et al. 1983
R=Me, R'=CHO	$C_{18}H_{18}O_6$	Kaneko et al. 1983
R=Me, R'=Me	$C_{18}H_{22}O_4$	Eriksson and Gierer 1985
R=Me, R'=CH_2OH	$C_{18}H_{22}O_6$	Kaneko et al. 1983
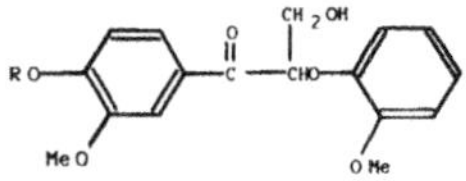		
R=H	$C_{17}H_{18}O_6$	Kaneko et al. 1983
R=Me	$C_{18}H_{20}O_6$	Kaneko et al. 1983
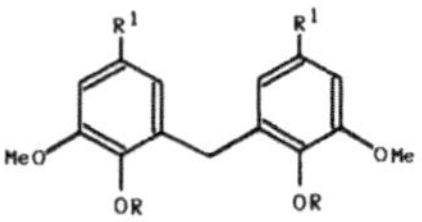		
R=H, R'=Me	$C_{17}H_{20}O_4$	Eriksson and Gierer 1985
R=Me, R'=Me	$C_{19}H_{24}O_4$	Eriksson and Gierer 1985
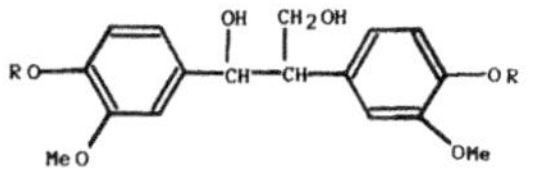		
R=H	$C_{17}H_{20}O_6$	Habu et al. 1990
R=Me	$C_{19}H_{24}O_6$	Tsutsumi et al. 1990

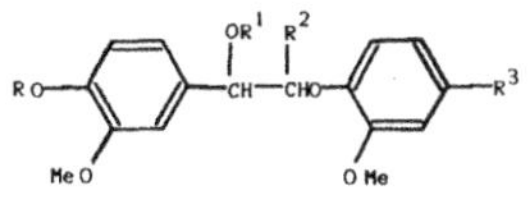		
R=H, R^1=H, R^2=CH_2OH, R^3=H	$C_{17}H_{20}O_6$	Kaneko et al. 1983, Habu et al. 1990
R=Me, R^1=H, R^2=CH_2OH, R^3=H	$C_{18}H_{22}O_6$	Kaneko et al. 1983, Habu et al. 1990
R=Me, R^1=H, R^2=Me, R^3=Me	$C_{19}H_{24}O_5$	Balousek et al. 1981
R=Me, R^1=Et, R^2=CH_2OH, R^3=H	$C_{20}H_{26}O_6$	Taneda et al. 1989
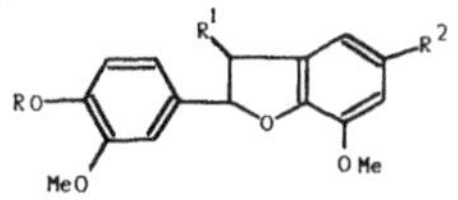		
R=H, R^1=Me, R^2=1-propenyl	$C_{20}H_{22}O_4$	Aulin-Erdtman and Tomita 1963
R=H, R^1=Me, R^2=propyl	$C_{20}H_{24}O_4$	Kaneko et al. 1983
R=H, R^1=CH_2OH, R^2=$(CH_2)_3OH$	$C_{20}H_{24}O_6$	Habu et al. 1990
R=Me, R^1=Me, R^2=propyl	$C_{21}H_{26}O_4$	Kaneko et al. 1983
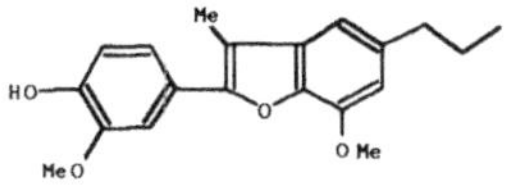	$C_{20}H_{22}O_4$	Tanahashi et al. 1975
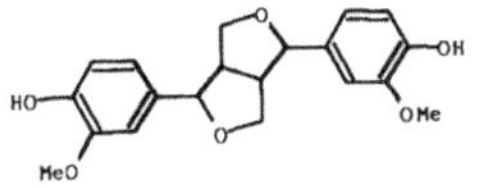	$C_{20}H_{22}O_6$	Matsumoto et al. 1989

ified products were derived from ozone attack on one or both of the aryl groups. One of these minor products (1.6% yield) was 2,3-dihydroxybutanoic acid (5) (stereochemistry unspecified). If a higher ratio of ozone to (4) had been used, the yield of (5) would likely have been much higher, because the other identified products would have been converted to (5) on further ozonation, as would many of the structural units in the 26% of unidentified GC-volatile products and in the 30–60% of nonvolatile products. Tsutsumi et al. (1990) obtained an 87–92% yield of threonic acid (11) and 0.9–1.2% yield of erythronic acid (10) on extensive ozonation of threo-veratrylglycerol.

Balousek et al. (1981) found that product (6) had a lower erythro/threo (E/T) ratio than did substrate (4), namely 0.81:0.19 for diol (6), compared with 0.92:0.08 for (4). They suggested that this could be a result of a higher rate of conversion for threo (4) relative to erythro (4) or a result of neighboring group participation by the α-hydroxyl group of (4). The protonated α, β-epoxide intermediate formed would be subject to attack by water at either the α- or β-carbon, thus permitting a change in stereochemistry. Taneda et al. (1989), on extensive ozonation of the very similar β-O-4 model compound (9) (Fig. 6.7.3), similarly observed a small reduction in the E/T ratio, from 0.90:1.00 for (9) to 0.80:1.00 for the derived tetronic acids (10) and (11). They concluded that the two ratios agreed within experimental error, and that the ratios between erythronic and threonic acids determined by GC are representative of the E/T

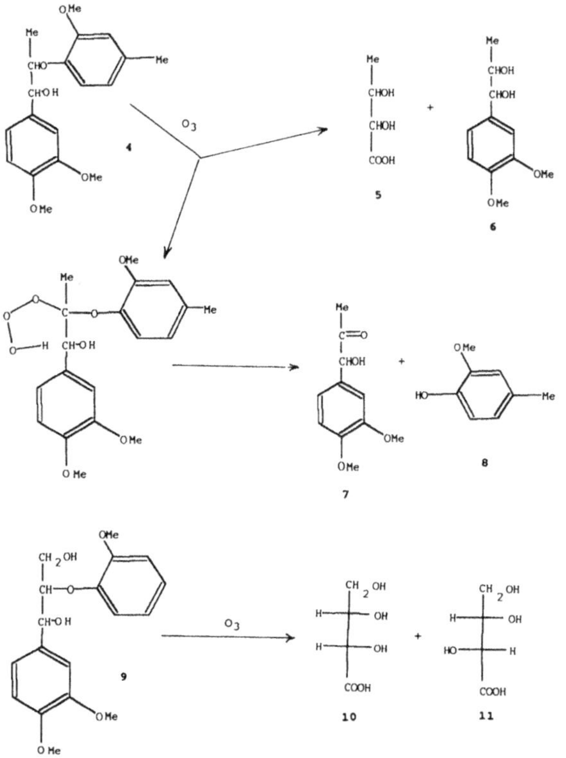

Fig. 6.7.3. Ozonation of β-O-4 lignin models. (Balousek et al. 1981, Taneda et al. 1989)

ratios of the substrates used. They also found that these ratios were unaffected by the duration of the ozonation. This has been confirmed by Tsutsumi et al. (1990), who ozonated a mixture of erythronic and threonic acids of known composition.

The most interesting result of the work of Balousek et al. (1981) was the detection of creosol (8) in an 8.0% yield based on consumption of (4), which apparently results from insertion of oxygen into the C—H bond of the β-carbon (Fig. 6.7.3), giving a hemiketal analog, from which the creosol is released on cleavage. Unfortunately, none of the predicted byproducts was identified, even though one of them, ketol (7), was shown to be "considerably more stable" to ozone than substrate (4). Matsumoto et al. (1986), on extensive ozonation of β-O-4 model (9), obtained acids (10) and (11) in 40–50% total yield, the highest so far reported, leaving considerable room for other simultaneous reactions such as direct ozone attack on the β-H (Balousek et al. 1981).

Additional evidence for the stereospecific conversion of β-O-4 and arylglycerol lignin models to tetronic acids under conditions that selectively oxidize aryl rings also comes from acidic permanganate oxidation (Tsutsumi et al. 1990). This method gives essentially the same products as ozonation.

Although the yields were lower, there was a high degree of stereospecificity. The erythro form of β-O-4 model compound (9) gave a 7.8% yield of erythronic acid (10) and a 0.4% yield of threonic acid (11). Similarly, threo-veratrylglycerol gave no detectable (10) and a 6.9% yield of (11). Finally, high stereospecificity has also been observed in the ozonation of phenylcoumaran (β-5) and diarylpropane (β-1) models, giving side chain-derived acids with largely the same stereochemistry (Aulin-Erdtman and Tomita 1963, Habu et al. 1990). This finding is relevant, since the mechanistic opportunities for scrambling of stereochemistry (e.g., via epimerization α to carbonyl intermediates) are much the same here.

The E/T ratio for lignin side chains involved in β-O-4 linkages is of practical as well as theoretical interest. The ratio has an effect on the rate of delignification during kraft pulping since etherified erythro β-O-4 structures hydrolyze more rapidly than threo structures (Miksche 1972). Under mildly acidic conditions, there is facile threo-erythro isomerization of such structures, with the erythro isomers being predominant at equilibrium (Adler et al. 1966). Thus, erythro isomers may become more abundant as the wood in a living tree ages, especially if it becomes increasingly acidic, e.g., as a result of the conversion of acetyl groups to acetic acid. E/T ratios for some lignocelluloses and lignins of different origin and processing history are given in Table 6.7.3. These data (Tsutsumi et al. 1990 and unpubl. results) are based on yields of the corresponding tetronic acids.

As noted above, etherified erythro β-O-4 structures undergo alkaline hydrolysis more rapidly than threo, and the latter should accordingly accumulate as alkaline pulping proceeds. Taneda et al. (1989) have observed decreasing E/T ratios during alkali pulping of spruce wood meal and during the corresponding alkaline hydrolysis of the model compound (9) (veratrylglycerol-β-guaiacyl ether). It is surprising that the E/T ratio is appreciably changed since the main reaction product is known, for model (9), to be the veratrylglycerol of the same erythro or threo form (Gierer and Noren 1962). An explanation for this seemingly incongruous finding has been proposed by Taneda et al. (1989).

Perhaps the most elegant application of ozonation to studies on the side chain in lignin deals with the frequency with which the glyceraldehyde-2-aryl ether unit occurs in lignin (Matsumoto et al. 1984a). After $NaBH_4$ reduction, ozonation, and saponification, model compound (12) gave a 95% yield of

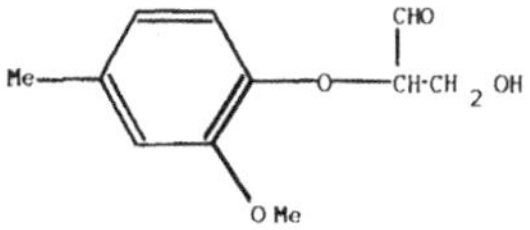

glycerol. Interestingly, ozonation for 60 min was required to achieve this maximum yield even though (12) was totally converted to some intermediate(s) within 5 min. Glycerol was completely stable on prolonged ozonation, its yield remaining constant for up to 3 h. Application of this analysis to wood meal revealed frequencies of 0.014 and 0.016 per C_6C_3 in spruce and beech, respectively. In keeping with the relative instability of glyceraldehyde-2-aryl ether units, their frequency in both spruce and white birch was found to decrease from

Table 6.7.3. Erythro/threo ratios of isolated and in situ lignins (Tsutsumi et al. 1990 and unpubl. work)

Substrate	Ozonation time (min)	E/T ratio
Lignins		
Kraft lignin	6	0.75
Klason lignin *Tsuga heterophylla*	18	0.76
Klason lignin *Populus deltoides*	18	0.78
Soda lignin *P. deltoides*	6	1.94
Native spruce Lignin	6	0.75
Native spruce Lignin	6	0.79
Native spruce Lignin	18	0.88
Native spruce Lignin	40	0.82
MWL *Corchorus capsularis* fiber	6	2.18
MWL *C. capsularis* wood	6	1.84
C. capsularis fiber lignin (enzymatic)	6	2.00
C. capsularis wood lignin (enzymatic)	6	2.06
Wood meals		
Norway spruce	40	1.05
Norway spruce (ozon. in gl. acetic acid)	40	1.03
Tsuga heterophylla	30	1.02
Tsuga heterophylla	40	1.27
Tsuga heterophylla	40	1.12
Tsuga heterophylla	60	1.01
Tsuga heterophylla (compression wood)	40	1.36
Tsuga heterophylla (compression wood)	40	1.11
Pseudotsuga menziesii	40	1.01
Pseudotsuga meziesii (compression wood)	40	1.39
Populus deltoides	20	2.72
Populus deltoides	40	3.03
Populus deltoides	60	2.42
Arbutus menziesii tension sapwood	40	3.69
Liquidambar styraciflua	40	2.86
Eucalyptus gummifera	40	2.60
E. grandis	40	2.45
Corchorus capsularis fiber	40	2.41
C. capsularis wood	40	2.02
Grasses		
Bagasse	40	1.92
Wheat straw	40	1.88
Wheat straw (saponified)	40	1.94
Pulps		
Soda pulp (cottonwood)	40	3.05
Bagasse pulp (1st stage delignif.)	40	1.94
Wheat straw pulp (1st stage delignif.)	40	2.25

about $0.020/C_6C_3$ in the outermost annual rings to about 0.012 in interior rings that were 20 to 40 years older.

Ozonation of the phenylcoumaran-type model (13) (Fig. 6.7.4) or the diarylpropane-type model (15) (Habu et al. 1990) yielded 2-hydroxy-3-hydroxymethylbutanedioic acid (14). The conversion of erythro (15) was highly

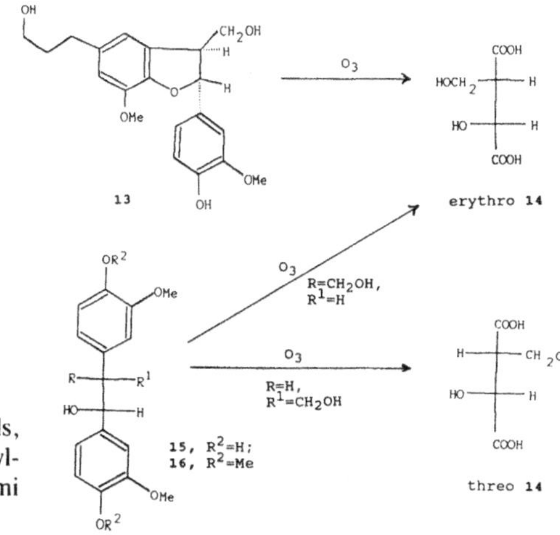

Fig. 6.7.4. β-5 and β-1 lignin models, ozonation to 2-hydroxy-3-hydroxymethyl-butanedioic acid (Habu et al. 1990, Tsutsumi et al. 1990)

stereospecific, giving nearly pure erythro (14), but that of erythro (13) was less so, affording 89% erythro (14) and 11% threo. The absolute yields of (14) were not stated, but (15) gave a yield 1.6 times higher than did (13). Compound (16), the veratryl analog of (15), gave a mixture of the two isomers of (14) in a combined yield of about 35% (Tsutsumi et al. 1990). Habu et al. (1990) used the yield of isomers of (14) from spruce MWL relative to the yield of tetronic acids in an argument based on the results of earlier analyses for β-O-4 and phenylcoumaran contents to show that there could be no more than 0.02 diarylpropane unit/monomer unit. This value is appreciably below previous estimates but in agreement with the content of glyceraldehyde-2-aryl ether units, which mechanistically are expected to be formed simultaneously in equal amounts.

Habu et al. (1990) also made further use of the ratio of threo to erythro (14). Since phenylcoumaran-type units exist only in the trans (erythro) form [see structure (13)], and since diarylpropane stereoisomers are known to be formed in more nearly identical amounts, any increase in the ratio of threo (14) to erythro (14) serves as an indication of an increased diarylpropane content in the lignin. Accordingly, mild hydrolysis of wood meal with dioxane and water at 180 °C gave a soluble lignin fraction having a much higher threo (14) to erythro (14) ratio relative to that found for the corresponding MWL. This explains why earlier diarylpropane frequencies, observed with isolated lignins, were too high.

Ozonation of the α-aryl structures found in Klason lignins should, by analogy, give the malonic acid (17) (Fig. 6.7.5), as suggested by Matsumoto et al. (1986). Using an ion exchange resin to separate the ozonation products into mono- and dicarboxylic fractions, they found that Klason lignin from yezo spruce (*Picea jezoensis*) gave an increased yield of a dicarboxylic acid fraction relative to the corresponding MWL (70% compared with 40%). No product derived from side chains was identified in this dicarboxylic acid fraction, but the Klason lignin itself gave a high yield of oxalic acid, as did also the MWL. No oxalic acid was detected, however, when a commercial thiolignin was ozonated. The authors noted that this observation was consistent with the fact that the

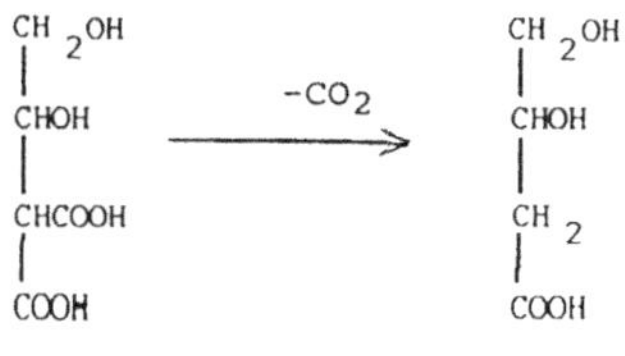

17 18

Fig. 6.7.5. Ozonation products from α-aryl condensed structures

1-alkylated and 1,6-dialkylated aromatic rings of uncondensed and condensed lignin units can be cleaved by ozone to produce two molecules of oxalic acid per ring, whereas the 1,5-dialkylated rings common in thiolignins yield only one molecule of oxalic acid per ring. Tsutsumi et al. (1990) identified 3,4-dihydroxybutanoic acid (18), which could be a decarboxylation product of malonic acid (17), among lignin ozonation products. On ozonation of hemlock and cottonwood Klason lignins, acid (18) was obtained in substantially increased yield relative to tetronic acids (10) and (11), thus suggesting that it was derived from the side chain of α-aryl units. The malonic acid (17) may be decarboxylated during ozonation or saponification.

The identification of pinoresinol units in lignins by ozonation has been studied by Matsumoto et al. (1989). They found that peaks at 82 and 83 ppm are unique in the C-13 NMR spectrum of the ozonation product mixture from pinoresinol (19). These peaks could not be detected in the ozonation products from beech, spruce, or eucalyptus MWL, although they were prominent at spiking levels as low as 0.03 unit of pinoresinol ozonation product per C_6C_3 unit of lignin substrate.

Matsumoto et al. (1986) found glyceric acid as a main ozonation product from a thiolignin and a Klason lignin, where its yield was much greater than that obtained with a MWL and a soda lignin. The greater yield of glyceric acid from the frequently dialkylated units of a thiolignin or Klason lignin, and its formation in only about 0.3% yield from vanillin (Tsutsumi et al. 1990), indicate its side chain origin. Tsutsumi et al. (1990) found that glyceric acid was one of many products obtained on ozonation of the tetronic acids, and that it was formed in about 14% yield after ozonation of threo-veratrylglycerol.

6.7.6 Conclusion

Ozonation has been demonstrated to be a superior method of oxidative degradation for lignins, and it offers special promise for studying in situ lignins. Refinements in the experimental procedure and more sophisticated correlation of known and yet-to-be-identified products with substrate structure can be expected.

References

Adler E (1977) Lignin chemistry – past, present, and future. Wood Sci Technol 11:169–218

Adler E, Delin S, Miksche GE (1966) Synthese des Isoeugenolglycol-β(2-methoxyphenyl)-äthers. Acta Chem Scand 20:1035–1037

Alen R, Niemelä K, Sjöström E (1984) Gas-liquid chromatographic separation of hydroxy monocarboxylic acids and dicarboxylic acids on a fused-silica capillary column. J Chromatogr 301:273–276

Aulin-Erdtman G, Tomita Y (1963) Studies on the degradation of lignin and model compounds. I. The configuration of dehydrodiisoeugenol. Acta Chem Scand 17:535–536

Bailey PS (1978) Ozonation in organic chemistry. I. Olefinic compounds. Academic Press, New York, 272 pp

Balousek PJ, McDonough TJ, McKelvey RD, Johnson DC (1981) The effects of ozone upon a lignin model containing the β-aryl ether linkage. Sven Papperstidn 84:R49–R54

Bes RS, Gas G, Molinier J, Vidal P, Mathieu J, Mora JC (1989) Enhancement of poplar cellulose susceptibility to cellulase enzyme hydrolysis by ozonation. Ozone Sci Eng 11: 217–226

Bonnet MC, de Laat J, Dore M (1989) Etude des sous-produits d'ozonation de la lignine et de glucides en milieu aqueux. Environ Technol Lett 10:577–590

Chan TH, Brook MA, Chaly T (1983) A simple procedure for the acetalization of carbonyl compounds. Synthesis 203–205

Chapman JR (1985) Practical organic mass spectrometry. John Wiley, Chichester, 197 pp

Criegee R (1975) Mechanism of ozonolysis. Angew Chem Int Ed Engl 14:745–751

Deslongchamps P, Atlani P, Frehel D, Malaval A, Moreau C (1974) The oxidation of acetals by ozone. Can J Chem 52:3651–3664

Dobinson F (1959) Ozonization of malonic acid in aqueous solution. Chem Ind 853–854

Doree C, Cunningham M (1913) The action of ozone on cellulose. III. Action on beech wood (lignocellulose). J Chem Soc 103:677–686

Dorland RM, Hawkins WL, Hibbert H (1939) Studies on lignin and related compounds. XLVI. The action of ozone on isolated lignins. J Am Chem Soc 61:2698–2701

Eriksson T, Gierer J (1985) Studies on the ozonation of structural elements in residual kraft lignins. J Wood Chem Technol 5:53–84

Eriksson T, Gierer J (1987) Ozonation of residual lignin. Proc Fourth Int Symp Wood Pulp Chem, Paris, April 27–30. Vol 1:293–294

Freudenberg K, Sohns F, Janson A (1935) Weitere Untersuchung des Lignins. 14. Mitteilung über Lignin. Justus Liebigs Ann Chem 518:62–85

Gierer J, Noren I (1962) Zur Spaltung von β-Hydroxyalkylaryläthern durch Alkali. II. Der stereochemische Verlauf. Acta Chem Scand 16:1976–1988

Godsay NP, Pearce EM (1984) Physico-chemical properties of ozone-oxidized kraft pulps. TAPPI Oxygen Delignification Symposium, San Francisco, Nov. 15–16. 55–69

Goodman SI, Markey SP (1981) Diagnosis of organic acidemias by gas chromatography-mass spectrometry. Alan R. Liss, New York, 158 pp

Grob RL (ed) (1985) Modern practice of gas chromatography. Wiley, New York, 897 pp

Habu H, Matsumoto Y, Ishizu A, Nakano J (1990) The role of the diarylpropane structure as a minor constituent in spruce lignin. Holzforschung 44:67–71

Haluk JP, Metche M (1986) Caractérisation chimique et spectrographique de la lignine de peuplier par acidolyse et ozonolyse. Cellul Chem Technol 20:31–50

Hayashi N, Shimizu K, Hosoya S (1989) Pretreatment of ozone for increasing the enzymic susceptibility of autohydrolyzed softwoods. Mokuzai Gakkaishi 35:521–529

Hyppanen T, Sjöström E, Vuorinen T (1983) Gas-liquid chromatographic determination of hydroxycarboxylic acids on a fused-silica capillary column. J Chromatogr 261:320–323

Kaneko H, Hosoya S, Iiyama K, Nakano J (1983) Degradation of lignin with ozone – reactivity of lignin model compounds toward ozone. J Wood Chem Technol 3:399–411

Kolsaker P, Bailey PS (1967) Ozonation of compounds of the type Ar—CH═CH—G; ozonation in methanol. Acta Chem Scand 21:537–546

Kondo T, Meshitsuka G, Ishizu A, Nakano J (1987) Preparation and ozonation of completely allylated and methallylated lignins. Mokuzai Gakkaishi 33:724–727

Kratzl K, Claus P, Reichel G (1976) Reactions of lignin and lignin model compounds with ozone. TAPPI 59(11):86–87

Matsumoto Y, Habu N, Minami K, Ishizu A, Nakano J (1989) Quantitative estimation of side-chain structure by ozonation. Proc Fifth Int Symp Wood Pulp Chem, North Carolina State University, Raleigh, NC, 365–368

Matsumoto Y, Ishizu A, Nakano J (1984a) Determination of glyceraldehyde-2-aryl ether type structure in lignin by the use of ozonolysis. Mokuzai Gakkaishi 30:74–78

Matsumoto Y, Ishizu A, Nakano J, Terasawa K (1984b) Residual sugars in Klason lignin. J Wood Chem Technol 4:321–330

Matsumoto Y, Ishizu A, Nakano J (1986) Studies on chemical structure of lignin by ozonation. Holzforschung 40:Suppl 81–85

Miksche GE (1972) Zum alkalischen Abbau der p-Alkoxyarylglycerin-β-arylätherstrukturen des Lignins. Versuche mit erythro-Veratrylglycerin-β-guajacyläther. Acta Chem Scand 26: 3275–3281

Miura M, Fujisaka T, Nojima M, Kusabayashi S, McCullough KJ (1985) Ozonolysis of 1-methylindenes. Solvent, temperature and substituent electronic effects on the ozonide exo/endo ratio. J Org Chem 50:1504–1509

Niemelä K (1988) GLC-MS studies on pine kraft black liquors. Part I. Identification of monomeric compounds. Holzforschung 42:169–173

Niemelä K, Sjöström E (1986) Simultaneous identification of aromatic and aliphatic low molecular weight compounds from alkali pulp liquor by capillary gas-liquid chromatography-mass spectrometry. Holzforschung 40:361–368

Phillips M, Goss MJ (1933) Chemistry of lignin VIII. The oxidation of alkali lignin. J Am Chem Soc 55:3466–3470

Sawada T, Motoi M, Nakamura Y (1989) Formation of organic acids from lignin by ozonolysis and biological degradation. AIChE For Prod Div Sess 1988:143–147

Soteland N (1971) Some attempts to characterize the oxidized lignin after ozone treatment of western hemlock groundwood. Part II. Nor Skogind 25(5):135–139

Tanahashi M, Nakatsubo F, Higuchi T (1975) Structural elucidation of bamboo lignin by acidolysis and ozonolysis I. Wood Res 58:1–11

Taneda H, Habu N, Nakano J (1989) Characterization of the side chain steric structures in the various lignins. Holzforschung 43:187–190

Tishchenko DV (1959) (Ozonation of lignin) Zh Priklad Khim 32:686–690

Tomita B, Kurozumi K, Takemura A, Hosoya S (1989) Ozonized lignin-epoxy resins, synthesis and use. In: Glasser WG, Sarkanen S (eds) Lignin, properties and materials. American Chemical Society, Washington, DC, 496–505

Tsutsumi Y, Islam A, Anderson CD, Sarkanen KV (1990) Acidic permanganate oxidations of lignin and model compounds: comparison with ozonolysis. Holzforschung 44:59–66

Wood BG (1923) Preparation of synthetic vanillin by the ozone process. Chem Met Eng 28: 806–808

Chapter 7
Functional Group Analysis

7.1 Determination of Total and Aliphatic Hydroxyl Groups

C.-L. Chen

7.1.1 Introduction

As discussed in Chapter 1.1, lignins are produced in vivo by an enzyme-initiated dehydrogenative polymerization of three monomeric precursors, p-hydroxycinnamyl alcohol (1), coniferyl alcohol (2) and sinapyl alcohol (3). These compounds all have a terminal aliphatic hydroxyl group at C-γ on the side chain in addition to a phenolic hydroxyl group at C-4 of the aromatic ring. In contrast to the phenolic hydroxyl group, approximately only 20–25 mol % of the aliphatic hydroxyl groups of the monomeric precursors is involved in the lignification process based on results from functional group analyses of lignins and investigations on the in vitro dehydrogenative polymerization of lignin (Freudenberg 1965, Freudenberg and Neish 1968, Lai and Sarkanen 1971, Adler 1977). This statement implies that lignins in wood tissue should contain a minimum total aliphatic hydroxyl content of approximately 0.75/C_9 unit. Indeed, as shown by the values in Table 7.1.1, milled wood and bamboo lignins (MWL and MBL) have total aliphatic hydroxyl contents of more than 1.1 mol/C_9 unit in contrast to total phenolic hydroxyl contents which are in the 0.2–0.4 mol/C_9 unit range (Freudenberg et al. 1964, Robert and Brunow 1984, Pan et al. 1990, Tai et al. 1990). Moreover, the total phenolic hydroxyl content of lignins in wood is usually less than 0.1 mol/C_9 unit (Yang and Goring 1980, Winston et al. 1986). Thus, it is evident that the role of the aliphatic hydroxyl group at C-γ is considerably less significant than that of the phenolic hydroxyl group in the dehydrogenative polymerization of the lignin precursors in the in vivo lignification process.

When coniferyl alcohol (2) undergoes abstraction of one hydrogen atom from the phenolic hydroxyl group by a one-electron transfer agent such as a peroxidase, a phenoxyl radical (R_a) is produced. As shown in Fig. 7.1.1, this radical is in resonance with radicals (R_b), (R_c) and (R_d). Coupling of these radicals results in the formation of dimeric quinonemethide intermediates; e.g., quinonemethide intermediate (4) is produced by the coupling of radicals (R_a) and (R_d). Nucleophilic addition of a mole of water or of the phenolic oxygen atom in precursor (2) to C-α of the intermediate (4) leads to the formation of β-O-4 substructures having an α-hydroxyl or α-aroxyl group, compound (5) or (6), respectively. Alternatively, addition of radical (R_a) to C-α of the intermediate (4) produces the trimeric radical (7) (Chen and Connors 1974). The fact that MWLs, MBLs and milled grass lignins (MGLs) have an aliphatic hydroxyl content of more than 1 mol/C_9 is significant not only in terms of the chemical

Springer Series in Wood Science
Methods in Lignin Chemistry
(Edited by S.Y. Lin and C.W. Dence)

Table 7.1.1. Total, phenolic and aliphatic hydroxyl contents of some representative milled wood and bamboo lignins (MWL and MBL) and technical lignins based on C_9 unit

Lignin preparation	C_9 Unit formula	Total OH	Phenolic OH	Aliphatic OH
		(mol/C_9 unit)		
Milled wood lignins (MWLs)[a]				
Spruce				
(*Picea glauca*)	$C_9H_{7.66}O_2(H_2O)_{0.48}(OCH_3)_{0.94}$	1.46	0.28	1.18
Zhong-Yang Mu				
(*Bischofia polycarpa*)	$C_9H_{6.49}O_2(H_2O)_{0.92}(OCH_3)_{1.13}$	1.33	0.22	1.11
Milled bamboo lignin (MBL)[a]				
Chinese bamboo				
(*Phyllostachys pubescens*)	$C_9H_{5.52}O_2(H_2O)_{0.84}(OCH_3)_{1.25}$	1.49	0.36	1.13
Willstätter (HCl) lignin				
Sweetgum				
(*Liquidambar styraciflua*)	$C_9H_{4.86}O_2(H_2O)_{1.15}(OCH_3)_{1.60}$	1.51	0.12	1.39
Technical lignins[b]				
Pine kraft lignin	$C_9H_{7.42}O_2(H_2O)_{0.28}S_{0.08}(OCH_3)_{0.77}$	1.35	0.58	0.77
Bamboo kraft lignin	$C_9H_{5.52}O_2(H_2O)_{1.60}S_{0.08}(OCH_3)_{1.00}$	1.00	0.44	0.56

[a] All data are corrected for carbohydrate and protein contents.

[b] These lignin preparations do not have a C_9 structure and are presented only for the purpose of comparison.

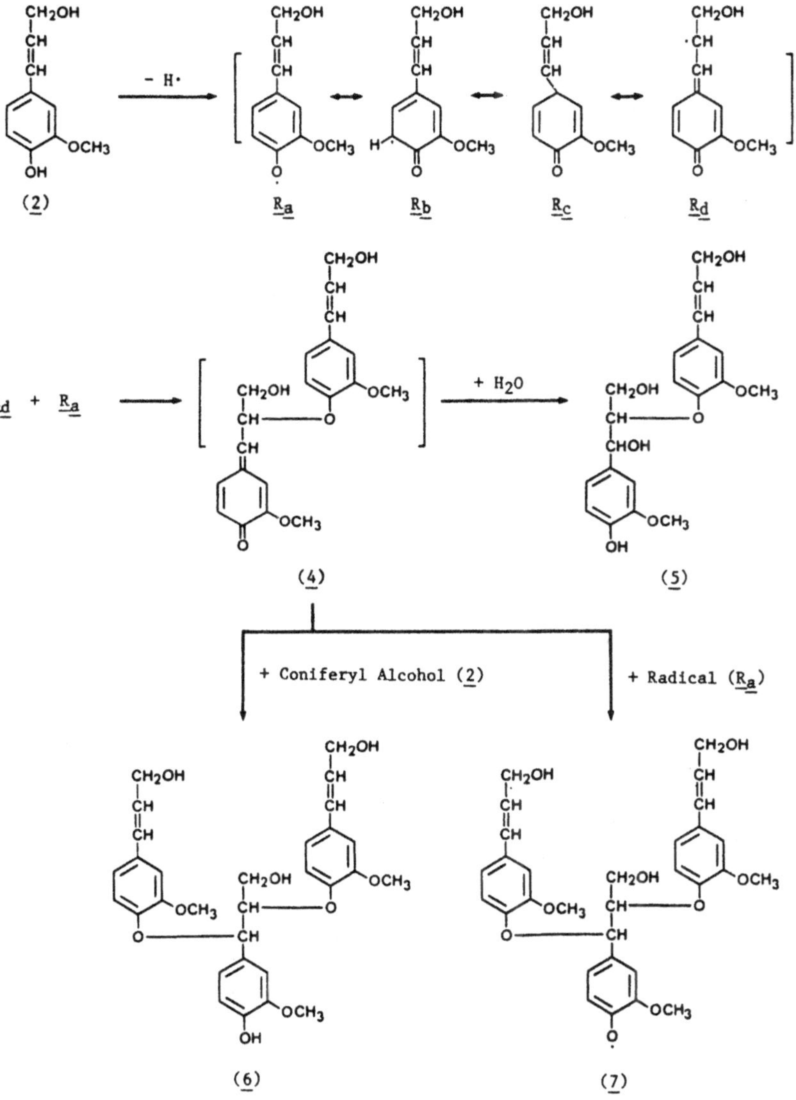

Fig. 7.1.1. Reaction sequences for the formation of α-hydroxyl and α-aroxyl groups in lignins

structure and biogenesis of lignins, but also with regard to changes occurring in the structure of lignins during the isolation process. For example, except for technical lignins, lignin preparations having an aliphatic hydroxyl content exceeding 1 mol/C_9 unit could also arise from acid-catalyzed hydrolysis and from homolytic cleavage of α-aryl ether bonds induced by ball-milling during the isolation of lignin from wood in addition to arising from the nucleophilic addition of water to C-α of a quinonemethide intermediate during the in vivo lignification process.

The contents of p-hydroxybenzyl alcohol-type hydroxyl and noncyclic α-aryl ether groups in Norway spruce (*Picea abies*) MWL, such as are present in structures (5) and (6) in Fig. 7.1.1, have been estimated to be approximately 0.06 and 0.02 mol/OCH_3, respectively (Freudenberg et al. 1964, Adler et al. 1966). Substructures of these types have primary roles in sulfidation reactions in kraft pulping and in sulfonation reactions in sulfite pulping (Gierer 1982a). These substructures are also of fundamental importance in bleaching processes. For example, they are directly or indirectly involved in side chain displacement reactions, some of them by way of quinonemethide intermediates, when pulp is delignified and/or bleached with chlorine, oxygen, and hydrogen peroxide (Gierer 1982b).

In general, the aliphatic hydroxyl content of a lignin preparation is determined by subtraction of the phenolic hydroxyl content (see Chap. 7.2) from the total hydroxyl content of the lignin. The total hydroxyl content of lignin (partial structure 11) can be determined by acetylation with acetic anhydride-pyridine reagent (10) followed by saponification of the acetate (partial structure 12) and titration of the resulting acetic acid (14) with a standard 0.05 N sodium hydroxide solution (Roth 1953).

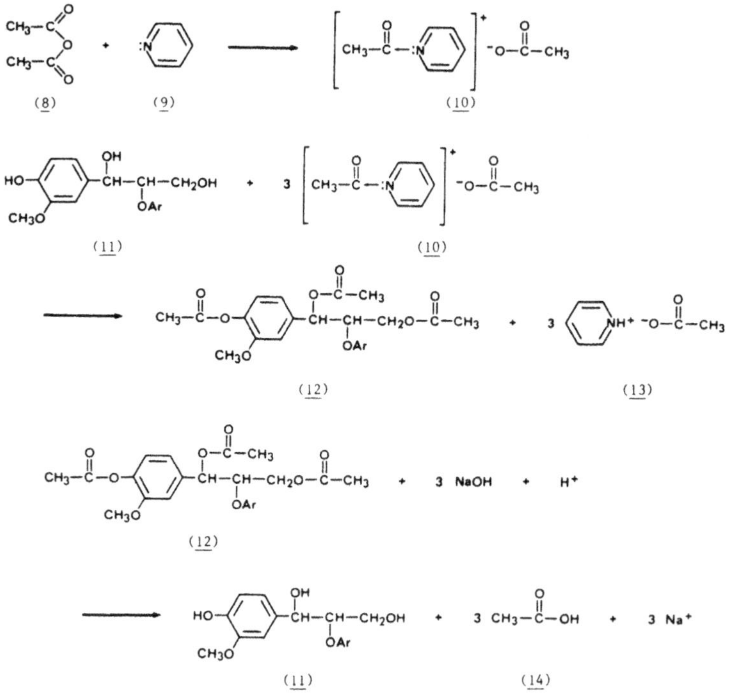

Since one O-acetyl group in the acetylated lignin (partial structure 12) produces 1 mol of acetic acid (14) on saponification, the titrimetric determination of acetic acid (14) in the resulting hydrolyzate provides a measure of the total O-acetyl content of the acetylated lignin (partial structure 12). This value, together with the results of elemental and methoxyl analyses, eventually permits calculation of the total hydroxyl content of the lignin preparation, since each hydroxyl group in the lignin is converted to one O-acetyl group. However, as some lignin preparations contain carboxylic acid groups, the direct titration of acetic acid in the hydrolyzate of such materials gives rise to a falsely high total hydroxyl content. Errors of this nature may be avoided by distilling the acetic acid (Kuhn and Roth 1933) from the hydrolyzate prior to titration with a standard alkali solution.

Alternatively, the acetic acid (14) in the hydrolyzate may be converted by treatment with tetrabutylammonium hydroxide (15) solution to tetrabutylammonium acetate (16). The latter is then transformed to benzyl acetate (18) by treatment with benzyl bromide (17):

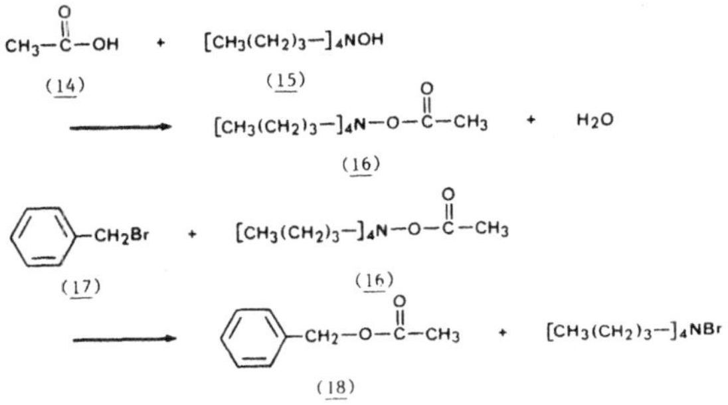

The benzyl acetate (18) is quantitatively determined by capillary column gas chromatography (GC). Benzyl butanoate (20), formed from butanoic acid (19) concurrently with the formation of benzyl acetate (18), serves as the internal standard (Månsson 1983). Since 1 mol of acetic acid (14) is converted to 1 mol of benzyl acetate (18), the total hydroxyl content of the lignin preparation can be determined by quantitative determination of the benzyl acetate (18). The Månsson procedure is a modification of the Bethge-Lindström method for determinating O-acetyl groups in acetylated carbohydrates (Bethge and Lindström 1973).

The presence of carbohydrates in lignin preparations significantly affects the accuracy in the determination of the total and aliphatic hydroxyl contents. Thus, it is necessary to remove such impurities from lignin preparations as completely as possible prior to determinating their total, aliphatic and phenolic hydroxyl, contents. Therefore, prior to performing hydroxyl group analysis, the carbohydrate content of the lignin sample must be determined (Note 1), so that the total and aliphatic hydroxyl contents of the lignin can be reported on a carbohydrate-free basis.

Note 1. See Chaps. 3 and 7.6.2 for lignin sample preparation.

7.1.2 Determination of Total Hydroxyl Groups

7.1.2.1 Acetylation of the Lignin Preparation

Reagents

Acetic anhydride, freshly distilled, b.p. 138–140 °C
Acetone, freshly distilled, b.p. 55–56 °C

Dichloromethane, b.p. 39–40 °C
Ethanol, absolute, b.p. 78–79 °C
Hydrochloric acid, 2 M in water (Note 2)
Methanol, b.p. 64–65 °C
Nitrogen, from a cylinder equipped with a pressure regulator and needle valve
Phosphorus pentoxide, 98%, powder
Pyridine, freshly distilled, b.p. 113–115 °C
Toluene, freshly distilled, b.p. 110–111 °C

Note 2. Preferably double-distilled or completely deionized water

Procedure

In a 50-ml beaker is placed 100 mg of the purified lignin preparation (Note 1) and 2 ml of acetic anhydride-pyridine reagent (1:1, v/v). The mixture is shaken gently until the lignin is dissolved. The beaker is covered with a watch glass and kept at room temperature for 48 h under a nitrogen atmosphere. To the reaction mixture is added 10 ml of ice-cold methanol-dichloromethane (1:8, v/v) and the mixture is allowed to stand for about 30 min with occasional stirring. The resulting solution is transferred to a 25-ml separatory funnel and washed several times with 5 ml of 2 M hydrochloric acid until free of pyridine. The solution is washed twice with 5 ml of water (Note 2), filtered, and the solvent is removed under reduced pressure. The resulting acetylated lignin is dried in a drying pistol at 50 °C under vacuum over phosphorus pentoxide. Yield, ~120–130 mg.

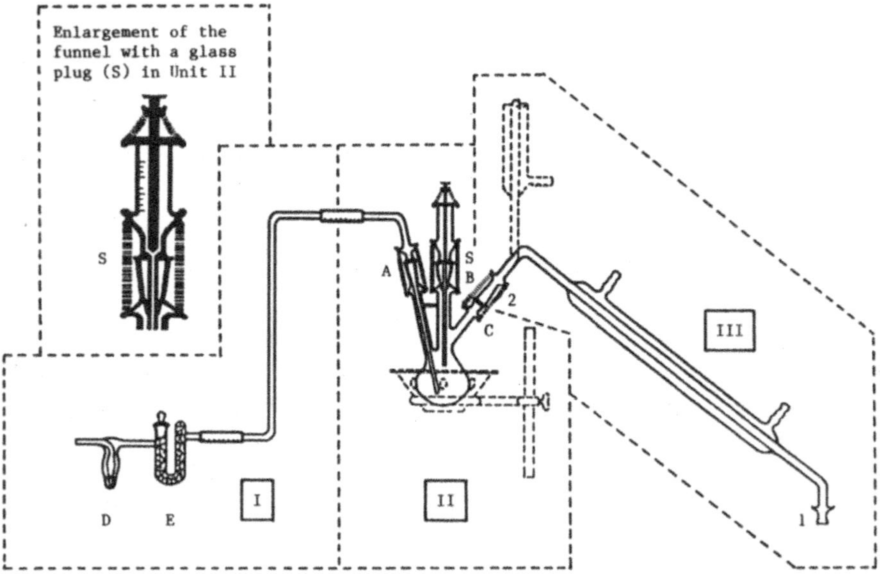

Fig. 7.1.2. The Kuhu-Roth apparatus

Alternatively, 100 ml to toluene is added to the reaction mixture, which is transferred to a 250-ml round-bottom flask. The acetic anhydride, acetic acid, and pyridine are removed as an azeotropic mixture at 40 °C by rotary evaporation (Gierer and Lindeberg 1980, Chum et al. 1985). A small amount of acetone is added to the solution to stabilize the acetylated lignin during removal of the toluene. If necessary, this operation may be repeated to insure complete removal of the acetylation reagents. The residual toluene is in turn similarly removed azeotropically from the mixture by adding 70 ml of ethanol-acetone (1 : 1, v/v). The product is dried in a drying pistol at 50 °C under vacuum. Yield, ~130 mg (Note 3).

Note 3. A blank test indicates that the product contains approximately 5 mg of nonlignin-related low-molecular weight (200–300 amu) impurities. These impurities are not detectable by ^{13}C NMR spectroscopy.

7.1.2.2 Determination of Total O-Acetyl Groups

7.1.2.2.1 The Kuhn-Roth Procedure (Kuhn and Roth 1933, Roth 1953)

Reagents

Barium chloride, crystals
Methanol, refluxed over solid potassium hydroxide (pellet form) for 15 min to remove residual acids and distilled, b.p. 64–65 °C
Metaphosphoric acid, powder
Potassium hydroxide solution, 50% in water
Sodium hydroxide solution, 1 M in methanol-water: dissolve 4 g of sodium hydroxide (pellet form) in 50 ml of water (Note 2) and 50 ml of methanol
Soda-lime pellets
Sulfuric acid, 6.5 M: add 100 ml of concentrated sulfuric acid to 200 ml of cold water (Note 2)
Phenolphthalein solution, 1%: dissolve 1 g of phenolphthalein in a mixture of 90 ml of ethanol and 10 ml of water
Standard sodium hydroxide solution, 0.01 N in water (Note 2)

Apparatus

The Kuhn-Roth apparatus for the determination of O-acetyl groups is shown in Fig. 7.1.2. The apparatus consists of a gas flow regulator (I), a reaction vessel (II), and a condenser (III). The bubble counter (D) in unit I is filled to an appropriate level with 50% potassium hydroxide solution, and the U-tube (E) with soda-lime. The reaction flask in unit II is made of Pyrex glass and has a capacity of approximately 45 ml. It is fitted with a 2-mm bore inlet tube inserted through neck A (80 mm × 6 mm i.d.) and extending to the bottom of the flask.

The ground joint of neck B (80 mm × 6 mm i.d.) is fitted with a funnel reaching just below the juncture with the flask. The funnel has an interior ground joint into which a glass plug (S) is fitted. The funnel has a capacity of 8 ml when the plug is in place and the 2 ml and 7 ml levels are marked. Neck C (65 mm × 5 mm i.d.) is attached to the reaction flask at an angle of 50° to the horizontal, and is connected to a condenser (360 mm overall length) in unit III through a vacuum-tight ground joint. As seen from Fig. 7.1.2, the condenser can be fitted to the reaction flask in two positions, one for a refluxing mode and the other for a distillation mode.

Prior to each determination, all glass parts of unit II should be carefully cleaned, rinsed with deionized water, and dried. The condenser in unit II is similarly cleaned; drying, however, is not necessary in this case.

Procedure

Nitrogen gas is passed through the apparatus via the bubble counter (Fig. 7.1.2) at a flow rate of approximately 50 bubbles/min. The condenser is tightly connected to the neck, C, of the reaction flask in the reflux position using a drop of water as a lubricant to seal the joint. In the reaction flask is placed 4 ml of 1 M sodium hydroxide solution in methanol-water. Approximately 10 mg of the dried, purified lignin sample is accurately weighed to the nearest 0.01 mg in a micro weighing bottle having an o.d. of 5 mm (Note 4; see also Chap. 7.6.3 and Fig. 7.6.1) and immediately dropped into the reaction flask. The gas inlet and funnel joints are wetted with metaphosphoric acid and firmly set in place. To the stoppered funnel is added 1 to 2 ml of water (Note 2). The solution in the reaction flask is gently refluxed for 1 h in a water bath and then cooled. While in place, the condenser is rinsed with 4 to 6 ml of water (Note 2) from a wash bottle after first removing the glass stopper from the funnel. The stopper is replaced in the funnel. The condenser is then removed, washed with 100 to 200 ml water, and firmly connected to the flask in the distilling position. About 5 ml of the reaction mixture, consisting mostly of methanol, is distilled, and the condenser is rinsed again with 100 to 200 ml of water and reconnected to the flask.

To the funnel is added 1 ml of 6.5 M sulfuric acid from a pipet. The acid is allowed to flow into the flask by carefully lifting the stopper; the funnel is rinsed with 2 to 3 ml of water (Note 2). The stopper is reinserted and water (Note 2) is added to the funnel up to the 7-ml mark. After adding a few boiling chips, the reaction mixture is distilled at a rate of 5 to 6 ml/5 min. The distillate is collected in a 75-ml Erlenmeyer flask. When the volume of the residue in the flask approaches 2 to 3 ml, without interrupting the distillation, the stopper of the funnel is lifted to allow the water to flow into the flask until the water level reaches the 2-ml mark on the funnel. The addition of 5 ml of water is repeated until the distillation is complete. When 20 ml of the first distillate is collected (Note 5), any distillate adhering to the condenser is removed by contacting the condenser tip with the Erlenmeyer flask. At the same time, another 50-ml Erlenmeyer flask is placed under the end of the condenser.

While the distillation is continuing, two or three crystals of barium chloride are added to the distillate in the first Erlenmeyer flask. This distillate is then boiled gently for several seconds to remove carbon dioxide as well as to ascertain if any turbidity appears (Note 6). If the distillate remains clear, it is titrated with standard 0.01 N sodium hydroxide solution after addition of four to five drops of the phenolphthalein solution (Note 7). All further distillates are treated as described above. If less than 4 ml of the standard alkali solution are required for the titration of the first distillate, the next 10 ml of distillate is similarly titrated. If the acetic acid in the initial distillate consumes more than 4 ml of standard alkali then three additional 5-ml distillates are collected and individually titrated as described above. When the indicator changes color upon the addition of the first drop of the alkaline solution, the analysis is complete.

Note 4. If an appropriate micro weighing bottle is not available, a small combustion boat may be used.

Note 5. The distillation is not complete at this point.

Note 6. Turbidity is caused by sulfuric acid and/or sulfonic acid carried over by entrainment. The presence of these acids introduces an experimental error. If turbidity occurs, the determination should be repeated.

Note 7. A calibrated 10-ml or 25-ml micro-buret is used.

Calculation

Since 1 ml of 1 N sodium hydroxide solution corresponds to 43.044 mg of O-acetyl, the total O-acetyl content of the sample is calculated as follows:

$$\%\ \text{Acetyl},\ -\overset{\overset{\displaystyle O}{\|}}{C}-CH_3,\ \text{(based on acetylated lignin)} = \frac{VN \times 4304.4}{W}, \qquad (1)$$

where V is the combined total volume in ml of standard sodium hydroxide solution required for titration, N is the normality of the sodium hydroxide solution, and W is the weight of the moisture-free sample in mg.

7.1.2.2.2 The Modified Bethge-Lindström Procedure (Månsson 1983)

Reagents

Acetone, b.p. 55–56 °C
Benzyl bromide, b.p. 198–199 °C
Butanoic acid solution, 0.05 M in water (Note 2)
Dioxane, refluxed over lithium aluminum hydride for 1 h (to decompose peroxides) and distilled, b.p. 100–102 °C
Methanol, refluxed over solid potassium hydroxide (pellet form) for 15 min to remove residual acids and distilled, b.p. 64–65 °C

Sodium methoxide solution, 1 M in methanol
Tetrabutylammonium hydroxide solution, 0.15 M in water (Note 2)

Procedure

Approximately 20 mg of a dried, purified acetylated lignin preparation is accurately weighed to the nearest 0.1 mg. The sample is placed in a 5-ml GC mini-vial together with 0.5 ml of dioxane and 0.5 ml of 1 M sodium methoxide solution in methanol. The mixture is shaken in an ultrasonic bath for 5 min and kept at room temperature for 30 min. To the mixture is added 2 ml of 0.05 M butanoic acid (0.1 mmol) (Note 8) which is converted to benzyl butanoate, the internal standard, in the subsequent GC analysis. The mixture is kept at room temperature for another 30 min and then acidified by passing it through an ion exchange column (80 mm × 6 mm o.d.) packed with 2 ml of wet Dowex 50 WX2-100 resin and prewashed with approximately 15 ml of a solution consisting of dioxane-methanol-water (1:1:4, v/v/v) (Note 9). The acidic constituents of the mixture are eluted from the column with about 20 ml of the foregoing solution. The collected effluent is concentrated to a syrup in a rotary evaporator under reduced pressure. This residue is dissolved in 1 ml of acetone and transferred to a 5-ml GC mini-vial to which is added approximately 40 μl of benzyl bromide. The reaction mixture is allowed to stand for 20 min at room temperature; acetone is then added until the total volume of the mixture is 5 ml. Approximately 0.5 μl of the resulting solution is injected into a gas chromatograph system equipped with a DB-5 bonded phase, fused silica capillary column (30 m × 0.32 mm i.d.; film thickness 0.25 μm) and a flame ionization detector. The GC conditions are as follows: carrier gas, helium (column flow rate, 2 ml/min); injection port temperature, 200 °C; temperature profile, hold for 1 min at 60 °C, 60–140 °C (15 °C/min), 140–250 °C (5 °C/min), final temperature 250 °C (10 min), post value 260 °C (10 min); detector temperature, 270 °C.

Note 8. The volume of 0.05 M butanoic acid must be adjusted so that the area ratio of the GC peaks corresponding to benzyl acetate to benzyl butanoate is in the 0.5–2 range. Compounds other than butanoic acid, e.g., hexanoic acid, can be used as internal standards.

Note 9. If at the end of this washing the eluent is not colorless, washing is continued until the eluent becomes colorless.

Calculation

Since 1 mol of acetic acid or butanoic acid reacts with 1 mol of benzyl bromide to produce 1 mol of benzyl acetate or benzyl butanoate, respectively, the O-acetyl content of the sample is calculated as follows:

$$\% \text{ Acetyl}, \ -\overset{\overset{\displaystyle O}{\|}}{C}CH_3, \text{ (based on acetylated lignin)} = \frac{4304.4 \times CIA_s}{WA_i}, \qquad (2)$$

where A_s and A_i are the areas of the GC peaks for benzyl acetate and benzyl butanoate (a derivative of the internal standard), I is the quantity of the internal standard (butanoic acid) in mmoles, C is the conversion factor (1.28) for benzyl acetate relative to benzyl butanoate (Note 10), and W is the sample weight in mg.

Note 10. The conversion factor C includes extraction and derivatization efficiencies and the response factor for the flame ionization detector. The value of C changes when a different GC system is used for the determination and when another compound is used as the internal standard.

7.1.2.3 Total Hydroxyl Content

Since the hydroxyl groups are converted to O-acetyl groups by acetylation on a one-to-one molar basis, the total hydroxyl content of the lignin is equal to the total O-acetyl content of the acetylated lignin. Provided the methoxyl content of the acetylated lignin is known (see Chap. 7.6.3), the total hydroxyl content of the lignin expressed in terms of methoxyl content can be calculated:

$$\text{Total } -OH/OCH_3 \text{ (moles/mole, based on lignin)}$$

$$= \text{Total } -\overset{\overset{\displaystyle O}{\|}}{C}CH_3/OCH_3 \text{ (Moles/mole, based on acetylated lignin)}$$

$$= \frac{31.034 \times (\% \text{ Total } -\overset{\overset{\displaystyle O}{\|}}{C}CH_3)}{43.044 \times (\% \ OCH_3)}. \qquad (3)$$

The total hydroxyl content of the lignin may be also expressed in terms of a C_9 unit, OH/C_9 (mol/C_9) as follows:

$$\text{Total } OH/C_9 (mol/C_9 \text{ unit}) = (\text{total } OH/OCH_3) \times Y, \qquad (4)$$

where C_9 is the C_9 building unit of the lignin and Y is the methoxyl content of the lignin expressed in mol/C_9 unit (Note 11).

The total hydroxyl content of the lignin/C_9 unit may also be calculated directly from the elemental composition of the acetylated lignin, i.e., % C, % H and the total O-acetyl and methoxyl contents (Freudenberg 1962).

Note 11. The values for total and aliphatic hydroxyl are meaningful only if the lignin preparations under study are sufficiently pure and have a C_9

structure. Otherwise, the total and aliphatic hydroxyl contents of lignin should be expressed in terms of methoxyl content (see Chap. 7.6.4 for the significance of the methoxyl analysis).

7.1.2.4 Total Aliphatic Hydroxyl Content

The aliphatic hydroxyl content of a lignin expressed on a methoxyl content and C_9 unit basis can be obtained by subtracting the phenolic hydroxyl content reported on a methoxyl or C_9 basis from the corresponding total hydroxyl value (see Chap. 7.2 for determination of phenolic hydroxyl groups).

$$\begin{aligned}&\text{Aliph OH/OCH}_3\text{ (moles/mole, based on lignin)}\\&= \text{Total OH/OCH}_3\text{ (moles/mole, based on lignin)}\\&\quad - \text{ArOH/OCH}_3\text{ (moles/mole, based on lignin).}\end{aligned} \qquad (5)$$

$$\begin{aligned}&\text{Aliph OH/C}_9\text{ (mol/C}_9\text{ unit, based on lignin)}\\&= \text{Total OH/C}_9\text{ (mol/C}_9\text{ unit, based on lignin)}\\&\quad - \text{ArOH/OCH}_9\text{ (mol/C}_9\text{ unit, based on lignin),}\end{aligned} \qquad (6)$$

where AliphOH and ArOH are, respectively, the aliphatic and phenolic hydroxyl contents of the lignin sample.

7.1.3 Discussion of the Methods

7.1.3.1 Precision and Accuracy of the Methods

The precision is generally about ±0.3% for the determination of O-acetyl groups in an acetylated lignin using the Kuhn-Roth procedure. The accuracy of the determination, however, depends largely on the amount of carbohydrate impurities present in the sample. In contrast to the Kuhn-Roth procedure, the modified Bethge-Lindström procedure has a precision of about ±3% for the determination of O-acetyl groups in acetylated lignins or about tenfold higher than that found using the Kuhn-Roth procedure. The accuracy of the modified Bethge-Lindström method is of the same order of magnitude as that found using the Kuhn-Roth procedure.

In addition to carbohydrates, milled bamboo lignins (MBLs) and milled grass lignins (MGLs) isolated from the corresponding saplings usually contain appreciable amounts of proteins (Tai et al. 1990) which also affect the accuracy of the results regardless of which procedure is followed. This statement is also probably true for milled wood lignins (MWLs) prepared from saplings. Therefore, it is essential to purify lignin preparations prior to performing the analysis by removing the contaminants as completely as possible. The removal of carbohydrates and low-molecular weight substances from lignin preparations can be accomplished using the procedure developed by Lundquist (Lundquist et al. 1977, Lundquist and Kirk 1980).

If the amounts of contaminants in the lignin sample are not negligible, e.g., the carbohydrate and protein contents are each more than 2% of the lignin, then the O-acetyl content of the lignin must be corrected (Pan et al. 1990, Tai et al. 1990). The correction for carbohydrate content is applied after performing a carbohydrate analysis of the lignin (Borchardet and Piper 1970, Fengel et al. 1977). With the exception of lignin isolated from saplings (Tai et al. 1990), a correction for protein content usually is not necessary because most lignin preparations do not contain proteins.

The precision found in the determination of aliphatic hydroxyl content depends on the combined precision observed for the determination of the O-acetyl content of the acetylated lignin and on the phenolic hydroxyl content in the original lignin sample (Chap. 7.2). Similarly, the accuracy of the aliphatic hydroxyl analysis depends on the accuracy obtained both in the determination of O-acetyl content in the corresponding acetylated lignin and in the determination of the phenolic hydroxyl content. The precision of the method for determining the aliphatic hydroxyl content of lignins is usually in the 0.8–6% range, depending on the method employed for the determination of the phenolic hydroxyl content. For example, the precision in the latter case is ±0.5% using the UV spectroscopic method developed by Goldschmid (1954). The overall precision in the determination of aliphatic hydroxyl content of lignin is ±0.8% when the Kuhn-Roth procedure is used to measure the O-acetyl content. In contrast, the precision for measuring the phenolic hydroxyl content by the Månsson procedure (Månsson 1983), is ±3%, and the overall precision for the determination of aliphatic hydroxyl groups in lignin is ±3.3% when the Kuhn-Roth procedure is used for the determination of O-acetyl content. This does not imply, however, that a combination of the Kuhn-Roth procedure and the UV spectroscopic ($\Delta\varepsilon_i$) method provides a more accurate value for the aliphatic hydroxyl content of a lignin than does the combination of Kuhn-Roth and Månsson procedures.

7.1.3.2 In-House as Opposed to Commercial O-Acetyl Analysis

Most commercial analytical laboratories have the capability of determining the O-acetyl content of organic compounds including acetylated lignins. The advantages and disadvantages of having the capability of conducting in-house O-acetyl determinations are the same as those listed for the methoxyl determination (Chap. 7.6.4).

References

Adler E (1977) Lignin chemistry – past, present and future. Wood Sci Technol 11:169–218

Adler E, Becker H-D, Ishihara T, Stamvik A (1966) Über die Benzylalkoholgruppen des Fichtenlignins. Holzforschung 20:3–11

Bethge PO, Lindström K (1973) Determination of O-acetyl groups in wood. Sven Papperstidn 76:645–649

Borchardet LG, Piper CV (1970) A gas chromatographic method for carbohydrates as alditol-acetates. Tappi 53:257–260

Chen C-L, Connors WJ (1974) New carbonyl compounds from the alkaline ferricyanide dehydrogenation of p-cresol. J Org Chem 39:3877–3880

Chum HL, Johnson DK, Ratcliff M, Black S, Schroeder HA, Wallace K (1985) Comparison between lignins produced by steam explosion and organosolv pretreatments. Proc 1985 Int Symp Wood Pulp Chem, Vancouver, BC, Canada, August 26–30, 1985, Tech Pap 223–226

Fengel D, Wegener G, Heizmann A, Przklent M (1977) Trifluoressigsäure zur schnellen und schonenden Hydrolyse von Cellulose und anderen Polysacchariden. Holzforschung 31: 65–71

Freudenberg K (1962) Forschung am Lignin. Fortschr Chem Org Naturst 20:41–72

Freudenberg K (1965) Lignin: its constitution and formation from p-hydroxycinnamyl alcohols. Science 148:595–600

Freudenberg K, Harkin JM, Werner H-K (1964) Das Vorkommen von Benzylaryläthern im Lignin. Chem Ber 97:902–920

Freudenberg K, Neish AC (1968) Constitution and biosynthesis of lignin. Springer, Berlin Heidelberg New York, 64–77

Gierer J (1980) Chemical aspects of kraft pulping. Wood Sci Technol 14:241–266

Gierer J (1982a) The chemistry of delignification: a general concept. Part I. Holzforschung 36:43–51

Gierer J (1982b) The chemistry of delignification: a general concept. Part II. Holzforschung 36:55–64

Gierer J, Lindeberg O (1980) Reactions of lignin during sulfate pulping. Part XIX. Isolation and identification of new dimers from a spent sulfate liquor. Acta Chem Scand B 34: 161–170

Goldschmid O (1954) Determination of phenolic hydroxyl content of lignin preparations by ultraviolet spectrophotometry. Anal Chem 26:1421–1423

Kuhn R, Roth H (1933) Mikro-Bestimmung von Acetyl-, Benzoyl- und C-Methylgruppen. Ber Dtsch Chem Ges 66:1274–1277

Lai YZ, Sarkanen KV (1971) Isolation and structural studies. In: Sarkanen KV, Ludwig CH (eds) Lignins: occurrence, formation, structure and reactions. Wiley-Interscience, New York, 165–240

Lundquist K, Kirk TK (1980) Fractionation-purification of an industrial lignin. Tappi 63(1): 80–82

Lundquist K, Ohlsson B, Simonson R (1977) Isolation of lignin by means of liquid-liquid extraction. Sven Papperstidn 80:143–144

Månsson P (1983) Quantitative determination of phenolic and total hydroxyl groups in lignins. Holzforschung 37:143–146

Pan D-r, Tai D-s, Chen C-L, Robert D (1990) Comparative studies on chemical composition of wood components in recent and ancient woods of Bischofia polycarpa. Holzforschung 44:7–16

Robert D, Brunow G (1984) Quantitative estimation of hydroxyl groups in milled wood lignin from spruce and in a dehydrogenation polymer from coniferyl alcohol using ^{13}C NMR spectroscopy. Holzforschung 38:85–90

Roth H (1953) Analytik der alkoholischen und phenolischen Hydroxylgruppe. In: Müller E, Bayer O, Meerwein H, Ziegler K (eds) Methoden der organischen Chemie (Houben-Weyl) 4. Aufl., Bd II Analytische Methoden, George Thieme, Stuttgart, Federal Republic of Germany, 330–380

Tai D-s, Chen C-L, Gratzl JS (1990) Chemistry of delignification during kraft pulping of bamboos. J Wood Chem Technol 10:75–99

Winston MH, Chen C-L, Gratzl JS, Goldstein IS (1986) Characterization of the lignin residue from hydrolysis of sweetgum wood with superconcentrated hydrochloric acid. Holzforschung 40(suppl):45–50

Yang J-M, Goring DAI (1980) The phenolic hydroxyl content of lignin in spruce wood. Can J Chem 58:2411–2414

7.2 Determination of Phenolic Hydroxyl Groups

Y.-Z. Lai

7.2.1 Significance of the Analysis

The phenolic hydroxyl group is one of the most important functionalities affecting the physical and chemical properties of lignin polymers (Adler 1977). It plays a prominent role in commercial pulping and bleaching processes by virtue of its ability to promote the base-catalyzed cleavage of interunitary ether linkages and the oxidative degradation of lignin (Gierer 1985, 1986). The chemical reactivity of lignin in various modification processes is also profoundly influenced by its phenolic hydroxyl content (e.g., in the reaction with formaldehyde for the production of lignin adhesives) (Olivares et al. 1988). On the other hand, the phenolic hydroxyl group significantly contributes to the poor brightness stability of lignin-containing pulps, which seriously limits their more widespread utilization (Gellerstedt and Pettersson 1977, Tschirner and Dence 1988). Quantitative measurement of phenolic hydroxyl groups thus provides pertinent information relating to the structure and reactivity of lignin and to the mechanism and extent of lignin degradation.

7.2.2 Survey of Analytical Approaches

Both physical and chemical methods or a combination of both have been used to estimate the phenolic hydroxyl content of lignin (Adler 1977). Common physical methods include potentiometric and conductometric titration (Chap. 7.5) ionization UV spectroscopy (Chap. 5.1), and NMR spectroscopy (Chap. 5.3).

The UV method (Aulin-Erdtman 1954, Wexler 1964), based on the difference in absorption at 300 nm of phenolic units in neutral and in alkaline solutions, is very suitable for soluble lignin preparations, but is not absolute, since the measurement must be calibrated by using appropriate lignin model compounds or a lignin of known phenolic hydroxyl content. The application of the UV method to the in situ measurement of the phenolic hydroxyl content of lignin in wood (Yang and Goring 1980) and in pulps (Yang and Goring 1981) has also been reported (Chap. 4.2).

The titration method is based on the acidity of phenolic hydroxyl groups and, under proper conditions, can be used to determine this group together with carboxyl or sulfonate groups that may be present coincidentally in lignin. An improved nonaqueous titration method in DMF developed by Pobiner (1983),

Springer Series in Wood Science
Methods in Lignin Chemistry
(Edited by S.Y. Lin and C.W. Dence)

using tetra-n-butylammonium hydroxide as the titrant and p-hydroxybenzoic acid as an internal standard, has significantly increased the accuracy and utility of the titration method. This improved method was shown to be more suitable than the UV method for bagasse lignins (Feng and Wang 1987).

In recent years, the use of ^{13}C NMR in estimating the phenolic hydroxyl content of lignin both in solution (Chap. 5.4) and in the solid state (Chap. 4.5) has continued to receive much attention. The determination is generally based on the methyl or carbonyl signals of phenolic acetyl groups (Robert et al. 1984), or on the signals corresponding to C_3, C_4, and C_5 in the aromatic ring (Hemmingson and Dekker 1987, Leary et al. 1988).

A pyrolytic gas chromatographic technique (Chap. 4.7) based on the difference in yield of phenol and guaiacol resulting from the thermal degradation of lignin before and after methylation has also been used to estimate the phenolic hydroxyl content of softwood lignin (Whiting and Goring 1982).

Chemical methods for analysis of phenolic hydroxyl groups include determination of the increase in methoxyl content resulting from diazomethane methylation (Björkman and Person 1957), the increase in phenolic acetyl group content after acetylation (Lenz 1968, Månsson 1983), and low-molecular weight compounds derived from the degradation of phenolic structures (Chap. 5.2). The phenolic acetyl group of acetylated lignin may also be determined by an NMR spectroscopic technique (Lenz 1968, Robert et al. 1986) or by a selective deacetylation in pyrrolidine (aminolysis) (Månsson 1983).

A degradation procedure based on the sequential oxidation of lignin after ethylation with potassium permanganate and hydrogen peroxide (Chap. 6.3) has been used in the estimation of phenolic hydroxyl contents of kraft lignins (Gellerstedt and Gustafsson 1987) and kraft pulps (Gellerstedt and Lindfors 1984a, Gellerstedt et al. 1988). Another degradation method based on the thioacidolysis (Chap. 6.4) of isolated lignin or wood after diazomethane methylation has been employed by Lapierre et al. (1988) to study the distribution of phenolic β-aryl ether structures among the p-hydroxyphenyl, guaiacyl, and syringyl units of straw lignins.

A method developed by Adler et al. (1958) is based on the oxidation of simple phenolic guaiacyl (Adler and Hernestam 1955) or syringyl (Chang et al. 1975) compounds with aqueous sodium periodate solution to ortho-quinone structures in which process nearly 1 mol of methanol per mol of phenolic hydroxyl group is released [Eq. (1)].

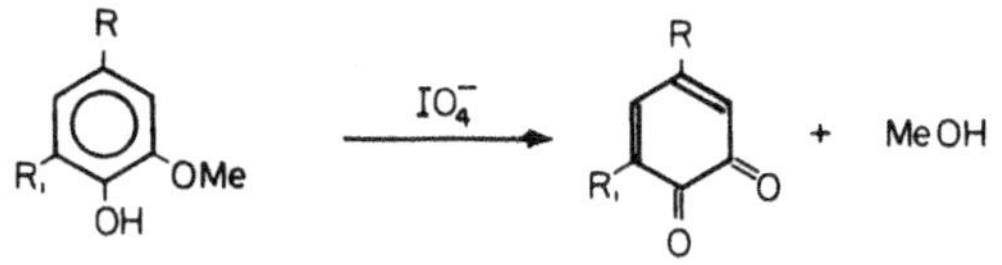

(1)

where R = lignin side chain
R_1 = H, OCH_3 or lignin unit.

Measurement of the methanol formed in the oxidation thus is approximately equivalent to the amount of adjacent phenolic hydroxyl groups. No methanol is formed when the oxidation is performed on etherified guaiacyl compounds, e.g., veratrole.

The periodate method, which has been applied extensively to isolated lignins from softwoods (Adler et al. 1958, Adler 1977), is also suitable for isolated hardwood lignins (Chang et al. 1975) and for wood samples (Lai et al. 1990). However, as this method is based on methanol formation alone, it cannot be applied to methoxyl-free phenolic units such as catechol-type structures or p-hydroxyphenyl units which are present in a large quantity in nonwood lignins and in compression wood lignin (Adler 1977, Feng and Wang 1987).

The aminolysis method is a direct procedure based on the finding that the rate of deacetylation of aromatic acetates in pyrrolidine under mild conditions is considerably higher than that of aliphatic acetates (Månsson 1983). Thus, the phenolic acetyl group of acetylated lignin can be selectively deacetylated in pyrrolidine which, as indicated by the formation of 1-acetylpyrrolidine [Eq. (2)], is a very rapid reaction. The method has been applied satisfactorily to acetylated lignin model compounds (Månsson 1982), to milled wood lignin (Månsson 1983), to kraft lignins (Månsson 1983, Robert et al. 1984), to wood (Gellerstedt and Lindfors 1984b), and to straw lignins (Liu and Li 1987).

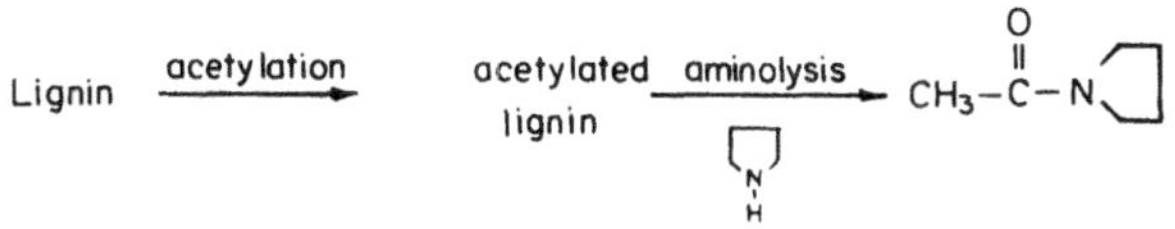

(2)

Of the various methods discussed above, the most widely used are those based on UV, titration, periodate oxidation, and aminolysis. These four methods generally give concordant results for relatively unmodified lignin preparations (e.g., milled wood lignin) from softwoods (Adler 1977), but they all (with the possible exception of aminolysis) have certain limitations when applied to chemically degraded lignins, lignins from hardwoods or nonwood species, or to lignocellulosic materials.

Detailed procedures for two driect chemical methods, periodate oxidation and aminolysis, for determination of the phenolic hydroxyl content of lignin or lignin-containing materials are described below.

7.2.3 Description of Procedures

7.2.3.1 Aminolysis

The procedure outlined below for the analysis of isolated lignin samples was originally developed by Månsson (1983), and consists of acetylation and aminolysis steps [Eq. (2)]; the procedure for lignocellulosic samples is essentially that of Gellerstedt and Lindfors (1984b) and includes a borohydride reduction step prior to acetylation.

7.2.3.1.1 Instrumentation and Chemicals

A gas chromatograph, equipped with a flame ionization detector and an electronic integrator for measuring peak areas, is required for analysis of the deacetylation product, 1-acetylpyrrolidine. The gas chromatograph is fitted with a 2.5 m × 0.32 cm glass column packed with 5% Castorwax on 60/80 mesh chromosorb G (Månsson 1983, Gellerstedt and Lindfors 1984b) or with a 1.83 m × 0.32 cm stainless steel column packed with 5% Carbowax 20 M on 60/80 mesh chromosorb G (Tschirner and Dence 1988, Lai et al. 1990). These columns are available from American Scientific Products McGaw Park, IL, or Chemical Research Supplies, Addison, IL; the Carbowax column is also available from Supelco Chromatography Products, Bellefonte, PA.

The column is heated isothermally at 180 °C with the injection port and detector temperatures set at 230 °C and 250 °C, respectively. The carrier gas, N_2, is delivered at a flow rate of 20 ml min^{-1}.

The pyrrolidine amides (1-acetylpyrrolidine and 1-propionylpyrrolidine) are prepared from the appropriate acid chloride and pyrrolidine (Månsson and Westfelt 1980).

Acetic anhydride, pyridine, pyrrolidine, acetyl or propionyl chloride, sodium borohydide, dioxane, and 1-methylnaphthalene are available from a number of suppliers, including Aldrich Chemical Co.

7.2.3.1.2 Calibration of Internal Standard

Stock solutions of 1-acetylpyrrolidine, and an internal standard (1-propionylpyrrolidine or 1-methylnaphthalene) in dioxane at a concentration of 10 mg ml^{-1} are prepared. A mixture of 1 ml each of the 1-acetylpyrrolidine and internal standard solutions is used to establish the GC response factor. The calibration is made by injecting a 1–2-μl aliquot of the mixture into the gas chromatograph and measuring the peak areas.

The calibration constant, k, is then calculated.

$$k = \frac{Aa \times Ws}{As \times Wa}, \quad (3)$$

where Aa = chromatographic area of product (1-acetylpyrrolidine in the aminolysis method or methanol in the periodate method), As = chromatographic area of internal standard (1-propionylpyrrolidine or 1-methylnaphthalene in the aminolysis method, or acetonitrile in the periodate oxidation method), Wa = weight of product, mg, Ws = weight of internal standard, mg.

7.2.3.1.3 Procedure

Borohydride reduction (Note 1)

Wood meals or ground pulp samples (Note 2) equivalent to about 25 mg of lignin are weighed accurately into a 20-ml glass centrifuge tube (Note 3). After the addition of 4 ml of distilled water and the calculated amount of sodium borohydride equivalent to one-half of the sample weight, the suspension is centrifuged at 1250 rpm for 60 min (Note 4).

The mixture, after the repeated addition of the same amount of sodium borohydride, is capped, centrifuged again for 2 h and then allowed to stand for 2 days with gentle stirring. After thorough mixing with 4 ml of water, the reduced mixture is centrifuged (5700 rpm for 15 min) and the aqueous phase is decanted. The residue is repeatedly washed with water (7 ml) in the tube followed by centrifugation until the washings are neutral.

The washed, reduced sample is dried by solvent exchange (Note 5) with acetone (7 ml) followed by pyridine (7 ml). Each solvent wash is repeated three times.

Acetylation

Lignin. A 20-mg sample in a 20-ml pear-shaped flask is acetylated with 1 ml of a pyridine-acetic anhydride (1:1, v/v) solution at ambient temperature for 15 h. The mixture, after the addition of 8 ml of 50% aqueous methanol, is evaporated to dryness under vacuum. The solvent treatment is repeated three times with toluene (3 × 8 ml), and once with methanol (8 ml). The acetylated sample is dried in vacuo at 50 °C overnight.

Wood. The borohydride-reduced wood or pulp sample in the centrifuge tube is acetylated by adding 4 ml of a pyridine-acetic anhydride (1:1, v/v) solution at ambient temperature for 3 days with gentle stirring. After being vigorously mixed with 5 ml of diethyl ether for 30 min, the mixture is centrifuged and the liquid phase is decanted. The residue is washed an additional three times with 7 ml of ether and dried in vacuo at 50 °C overnight.

Aminolysis

The acetylated lignin, dissolved in 1.0 ml of dioxane containing 5 mg of an internal standard, is treated with 1.0 ml of a dioxane-pyrrolidine (1:1, v/v) solution which initiates the aminolysis reaction. The rate of the 1-acetylpyrrolidine formation is followed by periodically injecting 1–2 μl of the reaction mixture into the gas chromatograph. The initial rapid phase is generally complete within 20 min, but a total of 60 min is required to complete the deacetylation process.

For an acetylated wood sample, 3 ml of dioxane and 1 ml of dioxane containing 5 mg of an internal standard are added to the sample. The mixture is homogenized in an ultrasonic bath for 10 min and then treated with 1 ml of a 1:1 (v/v) dioxane-pyrrolidine solution. With the cap firmly tightened on the reaction tube, the mixture is again homogenized by vigorous shaking and then allowed to stand with stirring.

The first sample is taken after approximately 30 min and subsequently, five to six samples are analyzed gas chromatographically at suitable intervals of about 30–60 min. The reaction mixture is centrifuged to obtain a clear solution prior to samping and homogenized again after samping.

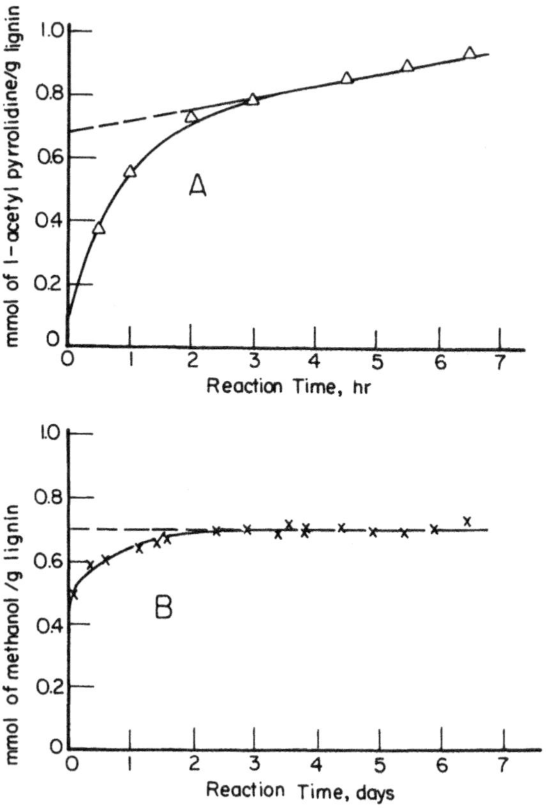

Fig. 7.2.1A,B. Rate of formation of 1-acetylpyrrolidine on **A** aminolysis of acetylated wood meal Norway spruce and of methanol on **B** periodate oxidation of wood meal. (Lai et al. 1990)

The rate of 1-acetylpyrrolidine formation is followed as in the case of acetylated lignin, but the overall process is much slower and a 6–7-h reaction period may be required as illustrated in Fig. 7.2.1A for Norway spruce wood meal.

The phenolic acetyl content (equivalent to the phenolic hydroxyl content) is calculated by extrapolating the linear region of the kinetic curve to zero time as illustrated in Fig. 7.2.1A

$$\text{mmol acetyl pyrrolidine/g lignin} = \frac{A \times Ws \times 1000}{As \times L \times k \times 113}, \qquad (4)$$

where A = chromatographic area of peak corresponding to 1-acetylpyrrolidine, As = chromatographic area of peak corresponding to internal standard, Ws = weight of internal standard, mg, L = weight of lignin, mg; for a wood sample it is equivalent to the product of the sample weight and its lignin content, 113 = the molecular weight of 1-acetylpyrrolidine, k = calibration constant as calculated using Eq. (3).

Table 7.2.1. A comparison of the aminolysis and periodate oxidation methods in the determination of the phenolic hydroxyl content of Norway spruce and aspen wood meals (Lai et al. 1990)

Method	Phenolic hydroxyl content mmol/g lignin			
	Spruce		Aspen	
	Av.	Std. Dev.	Av.	Std. Dev.
Aminolysis	0.649	0.029	0.445	0.031
Periodate oxidation	0.702	0.005	0.493	0.008

$$\text{Number of phenolic hydroxyl groups/100 } C_9 \text{ unit} = \frac{\text{mmol phenolic hydroxyl group} \times \text{unit weight}}{\text{g lignin} \times 10}, \quad (5)$$

where unit weight is that of an average phenylpropane unit (C_9) of lignin, e.g. 183 and 209 for spruce and birch MWL, respectively.

Note 1. The borohydride reduction step may be omitted for carbohydrate-free lignin samples.
Note 2. Wood meals or milled pulp samples ground to 40–60 mesh are generally sufficient.
Note 3. Centrifuge tubes with a Teflon screw cap are preferred.
Note 4. Caution should be taken to prevent part of the sample from escaping with the hydrogen gas liberated.
Note 5. The solvent exchange is necessary to preserve the swollen state of the wood samples and thereby assure high accessibility to the reagents during acetylation.

The standard deviation for a series of replicate determinations of Norway spruce and aspen wood meals is summarized in Table 7.2.1.

7.2.3.2 Periodate Oxidation

The procedure described below, a modified version of the method developed by Adler et al. (1958) and Gierer et al. (1964), is applicable to both lignin and lignocellulosic samples. The general procedure consists of oxidation of the sample with an aqueous sodium periodate solution and a direct measurement of the liberated methanol in the presence of an internal standard.

7.2.3.2.1 Instrumentation and Chemicals

The requirement for gas chromatography equipment is the same as that described in the aminolysis method.

Methanol and an internal standard (acetonitrile or benzene) can be determined with a 2 m × 6 mm stainless column packed with 30% Triton X-100 on chromosorb W (Gierer et al. 1964, Yang and Goring 1980) or with a 1.8 m × 0.32 cm stainless steel column packed with Tenax GC (Lai et al. 1990). The Triton column is available from Supelco Chromatography Products, Bellefonte, PA, while the Tenax column may be obtained from Alltech Associates, Inc., Deerfield, IL. Both columns are operated at 80 °C with an injection port temperature of 150 °C and a detector temperature of 250 °C. A nitrogen flow of 30 ml min^{-1} is maintained.

7.2.3.2.2 Calibration of Internal Standard

Stock solutions of methanol and acetonitrile (3 mg ml^{-1}) in distilled water are prepared.

The calibration constant (k) of acetonitrile and methanol is obtained and calculated as described in the section on aminolysis [Eq. (3)].

7.2.3.2.3 Procedure

A sample of lignin (~50 mg) or wood meal (~400 mg) (Note 6) in a 20-ml glass centrifuge tube (Note 3) is treated with sodium periodate, in an amount equivalent to the sample weight, dissolved in 6 ml of distilled water, and with 1 ml of distilled water containing 3 mg of an internal standard (Note 7). Both solutions are cooled to 4 °C prior to addition to the sample. The suspension is homogenized and kept in the dark at 4 °C in a refrigerator with occasional stirring.

The rate of methanol formation is followed gas chromatographically by periodically injecting a 1–2-μl aliquot of the reaction mixture. The mixture is homogenized and centrifuged to obtain a clear solution prior to sampling, and is homogenized again after sampling.

Methanol formation generally levels off after two days' reaction in the case of wood samples (Fig. 7.2.1B). For lignin preparations, the maximum formation of methanol is usually reached within 36 h (Adler et al. 1958).

The phenolic hydroxyl content of lignin is calculated from the maximum amount of methanol formed.

7.2.3.2.4 Calculation

$$\text{mmol } CH_3OH/\text{g lignin} = \frac{A \times Ws \times 1000}{As \times L \times k \times 32}, \tag{6}$$

where A, As, Ws, L, and k are defined as in Eq. (4); 32 is the molecular weight of methanol.

The phenolic hydroxyl frequency/100 C_9 units is similarly calculated as shown in Eq. (5).

Note 6. A wood or pulp sample (40–60 mesh) equivalent to about 100 mg of lignin is used.
Note 7. The oxidation of lignin may also be conducted in a homogeneous 80% acetic acid solution although the solvent reportedly causes problems during the GC analysis (Gierer et al. 1964).

The standard deviation for a series of replicate determinations of Norway spruce and aspen wood meals is shown in Table 7.2.1.

7.2.4 Discussion of Methods

Both aminolysis and periodate methods have been shown to be comparable in terms of the phenolic hydroxyl content obtained for milled wood lignin (MWL) and kraft lignin from Norway spruce (Månsson 1983) and for Norway spruce and aspen wood meals (Lai et al. 1990).

As indicated by the values in Table 7.2.1, the periodate oxidation method yields consistently higher values (8 and 10% for Norway spruce and aspen wood meals, respectively) for phenolic hydroxyl content than are obtained using the aminolysis procedure. The reasons for these variations are not clear. However, in a series of replicate determinations, the aminolysis method has a considerably higher standard deviation compared to the periodate method which may be partly attributable to the larger number of steps involved in the analysis.

Aminolysis is a laborious procedure, e.g., for wood samples, a 6–7-h reaction period may be required for the aminolysis step alone. The accuracy of this method is critically dependent upon both a quantitative acetylation of phenolic hydroxyl groups and a selective deacetylation of phenolic acetyl groups. Although these requirements may not represent a serious concern in the analysis of soluble or reagent-accessible lignin preparations, they could present a problem in the case of lignocellulosic materials.

In addition, the reducing end groups of carbohydrates behave like phenolic hydroxyl groups in the aminolysis method. Thus, a quantitative reduction of wood or pulp samples by borohydride prior to acetylation is essential.

In contrast, the periodate oxidation method is a relatively simple process and is unaffected by the presence of carbohydrates. Construction of a kinetic curve is not necessary in the periodate oxidation procedure, which requires only determination of the maximum amount of methanol formation for the calculation of phenolic hydroxyl content. Thus, for routine analysis, the periodate oxidation method may require only measurement of the amount of methanol liberated after 2 and 3 days' reaction for isolated lignin and wood samples, respectively (Gierer et al. 1964, Yang and Goring 1980).

Previous analysis of lignin samples by the periodate oxidation method was generally conducted in a homogeneous solution using 80% acetic acid (Adler et al. 1958, Gierer et al. 1964, Yang and Goring 1980). Under these conditions, methanol is partly converted to methyl acetate, thus giving rise to two gas chromatographic peaks corresponding to each of these two products. In addition,

Table 7.2.2. Phenolic hydroxyl contents of various lignins

Material	Phenolic hydroxyl content		Method	Reference
	mmol/g lignin	no./100 C_9 units		
Spruce wood	–	10	UV spectroscopy	Yang and Goring 1980
Spruce wood meal	0.70	13	Periodate ox.	Lai et al. 1990
	0.65	12	Aminolysis	Lai et al. 1990
Spruce MWL		18	Titration	Lindner and Wegener 1988
		20	Periodate ox.	Chang et al. 1975
		20	UV specroscopy	Chang et al. 1975
		20	Periodate ox.	Yang and Goring 1980
		20	^{13}C-NMR	Robert and Brunow 1984
		26	Periodate ox.	Lenz 1968
		27	^{1}H-NMR	Lenz 1968
		30	Periodate ox.	Adler et al. 1958
		33	Aminolysis	Månsson 1983
Spruce CEL[a]		20	Periodate ox.	Chang et al. 1975
		20	UV spectroscopy	Chang et al. 1975
Pine wood meal	0.73	13	Aminolysis	Gellerstedt and Lindfors 1984b
Pine kraft pulp[b]	0.84–1.47	15–26	Aminolysis	Gellerstedt and Lindfors 1984b
Pine kraft lignin[b]	2.9–4.1	55–70	Aminolysis	Robert et al. 1984
Aspen wood meal	0.49	10	Periodate ox.	Lai et al. 1990
	0.45	9	Aminolysis	Lai et al. 1989
Sweetgum MWL		14	Periodate ox.	Chang et al. 1975

[a] Cellulolytic enzyme lignin.
[b] Pulp yield varies in 45–90% range.

acetic acid passes through the Triton column slowly, exhibits marked tailing, and requires a proper conditioning of the column after each injection (Gierer et al. 1964). Interestingly, the rate of methanol formation in the oxidation of model compounds in water is significantly higher than that in 80% acetic acid.

The phenolic hydroxyl contents of a variety of lignins are shown in Table 7.2.2. For untreated wood, the value is generally less than 15%. It is noteworthy that the phenolic hydroxyl contents reported for MWLs show considerable variability. A large part of the variability can be attributed to lack of uniformity in the efficiency of the milling process used in the MWL preparation (Adler 1977).

From an inspection of the values in Table 7.2.2, it is clear that the phenolic content of lignin decreases in the order kraft lignin ≫ kraft pulp > wood, a trend which underscores the significance of aryl ether cleavage in delignification and of the effect of the phenolic hydroxyl groups on lignin solubility.

Although the periodate oxidation and aminolysis methods provide quantitative data on the frequency with which of phenolic hydroxyl groups occur in lignin, they do not reveal the structural environment in which they occur. Such information may be obtained by other means, e.g., UV spectroscopic or chemical degradation methods (Adler 1977).

References

Adler E (1977) Lignin chemistry – past, present, and future. Wood Sci Technol 11:169–218

Adler E, Hernestam S (1955) Estimation of phenolic hydroxyl groups in lignin. I. Periodate oxidation of guaiacol compounds. Acta Chem Scand 9:319–334

Adler E, Hernestam S, Walldén I (1958) Estimation of phenolic hydroxyl groups in lignin. Sven Papperstidn 61:64–647

Aulin-Erdtman G (1954) Spectrographic contributions to lignin chemistry. V. Phenolic groups in spruce lignin. Sven Papperstidn 57:745–760

Björkman A, Person B (1957) Studies on finely divided wood. Part 4. Some reactions of the lignin extracted by neutral solvents from Picea abies. Sven Papperstidn 60:285–292

Chang H-M, Cowling EB, Brown W, Adler E, Miksche G (1975) Comparative studies on cellulolytic enzyme lignin and milled wood lignin of sweetgum and spruce. Holzforschung 29:153–159

Feng JH, Wang ZH (1987) Chemical structure and utilization of bagasse lignin. In: Proc Fourth Int Symp Wood Pulp Chem, Paris, 1:61–65

Gellerstedt G, Gustafsson K (1987) Structural changes in lignin during kraft cooking. Part 5. Analysis of dissolved lignin by oxidative degradation. J Wood Chem Technol 7:65–80

Gellerstedt G, Gustafsson K, Northey RA (1988) Structural changes in lignin during kraft cooking. Part 8. Birth lignins. Nord Pulp Pap Res J 3:87–94

Gellerstedt G, Lindfors E-L (1984a) Structural changes in lignin during kraft pulping. Holzforschung 38:151–158

Gellerstedt G, Lindfors E (1984b) Structural changes in lignins during kraft cooking. Part 4. Phenolic hydroxyl groups in wood and kraft pulps. Sven Papperstidn 87:R115–R118

Gellerstedt G, Pettersson E-L (1977) Light-induced oxidation of lignin. Part 2. The oxidative degradation of aromatic rings. Sven Papperstidn 77:15–21

Gierer J (1985) Chemistry of delignification. Part 1. General concept and reactions during pulping. Wood Sci Technol 19:289–312

Gierer J (1986) Chemistry of delignification. Part 2. Reactions of lignins during bleaching. Wood Sci Technol 20:1–33

Gierer J, Lenz B, Noren I, Söderberg S (1964) Reactions of lignin during sulfate cooking. III. The splitting of aryl-alkyl ether bonds in milled wood lignin by alkali. Tappi 47:233–239

Hemmingson JA, Dekker FH (1987) A CP/MAS ^{13}C NMR study of residual lignin structure in autohydrolysis-exploded wood and bagasse. J Wood Chem Technol 7:229–244

Lai Y-Z, Guo X, Situ W (1990) Estimation of phenolic hydroxyl groups in wood by a periodate oxidation method. J Wood Chem Technol 10:365–377

Lapierre C, Monties B, Rolando C (1988) Thioacidolysis of diazomethane-methylated pine compression wood and wheat straw in situ lignins. Holzforschung 42:409–411

Leary GJ, Lloyd JA, Morgan KR (1988) A ^{13}C CP/MAS NMR studies of residual lignin in kraft pulps. Holzforschung 42:199–202

Lenz B (1968) Application of nuclear magnetic resonance spectroscopy to characterization of lignin. Tappi 51:511–519

Lindner A, Wegener G (1988) Characterization of lignins from organosolv pulping according to the organocell process. Part 1. Elemental analysis, nonlignin portions and functional groups. J Wood Chem Technol 8:323–340

Liu X, Li Z (1987) A new method for the determination of phenolic and total hydroxyl groups in straw lignins. Zhongguo Zaozhi 6:34–37

Månsson P (1982) Selective deacetylation of aromatic acetates by aminolysis. Tetrahedron Lett 23:1845–1846

Månsson P (1983) Quantitative determination of phenolic and total hydroxyl groups in lignins. Holzforschung 37:143–146

Månsson P, Westfelt L (1980) Selective aminolysis of mixed esters of cellulose. J Appl Polym Sci 25:1533–1540

Olivares M, Guzman JA, Natho A, Saavedra A (1988) Kraft lignin utilization in adhesives. Wood Sci Technol 22:157–165

Pobiner H (1983) Improved inflection points in the non-aqueous potentiometric titration of acid functionalities in lignin chemicals by using internal standardization and ion exchange. Anal Chim Acta 155:57–65 (1983)

Robert DR, Bardet M, Gellerstedt G, Lindfors EL (1984) Structural changes in lignin during kraft cooking. Part 3. On the structure of dissolved lignins. J Wood Chem Technol 4:239–263

Robert DR, Brunow G (1984) Quantitative estimation of hydroxyl groups in milled wood lignin from spruce and in a dehydrogenation polymer from coniferyl alcohol using ^{13}C NMR spectroscopy. Holzforschung 38:85–90

Robert DR, Gellerstedt G, Bardet M (1986) Carbon-13 NMR analysis of lignins obtained after sulfonation of steam exploded aspen wood. Nord Pulp Pap Res J 1:18–25

Tschirner U, Dence CW (1988) Attempts to photostabilize Norway spruce TMP by chemical modification. Pap Puu 70:338–347

Wexler AS (1964) Characterization of lignosulfonates by ultraviolet spectrometry. Anal Chem 36:213–221

Whiting P, Goring DAI (1982) Phenolic hydroxyl analysis of lignin by pyrolytic gas chromatography. Pap Puu 64:592–599

Yang J-M, Goring DAI (1980) The phenolic hydroxyl content of lignin in spruce wood. Can J Chem 58:2411–2414

Yang J-M, Goring DAI (1981) The phenolic hydroxyl content of the lignin residual in spruce wood cooked by sulphite and kraft. Cellul Chem Technol 15:337–342

7.3 Determination of Ethylenic Groups

C.W. DENCE

7.3.1 Ethylenic Structures in Lignin

Small amounts of ethylenic groups are present in in situ lignin mainly as components of unattached cinnamaldehyde and cinnamyl alcohol end groups (I and II, respectively, Table 7.3.1) (Lai and Sarkanen 1971). Additionally, the presence of carbonyl-conjugated ethylenic groups as elements of p-quinonoid (Pew and Connors 1971) and o-quinonoid (Harkin 1966) ring systems has been proposed.

On treatment of lignin with alkali or acid, stilbene and other ring-conjugated structures are generated through elimination reactions occurring at the propanoid side chains. Notable examples of such transformations have been provided by pulping studies where trans p,p′- and p,o′-dihydroxystilbenes (III and IV respectively, Table 7.3.1) have been detected following treatment of lignin or lignin model dimers with alkali under kraft or soda pulping conditions (Tanaka and Kondo 1937, 1959, Enkvist 1960, Falkehag 1961, Adler et al. 1964a). A ^{13}C-NMR study has provided additional evidence for the presence of stilbene units in kraft lignin (Kringstad and Mörck 1983). Trans 4,4′-dihydroxy-3,3′-dimethoxystilbene (III, Table 7.3.1) has also been detected in spent sulfite and bisulfite pulping liquors (Richtzenhain and van Hofe 1939, Kvasnicka and McLaughlin 1955, Hillis et al. 1966).

Other ring-conjugated ethylenic structures whose origins are associated with alkaline pulping are styryl aryl ethers, chalcones and 1,4-bis-(p-hydroxyphenyl)-buta-1,3-dienes (V-VII, respectively, Table 7.3.1). The presence of styryl aryl ethers in alkali lignins is well established (Adler et al. 1964b, Gierer et al. 1964); the existence of chalcones in alkali liquors is speculative and is based on the finding that vanillin and acetoguaiacone, compounds detected in black liquor, form such units when heated with alkali (Falkehag et al. 1966). The evidence for 1,4-bis-(p-hydroxyphenyl)-buta-1,3-diene consists of the finding that structures of this type are formed when pinoresinol lignin model dimers are heated with kraft, soda, or neutral sulfite pulping liquors (Gierer 1970).

Treatment of lignin with acid as is the situation, for example, in acidolysis (Chap. 6.1) gives rise to stilbene and phenylcoumarone structures (Adler et al. 1966). Stilbenes similar to those identified in acid sulfite pulping liquors have been detected in the steam hydrolyzate of aspen wood (Bardet et al. 1985), and a coniferyl aldehyde structure was formed when a β-aryl ether-type lignin model compound was heated with slightly acidified water (Kratzl et al. 1959).

Springer Series in Wood Science
Methods in Lignin Chemistry
(Edited by S.Y. Lin and C.W. Dence)

Table 7.3.1. Ring-conjugated ethylenic groups in lignin

	ACH=CHB	
Structure	A	B
I	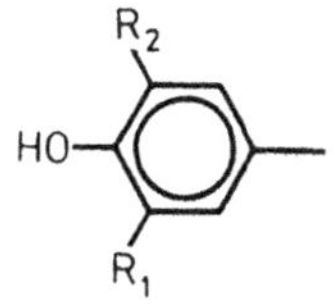	—CHO
II	R_1=R_2=H (p-hydroxyphenyl) R_1=OCH_3, R_2=H (guaiacyl) R_1=R_2=OCH_3 (syringyl)	—CH_2OH
III	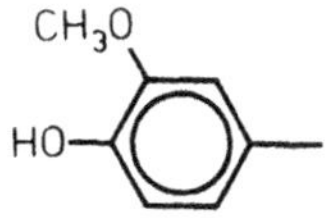	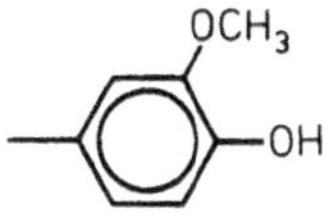
IV	"	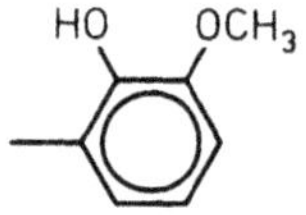
V	"	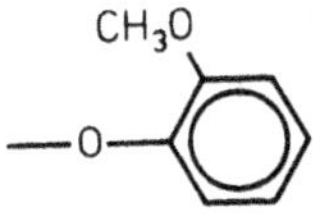
VI	"	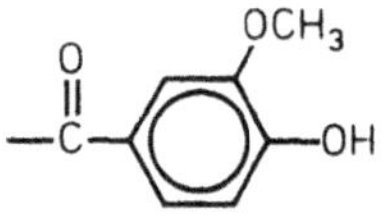
VII	"	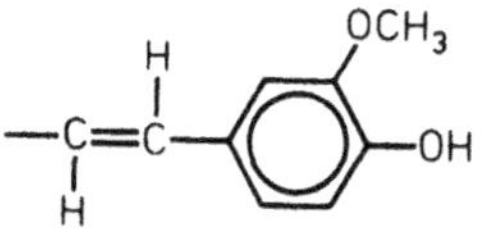

Oxidative treatment of lignin typically results in the conversion of phenolic nuclei to o- and p-quinonoid structures or in the rupture of phenolic rings with concomitant formation of enoic and dienoic acids (Chang and Allan 1971). Quinonoid structures, particularly the ortho type, are prone to undergo condensation reactions in aqueous acidic or alkaline media, forming complex products in which some ethylenic groups may be retained although in an altered structural environment.

7.3.2 Survey of Methods

At present there is no single procedure available for the determination of ethylenic groups in lignin which can be applied reliably to any type of sample without regard for the particular structural environment in which these units exist. Hence, most of the more widely accepted methods essentially consist of procedures used for the determination of larger structural elements comprising lignin, for example, cinnamaldehyde, cinnamyl alcohol, and stilbene units, in which the carbon-carbon double bond is incorporated.

Cinnamaldehyde and cinnamyl alcohol groups may be determined by visible spectroscopy after conversion to colored derivatives. In one approach, cinnamaldehyde groups are condensed with phloroglucinol (Wiesner reaction) (Chap. 2.1.3.1) yielding a chromophore absorbing in the vicinity of 550 nm (Browning 1967a, DeBaun and Nord 1951) or, alternatively, with aromatic amines such as p-aminobenzoic acid to form a Schiff base having an absorption maximum at approximately 445 nm (Adler and Marton 1959). In a similar approach, the ethylenic groups in cinnamyl alcohol structures may be transformed to p-dimethylamino-anilides of styrylglyoxylonitriles (λ max at approximately 475 nm) after reaction of the sample with tosyl chloride, p-nitroso-N,N-dimethylaniline, and potassium cyanide (Lindgren and Mikawa 1957).

Cinnamaldehyde and cinnamyl alcohol groups have been estimated as coniferaldehyde and coniferyl alcohol in softwood Brauns native lignins (Aulin-Erdtman and Hegbom 1958) and in softwood milled wood lignin (Marton and Adler 1961, Adler 1961) using a UV spectroscopic method based on the elimination of ethylenic groups through catalytic hydrogenation. In a similar approach, the effect on UV spectral characteristics produced by sulfonation with sodium bisulfite and methylation with diazomethane has served as the basis for the estimation of coniferaldehyde units in softwood Brauns native lignins (Aulin-Erdtman and Hegbom 1958).

An analytical method consisting of diazomethane methylation has also been used by Kratzl and Wittmann (1954) for the estimation of cinnamaldehyde groups in spruce native lignin. In this procedure diazomethane purportedly reacts with the ethylenic group in such units forming a pyrazoline derivative to an extent indicated by the gain in nitrogen content of the lignin. The reliability of this method is suspect, however, since it gives rise to an indicated cinnamaldehyde content for native lignin considerably higher than that obtained using other analytical approaches.

In another method based on the elimination of ring-conjugated ethylenic groups, the combined cinnamaldehyde and cinnamyl alcohol content of lignin in wood has been estimated through conversion of these structures to tetrahydrofuran derivatives by hydroformylation in the so-called oxo-reaction (Nahum 1969). In this particular analysis, the results apply only to the monomer fraction released during the hydroformylation reaction.

As indicated above, acidic and alkaline treatments of lignin may, under appropriate conditions, lead to the formation of stilbenoid structures. The amount of such groups in kraft Björkman lignin has been estimated by UV spectroscopy (Falkehag 1961, Falkehag et al. 1966). Specifically, the procedure involves the recording of the ionization difference spectrum of the lignin sample and comparison of the intensity of the long-wave $\Delta\varepsilon_i$ absorption maximum at approximately 378 nm with that of the appropriate model stilbene (Chap. 5.3). The amount of p,p′-dihydroxystilbene structures in kraft lignin has been estimated by oxidation with hydrogen peroxide in the presence of cupric ion. The stilbenequinone which is formed absorbs strongly in the 478 nm region (Adler and Häggroth 1950a,b). This procedure is not applicable to the determination of p,o′-dihydroxystilbenes.

Redinger (1961) has reported a value for the carbon-carbon double bond content of alkali lignin based on reaction with peroxyphthalic acid. The procedure is not described or referenced in the article.

The commonly used analytical approach based on the addition of a halogen, e.g., bromine, to unsaturated carbon-carbon bonds is not well suited for the determination of ethylenic groups in lignin mainly because of the occurrence of competing reactions such as ring bromination and oxidation. Nevertheless, Kürschner and Hostomsky (1962) claimed to have restricted bromine consumption to addition reactions by adding a trace of water to the bromine. Using this technique, the investigators determined the ethylenic group content of lignin in several wood samples from the bromine uptake and reported the results as the iodine number.

Attempts have been made to estimate ethylenic groups in lignin from ^{13}C-NMR measurements (Robert et al. 1986, Gellerstedt et al. 1986). However, the reliability of the results is questionable since the values reported are obtained indirectly by difference and because the calculations involved are based on the validity of several assumptions.

There is a dearth of reliable methods available for determining ethylenic groups as components of quinonoid ring systems in lignin. A sensitive and specific method developed by Lebo et al. (1990), based on an addition reaction with trimethyl phosphite, has been used to determine o-quinonoid structures in situ in the lignin component of refiner mechanical pulp. ^{31}P-NMR measurement capability is required for the analysis.

Of the methods described above, none can be regarded as being universally suited for the determination of ethylenic groups as the variety of structural environments in which this functional group exists in lignin realistically precludes such a possibility. Consequently, two methods, namely the determination of ethylenic groups present in the end units of native lignin and in stilbene structures present in chemically modified lignin, have been singled out as fulfilling the most common needs with respect to the determination of this group. The procedures for these analyses are outlined below.

7.3.3 Determination of Total Cinnamaldehyde and Cinnamyl Alcohol Units

7.3.3.1 Pre-Reduction of Carbonyl Groups with Lithium Aluminum Hydride (Note 1)

The following procedure is that described by Marton and Adler (1963). Finely powdered $LiAlH_4$ (4.4 g) is suspended in 50 ml of absolute dioxane in a hydrogen atmosphere and to the stirred mixture a solution of 3.0 g of lignin in 30 ml of absolute dioxane is added during 1 h (three-necked flask provided with a $CaCl_2$ drying tube). After complete addition, the reaction mixture is stirred for 1.5 h, cooled in an ice bath, and decomposed by the slow addition of a mixture of 50 ml of acetic acid and 30 ml of water. The resulting slurry is centrifuged and the solid material (A) obtained is washed with dioxane and the dioxane extract combined with the dioxane-acetic acid solution. The mixture is evaporated to dryness in vacuo, the residue dissolved in 20 ml of dioxane, and the solution is filtered. The filtrate is reprecipitated by addition of ether-petroleum ether (1:1, v/v; yield 0.7 g.

The solid material obtained after centrifuging the acidified mixture (material A) is further treated with dilute hydrochloric acid and the lignin residue thus obtained is centrifuged off, washed with water, dried, and purified by repreciptitating its dioxane solution as described above; yield 1.2 g. There is no apparent difference between the two lignin fractions.

7.3.3.2 Catalytic Hydrogenation

The following hydrogenation and isolation procedure is a composite of methods described by Adler and Marton (1961), Marton and Adler (1961), and Marton et al. (1961). The $LiAlH_4$-reduced lignin (0.1 g), dissolved in 15 ml of acetic acid containing 7% of water, is hydrogenated over Pd on $BaSO_4$ (Aldrich Chem. Co. and Fluka) (Note 2) at room temperature and atmospheric pressure for 6 h. The catalyst is centrifuged off and washed with acetic acid, the washings being combined with the solution. The solvent is removed under reduced pressure (bath temperature not exceeding 45 °C). The residue is redissolved in 93% acetic acid (5 ml for 0.1 g of lignin) and the solution is filtered and added dropwise to a 15-fold volume of water. The lignin precipitate is centrifuged off, washed with water, and dried in a vacuum desiccator over P_2O_5 and NaOH. This material is dissolved in a mixture of 1,2-dichloroethylene and ethanol (2:1, v/v), reprecipitated by pouring the filtered solution into ether-petroleum ether (1:1, v/v), and washed twice with the latter solvent mixture and finally with petroleum ether. After drying in a current of warm air or in a vacuum desiccator, the hydrogenated lignin is recovered in 90% yield.

7.3.3.3 Measurement of Hydrogenation Difference Spectrum

The principles and methodology relevant to the UV spectrophotometric determination of lignin and lignin model compounds have been discussed by Goldschmid 1954, Browning 1967b, and by Lin in Chapter 5.1.

The $LiAlH_4$-reduced lignin (X) and the hydrogenated $LiAlH_4$-reduced lignin (Y) are separately dissolved in 95% ethanol (Note 3) or in another suitable solvent (Table 5.3.2) to provide solutions containing 50 mg of X or Y/100 ml of solvent. The hydrogenation difference spectrum is conveniently measured in a recording spectrophotometer in which the solution of X is placed in the sample cuvette and the solution of Y in the reference cuvet. By this procedure the absorbance difference (ΔA) is obtained as a function of wavelength in the 210–400 nm spectral region. Calculation:

$$C = \frac{\Delta A_{295}}{(b)(\Delta\varepsilon_h)},$$

where C = molar concentration of ethylenic groups as the total of those initially present in the lignin as components of coniferaldehyde and coniferyl alcohol units, ΔA = absorbance difference, $\Delta\varepsilon_h$ = Molar absorptivity difference at 295 nm = $5100\,l\,mol^{-1}\,cm^{-1}$ (Marton and Adler 1961) (Note 4), b = Path length through the sample in cm.

The concentration of ethylenic groups in lignin is more commonly expressed as the number per phenylpropane (C_9) unit, per 100 phenylpropane units, or per methoxyl group. For softwood lignin the number per phenylpropane unit equals approximately the number per methoxyl group. With respect to the procedure outlined above, the following calculations apply:

$$\text{Ethylenic groups/MeO} = \frac{3100\,C}{(W)\ (\%\ \text{MeO in lignin})}$$

$$\text{Ethylenic groups/}C_9 = \frac{(C)\ (\text{unit weight of lignin})}{W}$$

W = sample weight of lignin in grams
Unit weight of spruce milled wood lignin = 185 (Marton 1971)

Note 1. The carbonyl groups also may be reduced with sodium borohydride using the procedure described by Marton and Adler (1961).
Note 2. The case for the use of freshly prepared catalyst has been discussed by Adler and Marton (1961).
Note 3. It is important that the sample be completely dissolved in the solvent to avoid underestimation of ethylenic group content.
Note 4. The absorption maximum and absorptivity are based on the hydrogenation difference spectrum of 4-0-isopropylconiferyl alcohol (Marton and Adler 1961).

7.3.4 Determination of Stilbenoid Units

7.3.4.1 Pre-Reduction of Carbonyl Groups with Lithium Aluminum Hydride ($LiAlH_4$)

The reduction is performed as described in Section 7.3.3.1.

7.3.4.2 Measurement of Ionization Difference Spectrum

Approximately 0.05 g of the $LiAlH_4$-reduced lignin accurately weighed to 0.001 g is dissolved in methylcellosolve-ethanol (1:4, v/v) or in another suitable solvent (Table 5.3.2) in a 100-ml volumetric flask and diluted to the mark with the same solvent.

Using a 1-ml pipet (0.01 ml subdivisions), a suitably sized aliquot (e.g., 0.3 ml) of the above solution is transferred to a 1-cm path quartz (reference) cuvette (~3-ml capacity) and diluted tenfold (i.e., to 3.0 ml) with the same solvent used to dissolve the sample. A second aliquot, identical in volume to the first, is similarly transferred to another 1-cm path quartz (sample) cuvette and diluted exactly tenfold with an aqueous sodium hydroxide solution of sufficient strength to raise the pH of the diluted solution to 12. The difference spectrum is measured immediately (Note 1) as the absorbance of the alkaline solution relative to that of the neutral solution in the 210–400 nm range. The difference in absorbance, ΔA, at approximately 378 nm should fall between 0.2 and 0.8 (Note 2).

Calculation:

Ethylenic groups in

$$\text{stilbenes/100 } C_9 \text{ units} = \frac{(\Delta A_{378})\ (\text{unit weight of lignin})\ (100)}{(\Delta\varepsilon_i)\ (W)},$$

where ΔA = absorbance difference, $\Delta\varepsilon_i$ = molar absorptivity difference at 378 nm = 24300 l mol cm^{-1} (Falkehag et al. 1966), W = sample weight of lignin in grams. Unit weight of pine kraft lignin = 178 (Marton 1971)

Note 1. The absorbance of alkaline solutions of certain stilbenes has been reported to decrease rapidly initially as a consequence of photochemical (Falkehag et al. 1966) and/or air oxidation (Lin et al. 1974) reactions. The latter authors have speculated on the underlying chemical changes responsible for the decreased absorbance.

Note 2. If the absorbance reading fails to fall within the prescribed range, the sample concentration should be adjusted accordingly and the spectrum re-measured.

7.3.5 Discussion of Procedures

The ethylenic group contents of several lignins are listed in Table 7.3.2 in terms of the structural type in which the group is present. Excluding o-benzoquinones, the values are uniformly low (<10%) irrespective of the method used for the determination and the material analyzed. The values included in Table 7.3.2 represent a major portion of the analytical data reported for the ethylenic content of lignin. Unfortunately, these figures encompass a very narrow range of lignin types; thus, for example, hardwood and grass lignins are poorly rep-

Table 7.3.2. Ethylenic group contents of various lignins

Material[a]	Structure	Method	Number/C_9 unit	Reference
Brauns native lignin	Cinnamaldehyde	Colorimetric	2.5	Adler and Ellmer 1948
Lignosulfonate	Cinnamaldehyde	Colorimetric	2	Adler and Ellmer 1948
Milled wood lignin[b]	Cinnamaldehyde	UV Spectroscopy ($\Delta\varepsilon_r$)	1–2	Klemola 1968
Milled wood lignin	Coniferyl alcohol	UV Spectroscopy ($\Delta\varepsilon_h$)	3	Marton and Adler 1961
Milled wood lignin	Coniferaldehyde	UV Spectroscopy ($\Delta\varepsilon_h$,$\Delta\varepsilon_i$)	3–4[c]	Aulin-Erdtman and Hegbom 1958, Marton and Adler 1961
Wood meal	Coniferaldehyde + Coniferyl alcohol	Hydroformylation	8	Nahum 1969
Alkali lignin	Unspecified	Perphthalic acid oxidation	10	Redinger 1961
Kraft milled wood lignin	Stilbenes	UV Spectroscopy ($\Delta\varepsilon_i$)	7–8	Falkehag et al. 1966
Refiner mech. pulp	o-Quinones	Formation of oxyphosphorane adduct	10–12	Lebo et al. 1988

[a] Except as indicated, all materials listed below were derived from softwoods.
[b] Prepared from birch wood.
[c] The majority of these are present in etherified coniferaldehyde units (Adler and Marton 1959, Aulin-Erdtman and Hegbom 1958).

resented in the table. This deficiency seriously restricts any attempt to identify trends based on comparisons of analytical data.

As was emphasized previously in this chapter, the procedures described above were developed for and are applicable only to the determination of ethylenic groups in a relatively small number of specified lignin structural types. Thus, ethylenic groups in several important structures such as styryl aryl ethers, p-quinonoid and phenylcoumarone units are excluded from analysis by the procedures described above. In addition to this limitation, neither method is particularly well suited for the routine analysis of ethylenic groups because each entails a step consisting of the time-consuming prior removal of α-carbonyl groups. This pre-reduction step is necessary since the spectra of conjugated carbonyl units partially overlap and hence interfere with those of structures containing ethylenic groups.

As is often the case when UV spectroscopy is used in lignin analysis, the procedures described above suffer from the disadvantage of involving calculations which are based on the assumption that the molar absorptivity of an appropriate model compound is identical with or closely approximates that of the lignin sample at the same wavelength. Selection of the most suitable wavelength at which to measure absorbance also constitutes a problem; for example p,p′- and p,o′-dihydroxystilbenes display absorption maxima at slightly different wavelengths and the absorptivity values corresponding to each absorption maximum are not identical (Falkehag et al. 1966).

The values obtained for total cinnamaldehyde and cinnamyl alcohol units in native lignins following the procedure outlined above can be safely assumed to be essentially free of contributions from ethylenic groups in other structural units. On the other hand, possible interference from styryl aryl ethers and other ring-conjugated groups in the determination of stilbene units in chemically modified lignins using the above procedure cannot as readily be ruled out. However, the fact that the long-wave absorption maximum in the ionization difference spectra of stilbenes is approximately 60 nm higher than that of other ring-conjugated ethylenic structures (Goldschmid 1971) suggests that an absorbance reinforcement resulting from spectral overlap will probably be small enough to be disregarded.

References

Adler E (1961) Recent studies on the structural elements of lignin. Pap Puu 43:634–643

Adler E, Ellmer L (1948) Coniferylaldehydgruppen in Holz und in isolierten Ligninpraparaten. Acta Chem Scand 2:839–840

Adler E, Falkehag S, Marton J, Halvarson H (1964b) The behavior of lignin in alkaline pulping. II. Model experiments with Arylalkyl-β-aryl ethers. Acta Chem Scand 18: 1313–1314

Adler E, Häggroth S (1950a) Über die Rötung des Sulfitzellstoffs. I. Studien über den Mechanismus und die Verhinderung der Rötung. Sven Papperstidn 53:287–294

Adler E, Häggroth S (1950b) Über die Rötung des Sulfitzellstoffs. II. Zur chemischen Natur der Rötung. Sven Papperstidn 53:321–326

Adler E, Lundquist K, Miksche GE (1966) The structure and reactivity of lignin. Adv Chem Ser 59:22–35

Adler E, Marton J (1959) Zur Kenntnis der Carbonylgruppen in Lignin. I. Acta Chem Scand 13:75–96

Adler E, Marton J (1961) Carbonyl groups in lignin. II. Catalytic hydrogenation of model compounds containing aryl carbinol, aryl carbinol ether, ethylene and carbonyl groups. Acta Chem Scand 15:357–369

Adler E, Marton J, Falkehag I (1964a) The behavior of lignin in alkaline pulping. I. Model experiments with phenylcoumarans. Acta Chem Scand 18:1311–1312

Aulin-Erdtman G, Hegbom L (1958) Spectrographic contributions to lignin chemistry. Sven Papperstidn 61:187–210

Bardet M, Robert D, Lundquist K (1985) On the reactions and degradation of the lignin during steam hydrolysis of aspen wood. Sven Papperstidn 88:R61–R67

Browning BL (1967a) Methods of wood chemistry, Vol. I. Wiley-Interscience, New York, 274–275

Browning BL (1967b) Methods of wood chemistry, Vol. II. Wiley-Interscience, New York, 762–764

Chang H-m, Allan GG (1971) Oxidation. In: Sarkanen KV, Ludwig CH (eds) Lignins. Occurrence, formation, structure and reactions. Wiley-Interscience, New York, 457–471

DeBaun RM, Nord FF (1951) A quantitative phloroglucinol-HCl test in the evaluation of lignin preparations. Tappi 34:71–73

Enkvist T (1960) New chemicals from cellulose waste liquids. Sven Kem Tidskr 72:93–100

Falkehag SI (1961) Stilbene structures in sulfate lignin. Pap Puu 43:655–656

Falkehag SI, Marton J, Adler E (1966) Chromophores in kraft lignin. Adv Chem Ser 59:75–89

Gellerstedt G, Gustavsson K, Lindfors E (1986) Structural changer in lignin during oxygen bleaching. Nord Pulp Pap Res J 3:14–17

Gierer J (1970) The reactions of lignin during pulping. A description and comparison of conventional pulping processes. Sven Papperstidn 73:571–595

Gierer J, Lenz B, Wallin N-H (1964) The reactions of lignin during sulfate cooking Part V. Model experiments on the splitting of aryl-alkyl ether linkages by 2 N sodium hydroxide and by white liquor. Acta Chem Scand 18:1469–1476

Goldschmid O (1954) Determination of phenolic hydroxyl content of lignin preparations by ultraviolet spectrophotometry. Anal Chem 26:1421–1423

Goldschmid O (1971) Ultraviolet spectra. In: Sarkanen KV, Ludwig CH (eds) Lignins. Occurrence, formation, structure and reactions. Wiley-Interscience, New York, 241–266

Harkin JM (1966) o-Quinonemethides as tentative structural elements in lignin. Adv Chem Ser 59:65–74

Hillis WE, Nelson P, Zadow G (1966) The cause of discoloration of Pinus Radiata bisulfite pulp. Appita 19:111–114

Klemola A (1968) Investigation of birchwood (Betula pubescens) lignin degraded by steam hydrolysis. Suom Kemistil B41:166–180

Kratzl K, Kisser W, Gratzl J, Silbernagel H (1959) 2-Guaiacyl ether of guaiacylglycerol and its conversion to coniferyl aldehyde and other arylpropane derivatives. Monatsh Chem 90: 771–782

Kratzl K, Wittmann E (1954) Reaktionen des Diazomethans mit α, β-ungesättigen Aldehyden. Monatsh Chem 85:7–22

Kringstad KP, Mörck R (1983) ^{13}C-NMR spectra of kraft lignins. Holzforschung 37:237–244

Kürschner K, Hostomsky G (1962) Über aliphatische Doppelbindungen in gewachsenen Ligninen. Holzforschung 16:180–189

Kvasnicka EA, McLaughlin RR (1955) Identification of spruce sulfite liquor components. Can J Chem 33:637–645

Lai Y-Z, Sarkanen KV (1971) Isolation and structural studies. In: Sarkanen KV, Ludwig CH (eds) Lignins. Occurrence, formation, structure and reactions. Wiley-Interscience, New York, 195–199

Lebo SE Jr, Lonsky WFW, McDonough TJ, Medvecz PJ and Dimmel DR (1990) The occurrence and light-induced formation of ortho-qûinonoid lignin structures in white spruce refiner mechanical pulp. J Palp Pap Sci 16:J139–J143

Lin SY, Falkehag SI, Gierer J, Szabo-Lin I (1974) A contribution to lignin spectroscopy. Paper presented at the 167th ACS National Meeting, Los Angeles, CA

Lindgren BO, Mikawa H (1957) The presence of cinnamyl alcohol groups in lignin. Acta Chem Scand 11:826–835

Marton J (1971) Reactions in alkaline pulping. In: Sarkanen KV, Ludwig CH (eds) Lignins. Occurrence, formation, structure and reactions. Wiley-Interscience, New York, 639–694

Marton J, Adler E (1961) Carbonyl groups in lignin. III. Mild catalytic hydrogenation of Björkman lignin. Acta Chem Scand 15:370–387

Marton J, Adler E (1963) Reactions of lignin with methanolic hydrochloric acid. A discussion of some structural questions. Tappi 46:92–98

Marton J, Adler E, Persson K-E (1961) Carbonyl groups in lignin. IV. Infrared absorption studies and examination of the volumetric borohydride method. Acta Chem scand 15: 384–397

Nahum LS (1969) Estimation of double bond content in lignin from the results of the oxo reaction of wood and lignin model compounds. Tappi 52:712–714

Pew JC, Connors WJ (1971) Color of coniferous lignin. Tappi 54:245–251

Redinger L (1961) Alkali lignin, its condensation products with phenols, and preparation of curable resins. Monatsber Dtsch Akad Wiss (Berl) 3:571–578

Richtzenhain H, van Hofe C (1939) A stilbene derivative from the sulfite waste liquor. Chem Ber 72:1890–1892

Robert E, Gellerstedt G, Bardet M (1986) Carbon-13 NMR analysis of lignins obtained after sulfonation of steam exploded aspen wood. Nord Pulp Pap Res J 3:18–25

Tanaka J, Kondo T (1957) Alkaline degradation products of protolignin. Parupu Kami Kôgyô Zasshi 11:29–33

Tanaka J, Kondo T (1959) Degradation of lignin. III. Alkaline oxidation products of hardowood lignin. Mokuzai Gakkaishi 5:105–108

7.4 Determination of Carbonyl Groups

C.-L. CHEN

7.4.1 Introduction

As early as 1922, Klason (1922) postulated the presence of carbonyl groups, in particular conjugated carbonyl groups, in coniferyl aldehyde-type structures, in spruce lignins on the basis of the chemical characteristics of lignosulfonates. Direct evidence for the presence of carbonyl groups of this type in lignins was, however, obtained only 25 years later by Adler and co-workers and by Pew (Adler et al. 1948, Adler and Ellmer 1948, Adler 1951, Pew 1951) during an investigation on the reaction mechanism of the Wiesner reaction. In this reaction, plant tissues containing lignins form a purple color when treated with phloroglucinol (4) in concentrated hydrochloric acid (Chap. 2.1.3). As shown in Fig. 7.4.1, the color is attributed to the formation of the cationic chromophore (6), a conjugate acid of the extended quinonemethide (5) produced by condensation of etherified p-hydroxycinnamaldehyde-type substructures, (1), (2), and (3), in lignins with phloroglucinol (4) (Adler et al. 1948, Adler 1951, Pew 1951). Using a procedure based on this reaction, the total content of etherified p-hydroxycinnamaldehyde-type substructures in spruce lignin was estimated to be approximately 2–3/100 C_9 units by UV spectroscopy (Adler and Ellmer 1948).

Three decades later, Geiger and Fuggerer (1979) further demonstrated that chromophore (6) is also formed from etherified p-hydroxycinnamyl alcohol substructures of types (7), (8), and (9) in lignins on treatment with phloroglucinol in concentrated hydrochloric acid in the presence of oxygen, i.e., by autoxidation of these three substructures to the corresponding etherified p-hydroxycinnamaldehyde intermediates (1), (2), and (3), and subsequent condensation with phloroglucinol. In addition to cinnamaldehyde-type conjugated carbonyls (1)–(3), spruce (*Picea abies*) milled wood lignin (MWL) contains conjugated carbonyl (α-carbonyl) groups typified by those present in structures (10) and (11) (see Fig. 7.4.2 for structures) and possibly non-conjugated (α-carbonyl) groups and quinone ketal (13) structures (Adler and Marton 1959, Larsson and Miksche 1972). Spruce MWL contains structure type (13) only in trace amounts. The α-carbonyl content increases with increasing ball-milling of the wood meal during the preparation of MWL indicating that some of α-carbonyl groups and possibly even β-carbonyl groups are produced artificially via homolytic cleavage of α- and β-aryl ether bonds during the isolation process.

Springer Series in Wood Science
Methods in Lignin Chemistry
(Edited by S.Y. Lin and C.W. Dence)

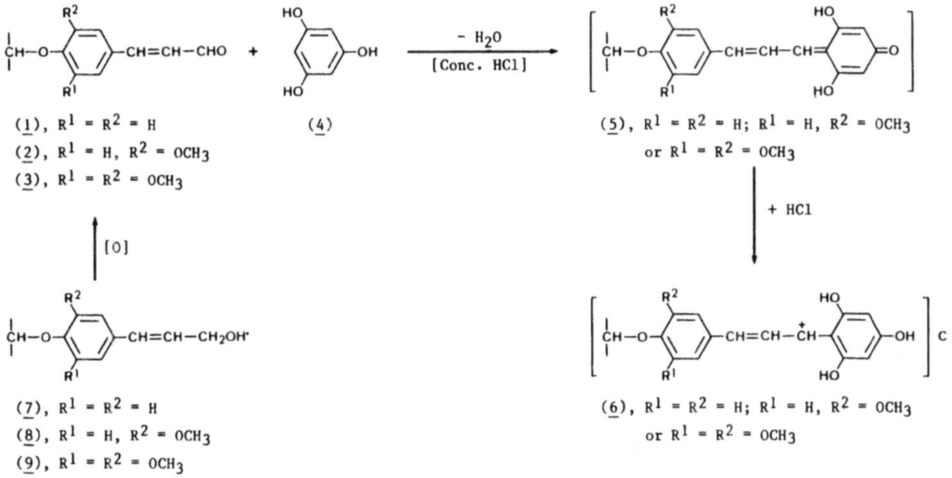

(1), $R^1 = R^2 = H$
(2), $R^1 = H$, $R^2 = OCH_3$
(3), $R^1 = R^2 = OCH_3$
(4)
(5), $R^1 = R^2 = H$; $R^1 = H$, $R^2 = OCH_3$ or $R^1 = R^2 = OCH_3$
(6), $R^1 = R^2 = H$; $R^1 = H$, $R^2 = OCH_3$ or $R^1 = R^2 = OCH_3$
(7), $R^1 = R^2 = H$
(8), $R^1 = H$, $R^2 = OCH_3$
(9), $R^1 = R^2 = OCH_3$

Fig. 7.4.1. Reaction mechanisms for Wiesner reaction

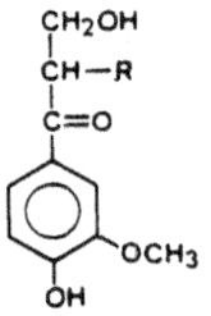
(10)

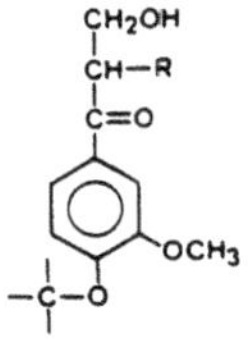
(11)

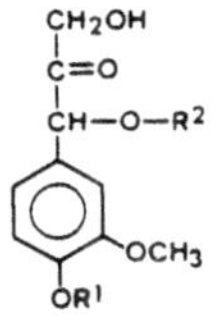
(12), R^1 = H or Alkyl, R^2 = H or Alkyl

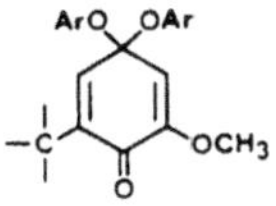
(13)

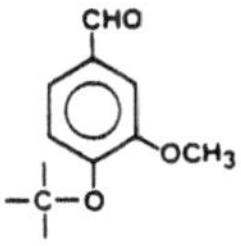
(14)

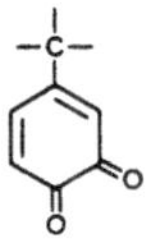
(15)

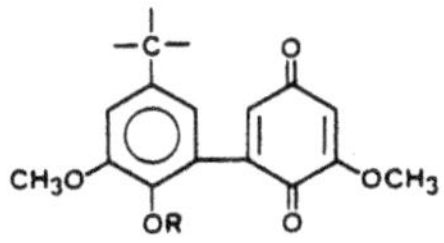
(16), R = H or Alkyl

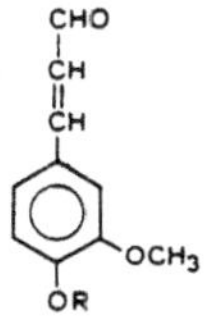
(17), R = CH_3
(18), R = $CH(CH_3)_2$

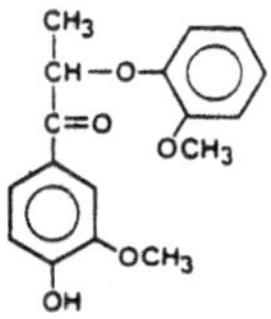
(19)

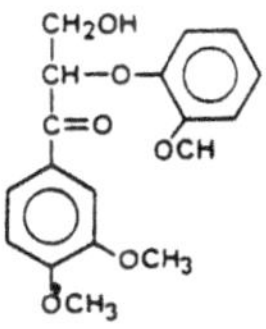
(20)

Fig. 7.4.2. Structures of lignin moieties with carbonyl groups and some lignin model compounds

Technical lignins, in particular alkali lignins, contain appreciable amounts of α-carbonyl groups in addition to benzaldehyde- and o-quinone-type carbonyl groups, (14) and (15), (see Fig. 7.4.2 for structures). Cinnamaldehyde-type carbonyl groups are not present in alkali lignins because such structures undergo reverse aldol condensation reactions, yielding p-alkoxybenzaldehyde structures via heterolytic cleavage of the C_α—C_β bond. Thus, the p-alkoxycinnamaldehyde-type structures (1)–(3) enhance, to some extent, the delignification rate during alkaline pulping. In contrast, arylpropane units, such as structure (11), which contain an α-carbonyl, undergo cleavage of C_β—C_γ bonds with elimination of formaldehyde via a reverse aldol reaction under pulping conditions. The cleavage of C_β—C_γ bonds does not contribute significantly to delignification during alkaline pulping.

Carbonyl groups in lignin, in particular those present as parts of o- and p-quinonoid, quinonemethide, and other extended conjugate enone systems, are contributors or potential contributors to the color observed in lignified plant tissues (Hon and Glasser 1979, Lebo et al. 1988). In addition, these structure types are recognized as being sensitizers in the photo-yellowing of lignocelluloses (Brunow and Eriksson 1971, Lin and Kringstad 1970). Oxidation of lignins with reagents such as molecular oxygen, hydrogen peroxide, and chlorine may further increase the carbonyl content (Chang and Allen 1971). The o- and p-quinone structures so generated in residual lignins of pulps may further alter the color of the pulps and promote their photosensitization capacity (Gierer and Lin 1972, Panchapakesan 1990).

7.4.2 Proposed Methods for the Determination of Carbonyl Groups

The carbonyl content of lignin preparations has been determined using such common reagents as phenylhydrazine (Schubert and Nord 1950), thiobenzylhydrazide (Björkman and Person 1957), and hydroxylamine hydrochloride (Adler and Gierer 1955, Gierer and Söderberg 1959, Adler and Marton 1959, Gierer and Lenz 1965). Other methods used for the determination of carbonyl groups in lignins consist of reduction with either sodium or potassium borohydride (Gierer and Söderberg, Marton et al. 1961), catalytic hydrogenation (Marton and Adler 1961) and UV spectroscopy (Adler and Marton 1959). Because thiobenzylhydrazide may also react with benzyl alcohol-type hydroxyl groups, it cannot be employed for the determination of the total carbonyl content in lignins.

Among the methods proposed for the determination of total carbonyl content in lignins based on reaction with carbonyl reagents, the reaction with hydroxylamine hydrochloride is the most effective and simplest method as it is conducted at room temperature and involves a relatively simple procedure. In addition, hydroxylamine hydrochloride reacts with almost all types of carbonyl groups, including those in diketones (Heuser 1953a) and p- and o-quinones (Heuser 1953b). In general, the carbonyl group in an aldehyde or ketone condenses with hydroxylamine hydrochloride, forming an oxime with liberation of hydrochloric acid:

$$>C{=}O + NH_2OH \cdot HCl \rightarrow \; >C{=}NOH + H_2O + HCl \qquad (1)$$

Since one carbonyl group quantitatively liberates 1 mol of hydrochloric acid, the total carbonyl content of a lignin preparation can be conveniently determined by titration of the hydrochloric acid liberated in the reaction (Heuser 1953a,b). Using this procedure, the total carbonyl content of spruce (*Picea abies*) milled wood lignin (MWL) was found to be approximately 0.19 mol/OCH_3 (Gierer and Soderberg 1959) or approximately 0.18 mol/C_9 unit, assuming that spruce MWL has an average methoxyl content of 0.96 mol/C_9 unit.

Alternatively, the total carbonyl content of a lignin preparation may be determined by reduction with either sodium or potassium borohydride. This method is based on the Lindberg-Misiorny method for the determination of carbonyl groups in carbohydrates (Lindberg and Misiorny 1952, Lindberg and Theander 1954). Both sodium and potassium borohydrides are insoluble in ether but soluble in water without decomposition. At room temperature, they effectively reduce aldehydes and ketones in aqueous and methanol solutions, in particular in weakly alkaline solutions, without simultaneously reducing olefinic double bones:

$$4 \; >C{=}O + NaBH_4 + 3H_2O \rightarrow 4 \; >CHOH + NaH_2BO_3 \qquad (2)$$

$$NaBH_4 + H_2SO_4 + 3H_2O \rightarrow NaHSO_4 + H_3BO_3 + 4H_2 \qquad (3)$$

Carbonyl groups in a lignin preparation are reduced with an excess of sodium borohydride in an alkaline solution [Eq. (2)]. After the reaction, excess sodium borohydride is decomposed by addition of dilute sulfuric acid solution [Eq. (3)]. The volume of molecular hydrogen thus liberated is measured volumetrically. The same experiment is repeated in the absence of the lignin sample (blank test). The total carbonyl content of the lignin preparation is determined from the difference in the volumes of hydrogen liberated in the presence and absence of sample, since each mole of hydrogen consumed corresponds to a mole of carbonyl group, as Eqs. (2), and (3) indicate (Gierer and Soderberg 1959, Marton et al. 1961).

The UV spectra of p-hydroxycinnamaldehyde, 1-(p-hydroxyphenyl)-1-propanone, and their derivatives have intense adsorption bands near 400 and 350 nm, respectively, in an alkaline solution. Similarly, p-alkoxycinnamaldehyde and 1-(p-alkoxyphenyl)-1-propanone derivatives absorb near 340 and 310 nm in neutral and alkaline solutions, respectively. These bands disappear when the carbonyl groups in compounds of these types are reduced to the corresponding benzylic alcohol groups with sodium borohydride. Thus, the content of carbonyl groups in lignin preparations can be directly determined by a combination of borohydride reduction and UV spectroscopy if appropriate lignin models are selected as reference compounds for determining the change in molar absorptivity of the absorption band resulting from reduction of a particular carbonyl group to the corresponding benzylic alcohol (Adler and Marton 1959). The differential molar absorptivity for the reduction of carbonyl groups is a function of time, since the change of molar absorptivity during the reduction is a kinetic measure of the reaction.

The carbonyl contents of the major carbonyl-containing substructures in spruce (*Picea abies*) MWL, typified by (2) in Fig. 7.4.1 and by (10), (11), and (12) in Fig. 7.4.2, have been determined by combined sodium borohydride reduction and UV spectroscopic techniques (Adler and Marton 1959, Marton and Adler 1961, Marton et al. 1961). By this combination of techniques, the quantities of conjugated carbonyls of structure types typified by (2), (10), and (11) in the MWL were found to be approximately 0.03, 0.01, and 0.05–0.06 mol/OCH_3, respectively, for a total carbonyl content of 0.09–0.10 mol/OCH_3. The difference between the total carbonyl content, 0.2 mol/OCH_3 as determined by the hydroxylamine method, and the total quantity of conjugated carbonyls, was assumed to be the quantity of carbonyls present in aryl-2-oxopropane-type substructures (12), i.e., β-carbonyl groups. However, no direct evidence has been presented to support the existence of such groups in lignins present in wood.

7.4.3 Determination of Total Carbonyl Content of Lignins by Reaction with Hydroxylamine Hydrochloride (Gierer and Söderberg 1959, Gierer and Lenz 1965)

Reagents

Ethanol, 95%
Hydroxylamine hydrochloride solution, 0.05 M in 80% ethanol (3.475 g in 1000 ml of 80% ethanol)
Nitrogen gas
Standard sodium hydroxide solution: 0.01 N

Procedure

Approximately 200 mg of the lignin sample (Note 1), accurately weighed to the nearest 0.1 mg, is placed in a 100-ml three-necked round-bottomed flask fitted with a reflux condenser and an inlet for nitrogen. The sample is dissolved in 10 ml of 95% ethanol with addition of water if necessary. The pH of the solution is adjusted to 4.0 with dilute hydrochloric acid. The flask with the solution is placed in a constant temperature bath at 25 °C. After thermal equilibrium is attained, 40 ml of 0.05 M hydroxylamine hydrochloride solution (2 mmol) in 80% ethanol, adjusted to pH 4 with dilute hydrochloric acid (Note 2) and to a temperature of 25 °C, is added to the sample solution in the flask at reaction time zero. A stream of nitrogen saturated with ethanol (Note 3) is passed through the mixture, which is then allowed to react at 25 °C under mechanical stirring for specified reaction times up to a maximum of 30 h (Note 4).

Note 1. See Part 3 and Chapter 7.1.2 for lignin sample preparation.
Note 2. Hydroxylamine hydrochloride is a salt of weak base and strong acid and undergoes hydrolysis at pH 5–7. Thus, the reaction mixture is adjusted to pH 4 with hydrochloric acid to prevent the hydrolysis of this salt.

Note 3. Nitrogen gas is first passed through a trap containing ethanol.
Note 4. Reaction periods are 1, 2, 4, 7, 10, 20, and 30 h.

Titration

At the end of each reaction period (Note 4), the liberated hydrochloric acid in the reaction mixture is titrated with standard 0.01 N sodium hydroxide solution, preferably employing an automatic titrator, to the pH of a blank solution treated in the same manner but in the absence of sample. Alternatively, the liberated hydrochloric acid is concurrently titrated with standard 0.01 N sodium hydroxide solution using an autotitrator to maintain the pH of the solution constant at 4. Buret readings (Note 5) are taken at various reaction periods during the 30-h reaction time.

Note 5. A calibrated 10- or 25-ml microburet is used.

Calculation of the Total Carbonyl Content

Since 1 ml of 1 N sodium hydroxide solution is equivalent to 28.01 mg of carbonyl group, the total of carbonyl groups reacted at reaction time t is calculated as follows:

$$(\text{Percent C=O})_t \text{ (based on lignin sample)} = \frac{2801(V_s - V_b)N}{W} \quad (4)$$

$$(\text{C=O/OCH}_3)_t \text{ (mol C=O/mol OCH}_3) = \frac{31.034(\%\ \text{C=O})_t}{28.01(\%\ \text{OCH}_3)}, \quad (5)$$

where V_s is the ml of standard sodium hydroxide solution required for titration of the reaction mixture at reaction time t, V_b is the ml of standard sodium hydroxide solution required for titration of the blank, N is the normality of the standard sodium hydroxide solution, W is the weight of the moisture-free sample in mg (Note 1), and % OCH_3 is the methoxyl content of the sample expressed as a weight percentage (Note 6). The values for $(C{=}O/OCH_3)_t$ are plotted against time. The value for the total carbonyl content of the lignin, $(C{=}O/OCH_3)_0$, is determined by extrapolation of the straight line portion of the curve to zero time (Note 8). The total carbonyl content of the lignin may also be expressed as $C{=}O/C_9$:

$$\text{C=O/C}_9 \text{ (mol/C}_9 \text{ unit)} = (\text{C=O/OCH}_3)_0\, X, \quad (6)$$

where C_9 is the phenylpropane unit of the lignin and X is the methoxyl content of the lignin expressed in mol/C_9 unit (Note 8).

Note 6. See Chapter 7.6.3 for determination of methoxyl groups in lignins.
Note 7. Although the reaction is conducted under nitrogen atmosphere by passing nitrogen gas through the reaction mixture, a small amount of

oxygen may be present in the reaction vessel. This oxygen may initiate autoxidation of the lignin sample, resulting in formation of new carbonyl groups. In addition, hydroxylamine hydrochloride reacts with vinyl ethers in the same manner as it reacts with carbonyls. These reactions result in the liberation of extraneous hydrochloric acid with increasing reaction time. Thus, extrapolation of the $C{=}O/OCH_3$ value to zero time is necessary.

Note 8. The value $C{=}O/C_9$ is meaningful only if the lignin preparations under study are sufficiently pure and have C_9 structure (some technical lignins may not have a true C_9 structure). In such a case, results for the total carbonyl content of lignin should be expressed as $C{=}O/OCH_3$. See Chapter 7.6.3 for the significance of the methoxyl content of lignin.

7.4.4 Determination of Conjugated Carbonyl Groups by UV Spectroscopy (Adler and Marton 1959)

Reagents
2-Methoxyethanol
Ethanol. 95%
Sodium hydroxide solution, 0.03 M
Sodium borohydride solution, 0.05 M in 0.03 M sodium hydroxide solution (189 mg in 100 ml of 0.03 M sodium hydroxide solution)
Several procedures are described in the following sections.

7.4.4.1 Preparation of Lignin Sample Stock Solution

Approximately 10 mg of the lignin (Note 1), accurately weighed to the nearest 0.01 mg, is placed in a 100-ml volumetric flask containing approximately 80 ml of a mixture of 2-methoxyethanol-ethanol (2:3, v/v). When the sample is completely dissolved, the solution is diluted to 100 ml with the 2-methoxyethanol-ethanol mixture. This sample solution is the stock solution.

7.4.4.2 UV Spectrum of Lignin in Alkaline Solution (Chap. 5.3.4.3)

A UV spectrum (200–400 nm) of the sample in an alkaline solution is obtained by placing, in a 10-mm cell, a solution consisting of 2 ml of the stock solution and 1 ml of 0.03 M sodium hydroxide solution. A solution prepared from the 2-methoxyethanol-ethanol mixture and 0.03 M sodium hydroxide solution (2:1, v/v) is used as the reference solution. The UV spectrum of the sample is obtained by plotting molar absorptivity/OCH_3 against wavelength. The molar absorptivity/OCH_3 at a wavelength i, $(\Delta')_i$, is obtained from the following relationship:

$$(\Delta')_i = \frac{A_i(C_9\text{–}UW)}{bc}\,(\ell\,mol^{-1}\,cm^{-1}), \tag{7}$$

where A_i is the absorbance at wavelength i, b is the length of the light path in cm (b = 1), c is the concentration of the sample solution in g ℓ^{-1}, and C_9–UW is the C_9 unit formula weight of the lignin sample in g.

7.4.4.3 UV Spectrum of Sodium Borohydride-Reduced Lignin in Alkaline Solution

An alkaline solution of the sodium borohydride-reduced lignin sample, prepared by mixing 2 ml of the stock solution and 1 ml of the 0.05 M sodium borohydride solution, is placed in a 10-mm cell and the UV spectrum (200–400 nm) is recorded against a reference solution composed of the 2-methoxyethanol-ethanol mixture and 0.03 M sodium hydroxide solution (2:1, v/v). UV spectra of the sodium borohydride-reduced lignin sample are obtained after reduction times of 100 min and 45 h (Note 9) at room temperature in the same manner as described for the corresponding lignin sample in Sect. 7.4.4.2.

Note 9. At 100 min, reduction of the p-alkoxycinnamaldehyde-type carbonyl group in structures (1)–(3) and 1-(p-alkoxyphenyl)-1-propanone-type carbonyl groups (α-carbonyl groups conjugated with nonphenolic rings) in structure (11) is virtually complete since the half-life time for the sodium borohydride reduction of the carbonyl group in 4-O-methylconiferyl aldehyde (17), 4-O-isopropylconiferyl alcohol (18), and 2-(2-methoxyphenoxy)-1-(3,4-dimethoxyphenyl)-3-hydroxy-1-propanone (20) is in the range of 0.5–2.0 min (see Fig. 7.4.1 and 7.4.2 for structures). In contrast, at 100 min, reduction of a 1-(p-hydroxyphenyl)-1-propanone-type carbonyl group (α-carbonyl groups conjugated with phenolic rings) typified by structure (10) is only 5 mol % complete. At 45 h, the reduction of the phenolic α-carbonyl group in structure (10) is almost complete since the half-life time for sodium borohydride reduction of carbonyl group in 2-(2-methoxyphenoxy)-1-(4-hydroxy-3-methoxyphenyl)-1-propanone, (19), is 10.0 h. Extending the reaction to 60 h does not result in an increase of the $\Delta\varepsilon_r$ value at 356 nm.

7.4.4.4 Reduction Difference ($\Delta\varepsilon_r$) Spectrum (see Chap. 5.3.4.4)

The $\Delta\varepsilon_r$ spectra, A and B, of a lignin can be obtained by subtracting the UV spectrum of the original lignin from that of the corresponding sodium borohydride-reduced lignin after reduction times of 100 min and 45 h, respectively. Alternatively, the $\Delta\varepsilon_r$ spectra, A and B, can be obtained by directly measuring the absorbance of the sodium borohydride-reduced lignin sample solution after reduction times of 100 min and 45 h (Sect. 7.4.4.3) against that of a solution of the corresponding original lignin placed in the reference cell (Sect. 7.4.4.3). In the latter case, the $(\Delta\varepsilon_r)_i$ value at wavelength i is calculated as follows:

$$(\Delta\varepsilon_r)_i = \frac{(\Delta A)_i(C_9\text{-}UW)}{bc} \; (\ell\,mol^{-1}\,cm^{-1}), \qquad (8)$$

where $(\Delta A)_i$ is the absorbance difference at wavelength i, b is the length of the light path in cm (b = 1), c is the concentration of lignin in g ℓ^{-1}, and C_9–UW is the C_9 unit formula weight of the lignin in g.

7.4.4.5 Calculation of the Conjugated Carbonyl Contents

As shown in Table 7.4.1, 4-O-isopropylconiferyl aldehyde (18) and 2-(2-methoxyphenoxy)-1-(3,4-dimethoxyphenyl)-3-hydroxy-1-propanone (20) have $\Delta\varepsilon_r$ maxima at 342 and 310 nm with corresponding $\Delta\varepsilon_r$ values of 19 800 and 9400 $\ell\,mol^{-1}\,cm^{-1}$, respectively, at the reduction time of 100 min (see Table 7.4.1 for $\Delta\varepsilon_r$ values and Fig. 7.4.2 for structures of model compounds). In addition, compound (18) has a $\Delta\varepsilon_r$ value of 2020 $\ell\,mol^{-1}\,cm^{-1}$ at 310 nm, while compound (20) has a $\Delta\varepsilon_r$ value of 10 500 $\ell\,mol^{-1}\,cm^{-1}$ at 342 nm (Adler and Marton 1959). Thus, the quantities of etherified coniferyl aldehyde-type (2) and etherified 1-(guaiacyl)-1-propanone-type (11) structures in lignin can be estimated by simultaneously solving Eqs. (9) and (10):

$$\text{At 310 nm: } 10\,500\,x + 9400\,y = (\Delta\varepsilon_r)_1. \qquad (9)$$

$$\text{At 342 nm: } 19\,800\,x + 2020\,y = (\Delta\varepsilon_r)_2, \qquad (10)$$

where x is the quantity of structure type (2) in mol/C_9, y is the quantity of structure type (11) in mol/C_9, $(\Delta\varepsilon_r)_1$ is the $\Delta\varepsilon_r$ value at 310 nm for a reduction time of 100 min, and $(\Delta\varepsilon_r)_2$ is the $\Delta\varepsilon_r$ value at 342 nm for a reduction time of 100 min (Note 10).

Since 2-(2-methoxyphenoxy)-1-(4-hydroxy-3-methoxyphenyl)-1-propanone (19) has a $\Delta\varepsilon_r$ value of only 1200 $\ell\,mol^{-1}\,cm^{-1}$ at 356 nm for a reduction time of 100 min and a $\Delta\varepsilon_r$ value of 24 300 $\ell\,mol^{-1}\,cm^{-1}$ at 356 nm for a reduction time of 45 h, it follows that the contribution of 1-(guaiacyl)-1-propanone-type (10)

Table 7.4.1. UV spectral properties of carbonyl-containing lignin model compounds on reduction with sodium borohydride. (Adler and Marton 1959)

Model compounds	$\Delta\varepsilon_r$-Maximum at λ (nm)	$\Delta\varepsilon_r$	Reduction time	Half-life time ($t_{\frac{1}{2}}$) for reduction
4-O-Methyl-coniferyl aldehyde (17)	340	24 000	100 min	1.0 min
4-O-Isopropyl-coniferyl aldehyde (18)	342	19 800	100 min	0.5 min
2-(2-Methoxyphenoxy)-1-(4-hydroxy-3-methoxyphenyl)-1-propanone (19)	356	1200	100 min	10.0 h
	356	24 300	45 h	
2-(2-Methoxyphenoxy)-1-(3,4-dimethoxyphenyl)-3-hydroxy-1-propanone (20)	310	9400	100 min	2.0 min

structures to the $\Delta\varepsilon_r$ spectrum of lignin at 356 nm is insignificant after a reduction time of 100 min. Thus, the quantity of 1-(guaiacyl)-1-propanone-type (10) structures in lignin can be estimated from the difference in $\Delta\varepsilon_r$ values at 356 nm for reduction times of 100 min and 45 h as follows:

$$\text{At 356 nm: } z = \frac{(\Delta\varepsilon_r)_b - (\Delta\varepsilon_r)_a}{24\,300 - 1200} \text{ (mol/C}_9\text{ unit)}, \tag{11}$$

where z is the quantity of structure type (10) in mol/C_9, $(\Delta\varepsilon_r)_a$ is the $\Delta\varepsilon_r$ value at 356 nm for a reduction time of 100 min, and $(\Delta\varepsilon_r)_b$ is the $\Delta\varepsilon_r$ value at 356 nm for a reduction time of 45 h (Note 10).

Note 10. The values x, y, and z may also be expressed as mol/OCH_3 by using Eq. (6).

7.4.5 Discussion of the Methods

7.4.5.1 Comparison of the Hydroxylamine Hydrochloride and Borohydride Methods

As mentioned previously (Sect. 7.4.1), the total carbonyl content of spruce (*Picea abies*) milled wood lignin (MWL) was reported to be approximately 0.19 mol/OCH_3 using hydroxylamine hydrochloride compared to approximately 0.43 mol C=O/OCH_3 for the same sample analyzed by the borohydride method (Gierer and Soderberg 1959). Employing the same methods, Marton and co-workers (1961) obtained total carbonyl contents of 0.2 and 0.4 mol/OCH_3, respectively, for spruce MWL. Based on these values, the total carbonyl content of spruce MWL determined by the borohydride method is about twice the value determined by the hydroxylamine hydrochloride method, although the precision and accuracy of both methods are rather poor, showing errors of approximately ±3% or higher. The errors are particularly large in the determination of the total carbonyl content by the borohydride method.

It the results obtained using the borohydride method are correct, the total carbonyl content of spruce MWL (0.4–0.43 mol C=O/OCH_3) corresponds to approximately one carbonyl group/2.5 C_9 units. This would imply that the carbonyl group is one of the major functional groups in spruce lignin and possibly in other softwood lignins, and even in hardwood lignins as well. This presumption led Marton and co-workers (1961) to conduct further investigations on the structural character of carbonyl groups in spruce lignin.

As stated above, spruce MWL contains conjugated carbonyl groups of structure types (2), (10), and (11) (see Figs. 7.4.1 and 7.4.2 for structures) in concentrations of approximately 0.03, 0.01, and 0.05–0.06 mol/OCH_3, respectively, or a total of approximately 0.09–0.10 mol C=O/OCH_3. The total volume of hydrogen consumed in the hydrogenation of spruce MWL over a nickel boride catalyst indicates that the sum of carbonyl and ethylene groups in this lignin does not exceed 0.30 group/OCH_3 (Marton and Adler 1961). Since the number of ethylene groups was found to be approximately 0.06–

0.08 mol/OCH_3, the total carbonyl content of the lignin would be approximately in the 0.22–0.24 mol/OCH_3 range by the catalytic hydrogenation method. This value differs by only 0.02–0.04 mol/OCH_3 from the value of approximately 0.2 mol/OCH_3 obtained by the hydroxylamine hydrochloride method. This difference could be well within the limit of experimental error for the two methods. In addition, hydroxylamine hydrochloride reacts with almost all types of carbonyl groups (Heuser 1953a,b). Thus, the hydroxylamine hydrochloride method is more appropriate for the determination of total carbonyl content in lignins than is the borohydride method.

7.4.5.2 Accuracy of the Procedure for Determining Conjugated Carbonyl Groups

The conjugated carbonyl contents of structure types (2), (10), and (11) in spruce MWL are determined by measurement of the UV $\Delta\varepsilon_r$ spectra of the lignin using the $\Delta\varepsilon_r$-maxima in the $\Delta\varepsilon_r$ spectra of model compounds (18), (19), and (20) as standards, respectively (Marton et al. 1961) (see Figs. 6.4.1 and 6.4.2 for structures). Within the limit of Beer's law, the precision of the method is approximately ±2% for the determination of total conjugated carbonyl groups in lignin preparations. However, the accuracy of the method cannot be estimated because the chemical structure of lignin containing structure types (2), (10), and (11) is not known exactly, and because the model compounds are arbitrarily selected as standards for the determination. For example, as shown in Table 7.4.1, 4-O-methylconiferyl aldehyde (17) and 4-O-isopropylconiferyl aldehyde (18) have $\Delta\varepsilon_r$-maximum values of 24000 and 19800 $\ell\,mol^{-1}\,cm^{-1}$ at 340–342 nm, respectively. Consequently, if compounds (17) and (19) are chosen as the standards for the determination of conjugated carbonyl groups in lignins instead of compounds (18) and (19), the estimated value for structure type (2) in the spruce MWL would be approximately 15% less and the value for the structure type (11) approximately 10% more than the corresponding values obtained using the method described in Sect. 7.4.4.5.

References

Adler E (1951) Sulfite pulping properties of spruce wood from unpeeled, floated logs. Part 1. Catechol tannins in the surface layer of the sapwood. Sven Papperstidn 54:445–450

Adler E, Björkqvist KJ, Häggroth S (1948) Über die Ursache der Farbreaktionen des Holzes. Acta Chem Scand 2:93–94

Adler E, Ellmer L (1948) Coniferylaldehydgruppen im Holz und in isolierten Ligninpraparaten. Acta Chem Scand 2:839–840

Adler E, Gierer J (1955) The alkylation of lignin with alcoholic hydrochloric acid. Acta Chem Scand 9:84–93

Adler E, Marton J (1959) Zur Kenntnis der Carbonylgruppen im Lignin. I. Acta Chem Scand 13:75–96

Björkman A, Person B (1957) Studies on finely divided wood. Sven Papperstidn 60:285–292

Brunow G, Eriksson B (1971) α-Carbonyl groups as sensitizers in the photo-dehydrogenation of phenolic structures in Lignin. Acta Chem Scand 25:2779–2781

Chang H-m, Allen GG (1971) Oxidation. In: Sarkanen KV, Ludwig CH (eds) Lignins. Occurrence, formation, structure and reactions. Wiley-Interscience, New York, 433–485

Geiger H, Fuggerer H (1979) Über den Chemismus der Wiesener-Reaktion auf Lignin. Z Naturforsch 34b:1471–1472

Gierer J, Lenz B (1965) Reaction of lignin during sulphate cooking. Sven Papperstidn 68: 334–338

Gierer J, Lenz B, Noren I, Soderberg S (1964) Reactions of lignin during sulfate cooking III. The splitting of aryl-alkyl ether bonds in milled wood lignins by alkali. Tappi 47:233–239

Gierer J, Lin SY (1972) Photodegradation of lignin. A contribution to the mechanism of chromophore formation. Sven Papperstidn 75:233–239

Gierer J, Söderberg S (1959) Über die Carbonylgruppen des Lignins. Acta Chem Scand 13:127–137

Green JW (1963) Determination of carbonyl group. In: Whistler RI (ed) Methods in carbohydrates Chem Vol. III, Academic Press, New York, 49–54

Heuser E (1953a) Analytik der Carbonylgruppe. In: Müller E, Bayer O, Meerwein H, Ziegler K (eds) Methoden der organischen Chemie (Houben-Weyl) 4. Aufl., Bd II Analytische Methoden. George Thieme, Stuttgart, Federal Republic of Germany, 434–472

Heuser E (1953b) Analytik von Chinonen. In: Müller E, Bayer O, Meerwein H, Ziegler K (eds) Methoden der organischen Chemie (Houben-Weyl) 4. Aufl., Bd II Analytische Methoden. George Thieme, Stuttgart, Federal Republic of Germany, 434–472

Hon D N-S, Glasser W (1979) On possible chromophore structures in wood and pulps – a survey of the present state of knowledge. Polym-Plast Technol Eng 12:157–179

Klason P (1922) Beitrag zur Kenntnis der Konstitution des Fichtenholz-Lignins (II). Ber Dtsch Chem Ges 55:448–455

Larsson S, Miksche GE (1972) Gaschromatographische Analyse von Ligninoxidations-produkten. VI. 4,4-Bis-aryloxy-cyclohexa-2,5-dienonestrukturen im Lignin. Acta Chem Scand 26:2031–2038

Lebo SE, Lonsky WFW, McDonough TJ, Medvecz PJ (1988) The occurrence and light induced formation of ortho-quinonoid lignin structures in white spruce refiner mechanical pulps. Proc Int Pulp Bleach Conf, Orlando, FL, June 6–9, 277–251

Lin SY, Kringstad KP (1970) Photosensitive groups in lignin and lignin model compounds. Tappi 53:658–663

Lindberg B, Misiorny A (1952) Quantitative analysis of reducible carbohydrates by means of sodium borohydride. Sven Papperstidn 55:13–14

Lindberg B, Theander O (1954) Quantitative determination of carbonyl groups in oxycellulose by means of sodium borohydride. Sven Papperstidn 57:83–85

Marton J, Adler E (1961) Carbonyl groups in lignin III. Mild catalytic hydrogenation of Björkman lignin. Acta Chem Scand 15:370–383

Marton J, Adler E, Persson K-I (1961) Carbonyl groups in lignin IV. Infrared absorption studies and examination of the volumetric borohydride method. Act Chem Scand 15: 384–392

Panchapakesan B (1990) Photo-induced reactions of lignin and lignin fragments: Photo-oxidation of chlorinated organics in bleach plant effluents and brightness stability of high yield pulps. Ph.D. Thesis, North Carolina State University, Raleigh, North Carolina, USA

Pew JC (1951) Structural aspect of the color reaction of lignin with phenols. J Am Chem Soc 73:1678–1685

Schubert WJ, Nord FF (1950) Investigation on lignin and lignification I. Studies on softwood lignin. J Am Chem Soc 72:977–981

7.5 Determination of Carboxyl Groups

C.W. DENCE

7.5.1 Introduction

Carboxyl groups reportedly have been detected in native lignin, specifically in milled wood lignin, by infrared spectroscopy (Ekman and Lindberg 1960). However, the collective findings of a large number of structural studies indicate that if such groups are in fact present in native lignin, their concentration is extremely small. On the other hand, when native lignin is subjected to chemical (e.g., pulping and bleaching) or biological treatments, significant amounts of carboxyl groups are frequently detected. This effect is particularly pronounced in oxidative treatments where the resulting rupture of phenolic rings in the lignin gives rise to carboxyl-containing entities. Quantitative measurement of carboxyl groups may therefore provide information regarding the degree to which the lignin has been degraded or otherwise modified as the result of a specific chemical or biological treatment. The carboxyl content of a modified lignin may also provide information relating to its solubility characteristics and, for pulps, to strength properties.

The most widely accepted methods for determining carboxyl groups in lignin are based directly or indirectly on the acidity of this group. The approach followed most consistently in the analysis involves neutralization of the carboxylic acid using potentiometry or conductivity to detect the titrimetric end point. Under optimal conditions, potentiometric and conductometric titrations provide the advantage of determining not only the carboxyl groups in lignin but, concurrently, the weakly acidic phenolic hydroxyl groups; when combined with an ion exchange treatment, the aforementioned titrimetric procedures can also be used to determine the strongly acidic sulfonic acid groups in lignosulfonates.

Conductometric titrations have been employed in the measurement of carboxyl groups in lignosulfonic acids (Mikawa et al. 1956, James and Tice 1965), in alkali lignins (Gaslini and Nahum 1959), and in sulfite pulps (Katz et al. 1984). When applied to pulps, this method does not differentiate between carboxyl groups in the lignin and those present in the polysaccharide component and hence a value corresponding to the total is obtained. In this connection, the procedure described by Katz et al. (1984) for the determination of carboxyl groups in sulfite pulps may, in all likelihood, be adapted to isolated lignins provided the carboxyl groups are in, or can be readily converted to, their hydrogen form. Using potentiometric titrations, the carboxyl contents of such diverse materials as dioxane lignin (Crozier et al. 1979), wheat straw lignin (Scalbert and Monties 1986), enzymatic lignin (Kirk and Chang 1975), and lignin

Springer Series in Wood Science
Methods in Lignin Chemistry
(Edited by S.Y. Lin and C.W. Dence)

obtained from the ethanol extract of steam-exploded aspenwood (Chum et al. 1984) have been determined.

The carboxyl groups in lignin have also been determined using the so-called calcium acetate method (Ekman and Enkvist 1955) which is essentially an ion exchange method. In this procedure the acidic hydrogens of the carboxyl groups are selectively replaced by calcium ions supplied by calcium acetate and the acetic acid produced by it is subsequently determined by titration with base. In another approach, the carboxyl content of lignin isolated from wood by means of white-rot fungi has been determined by converting the carboxyl groups to their corresponding methyl esters by methylation with diazomethane and subsequently measuring the methanol liberated after saponification (Kirk and Chang 1975).

The procedure described below was developed by Pobiner (1983) and consists of the nonaqueous potentiometric titration of lignin with tetra-n-butylammonium hydroxide in the presence of an internal standard, p-hydroxybenzoic acid. As is true also for conductometric titrations of lignin (see Chap. 7.7), phenolic hydroxyl and sulfonic acid groups may be coincidentally determined by this procedure.

7.5.2 Determination of Carboxyl Groups by Nonaqueous Potentiometric Titration

7.5.2.1 Instrumentation and Chemicals

An automatic titrator is used in conjunction with a glass indicating electrode and a calomel reference electrode. The saturated KCl electrolyte in the calomel electrode must be replaced by 1 M tetra-n-butylammonium chloride (TnBACl) in water which is available as polarographic grade material (Southwestern Analytical Chemicals, Austin, TX). The electrode is stored in the TnBACl solution when not in use. A nitrogen flow of 50 ml min^{-1} is maintained over the solution in the titration vessel.

To prepare 0.05 M TnBAH, transfer 100 g of tetra-n-butylammonium hydroxide, available as 25% (w/w) in methanol in 100-g quantities (Baker, Aldrich, Fluka and others), into a 3-l glass-stoppered reagent bottle. Add 2 l of anhydrous 2-propanol, blanket with nitrogen, close the bottle, and mix the contents thoroughly. The solution is stored under nitrogen in the refrigerator when not in use. p-Hydroxybenzoic acid (99+%) is available from a number of suppliers including Aldrich Chemical Co.

7.5.2.2 Standardization of Titrant

The titrant, TnBAH solution, requires standardization at least twice daily when in use over an 8-h period. Duplicate samples of primary standard grade benzoic acid are weighed into the titration vessel and 60 ml of DMF (N,N′-dimethylformamide) is added. The glass and TnBACl-filled calomel electrodes are immersed in the solution, the titrant delivery tip is placed in the vessel, and the surface of the solution is blanketed with a stream of nitrogen delivered at 50–100 ml/min. The solution is titrated potentiometrically with the TnBAH solution to a sharp inflection break (24–25 ml is usually required to titrate 0.15 g of benzoic acid). After determining the titrant volume corresponding to the inflection point (Note 1), the normality of the TnBAH, N, is calculated.

$$N = \frac{\text{g benzoic acid}}{(\text{ml titrant})\ (0.12212)}.$$

7.5.2.3 Procedure for Carboxyl Group Determination

The lignin sample (0.35 g) and p-hydroxybenzoic acid (0.05–0.08 g) are weighed accurately to 0.1 mg into the titration vessel (Note 2) and 2 ml of distilled water, 0.2 ml of concentrated, reagent grade hydrochloric acid accurately delivered by pipet (Note 3), and 60 ml of DMF are added (Note 4). The titration vessel is covered (with a Fleaker cap if a Fleaker beaker is used) and stirred magnetically for 5 min to insure complete sample dissolution. The electrodes are immersed in the solution which is then titrated with a 0.05 N TnBAH under a nitrogen atmosphere as described above in the standardization procedure. A blank is run on a solution consisting of 0.050 g of p-hydroxybenzoic acid, 2 ml of distilled water, 0.2 ml of concentrated hydrochloric acid, and 60 ml of DMF.

There will be three inflections in the titration curve as illustrated in Fig. 7.5.1: the first inflection, near +200 to +100 mV, is assignable to excess hydrochloric acid and strong acids (for example, sulfonic) present in the sample; the second inflection, near −350 mV, corresponds to carboxylic acids; and the third, near −480 to −520 mV, to phenolic hydroxyl groups. These inflections can be resolved in the 500–200 mV (full-scale deflection) ranges.

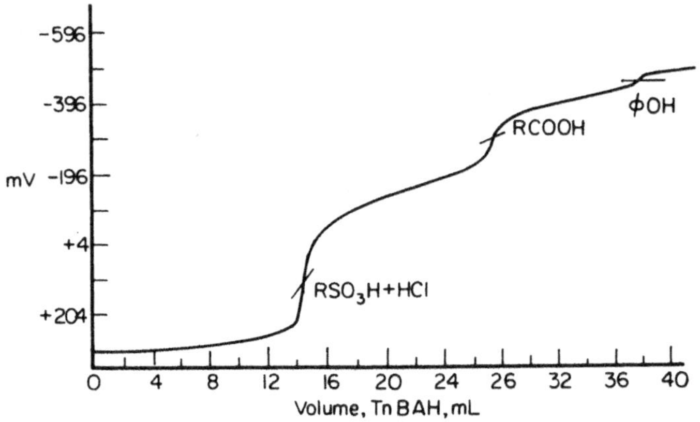

Fig. 7.5.1. Potentiometric titration curve of a lignosulfonate in the presence of an internal standard, p-hydroxybenzoic acid

7.5.2.4 Calculation

$$\text{mEq carboxyl (RCOOH)/g sample} = \frac{[(y) - (x) - (c) - (a)](N)}{w},$$

where y = titration volumes up to and including the second inflection point, x = titration volume up to and including the first inflection point, c = correction factor for HCl interference, a = theoretical titer for the p-hydroxybenzoic acid added to the sample, N = normality of TnBAH, w = weight of original lignin sample taken for analysis.

The theoretical titer, a, assignable to the internal standard, p-hydroxybenzoic acid, is calculated from the expression

$$a(\text{mL}) = \frac{\text{g p-hydroxybenzoic acid}}{0.13812\ (\text{normality of TnBAH})}.$$

The correction factor, c, for hydrochloric acid interference is calculated with reference to the titration curve obtained in the blank run described above. The titer corresponding to the second inflection point, near −320 mV, is subtracted from that corresponding to the first inflection point (near +130 mV) to obtain the titer for the carboxyl group in p-hydroxybenzoic acid. Subtraction of the theoretical titer for p-hydroxybenzoic acid, a, from this value affords the HCl interference correction factor, c:

c(ml) = (measured titer for RCOOH) − (theoretical titer for RCOOH of int. standard).

The phenolic hydroxyl content is calculated as follows:

$$\text{mEq phenolic hydroxyl/g sample} = \frac{[(z) - (y) - (a)](N)}{w},$$

where z = titration volume up to and including the third inflection point; the other terms are as defined above

Note 1. The inflections may be determined by the use of a Metrohm endpoint locator ruler (Brinkmann Instruments, Westbury, NY)

Note 2. The titration is conveniently conducted in a Fleaker beaker (Fisher Scientific) (Pobiner 1983).

Note 3. Addition of hydrochloric acid may be omitted when the carboxyl groups are known to be present entirely in the hydrogen form. In this situation the expression for calculating carboxyl content becomes

$$\text{mEq RCOOH/g sample} = \frac{(y - x - a)(N)}{w}.$$

In the absence of any strongly acidic (for example, sulfonic) groups in the lignin sample, as indicated by the lack of an inflection near +130 mV, the equation for calculating the carboxyl content is further contracted to

$$\text{mEq RCOOH/g sample} = \frac{(\text{titration volume at RCOOH infl.} - a)(N)}{w}.$$

Note 4. Scalbert and Monties (1986) have titrated DMF solutions (20 ml) of lignin in concentrations ranging from 0.25 mg to 1.25 mg ml^{-1} with 0.05 M TnBAH in isopropyl alcohol solution (Fluka Chemical Corp.). p-Hydroxybenzoic acid (2.5 mg in each lignin solution titrated) was used as an internal standard.

The standard deviation in a series of replicate determinations of carboxyl group concentrations at the 1.05 mEq g^{-1} level in a lignosulfonate was found to be 0.060 mEq g^{-1}; for a lignosulfonate having a carboxyl content at the 2.99 mEq g^{-1} level, the standard deviation was 0.043 mEq g^{-1} (Pobiner 1983).

7.5.3 Discussion of Method

Most isolated lignins usually contain at least two types of acidic functionalities, namely carboxyl and phenolic hydroxyl groups. Additionally, sulfonic acid groups are present in sulfonated lignins. Under the appropriate titration conditions, lignins containing such groups should therefore give rise to titration curves displaying two or three inflection points assuming the pK_a's corresponding to the acid groups being titrated are sufficiently dissimilar.

The problem frequently encountered in performing potentiometric titrations of lignin is that the inflections in the titration curve corresponding to the various acidic groups are often indistinct, making detection of the end points of the weaker acids difficult, particularly when lignosulfonates are titrated (Mikawa et al. 1956, James and Tice 1965, Forss and Fremer 1978, Pobiner 1983, Katz et al. 1984, Scalbert and Monties 1986). Possible explanations for the observed lack of definition in the inflection points have been offered by Pobiner (1983) and Scalbert and Monties (1986).

In the procedures outlined above, the problem of indistinct end points has been circumvented by the addition of an internal standard, p-hydroxybenzoic acid, which markedly improves the instrumental response, thereby promoting resolution of the individual inflections in the titration curves.

The criteria used by Pobiner (1981, unpubl. results) for the selection of an internal standard are as follows: (1) the model compound should bear the two functionalities of interest, namely carboxylic acid and phenolic hydroxyl, (2) the titration curve of the internal standard should display two cleanly resolved inflections that significantly overlap the poorly resolved slope changes evident in the titration curve of the lignins or the lignosulfonate, (3) the model compound should be soluble in the titration medium (DMF to which some HCl and H_2O are added), and not react with the lignin, and (4) the pK_a of the internal standard should be close to that of the acidic groups in the lignin so that in a mixture their titration volumes in the acid-base titration are additive at the inflection points. Other internal standards that meet the above criteria and give results

Table 7.5.1. Carboxyl contents of various lignins

Lignin	COOH, mEq/g	Method	Reference
Hardwood kraft	1.44	Potent. titr.	Pobiner 1981, Unpubl. results
Hardwood native	0.92	Potent. titr.	Pobiner 1981, Unpubl. results
Hardwood + softwood LSA[a]	0.31–2.08	Potent. titr.	Pobiner 1981, Unpubl. results
Wheat straw milled wood	0.81	Potent. titr.	Pobiner 1981, Unpubl. results
Spruce milled wood	0.12	Potent. titr.	Scalbert and Monties 1986
Decayed spruce	0.55	Potent. titr.	Kirk and Chang 1975
Birch hydrolysis	0.41	Ca acetate	Klemola 1968
Spruce LSA	0.61–1.26	Cond. titr.	James and Tice 1965
Softwood kraft	0.8	Cond. titr.	Mörck et al. 1986
Softwood kraft	0.89	Methylation	Marton and Adler 1963

[a] LSA = lignosulfonic acid.

equivalent to those obtained with p-hydroxybenzoic are p-hydroxycinnamic acid and 3,5-dimethoxy-4-hydroxycinnamic acid (Pobiner 1983).

The presence of sodium hydroxide as an impurity in a lignosulfonate in concentrations up to 15% was found not to interfere with the weak acid titration since an excess of hydrochloric acid was available to neutralize the alkali. However, sodium sulfate interfered with the titration when present in lignosulfonates in levels exceeding 5% (Pobiner 1983).

The carboxyl contents of different lignins are shown in Table 7.5.1. In the case of lignins isolated or modified without recourse to strong oxidation procedures, the carboxyl content is generally less than 10%. However, when lignin is subjected to the action of strong oxidants, as is the case, for example, in the bleaching of chemical pulps, the carboxyl content may substantially exceed this value.

References

Chum HL, Ratcliff M, Schroeder HA, Sopher DW (1984)Electrochemistry of biomass-derived materials. I. Characterization, fractionation, and reductive electrolysis of ethanol-extracted explosively depressurized aspen lignin. J Wood Chem Technol 4:505–532

Crozier TE, Johnson DC, Thompson NS (1979) Changes in a southern pine dioxane lignin on oxidation with oxygen in sodium carbonate media. Tappi 62:107–111

Ekman K, Enkvist T (1955) Some determinations of weak and strong acids in various lignin preparations and pulps. Pap Puu 37:369–382

Ekman KH, Lindberg JJ (1960) On the origin of the infra-red bands in the 1720 cm^{-1} region in lignins. Pap Puu 42:21–22

Enkvist T, Alm B, Holm B (1956) Some experiments concerning the acidic groups of lignin. Pap Puu 38:1–8, 12

Forss K, Fremer K-E (1978) Potentiometric titration of the acidic groups in lignosulfonic acids. Pap Puu 60:121–127

Gaslini F, Nahum LZ (1959) Conductometric titration of alkali lignin. Sven Papperstidn 62:520–523

James AN, Tice PA (1965) The presence of carboxyl groups in lignosulfonate preparations. Tappi 48:239–244

Katz S, Beatson RP, Scallan AM (1984) The determination of strong and weak acidic groups in sulfite pulps. Sven Papperstidn 87:R48–R53
Kirk TK, Chang H-m (1975) Decomposition of lignin by white-rot fungi. II. Characterization of heavily degraded lignins from decayed spruce. Holzforschung 29:56–64
Klemola A (1968) Steam hydrolysis of birchwood. Part V. An investigation of the carbonyl and carboxyl groups in the degraded lignin. Suom Kemistil B41:152–160
Marton J, Adler E (1963) Reactions of lignin with methanol hydrochloric acid. A discussion of some structural questions. Tappi 46:92–96
Mikawa H, Sato K, Takasaki C, Ebisawa K (1956) Studies on the cooking mechanism of wood. XIII. On the possible hydroxyl group of lignosulfonic acid. Bull Chem Soc Jpn 29:245–254
Mörck R, Yoshida H, Kringstad KP, Hatakeyama H (1986) Fractionation of kraft lignin by successive extraction with organic solvents. I. Functional groups, ^{13}C-NMR-spectra and molecular weight distributions. Holzforschung 40 (Suppl):51–60
Pobiner H (1983) Improved inflection points in the non-aqueous potentiometric titration of acid functionalities in lignin chemicals by using internal standardization and ion exchange. Anal Chim Acta 155:57–65
Scalbert A, Monties B (1986) Comparison of wheat straw preparations. II. Straw lignin solubilization in alkali. Holzforschung 40:249–254

7.6 Determination of Methoxyl Groups

C.-L. Chen

7.6.1 Introduction

Modified Zeisel procedures have been used for quantitative determination of alkoxyl groups such as methoxyl (Roth 1953). In the original Zeisel procedure (Zeisel 1885, Zeisel and Fanto 1902, 1903), the treatment of methoxyl- or ethoxyl-containing compounds with concentrated hydriodic acid (HI) at refluxing temperature leads to cleavage of the alkoxyl groups with formation of methyl or ethyl iodide. The volatile iodide, when passed through a solution of silver nitrate, yields a precipitate of silver iodide which is used to determine quantitatively the methoxyl or ethoxyl content of the compound. However, any sulfur-containing groups are converted quantitatively to hydrogen sulfide during the hydriodic acid treatment. The hydrogen sulfide, in turn, reacts with silver nitrate, producing a precipitate of silver sulfide. Thus, the original Zeisel procedure is not appropriate for the determination of alkoxyl groups in sulfur-containing compounds. In order to circumvent this difficulty, Vieböck and co-workers converted the alkyl iodide to iodic acid by oxidation with bromine while simultaneously oxidizing hydrogen sulfide to sulfuric acid (Vieböck and Brecher 1930, Vieböck and Schwappach 1930). This step eliminates the interference of sulfur-containing groups and makes the Vieböck and Schwappach procedure suitable for the determination of methoxyl and ethoxyl groups in compounds containing sulfur. This is of primary importance in the characterization of technical lignins, such as kraft lignins and lignosulfonates, which contain varying amounts of sulfur-containing groups.

When a compound with an alkoxyl group is heated with concentrated hydriodic acid, the alkoxyl group is cleaved and converted to the corresponding alkyl iodide as follows:

$$
\begin{array}{llllll}
\text{R—O—R}' & + & \text{HI} & \rightarrow & \text{R—OH} & + \quad \text{R}'\text{I} \\
\text{Ar—O—R}' & + & \text{HI} & \rightarrow & \text{AR—OH} & + \quad \text{R}'\text{I} \\
\text{RC(=O)—OR}' & + & \text{HI} & \rightarrow & \text{RCOOH} & + \quad \text{R}'\text{I} \\
\text{R—S—R}' & + & \text{HI} & \rightarrow & \text{R—SH} & + \quad \text{R}'\text{I} \\
\text{R—SH} & + & \text{HI} & \rightarrow & \text{H}_2\text{S} & + \quad \text{RI}
\end{array}
$$

When the alkyl iodide (R′I) produced is volatile, as is the case with methyl or ethyl iodide, it is quantitatively stripped from the reaction mixture by purging with an inert gas such as carbon dioxide. The methyl or ethyl iodide in the

Springer Series in Wood Science
Methods in Lignin Chemistry
(Edited by S.Y. Lin and C.W. Dence)

carrier gas is collected in a trap containing a solution of sodium acetate in glacial acetic acid to which is added a few drops of bromine. The bromine reacts with the volatile alkyl iodide to produce the corresponding alkyl bromide (RBr) and iodine bromide (IBr). The excess bromine further oxidizes the iodine bromide to iodic acid (HIO_3). The remaining bromine is then reduced by formic acid (HCOOH) to hydrogen bromide, which is neutralized by the sodium acetate solution:

$$CH_3I + Br_2 \rightarrow CH_3Br + IBr$$
$$IBr + 3H_2O + 2Br_2 \rightarrow HIO_3 + 5HBr$$
$$HIO_3 + 5KI + 5H^+ \rightarrow 3I_2 + 3H_2O + 5K^+$$
$$I_2 + 2Na_2S_2O_3 \rightarrow 2NaI + Na_2S_4O_6$$

Finally, the iodine thus produced is titrated with a dilute standard sodium thiosulfate solution using 1% starch solution as indicator (Leipert 1929). Since six atoms of iodine are liberated for each alkoxyl group, the alkoxyl content can be quantitatively determined with very high accuracy, even on a micro scale.

In the original Zeisel procedure, the methyl iodide produced is absorbed in an alcoholic solution of silver nitrite. The resulting solution is diluted with chloride ion-free distilled water, acidified with nitric acid, and boiled. The silver iodide is removed by filtration, washed, and weighed in the same fashion as for the determination of halides by gravimetry. Alternatively, the methyl iodide is absorbed in pyridine and the resulting 1-methylpyridinium iodide is titrated with either silver nitrate solution, using sodium chromate as indicator (Kirpal and Buhn 1914) or tetrabutylammonium hydroxide (Cundiff and Markunas 1961). The 1-methylpyridinium iodide may be also determined by UV spectroscopy (Sobue et al. 1959). Sulfur-containing groups in the sample cause no interference in these procedures. The methyl iodide can be also quantitatively determined by gas chromatography after absorption by an organic solvent such as chloroform containing 0.5% of methylene chloride (Vertailer and Martin 1958) or after extraction of the reaction mixture by carbon tetrachloride (Sporek and Danyi 1962). The precision and accuracy of these methods in the determination of methoxyl in lignins are, however, appreciably lower than in the modified Vieböck and Schwappach procedure (Silbernagel 1980, pers. commun.) described below.

7.6.2 Sample Preparation

To express the methoxyl content of lignin on a particular basis, for example as equivalents per C_9 unit, the lignin must be isolated and purified prior to the methoxyl analysis. Useful isolation and purification procedures are described by Björkman (1956) Lundquist et al. (1977), and Lundquist and Kirk (1980) (Chap. 3.1). In addition to the methoxyl analysis, other analyses of the sample should be conducted: (1) moisture content, (2) Klason lignin content, (3) carbohydrate analysis, and (4) elemental analyses for carbon, hydrogen, and oxygen. If the Klason lignin content of the sample is less than about 95%, the sample should be further purified.

Since many lignin preparations are hygroscopic, it is imperative to dry the lignin samples completely such as at 50 °C over phosphorous pentoxide (P_2O_5) (<20 mm Hg) for 48 h. The dried samples are kept in a desiccator over P_2O_5.

7.6.3 Determination of Methoxyl Content by the Vieböck and Schwappach Procedure (Silbernagel 1980, pers. commun.)

Reagents

Bromine, iodine-free, stored in a brown dropping bottle of appropriate volume
Carbon dioxide, obtained from a cylinder equipped with pressure regulator and needle valve
Hydriodic acid, freshly distilled at constant boiling point (127 °C); concentration, 57%, density, 1.7 (Note 1)
Formic acid, 90%
Phenol, freshly distilled
Red phosphorus, prepared by digesting 5 g of commercial red phosphorus in 50 ml of 5% ammoniacal water on a water bath for 30 min. The precipitate is washed with distilled water (Note 2) until the washings are neutral, then with ethanol. Finally, the purified red phosphorus is dried in a forced air oven at 100 °C and stored in a desiccator.
Potassium iodide
Sodium acetate solution, 20% in 96% acetic acid by dissolving 20 g of pure sodium acetate in 80 g of 96% acetic acid
Sodium acetate, powder
Saturated sodium bicarbonate solution (Note 2)
Standard sodium thiosulfate solution, 0.05 N (Note 2)
Sulfuric acid solution, 10% (Note 2)
Starch solution, 1%: grind 0.5 g of starch thoroughly in a small mortar with a small amount of water until a smooth paste is formed, then pour slowly into 50 ml of boiling water. Boil for 1 or 2 min, cool, and decant into a small glass-stoppered bottle. The starch solution must be prepared fresh every 6 days (Note 3).
Thyodene solution, 5% (Note 3)
o-Methoxycinnamic acid (m.p. 119–120 °C), a test substance used occasionally to verify the reliability of the procedure (theoretical methoxyl content, 17.37%)

Note 1. A mixture of 100 ml of hydriodic acid and 1 g of purified red phosphorous is refluxed gently for 2 h under a carbon dioxide atmosphere and then distilled. Distillate, having a constant boiling point of 126–127 °C, is collected. It is convenient to divide the distillate into 5-ml portions and transfer each portion to a 10-ml brown, general-purpose ampule. The ampules are flushed with nitrogen and sealed with a flame-sealing torch. One ampule of hydriodic acid is used for each methoxyl analysis. The distillation must be conducted in a hood and behind a safety shield.

Note 2. Preferably, use doubly distilled water or completely deionized water.

Note 3. Thyodene (Fisher Scientific) may be used as the indicator instead of the 1% starch solution.

Apparatus for Determination of Methoxyl Groups

Figure 7.6.1 shows a diagram of the Vieböck and Schwappach apparatus modified to improve the efficiency and accuracy of the procedure (Silbernagel 1980, pers. commun.).

The apparatus consists of a reaction flask (A), a condenser-scrubber trap (B), a series of traps (C) and a micro weighing bottle (D). The reaction flask in A is a round-bottomed flask having a volume of approximately 13 ml fitted with a gas inlet tube in the form of a capillary tube extending to the bottom of the flask. The scrubber trap in B has a side arm with a ground glass stopper in the upper part and is capable of holding a maximum of about 4 ml of wash solution.

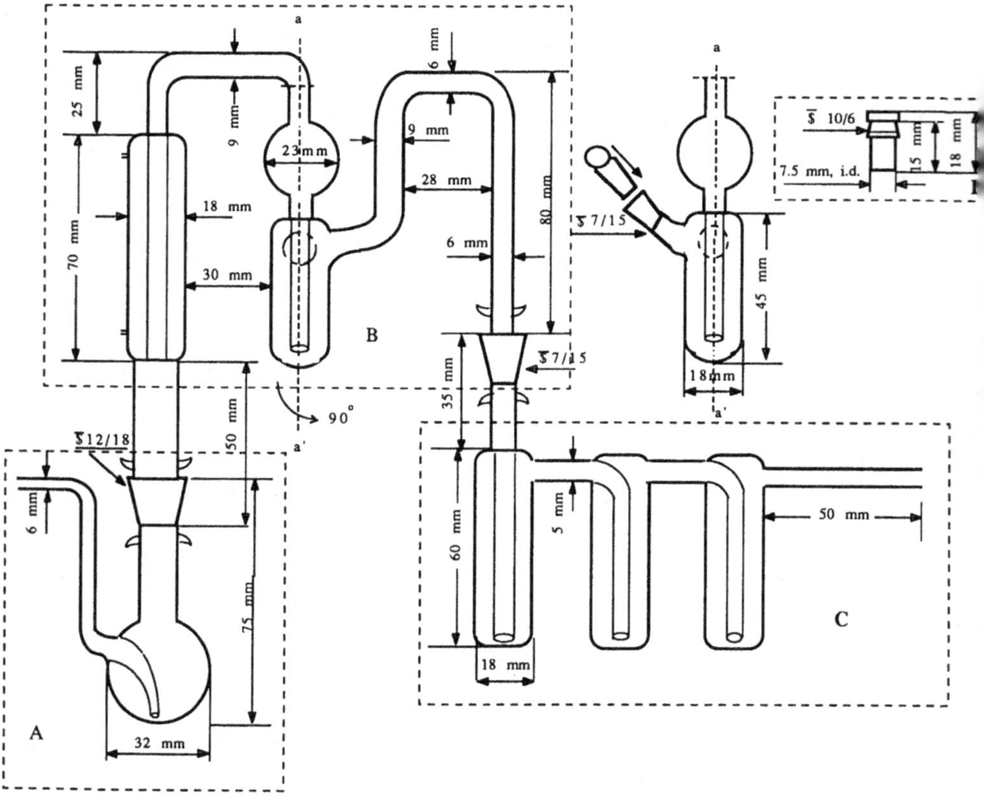

Fig. 7.6.1. The modified Vieböck-Schwappach apparatus (for explatation see text)

Trap C consists of three gas-trapping bottles of equal volume joined in series, each capable of holding a maximum of about 4 ml of trapping solution. Prior to the analysis, all parts of the apparatus must be cleaned with cleaning solution.

Procedure

In flask A (Fig. 7.6.1) are placed 5 ml of freshly distilled, constant boiling hydriodic acid, several crystals of phenol, and 25 mg of red phosphorus. The phenol is added to the mixture to increase the solubility of the lignin. The scrubber trap in unit B is filled with 3 ml of distilled water (Note 3). The first gas trapping bottle in C is filled with about 1.5–2 ml of a 20% solution of sodium acetate in 96% acetic acid. Approximately 10–12 drops of bromine are added to the solution with gentle shaking, and the first trapping bottle is finally filled with 20% sodium acetate solution. The resulting solution is distributed evenly among the three gas trapping bottles by inclining C with the third trapping bottle tilted downward. A and B are connected and the stopper is inserted in the side arm of the scrubber trap B. Carbon dioxide gas is passed slowly through the gas inlet tube into the reaction flask at the rate of six bubbles/10 s. The mixture in the reaction flask is gently refluxed for 2 h with a heating mantle or oil bath (bath temp. 145–150 °C). Any free iodine initially present in the reaction mixture or produced from the reactions is converted by the phosphorus into hydriodic acid as follows:

$$2\,P + 3\,I_2 \rightarrow 2\,PI_3$$
$$2\,PI_3 + 3\,H_2O \rightarrow 3\,HI + H_3PO_3$$

When the refluxing is completed, the distilled water in the scrubber is replaced by saturated sodium bicarbonate solution (Note 4), and C, filled with the absorption solution, is connected to B. About 10 mg of the dried sample is accurately weighed to the nearest 0.01 mg into a constant weight micro weighing bottle. The cap of the micro weighing bottle is removed and the weighing bottle containing the sample is dropped into the reaction mixture through the top of the reaction flask. A is connected to B and the reaction mixture is again gently refluxed for exactly 1 h.

Note 4. The distilled water is conveniently removed with a 10-ml syringe fitted with a hypodermic needle to which is connected Teflon tubing of appropriate length (about 10 cm) and diameter. The scrubber trap is then filled with the saturated sodium bicarbonate solution.

Titration

The reaction mixture in the trapping bottles in C is transferred to a 250-ml Erlenmeyer flask containing 1.5 g of sodium acetate. The traps in C are thoroughly washed three times with 20 ml of distilled water, and the washings are added to the reaction mixture in the Erlenmeyer flask. Approximately five to ten drops of formic acid are added to the combined reaction mixture and washings with gentle shaking until the brown color of the bromine is completely discharged. To the resulting solution are added 3 ml of dilute (10%) sulfuric acid and 1.5 g of potassium iodide. After standing for 5 min, the solution is titrated (Note 5) with standard 0.05 N sodium thiosulfate until the color of the liberated iodine nearly disappears (a light straw color); 1–2 ml of the 1% starch or 5% Thyodene solution (Note 6) is added and the mixture is further titrated until the iodo-starch or iodo-Thyodene blue color is completely discharged. Since 1 ml of 1 N sodium thiosulfate solution corresponds to 5.1706 mg of methoxyl group, 1 ml of 0.05 N sodium thiosulfate solution is equivalent to 0.25853 mg of methoxyl group. More accurate results may be obtained when a blank methoxyl determination is conducted (Note 7). The methoxyl content of the sample is calculated as follows:

$$\%\ CH_3O\ (\text{lignin basis}) = \frac{(V_s - V_b)N \cdot 517.06 \cdot f}{W}, \tag{1}$$

where V_s is the ml of standard sodium thiosulfate solution required for titration of a sample, V_b is the ml of sodium thiosulfate solution required for titration of the blank, N is the normality of the sodium thiosulfate solution, f is the factor for the standard sodium thiosulfate solution (Note 8), and W is the weight of the moisture-free sample in mg.

Note 5. A calibrated 25-ml micro-buret is used.

Note 6. Thyodene powder (50–100 mg) can be directly used as the indicator instead of a 5% Thyodene solution (1–2 ml).

Note 7. Potassium iodide sometimes contains an appreciable amount of potassium iodate as an impurity. In the presence of acid, iodine would be liberated from the potassium iodate as stated in Section 7.6.1. This iodine would consume some of the thiosulfate and thus give too low a value for the methoxyl content. Unless the iodide is known to be free of iodate, a blank test should be conducted.

Note 8. The standardization of a thiosulfate solution is usually conducted against pure potassium iodate. For example, a 0.1 N potassium iodate solution (3.6667 g/l) is titrated with the sodium thiosulfate solution to be standardized. The factor for the standard sodium thiosulfate solution is calculated by Eq. (2):

$$f = \frac{N_i V_i}{N_s V_s}, \tag{2}$$

where N_i is normality of the potassium iodate solution, V_i is ml of the potassium iodate solution used, N_s is normality of the sodium thiosulfate solution to be standardized, and V_s is ml of the sodium thiosulfate solution required to titrate the potassium iodate solution (V_i ml). Usually, f is approximately 1.

7.6.4 Discussion of Method

In general, the precision and accuracy for the determination of methoxyl content of the test substances (e.g., o-methoxycinnamic acid) are about 0.2% and 0.3%, respectively, expressed as average deviation. For the determination of methoxyl content in lignin preparations, the precision is of the same order as that of the test substances, whereas the accuracy deviates by approximately 0.6% from that of the test substances because of the hygroscopicity of lignins and the presence of impurities such as carbohydrates. Thus, on treatment with hydriodic acid, glucose and the other carbohydrates give rise to "apparent" methoxyl contents of 0.1–0.4%, although methyl iodide is not formed (Gran 1953a,b). For lignin samples with carbohydrate contents of less than 5%, the "apparent" methoxyl content is less than 0.02%, or within the limit of experimental error, and may therefore be neglected. In addition to ethers and esters, lower alcohols such as methanol and ethanol also produce the corresponding volatile alkyl iodides on treatment with hydriodic acid and hence are potential sources of experimental error in the determination of methoxyl content. Residual quantities of dioxane in lignin preparations may also lead to high methoxyl values by giving rise to ethyl iodide on reaction with hydriodic acid (Kaneko et al. 1981). The presence of dioxane in a lignin sample can be verified by running a routine ^{13}C-NMR spectrum of the sample (solvent, DMSO-d_6). A signal at δ 66.1–66.3 ppm indicates the presence of dioxane in the sample (Griggs 1985). Thus, to avoid such errors, it is important to purify and dry lignin preparations as thoroughly as possible prior to determining their methoxyl content.

Technical lignins do not have a true C_9 unit structure, because the side chains of lignins in plant tissues undergo rather extensive degradation during pulping or other industrial processes. However, it is common practice to express the chemical composition of technical lignins in terms of a C_9 unit in order to evaluate and compare the complex structural characteristics of these materials. Such data, although not very informative in terms of revealing the chemical composition and properties of lignins, are convenient for comparing the structure and properties of technical lignins and their modified products.

References

Björkman AJ (1956) Studies on finely divided wood. Part I. Extraction of lignin with neutral solvent. Sven Papperstidn 59:477–485

Cundiff RH, Markunas PC (1961) Determination of alkoxyl groups by nonaqeuous titration. Anal Chem 33:1028–1030

Gran G (1953a) Determination of alkoxyl groups III. An investigation on the determination of methoxyl in carbohydrate-containing substances. Sven Papperstidn 56:179–180

Gran G (1953b) Determination of alkoxyl groups IV. An investigation on absorption solutions. Sven Papperstidn 56:202–203

Griggs BF (1985) Modification of kraft lignins by sulfonation: structure and mechanism. Ph.D. Thesis, North Carolina State University, Raleigh, NC, 54–84

Kaneko H, Tokasuka C, Nakano J (1981) Effect of trace amounts of dioxane in lignin on determination of the methyoxyl groups. Mokuzai Gakkaishi 27:523–525

Kirpal A, Buhn T (1914) Methoxyl-Bestimmung schwefelhaltiger Verbindungen. Ber Dtsch Chem Ges 47:1084–1087

Leipert T (1929) Über die quantitative mikroanalytische Bestimmung von Jod in organischen Substanzen. Mikrochem J 7 (Sonderband: Pregel-Festschrift): 266–271

Lundquist K, Kirk TK (1980) Fractionation-purification of an industrial kraft lignin. Tappi 63:80–82

Lundquist K, Ohlsson B, Simonson R (1977) Isolation of lignin by means of liquid-liquid extraction. Sven Papperstidn 80:143–144

Roth H (1953) Analytik der Alkoxylgruppe. In: Muller E, Bayer O, Meerwein H, Ziegler K (eds) Methoden der organischen Chemie (Houben-Weyl) 4. Aufl., Band II Analytische Methoden. George Thieme, Stuttgart, 401–420

Sobue H, Hatano A, Arai T (1959) Determination of methoxyl group. Sen-i Gakkaishi (J Soc Textile Cellul Ind Jpn) 15:21–23

Sporek KF, Danyi MD (1962) Detection and identification of alcohols, alkoxy groups, lignin, and wood by gas liquid chromatography. Anal Chem 34:1527–1529

Vertailer S, Martin F (1958) Microdetermination séléctive des groupes alcoxy chromatographie gaz-liquide. Chim Anal 40:80–86

Vieböck F, Brecher C (1930) Eine neue Methode zur massanlytischen Bestimmung der Methoxyl- und Athoxylgruppe II. Mikro-Analyse. Ber Dtsch Chem Ges 63:3207–3210

Vieböck F, Schwappach A (1930) Eine neue Methode zur massanalytischen Bestimmung der Methoxyl- und Athoxylgruppe I. Makro-Analyse. Ber Dtsch Chem Ges 63:2818–2823

Zeisel S (1885) Über ein Verfahren zum quantitativen Nachweis von Methoxyl. Monatsh Chem 6:989–996

Zeisel S, Fanto R (1902) Neues Verfahren zur Bestimmung des Glycerins. Z Landw Vers-Wes Öst 5:729–745

Zeisel S, Fanto R (1903) Bestimmung des Rohglycerins im Weine mittelst der Jodidmethode. Z Anal Chem 42:549–578

7.7 Determination of Sulfonate Groups and Total Sulfur

R. P. BEATSON

7.7.1 Significance of the Analysis

The reactions between sulfurous acid and lignin have been important since the beginning of sulfite pulping. In 1890, Pedersen realized that some of the sulfur found in spent sulfite liquors was chemically bound to lignin. Shortly after, Lindsey and Tollens (1892) deduced that this sulfur was in the form of sulfonic acid groups. These discoveries led to extensive research on the role of sulfonate groups in sulfite pulping, involving many determinations of the sulfur and sulfonic acid content of lignin. The methods used are described in the references contained in several excellent reviews of the earlier literature (Brauns 1952, Brauns and Brauns 1960, Rydholm 1965, Glennie 1971).

New significance has been given to sulfonate groups in lignin with the advent of sulfite mechanical pulping and demonstration of a relationship between sulfonate content and pulp properties (Kvisgaard 1965, Atack et al. 1980). As a result, it became important to have a relatively simple and reliable method for the determination of sulfonate group content of lignin in pulp.

7.7.2 Survey of Available Methods

Several methods of estimating the sulfonate content of lignin in pulp simply measure total sulfur content and assume that all the sulfur is present in sulfonate groupings.

Such methods are wet oxidation of pulp followed by estimation of sulfate by precipitation of barium sulfate (Canadian Pulp and Paper Association Standard G28 1970), X-ray fluorescence spectroscopy (Rivington 1988, Kibblewhite et al. 1987), and combustion of pulp followed by analysis of sulfur as sulfur dioxide or as sulfate. The sulfur dioxide evolved is determined by iodometric titration (Canadian Pulp and Paper Association useful method G.7U, March 1959). Sulfate can be determined by titration with barium chloride (Ora 1960), back-titration with sulfuric acid after addition of barium perchlorate (Aldrich 1974), potentiometric titration with lead perchlorate using an ion-selective electrode (Ross and Frant 1969), or ion chromatography (Douek and Ing 1989).

Methods that attempt to measure the sulfonate content directly are based on retention of benzidinium ions by sulfonate groups (Sjöström and Enström 1966), titration of the hydrogen ions released when a sulfite pulp in its acid form is washed with potassium chloride adjusted to pH 4 (Cappelen and Schöön

Springer Series in Wood Science
Methods in Lignin Chemistry
(Edited by S.Y. Lin and C.W. Dence)

1966), selective elution of magnesium from carboxylate and sulfonate groups and conversion of pulp to the acid form followed by conductometric titration of sulfonic acid groups (Katz et al. 1984). Potentiometric titration is capable of giving the total acid content and, in the presence of an internal standard, may be used to determine sulfonic acid, carboxylic acid, and phenolic hydroxyl functionalities in soluble lignins (Pobiner 1983). The procedure developed by Pobiner for the determination of carboxylic acid groups in lignin is described in Chapter 7.5.

For both direct and indirect methods of estimating the sulfonate content of lignin in pulp, the sulfonate content of the lignin is obtained by dividing the sulfonate content of the pulp by the sum of its Klason and acid-soluble lignin contents (Lorås and Loschbrandt 1961).

The most convenient method of obtaining a direct measure of the sulfonate content of lignin in pulp has proven to be conductometric titration (Katz et al. 1984). The preferred method for analysis of total sulfur in lignin in pulp is combustion followed by ion chromatography (Douek and Ing 1989).

7.7.3 Sulfonate Group Determination

7.7.3.1 Principle of Conductometric Titration

In conductometric titrations the electrical conductivity of the solution is measured after successive additions of reagent, and the values are plotted against the volume of reagent added (Kolthoff and Sandell 1952). All ions present contribute to the electrical conductivity of the solution.

In the titration of a strong acid with a strong base as typified by the reaction

$$H^+ + Cl^- + Na^+ + OH^- \rightarrow H_2O + Na^+ + Cl^-,$$

the mobility of H^+ is much greater than Na^+ and thus conductivity decreases during the titration. After neutralization of the acid, addition of excess OH^- leads to a rapid increase in conductivity resulting from the high mobility of the hydroxyl ion. The resulting titration curve has the form shown in Fig. 7.7.1.

When a weak acid is titrated with a strong base as illustrated by the reaction

$$CH_3COOH + Na^+ + OH^- \rightarrow CH_3COO^- + Na^+ + H_2O,$$

the mobility of the product is greater than that of the weakly dissociated acid and the conductivity increases slowly until the end point is reached. Again, an excess of the highly mobile hydroxyl ion causes a rapid increase in conductivity after the end point is reached. The resulting titration curve is shown in Fig. 7.7.2.

The end point is determined by extrapolation of the linear sections of the conductometric plot to the point of intersection.

In the conductometric titration of sulfonic acid groups in pulp (Katz et al. 1984), it is first necessary to convert all the acid groups to their protonated form. The pulp is then dispersed in water containing 0.001 M sodium chloride and the conductivity monitored during titration with 0.1 M sodium hydroxide. The

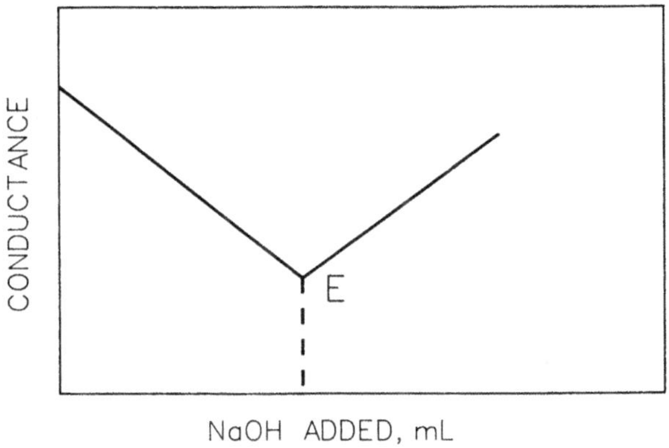

Fig. 7.7.1. Conductometric titration curve of a strong acid with a strong base. *E* designates the end point

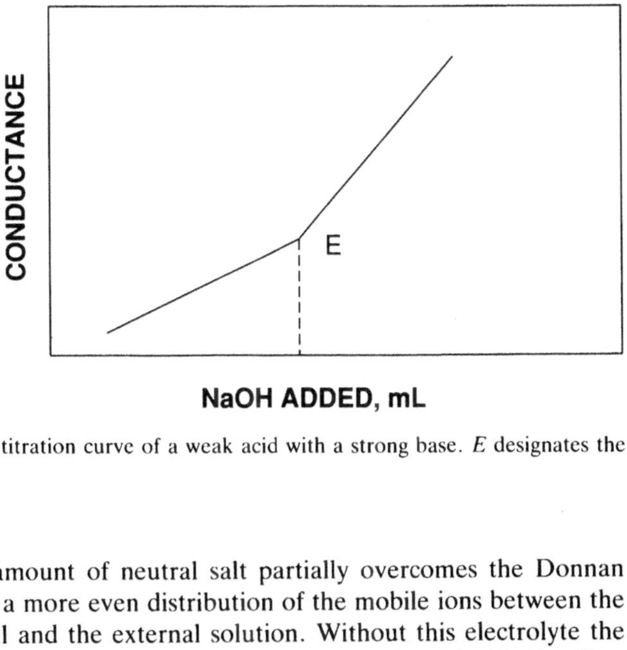

Fig. 7.7.2. Conductometric titration curve of a weak acid with a strong base. *E* designates the end point

presence of the small amount of neutral salt partially overcomes the Donnan equilibrium and creates a more even distribution of the mobile ions between the interior of the fiber wall and the external solution. Without this electrolyte the hydrogen ions from the acid groups would be largely confined inside the fiber during titration and, as a result, the end points would be ill-defined (Scallan et al. 1989). A larger amount of salt creates too great a background conductivity for observation of the small changes that occur during the titration.

The shape of the titration curve is shown in the Fig. 7.7.3.

The initial high conductivity is attributable to the protons from the highly dissociated sulfonic acid groups. As alkali is added, the conductivity drops linearly as H^+ is replaced by the less conducting Na^+ ion. The horizontal portion of the curve corresponds to the titration of weaker carboxylic acid

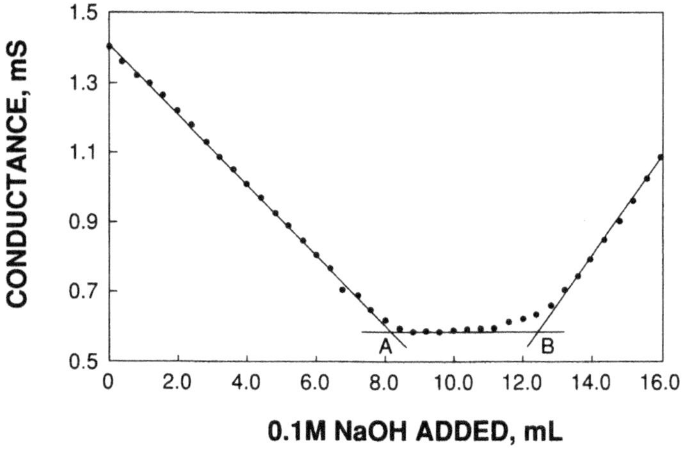

Fig. 7.7.3. Conductometric titration curve of a refiner mechanical pulp made from sulfonated spruce chips (3.47 g). End point *A* corresponds to titration of the sulfonic acid groups and end point *B* coresponds to titration of the sulfonic and carboxylic acid groups. (Katz et al. 1984)

groups. As the carboxylic acids are weakly dissociated, they provide few free protons to the external solution and thus, the conductivity does not drop as they are titrated. The slow rise in conductivity expected from the titration of weak acids in solution is not observed in this case, as the added sodium is associated with the carboxylate groups in the fiber wall. The final steeply rising linear portion of the curve corresponds to the accumulation of excess sodium hydroxide.

The end points, like those for conductometric titration of soluble acids, are found at the intersection of the extrapolated linear portions of the titration curves. The first end point corresponds to sulfonic acid sulfur and the second to the total acid content, i.e., sulfonic plus carboxylic.

7.7.3.2 Description of Conductometric Titration of Pulp

Acidic groups are converted to their hydrogen form by twice soaking approximately 3 g (o.d. basis) of the pulp in 0.1 M hydrochloric acid for 45 min and then washing it with deionized water to constant conductance. The pulp is drained and dispersed in 450 ml of 0.001 M sodium chloride prepared with deionized water. The pulp suspension is stirred with a magnetic stirrer and the conductivity is measured using a dip-type electrode and a conductivity meter. The conductivity meter should have a resolution of at least 0.01 mS in the 0-10 mS range (for example, the CBM 83 available from Radiometer). A dip-type electrode with a cell constant of 1 cm^{-1} can be used (for example, the CDC 304 available from Radiometer). The protective sheath is removed from the electrode to prevent accumulation of pulp fines. An atmosphere of nitrogen is maintained above the surface of the water to prevent interference from carbon dioxide absorption.

Alkali, 0.1 M sodium hydroxide, is added in 0.5 ml aliquots. The suspension is stirred for 5 min after each addition to allow equilibrium to be reached. The conductivity is noted and the process repeated. Following titration, the pulp is washed and dried at 105 °C for 24 h to obtain an accurate value for the o.d. weight. The conductivity is plotted against the volume of sodium hydroxide added to provide plots of the type shown in Fig. 7.7.3.

If the initial rapidly dropping portion of the plot is too short to allow an accurate extrapolation to the first end point, as is the case for pulps having low sulfonate contents, an accurately measured volume of 0.1 M hydrochloric acid may be added to the pulp prior to the titration with base.

Sulfonic acid contents are expressed by the following equation:

$$\text{Sulfonic acid content} = \frac{AM_1 - VM_2}{W} \ \text{mmol kg}^{-1},$$

where A is the first end point for titration of sulfonic acid groups in ml, V is the volume of hydrochloric acid added in ml, M_1 is the molarity of sodium hydroxide, M_2 is the molarity of hydrochloric acid, W is the o.d. weight of pulp in kg.

The total acid group content of the pulp is calculated in the same manner using the second end point B in the above equation. The difference between total and sulfonic acid contents corresponds to the carboxylic acid content of the pulp. The sulfonic acid content of lignin in pulp is obtained by dividing the sulfonic acid content of pulp by the sum of Klason and acid-soluble lignin contents expressed as a percentage.

If the total acid content of the pulp is in the vicinity of 300 mmol kg^{-1}, approximately 14 ml of 0.1 M sodium hydroxide will be required to complete the titration. At a rate of 0.5 ml added every 5 min, this process will take 140 min. Thus, if many titrations are anticipated, it is worthwhile to automate the titration as has been done using a personal computer interfaced with an automatic titrator and conductivity meter.

The procedure can be modified to determine the sulfonate content of sulfonated kraft lignin (Öster et al. 1988).

7.7.4 Total Sulfur Determination

7.7.4.1 Principle of Combustion/Ion Chromatography

The first step in determination of total sulfur in pulp by combustion/ion chromatography entails burning the pulp in a Schöniger combustion flask (Douek and Ing 1989). Combustion of the pulp in an oxygen atmosphere in a platinum boat oxidizes the sulfur to sulfate which dissolves in deionized water at the bottom of the flask. The presence of a few drops of hydrogen peroxide in the deionized water ensures that oxidation to sulfate is complete.

In the second step, a sample of the sulfate solution, after appropriate dilution, is analyzed by ion chromatography. The separator column contains an ion exchange resin consisting of a sulfonated polystyrene/divinylbenzene core with an outer layer of latex particles bearing quaternary ammonium groups as the ion exchange functions. The different affinities of the anions for the resin causes them to pass through the column at different rates and, under controlled conditions, to exit the column at characteristic times.

After the separator column, the sulfate is carried by the eluent into a suppressor column containing a strongly acidic cation exchange resin. This resin reduces the conductivity of the carbonate/bicarbonate eluent by conversion to weakly ionized carbonic acid and increases the conductivity of the sulfate by conversion to highly ionized sulfuric acid. In this manner, the sulfate can be easily detected by a conductivity detector. The peak height corresponding to sulfate is proportional to the concentration of sulfate in the original solution and can be converted into concentration using a previously determined calibration curve.

7.7.4.2 Description of Combustion/Ion Chromatography

The pulp is thoroughly washed with deionized water to remove any free soluble sulfur. Two to three grams of each pulp sample is ground in a Wiley mill in order to ensure homogeneous sampling. Approximately 100 mg of the ground pulp is pressed into a pellet, using a KBr pellet die, with a force of 7 tons in a laboratory press (effective pressure on the pulp of approximately 530 MPa) for approximately 10 s The pellet is removed from the die, dried in an oven at 105 ± 2 °C for at least 2 h, cooled in a desiccator, and weighed. The pulp is then wrapped in a black sample wrapper, with the narrow tail extending for ignition, and placed in the platinum sample carrier of a 1000-ml Schöniger combustion flask (available from A. H. Thomas Co.).

Deionized water (20 ml) containing a few drops of 30% hydrogen peroxide are placed in the bottom of the Schöniger flask and the flask is filled with oxygen. The combustion flask is introduced into an infrared igniter (available from A.H. Thomas Co.) and the wrapper is ignited. After combustion is complete, the flask is removed and allowed to cool for several minutes. The contents are shaken for 30 s and allowed to stand for about 1 h to ensure complete absorption. The solution is then transferred into a 100-ml volumetric flask and diluted to the mark with deionized water prior to ion chromatographic analysis.

Analyses for sulfate can be performed on a Dionex ion chromatograph or equivalent equipment. The sample is injected, after appropriate dilution, onto an AS4A ion-exchange separator column. A prefilter (0.2 μm, 25 mm diameter from John Scientific Co.) is placed ahead of the injection port to remove particulates from the sample solution. The eluent is a 2.8 mM $NaHCO_3$, 2.25 mM Na_2CO_3 aqueous solution pumped at a rate of 1.5 ml min^{-1}. A filter combination (eluent filter kit from Dionex) consisting of two superimposed filters (10 and 35 μm) installed ahead of the separator column removes trace impurities from the eluent and sample. The eluent is passed into the fiber suppressor column and then to a conductivity detector. The output from the detector may be measured on a Hewlett-Packard 3390A integrator. The sulfate concentration is determined from the peak height by reference to a calibration curve generated using standard sodium sulfate solutions. A blank correction is provided by repeating the entire combustion using the wrapper with no sample added.

The percent sulfate in the pulp sample is given by equation,

$$\text{Sulfate, \%} = \frac{(C_s - C_b)\,DV}{W_s \times 10},$$

where C_s = concentration of sulfate in sample solution ($mg\,l^{-1}$), C_b = concentration of sulfate in blank solution ($mg\,l^{-1}$), D = dilution factor, V = volume of flask (ml) into which the solution containing the combustion products is transferred. A 100-ml flask was found to be convenient for this work, W_s = weight of sample, mg.

To convert from percent sulfate to sulfonate content in $mmol\,kg^{-1}$ the result is multiplied by 104.2. As in the conductometric titration, the sulfonate content of lignin in the pulp is obtained by dividing the sulfonate content of the pulp by the sum of Klason and acid-soluble lignin expressed as a percentage.

This procedure should be equally applicable to the determination of sulfur in lignin sulfonates or sulfonated kraft lignins.

7.7.5 Discussion of Methods

Methods of determining the sulfonate content of lignin fall into two main classes: those typified by conductometric titration, in which sulfonate content is measured directly and those, as represented by combustion/ion chromatography, which measure the sulfur content and assume that all the sulfur is present as sulfonate groups is approximate in many ultra-high-yield sulfite pulps (Katz et al. 1984) but may lead to large errors (Samuelson and Westlin 1947, 1948, Regestad and Samuelson 1958, Schöön 1961, Janson and Sjöström 1966, Katz et al. 1984).

Methods that attempt to directly measure the sulfonate content of lignin in pulps must overcome the challenge of distinguishing sulfonate from accompanying carboxylate groups. One such method (Sjöström and Enström 1966, Sjöström et al. 1965) is based on the concept that, under acidic conditions, benzidinium ions form salts only with the sulfonate groups. The pulp is saturated with an aqueous solution of benzidine hydrochloride and washed with water to remove excess reagent; the benzidinium ions are eluted with hydrochloric acid and determined by UV spectrophotometry.

A problem inherent in the foregoing method is that benzidinium ions are not entirely specific for sulfonate groups and a correction is necessary for the extent of their reaction with carboxylate groups. The correction is determined by calculating the difference between the magnesium and benzidinium ion capacities of the pulp and assuming that this value is equal to the carboxylate content. The amount of benzidinium ion bound to the carboxylate groups is estimated from the amount that binds to carboxylate groups in bleached (i.e., delignified) pulps. Another drawback of the method is nonionic binding of benzidine to lignin, which leads to an overestimation of the sulfonate content in high lignin content pulps. Thus, the method cannot be safely applied to high-yield sulfite mechanical pulps.

The second direct method depends on the ability of aqueous potassium chloride, adjusted to pH 4 with carbon dioxide, to selectively elute hydrogen ions from sulfonic acid groups in pulps that first have been converted to their hydrogen form with 0.1 M hydrochloric acid (Cappelen and Schöön 1966). Carbon dioxide is removed from the eluent by sparging with nitrogen and the remaining acid is titrated with 0.1 M sodium hydroxide. Again, a correction factor for interference from carboxylic acids is required. This factor, as before, is based on the protons eluted from bleached pulps by the eluent. As the results depend on the concentration of potassium chloride used, the letter is adjusted so that the sulfonate content corresponds to the sulfur content of pulps assumed to contain only acidic sulfur.

The third direct method (Katz et al. 1984) distinguishes between sulfonate and carboxylate groups by the ability of 0.01 M acetic acid to selectively elute magnesium from carboxylate groups. The sulfonate and carboxylate groups in sulfite pulp are first converted to their magnesium forms by treatment with magnesium chloride solution. After washing the pulp with water, magnesium is eluted from the carboxylate groups using 100-ml aliquots of 0.01 M acetic acid until a constant titer with 0.001 M EDTA is obtained in three successive elutions. This constant titer is assumed to correspond to the amount of magnesium eluted from sulfonate groups by each aliquot of acetic acid and is used as the correction factor. The elution is continued with 0.1 M hydrochloric acid until no further magnesium is released. The content of sulfonate groups is determined by subtracting from the total titer the corrected titer for the carboxylic acid. Although, as is the case for the other two methods, a correction is required to compensate for interference from carboxylate groups, the necessary corrections are assessed on the pulp under investigation at the same time as the sulfonate content is determined.

The method of choice, conductometric titration (Katz et al. 1984), through its ability to easily distinguish between carboxylate and sulfonate groups, provides the most direct and convenient measure of sulfonate content of lignin in pulp. This ability to distinguish between carboxylate and sulfonate groups is inherent in the way end points are located in conductometric titrations. The end points are found by extrapolation of the straight lines to the equivalence point and hence, in contrast to potentiometric titrations, measurements near the equivalence point have no special significance.

Conductometric titrimetry can be applied to a wide range of pulp types with a wide range of sulfonate contents. The method has been applied to softwood chips treated with sulfite, softwood sulfite mechanical pulps, high-yield bisulfite pulps, low-yield acid sulfite pulp (Katz et al. 1984, Gummerus 1985, Heitner and Hattula 1988) and sulfite treated aspen wood (Beatson et al. 1985) with sulfonate contents ranging from 20 to 300 $mmol\,kg^{-1}$. In addition, the method has been adapted to measure sulfonate content of sulfonated kraft lignin (Öster et al. 1988).

Table 7.7.1. Comparison of sulfonic acid contents determined by various methods

Material[a]	Yield %	Sulfur mmol kg^{-1} (a)[b]	Sulfonic acid mmol kg^{-1} (b)[b]	(c)[b]
Wood, sulfonated for 12.5 min at pH 7	99	128	120	107
Wood, sulfonated for 35 min at pH 7	97	216	181	179
Wood, sulfonated for 110 min at pH 7	95	250	227	213
CMP from chips sulfonated at pH 7	ca. 92	269	236	246
High-yield bisulfite	ca. 70	219	208	196
Low-yield bisulfite	47	40	28	20
Low-yield acid sulfite	42	24	6	6
Oxidatively sulfonated kraft lignin ($KMnO_4$)	–	1250	1250	–
Lignin sulfonic acid	–	(1811)[c]	1503	–

[a] Results on wood and pulp from Katz et al. 1984. Sulfonated lignin Öster et al. 1988. Lignin sulfonic acid Samuelson 1948.
[b] (a) Gravimetric (b) Conductometric (c) Magnesium elution.
[c] Sulfur content measured by combustion.

Analogous conductometric titrations have been used to determine the sulfonic acid content of isolated lignin sulfonates (James and Tice 1965, Kojima et al. 1960, Peniston and McCarthy 1948, Samuelson 1948) and sulfonated lignin model compounds (Mikawa and Sato 1958).

Provided the described experimental procedure is followed, there seems to be little restriction in the use of the analysis. Typical data obtained by conductometric titration for the sulfonate contents of a variety of pulps and lignin are shown in Table 7.7.1 along with data obtained by other methods.

It is apparent that, for lignin in pulp or wood, gravimetric analysis (CPPA G28) gives sulfonate group contents about 20 mmol kg^{-1} higher than those obtained by conductometric titration. Figure 7.7.4 shows that magnesium exchange and conductometric titration afford comparable results which differ on average by about 5%. Results from conductometric measurements on a typical high-yield sulfite pulp gave a mean of 208 mmol kg^{-1} with a standard deviation of 3 mmol kg^{-1}.

Of the methods that measure total sulfur content, combustion/ion chromatography has several advantages over other methods. It is much simpler and faster than the standard gravimetric method and, when compared to titrimetric methods, is less prone to interference by other ions. The procedure has been applied to chemimechanical pulps with sulfur contents ranging from 0.1 to 1%. Measurements on a typical sulfite mechanical pulp gave a mean of 0.612%, with a standard deviation of 0.013%.

The accuracy of the method has been assessed by analysis of a standard sample of bovine liver from the National Bureau of Standards and a standard sulfonated hydrocarbon. For the bovine liver the sulfur content by combustion/ion chromatography was about 1.2% below the accepted value whereas for the standard hydrocarbon, with a reported sulfur content of 0.97%, combustion/ion

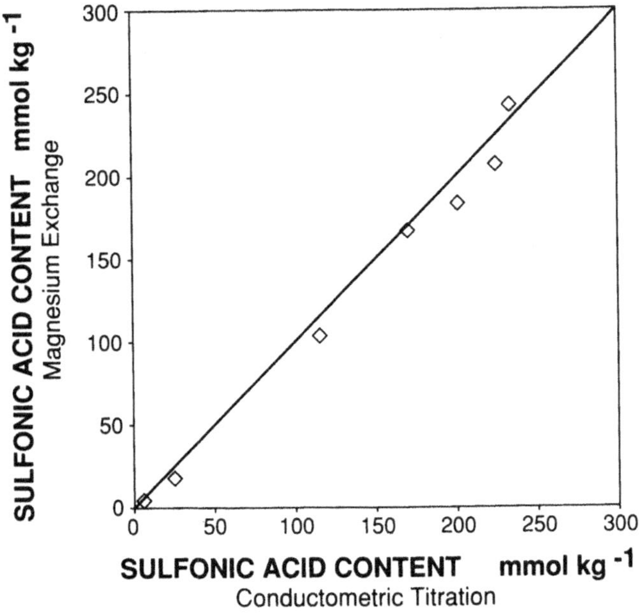

Fig. 7.7.4. Relationship between the sulfonate contents of chemimechanical pulps obtained by conductometric titration and those obtained by magnesium exchange. (Katz et al. 1984)

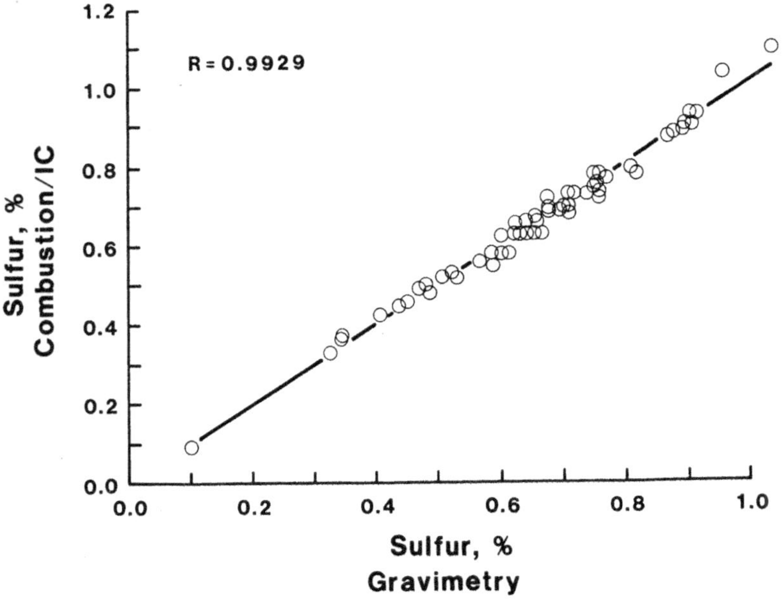

Fig. 7.7.5. Relationship between sulfur content of chemimechanical pulps obtained by combustion/ion chromatography and by CPPA standard gravimetric method G28. (Douek and Ing 1989)

chromatography gave a value 0.978%. In addition, good agreement has been obtained between the results from combustion/ion chromatography and from gravimetric analysis as shown in Fig. 7.7.5 (Douek and Ing 1989).

Although the reproducibility and accuracy of the measurements are good for both analyses, other errors may arise when estimating the sulfonate content of the lignin in the pulp. For example, uncertainty exists in estimating the lignin content of the pulp (Lorås and Loschbrandt 1961). Also, bisulfite may add to carbonyl groups in the cellulosics to form alpha-hydroxysulfonates (Rydholm 1965). However, for conductometric titration, any error from this source is unlikely to be major, as alpha-hydroxysulfonates should not survive the acid soaks performed prior to conductometric titration. In addition, it must be remembered that as combustion/ion chromatography measures total sulfur content, in many instances it will overestimate sulfonate content.

References

Aldrich LC (1974) A volumetric method for the determination of sulfates. Tappi 57:122–123

Atack D, Heitner C, Karnis A (1980) Ultra-high yield pulping of eastern black spruce. Part 2. Sven Papperstidn 83:133–141

Beatson RP, Heitner C, Rivest M, Atack D (1985) Sulphite treatment of aspen. Factors affecting the formation of carboxylate and sulphonate groups. Pap Puu 11:702–708

Brauns FE (1952) The sulfonation of lignin. In: Brauns FE (ed) The chemistry of lignin. Academic Press, New York, 359–393

Brauns FE, Brauns DA (1960) The sulfonation of lignin. In: Brauns FE, Brauns DA (eds) The chemistry of lignin. Supplement volume. Academic Press, New York, 313–386

Canadian Pulp and Paper Association (1970) Standard Method G28. In: Standard Testing Methods, Technical Section, Canadian Pulp and Paper Association, Montreal, Canada

Canadian Pulp and Paper Association Useful Method G.7U. In: Standard Testing Methods, Technical Section, Canadian Pulp and Paper Association, Montreal, Canada

Cappelen H, Schöön N-H (1966) Determination of sulfonic acid groups in unbleached sulfite pulps. Sven Papperstidn 69:322–325

Douek M, Ing J (1989) Determination of sulphur and chlorine in pulp and paper samples by combustion/ion chromatography. J Pulp Pap Sci 15:J72–J78

Glennie DW (1971) Reactions in sulphite pulping. In :Sarkanen KV, Ludwig CH (eds) Lignins. Occurrence, formation, structure and reactions. Wiley-Interscience, New York, 597–637

Gummerus M (1985) Sulphite treatment of TMP rejects. Part 1. Properties of fibre fractions at various freeness levels and their influence on the reject pulp. Pap Puu 11:635–647

Heitner C, Hattula T (1988) Ultra-high-yield pulping. Part VI: The effect of sulphonation on the development of fibre properties. J Pulp Pap Sci 14:J6–J11

James AN, Tice PA (1965) The presence of carboxyl groups in lignosulfonate preparations. Tappi 48:239–244

Janson J, Sjöström E (1966) Sulphur content in relation to the degree of sulphonation in birch and spruce sulphite pulps. Sven Papperstidn 69:107–110

Katz S, Beatson RP, Scallan AM (1984) The determination of strong and weak acidic groups in sulfite pulps. Sven Papperstidn 87:R48–R53

Kibblewhite RP, Corson SR, Graham KL (1987) Chemimechanical and thermomechanical pulps of radiata pine corewood and slabwood. Part 3. Factors determining paper quality. Appita 40:121–131

Kojima Y, Hayashi A, Tachi I (1960) Studies on lignosulphonic acid (VI). Conductometric titration with weak base. 7pn Tappi 14:831–834

Kolthoff IM, Sandell EB (1952) Textbook of quantitative inorganic analysis. Macmillan, New York, 759 pp

Kvisgaard HJ (1965) Postsulfonation. Nor Skogind 19:155–163

Lindsey JB, Tollens B (1892) Über Holz-Sulfitflüssigkeit und Lignin. Ann Chem 267:341–366

Lorås V, Loschbrandt F (1961) Determination of lignin in sulphite pulps. Nor Skogind 7: 302–309

Marth DE (1959) Studies on the lignin fraction of aspenwood pulps produced by sulphite-bisulphite cooking liquor systems. Tappi 42:301–308

Mikawa H, Sato K (1958) On the cooking mechanism of wood. XVII. Further investigations on the conductometric titrations of lignin model compounds with phenolic hydroxyl groups and a brief theory of the conductometric titration of phenols. Bull Chem Soc 7pn 31: 628–634

Ora A (1960) Determination of the total sulphur content of sulphite and sulphate waste liquor. Pap Puu 42:299–302

Öster R, Kringstad KP, Hirose S, Hatakeyama H (1988) Oxidative sulfonation of kraft lignin. Nord Pulp Pap Res J 3:68–74

Peniston QP, McCarthy JL (1948) Lignin. II. Liberation of phenolic hydroxyl groups by alkaline cleavage of lignin sulfonic acids. J Am Chem Soc 70:1329–1332

Pobiner H (1983) Improved inflection points in the non-aqueous potentiometric titration of acid functionalities in lignin chemicals by using internal standardization and ion exchange. Anal Chim Acta 155:57–65

Regestad SO, Samuelson O (1958) Side-reactions during sulphite cooking. Sven Papperstidn 61:735–740

Rivington DE (1988) Total sulphur in pulp by X-ray fluorescence spectrometry. Proc 74th annual meeting of the Tech Section CPPA A63–A68. Technical Section Canadian Pulp and Paper Association, Montreal, Canada

Ross JW Jr, Frant MS (1969) Potentiometric titrations of sulphate using an ion-selective lead electrode. Anal Chem 41:967–979

Rydholm SA (1965) Chemical pulping I., Sulfite process. In: Rydholm SA (ed) pulping processes. Kreiger, Malabar, Florida, 439–576

Samuelson O (1948) A contribution to the chemistry of the sulphite process. III. A comparison between the sulphur content and the amount of sulphonic acid groups in lignosulphonic acid. Sven Kem Tidskr 60:128–130

Samuelson O, Westlin A (1947) Chemistry of the sulphite process. Sven Papperstidn 50: 149–154

Samuelson O, Westlin A (1948) A contribution to the chemistry of the sulphite process. Part II. Sven Papperstidn 51:438–444

Scallan AM, Katz S, Argyropoulos DS (1989) The conductometric titration of cellulosic fibres. In: Schuerch C (ed) Cellulose and wood-chemistry and technology. Wiley, New York, 1457–1471

Schöön N-H (1961) The reaction between thiosulphate and wood during sulphite cooking. Sven Papperstidn 64:624–633

Sjöström E, Enström B (1966) A method for the separate determination of sulphur and carboxyl groups in sulphite pulps. Sven Papperstidn 69:55–59

Sjöström E, Janson J, Haglund P, Enström B (1965) The acidic groups in wood and pulp as measured by ion exchange. J Polym Sci C11:221–241

Chapter 8

Determination of Molecular Weight, Size, and Distribution

8.1 Gel Permeation Chromatography

G. GELLERSTEDT

8.1.1 Introduction

Gel permeation chromatography (GPC), sometimes referred to as gel filtration chromatography (GFC) or size exclusion chromatography (SEC), entails the chromatographic fractionation of macromolecules according to molecular size.

In the early development of this technique, cross-linked polydextran gels having varying pore sizes (Sephadex, Pharmacia, Sweden) were used as the stationary phase (Porath and Flodin 1959, Flodin 1962). When swollen in water, such gels form a three-dimensional network that acts as a molecular sieve. When an aqueous solution of macromolecules is allowed to move through a column containing the gel, a chromatographic separation takes place; molecules of low molecular weight (size) (i.e., molecules having a small hydrodynamic radius) are able to penetrate into the gel particle pores but large molecules are excluded from the pores and pass directly through the column. Consequently, the largest molecules elute first and the smallest last.

8.1.1.1 Chromatography of Lignins on Sephadex Gels

The technique of fractionating macromolecules on Sephadex has been applied to lignin materials since the early 1960s. Thus, in a series of papers, Forss and co-workers reported the use of gel filtration to determine the molecular weight distribution of lignosulfonates obtained from different types of sulfite cooks (acid sulfite and bisulfite) (Jensen et al. 1962, Forss and Fremer 1964, Forss et al. 1976, Alekseev et al. 1978). The fractionation curves were monitored by UV detection. Isolated lignosulfonate fractions, whose molecular weight had been previously determined by light scattering, were used to calibrate the column, and water, with or without the addition of an electrolyte, was the eluent.

In later work, Forss et al. also analyzed kraft lignins by the same technique, using aqueous sodium hydroxide as the eluent (Forss et al. 1976, 1984). Fractionation of kraft lignins on Sephadex gels has also been carried out by other investigators (Connors 1978, Connors et al. 1978, 1980, Sarkanen et al. 1982, Yan et al. 1984, Kondo and McCarthy 1985, Dolk et al. 1986).

As seen in Fig. 8.1.1, lignosulfonates, as well as kraft lignins, give rise to elution profiles consisting of an excluded peak (i.e., molecules which are larger than the largest pores of the swollen gel) followed by a generally uniform distribution curve when chromatographed on Sephadex gels. It should be noted,

Springer Series in Wood Science
Methods in Lignin Chemistry
(Edited by S.Y. Lin and C.W. Dence)

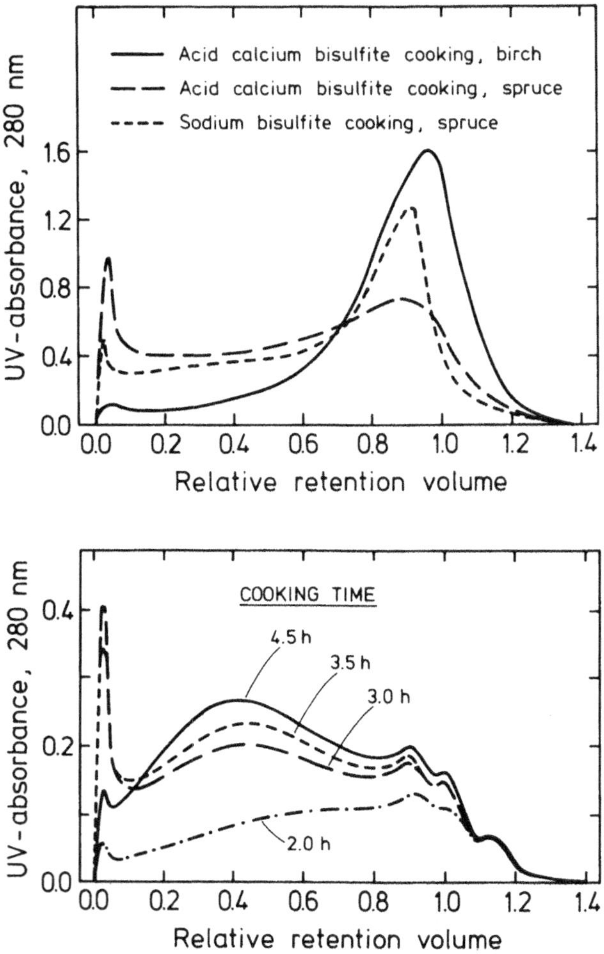

Fig. 8.1.1. Elution of lignosulfonates on Sephadex G-75 with 0.25 M calcium chloride as eluant (*upper curve*). Elution of kraft lignins on Sephadex G-50 with 0.5 M sodium hydroxide as eluant (*lower curve*). (Forss et al. 1976)

however, that different Sephadex gels were used in the experiments referred to in Fig. 8.1.1. In the analysis of kraft lignin samples on Sephadex G-50 in 0.5 M sodium hydroxide, a weight-average molecular weight exclusion limit of approximately 18000 was found. On the other hand, chromatography of lignosulfonates on Sephadex G-75 in 0.25 M calcium chloride demonstrated that, in this system, the $\bar{M}_w$ exclusion limit was above 50000 (Forss et al. 1976). These features illustrate one difficulty encountered in performing gel filtration, namely, the exclusion limit of the gel. Gels must be chosen such that their pore size distribution includes the total molecular size range of the polymer under analysis. Otherwise, poor or negligible separation will result in certain molecular size ranges.

8.1.1.2 Modified Sephadex Gels

More recently, modified polydextran gels, as well as completely new gel materials, have been employed in GPC. Derivatization of the original Sephadex gels, for example, by alkylation of the hydroxyl groups, has led to the creation of gels suitable for use in organic solvents. Connors (1978) and Connors et al. (1980) used such gels to analyze the apparent molecular weight distribution of DHP and native, kraft, and cellulase lignins. Column calibration was performed with lignin model compounds and polystyrene fractions of known molecular weight. Therefore, apparent molecular weights and not absolute molecular weights were reported. Such values are nevertheless very useful for comparing the molecular weight distribution in different lignin samples. In this work, it was also found that completely different elution curves were obtained for lignin samples analyzed in pure dimethylformamide (DMF) and in DMF containing LiCl. Lignin association complexes were assumed to be responsible for the higher apparent weights observed in DMF alone. These effects decreased in intensity or disappeared completely as the ionic strength of the eluant was increased by addition of LiCl.

8.1.2 High-Performance Size Exclusion Chromatography (HPSEC)

The introduction of cross-linked, semi-rigid polystyrene gels of uniform particle size has made it possible to perform GPC in high pressure systems using small, high-efficiency columns. With such systems, the chromatographic time can be substantially reduced and the resolution power enhanced. A variety of organic solvents can be used. The technique is referred to as HPSEC (High-Performance, Size-Exclusion Chromatography) and has been described as "a system in which the number of theoretical plates generated per second is about ten-fold greater than that of traditional SEC systems" (Yau et al. 1979). Apart from polystyrene gels, other available packing materials suitable for HPSEC include a variety of completely rigid inorganic particles based usually on porous silica.

HPSEC performed on polystyrene and sometimes on silica gels (Plastre 1983), as well as GFC with Sephadex gels, have been utilized in lignin analysis. Though aqueous or sometimes polar organic solvents were used for the latter gels as described above, polystyrene gels should only be used with organic solvents such as tetrahydrofuran (THF), methylene chloride, or toluene (Yau et al. 1979). This constitutes a problem since lignins, whether extracted directly from plant tissue of isolated after various pulping or bleaching processes, have a diverse range of solubilities. Thus, plant source, isolation procedure, functionality, and molecular weight distribution are among the factors that, to a great extent, determine the solubility of the lignin sample in a particular solvent. This also means that no universal GPC method exists that is suitable for the analysis of all types of lignins.

In the following subsection, HPSEC on polystyrene gels using THF as solvent is emphasized since this system has been found to give the most reliable results with a variety of lignins (Froment and Robert 1977, Kaneko et al. 1980, Kringstad et al. 1981, Månsson 1981, Yamasaki et al. 1981, Gellerstedt and Lindfors 1984, 1987, Robert et al. 1984, Mörck et al. 1986, Chum et al. 1987). Possible chromatographic media other than Sephadex for analyzing water-soluble lignins from sulfite pulping liquors and different bleaching effluents have been described (Pellinen and Salkinoja-Salonen 1985a, 1986) and will not be discussed here.

8.1.3 Experimental

8.1.3.1 Equipment

A HPLC system consisting of a solvent delivery system, pump, injection port, column(s) and detector(s) should be used. Such chromatographic systems, usually digitally controlled, are available from a large number of manufacturers. In addition, appropriate data-handling systems for calculating molecular weight averages are available. The solvent (THF) should be of high purity and further purified before use by distillation, degassing, and filtration. The pump should be able to deliver a precise, pre-determined flow-rate with a minimum of pulsation to obviate detector noise. Columns packed with a semi-rigid, cross-linked styrene/divinylbenzene copolymer are commercially available as μ-Styragel (Waters Assoc.), Mikro Pak Bk (Varian Assoc.) or μ-Spherogel (Altex Scientific). These columns have different pore sizes and thus different molecular weight fractionation ranges. They may be coupled in series, thus enhancing considerably the area of maximum resolution power. However, such an arrangement makes it more difficult to compare directly chromatograms obtained in different laboratories.

Detection is usually achieved by ultraviolet light (UV) or by refractive index (RI), although other types of detectors, e.g., those based on low-angle laser light-scattering (LALLS) (Kolpak et al. 1983) and differential viscometry (Himmel et al. 1989, Siochi et al. 1989) are becoming available. The choice of UV detection for lignin analysis is obvious but also has the further advantage of being highly sensitive and of good linearity. Furthermore, UV detectors are relatively insensitive to changes in solvent flow rate and temperature. The latter feature is a definite advantage since, by comparison, RI detectors are extremely solvent-sensitive and require strict temperature control of the chromatograpic system. One disadvantage of UV detection is, however, that the response factor for lignins may vary with the molecular weight. Thus, Lange et al. (1983) found absorption coefficients of lignin to differ by 5% on proceeding from samples of low to high molecular weight. Chum et al. (1987) found that UV absorptivity of low-molecular weight fractions to be higher at wavelengths >300 nm, indicating the presence of vinyl structures and/or carbonyl groups conjugated to aromatic rings.

8.1.3.2 Sample Preparation

The lignin sample should be completely soluble in THF. For the vast majority of lignin preparations, this condition is not automatically fulfilled. Consequently, HPSEC of lignins, using the system described here, should be conducted only after appropriate derivatization. Acetylation, as well as methylation, may be used. As an additional derivatization procedure, silylation has been tried. A comparison of acetylated, silylated and unmodified samples has been made by Pellinen and Salkinoja-Salonen (1985b). Despite simplicity of preparation, silylated derivatives may be less suited for GPC analysis than methylated or acetylated samples since they are very unstable and may decompose before or during the analysis. As a consequence, a gradual deterioration of the chromatographic system may occur.

For lignin samples containing different types of hydroxyl groups, but with no or minor amounts of carboxylic acid groups, acetylation or methylation can be used to provide derivatives of good solubility in THF. Samples of this type include different native lignin preparations and kraft, soda, and organosolv lignins. Lignins that have been exposed to oxidative conditions, such as those from different types of bleaching operations, contain carboxylic acid groups. For such lignin samples, methylation with diazomethane is the preferred mode of derivatization. Diazomethane converts all carboxylic acid groups to methyl esters while phenolic (and enolic) hydroxyl groups are simultaneously etherified.

8.1.3.3 Acetylation

A weighed amount of lignin is acetylated overnight at room temperature with a mixture of purified pyridine-acetic anhydride (1:1, v/v). Excess acetic anhydride is decomposed by the addition of methanol-ice water. After evaporation of the mixture to dryness, the residue is suspended in toluene and again evaporated (repeated three times). In this process, the remaining traces of acetylation mixture are removed by azeotropic distillation. Subsequently, the toluene can be removed by addition of methanol and evaporation. This method allows complete recovery of the sample with a minimum of added impurities.

8.1.3.4 Methylation

A weighed amount of lignin is dissolved or suspended in dioxane-methanol (9:1, v/v) and one drop of water/2 ml solvent (approximately 5 mg of lignin ml^{-1} of solvent). Diazomethane is added to the solution under a stream of nitrogen over a 5-min period at room temperature. (The reaction should be conducted in a well-ventilated hood.) The diazomethane is conveniently prepared from N-methyl-N-nitroso-p-toluenesulfonamide (Merck-Schuchardt, Aldrich, Fisher, Fluka, and others) in a special reaction flask as described by Schlenk and Gellerman (1960). After a reaction time of approximately 20 min, the procedure is repeated and diazomethane is again added over a 5-min period. The procedure is repeated a third time to ensure complete methylation and solubilization of the sample. The reaction is terminated after a further 20 min by bubbling nitrogen through the solution. After being dried over magnesium sulfate, the solution is carefully evaporated and the residue is dissolved in THF. A 5-10-mg amount of sample ml^{-1} of THF is suitable for analysis. The injection volume should be in the order of 20 μl.

8.1.4 Discussion of Method

8.1.4.1 Problems Encountered in Gel Permeation Chromatography

A difficulty encountered when using GPC is that different solvent systems may swell the gel to different degrees with the result that pore size distribution is altered. Analysis of polyelectrolytes such as lignins, which contain acidic groups of various types (phenolic hydroxyl, carboxyl, and sulfonic acids in lignosulfonates), constitutes a special problem. Thus, chromatographic separation may be affected by expansion of the molecules stemming from electrostatic repulsion of the charged groups and solvation. Such effects lead to a separation behavior that is strongly dependent on the ionic strength of the eluant.

Furthermore, the separation behavior can be influenced by the presence of charged groups in the gel. If these groups have the same charge as those of the sample, the latter may be excluded from the pores and hence eluted from the column at an erroneous (higher) rate (ion exclusion). If the gel and the sample are oppositely charged, the reverse effect can occur (ion exchange).

A third ionic effect may also influence the separation behavior, namely, ion inclusion. In this instance, small, charged molecules from the sample become concentrated in the gel pores and thus are eluted at a lower rate than expected. The described ionic effects are usually diminished by the addition of an electrolyte to the solvent.

Finally, adsorption phenomena between gel and sample, caused by hydrogen bonding or hydrophobic interactions, may occur, particularly when hydrophilic solvents such as DMF are used in conjunction with polystyrene-divinylbenzene gels. This also results in a slower (or sometimes negligible) elution. If adsorption on the gel is irreversible, a gradual deterioration of the gel and the separation power of the column takes place.

8.1.4.2 HPSEC on Styragel

High performance size exclusion chromatography on Styragel can be used for rapid determination of the apparent molecular weight distribution in lignin resulting from, for example, a chemical reaction. Thus, ozonation of lignin samples and its effect on the molecular size distribution have been investigated by Kaneko et al. (1980) and Kringstad et al. (1981). The latter investigators also analyzed lignin treated with oxygen/alkali, chlorine dioxide, phenol/acid, kraft cooking liquor, and with acid alone. A comparison of milled wood lignin

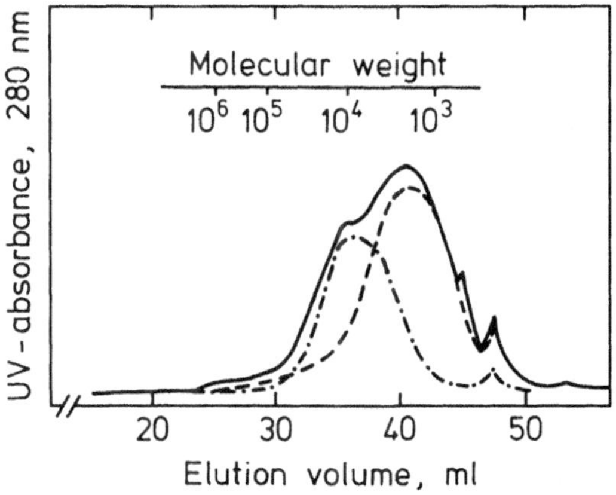

Fig. 8.1.2. High-performance size-exclusion chromatography of silylated samples of pine kraft lignin (—), birch kraft lignin (---) and milled wood lignin from spruce (-·-·). (Månsson 1981)

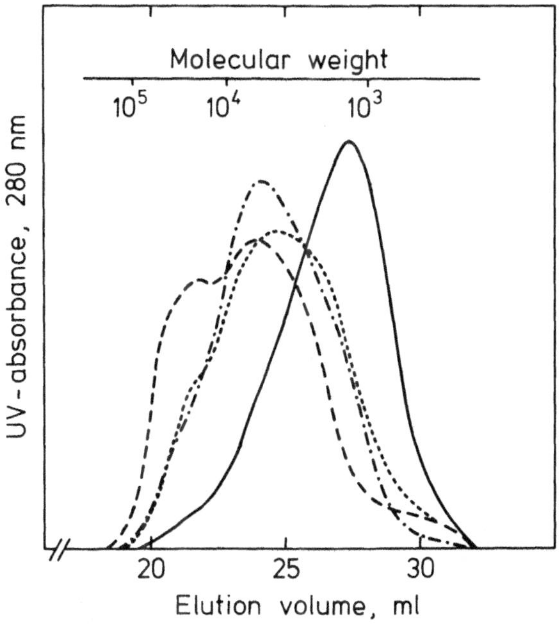

Fig. 8.1.3. High-performance size-exclusion chromatograms of acetylated samples of milled wood lignin from pine (-·-·), pine kraft lignin (—) and two samples of residual lignin isolated from pine kraft pulp (–––, -----). (Yamasaki et al. 1981)

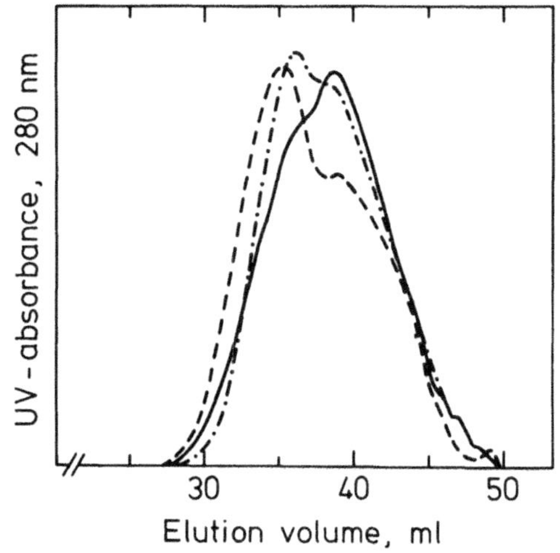

Fig. 8.1.4. High-performance size-exclusion chromatograms of methylated samples of pine kraft lignin (—) and of two lignins obtained from oxygen bleaching effluents (---, -·-·). (Gellerstedt and Lindfors 1987)

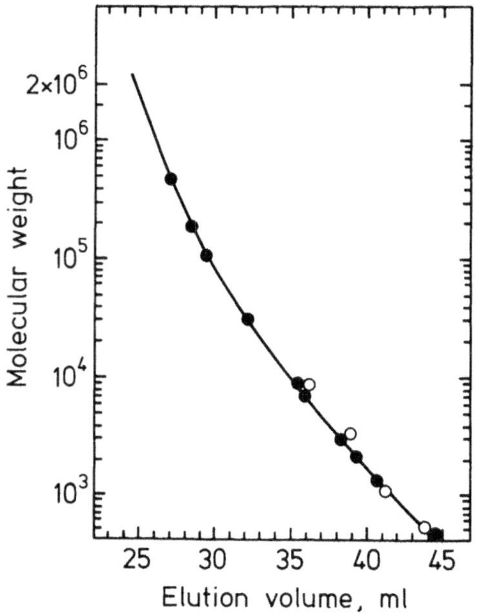

Fig. 8.1.5. Calibration of styragel columns with polystyrene standards (•) and with lignin samples (○). The molecular weights of the latter were determined by vapor phase osmometry. (Månsson 1981)

(MWL) with kraft lignins has been made by Månsson (1981) (Fig. 8.1.2). Similarly, Yamasaki et al. (1981) have compared the molecular-weight distribution of milled wood lignin, kraft lignin, and two samples of lignin isolated from kraft pulp (Fig. 8.1.3). Lignin obtained by steam explosion or organosolv pulping has been compared to MWL by Chum et al. (1987).

Gellerstedt and Lindfors (1984) and Robert et al. (1984) have used HPSEC to compare the molecular weight distribution in a series of lignins taken from different stages of kraft pulping. Gellerstedt and Lindfors (1987) have also used HPSEC to compare a kraft lignin with the lignin dissolved in an oxygen bleaching stage (Fig. 8.1.4). Kraft lignin fractions have been similarly analyzed by Mörck et al. (1986).

In most of the studies mentioned above, polystyrene fractions of known molecular weight, together with low-molecular weight lignin fractions and/or lignin model compounds, were used to calibrate the chromatographic system. It has been shown that low-molecular weight polystyrene and lignin fractions are eluted at approximately the same elution volume (Fig. 8.1.5) (Månsson 1981). The same good correlation cannot, however, be expected for lignin fractions of higher molecular weight owing to increased branching and/or cross-linking. The use of polystyrene standards for determining molecular weight distribution of lignin is not totally satisfactory. However, a molecular weight scale based on polystyrene standards can facilitate comparison of chromatograms obtained in different laboratories and when different combinations of columns are used.

The possibility of using GPC for determining the molecular weight of polymers is widely recognized and, as mentioned above, data-handling programs are available for direct calculation of $\bar{M}_n$ and $\bar{M}_w$ ($\bar{M}_n$ = apparent number-average molecular weight; $\bar{M}_w$ = apparent weight-average molecular weight) from the chromatograms. To give accurate and reproducible results, the GPC method requires that the baseline be well defined on both sides of the chromatographic curve. This is often not the case in the analysis of lignins owing to their heterogeneity. Thus, in addition to the "bulk" polymer, lignin samples may frequently contain low-molecular weight materials and/or materials in the very high-molecular weight range. These features can be seen in Fig. 8.1.2. From this chromatogram, it is not possible to define a true "starting point", i.e., a specific retention time (or elution volume) or a good end point in the low-molecular weight range. Occasionally, the end point may not be regained even after elution of more than one theoretical column volume. In such cases, calculation of the molecular weight becomes highly uncertain. The problems encountered in determining molecular weight by GPC have been treated comprehensively by Yau et al. (1979).

For lignin samples that have been "purified", for example, by chromatographic fractionation or solvent partition, $\bar{M}_n$ and $\bar{M}_w$ values probably can be calculated with some certainty. In such samples, it is assumed that both high- and low-molecular weight material have been removed in the purification step. Therefore, in such instances, the molecular weight distribution reflects only the isolated fraction and not the overall sample.

References

Alekseev A, Forss K, Johanson M, Stenlund B (1978) On the polymerization of spruce lignosulphonates in acid bisulphite pulping. Pap Puu 60:195–198

Chum HL, Johanson DK, Tucker MP, Himmel ME (1987) Some aspects of lignin characterization by high performance size exclusion chromatography using styrene divinylbenzene copolymer gels. Holzforschung 41:97–108

Connors WJ (1978) Gel chromatography of lignins, lignin model compounds and polystyrene using Sephadex LH-60. Holzforschung 32:145–147

Connors WJ, Lorenz LF, Kirk TK (1978) Chromatographic separation of lignin models by molecular weight using Sephadex LH-20. Holzforschung 32:106–108

Connors WJ, Sarkanen S, McCarthy JL (1980) Gel chromatography and association complexes of lignin. Holzforschung 34:80–85

Dolk M, Pla F, Yan JF, McCarthy JL (1986) Lignin 22. Macromolecular characteristics of alkali lignin from western hemlock wood. Macromolecules 19:1464–1470

Flodin P (1962) Dextran gels and their applications in gel filtration. Thesis, University of Uppsala, Uppsala, Sweden

Forss K, Fremer KE (1964) The dissolution of wood components under different conditions of sulfite pulping. Tappi 47:485–493

Forss K, Janson J, Sagfors p (1984) Influence of anthraquinone and sulfide on the alkaline degradation of the lignin macromolecule. Pap Puu 64:77–79

Forss KG, Stenlund BG, Sagfors P (1976) Determination of the molecular-weight distribution of lignosulfonates and kraft lignin. Appl Polym Symp 28:1185–1194

Froment P, Robert A (1977) Importance de l'étalonnage dans la détermination par GPC des masses moleculaires moyennes de la lignine. Cell Chem Technol 11:691–696

Gellerstedt G, Lindfors E (1984) Structural changes in lignin during kraft pulping. Holzforschung 38:151–158

Gellerstedt G, Lindfors E (1987) Hydrophilic groups in lignin after oxygen bleaching. Tappi J 70(6):119–122

Himmel ME, Tatsumoto K, Oh KK, Grohmann KG, Johnson DK, Chum HL (1989) Molecular weight distribution of aspen lignins estimated by universal calibration. In: Glasser WG, Sarkanen S (eds) Lignin, properties and materials. ACS Symp. Ser. Vol. 397, American Chemical Society, Washington, DC, 82–99

Jensen W, Fremer K, Forss K (1962) The separation of the components in spent sulfite liquor. Tappi 45:122–127

Kaneko H, Hosoya S, Nadano J (1980) Degradation of lignin with ozone. Mokuzai Gakkaishi 26:752–758

Kolpak FJ, Cietek DJ, Fookes W, Cael JJ (1983) Analysis of lignins from spent alkaline pulping liquors by gel permeation chromatography/low-angle laser light scattering (GPC/LALLS). Appl Polym Symp 37:491–507

Kondo R, McCarthy JL (1985) Incremental delignification of hemlock wood and characterization of lignin products. Holzforschung 39:231–234

Kringstad KP, Månsson P, Mörck R (1981) Changes in the molecular weight distribution of kraft lignins resulting from various chemical treatments. The Ekman Days 1981. Proc Int Symp Wood Pulp Chem, SPCI, Stockholm, 5:91–93

Lange W, Faix O, Beinhoff O (1983) Über Eigenschaften und Abbaubarkeit von mit Alkohol-Wasser-Gemischen isolierten Ligninen. VIII. The inhomogeneity of the lignins from birch and spruce wood. Holzforschung 37:63–67

Månsson P (1981) GPC of kraft lignins. The Ekman Days 1981. Proc Int Symp Wood Pulp Chem, SPCI, Stocckholm, 5:94–95

Mörck R, Yoshida H, Kringstad KP, Hatakayama H (1986) Fractionation of kraft lignin by successive extraction with organic solvents. 1. Functional groups, ^{13}C-NMR-spectra and molecular weight distributions. Holzforschung 40:51–60

Pellinen J, Salkinoja-Salonen M (1985a) Aqueous size exclusion chromatography of industrial lignins. J Chromatogr 322:129–138

Pellinen J, Salkinoja-Salonen M (1985b) High-performance size exclusion chromatography of lignin and its derivatives. J Chromatogr 328:299–308
Pellinen J, Salkinoja-Salonen M (1986) Characterization and biodegradability of lignin-containing wastewater from forest industry. In: Bjorseth A, Angeletti G (eds) Organic micropollutants in the aquatic environment. D Reidel, Dordrecht, 354–362
Plastre D (1983) Determination des proprietes macromoleculaires de la lignine mise au point de techniques d'analyse. Applications à la caracterisation de lignines kraft et a l'étude de cuissons alcalines avec additifs. Thesis, l'Institut National Polytechnique de Grenoble, Grenoble, France
Porath J, Flodin P (1959) Gel filtration: A method for desalting and group separation. Nature 183:1657–1659
Robert DR, Bardet M, Gellerstedt G, Lindfors E (1984) Structural changes in lignin during kraft cooking. Part 3. On the structure of dissolved lignins. J Wood Chem Technol 4:239–263
Sarkanen S, Teller DC, Abramowski E, McCarthy JL (1982) Lignin 19. Kraft lignin component. Conformation and associated complex configuration in aqueous alkaline solution. Macromolecules 15:1098–1104
Schlenk H, Gellerman JL (1960) Esterification of fatty acids with diazomethane on a small scale. Anal Chem 32:1412–1414
Siochi EJ, Haney MA, Mahn W, Ward TC (1989) Molecular weight determination of hydroxypropylated lignins. In: Glasser WG, Sarkanen S (eds) Lignin. Properties and materials. ACS Symp. Ser. Vol. 397. American Chemical Society, Washington, DC, 100–108
Yamasaki T, Hosoya S, Chen C, Gratzl JS, Chang H-m (1981) Characterization of residual lignin in kraft pulp. The Ekman Days 1981. Proc Int Symp Wood Pulp Chem, SPCI, Stockholm, 2:34–42
Yan JF, Pla F, Kondo R, Dolk M, McCarthy JL (1984) Lignin 21. Depolymerization by bond cleavage reactions and degelation. Macromolecules 17:2137–2142
Yau WW, Kirkland JJ, Bly DD (1979) Modern size-exclusion liquid chromatography. Practice of gel filtration and gel filtration chromatography. Wiley, New York, 476 pp

8.2 Light Scattering

F. PLA

8.2.1 Introduction

Among the physicochemical methods used to study the macromolecular properties of lignins, light scattering has been considered as one of the most useful. It is a convenient method for deriving several parameters such as the weight-average molecular weight, $\bar{M}_W$, the second virial coefficient, A2, and the z-average root mean square radius of gyration $\langle s^2 \rangle z$, of polymers in solution. The method is also probably the most suitable for determining the actual shapes of molecules. Furthermore, light scattering data have often been used for the estimation of polydispersity and also as a means of determining microtacticity.

This analytical procedure has been used to determine the $\bar{M}_W$ of soda and kraft lignins (Gupta and Goring 1960, Lindberg et al. 1964) and lignosulfonates (Moacanin et al. 1955, Yean et al. 1964, Forss et al. 1967, Forss and Stenlund 1969). However, with lignins, several important and characteristic difficulties must be overcome to obtain reproducible and accurate results since the relatively small dimensions of the molecules give low intensity scattered light, the specific optical properties of lignins, i.e., absorbance, fluorescence, and anisotropy, may interfere with the measurement, and because of the possible presence of lignin aggregates in the solutions. The classical instruments (Sofica and Brice Phoenix light scattering photometers) appear to be unsuitable for lignin studies because of their poor performance in the range of molecular weights under investigation (Pla and Robert 1974). Furthermore, with these instruments, the measurements are extremely sensitive to traces of colloidal debris in the solutions and require extensive solution clarification.

Nevertheless, these problems have recently been overcome, at least in part, by the use of two new and sophisticated instruments. Both are equipped with a laser source, an interference filter, and an analyzing polarizer. The first is a prototype operating at various angles. It has been used with lignins (Pla et al. 1977, Pla 1980) at 90°. The second commercially available instrument (Plastre 1983, Kolpak et al. 1983, Kim 1985, Dolk et al. 1986, Pla et al. 1986) is a low-angle laser light scattering photometer (Kromatix Model KMX-6).

8.2.2 Principle

The determination of molecular weights by the light scattering method is based on the principle that the intensity of light scattered at some angle, θ, is a

Springer Series in Wood Science
Methods in Lignin Chemistry
(Edited by S.Y. Lin and C.W. Dence)

function of the size of the scattering particles. The coefficient commonly used to describe the light scattering property of materials is the Rayleigh factor $R(\theta)$, expressed in the following form (Stacey 1956):

$$R(\theta) = J(\theta)/(I_o V), \tag{1}$$

where $J(\theta)$ is the radiant intensity per unit irradiation, and I_o is the illuminating beam scattered at the angle θ, per unit volume V of the scattering medium. The scattering volume, V, is the portion of the illuminated volume that is sensed by the detector. Using the general definitions of the radiant intensity and irradiance and the geometry of the instrument, Eq. (1) becomes (Kaye and Havlik 1973):

$$R(\theta) = G(\theta)/G_o \cdot D \cdot (\sigma' l')^{-1}, \tag{2}$$

where $G(\theta)$ and G_o are the photomultiplier signals from the scattered and incident beams, respectively. D is the transmittance of the measuring attenuators used for the measurement of G_o, σ' is the solid angle over which $J(\theta)$ is viewed and l', the length parallel to the incident beam of the scattering volume. The value of σ' and l' are both determined by the geometry of the instrument and are referred to as instrument constants that depend on the angle of measurement and on the solvent refractive index (Kaye and McDaniel 1974).

According to the fluctuation theory of light scattering (Flory 1953), the relationship between the Rayleigh factor and the physical characteristics of the macromolecules in solution is expressed in the following general form:

$$Kc/\Delta R(\theta) = 1/M_W P(\theta)^{-1} + 2A_2 c + 3A_3 c^2 + \ldots, \tag{3}$$

where c is the concentration of the polymer in $g\,ml^{-1}$, $P(\Theta)$, the particle scattering function, A_2 and A_3, respectively, the second and third virial coefficients, $\Delta R(\theta)$, the difference in the Rayleigh factors for the solution and pure solvent. K is an optical constant which, when the primary beam is vertically polarized, is given by:

$$K = [4\pi^2 n_o^2/\lambda_o^4 N] \cdot (dn/dc)^2, \tag{4}$$

where n_o is the refractive index of the solvent, λ_o the wavelength of the incident light in a vacuum, N is Avogadro's number, and dn/dc the specific refractive index increment.

The term $P(\theta)$ describes the angular variation of light scattered at constant concentration (Kratochvil 1972). The form of $P(\theta)^{-1}$ is dependent upon the size and shape of the scattering particles and provides the basis for their determination. Under the limiting condition of small angles, $P(\theta)^{-1}$ may be expressed in the form of Eq. (5):

$$\lim_{\theta \to 0,} P(\theta)^{-1} = 1 - 16\pi^2/3\lambda^2 \langle s^2 \rangle z \sin^2(\theta/2), \tag{5}$$

where λ is the wavelength of the incident light in the media and $\langle s^2 \rangle z$ is the mean square radius of gyration of the particles.

This equation shows that for particles such as lignin molecules, generally much smaller than the wavelength ($\langle s^2 \rangle z \ll \lambda$), $P(\theta)^{-1}$ approaches unity. Thus, for dilute solutions, Eq. (3) becomes:

$$Kc/\Delta R(\theta) = 1/\bar{M}_W + 2A_2c. \tag{6}$$

As mentioned above, with lignins, optical anisotropy, absorbance, and fluorescence must be corrected for or minimized (Pla et al. 1977).

8.2.2.1 Optical Anisotropy Correction

Lignin molecules are relatively small and cause depolarization of the incident beam, giving rise to an enhancement of scattered light intensity. The effect is due to electrical anisotropy within the molecules and cannot be disregarded (Stacey 1956). Its correction is made from the resulting vertical and horizontal components of the scattered light, measured with an analyzing polarizer inserted between the sample and the detector. These two components are complex functions of $\sin\theta$ and $\cos\theta$ (Russo et al. 1984). For small values of θ these functions are simplified and the excess Rayleigh factor, corrected for anisotropy, is then given by the following expression:

$$\Delta R(\theta) = \Delta Rv(\theta) - 4/3\ \Delta Rh(\theta), \tag{7}$$

where $\Delta Rv(\theta)$ and $\Delta Rh(\theta)$ are, respectively, the vertical and horizontal excess Rayleigh factors.

Another method for correcting for optical anisotropy has been described by Stacey (1956) and applied to lignosulfonates (Forss and Stenlund 1969). It involves essentially the multiplication of the measured $Kc/\Delta R(\theta)$ values by the Cabannes factor, expressed for the case in which the incident light is vertically polarized (Armizadeh and McDonnel 1982), as:

$$Cv(\theta) = [1 - 7/3(\rho_v/1 + \rho_v)]^{-1}, \tag{8}$$

where

$$\rho_v = \frac{\bar{\rho}(1 + \cos^2\theta)}{2 - \bar{\rho}(1 - \cos^2\theta)}, \tag{9}$$

and $\bar{\rho}$ is the ratio of the horizontally to the vertically polarized scattered light at an angle, θ, when the incident light is vertically polarized.

8.2.2.2 Absorbance Correction

It is well known that lignins exhibit an absorption spectrum with a maximum in the near ultraviolet light tailing into the visible region. Thus, both incident and scattered light intensities are attenuated and, without correction, lead to a low estimation of molecular weights.

The correction of absorption depends on the geometry of the cell and the scattering volume. Different relationships have been proposed to correct the Rayleigh factor measured with conventional goniometers, in which the detector rotates around the sample in a horizontal plane (Brice et al. 1953, Forss et al. 1967, Pla et al. 1977). With LALLS (low angle laser light scattering) photometers, e.g., the KMX-6 instrument, the scattering volume is located at the center of the cell, and the incident and scattered beams are equally attenuated. Consequently, the correction for absorption is reduced to the measurement of the intensity of the transmitted beam, G_o, for each concentration.

8.2.2.3 Influence of Fluorescence

Previous light scattering studies (Moacanin et al. 1955, Gupta and Goring 1960, Forss et al. 1967, Pla et al. 1977) indicated that lignins fluoresce when irradiated at the working wavelength (546 nm). For kraft lignins, the maximum fluorescence, as measured with a fluorescence spectrophotometer, occurs at λ_{exc} = 335 nm (Lundquist et al. 1981) while no significant fluorescence is observed at λ_{exc} = 633 nm (Kolpak et al. 1983). However, in LALLS measurements, an important decrease of the scattered light intensity is observed when an interference filter, which removes the effects of fluorescence, is placed between the scattering cell and the photomultiplier (Kolpak et al. 1983, Plastre 1983, Kim 1985). Consequently, this causes a drastic change in the calculated value of $\bar{M}_W$. The phenomenon is a result of the fact that the xenon lamp used in exciting the sample in a fluorescence spectrophotometer is considerably less powerful than the 2 mW He-Ne laser used in the LALLS photometer, and the power of this laser radiation is high enough to induce lignin fluorescence in contrast to the xenon lamp which does not cause a measurable excitation.

8.2.3 Method

8.2.3.1 Instrument Description

The only low-angle laser light scattering photometer in use today for lignin studies is the Chromatix KMX-6 model. Its optical design allows the measurements of $G(\theta)$ and G_o with minimum scattering contribution from air/glass and glass/liquid interfaces and permits visual examination of the scattering sample. Figure 8.2.1 schematically illustrates its main components.

The light source (1) is a 2 mW vertically polarized helium-neon laser which provides a small-diameter, coherent incident beam of 633 nm wavelength which is folded back on itself via a two-prism system (2). This directs the beam toward three measuring attenuators (3, 4, 5) a calibrating/shutter attenuator (6), a condensing lens (7), the sample compartment (8) and finally, through an optical system, to the detector.

The condensing lens focuses the incident light to a beam width of about 0.08 mm inside the sample, contained between two highly polished fused silica windows and a Teflon spacer. The volume of the measuring cell is 150 μl, but the sample volume participating effectively in the scattering amounts to only about 0.1 μl.

The light scattered at small angles, as determined by a series of five annuli (9), is detected by a highly linear photomultiplier (16). These annuli are located

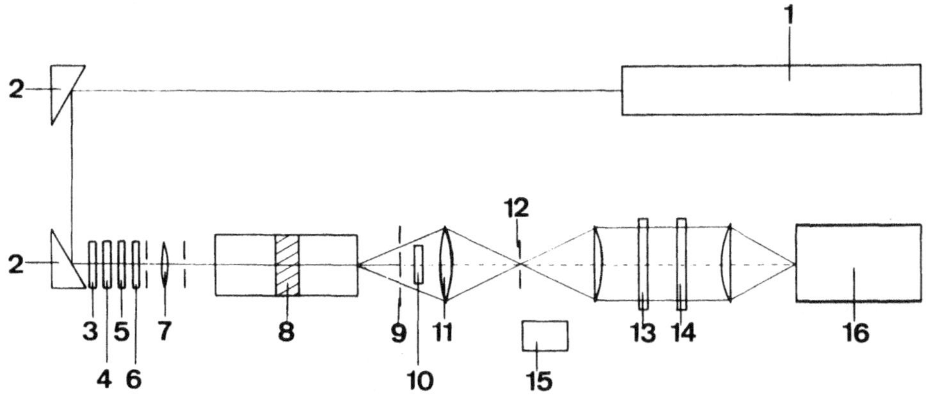

Fig. 8.2.1. Low angle laser light scattering photometer (Chromatix KMX-6): simplified optical diagram. *1* Helium-neon laser; *2* prism system; *3,4,5* measuring attenuators; *6* calibrating/shutter attenuator; *7* condensing lens; *8* sample compartment; *9* annuli; *10* safety attenuator; *11* relay lens; *12* field stops; *13* interference filter; *14* analyzing polarizer; *15* microscope objective; *16* photomultiplier

on an indexed wheel and offer a choice of observation angles which range from 2 to 7°. In addition to the five annuli, a small aperture on the wheel allows the measurement of G_0. Beam stops situated within the annuli impede the direct laser beam from reaching the photomultiplier. A safety attenuator (10) located behind the annuli wheel protects the photomultiplier tube from excessively high light levels.

The scattered light is then imaged by a relay lens (11) onto the field stops (12), which are contained on a wheel and vary in size from 1.5 to 0.005 mm. Beyond the field stops are two lenses between which are inserted an interference filter (13) centered on 633 nm and with a bandwidth of 4 nm to eliminate the fluorescence and an analyzing polarizer (14) which measures both the vertical and horizontal components of the scattered light. Immediately behind the field stops lies a microscope objective (15) which allows observation of the light scattered from the sample solution so that the system can be aligned appropriately.

8.2.3.2 Procedure

The procedure employed is thoroughly described in the instrument instructional manual. Thus, the following sections will deal only with the general operations and key operating parameters.

8.2.3.2.1 Solvents and Solutions

The common solvents employed and their refractive indices as measured at 633 nm wavelength are given in Table 8.2.1. The solutions are prepared by

Table 8.2.1. Common lignin solvents of lignins employed in light scattering measurements

Solvent	THF (a)	Dioxane (a)	2-Methoxyethanol (a)	DMF (b)
Refractive index	1.405	1.420	1.400	1.4066

Temperature = (a): 20 °C; (b): 69.6 °C

dissolving lignin samples in 50 or 100 ml of solvent to give 2×10^{-3} to $3 \times 10^{-3}\,g\,ml^{-1}$ of stock solutions from which lower concentrations are obtained by volumetric dilution.

Although the scattering volume is very small and the effect of dust is minimal in the solvents and lignin solutions, it is, however, necessary to remove any dust from them prior to making light scattering measurements. This involves filtration through a $0.2\,\mu m$ Fluoropore membrane filter (Millipore Corporation) using a syringe whose plunger may be slightly forced either by hand or by a syringe pump. The syringe is connected directly to the light scattering cell to prevent the reintroduction of dust into the filtered solutions through contact with air.

8.2.3.2.2 Measurements

These consist essentially of the following steps:

1. Preparation and insertion of the sample cell into the cell compartment and introduction of the solvent (or solution)
2. Focusing the incident light by rotating the condensing lens until the glow has a minimum diameter as seen through the telescope
3. Selection of the annulus and field stop; this allows determination of σ' and l'
4. Selection of the gain and filter time constant and measurement of $Gv(\theta)$ and $Gh(\theta)$
5. Measurement of G_o using the adequate set of attenuators (with the corresponding transmittance, D)

8.2.3.2.3 Determination of Rayleigh Factors

The vertical, $Rv(\theta)$, and horizontal $Rh(\theta)$, Rayleigh factors are calculated from Eq. (2), and the excess Rayleigh factor, corrected for anisotropy, is finally obtained from Eq. (7).

8.2.3.2.4 Specific Refractive Index Increment, dn/dc

Accurate values of the specific refractive index increment are required in light scattering equation because dn/dc appears in a quadratic form in Eq. (4).

They are determined at constant temperature with a differential refractometer employing the same wavelength as used in the light scattering measurement (λ = 633 nm). Usually the lignin concentration varies from 0.2×10^{-3} to 3×10^{-3} g ml^{-1}.

The differential refractometer is generally calibrated with aqueous salt solutions, e.g., KCl, NaCl, NH_4NO_3 of known concentration. The difference in refractive index at 633 nm between the salt solutions and distilled or deionized water is represented by the following equations:

$$\text{KCl solutions: } \Delta n/c = [137.14 - 1.178\,c^{1/2}] \times 10^{-6} \quad (10)$$
$$\text{NaCl solutions: } \Delta n/c = [123.86 - 1.217\,c^{1/2}] \times 10^{-6} \quad (11)$$
$$\text{NH}_4\text{NO}_3 \text{ solutions: } \Delta n/c = [176.58 - 1.985\,c^{1/2}] \times 10^{-6}, \quad (12)$$

where c is the salt concentration in g kg^{-1}.

Furthermore, the solvent refractive index, n_o, required in Eq. (4) must be determined under the same wavelength and temperature conditions.

8.2.3.3 Data Processing

The value of dn/dc and n_o allows the calculation of the optical constant, K, from Eq. (4). A graph of $Kc/\Delta R(\theta)$ against the lignin concentration, c, gives the weight-average molecular weight from the intercept and the second virial coefficient from the slope.

8.2.4 Discussion

As stated in Section 8.2.1, only limited results concerning $\bar{M}_W$ determinations of lignins with the use of LALLS photometry have been published. This radically new and sophisticated technique constitutes an improvement over other methods, such as sedimentation equilibrium and size exclusion chromatography, for the evaluation of $\bar{M}_W$. Its low angle capability, combined with high sensitivity, small sample size, and simplified clarification procedures, allow simple and accurate determinations to be made even on lignin samples which have been considered very troublesome. Moreover, as stated in Section 8.2.2, additional difficulties, inherent lignin properties, i.e., absorbance, fluorescence and anisotropy, may also be overcome.

8.2.4.1 Clarification of Solvents and Solutions

As described in "Solvents and Solutions" under Section 8.2.3.2, solvents and solutions are filtered and allowed to flow slowly through the sample cell during the measurements. Accordingly, the residence time of dust particles inside the scattering volume (<0.1 μl) is short. Thus, the sharp and intense signals resulting from dust or other contaminant particles may be easily disregarded when the output signal corresponding to the scattered light is displayed on a strip chart.

8.2.4.2 *Influence of Fluorescence*

The use of an interference filter to eliminate undesirable emissions such as Raman and fluorescence radiation from the lignin solutions causes a drastic change in the scattered light intensity. Thus, for instance, the weight-average molecular weight of a kraft lignin fraction was found to be 24 500 in the absence of a fluorescence correction instead of the actual value of 5700 obtained when fluorescence was taken into account (Plastre 1983). Similar results were observed by Kim (1985) with purified kraft lignins.

8.2.4.3 *Influence of Anisotropy*

Since anisotropic molecules enhance the intensity of scattered light and, consequently, lead to higher Rayleigh factors, molecular-weight values determined without correcting for anisotropy are higher than the actual values. This trend was observed for kraft lignins (Plastre 1983, Kim 1985), organosolv, and alkali lignins (Pla et al. 1986, Dolk et al. 1986). In all cases, if anisotropy is not accounted for, there is a significant error (20–50%) in the determination of $\bar{M}_W$ by light scattering.

Table 8.2.2. Weight-average molecular weights of acidic organosolv lignin fractions from black cottonwood determined with a LALLS photometer. (Pla et al. 1986)

Parameter	Fraction						
	1	2	3	4	5	6	7
$\bar{M}_w$	1500	2550	4000	9000	19 000	29 000	59 500
dn/dc ml g^{-1}	0.168	0.159	0.157	0.157	0.156	0.157	0.159
$A_2 \times 10^2$ mol ml g^{-2}	2.1	1.5	2.8	1.7	0.15	0.11	0.18

Table 8.2.3. Weight-average molecular weights of alkaline lignin fractions from western hemlock determined with a LALLS photometer. (Dolk et al. 1986)

Parameter	Fraction						
	1	2	3	4	5	6	7
$\bar{M}_w$	2250	5200	7850	12 450	28 500	32 200	38 500
dn/dc ml g^{-1}	0.260	0.260	0.240	0.190	0.210	0.230	0.200
$A_2 \times 10^3$ mol ml g^{-2}	−3.0	−2.2	0	−4.1	0	0	−0.05

Table 8.2.4. Weight-average molecular weights of fractions of spruce dioxane lignin. (Pla et al. 1977)

Parameter	Fraction				
	D3	D4	D5	D6	D7
$\bar{M}_w$ LALLS	1600	2100	3350	5400	7530
$\bar{M}_w$ SEC	1550	2100	3100	5150	7250

Table 8.2.5. Weight average molecular weights of different lignins determined by light scattering

Lignin	Molecular weight range ($\bar{M}_w$)	Reference
Soda	28000–146000	Lindberg et al. (1964)
Kraft	5600–25000	Plastre (1983)
Kraft	2000–17500	Plastre (1983)
Kraft	13700–48300	Kim (1985)
Lignosulfonates	7900–126000	Forss and Stenlund (1969)

8.2.4.4 Typical Analytical Data

Tables 8.2.2 and 8.2.3 give typical results for two series of polydisperse lignin fractions obtained from acidic organosolv delignification of black cottonwood (Pla et al. 1986) and from alkaline delignification of western hemlock (Dolk et al. 1986). In both cases, LALLS allows accurate determination of low molecular weight values. The nearly identical dn/dc values for a given series of lignin fractions indicate the good reproducibility and accuracy of the technique. However, the second virial coefficients, A2, vary considerably depending upon the fraction measured.

Negative or near zero A2 values have been obtained showing a behavior typically exhibited by polymers dissolved in thermodynamically poor or theta solvents. Negative values may indicate an association effect, but light scattering alone is insufficient to ascertain association of lignin. The results listed in Table 8.2.4 show very similar values of $\bar{M}_W$ determined by laser light scattering and size exclusion chromatography (Pla et al. 1977). Table 8.2.5 also shows other $\bar{M}_W$ values for kraft lignins and lignosulfonates.

8.2.4.5 Accuracy of the Method

Different sources of error must be considered in estimations of the accuracy of weight-average molecular weights of lignins, namely, those inherent in the Rayleigh factor, $R(\theta)$, measurements and those arising from the light scattering constant, K. Errors related to concentration are generally assumed to be negligible.

The factor, $D(\sigma' l')^{-1}$, used for the determination of $R(\theta)$ is obtained from very accurate and independent optical and geometrical measurements. Thus, owing to the high sensitivity of the photomultiplier, the error in $R(\theta)$ is normally lower than 2%. The light scattering constant contains measured quantities (solvent refractive index, wavelength, and specific refractive index increment) of which only the latter (dn/dc) is susceptible to any real error, usually ranging from 2 to 3%. Hence, an overall accuracy of about 5% in lignin $\bar{M}_W$ measurement is realistic.

References

Amirzadeh J, McDonell ME (1982) Scaling analysis of static properties for semidilute solutions. Macromolecules 15:927–933

Brice BA, Nutting GC, Halwer M (1953) Correction of absorption and fluorescence in the determination of molecular weights by light scattering. J Am Chem Soc 75:824–828

Dolk M, Pla F, Yan JF, McCarthy JL (1986) Lignin. 22. Macromolecular characteristics of alkali lignin from western hemlock wood. Macromolecules 19:1464–1470

Flory PJ (1953) Principles of polymer chemistry. Cornell University Press, Ithaca, New York, 672 pp

Forss K, Schott O, Stenlund B (1967) Light absorption and fluorescence of lignosulfonates dissolved in water and dimethylsulfoxide. Pap Puu 49:525–530

Forss K, Stenlund B (1969) Molecular weights of lignosulfonates fractionated by gel permeation chromatography. Pap Puu 51:93–105

Gupta PR, Goring DAI (1960) Physicochemical studies of alkali lignins. III. Size and shape of the macromolecule. Can J Chem 38:270–279

Kaye W, Havlik AJ (1973) Low-angle laser light scattering – absolute calibration. Appl Opt 12:541–550

Kaye W, McDaniel JB (1974) Low-angle laser light scattering – Rayleigh factors and depolarization ratios. Appl Opt 13:1934–1937

Kim H (1985) The effect of kraft pulping conditions on molecular weights of kraft lignins. Ph.D. Thesis, University of Maine, Orono, 179 pp

Kolpak FJ, Cietek DJ, Fookes W, Cael JJ (1983) Analysis of lignins from spent liquors by gel permeation chromatography/low-angle laser light scattering (GPC/LALLS). Appl Polym Symp 37:491–507

Kratochvil P (1972) Particle scattering functions. In: Huglin MB (ed) Light scattering from polymer solutions. Academic Press, New York, 333–384

Lindberg JJ, Tilly H, Majani C (1964) Notes on the molecular weights and the fractionation of lignins with organic solvents. Pap Puu 46:521–526

Moacanin JL, Felicetta VF, Haller W, McCarthy JL (1955) Lignin VI. Molecular weight of lignin sulfonates by light scattering. J Am Chem Soc 77:3470–3475

Pla F (1980) Etude de la structure macromoléculaire des lignines. Thesis, University of Grenoble, France, 167 pp

Pla F, Dolk M, Yan JF, McCarthy JL (1986) Lignin. 23. Macromolecular characteristics of alkali lignin and organosolv lignin from black cottonwood. Macromolecules 19:1471–1477

Pla F, Froment P, Capitini R, Tistchenko AM, Robert A (1977) Etude d'une lignine d'extraction par diffusion de la lumière avec source laser. Cellul Chem Technol 11: 711–718

Pla F, Robert A (1974) Etude du comportement hydrodynamique d'une lignine d'extraction. Cellul Chem Technol 8:3–10

Plastre D (1983) Determination des propriétés macromoléculaires de la lignine, mise au point de techniques d'analyse; application à la characterisation de lignines kraft et à l'étude de cuissons alcalines avec additifs. Thesis, University of Grenoble, France, 165 pp

Russo PS, Bishop M, Langley KH, Karash FE (1984) Comment on the correct polarization factor for light scattering photometers employing annular detection. Macromolecules 17:1289–1291

Stacey KA (1956) Light scattering in physical chemistry. Academic Press, New York, 230 pp

Yean WQ, Rezanowich A, Goring DAI (1964) The molecular weight and configuration of sodium lignin sulphonate from spruce wood. Symposium international de Grenoble France. Chimie et biochimie de la lignine, de la cellulose et des hémicelluloses. Les imprimeries réunies de Chambéry (ed) 327–343

8.3 Vapor Pressure Osmometry

F. Pla

8.3.1 Introduction

In earlier physicochemical characterizations of lignins, several absolute procedures were employed to determine number-average molecular weights, $\bar{M}_n$. All of these methods were based on colligative properties such as freezing point depression – cryoscopy – (Gross et al. 1958, Doucet et al. 1970), boiling point elevation – ebulliometry – (Grushnikov et al. 1969), osmotic pressure – membrane osmometry – (Hess 1952, Karpovskaya et al. 1964, Rinaudo and Pla 1967), and vapor pressure depression – vapor pressure osmometry – (Marton and Marton 1964, Brown 1967, Froment et al. 1971, Pla and Robert 1974a,b, 1984, Dolk et al. 1986, Pla et al. 1986).

However, with the exception of vapor pressure osmometry, these methods have not been used in recent years because of many experimental difficulties. For instance, the choice of the appropriate solvent for the cryoscopic method is difficult since the lignin may precipitate from solution before solvent freezing is achieved. Ebulliometry requires a low boiling point solvent to avoid lignin degradation during the measurement. Moreover, this method suffers from a more practical limitation, namely, the tendency of lignin solutions to foam on boiling. Lignins concentrate in the foam thus rendering uncertain the actual concentration of the lignins in solution, a key parameter for the assessment of $\bar{M}_n$. Membrane osmometry is generally used for molecular weights above 25000 and is limited by the permeability of the membranes to the low-molecular weight species. Thus, true number-average molecular weights can only be obtained after a very careful selection of the membrane material. The perturbing effect of relative permeability increases for a given $\bar{M}_n$ value with the broadness of the molecular weight distribution.

Vapor pressure osmometry (VPO) is commonly used to determine $\bar{M}_n$ in the range of 100 to 10000. Beyond the upper limit, the sensitivity becomes inadequate. Nevertheless, VPO appears now to be the preferred method for determining number-average molecular weights of lignins and lignin derivatives. It should be pointed out that in the range 10000 to 25000, all colligative methods are uncertain and average molecular weights within this range are therefore difficult to determine accurately.

Springer Series in Wood Science
Methods in Lignin Chemistry
(Edited by S.Y. Lin and C.W. Dence)

8.3.2 Principle

As mentioned above, vapor pressure osmometry is a suitable and practical method for measuring the number-average molecular weights of lignins. Because of recent improvements in the technique, the measurements are now easily carried out with a high precision using a wide range of temperatures and solvents. As with all colligative properties, vapor pressure depression is based on the chemical potential decrease, $\Delta\mu_1$, of the solvent on addition of the solute that is

$$\Delta\mu_1 = -RT[(V_1/M_2)c_2 + (1 - \chi_1)/2 \cdot (c_2/d_2)^2 + \ldots], \tag{1}$$

where V_1 is the molar volume of the solvent, M_2 the molecular weight of the solute, c_2 the concentration of the solution (gml^{-1}), d_2 the density of the solute, χ_1 the Flory's polymer-solvent interaction coefficient, R the gas constant, and T the absolute temperature.

If the concentration, c_2, is replaced by c_w, expressed in g per kg of solvent, i.e., $c_2 = d_1c_w/1000$, where d_1 is the density of the solvent, then $\Delta\mu_1$ becomes

$$\Delta\mu_1 = -RT10^{-3}[(M_1/M_2)c_w - (1/2 - \chi_1)10^{-3}(d_1/d_2)^2c_w^2 + \ldots], \tag{2}$$

where M_1 is the molecular weight of the solvent.

The principle of the method is based on the measurement, at a given temperature, of the vapor pressure depression, Δp, of the solvent for dilute polymer solutions. Two matched thermistor beads, suspended in a thermostated cell saturated with solvent vapor, are connected with a Wheatstone bridge. The cell temperature is adjustable over a wide range. By means of guided syringes, a drop of solvent is placed on one thermistor (the reference bead) and a drop of solution on the other thermistor (the measuring bead). Initially, solvent vapor condenses on each drop given their lower temperature, but with a condensation rate which, as a result of the lower vapor pressure of the solution, is proportional to the solute concentration. The heat resulting from condensation enthalpy raises the temperature of the solution drop, thereby raising its vapor pressure. Condensation of solvent vapor will cease when the vapor pressure of the solution drop reaches equilibrium with the vapor pressure of the pure solvent. The difference in temperature, ΔT, between the two beads causes the Wheatstone bridge to become unbalanced; thus ΔT is measured as a resistance difference.

At steady state, according to the hypothesis that ΔT is proportional to the chemical potential decrease, $\Delta\mu_1$, of the solvent, one can write

$$\Delta T = -(T/\Delta H_v)\Delta\mu_1, \tag{3}$$

Where T is the cell temperature, and ΔH_v the molar latent heat of vaporization of the solvent.

Using Eq. (2), Eq. (3) becomes

$$\Delta T = (RT^2/\Delta H_v)\,10^{-3}[M_1/M_2c_w + (1/2 - \chi_1)\,10^{-3}(d_1/d_2)^2c_w^2 + \ldots]. \tag{4}$$

In the actual determination, ΔT is measured as a resistance difference, ΔR, (Bonnar et al. 1958, Lewis and Randall 1961), and can be expressed as

$$\Delta R/c_w = K/M_2 + A_2c_w + A_3c_w^2 + \ldots \tag{5}$$

where K is the apparatus constant for a given solvent temperature and thermistor assembly and A_2 and A_3 are the second and third virial coefficients, respectively. In general, at concentrations below 1%, the term $A_3c_w^2$ becomes negligible, and the expression simplifies to

$$\Delta R/c_w = K/M_2 + A_2c_w. \tag{6}$$

For polydisperse solutes, M_2 is the number-average molecular weight. The calibration constant is commonly determined using a substance of known molecular weight at various concentrations. As a rule, ΔR is measured at several lignin concentrations; a plot of $\Delta R/c_w$ against c_w is a straight line. If K is known under the same experimental conditions, the intercept at infinite dilution is $\bar{M}_n$.

8.3.3 Method

8.3.3.1 Instrument Description

Several vapor pressure osmometers are now commercially available. Although they are mainly used for determining number-average molecular weights in aqueous and organic solvents, they can also be employed to evaluate the total osmolality of biological solutions or dissociation and activity coefficients. Each model has its own technical characteristics. However, all are comparable in terms of general measurement procedure and sensitivity.

In principle, the operating temperature is continuously adjustable and electronically controlled between about 5 °C above room temperature and up to 120 or 130 °C. The accuracy in temperature control is approximately 10^{-3} °C per hour.

As illustrated schematically in Fig. 8.3.1, a thermostat (1) surrounds the beaker-shaped measurement cell (2) which can be enclosed by an appropriate lid (3). In operation, the cell atmosphere is saturated with solvent vapor from porous wicks (4) positioned in a glass solvent reservoir. Above the measurement cell there is a second thermostated heating block (5) in which a series of hypodermic syringes (6) are preheated. The temperature of this heating block follows automatically that of the cell. Two bead thermistors (7) are maintained within the measurement cell through the heating block and are connected electrically to a Wheatstone bridge circuit. After amplification, the bridge unbalance is measured by a detector circuit which usually has, at the highest sensitivity, a temperature resolution of less than 10^{-4} °C. The amplifier output can be connected, as desired, to a digital meter and/or to a recorder.

The two thermistors are generally extended from a stainless steel tube to form a thermistor probe (8). In existing instruments, several kinds of thermistor probes are used. The most convenient appear to be those with thermistors fused into the ends of stainless steel tubes (e.g., Knauer Model 11.00). In this case, the steel tubes provide both mechanical and chemical protection to the thin

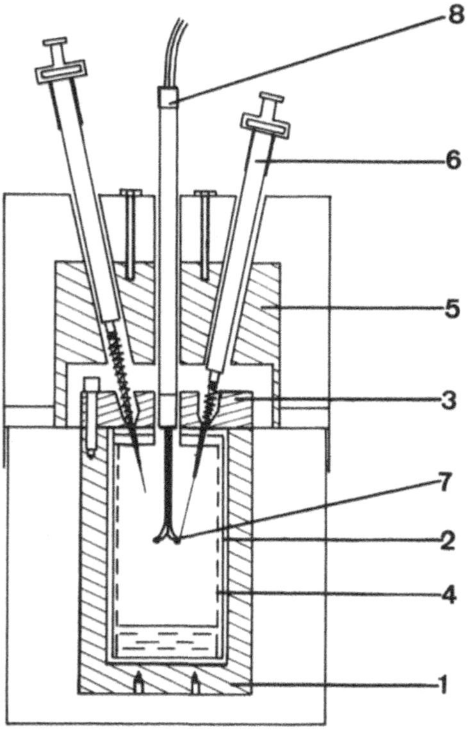

Fig. 8.3.1. Simplified representation of the Knauer Model 11.00 Vapor pressure osmometer *1* Thermostat; *2* measurement cell; *3* lid; *4* porous wicks; *5* thermostated heating block; *6* hypodermic syringes; *7* bead thermistors; *8* thermistor probe

conductors passing through them. Furthermore, the tips of the probes carry a short spiral of thin stainless steel wire on which the solution and solvent drops fall; these feed the drops and ensure a constant drop size on the probes. Two other types of thermistors are also often used. The first is a classical pendant-drop type (Mechrolab Model 302, Hewlett-Packard Co.) while the second (Corona/Wescan, Model 232A or Hitachi Model 115) employs inversely arranged thermistors mounted at the bottom of the measurement cell to keep the volume of the sample drop constant.

8.3.3.2 Procedure

The following section is devoted exclusively to the description of the main operations and key operating parameters.

8.3.3.2.1 Selection of Solvents

As for all polymers, the molecular weight of a lignin sample is a characteristic of isolated lignin molecules and all methods for its determination require that the interactions between lignin molecules be minimized. This calls for the use of thermodynamically good solvents. These solvents must be extremely pure and thermally stable. Mixed solvents are strictly prohibited. Furthermore, solvents with high ebullioscopic constants are recommended since the measuring sensitivity is thereby increased.

The solvents used for lignins are mainly tetrahydrofuran (Pla and Robert 1974a, 1984), dioxane (Pla 1980), N,N-dimethylformamide (Kim 1985), and 2-methoxyethanol (Plastre 1983, Pla et al. 1986, Dolk et al. 1986).

8.3.3.2.2 Measurement Steps

1. Adjust the thermistor probe in the cell lid so that one of the thermistors can be loaded only with solvent, while the second can receive either solvent or lignin solutions.
2. Insert the vapor wicks in the glass solvent reservoir and introduce the solvent into the cell.
3. Assemble and activate the instrument.
4. Set the operating temperature: the desired cell temperature should be in a range extending up to about 5 °C below the boiling temperature in order to prevent decomposition of the solvent or the sample. Table 8.3.1 lists the temperature generally employed with the usual lignin solvents.
5. Prepare the syringes: two syringes are filled with solvent and the others with sample solutions (lignins or calibrating compound). All are inserted into the head.
6. Set the bridge to zero: the zero balance is adjusted with solvent on both thermistors by means of a potentiometer. For good reproducibility, the drop size should remain as uniform as possible.
7. Solution measurements: the measurements are performed with different concentrations, starting preferably with the lowest and proceeding to the highest.

8.3.3.3 Evaluation of Results

8.3.3.3.1 Calibration Factor

As mentioned in Section 8.3.2, the instrument must be calibrated with a substance of known molecular weight. The experimental conditions, i.e., solvent and temperature, must be strictly identical with those used for lignin samples. The calibrating substances usually employed are benzil ($M = 210.2\,g\,mol^{-1}$) and sucrose octaacetate ($M = 678.6\,g\,mol^{-1}$) with concentrations, m, in the range between 10^{-2} and $10^{-1}\,mol\,kg^{-1}$.

A plot of $\Delta R/m$ against m gives a straight line whose intercept at infinite dilution is K ($Ohm\,mol^{-1}\,kg$).

Table 8.3.1. Primary solvents and operating-temperature ranges used for lignin experiments by vapor pressure osmometry

Solvent	Boiling temperature (°C)	Operating temperature range (°C)
Tetrahydrofuran	65.5	45
Dioxane	101	45–60
N-N-dimethylformamide	155	80–120
2-Methoxyethanol	124	45–90

8.3.3.3.2 Number-Average Molecular Weight

The lignin solution is measured at various concentrations under the same experimental conditions as the calibration solution. The concentrations, c_w, expressed in $g\,kg^{-1}$, are in the range of 0.5 and $10\,g\,kg^{-1}$. $\bar{M}_n$ is calculated as described in Section 8.3.2.

8.3.4 Discussion

The effect of several operating parameters must be emphasized. They are the size of the drops on the thermistor beads, the response time, and the purity of both solvents and lignin samples. Furthermore, it must be noted that conflicting results have been reported on the constancy of the calibration factor (Bersted 1973, Brzezinski et al. 1973, Kamide et al. 1976, Burge 1979, Marx-Figini and Figini 1980, Kim 1985, Froment and Pla, unpubl. results, 1988).

In this section, all these factors are briefly examined together with the influence of other experimental parameters.

8.3.4.1 Drop Size Effects

The drop size appears to have a significant effect on the results, particularly with the classical pendant drop-type instruments. For example, using a Mechrolab Model 302B, Morris (1977) showed that the drop size effect was related to the concentration of the solution and became important for concentrations greater than about $6\,g\,kg^{-1}$. For concentrations of $6–12\,g\,kg^{-1}$, the shape of the bridge imbalance-against-time curves showed a downward curvature with small drops (of vertical diameter, b, ~1.4–1.5 mm) but not with larger drops. For concentrations higher than $12\,g\,kg^{-1}$, this curvature appeared for drops with a vertical diameter, b, up to about 1.8 mm.

To minimize these effects, it is recommended therefore that medium size drops (b = 1.8–1.9 mm) and solution concentrations in the range of $2–10\,g\,kg^{-1}$ be employed. However, these effects are reduced or eliminated using the type of thermistor beads described in Section 8.3.3.1 which maintain the drop volume constant.

8.3.4.2 Response Time

In most operating manuals, it is claimed that the time to reach the equilibrium values must remain, for all measurements, constant and equal to that required to set the bridge to zero. However, for a given drop size, the direction and rate of approach to the equilibrium value were found to vary (Morris 1977). The use of a recorder to show the shape of the bridge imbalance-time trace is therefore strongly recommended. Readings must only be taken when the system has reached a steady state, rather than after a fixed time interval.

8.3.4.3 Purity of Solvents and Lignin

As mentioned in Section 8.3.3.2, the solvents must be of high purity and thermally stable. If not, the equilibrium values are difficult to reach and the results are not reproducible. This was observed, for example, with chloroform and tetrahydrofuran, particularly if the instrument was previously contacted with another solvent (e.g., toluene or water) at a higher temperature. As a general rule, this effect was eliminated by replacing the solvent in the reservoir.

Furthermore, since colligative properties are dependent on the number, but not on the nature, of the dissolved molecules, vapor pressure osmometry experiments may give rise to large errors if low-molecular weight impurities are present, even in small amounts, in the solution. The lignins must therefore be carefully dried and purified by removing all phenolic degradation products, sugars, and inorganic salts. For example, after a simple extraction with diethyl ether in a Soxhlet apparatus for eight hours (Pla 1980), the number-average molecular weight of a spruce dioxane lignin increased from 1200 to 3000, in spite of a weight loss of about 5%. Moreover, when traces of diethyl ether remained present in the sample, no equilibrium value was reached in the normal reading time because of volatility problems.

8.3.4.4 Constancy of the Calibration Factor

It is commonly assumed that the calibration factor, K, by means of which, according to Eq. (8), the measurement data are converted into the number-average molecular weights, depends only on the solvent, the temperature, and an "apparatus constant."

Recently, however, (Bersted 1973, Brzezinski et al. 1973, Morris 1977, Marx-Figini and Figini 1980) obtained results with pendant drop-type instruments which showed that K may also exhibit a significant dependence on the molecular weight of the calibrating compound. However, with an apparatus equipped with thermistor beads which maintained the drop size constant, this dependence was not observed (Kamide et al. 1976, Burge 1979, Kim 1985, Froment and Pla, unpubl. results, 1988). For accurate molecular weight measurements, it is therefore strongly recommended that K be determined, using the same instrument, if it shows a dependence on molecular weight.

Table 8.3.2. Number-average molecular weights of spruce dioxane lignin fractionated by preparative size exclusion chromatography. (Pla and Robert 1974a, Froment and Robert 1977)

Method	Solvent	Fraction								
		1	2	3	4	5	6	7	8	9
$\bar{M}_n$ (VPO)	THF	820	970	1250	1650	2250	3100	4000	5000	6050
	Dioxane	800	970	1300	1700	2250	3100	4000	5000	6050
$\bar{M}_n$ (SEC)	THF	820	980	1250	1650	2300	3150	4300	5100	5900

Table 8.3.3. Number-average molecular weights of black cottonwood alkali lignins measured at various temperatures (solvent: 2-methoxyethanol). (Pla et al. 1986)

Fraction	Number-average molecular weight Temperature, °C 25	37	45	60
1	3700	3750	3700	3700
2	5000	5000	5050	5000
3	5900	5950	5860	5900
4	6400	6400	6400	6450

Table 8.3.4. Number-average molecular weights of different kraft lignins determined by vapor pressure osmometry

Wood species	Range of molecular weight ($\bar{M}_n$)	Reference
Pine	1600–2700	Marton and Marton (1964)
Hardwood	1000–1100	Marton and Marton (1964)
Pine	1400–2500	Plastre (1983)
Southern slash pine	2300–6100	Kim (1985)

8.3.4.5 Typical Analytical Data

Some typical results for fractions obtained from preparative-scale exclusion chromatography of a purified spruce dioxane lignin (Pla and Robert 1974a, Froment and Robert 1977), are listed in Table 8.3.2. The corresponding $\bar{M}_n$ (VPO) values are generally independent of the solvent (Pla 1980) and similar to those determined by size exclusion chromatography (SEC) (Froment and Robert 1977). The molecular weight of alkali lignins in 2-methoxyethanol is also independent of temperature as shown in Table 8.3.3 (Pla et al. 1986). Table 8.3.4 also shows other $\bar{M}_n$ data for kraft lignins.

References

Bersted BH (1973) Molecular weight determination of high polymers by mean of vapor pressure osmometry and the solute dependence of the constant of calibration. J Appl Polym Sci 17:1415–1430

Bonnar RU, Dimbat M, Stross FH (1958) Number average molecular weight. Interscience, New York, 320 pp

Brown W (1967) Solution properties of lignin: thermodynamic properties and molecular weight determination. J Appl Polym Sci 11:2381–2396

Brzezinski J, Glowala H, Kornas-Calka A (1973) Note on the molecular weight dependence of the calibration constant in vapor pressure osmometry. Eur Polym J 9:1251–1253

Burge DE (1979) Calibration of vapor pressure osmometers for molecular weights measurements. J Appl Polym Sci 24:293–299

Dolk M, Pla F, Yan JF, McCarthy JL (1986) Lignin. 22. Macromolecular characteristics of alkali lignin from western hemlock wood. Macromolecules 19:1464–1470

Doucet J, Porta G, Finiels G (1970) Sur la détermination de la masse moléculaire d'une lignine. CR Acad Sci Ser C 270:1208–1211

Froment P, Pla F, Robert A (1971) Determination par GPC de la polydispersité de lignines d'extraction. Remarques sur les possibilitiés d'étalonnage. J Chim Phys 68:203–206

Froment P, Robert A (1977) Importance de l'étalonnage dans la détermination par GPC des masses moléculaires moyennes de la lignine. Cellul Chem Technol 11:691–696

Gross SK, Sarkanen KV, Schuerch C (1958) Determination of molecular weight of lignin degradation products by three methods. Anal Chem 30:518–521

Grushnikov OP, Shorygina ND, Mikhailov NP (1969) (Molecular weights of certain nitrolignins). Izv Akad Nauk USSR Ser Khim 11:2447–2453

Hess CL (1952) An investigation of the homogeneity of isolated native black spruce lignin. Tappi 35:312–320

Kamide K, Terakawa T, Uchiki H (1976) Molecular weight determination of macromolecules by vapor pressure osmometry. Makromol Chem 177:1447–1464

Karpovskaya RL, Ledviokova VL, Dorzet NM, Reznikov VM (1964) (The chemical and physical heterogeneity of dioxane lignin). Zh Prikl Khim 37:1318–1324

Kim H (1985) The effect of kraft pulping conditions on molecular weights of kraft lignins. PhD Thesis University of Maine, Orono, 179 pp

Lewis GN, Randall M (1961) Thermodynamics. 2nd edn, McGraw-Hill, New York, 735 pp

Marton J, Marton T (1964) Molecular weight of kraft lignins. Tappi 47:471–476

Marx-Figini M, Figini RV (1980) On the molecular weight determination by vapour pressure osmometry. 1. Consideration of the calibration function. Makromol Chem 181:2401–2407

Morris CEM (1977) Aspects of vapor pressure osmometry. J Appl Polym Sci 21:435–448

Pla F, Robert A (1974a) Etude du comportement hydrodynamique d'une lignine d'extraction. Cellul Chem Technol 8:3–10

Pla F, Robert A (1974b) Etude du caractère ramifié des lignines d'extraction. Cellul Chem Technol 8:11–19

Pla F (1980) Étude de la structure macromoléculaire des lignines. Thesis, University of Grenoble, France, 167 pp

Pla F, Robert A (1984) Etude des lignines d'extraction par GPC viscosimétrie et ultracentrifugation: détermination du degré de ramification. Holzforschung 38:37–42

Pla F, Dolk M, Yan JF, McCarthy JL (1986) Lignin 23. Macromolecular characteristics of alkali lignin and organosolv lignin from black cottonwood. Macromolecules 19:1471–1477

Plastre D (1983) Détermination des propriétés macromoléculaires de la lignine; mise au point de techniques d'analyse; application à la caractérisation de lignines kraft et à l'étude de cuissons alcalines avec additifs. Thesis, University of Grenoble, France, 165 pp

Rinaudo M, Pla F (1967) Fractionnement par filtration sur gel et distribution en masses moléculaires sur un haut polymère naturel: la lignine. Chim Anal 49:320–326

8.4 Ultrafiltration

S.Y. Lin

8.4.1 Introduction

Membrane separation is a relatively new technology, but already is playing an increasingly important role in many industries, including the chemical, pulp and paper, pharmaceutical, dairy, and water treatment industries. Ultrafiltration (UF) is a pressure-driven membrane separation technique. UF, usually conducted at low to medium pressure, has separation capabilities spanning the upper limits of the ionic range through the macromolecular range and into the micron-particle range, that is, from 20 to 100 000 Å (Moes 1987). It has been applied successfully to fractionation of kraft lignin (Forss and Fuhrman 1978, Hill and Fricke 1984), and lignosulfonates (Collins et al. 1977).

UF is an important laboratory as well as industrial method for separating lignins on the basis of molecular size. Although not a standard method for determining the molecular weight distribution of lignins, UF has potential that merits greater use in laboratories. This section describes a procedure for measuring the relative molecular size/weight distribution of water-soluble lignins.

8.4.2 Principle

The first membrane separation was performed with nitrocellulose in 1855. During the following 100 years, the technology played a limited role as a research tool in analytical chemistry. A major breakthrough occurred in 1958–1961 when the anisotropic or asymmetric membrane was developed. While membranes employed previously were uniform throughout, the upper portion of anisotropic membranes represents only 1% of the total film and is the actual filter, the other 99% acting as a support. The thinness of the membrane and the very fine pore structure promote excellent throughput for UF.

A wide range of membrane materials is available for UF. In 1970, there were only two types of membrane materials, polyamide and cellulose acetate. Today, polysulfone and ceramic membranes offer greater resistance to oxidizing chemicals, better pH stability, and other distinctive advantages. Spiral-wound, hollow-fiber and plate devices are used to house the membranes for UF systems. Newer devices employ a tangential flow (cross-flow) concept: the process stream flows across the membranes, keeping the surfaces clean and allowing inorganic and small organic molecules to pass through the membrane pores with water

Springer Series in Wood Science
Methods in Lignin Chemistry
(Edited by S.Y. Lin and C.W. Dence)

while retaining large molecules. The pressure applied to the UF system to achieve such molecular separation normally ranges between 200 and 1500 kPa (30–220 psi).

The term molecular weight cutoff (MWCO) is used to describe separation capabilities. If a membrane has a nominal MWCO of 10 000, at least 90% of the soluble macomolecules that have a molecular weight of 10 000 or more will be rejected by the membrane.

The amount of material that will pass through a given area of UF membrane in a given period is commonly called membrane flux, which may be expressed by the equation

$$J = Kp/t,$$

where J is the membrane flux, k the permeability of membrane, p the pressure difference across the membrane, and t the thickness of the membrane. Because of membrane compaction, a nonlinear flux response may be obtained at high pressure. This effect, however, is irreversible. For this reason, a membrane should never be exposed to a pressure exceeding the recommended operating limit, and if it is, the pressure should be reduced. The composition, structure, and thickness of membranes, and the pressure across the membrane, also affect flux.

Factors such as concentration polarization, shear rate, and temperature also have a pronounced effect on UF performance. In general, high shear rate and temperature tend to increase membrane flux. In contrast, high concentration and low flow rate reduce the flux. For more details of UF as a separation process, see Klinkowski (1978), and Claussen (1978), Lonsdale (1982), Woerner and McCarthy (1986).

8.4.3 Method

8.4.3.1 UF Equipment and Membranes

An Amicon Model TCE thin channel multiple disc system is suitable for laboratory fractionation work of lignin. It has an UF surface area of 137 cm^2 and a

Table 8.4.1. Diaflo ultrafilters

Series no.	MWCO	Maximum recommended pressure (kPa)[a]	Maximum operating temperature (°C)
UM 2	1 000	678	50
UM 10	10 000	678	50
UM 20	20 000	678	50
PM 30	30 000	476	121
XM 50	50 000	476	70
XM 100A	100 000	476	70
XM 300	300 000	476	70

[a] 476 kPa (69 psi), 678 kPa (98 psi).

hold-up volume of 5.5 ml. The system is connected to a pump for delivering the feed solution. The flow rate of the solution may be adjusted by controlling pump speed.

A series of Diaflo ultrafilters (Amicon, Inc.), cellulose acetate membranes, are used with the UF equipment. Table 8.4.1 lists, for each membrane, the series number, nominal MWCO, recommended maximum operating pressure, and temperature. The membranes should be stored in a 10% ethanol-water solution and refrigerated. For details, Publication 1-101L, Instruction for Diaflo Ultrafilters, (Amicon, 21 Hartwell Avenue, Lexington, MA) may be consulted.

8.4.3.2 Preparation of Feed Solution

Prepare a 5% aqueous lignin solution by dissolving a water-soluble lignin sample in water and adjust the pH to 7 with acid or alkali. With water-insoluble lignins (e.g., kraft and organosolv lignins), presulfonation with 4-chloro-2-hydroxypropylsulfonate (Lin and Detroit 1981) is recommended to solubilize samples prior to ultrafiltration. Alternatively, water-insoluble lignins may be dissolved in aqueous alkali; however, this route is not favored because of the possible association of nonsulfonated lignin molecules during UF. The feed should be prefiltered through a Whatman filter paper to eliminate any particulate material.

8.4.3.3 Procedure

The prefiltered feed solution is circulated through the UF system equipped with XM 300 membranes, using a pump to maintain a suitable flow rate to provide maximum liquid permeation. This condition normally is met at a pressure of about 170 kPa (25 psi). Pressures higher or lower than the optimum usually cause a drop in flux. Because factors such as membrane type, temperature, and concentration of feed solution affect the permeation of lignin, the optimal operating pressure needs to be determined empirically by adjusting the pump speed until the maximum permeation is obtained. Ultrafiltration is stopped when the volume of feed solution is reduced to one-tenth of the original amount. The concentrate (Fraction 1) is saved for analysis.

The permeate is then filtered as described above using an XM 100A membrane to obtain a second concentrate (Fraction 2). This procedure is repeated in turn using XM50, PM30, UM20, UM10, and UM2 membranes to obtain six additional fractions. Yields of the fractions are determined by weighing if the sample is a purified lignin, or by ultraviolet light absorption if the sample is a spent pulping liquor.

8.4.3.4 Data Presentation

Yields of the various lignin fractions may be presented in a tabular form, showing the distribution of lignin in relation to molecular weight cutoffs. Tables

Table 8.4.2. Molecular weight distribution of spruce lignosulfonate in spent sulfite liquor measured by ultrafiltration

Fraction	MWCO	Weight fraction	GPC $\bar{M}_w$
1	>300 000	0.02	33 250
2	100 000–300 000	0.20	22 810
3	50 000–100 000	0.06	10 880
4	30 000–50 000	0.05	7 500
5	20 000–30 000	0.18	4 290
6	10 000–20 000	0.29	2 460
7	1 000–10 000	0.15	750
8	<1000	0.05	240

Table 8.4.3. Molecular weight distribution of pine kraft lignin in black liquor measured by ultrafiltration

Fraction	MWCO	Weight fraction	GPC $\bar{M}_w$
1	>300 000	0.01	18 500
2	100 000–300 000	0.02	13 100
3	50 000–100 000	0.06	7 200
4	30 000–50 000	0.28	3 480
5	20 000–30 000	0.42	2 150
6	10 000–20 000	0.16	920
7	1 000–10 000	0.03	670
8	<1000	0.01	310

8.4.2 and 8.4.3 illustrate the tabular presentation of ultrafiltration data for a spruce spent sulfite liquor and a sulfonated kraft lignin. From these data, the weight-average molecular weight ($\bar{M}_w$) of lignin is estimated by the equation

$$\bar{M}_w = \Sigma M_i \times f_i,$$

where M_i is mid-point MWCO and f_i the weight fraction of the ith fraction. The caculated $\bar{M}_w$ values for the lignosulfonate and kraft lignin were 62 200 and 35 700, respectively.

The $\bar{M}_w$ values determined by analytical gel permeation chromatography using Sephadex G-100 gel according to Lin and Detroit (1981) are also shown in Tables 8.4.2 and 8.4.3. These GPC values deviate greatly from the corresponding nominal MWCOs of the membranes. Calculations based on the GPC data give $\bar{M}_w$ values of 7860 and 2920 for the lignosulfonate and kraft lignin, respectively, which are about one-tenth of the values estimated from the nominal MWCOs. This discrepancy can be partially attributed to differences in the shape of lignins (branched, three-dimensional) and in the standards (spherical) used for calibration of the membrane MWCOs.

8.4.4 Discussion of Method

As a tool for measuring the molecular weight distribution of polydisperse lignins, ultrafiltration has some distinctive advantages over classical physical techniques such as vapor pressure osmometry and light scattering. Thus, while physical techniques require ultrapure lignin samples, UF is usually not influenced by impurities. Lignin permeates through UF membranes practically unhindered by sugars, sugar acids, and inorganic salts that occur with lignin in the spent pulping liquor. Comtamination of lignin with these impurities, however, renders the physical techniques unusable. Information obtained by the UF method is particularly useful in the industrial ultrafiltration of pulping liquors.

The advantage of the UF method over gel permeation chromatography is related to the retardation effect commonly associated with GPC that causes underestimation of the molecular weight of lignin (Chap. 8.1). By contrast, there is little interaction between lignin macromolecules and the thin UF membrane; hence, in UF, true fractionation of the molecules based on molecular weight is obtained.

An order of magnitude difference between the $\bar{M}_w$ values estimated by GPC and UF as shown in Tables 8.4.2 and 8.4.3 has been reported by Wilson et al. (1986). The disparity can be partly accounted for by the difference in molecular configuration between lignin and the proteins used by membrane manufacturers to measure nominal MWCOs. In addition, the disparity may be due to the retardation effect in GPC that results in underestimation of molecular weight.

Results obtained in the ultrafiltration experiment described in this section are reproduced only by strictly following the procedure specified in Section 8.4.3.3. Any departure from the specified solution concentration in the feed solution, time of filtration, and other conditions will give a different fractionation pattern. Since UF is a nonequilibrium process, the method outlined in this section obviously is a qualitative method for measuring a relative molecular weight distribution, instead of the precise molecular weights obtained by the classical physical techniques. For quantitative information, the UF method should be used in conjunction with other methods for determining molecular weights of polymers, such as light scattering or osmometry.

References

Claussen P (1978) Membrane filtration of SSL for byproduct recovery and pollution control. Pulp Pap Can 79(3):41–45

Collins JW, Torkelson JM, Webb AA (1977) Viscosity properties of lignosulfonate isolated by ultrafiltration. J Agric Food Chem 25:743–746

Forss KG, Fuhrman AGM (1978) US Pat 4,105,606

Hill M, Fricke A (1984) Ultrafiltration studies on a kraft black liquor. Tappi J 67(6):100–103

Klinkowski PR (1978) An emerging unit operation. Chem Eng 8:165–173

Lin SY, Detroit WJ 1981 Heterogeneity of technical lignins – its significance in lignin utilization. Proc 1981 Int Symp Wood Pulp Chem. Ekman-Days 1981 (Stockholm). SPCI Rep 38,4:44–52

Lonsdale HK (1982) The growth of membrane technology. J Membr Sci 10:81–181

Moes H (1987) The expanding role of UF and RO separations in CPI applications. Chem Proc 2:35–40

Wilson KP, Laver ML, Frederick WJ (1986) Ultrafiltration of kraft black liquor: membrane performance and liquor characteristics. Proc AIChE Summer Natl Meet (Boston) Aug 24–27, 1986, Preprint 67c: 26p

Woerner DL, McCarthy JL (1986) The effect of manipulatable variables on fractionation by ultrafiltration. AIChE Sym Ser 248, 82:77–86

Chapter 9

Separation of Identification of Low-Molecular Weight Fragments and Model Compounds

9.1 Gas Chromatography-Mass Spectrometry (GS-MS)

C.-L. CHEN

9.1.1 Introduction

Gas chromatography-mass spectrometry (GC-MS) is one of the most valuable analytical techniques for the separation and identification of low-molecular weight fragments such as are found in product mixtures after the application of chemical degradation methods to lignin (Chap. 6) and in spent pulping liquors (Gierer and Petterson 1977, Gierer and Lindeberg 1980) and bleaching effluents (Lindström and Nordin 1976, Lindström and Österberg 1980, 1984).

In the past three decades, significant progress has been made in the field of instrumental analysis. Instruments for gas chromatography (GC) and mass spectrometry (MS) have been constantly improved and have attained a high degree of sophistication, comparable to that achieved in ^{1}H and ^{13}C NMR spectroscopy and other instrumental analytical techniques with computer-derived data acquisition and processing capabilities. Obviously, the combining of GC and MS methodologies into one package provides an analytical method of great versatility and power. The noteworthy advantages of GC-MS techniques are (1) short time requirement for sample analysis, (2) greater sensitivity in the identification of sample constituents, (3) elimination of classical isolation steps, resulting in decreased risk of sample contamination by impurities, (4) suitability for application in routine analysis through automation once procedures for the analyses are standardized, and (5) very short time requirement for the acquisition and processing of data produced by a computer-derived work station.

In general, the objectives of GC-MS analysis are twofold, namely, to confirm the identity of compounds suggested by GC analysis and to determine the possible structures of unidentified components in the sample mixture. The disadvantage of GC-MS is that assignment of chemical structures to sample components is only tentative unless each structural assignment is further verified by comparison of the GC and mass spectral characteristics of the tentatively identified compound with those of the corresponding authentic compound, and by application of other analytical and/or chemical methods. When GC-MS data suggest that the compounds in question are new, these compounds must be isolated and/or synthesized, and their GC and mass spectral characteristics and, if possible, other physical and chemical properties, must be compared to establish their identity rigorously.

A gas chromatography-mass spectrometer consists of gas chromatograph, mass spectrometer and computer-based data acquisition systems. Since success

Springer Series in Wood Science
Methods in Lignin Chemistry
(Edited by S.Y. Lin and C.W. Dence)

in counducting a GC-MS analysis is especially dependent on the GC component of the combination, it is of primary importance that the GC operating conditions be optimized. The conditions and equipment to be optimized for GC analysis include sample concentration (Sect. 9.1.2.4), type of capillary column, column flow rate of carrier gas, split ratio of the sample if the injection port is fitted with a split liner, and oven temperature profile.

An appropriate amount (usually in the range of 2–5 μl) of a sample solution of suitable concentration (Sect. 9.1.2.4) is injected into the GC system of a GC-mass spectrometer fitted with an appropriate capillary column. The effluent emerging from the GC column is continuously introduced into the ion source of the mass spectrometer at a pressure in the order of 10^{-6} to 10^{-7} Torr. In the ion source, the energy of the bombarding electrons can be adjusted between approximately 10 and 100 eV, although the normal operating energy is 70 eV. In general, a mass spectrometer has the capability of producing mass spectra by the electron impact (EI) (Budzikiewicz et al. 1967), chemical ionization (CI) (Munson and Field 1966a,b) or negative chemical ionization (NCI) mode (Harrison 1983). Whatever the operational mode used in MS analysis, several mass spectra (from five to ten) are usually obtained from a single compound emerging from the column. The numerous mass spectra thus produced from GC-MS analysis of a sample are scanned quickly (approximately 45–80 scans/min depending on the type of gas chromatograph-mass spectrometer), and the spectral data are recorded simultaneously with equal rapidity in a computer system.

The resulting gas chromatogram is plotted as ion current against retention time and is referred to as the "total ion chromatogram" (TIC). Since the ion current is dependent on both the concentration of sample solution and the amount of the sample solution injected into the gas chromatograph, the intensity of a GC peak in a TIC is proportional to the ion current produced, i.e., the intensity is not an absolute value but is relative to the ion current produced. The mass spectrum of a particular component is obtained by retrieval and concomitant editing of the mass spectral data stored in the computer. The editing process involves subtraction of background spectra from the spectrum obtained at the scan number corresponding to the maximum of the peak of interest. Most GC-MS facilities subscribe to one or more mass spectra data centers where standard mass spectra of over 10000 compounds, recorded mostly in the EI mode, are stored in a computer system. These facilities provide a library computer search service for finding a matching spectrum.

9.1.2 Sample Preparation

9.1.2.1 Extraction of Mixtures of Chemical Degradation Products

Extraction of products arising from the chemical degradation is described as part of several procedures detailed in Chapter 6.

9.1.2.2 Extraction of Spent Pulping Liquors and Bleaching Effluents

Approximately 400 ml of spent pulping black liquor or bleaching effluent, e.g., E_1 effluent, is adjusted to pH 14, placed in an ether extractor with a 500-ml capacity extractor tube (Note 1) and extracted continuously with 250 ml of diethyl ether for 8 h, giving the neutral fraction. Carbon dioxide gas is slowly passed through the cooled liquor in the extractor until the pH becomes constant (pH of approximately 9) (Note 2). The resulting mixture is extracted continuously with 250 ml of fresh diethyl ether for 24 h to remove phenolic substances. The liquor is adjusted to pH 2 by adding conc. hydrochloric acid dropwise with gentle shaking (Note 2). The resulting mixture is again extracted continuously with 250 ml of fresh diethyl ether for 24 h to obtain the acidic fraction. The solvent is removed from the neutral, phenolic, and acidic fractions under reduced pressure. The phenolic fraction is dissolved in 20 ml of acetone, and the resulting solution is filtered and transferred to a 50-ml round-bottomed flask where the solvent is removed at 40 °C by rotary evaporation. Approximately 10 ml of benzene is added to the residue, and any moisture in the residue is removed as an azeotropic benzene-water mixture by rotary evaporation. The neutral and acidic fractions are dried at 40 °C in the same manner as described for drying the phenolic fraction.

Note 1. The volume of black liquor should be a maximum of 80% of the volume of extractor tube so that it does not overflow into the solvent flask.

Note 2. Some technical lignins will precipitate at this pH. It is important to prevent their deposition on the fritted disc at the base of the inner tube to insure that extraction will not cease.

9.1.2.3 Derivatization

In general, GC and GC-MS analysis of phenolics having molecular weights of less than 300 does not require derivatization when a gas chromatograph equipped with a DB-5 capillary column is used for the analysis, provided the major constituents of the sample do not contain an aromatic carboxylic acid group. The sample must be derivatized when the objective of the analysis is to determine aromatic acid components in the sample, even if they are present in the sample in insignificant amounts. Before derivatization, an appropriate gel permeation chromatographic analysis of the sample should be conducted to determine the molecular weight distribution pattern of the specimen (see Chap. 9.3). When more than two-thirds of the sample contains relatively high-molecular weight material (>800), it should first be fractionated to remove as much this material as is possible (Note 3). Of the various available procedures, the following four are most frequently used for the derivatization of lignin-related samples in GC and GC-MS analyses.

Note 3. The high-molecular weight fraction can be removed by the following procedure (Tai et al. 1990a): the sample is dissolved in the minimum volume of chloroform and any insoluble material is removed by filtration. The resulting solution is added dropwise to ten times the volume of diethyl ether to precipitate high-molecular weight contaminants. Any precipitate is centrifuged off, suspended in an appropriate volume of diethyl ether under mechanical stirring, and again centrifuged off. The diethyl ether solution and washings are combined and dried over anhydrous magnesium sulfate; subsequently, the solvent is removed under reduced pressure. The ether-soluble low-molecular weight fraction is dried over phosphorous pentoxide in a vacuum desiccator.

O-Acetylation (Tai et al. 1990b,c) (Note 4)

Approximately 50 mg of a dried sample is weighed accurately to the nearest 0.1 mg and placed in a 2-ml micro-reaction vial fitted with a Teflon-lined screwcap. After adding 1 ml of acetic anhydride-pyridine reagent (1:1, v/v), the reaction vial is tightly sealed with the screwcap and allowed to stand at room temperature for 48 h. The reaction mixture is poured into a 25-ml separatory funnel containing 10 ml of diethyl ether. The reaction vial is washed thoroughly with 6 ml of diethyl ether in three portions, and the washings are combined and added to the solution in the separatory funnel. The combined solution is washed with 2 M hydrochloric acid until free of pyridine, twice with 5 ml of double distilled or completely deionized water and filtered after which the solvent is removed. The resulting acetylated sample is dried over phosphorus pentoxide in a vacuum desiccator.

Alternatively, approximately 10 mg of a dried sample is accurately weighed and placed in a 1-ml micro-reaction vial fitted with a Teflon-lined screwcap. Acetic anhydride (0.1 ml) and 20 μl of pyridine are added to the sample. For quantitative GC analysis, a measured amount of an appropriate internal standard is added to the mixture. The reaction vial is tightly sealed with the screwcap, and the resulting mixture is heated at 60 °C for 2 h with occasional shaking. The sample solution is now ready for GC and GC-MS analyses (Schultz et al. 1981).

Note 4. See Chapter 7.1.2.1 for alternative methods. O-Acetylation modifies both aliphatic and phenolic hydroxyl groups. In the event subsequent O-methylation of carboxylic acid groups by ethereal diazomethane is required, it is essential to carry out the O-acetylation of phenolic hydroxyl groups as completely as possible as unacetylated phenolic hydroxyl groups also undergo O-methylation with diazomethane.

O-Acetylation and Subsequent Methylation (Tai et al. 1990b,c)

Approximately 10 mg of an O-acetylated sample, prepared as described in the proceeding section, is weighed accurately to the nearest 0.1 mg and placed in a 5-ml micro-reaction vial fitted with a Teflon-lined screwcap, and 0.5 ml of ethereal diazomethane (Vogel 1978) is added (Note 5). If the yellow color of diazomethane is discharged with evolution of nitrogen immediately after the addition of diazomethane, further portions of the diazomethane solution are added to the reaction mixture until the yellow color persists. The reaction vial is then sealed tightly with the screwcap and allowed to stand overnight at room temperature. The solvent is stripped from the reaction mixture under a slow stream of nitrogen gas at room temperature. The residue is dried over phosphorus pentoxide in a vacuum desiccator and weighed. The dried product is now ready for use in the preparation of the sample solution for GC and GC-MS analyses.

Note 5. Diazomethane and diazoethane are not only very toxic gases but also may explode when heated to 100 °C on contact with a rough glass surface. The methylation or ethylation of organic compounds with diazomethane or diazoethane therefore must be conducted in a hood equipped with a powerful exhaust system. The use of a safety glass screen is strongly recommended and safety glasses should be worn.

O-Ethylation

Approximately 10 mg of a dried sample is accurately weighed to the nearest 0.1 mg and placed in a 5-ml micro-reaction vial fitted with a Teflon-lined screwcap, and 0.5 ml of ethereal diazoethane (Adamson and Kenner 1937, Eistert et al. 1968) is added (Notes 5 and 6). If the orange-red color of diazoethane is discharged with evolution of nitrogen gas immediately after the addition of diazoethane, additional portions of the diazoethane solution are added to the reaction mixture until the orange-red color persists. Subsequent reaction and work-up are performed as described above for diazomethane methylation.

Note 6. Ethereal diazoethane ethylates both phenolic and alkyl hydroxyl groups.

O-Trimethylsilylation (Lapierre et al. 1985)

Approximately 10 mg of a dried sample is accurately weighed to the nearest 0.1 mg and placed in a 5-ml micro-reaction vial fitted with a Teflon-lined screwcap, and 0.5 ml of methylene dichloride, 2 ml of N,O-bis(trimethylsilyl)trifluoroacetamide (BSTFA) and 0.25 ml of pyridine are added. The vial is tightly sealed, shaken gently, and allowed to stand at room temperature for 2 h. The trimethylsilyl ether derivatives of aliphatic, phenolic, and carboxylic hydroxyl groups thus obtained are stable for at least 24 h. The reaction mixture is used directly in GC and GC-MS analyses.

9.1.2.4 Preparation of Sample Solution

For GC analysis of a sample by a GC system equipped with a capillary column and flame ionization detector (FID) (Note 7), the minimum amount of a constituent that can be detected by a FID is, on the average, approximately 1 ng. Conservatively, the solute concentration of a GC sample should be approximately 50 ng/l solvent [=0.05 mg/ml] per major constituent. Thus, for a sample with ten major constituents, the sample concentration would be approximately 1 mg/ml solvent (=10 mg/10 ml) after taking into account the presence of minor constituents. The sample solution thus prepared is usually divided into two equal parts. One part is used for determination of the optimum conditions for GC analysis of the sample, i.e., optimum conditions for GC-MS analysis. After the identity of the major constituents of the sample has been established, the remaining half of the solution is used for the quantitative determination of the major constituents.

Note 7. In the GC analysis of halogenated organic compounds, an electron capture detector (ECD) is usually used instead of a FID, as the latter is not sufficiently sensitive. The sensitivity of ECD in the detection of halogenated compounds is approximately 10000 times that of a FID. However, the ECD is not necessarily more sensitive than FID for the detection of nonhalogen-containing compounds, and this factor must be taken into account when preparing a sample containing halogenated organic compounds for GC-MS analysis. Thus, a large area for a GC peak corresponding to a halogenated organic compound relative to that of a nonhalogen-containig compound would not necessarily be indicative of a correspondingly higher concentration of the halogenated compound in the sample.

9.1.3 GC-MS Analysis

9.1.3.1 Selection of Capillary Column for GC Analysis

As with any form of chromatography, the separating efficiency of capillary columns in gas chromatography is strongly dependent on the column stationary phase, carrier gas flow rate, and temperature. Because of the high separation efficiency of capillary columns, only a limited number of stationary phases can be substituted for the numerous phases used in most packed column applications. The choice of a stationary phase is commonly dictated by experience. A phase that has been successfully used by others is usually a good choice. Fre-

Table 9.1.1. Column selection guide for stationary phases used in capillary columns

Group	Stationary phases		Polarity[a]	Classes of compounds effectively separated by columns with stationary phases of the same group
1	Apiezon L CP SIL-5 DB-1 DC-200 007-1 OV-1 OV-101	RSL-150 Rt_x-1 SE-30 SF-90 SP-2100 Squalane UCW-982	Nonpolar	Aliphatic and aromatic hydrocarbons Lower aliphatic alcohols Aliphatic acid methyl esters
2	CP SIL-8 DB-5 DC-200 DC-560 Dexsil 300 Fluorolube	007-2 OV-5 RSL-200 Rt_x-5 SE-52 SE-54	Nonpolar	Aliphatic and aromatic hydrocarbons Aliphatic alcohols and their acetates Phenols and their acetates Halophenols and their acetates Aliphatic acids and their esters Aromatic acid esters
3	DB-17 DC-710 Hallcomid OV-11 OV-17	OV-22 OV-25 RSL-300 Rt_x-50 SP-2250	Intermediate	Aliphatic and aromatic hydrocarbons Aliphatic alcohols Lower aliphatic acids Aliphatic and aromatic acid esters Mono- and sesquiterpenes and their alcohols
4	DB-210 OV-210 QF-1 RSL-400	SP-2401 Squalene Triton X-100 UCON HB 280X	Polar	Aliphatic and aromatic halohydrocarbons Lower aliphatic alcohols Aliphatic and aromatic halohydrocarbons
5	DB-624 OV-624 Rt_x-502.2	Rt_x-Volatiles VOCOL	Special phases	Aliphatic and aromatic hydrocarbons Aliphatic and aromatic halohydrocarbons Halophenols

[a] Except for group 5, the polarity of the stationary phases increases with increasing group number.

quently, a phase that is useful in packed column GC can be applied successfully in capillary column GC. Table 9.1.1 may be used as a guide in selecting the proper stationary phase for capillary GC columns frequently used in wood and pulping chemistry. The phases are listed in order of increasing polarity.

9.1.3.2 Selection of Optimum Conditions for GC Operation

The optimum operating conditions for GC analysis must be determined prior to the GC-MS analysis. In general, this is accomplished by injecting the sample solution several times while varying such parameters as carrier gas, splitting ratio of carrier gas at the injection port fitted with a split liner, carrier gas flow rate, and oven temperature profile. In addition, the type of capillary column used and the concentration and volume of the sample to be injected also have to be optimized.

The following GC operating conditions are optimal for phenolic compounds, including those present in spent pulping liquor or bleaching effluent with or without prior O-acetylation and subsequent O-methylation or O-ethylation:

1. Column: DB-5 bonded phase, fused silica capillary column (30 m × 0.32 mm ID; film thickness 0.25 μm) or similar capillary column.
2. Carrier gas: helium.
3. Carrier gas splitting ratio: 20 : 1.
4. Carrier gas column flow rate: 2 ml/min.
5. Injection port temperature: 200 °C.
6. Oven temperature profile: hold 1 min at 60 °C; 60-140 °C (15 °C/min); 140-260 °C (5 °C/min); final temperature 260 °C (10 min); post value 270.007C (10 min).
7. Detector: Flame Ionization Detector (FID) for non-halogenated compounds Electron Capture Detector (ECD) for halogenated compounds

9.1.3.3 Selection of Most Suitable Mode for MS Operation

Together with determination of the optimum GC operating conditions, selection of an apropriate MS mode is of primary importance in conducting a successful GC-MS analysis.

In the electron impact (EI) mode, the effluent from a GC unit is introduced into the ion source of the MS unit and directly bombarded with electrons at very low pressure (approximately 10^{-7} Torr). Electron impact causes the removal of an electron form a compund, thereby producing a radical cation-type molecular ion at a relatively high ionizing energy (approximately 50 eV) according to the reaction $M + e \rightarrow [M^{+}] + 2e$. An electron may be removed from any bond in the molecule, M, producing an electronically excited radical cation species, which does not decompose immediately. In a molecule containing hetero atoms or π-bonds, the charge frequently resides on the hetero atom or the π-bond system. In general, the molecular ion has sufficient time to transfer much of its excess electronic energy through vibrational energy and form chemical species of lower-lying excited energy states. The process of molecular ion decomposition is called fragmentation.

In the chemical ionization (CI) mode, the substance to be investigated is introduced into the ion source with an excess of methane at a higher pressure than is used in the EI mode. In general, the partial pressure of the substance is approximately 10^{-3} Torr and that of methane is approximately 1 Torr. Under these conditions, the component subjected to electron impact ionization is predominantly methane, which gives rise to CH_2^+, CH_3^+ and CH_4^+ ions. These ions in turn react with the excess of neutral methane molecules forming mainly CH_5^+ and $C_2H_5^+$, strong Lewis acids, which react with the substance through transfer of a proton to give (M + 1) cation species with concomitant formation of methane and ethane as shown in reactions (1) and (2), respectively, using an alkyl propionate as example (Munson and Field 1966a,b). By way of comparison, reaction (3) shows the formation of a radical cation-type molecular ion (M^{+}) from an alkyl propionate according to the EI mode. Since the (M + 1) cation species formed in the CI mode are conjugate acids of the molecules being investigated and the reaction involves relatively low energy, they are more stable than the corresponding M^{+} species formed in the EI mode. Subsequent fragmentation of a (M + 1) ion species produces a CI mass spectrum of the compound.

$$CH_3{-}CH_2{-}\overset{\overset{\displaystyle O}{\|}}{C}{-}OR + CH_5^{+} \rightarrow CH_3{-}CH_2{-}\overset{\overset{\displaystyle ^{+}OH}{\|}}{C}{-}OR + CH_4 \quad (1)$$

$$CH_3{-}CH_2{-}\overset{\overset{\displaystyle O}{\|}}{C}{-}OR + CH_2 \cdot CH_5^{+} \rightarrow CH_3{-}CH_2{-}\overset{\overset{\displaystyle ^{+}OH}{\|}}{C}{-}OR + CH_2{=}CH_2 \quad (2)$$

$$CH_3{-}CH_2{-}\overset{\overset{\displaystyle O}{\|}}{C}{-}OR + e \rightarrow CH_3{-}CH_2{-}\overset{\overset{\displaystyle O^{+}}{\|}}{C}{-}OR + 2\,e \quad (3)$$

Under the operating conditions employed in the CI mode, the reagent gas, e.g., methane, can also act only as a moderating gas to produce a high population of near-thermal energy electrons which are captured by the sample molecules (Harrison 1983). Formation of negative ions by interaction of these electrons and sample molecules (AB) can occur by three different mechanisms (Hunt et al. 1976):

a) Resonance Capture
$$AB + e \rightarrow AB^{\overline{\cdot}} \quad (4)$$
b) Dissociative Resonance Capture
$$AB + e \rightarrow A\cdot + B^{-} \quad (5)$$
c) Ion-Pair Production
$$AB + e \rightarrow A^{+} + B^{-} + e \quad (6)$$

Each of these processes strongly depends on electron energy. Resonance capture affords molecular anions and involves electrons with energy near O eV. Dissociative resonance capture is observed with electrons in the energy range of 0–15 eV, while ion-pair production usually requires electron energies above 10 eV. Collectively, these processes are referred to as the negative chemical ionization (NCI) mode. In the NCI mode, methane or isobutane at pressures near 1 Torr is commonly used as the reagent gas (Dougherty et al. 1972, Hunt et al. 1976).

In general, the EI mode is more suitable for determining chemical structure, because mass spectra obtained by GC-MS analysis using this mode have considerably more fragment ion peaks, i.e., a more detailed fragmentation pattern, than the corresponding CI and NCI mass spectra. The fragmentation is diagnostically important for identifying a particular partial structure of the compound under investigation. Mass spectra obtained by the EI mode frequently do not contain a molecular ion (M^{+}) peak. In contrast, mass spectra obtained by the CI or NCI mode usually contain prominent (M + 1) or M ion peaks compared to the M ion peak in the corresponding EI mass spectra. Thus, whenever the EI mode fails to produce molecular ion peaks for the constituents in the sample under GC-MS analysis, the CI or NCI mode may be applied. The NCI mode is particularly effective in the molecular weight determination of halogenated organic compounds and compounds having functional groups abundant in ion-pair electrons.

9.1.3.4 GC-MS Operation

After the optimum GC conditions and MS mode have been selected, the operating parameters of the GC-MS spectrometer are set accordingly. When the initial oven temperature attains an equilibrium state, a sample solution in the range of 2–5 μl is injected into the gas chromatograph. The amount of sample injected depends on the number and the concentration of the constituents in the sample solution. The constituents are separated in the capillary column and continuously introduced into the ion source of the MS system at a pressure in the order of 10^{-6} to 10^{-7} Torr, whereupon the mass spectra corresponding to the individual components in the sample are generated. Irrespective of the MS operational mode used, the mass spectra thus produced are scanned

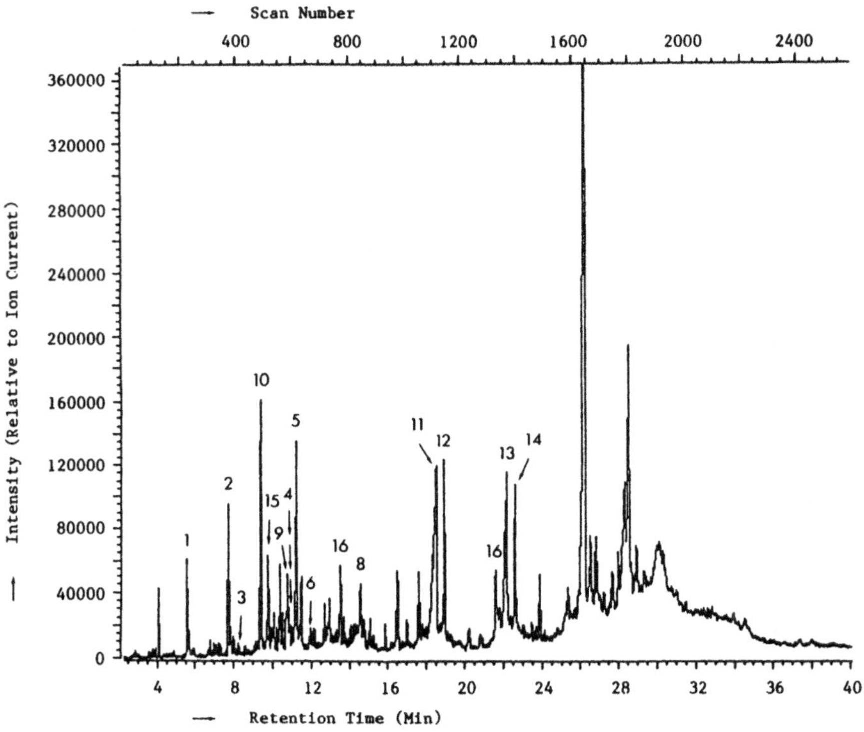

Fig. 9.1.1. Total ion chromatogram (TIC) of the phenolic fraction of a bleach plant E_1 effluent by gas chromatography-mass spectrometric (GC-MS) analysis

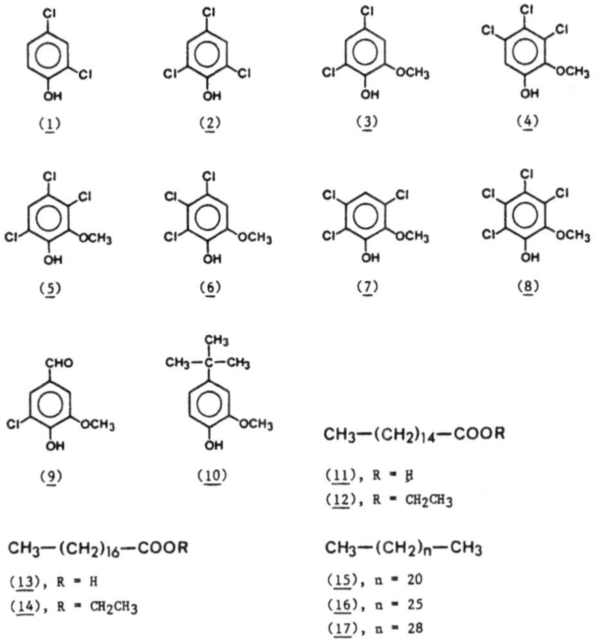

Fig. 9.1.2. Compounds identified in the phenolic fraction of a bleach plant E_1 effluent by gas chromatography-mass spectrometric (GC-MS) analysis

quickly, approximately 45–80 scans/min depending on the type of GC-mass spectrometer, and the spectral data are stored in a computer system. The ion current produced, which is proportional to signal intensity at time t_i, is plotted against retention time to produce a chromatogram which is referred to as a "total ion chromatogram" (TIC). Figure 9.1.1 shows a total ion chromatogram of the phenolic fraction of an E_1 effluent from a conventional chlorine bleach plant. The compounds identified in the phenolic fraction of this effuent are shown in Fig. 9.1.2.

9.1.3.5 Retrieval and Editing of Mass Spectral Data

The mass spectrum of a particular constituent in a sample is obtained by retrieval and concomitant editing of the mass spectral data stored in the computer system. The editing process consists of subtracting the background spectra from the spectrum obtained at a scan number corresponding to the maximum of a TIC peak of interest. For example, if the mass spectrum at scan number 374 corresponds to the maximum of a TIC peak, the mass spectrum of the compound giving rise to this peak is obtained by subtracting background spectra at scan numbers 371 and 377 from the spectrum at scan number at 374. Any unedited mass spectrum obtained from GC-MS analysis contains background spectra which originate from trace amounts of impurities. Most GC-MS systems

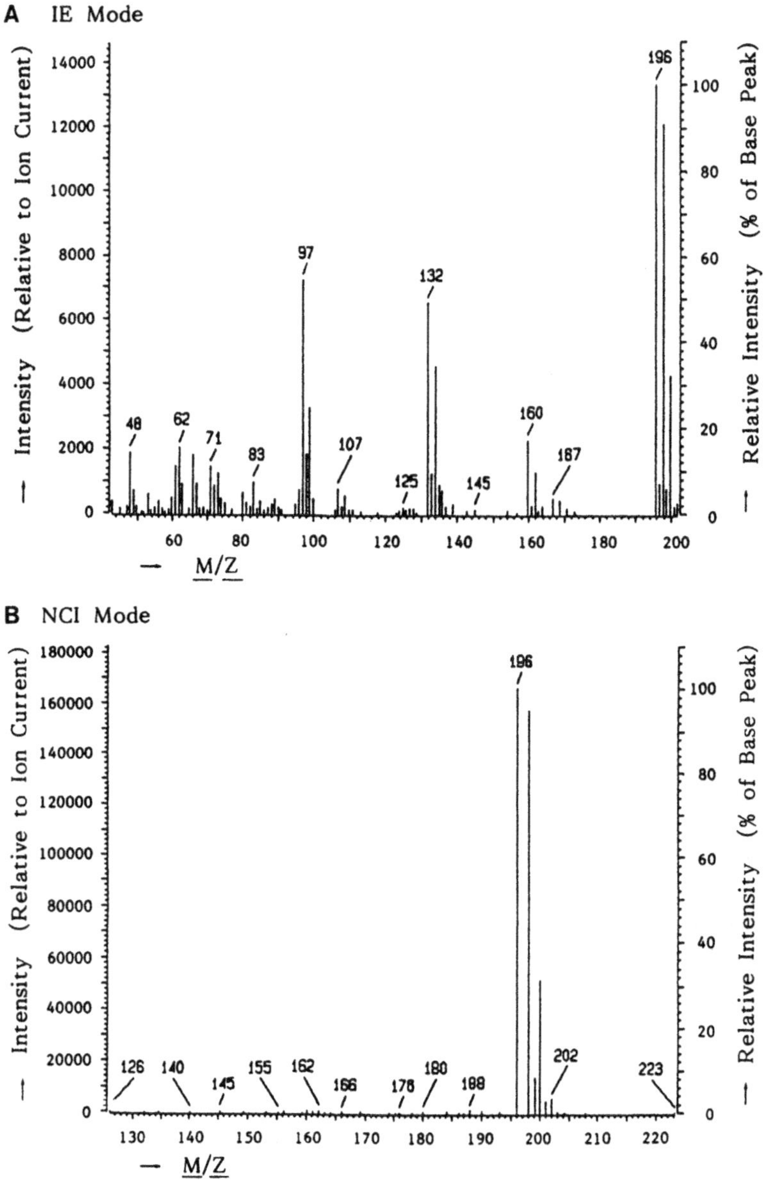

Fig. 9.1.3A,B. Mass spectra of 2,4,6-trichlorophenol (2) (TIC retention time 7.74 min, Scan No. 374) obtained by GC-MS analysis. **A** EI spectrum, **B** NCI spectrum

are fitted with an auto-editing device that automatically edits the mass spectrum of the constituent of interest on command.

Figures 9.1.3–9.1.7 show the EI mass spectra of the major constituents and Fig. 9.1.8, the mass spectrum of a minor constituent in the phenolic fraction of an E_1 effluent. An inspection of the spectra indicates that, except for the

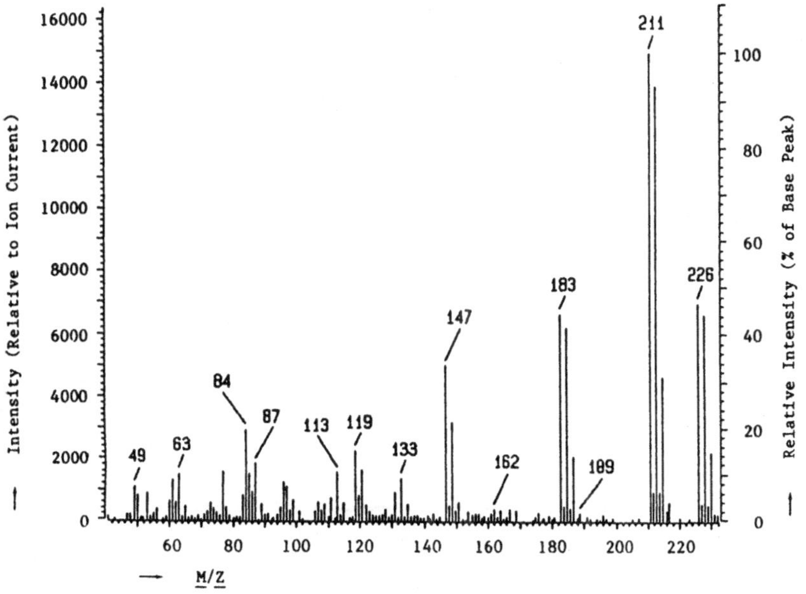

Fig. 9.1.4. EI mass spectrum of 3,4,6-trichloroguaiacol (5) (TIC retention time 11.25 min, Scan No. 616) from GC-MS analysis

spectrum in Fig. 9.1.8, all of the spectra are almost free from contaminating ion peaks and have fragmentation patterns that can be interpreted. In contrast, the mass spectrum in Fig. 9.1.8 contains several ion peaks (e.g., at m/z 218 and 203) originating from contaminants. Evidently, the ion species giving rise to these peaks have no meaningful relationship with the species generating the ion peaks at m/z 192, 194, and 196. The relative intensities of these peaks are in the approximate ratio of 3:2:0.5 (Table 9.1.2), which is characteristic of dichloro compounds. Since the ion species at m/z 218 and 203 do not contain chlorine atoms as evidenced by the absence of isotope ion peaks at m/z 220 and 205, these ion species are structurally unrelated to the compound giving rise to the ion species at m/z 192.

To examine the spectrum in Fig. 9.1.8 more precisely, the part of the TIC (see Fig. 9.1.1) encompassing retention times 7.80 to 8.80 min, has been expanded (Fig. 9.1.9, I). In addition, the relative intensity of the ion species at m/z 218 and 192 has been individually plotted against retention time (Fig. 9.1.9, II and III, respectively). These results clearly indicate that the spectrum of material corresponding to retention time 8.15 min (Scan No. 443) is composed of at least three compounds, i.e., compounds A, B, and C (Fig.9.1.9) having retention times of 8.18, 8.10, and 8.06 min, respectively, in addition to substances contributing to the background spectra. The major compound, A, has a retention time of 8.18 min (Scan No. 445). Subtraction of the background spectra at scan numbers 442 and 449 from the spectrum at scan number 445 yields a mass spectrum identical to that of authentic 4,6-dichloroguaiacol (3) (see Fig. 9.1.2 for structures).

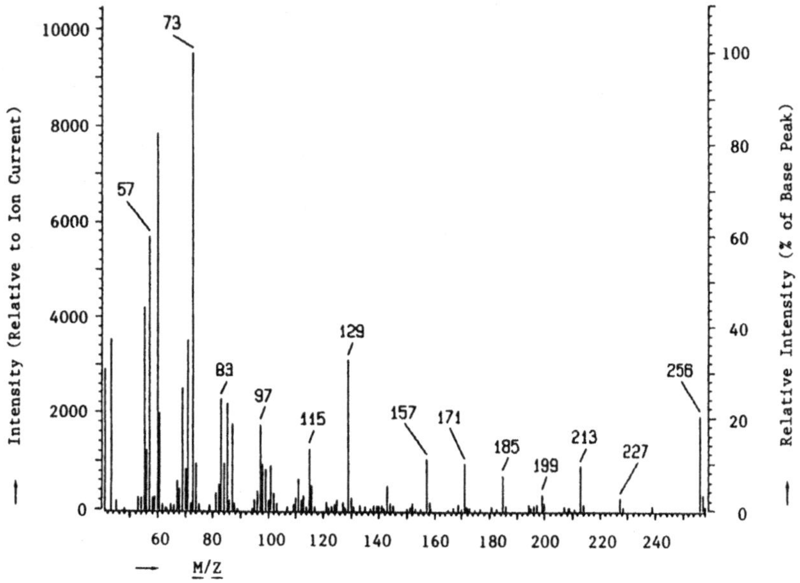

Fig. 9.1.5. EI mass spectrum of palmitic acid (11) (TIC retention time 18.52 min, Scan No. 1116) from GC-MS analysis

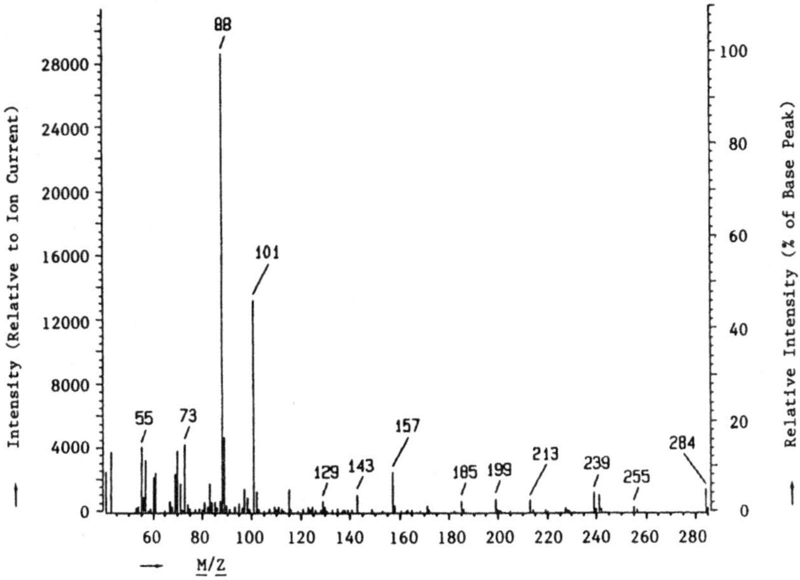

Fig. 9.1.6. EI mass spectrum of palmitic acid ethyl ester (12) (TIC retention time 18.94 min, Scan No. 1145) from GC-MS analysis

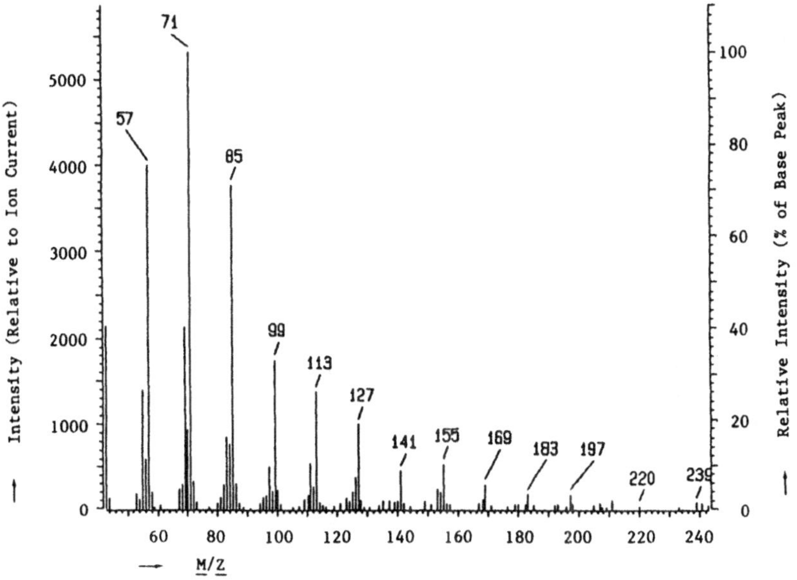

Fig. 9.1.7. EI mass spectrum of triacontane (17) (TIC retention time 21.62 min, Scan No. 1329) from GC-MS analysis

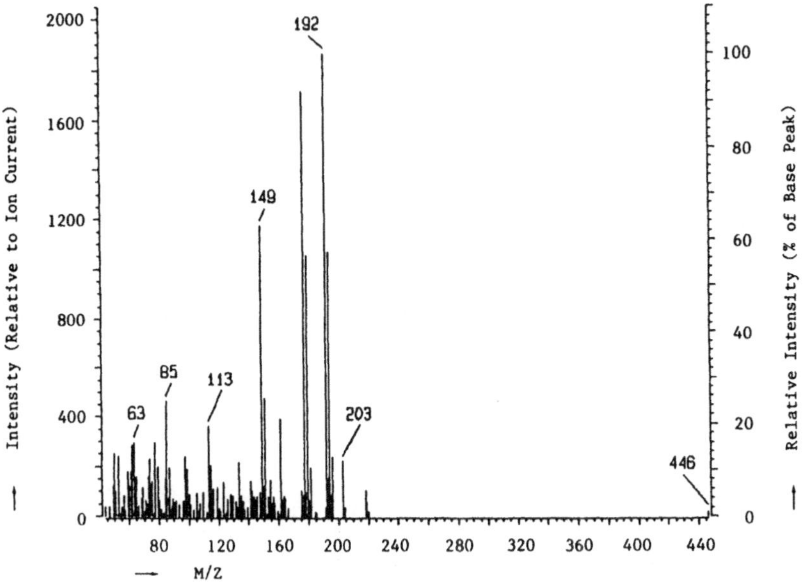

Fig. 9.1.8. EI mass spectrum of 4,6-dichloroguaiacol (3) with some contaminants (TIC retention time 8.15 min, Scan No. 443) from GC-MS analysis

Table 9.1.2. Relationship between the number of chlorine atoms present in a compound and the relative probability of peaks occurring at mass numbers corresponding to ^{35}Cl isotopes

Atoms present	Relative probability of occurrence of peaks at masses[a]							
	x	x + 2	x + 4	x + 6	x + 8	x + 10	x + 12	x + 14
Cl	0.7540	0.2460						
Cl_2	0.5685	0.3710	0.0605					
Cl_3	0.4287	0.4196	0.1369	0.0149				
Cl_4	0.3232	0.4218	0.2064	0.0449	0.0037			
Cl_5	0.2437	0.3976	0.2594	0.0846	0.0138	0.0009		
Cl_6	0.1838	0.3597	0.2934	0.1276	0.0312	0.0041	0.0002	
Cl_7	0.1386	0.3164	0.3097	0.1684	0.0549	0.0108	0.0012	0.0001
Cl_8	0.1045	0.2727	0.3114	0.2032	0.0829	0.0216	0.0035	0.0003

[a] x is the mass corresponding to molecules containing only ^{35}Cl.

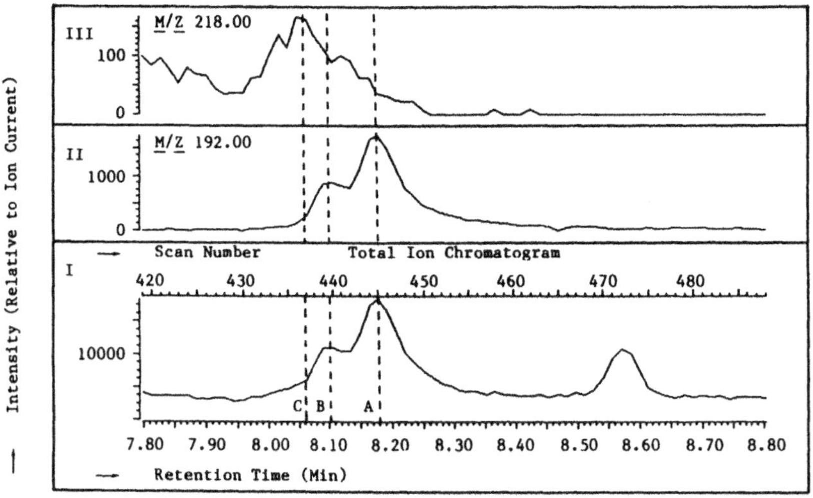

Fig. 9.1.9. **I** Expansion of the total ion chromatogram (TIC) shown in Fig. 9.1.1 for retention time interval 7.80–8.80 min. **II** Selective plotting of ion species with mass number (m/z) 192.00 against retention time. **III** Selective plotting of ion species with mass number (m/z) 218.00 against retention time

9.1.4 Discussion of Method

9.1.4.1 Validity of Structural Assignments

Structural assignments based on mass spectral data obtained by GC-MS analysis are only tentative. Tentative structural assignments can be verified by comparison of gas and/or other chromatographic retention times after co-chromatographing the sample solution with authentic samples of tentatively identified compounds and/or by high resolution GC-MS analysis. Alternatively, the identity of tentatively identified compounds can be established using additional analytical methods, provided reasonably large amounts of the sample are available (approximately 50 mg per component). These methods include, after isolation and purification of the compound, elemental analysis, IR, UV, and ^{1}H NMR spectrometry and, if possible, ^{13}C NMR spectrometry.

9.1.4.2 Mass Spectra Quality

Mass spectral quality is an important consideration in performing a successful GC-MS analysis. The quality of the mass spectra depends on (1) the concentration of the constituents in the sample solution, (2) the GC operating conditions used to resolve the peaks, and (3) excessive pressure fluctuation in the MS unit of the GC-MS system leading to distortion of the mass spectrum, especially an EI mass spectrum, as reflected in the relative abundance of the ion peaks. The implication of (3) is that distortions of this type could lead to misinterpretation of the spectrum even though a library search is performed.

Furthermore, good mass spectra of the minor constituents in a sample are difficult to obtain by GC-MS analysis if the relative intensity of these constituents is less than 2% of the area above the base line in the corresponding total ion chromatogram (TIC) because of the presence of background spectra. However, this problem can be circumvented by applying specific mass number plotting as described in Section 9.1.3.4.

9.1.4.3 Interpretation of Mass Spectra

In contrast to the mass spectrum of a compound obtained by direct probe measurement, the spectrum of the same compound obtained by GC-MS analysis does not contain meta-stable ions. The absence of these ions makes the interpretation of mass spectra obtained by GC-MS analysis more difficult than those obtained by direct probe measurement. In addition, during editing of mass spectral data (see Sect. 9.1.3.4), the ratio of the relative intensity of an ion fragment at m/z p to its naturally occurring ^{13}C isotope ion peak at m/z (p + 1) may be distorted.

In addition to a library computer search for matching spectra, the usual procedure for interpretation of a mass spectrum involves (1) identification of the molecular ion peak and, if possible, determination of the elemental composition of the molecule based on the observed mass number (m/z) for the molecular ion, (2) establishment of the fragmentation patterns of the molecular ion, and finally (3) reduction of the number of possible structures for the compound, based on the interrelationships of the observed fragmentation patterns. The absence of meta-stable ions mass spectra obtained by GC-MS analysis makes (2) very difficult to achieve.

It is safe to assume that lignin-related degradation products usually do not contain nitrogen, although they may contain chlorine and sulfur in addition to

Table 9.1.3. Relationship between the number of sulfur atoms present in a compound and the relative probability of peaks occurring at mass numbers corresponding to ^{32}S isotopes

Atoms present	Relative probability of occurrence of peaks at masses[a]								
	x	x + 1	x + 2	x + 3	x + 4	x + 5	x + 6	x + 7	x + 8
S	0.9503	0.0076	0.0420		0.0001				
S_2	0.9030	0.0144	0.0799	0.0006	0.0020				
S_3	0.8581	0.0206	0.1138	0.0018	0.0054		0.0001		
S_4	0.8154	0.0261	0.1445	0.0035	0.0101	0.0002	0.0004		
S_5	0.7748	0.0310	0.1717	0.0055	0.0158	0.0004	0.0008		
S_6	0.7363	0.0353	0.1960	0.0078	0.0223	0.0007	0.0014		0.0001
S_7	0.6997	0.0392	0.2174	0.0104	0.0296	0.0012	0.0023	0.0001	0.0001
S_8	0.6649	0.0425	0.2363	0.0132	0.0375	0.0018	0.0035	0.0001	0.0002

[a] x is the mass corresponding to molecules containing only ^{32}S.

carbon, hydrogen, and oxygen. This implies that, for sulfur-free products, the mass number corresponding to the molecular ions of the products must be an even number. By contrast, when the products do contain sulfur, the mass number for molecular ions of such products must be an odd number when an odd number of sulfur atoms is present and an even number when an even number of sulfur atoms is present in the molecule.

Since naturally occurring chlorine consists of two stable isotopes, ^{35}Cl and ^{37}Cl, in the approximate ratio of 3:1, an ion fragment containing carbon, hydrogen, and oxygen atoms and one chlorine atom always has ion peaks at m/z x and (x + 2) in the approximate ratio 3:1, where x is the ion fragment containing several ^{12}C, ^{1}H and ^{16}O atoms and one ^{35}Cl atom. Ion peak (x + 2), the corresponding isotope peak of ion peak x, contains one ^{37}Cl atom instead of a ^{35}Cl atom. A similar ion fragment containing two chlorine atoms has ion peaks at m/z x, (x + 2), and (x + 4) in the approximate ratio 9:6:1. Table 9.1.2 summarizes the relationship between the number of naturally occurring chlorine atoms (up to 8) and the number of possible isotope ion peaks in a molecular ion or ion fragment containing carbon, hydrogen, oxygen, and chlorine atoms (Beynon et al. 1968). Similarly, naturally occurring sulfur is composed of four isotopes, ^{32}S, ^{33}S, ^{34}S, and ^{36}S, in the approximate ratio 6800:55:300:1. Table 9.1.3 shows the relationship between the number of naturally occurring sulfur atoms and the number of possible isotope ion peaks in a molecular ion or ion fragment containing carbon, hydrogen, oxygen and sulfur atoms (Beynon et al. 1968).

Figure 9.1.3 shows the EI and NCI mass spectra of 2,4,6-trichlorophenol (2) (TIC retention time 7.74 min, Scan No. 374) obtained by GC-MS analysis of the phenolic fraction of a bleach plant E_1 effluent. Spectrum A exhibits ion peaks at m/z 196, 198, and 200 in an intensity ratio of approximately 10:9:3 and several rather prominent ion fragment peaks originating from these peaks, whereas spectrum B contains only ion peaks at m/z 196, 198, and 200 in an intensity ratio of approximately 10:9.5:3. Taking into account of possible experimental error, these sets of values are practically identical. A comparison of these values with the data in Table 9.1.2 reveals that the compound in question most likely contains three chlorine atoms. Based on this information, one can safely assume that (1) the compound consists of carbon, hydrogen, oxygen and three chlorine atoms, (2) the molecular weight of the compound is 196, and (3) the ion peaks at m/z 198 and 200 are the corresponding isotope peaks since the ion peaks at m/z 196, 198, and 200 are the highest even mass number ion peaks in the spectra (Note 8).

A calculation based on these data leads to an empirical formula of $C_6H_3OCl_3$ or $C_7H_7Cl_3$ for the compound. The former empirical formula is consistent with that of an aromatic chloride, while the latter formula is compatible with that of a poly-unsaturated aliphatic chloride. A close inspection of the EI mode spectrum, A, reveals the presence of ion peaks at m/z 160, 162, and 164 in an intensity ratio of approximately 9:5:1, and a cluster of ion peaks at m/z 132, 134, and 136 in an intensity ratio of approximately 8:5:1, showing that the m/z 160 and 132 ion species both contain two ^{35}Cl atoms. In addition, the intensity ratio of the ion peaks at m/z 97 and 99 is approximately 5:2, indicating that the m/z 97 ion species contains one ^{35}Cl atom. Thus, it is evident that the molecular ion at m/z 196 is converted to ion fragment m/z 160 by elimination of HCl. Loss of CO from ion fragment m/z 160 in turn generates ion fragment m/z 132. Elimination of a Cl radical from the latter ion fragment affords ion fragment m/z 97.

Thus, the major fragmentation pattern of the molecular ion is deduced by examining the intensity ratio of the isotope peaks of chlorine-containing ion species. Analysis of the fragmentation pattern not only verifies the presence of three chlorine atoms in the compound, but also effectively eliminates from consideration a compound having elemental composition, $C_7H_7Cl_3$, since formation of ion fragment m/z 132 from ion fragment m/z 160 requires an elimination of CO, i.e., the compound must contain at least one oxygen atom. It follows that the compound in question has the empirical formula $C_6H_3OCl_3$ and is possibly a trichlorophenol. Co-chromatography of the sample solution with an authentic sample of 2,4,6-trichlorophenol (2) using TLC, HPLC and GC establishes that the compound is indeed (2).

Figure 9.1.4 shows the mass spectrum of 3,4,6-trichloroguaiacol (5) (TIC retention time 11.25 min, Scan No. 616) obtained by GC-MS analysis of the phenolic fraction of E_1 effluent. The spectrum exhibits clusters of ion peaks at m/z 226, 228, and 230, at m/z 211, 213, and 215, and at m/z 183, 185, and 187. The intensity ratios for these three groups of isotope ion peaks are the same and approximately 10:9.5:3, indicating that ion species m/z 226, 211, and 183 contain three ^{35}Cl atoms (see Table 9.1.2) in addition to carbon, hydrogen, and oxygen atoms. Assuming that the highest even mass number ion peak at m/z 226 is the molecular ion peak (Note 8), the molecular ion must have three ^{35}Cl atoms and the ion peaks at m/z 228 and 230 must be its chlorine isotope peaks. In addition, the ion peaks at m/z 147, 149, and 151 show an intensity ratio of approximately 8:5:1, indicating that the m/z 147 ion contains two ^{35}Cl atoms. Evidently, the molecular ion fragments decompose to form the very prominent m/z 211 ion by loss of a methyl radical. The m/z 211 ion in turn is converted to ion fragment m/z 183 by loss of CO. Elimination of HCl from the m/z 183 ion

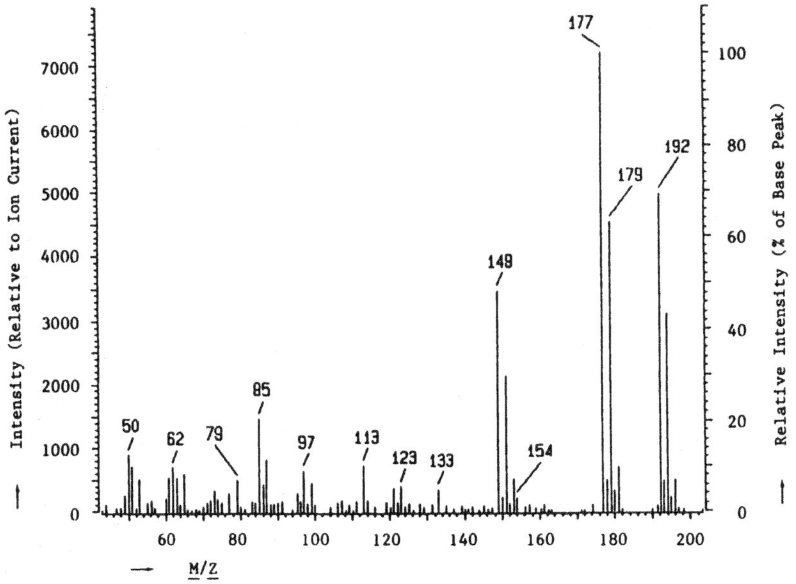

Fig. 9.1.10. EI mass spectrum of 4,6-dichloroguaiacol (3) (TIC retention time 8.18 min, Scan No. 445) from GC-MS analysis

affords ion fragment m/z 147. Thus, the first two steps of the fragmentation pattern of the molecular ion do not involve elimination of chlorine as a radical or as HCl. The prominence of methyl radical elimination from the m/z 226 molecular ion and the subsequent loss of CO from the resulting m/z 211 ion indicate that the process is the result of the so-called ortho effect and involves a methoxyl group substituted ortho to a phenolic hydroxyl group on an aromatic ring. This observation and the assumption that the molecular weight of the compound is 226 lead to the conclusion that the compound is a trichloroguaiacol, i.e., 3,4,5-trichloroguaiacol (4), 3,4,6-trichloroguaiacol (5), 4,5,6-trichloroguaiacol (6), or 3,5,6-trichloroguaiacol (7). The compound is identified as compound (5) through co-chromatography of the sample solution with authentic samples of the trichloroguaiacols in question using TLC, HPLC, and GC.

In general, mono-, di-, tri-, and tetrachloroguaiacol derivatives have the same major MS fragmentation patterns, i.e., loss of a methyl radical from the molecular ion, elimination of CO from the resulting (M-15) ion, and subsequent loss of HCl from the resulting (M-43) ion to give the (M-73) ion. This fragmentation pattern can be also observed in the mass spectrum of 4,6-dichloroguaiacol (3) (TIC retention time 8.18 min, Scan No. 445) as shown in Fig. 9.1.10.

The mass spectrum of palmitic acid (11) (TIC retention time 18.52 min, Scan No. 1116) is shown in Fig. 9.1.5. The saturated fatty acid nature of this compound is established by the presence of prominent m/z 60, 73, and 129 ion peaks and by the presence of $(73 + 14 \times n)$ ion peaks where n = 0, 1, . . . , 10, 11 (Odham and Stenhagen 1972). The m/z 256 ion is shown to be the molecular ion by the presence m/z 213 and 227 ion species which correspond to the elimination of propyl and ethyl radicals, respectively. The identity of the compound is verified through co-chromatography of the sample solution with an authentic sample of the acid.

Figure 9.1.6 shows the mass spectrum of palmitic acid ethyl ester (12) (TIC retention time 18.94 min, Scan No. 1145). The major MS fragmentation patterns of the compound are similar to those of palmitic acid, except that all of the important ion species in the spectrum have mass numbers 28 units higher than the corresponding ion species in the spectrum of palmitic acid (11); for example, ion species m/z 88, 101, 157, and 284 compared to ion species m/z 60, 73, 129, and 256 (Odham and Stenhagen 1972).

Figure 9.1.7 shows the mass spectrum of triacontane (17) (TIC retention time 21.62 min, Scan No. 1329). This spectrum exhibits prominent ion peaks corresponding to m/z $(43 + 14 \times n)$ where $n = 0, 1, \ldots, 13, 14$, indicating that the compound is a n-alkane. Although the molecular ion peak is not present, the spectrum coincides with the standard spectrum of docosane with 70% probability as determined by a library computer search. Therefore, this compound is only tentatively identified.

As shown in Fig. 9.1.2, the phenolic fraction of a bleach plant E_1 effluent contains not only chlorinated phenols but also fatty acids and their esters, and even n-alkanes. The latter two classes of compounds probably are carry-over extractives from the wood during pulping or, in part, originate from defoaming agents added to the pulping liquor.

Note 8. The ion species with the highest even mass number is not necessarily that of the molecular ion since such a species could be an ion fragment formed by loss of a neutral fragment with an even number mass unit from the molecular ion.

References

Adamson DW, Kenner J (1937) Improved preparation of aliphatic diazo-compounds, and certain of their properties. J Chem Soc 1551–1556

Beynon JH, Saunders RA, Williams AE (1968) The mass spectra of organic molecules. Elsevier, Amsterdam, 352–394

Budzikiewicz H, Djerassi C, Williams DH (1967) Mass spectrometry of organic compounds. Holden-Day, San Francisco, 9–27

Dougherty RC, Dalton J, Biros FJ (1972) Negative chemical ionization mass spectra of polycyclic chlorinated insecticides. Org Mass Spectrom 6:1171–1181

Eistert B, Regitz M, Schwall H (1968) Methoden zur Herstellung und Umwandlung von aliphatischen Diazoverbindungen. In: Müller E, Bayer O, Meerwein H, Ziegler K (eds) Methoden der organischen Chemie (Houben-Weyl) 4. Aufl., Band X/4 Stickstoff-Verbindungen I, Teil 4, George Thieme, Stuttgart, 554–557

Gierer J, Lindeberg O (1980) Reactions of lignin during sulfate pulping. Part XIX. Isolation and identification of new dimers from a spent sulfate liquor. Acta Chem Scand B 34: 161–172

Gierer J, Petterson I (1977) Study on the condensation of lignins in alkaline media (2). Formation of stilbene and arylcoumaran structures through neighboring group participation reactions. Can J Chem 55:593–599

Harrison A (1983) Chemical ionization mass spectrometry. CRC Press, Boca Raton, 22–45

Hunt DF, Stafford GC, Crow FW, Russell JW (1976) Pulsed positive negative ion chemical ionization mass spectrometry. Anal Chem 48:2098–2105

Lapierre C, Monties B, Rolando C (1985) Thioacidolysis of lignin: comparison with acidolysis. J Wood Chem Technol 5:277–292

Lindström K, Nordin J (1976) Gas chromatography-mass spectrometry of chlorophenols in spent bleach liquors. J Chromatogr 128:13–26

Lindström K, Österberg F (1980) Synthesis, X-ray structure determination, and formation of 3,4,5-trichloroguaiacol occurring in kraft pulp spent liquors. Can J Chem 58:815–822

Lindström K, Österberg F (1984) Characterization of the high-molecular-mass chlorinated matter in spent bleach liquor (SBL). Part 1. Alkaline SBL. Holzforschung 38:201–212

Munson MSB, Field FH (1966a) Chemical ionization mass spectrometry I. General introduction. J Am Chem Soc 88:2621–2630

Munson MSB, Field FH (1966b) Chemical ionization mass spectrometry II. Esters. J Am Chem Soc 88:4337–4345

Odham G, Stenhagen E (1972) Fatty acids. In: Waller GR (ed) Biochemical applications of mass spectrometry. Wiley-Interscience, New York, 211–228

Schultz TP, Chen C-L, Goldstein IS, Scaringelli FP (1981) Analysis of lignin hydrogenation products by gas chromatography. J Chromatogr Sci 19:235–239

Tai D-S, Terazawa M, Chen C-L, Chang H-m (1990a) Lignin biodegradation products from birch wood decayed by Phanerochaete chrysosporium. Part 1. Fractionation of methanol-extractive and characterization of ether-insoluble low-molecular weight fractions. Holzforschung 44:185–190

Tai D-S, Terazawa M, Chen C-L, Chang H-m (1990b) Lignin biodegradation products from birch wood decayed by Phanerochaete chrysosporium. Part 2. The constituent of ether-soluble low-molecular weight fractions. Holzforschung 44:257–262

Tai D-S, Chen C-L, Gratzl JS (1990c) Chemistry of delignification during kraft pulping of bamboos. J Wood Chem Technol 10:75–99

Vogel A (1978) A textbook of practical organic chemistry, 4th edn, Longmans, London, 289–291

9.2 High Performance Liquid Chromatography (HPLC)

N.G. Lewis

9.2.1 Significance of the Analysis

The general principles of liquid chromatography were first outlined in 1903 by the botanist Tswett in his description of adsorption chromatography for the separation of natural pigments, such as chlorophylls (Tswett 1903). Since then remarkable progress has been made, as evidenced by the multitude of techniques employed today in thin layer, low pressure, flash, and high performance liquid chromatographic (HPLC) methods (Heftmann 1983). Without these developments, the separation and isolation of chemical constituents within complex, and even simple, mixtures would not be possible. To underscore the general importance of chromatography, two examples will suffice: our knowledge of mechanisms in organic chemistry (including that pertaining to lignin) would be very limited if products could not be isolated and quantified by chromatographic methods, and their structures subsequently determined. Similarly, without chromatography, our detailed understanding of biochemical pathways and metabolic products would not be as highly developed as it is today.

Currently, HPLC is the fastest-growing technique in chromatography, even though it is used mainly for analytical (nonpreparative) separations. Indeed, it is of such general applicability that the interested reader is referred to several excellent reviews which more adequately illustrate the broad range of applications available (e.g., see volume edited by Heftmann 1983). As regards its particular significance to lignin chemistry, biochemistry, and structure, the pace of development lags behind that of other fields of research. In part, this can be attributed to the complexity of lignin-derived mixtures which are obtained upon isolation of lignin from lignocellulosic materials by either chemical or biochemical means. Nevertheless, several procedures have been reported which are suitable for the HPLC separation of lignin-related substances (Table 9.2.1).

The developments represented by these studies now permit the facile analysis, identification, and quantification of constituents of the type described above in plant extracts, and the associated methods may also be used to help define reaction pathways and conditions for both delignification and lignin modification. To illustrate the utility of HPLC, selected examples will be discussed, as appropriate, in the accompanying text; chosen examples simply reflect the author's interests and should not be misconstrued as being comprehensive.

Springer Series in Wood Science
Methods in Lignin Chemistry
(Edited by S.Y. Lin and C.W. Dence)

Table 9.2.1. HPLC applications for separation of lignin-related compounds

Compound type	Reference
Monolignols	Mortimer 1982, 1985, Morelli et al. 1986, Pecina et al. 1986, Lewis et al. 1989
Monolignol glucosides	Lewis et al. 1989
Hydroxycinnamic acids	Wulf and Nugel 1976, Nagels and Parmentier 1976, Murphy and Stutte 1978, Hartley and Buchan 1979, Hardin and Stutte 1980, Villeneuve et al. 1982
Lignin-derived fragments	
1. Sulphonomethyl derivatives of lignosulphonates	Bialski et al. 1986, Luthe and Lewis 1986
2. Thioacidolysis lignins	Lapierre and Monties 1986
3. Aromatic aldehydes (vanillin, syringaldehyde, *p*-hydroxybenzaldehyde)	Bourbonnais and Paice 1987
4. Aromatic alcohols (vanillyl, coniferyl, and cinnamyl alcohols)	Bourbonnais and Paice 1987
5. Aromatic acids and aldehydes (e.g., vanillic acid/vanillin)	Hartley and Buchan 1979, Villeneuve et al. 1982, Pecina et al. 1986, Bourbonnais and Paice 1987, Pometto and Crawford 1988
6. Lignin model compounds (e.g., guaiacyl- and veratryl-glycerol-*β*-guaiacyl ethers, substituted phenylcoumarans, and acetophenone dimers	Bourbonnais and Paice 1987
Lignans	
1. Podophyllotoxin	Buchardt et al. 1986
2. Arctiin, phillyrin, epipinoresinol-4′-*β*-D-glucoside, matairesonide, and pinoresinol 4-*β*-D-glucoside	Nishibe et al. 1988
(+) and (−) Kadsurenone	Pondipom et al. 1986
(+) and (−) Syringaresinol	Tanahashi et al. 1987
(+) and (−) Todolactol	Ozawa and Sasaya 1988
(+) and (−) Pinoresinol	Umezawa et al. 1990, Davin et al. 1991
(+) and (−) Matairesinol, secoisolariciresinol	Umezawa et al. 1990a, Davin et al. 1991

9.2.2 Survey of Available Methods

Until the mid 1970s, almost all low-molecular weight phenolic materials, many of which were lignin-related, were separated either by paper chromatography, or on silica or cellulose thin layer plates and columns. In some cases, ion exchange chromatography was also used. This situation has changed dramatically ever since Charalambous et al. (1973) described the use of high performance liquid chromatography (HPLC) for the separation of low-molecular weight phenolics. The advantages of HPLC over conventional methods are speed of analysis, ease of quantification, reliability, and automation. Additionally, separations by HPLC do not demand the same technical skills as are required for column and thin layer chromatography.

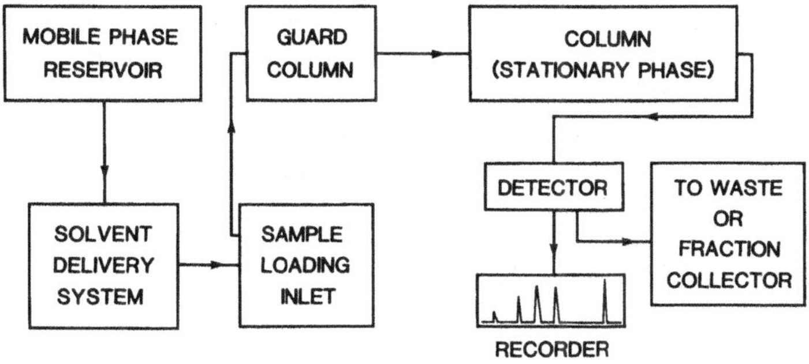

Fig. 9.2.1. Flow diagram of essential equipment requirements for high performance liquid chromatography

Figure 9.2.1 shows the essential equipment requirements for high performance liquid chromatographic separations. This diagram does not take into account the different levels of sophistication in both soft- and hardware currently available. The simplest systems consist of a pump for precise (isocratic) solvent delivery, a built-in sample injection port (or some equivalent attachment loop), an HPLC column (and accompanying guard column), a detector (normally UV), and strip-chart recorder. Such configurations are normally described as manual, since their operation (including introduction of sample into the injection loop) cannot occur unattended.

More advanced systems employ some form of computerized programming whereby mobile phase (composition, flow rate), size of sample (e.g., 10, 50, 100 μl) to be applied to the column, choice of sample for analysis, and wavelength of detection can be predetermined and automated. Such equipment can be programmed to operate overnight, hence permitting unattended routine analysis of large numbers of samples. In such cases, chromatograms can be acquired either during individual run times, or stored on a disket for subsequent printout.

HPLC separation of mixtures of lignin-derived (or lignin-like derived) components is normally achieved by normal or reversed phase chromatography with the latter predominating. In normal phase chromatography, the stationary phase is more polar than its mobile counterpart, whereas with reversed phase chromatography the opposite is true. The stationary bed in normal phase (adsorption) chromatography is most often particulate silica gel with a functional surface described as —Si(OH)—O—Si(OH)— (Yost et al. 1980). In some instances, even when the surface of the silica gel is covalently modified to carry $—NH_2$, —CN or diol functionalities, the general principles of adsorption chromatography still apply. This is not the case for nonpolar hydrocarbon coatings (C_2 and C_{18} chains) where separations are more dependent upon the principles of partition chromatography. The latter is often referred to as reversed phase chromatography and, in HPLC, separations based on this approach are those most often encountered.

With respect to the stationary (bed) phase, it should be understood that both the volume and size of the packing material may vary depending upon the application, e.g., from being very fine (4–10μ) to coarse (~40–60μ) (Braithwaite and Smith 1985). Coarse particulate materials are normally used for preparative applications (milligram to gram quantities of sample) where resolution may sometimes be compromised. As with all forms of chromatography, the mobile phase may be isocratic (composition remains constant during separation) or gradient (composition changes during chromatographic separations). Following separation(s), detection of eluted components is most often achieved by continuously monitoring the eluate by means of a UV detector (fixed or variable wavelength).

9.2.3 Principle of the Preferred Method

As discussed briefly earlier, the stationary phases commonly encountered in high performance liquid chromatography are either polar adsorbents (silica, alumina) or those which contain bonded phases (Yost et al. 1980, Braithwaite and Smith 1985). The latter are most likely to be silica coated with either polar (—NH_2, —CN, or diol) or nonpolar (hydrocarbonaceous) ligands covalently attached via siloxane bonds.

9.2.3.1 Adsorption Chromatography

For adsorbents such as silica, the lattice is terminated with polar hydroxyl groups (or other polar ligands in bonded phases) which are responsible for surface interactions with the sample (solute) and mobile phase. The polar functional groups on the applied sample (i.e., compound or mixture of compounds) are then attracted to the polar groups on the stationary phase by adsorption (Braithwaite and Smith 1985). Eventually, these forces of attraction are subsequently overcome at that particular site by the percolating mobile phase. The sample is then desorbed and subsequently readsorbed onto fresh sites further down on the stationary bed. This adsorption/desorption process continues until the compound (or compounds) of interest is (are) eluted from the stationary phase. In most applications of normal phase (adsorption) chromatography, the mobile phase is usually nonpolar (e.g., hexane) with a polar modulator (e.g., ethanol). The more polar the mobile phase, the faster is the elution of the adsorbed components. Elution profiles of samples are thus in the order of increasing polarity of the constituents within the mixture. In general, the separation of nonpolar to moderately polar compounds by adsorption HPLC is achieved most effectively with moderately polar eluents (e.g., hexane/isopropyl alcohol) and least successfully with strongly polar mobile phases (e.g., H_2O).

9.2.3.2 Reversed Phase Chromatography

In reversed phase chromatography, the stationary phase is normally silica coated with covalently attached nonpolar hydrocarbonaceous ligands (normally alkyl

C_{18} chains). Consequently, the mobile phases need to be more polar (e.g., $CH_3CN—H_2O$, $MeOH—H_2O$) than the coated surface. During elution, separation of constituents of an applied sample mainly relies upon the differences in distribution coefficients between the mobile phase and the hydrocarbon coatings of the stationary phase. For all these reasons, the more polar constituents of a mixture in reversed phase chromatography are eluted before nonpolar components. Note also that although water is more polar than acetonitrile or methanol, it is a poorer eluent in reversed phase chromatography because of the alkyl surface character of the stationary phase.

9.2.3.3 Chromatographic Theory

The following treatment of chromatographic theory deals only with the bare essentials needed for the separations described later. It is highly recommended that the reader becomes acquainted with more comprehensive treatments (e.g., Yost et al. 1980).

For practical purposes, several terms need to be defined. These are capacity factor (k'), theoretical plate number (N), height equivalent to one theoretical plate (HETP), selectivity (α) and peak asymmetry (b/a). As will be discussed later in specific examples (Sects. 9.2.4 and 9.2.5), these parameters are of crucial importance in monitoring and maintaining HPLC column efficiency.

Capacity Factor (k')

Regardless of the stationary phase, most applied solutes are eluted at larger retention volumes (V_R) than that required for solvent passage through the

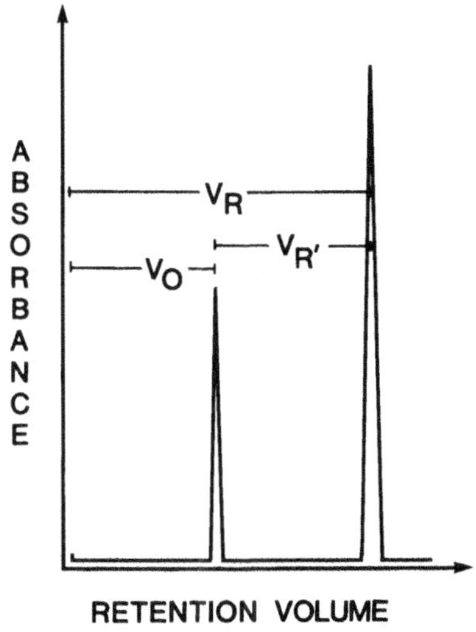

Fig. 9.2.2. An HPLC chromatogram showing components eluted at V_0 (the interstitial volume) and $V_{R'}$, respectively. $V_{R'}$ = adjusted retention volume

column (V_O, the interstitial volume) (Fig. 9.2.2). The extent to which a solute is retarded behind V_O is called the adjusted retention volume (V_R'), and is given by the equation:

$$V_R' = V_R - V_O.$$

From this equation, an important parameter termed the capacity factor, k', can be calculated. This factor is a comparison of the adjusted retention volume of a particular constituent on the HPLC column relative to the interstitial volume, V_O, and is given by:

$$\text{Capacity factor, } k' = \frac{V_R'}{V_O} = \frac{V_R - V_O}{V_O}.$$

For practical purposes, values ranging from $k' > 1$ to 10 are used; higher values tend to be accompanied by severe line-broadening of the peaks of eluted components.

Theoretical Plate Number (N) and Height Equivalent to One Theoretical Plate (HETP)

Theoretically, any eluted component should have a Gaussian (or bell-shaped peak) distribution. For practical purposes, sharp rather than broad peaks are required simply because the sharper the peak the greater the resolving capacity of the particular column.

As shown in Fig. 9.2.3, peak width (W) can be expressed as a multiple of the standard deviation σ (Yost et al. 1980). From the value of σ and the

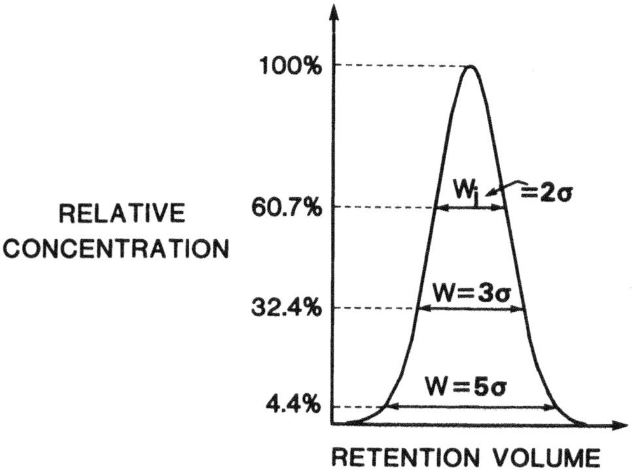

Fig. 9.2.3. Schematic representation of standard deviation (σ) for a Gaussian peak as a function of peak width (W) at different peak heights: at 60.7% (point of inflection), $W = 2\sigma$, at 32.4%, $W = 3\sigma$, and at 4.4%, $W = 5\sigma$. (Yost et al. 1980)

retention volume, V_R, the number of partitioning steps that the solute undergoes during passage through the column can be determined. This is defined as the theoretical plate number, N, which is a term taken from solvent distillation theory. It can be calculated as shown below:

$$\text{Plate number, } N = \left(\frac{V_R}{\sigma}\right)^2.$$

As mentioned, the standard deviation can readily be determined by measuring the peak width. Thus, as shown in Fig. 9.2.3, at 4.4% peak heights the standard deviation corresponds to 5σ, at 32.4% peak height to 3σ, at the point of inflection (Wi) to 2σ and so on.

$$\text{Since } N = \left(\frac{V_R}{\sigma}\right)^2,$$

Then, when

$$W_{4.4\%} = 5\sigma,\ N = 25\left(\frac{W_R}{W_{5\sigma}}\right)^2,$$

$$W_{32.4\%} = 3\sigma,\ N = 9\left(\frac{W_R}{W_{3\sigma}}\right)^2,$$

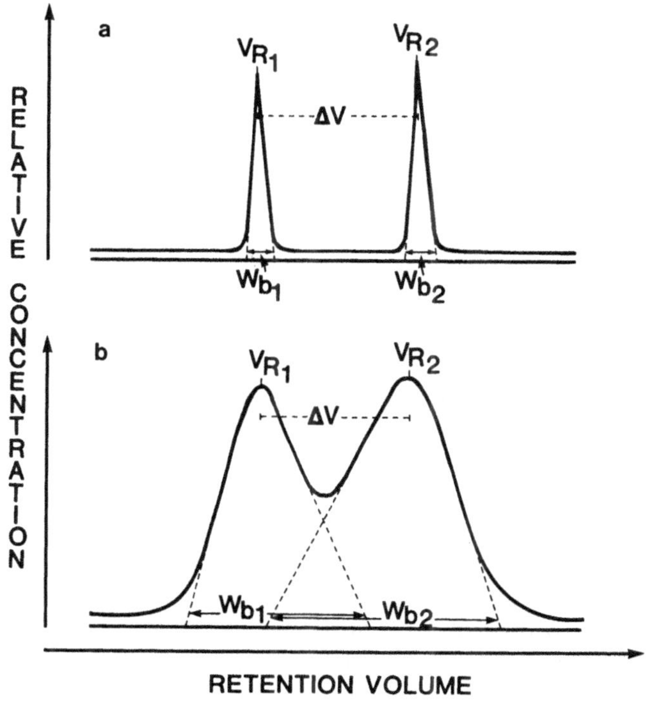

Fig. 9.2.4a,b. Schematic representation of the separation of two components by HPLC with baseline separation (*above*) and partial separation (*below*)

$$W_i = 2\sigma,\ N = 4\left(\frac{V_R}{W_{2\sigma}}\right)^2.$$

Since the value of N depends on column length, a more useful term is plate height, (h), or height equivalent to one theoretical plate (HETP). This term is derived as shown below:

$$h = \frac{L}{N},\ \text{where L = length of column, usually in millimeters (mm)}$$

This is an important term, since the lower the value of h, or the greater the value of N, the more efficient is the column.

Separation Factor (α)

As shown in Fig. 9.2.4, most chromatographic analyses involve the separation of more than one component. The efficiency of the separation is given by the separation factor, α, which is a measure of the relative position of each eluted component.

$$\alpha = \frac{V'_{R_2}}{V'_{R_1}} = \frac{k'_2}{k'_1}.$$

This term, while useful, gives no information about the actual degree of separation of the components. This depends upon the peak resolution, R, which is simply the ratio of the distance between the peak maxima (ΔV) to that of the mean of the peak widths at the baseline (Wb).

$$R = \frac{\Delta V}{1/2\,(Wb_1 + Wb_2)} = \frac{2\Delta V}{Wb_1 + Wb_2}.$$

A value of $R > 1.5$ means that baseline separation has occurred and $R = 1$ indicates ~90% seqaration.

Peak Asymmetry (b/a)

In ideal situations, all peak shapes are Gaussian. However, in certain cases, many eluted components do not show true Gaussian behavior. A measure of this nonideal behavior is the asymmetry factor, (b/a). This factor can be determined by drawing a line from the peak maxima perpendicular to the baseline and then measuring the widths (b and a) at one-tenth peak height (see Fig. 9.2.5). Values of ~1.0 for b/a are recommended; higher values indicate considerable skewing of the eluted component.

The effects of flow velocity leading to band spreading are not discussed here, since a fairly comprehensive treatment has been given by Yost et al. (1980).

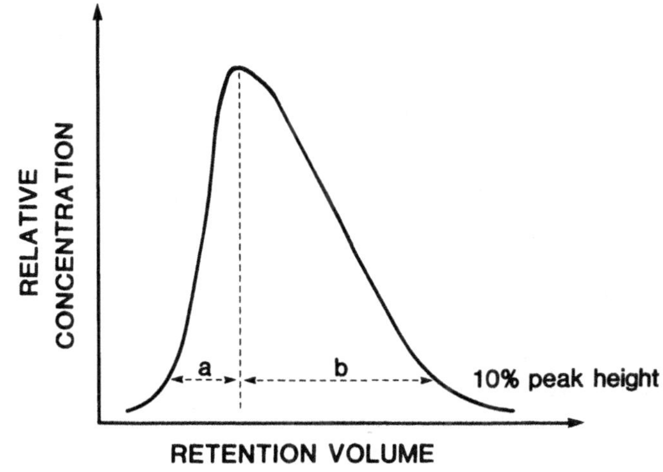

Fig. 9.2.5. Elution of a components (peak) showing non-Gaussian behavior (tailing)

9.2.4 Description of the Preferred Method

The HPLC methodology currently employed may be conveniently discussed using as a specific example, the separation of E/Z *p*-coumaryl 1, 5, coniferyl 2, 6, and sinapyl 3, 7 alcohols (see Fig. 9.2.6). This separation can be achieved using a Waters Novapak column (4 μ; 3.9 × 150 mm) eluted with MeOH—H_2O (15:85; v/v) as described below (Lewis et al. 1989).

1, $R_1, R_2, R_3 = H$
2, $R_1 = OCH_3$, $R_2, R_3 = H$
3, $R_1, R_2 = OCH_3$, $R_3 = H$

4, $R_1 = OCH_3$, $R_2 = H$, $R_3 = Glc$
5, $R_1, R_2, R_3 = H$
6, $R_1 = OCH_3$, $R_2, R_3 = H$
7, $R_1, R_2 = OCH_3$, $R_3 = H$

Fig. 9.2.6. E and Z monolignols and glucoside, Z-coniferin *4*

9.2.4.1 Mobile Phase and Pump Considerations

All mobile phases (whether organic or aqueous) must be of high purity (redistilled or HPLC grade), filtered and degassed prior to use. Such manipulations are necessary to circumvent operative problems such as clogging of HPLC filters (frits) or columns. These problems may arise from the introduction of foreign particulate materials, trapping of air bubbles in either the pumphead (leading to severe pressure fluctuations) or in the detector cell (leading to erratic baseline responses). Currently, the trend in most laboratories is to purchase prepurified HPLC-grade solvents rather than experience the inconvenience of redistillation. Whatever the source, the mobile phase must be filtered and degassed prior to use.

Filtering and degassing are presently most conveniently carried out by use of a preassembled HPLC solvent delivery system. In such systems, the mobile phase is transferred to the HPLC mobile phase reservoir via suction through a filter (normally 0.4μ) using a water aspirator or diaphragm pump (610–640 mm Hg). The mobile phase is then held under reduced pressure for several minutes longer until degassing is complete, following which atmospheric pressure is restored by introduction of either an inert gas (He, Ar) or air. After these manipulations, appropriate attachments for mobile phase delivery to the HPLC metering pump are connected. The HPLC pump is then ready for priming prior to use.

Priming is achieved by attaching a hypodermic syringe (normally 50–100 ml) to the draw-off valve and manually withdrawing solvent until no air bubbles are evident. Following this priming step, the HPLC pump is activated initially at a mobile phase flow rate of $\sim 1\,\text{ml}\,\text{min}^{-1}$. The flow rate is then stepped up in 1-ml increments until a high flow rate ($\sim 9.0\,\text{ml}\,\text{min}^{-1}$) is obtained. If the mobile phase is discharged for 1–2 min with no obvious aberration in flow, the pumping rate is then dropped down sequentially to $1\,\text{ml}\,\text{min}^{-1}$, or to the flow rate required for the analysis. When priming has been completed, the sample injection loops are purged with mobile phase to remove any impurities that may contaminate the samples to be analyzed.

9.2.4.2 HPLC Column: Attachment, Equilibration, and Performance Checks

The next step involves attachment of both guard and HPLC columns and their subsequent equilibration with mobile phase. Vendors recommend upper pressure limits for their products and these recommendations must not be ignored.

The guard column is normally dry-packed with a stationary phase comparable to that present in the HPLC column (e.g., 37–50 μ Bondapak C_{18}/Corasil when using 4μ Novapak C_{18} HPLC columns). The guard column serves the important purpose of extending the lifetime of the more expensive HPLC column by absorbing "baseline" materials. Following dry-packing, the guard column is wetted with mobile phase and connected to the sample-injection outlet. Mobile phase is then slowly introduced into the guard column at $\sim 1\,\text{ml}\,\text{min}^{-1}$ (or the analysis flow rate) until equilibration of mobile and stationary phases is achieved (~ 5 min).

As far as HPLC columns are concerned, they are normally not supplied in dry-packed form but instead are received equilibrated with a recommended solvent composition for storage. Prior to use, the column(s) must then be equilibrated with the required mobile phase for chromatographic separations.

This is a rather trivial procedure if both storage solvent and mobile phase compositions are miscible; equilibration can be achieved by slowly changing the composition of one mobile phase to the other (~5% change in composition per minute) until the desired composition is obtained. The column is then equilibrated with at least seven bed volumes of mobile phase prior to chromatographic use. This operation can be carried out by either manually changing the mobile phase in the reservoir, or by selecting (programming) an appropriate gradient. It should be self-evident that if both storage solvent and mobile phase compositions are immiscible, other intermediate mobile phases miscible with both storage and desired mobile phases must be employed.

When the final mobile phase composition has been achieved, the detector is attached in series as shown in Fig. 9.2.1. Mobile phase elution is then continued until, at the wavelength under investigation (normally 254 or 280 nm for phenylpropanoids), the baseline is stable and fluctuation- or drift-free. One important feature is to ensure that there are no leaks at any attachment points, since these will adversely affect both pressure and elution behavior.

The next step is to ensure that the column meets the specifications of the manufacturer's claims. This requires determination of the column's efficiency, capacity factor, selectivity, and peak asymmetry. For example, when the Waters Novapak C_{18} column (3.9 × 150 mm) or equivalent is equilibrated with CH_3CN—H_2O (60:40, v/v), column characteristics can be determined by examining the separation of naphthalene and acenaphthene (Fig. 9.2.7). Thus, naphthalene (20.39 mg) and acenaphthene (40.35 mg) are individually dissolved in 100 ml of CH_3CN (or another solvent miscible system), and the solutions are individually filtered. This operation is normally carried out using a 0.45 μ filter hypodermic syringe attachment, with the filtrate then being subjected to HPLC analysis. Note that different filters are required depending upon whether

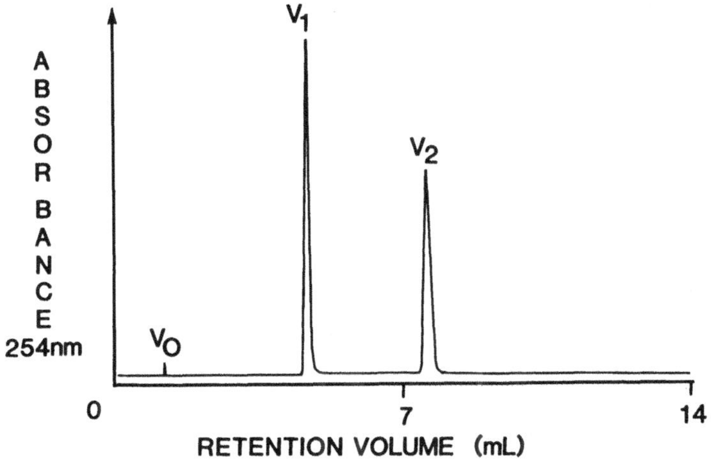

Fig. 9.2.7. HPLC separation of naphthalene (V_I) and acenaphthene (V_2). V_0 = interstitial volume. Conditions: Waters Novapak C_{18} (13.9 × 150 mm) eluted with CH_3CN—H_2O), 60:40, v/v. Flow rate = 0.7 ml min^{-1}

aqueous or organic solutions are used. A 2-μl aliquot of a 1:1 sample of acenapthene and naphthalene is then applied to the HPLC column via the sample injection loop (or some equivalent). Following chromatographic separation, the column efficiency (N), capacity factor (k'), selectivity (α) and peak asymmetry (b/a) are determined for each component (see below). Such evaluations are carried out periodically to ensure that column operating parameters are acceptable.

Parameter	Formula	Value Obtained
Theoretical plates or efficiency, N	$N = 25\left(\frac{V_R}{W_{5\sigma}}\right)^2$	$N_1 = 6700$ $N_2 = 8130$
Capacity factor, k'	$k'_1 = \frac{V_1 - V_0}{V_0}$	3.9
	$k'_2 = \frac{V_2 - V_0}{V_0}$	7.1
Selectivity, α	$\alpha = \frac{k'_2}{k'_1}$	1.8
Asymmetry, (b/a)	b/a	0.8

9.2.4.3 E/Z Monolignols – Separation and Calibration

Development of any HPLC system for the separation of a mixture requires the correct choice of mobile phase composition (isocratic or gradient) and HPLC column. Following column selection, it is best to first identify conditions for the rapid elution of the components being studied. Subsequently, the conditions necessary to separate the individual constituents may be identified by slowly changing the mobile phase composition. The purpose of this exercise is to slowly increase the retention volume of the components of interest to enhance their separation from one another. For practical purposes, separations should be achieved within 5–60 min to minimize peak broadening. For the separation of monolignols, a system was first developed for the separation of E- and Z-coniferyl alcohols 2 and 6 (Morelli et al. 1986). Thus, solutions of E- and Z-coniferyl alcohols 2 and 6 in methanol (concentrations ~9 mg ml^{-1}) were individually filtered (0.45 μ), and the resulting filtrates placed in an HPLC vial (~1 ml). An aliquot (10 μl) of each monolignol solution was then applied to a Novapak C_{18} column previously equilibrated with MeOH—H_2O (15:85, v/v). Eluate detection was carried out at λ = 262 nm. Two strong absorbances were observed at retention volumes of 17.15 and 20.64 ml, which corresponded to the E and Z isomers, respectively. However, it cannot be overemphasized that the appearance of an absorbance (i.e., a peak) does not necessarily prove that the component under investigation is being eluted, since any minor impurity within the sample may be eluted first. Consequently, additional product verification should be employed in all investigations using, for example, NMR, mass, infrared and ultraviolet spectroscopic analyses. In this case, confirmation

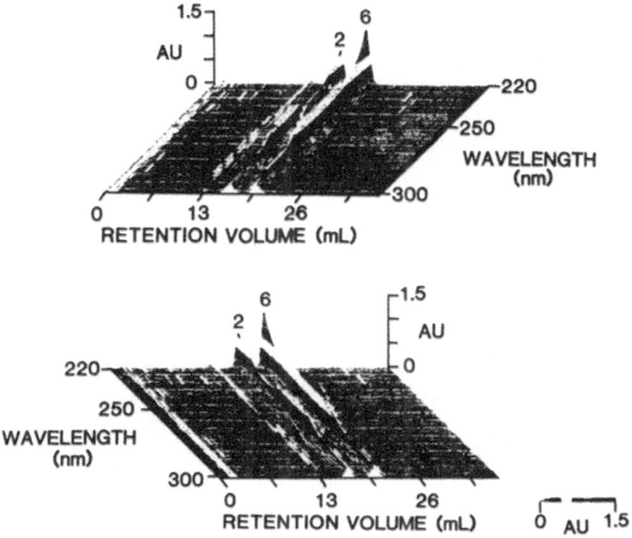

Fig. 9.2.8. Three-dimensional plots of UV-spectrum of E-coniferyl *2* and Z-coniferyl *6* alcohols during HPLC separation. Detection: Waters 990 Photodiode array detector equipped with NEC Power Mate 2 and NEC pinprinter

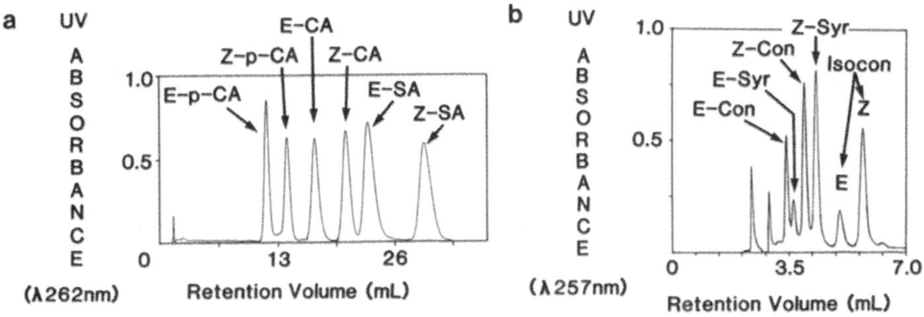

Fig. 9.2.9a,b. HPLC separation of **a** E/Z monolignols, using Waters' Novapak C_{18} column eluted with MeOH—H_2O (15:85) and **b** monolignol glucosides, using two Waters' Novapak C_{18} columns eluted with CH_3CN—H_2O (15:85). **a** E-*p*-CA, Z-*p*-CA. E- and Z-*p*-coumaryl alcohols *1*, *5*; *E-CA*, *Z-CA*. E- and Z-coniferyl alcohols *2*, *6*; *E-SA*, *Z-SA* E- and Z-sinapyl alcohols *3*, *7*. **b** *E-Con/Z-Con*. E- and Z-coniferins; *E/Z Syr*. E- and Z-syringins; *E/Z Isocon*. E- and Z-isoconiferins

was achieved by analysis of (1) the full UV spectrum (220–300 nm) of each eluted component during chromatographic separation using a photodiode array spectrometer (Fig. 9.2.8) and (2) the ^{1}H NMR spectrum of each eluted component. Subsequently, conditions were determined for the facile separation of all six E/Z monolignols and their corresponding glucosides (Lewis et al. 1989) (Fig. 9.2.9).

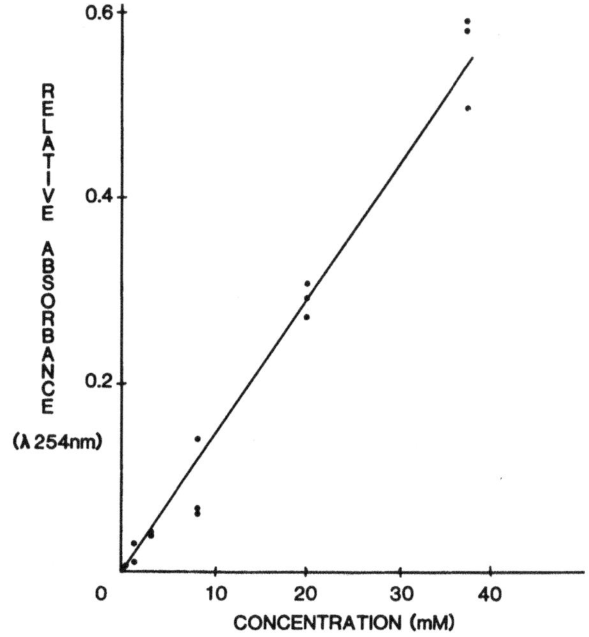

Fig. 9.2.10. HPLC calibration curve for Z-coniferin *4*

Having satisfactorily determined elution conditions for a particular compound or group of compounds, calibration curves must then be obtained to ensure the linearity (or lack thereof) of each integrated signal at different concentrations. An example of such a calibration curve is shown for Z-coniferin 4 (Fig. 9.2.10) as a function of absorbance against concentration. As expected, a linear response was noted over the concentrations studied. This would not have been the case had the column been overloaded or if nonlinear adsorption had occurred. Such graphs are particularly useful when quantification of a particular constituent is required.

9.2.5 Discussion of Methods

Table 9.2.1 summarizes some of the numerous analyses of low molecular weight lignin precursors, lignin fragments, and model compounds that can be performed with the use of HPLC.

9.2.5.1 Monomers

These are subclassified according to type, i.e., monolignols and their glucosides, hydroxycinnamic acids and aldehydes, lignosulphonates, thioacidolysis and acidolysis lignins, aromatic acids, lignin, and lignin model compound biodegradation products.

Monolignols and Their Glucosides

Many studies of plant extracts for monolignol and monolignol glucosides have relied heavily upon TLC comparison with authentic E-standards. Unfortunately, these methods normally only gave partial or no separation of the E and Z isomers. Consequently, development of the aforementioned HPLC methodology (Lewis et al. 1989) now permits the facile analysis of plant extracts, pulping liquors and the like for both E and Z isomers. As previously shown in Fig. 9.2.8, the combined use of a photodiode array detector in such studies can determine whether any other UV-absorbing impurities co-elute with the components of interest.

Hydroxycinnamic Acids and Aldehydes

Hydroxycinnamic acids, e.g., ferulic acid (Nagels and Parmentier 1976, Wulf and Nagel 1976, Murphy and Stutte 1978, Hartley and Buchan 1979, Hardin and Stutte 1980, Villeneuve et al. 1982), and hydroxycinnamyl aldehydes, e.g., coniferaldehyde (Dubelsten 1987), can be analyzed by reversed phase HPLC. In the case of the hydroxycinnamic acids, this methodology is also suitable for the separation of both E and Z isomers, e.g., ferulic acid (Hartley and Buchan 1979, Dubelsten 1987).

Lignosulphonates

Methodology has been described for the separation and identification of mono-, di- and tri-lignosulphonic acids obtained by sulphonation of lignin (Bialski et al. 1986, Luthe and Lewis 1986). This analysis first relies upon converting the mixture of lignosulphonic acids to their corresponding methyl derivatives with diazomethane, and then separating these derivatives on normal phase (μ-porasil) HPLC columns using 0.25% MeOH in CH_2Cl_2 as eluent.

Thioacidolysis and Acidolysis Lignins

Recently, Lapierre and Monties (1986) described a method whereby lignin samples may be partially degraded, using ethanethiol/BF_3 etherate, to afford the diastereomeric monomeric adducts 8–10 (see Fig. 9.2.11 and Chap. 6.4). Since each diastereomer can be separated by reversed phase HPLC, this method is used to estimate the relative amounts of each monomer in the lignin polymer. Additionally, the extent of β-O-4 linkages in the lignin polymer can also be estimated. This objective is achieved by comparing the results of thioacidolysis from a particular lignin preparation before and after methylation. In the case of premethylation with diazomethane, any free phenolic moiety in the lignin preparation is converted to its methyl ether. Subsequent treatment with ethanethiol/BF_3 etherate, as before, results in the liberation of the methylated adducts

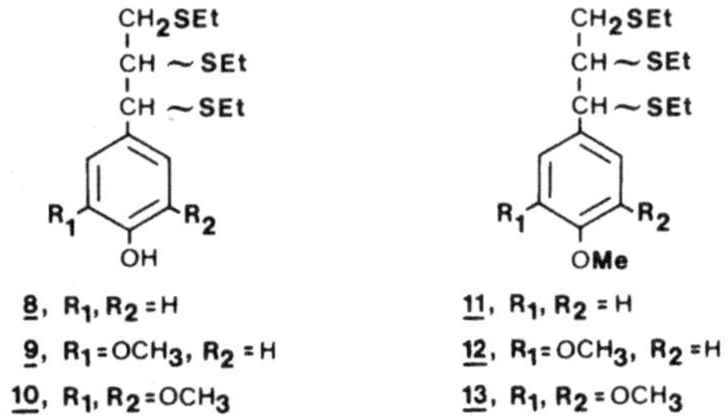

Fig. 9.2.11. Monomeric thioacidolysis derivatives from lignins and premethylated lignins

11–13. Additionally, cleavage of any β-O-aryl linkage in the premethylated lignins releases adducts 8–10. Hence an estimate of β-O-aryl linkages can be obtained by comparing the relative amounts of monomers 8–10 and 11–13 liberated following thioacidolytic treatment of lignin samples before and after methylation. This approach relies upon the assumption that all phenolic moieties, released after premethylation/thioacidolysis, result only from cleavage of the β-O-aryl linkage. Note also that HPLC can be used to analyze monomeric acidolysis lignins (see Chap. 6.1 on acidolysis).

Aromatic Aldehydes

Treatment of lignins with alkaline nitrobenzene at elevated temperatures results in the release of aldehydes such as *p*-hydroxybenzaldehyde, vanillin, and syringaldehyde (Scalbert et al. 1986) (Chap. 6.2). These can also be separated and quantified by HPLC; this technique is widely used to estimate monomer ratios in the noncondensed units of lignin. While this is a rather simple analysis, it should not be forgotten that the ratios so obtained do not adequately reflect the true monomer content of the polymer (Manders 1987).

Lignin and Model Compound Biodegradation Products

A substantial usage of HPLC methodology has been employed to examine the effects of biodegradative processes on both lignin-derived products and lignin model compounds, with the latter predominating. For the most part, this includes the analysis of degradation products following treatment of substrate with fungi such as *Phanerochaete chrysosporium* and *Aureobasidium pullulans* (Bourbonnais and Paice 1987), or bacteria such as streptomyces (Pometto and Crawford 1988). These HPLC techniques have mainly been developed for simple biodegradation substrates, such as guaiacylglycerol-β-guaiacyl ether or veratrylglycerol-β-guaiacyl ether, and their degradation products. The products include (among others) the hydroxycinnamic acids (e.g., ferulic acid) previously mentioned, substituted benzaldehydes (e.g., vanillin, syringaldehyde) and benzoic acids (e.g., vanillic and syringic acids). These fragments can be separated satisfactorily by reversed phase HPLC systems (Hartley and Buchan 1979, Villeneuve et al. 1982, Pecina et al. 1986, Bourbonnais and Paice 1987, Pometto and Crawford 1988).

9.2.5.2 Dimers

These comprise an abundant group of compounds present in both angiosperms and gymnosperms, and are principally found as optically active lignans and neolignans (Rao 1978). Again, reversed phase HPLC is the method of choice for their separation and analysis, e.g., for arctiin, epipinoresinol, matairesinol, matairesonide, and phyllyrin, (Nishibe et al. 1988). Additionally, the facile enantioselective separation of optically active (+ and −) forms of lignans and neolignans is now possible (Davin et al. 1991). This was first reported using poly (methyl methacrylate) coated silica (Chiralpak, Daicel, Japan) for resolution of enantiomers of kadsurenone (Pondipom et al. 1986). However, more recent applications include cellulose triphenylcarbamate supported on silica gel (Chiralcel OC, Daicel, Japan) for separation of (+)- and (−)- forms of syringaresinol (Tanahashi et al. 1987) and todolactol (Ozawa and Sasaya 1988), and cellulose tris (3,5-dimethylphenylcarbamate) coated on silica (Chiralcel OD, Daicel, Japan) for resolution of (±)-pinoresinol (Umezawa et al. 1990b), and (±)-secoisolariciresinol and (±)-matairesinol (Umezawa et al. 1990a, Umezawa et al. 1991); such separations are discussed in detail elsewhere (Davin et al. 1991).

Another important advance has also occurred in the thioacidolysis of lignins which, following Raney Ni desulphuration, now affords a group of dimeric lignin-derived compounds. The identity of these compounds is beginning to shed more light on different bonding patterns between monomeric units lignins (Lapierre et al. 1991). As described above, HPLC is also used widely to study biodegradation of synthetic dimers, such as guaiacylglycerol-β-guaiacyl and veratrylglycerol-β-guaiacyl ethers (Bourbonnois and Paice 1987).

9.2.5.3 Trimers and Higher Oligomers (Lignin Model Compounds)

Very little work has been carried out on the HPLC separation of higher molecular weight lignin oligomers, such as trimers and tetramers, using normal or reversed phase chromatography. However, synthetic routes to β-O-4 linked tetramers, and higher lignin model oligomers have now been described (Hyatt 1987). These substances have been analyzed using HPLC, but not with either normal or reversed phase chromatography. Instead, high performance size exclusion chromatography (HPSEC) was successfully employed using polystyrene-divinylbenzene columns eluted with tetrahydrofuran (Johnson et al. 1989). It remains to be demonstrated whether these oligomers can also be satisfactorily separated by conventional reversed or normal phase HPLC systems.

References

Bialski AM, Luthe CE, Fong JL, Lewis NG (1986) Sulphite-promoted delignification of wood: identification of paucidisperse lignosulphonates. Can J Chem 64:1336–1344

Bourbonnais R, Paice MG (1987) Oxidation and reduction of lignin-related aromatic compounds by *Aureobasidium pullulans*. Appl Microbiol Biotechnol 26:164–169

Braithwaite A, Smith FJ (1985) Chromatographic methods. Chapman and Hall, New York, 258–266

Buchardt O, Jensen RB, Hansen HF, Nielsen PE, Andersen D, Chinoin I (1986) Thermal chemistry of podophyllotoxin in ethanol and a comparison of the cytostatic activity of the thermolysis products. J Pharm Sci 75:1076–1080

Charalambous G, Bruckner KJ, Hardwick WA, Linneback A (1973) Determination of beer flavor compounds by high pressure liquid chromatography. Tech Quart, Master Brew Assoc Am 10(2):74–78

Davin LB, Umezawa T, Lewis NG (1991) Enantioselective separations in phytochemistry. Recent Adv Phytochem 25:75–112

Dubelsten P (1987) The E/Z Isomerization step in the biosynthesis of Z-coniferyl alcohol in *Fagus grandifolia* Ehrh, M.S. Thesis, Virginia Tech, Blacksburg, USA, 67 pp

Hardin JM, Stutte CA (1980) Analysis of phenolic and flavonoid compounds by high-performance liquid chromatography. Anal Biochem 102:171–175

Hartley RD, Buchan H (1979) High performance liquid chromatography of phenolic acids and aldehydes derived from plants or from the decomposition of organic matter in soil. J Chromatogr 180:139–143

Heftmann E (ed) (1983) Chromatography: fundamentals and applications of chromatographic and electrophoretic methods. Part B: Applications. Elsevier Scientific, Amsterdam. J Chromatogr Libr 22B:B1–B564

Hyatt JA (1987) Synthesis of some tetrameric lignin model compounds containing β-O-4 and 5,5′-interunit linkages. Holzforschung 41:363–370

Johnson DK, Chum HL, Hyatt J (1989) Molecular weight distribution studies using lignin model compounds. In: Glasser WG, Sarkanen S (eds) Lignin: properties and materials. ACS Symp Ser 397:109–123

Lapierre C, Monties B (1986) Preparative thioacidolysis of spruce lignin: isolation and identification of main monomeric products. Holzforschung 40:47–50

Lapierre C, Pollet B, Monties B (1991) Heterogenous distribution of diarylpropane structures in spruce lignins. Phytochemistry 30:659–662

Lewis NG, Inciong MaEJ, Dhara KP, Yamamoto E (1989) High performance liquid chromatographic separation of E and Z-monolignols and their glucosides. J Chromatogr 979: 345–352

Luthe CE, Lewis NG (1986) Identification and characterization of paucidisperse lignosulphonates. Holzforschung 40 (Suppl): 153–157

Manders WF (1987) Solid state ^{13}C nmr determination of the syringyl/guaiacyl ratio in hardwoods. Holzforschung 41:13–18

Morelli E, Rej RN, Just G, Lewis NG, Towers GHN (1986) The occurrence of cis-monolignols in *Fagus grandifolia*. Phytochemistry 25:1701–1705

Mortimer RD (1982) The formation of coniferyl alcohol during alkaline delignification. J Wood Chem Technol 2:383–415

Mortimer RD (1985) The formation of coniferyl alcohol during alkaline delignification with anthraquinone. Part II. J Wood Chem Technol 5:347–361

Murphy JB, Stutte CA (1978) Analysis for substituted benzoic and cinnamic acids using high-pressure liquid chromatography. Anal Biochem 86:220–228

Nagels L, Parmentier F (1976) Kinetic study of possible intermediates in the biosynthesis of chlorogenic acid in *Cestrum poeppigii*. Phytochemistry 15:703–706

Nishibe S, Sakushima A, Kitagawa S, Klimek B, Benecke R, Thieme H (1988) Phenolic compounds from Forsythia leaves (III). On the comparison of constituents between hybrids and parents. Shoyakugaku Zasshi 42:324–328

Ozawa S, Sasaya T (1988) Extractives of Todomatsu *Abies sachalinensis* masters V. A novel dibenzylbutyrolactol lignan from the wood of *Abies sachalinensis*. Mokuzai Gakkaishi 34:851–857

Pecina R, Burtscher P, Bonn G, Bobleter O (1986) GC-MS and HPLC analyses of lignin degradation products in biomass hydrolyzates. Fresenius Z Anal Chem 325:461–465

Pometto III AL, Crawford DC (1988) High performance liquid chromatography of aromatic fragments from lignin degradation. Methods Enzymol 161:183–190

Pondipom MM, Yue BZ, Bugianesi RL, Brooker DR, Chuang MN, Shen TY (1986) Total synthesis of kadsurenone and its analogs. Tetrahedron Lett 27:309–312

Rao CBS (ed) (1978) Chemistry of lignans. Andhra University Press, Series No. 149, Waltair, Visakhapatnam, India, 377 pp

Scalbert A, Monties B, Guittet E, Lallemand JY (1986) Comparison of wheat straw lignin preparations. I. Chemical and spectroscopic characterizations. Holzforschung 40:119–127

Tanahashi M, Karina M, Higuchi T (1987) Cleavage of lignin in wood by steam-explosion. Proc 4th Int Symp on wood and pulping chemistry, April 27–30, (1987), Paris. EUCEPA-TAPPI, 343–347

Tswett MS (1903) (publ. 1905) TR Protok, Varshav, Obschch, Entestvoispyt Odt Biol *14* (English translation: Hesse G, Weil H (1954) Michael Tswett's first paper on chromatography. M. Woelm Eschwege 35 pp

Umezawa T, Davin LB, Lewis NG (1990a) Formation of the lignan, (−)-secoisolariciresinol, by cell-free extracts of *Forsythia intermedia*. Biochem Biophys Res Commun 171:1008–1014

Umezawa T, Davin LB, Lewis NG (1991) Formation of the lignans (−)-secoisolariciresinol and (−)-matairesinol with *Forsythia intermedia* cell-free extracts. J Biol Chem 266(16): 10210–10217

Umezawa T, Davin LB, Yamamoto E, Kingston DAI, Lewis NG (1990b) Lignan biosynthesis in Forsythia species. J Chem Soc Chem Commun 1405–1408

Villeneuve F, Abravanel G, Moutounet M, Alibert G (1982) General scheme of analysis of phenolic compounds in plant extracts by reversed phase high-performance liquid chromatography. J Chromatogr 234:131–140

Wulf LW, Nagel CW (1976) Analysis of phenolic acids and flavonoids by high-pressure liquid chromatography. J Chromatogr 116:271–279

Yost RW, Ettre LS, Conlon RD (1980) Practical liquid chromatography: an introduction. Perkin-Elmer, Norwalk, CT., 257 pp

Subject Index

GPSR Compliance

The European Union's (EU) General Product Safety Regulation (GPSR) is a set of rules that requires consumer products to be safe and our obligations to ensure this.

If you have any concerns about our products, you can contact us on ProductSafety@springernature.com

In case Publisher is established outside the EU, the EU authorized representative is:

Springer Nature Customer Service Center GmbH
Europaplatz 3
69115 Heidelberg, Germany

Batch number: 10798362

Printed by Printforce, the Netherlands